THE BILE ACIDS

Chemistry, Physiology, and Metabolism

VOLUME 4: METHODS AND APPLICATIONS

THE BILE ACIDS

Series Editors: Padmanabhan P. Nair and
David Kritchevsky

Volume 1: Chemistry
Volume 2: Physiology and Metabolism
Volume 3: Pathophysiology
Volume 4: Methods and Applications

A Continuation Order Plan is available for this series. A continuation order will bring delivery of each new volume immediately upon publication. Volumes are billed only upon actual shipment. For further information please contact the publisher.

THE BILE ACIDS

Chemistry, Physiology, and Metabolism

VOLUME 4: METHODS AND APPLICATIONS

Edited by

Kenneth D. R. Setchell

Children's Hospital Medical Center
Cincinnati, Ohio

David Kritchevsky

The Wistar Institute
Philadelphia, Pennsylvania

and

Padmanabhan P. Nair

Lipid Nutrition Laboratory
Beltsville Human Nutrition Research Center
ARS, USDA
Beltsville, Maryland
and The Johns Hopkins University School of Hygiene and Public Health
Baltimore, Maryland

PLENUM PRESS • NEW YORK AND LONDON

Library of Congress Cataloging in Publication Data
(Revised for vol. 4)

Nair, Padmanabhan P., 1931–
  The bile acids; chemistry, physiology, and metabolism.

  Includes bibliographical references.
  Contents: v. 1. Chemistry. — v. 2. Physiology and metabolism. —   — v. 4. Methods
and applications.
  1. Bile acids — Collected works. 2. Bile acids — Metabolism — Collected works. 3. Bile
salts — Collected works. I. Title. II. Kritchevsky, David, 1920-   , joint author.
QP752.B54N34   599'.01'9243   71-138520

ISBN-13: 978-1-4612-8236-5   e-ISBN-13: 978-1-4613-0901-7
DOI: 10.1007/978-1-4613-0901-7

CONTRIBUTORS

Peter Back
Hospital of the University of Freiburg
Freiburg im Breisgau, West Germany

Stephen Barnes
Departments of Pharmacology and
* Biochemistry*
Comprehensive Cancer Center
University of Alabama at Birmingham
Birmingham, Alabama

C. E. Bugg
Departments of Pharmacology and
* Biochemistry*
Comprehensive Cancer Center and Center
* for Macromolecular Crystallography*
University of Alabama at Birmingham
Birmingham, Alabama

C. Colombo
Department of Pediatrics and Obstetrics
University of Milan
Milan, Italy

M. DeLuca
Department of Chemistry
University of California
San Diego, California

Nina K. Dodd
Department of Food and Nutrition
SVT College of Home Science
Juho, Bombay, India

Jacqueline Dupont
Department of Food and Nutrition
Iowa State University
Ames, Iowa

S. E. Ealick
Departments of Pharmacology and
* Biochemistry*
Comprehensive Cancer Center and Center
* for Macromolecular Crystallography*
University of Alabama at Birmingham
Birmingham, Alabama

D. Festi
Institute of Chemical Sciences and
* Departments of Internal Medicine and*
* Gastroenterology*
University of Bologna
Bologna, Italy

Pilar A. Garcia
Department of Food and Nutrition
Iowa State University
Ames, Iowa

Satindra K. Goswami
Department of Food and Nutrition
Iowa State University
Ames, Iowa
Present address: Institute for Basic
* Research*
Staten Island, New York

v

Junichi Goto
Pharmaceutical Institute
Tohoku University
Sendai, Japan

Bernhard Hennig
Departments of Nutrition and Food Science
University of Kentucky
Lexington, Kentucky

Walter G. Hyde
Veterinary Diagnostic Laboratory
Iowa State University
Ames, Iowa

David N. Kirk
Department of Chemistry
Queen Mary College
University of London
London, England

A. M. Lawson
Section of Clinical Mass Spectrometry
Clinical Research Centre
Harrow, Middlesex, England

G. M. Murphy
Gastroenterology Unit
Department of Medicine
Guy's Campus
United Medical and Dental Schools
London, England

Padmanabhan P. Nair
Lipid Nutrition Laboratory
Beltsville Human Nutrition Research
Center
Agricultural Research Service
U.S. Department of Agriculture
Beltsville, Maryland and
Department of Biochemistry
The Johns Hopkins University School
of Hygiene and Public Health
Baltimore, Maryland

Toshio Nambara
Pharmaceutical Institute
Tohoku University
Sendai, Japan

Suk Yon Oh
Division of Food and Nutrition
College of Health
University of Utah
Salt Lake City, Utah

A. Roda
Institute of Chemical Sciences and
Departments of Internal Medicine and
Gastroenterology
University of Bologna
Bologna, Italy

E. Roda
Institute of Chemical Sciences and
Departments of Internal Medicine and
Gastroenterology
University of Bologna
Bologna, Italy

J. Schölmerich
Department of Internal Medicine
University of Freiburg
Freiburg, West Germany

K. D. R. Setchell
Department of Pediatric Gastroenterology
and Nutrition
Clinical Mass Spectrometry Laboratories
Children's Hospital Medical Center
Cincinnati, Ohio

Charles E. Sizer
Department of Food and Nutrition
Iowa State University
Ames, Iowa
Present address: Brikpak Inc.
Dallas, Texas

J. Sjövall
Department of Physiological Chemistry
Karolinska Institute
Stockholm, Sweden

J. M. Street
Department of Pediatric Gastroenterology
  and Nutrition
Clinical Mass Spectrometry Laboratories
Children's Hospital Medical Center
Cincinnati, Ohio

Nabila Turjman
Lipid Nutrition Laboratory
Beltsville Human Nutrition Research
  Center
Agricultural Research Service
U.S. Department of Agriculture
Beltsville, Maryland
Present address: Department of
  Microbiology and Immunology
Duke University
Durham, North Carolina.

PREFACE

Over a decade has elapsed since the last volume in this series was published. At that time we considered that we had comprehensively covered all aspects relating to bile acid chemistry and physiology. However, major strides have been made in our understanding of the physiology and pathophysiology of bile acids, due largely to the great advances which have taken place in analytical technology. As a result, the need to document these advances was felt acutely, and therefore this volume is devoted to methodologies in bile acid analysis and their applications.

This volume includes twelve chapters written by prominent scientists in the field of bile acid research. The initial chapter discusses techniques of extraction and isolation of bile acids from biological fluids. It is followed by descriptions of physical methods of analysis and discussions of the way these techniques have been applied in the field of bile acid research. Of practical value is the inclusion of a comprehensive list of spectra obtained for references by nuclear magnetic resonance spectroscopy and mass spectrometry. These chapters are followed by reviews of biological methods of immunoassay and bioluminescence. Specific applications of these techniques are then addressed in contributions relating to bile acid analysis of tissue, serum, urine, and feces.

With this integrated approach we have attempted to provide a volume which represents a comprehensive review of the analytical field of bile acids, while also serving as a useful reference book for those workers involved in bile acid analysis.

Inevitably there are several areas not covered in this volume and the description of applications has been largely restricted to the analysis of biological fluids from man. As editors we take responsibility for these omissions, our excuse being that the content is already at the limit of what can be reasonably accommodated in a single volume.

We wish to express our sincere thanks to Mary Jo McCarthy for typing

assistance, and we are indebted to Frances M. Murray and Shirley A. Tepper
for their expert help in the preparation of the index.

K.D.R.S.
Cincinnati, Ohio

D.K.
Philadelphia, Pennsylvania

P.P.N.
Beltsville, Maryland

CONTENTS

Chapter 3
Nuclear Magnetic Resonance **65**
Stephen Barnes and David N. Kirk

Chapter 6
Immunological Methods for Serum Bile Acid Analysis 269
A. Roda, E. Roda, D. Festi, and C. Colombo

Chapter 7
Bioluminescence Assays Using Immobilized Enzymes in
Bile Acid Analysis 315
J. Schölmerich, A. Roda, and M. DeLuca

Chapter 8
Bile Acids in Extrahepatic Tissues **341**
Jacqueline Dupont, Pilar A. Garcia, Bernhard Hennig, Nina K. Dodd,
Suk Yon Oh, Charles E. Sizer, Satindra K. Goswami, and Walter G. Hyde

Chapter 9
Tissue-Bound Bile Acids **373**
Nabila Turjman and Padmanabhan P. Nair

THE BILE ACIDS

Chemistry, Physiology, and Metabolism

VOLUME 4: METHODS AND APPLICATIONS

Chapter 1

TECHNIQUES FOR EXTRACTION AND GROUP SEPARATION OF BILE ACIDS

J. Sjövall and K. D. R. Setchell

1. INTRODUCTION

When bile acids in biological materials are to be analyzed and identified, there is usually a need for extraction and prepurification procedures. Bile and duodenal fluid are exceptions; when the concentration of bile acids is sufficiently high, these materials may be directly applied to the starting line of paper and thin-layer (TLC) chromatograms (see Ref. 1). There is a continuing development in this area [2] also by application of high-performance [3], reversed-phase [4], and overpressured TLC [5]. Direct analysis of the major bile acids by electrophoresis is possible [6,7], and isotachophoresis in organic solvents is a recent advance [8]. Bile may be directly injected on high-performance liquid chromatography columns (e.g., Ref. 9; see this volume, Chapter 2), but column-switching techniques with automated sample extraction and cleanup are likely to be more widely used.

Radioimmunoassay and enzymatic methods are often claimed to be sufficiently specific for analysis of crude biological fluids. Although this may be true for major components or particular needs (see Chapters 6 and 7), it is not the case for minor bile acids in complex mixtures and different states of conjugation. Although the gross correlation between results obtained with these indirect methods and methods based, for example, on mass spectrometry may be satisfactory, it is difficult to evaluate the specificity of the indirect methods for individual samples, and major abnormalities in bile acid profiles may escape detection. There are also examples where grossly misleading results were obtained by

J. Sjövall Department of Physiological Chemistry, Karolinska Institute, Stockholm, Sweden
K. D. R. Setchell Department of Pediatric Gastroenterology and Nutrition, Clinical Mass Spectrometry Laboratories, Children's Hospital Medical Center, Cincinnati, Ohio

indirect (enzymatic) methods of analysis [10–12]. However, improved indirect methods are being developed [13] which can be expected to find important use in routine analyses of selected bile acids.

Methods of extraction and purification were reviewed in the first volume of this series [1]. Since that time many important advances have been made in the analysis of bile acids. Capillary column gas–liquid chromatography (GLC) alone and in combination with mass spectrometry (GC/MS) has increased the need for procedures that minimize sample load on the column (see this volume, Chapter 5). High-performance liquid chromatography (HPLC) also in combination with mass spectrometry has developed greatly (see this volume, Chapter 2), and extraction/purification procedures are needed for protection of the columns and to increase the specificity of the analyses. Fast atom bombardment and secondary ion mass spectrometry have been introduced for analysis of conjugated bile acids and crude samples (see this volume, Chapter 5), and appropriate extraction and purification methods enhance the sensitivity and increase the amount of information obtained from such analyses. Thus, extraction and group separation methods are required for all types of analyses and studies of bile acids in biological materials as well as for isolation of products from organic syntheses. While standard methods of liquid extraction and chromatography [1] continue to have important applications, some new and useful general procedures have been developed which will be covered in this chapter. Aspects of bile acid analysis and extraction and group fractionation procedures have been discussed in some recent reviews [14–19].

2. EXTRACTION

2.1. Solvent Extraction

2.1.1. Biological Fluids

Although solid-phase extractions have largely replaced solvent extractions of biological fluids, solvent extraction methods continue to be used and are important for validation of the efficiency of solid-phase extraction procedures.

Extraction of bile acids in urine can be achieved with n-butanol at acidic pH [1,20]. It is of interest that taurine-conjugated and -sulfated bile acids can be extracted with butanol at an alkaline pH [21], as originally observed by Bremer [22]. This is of practical importance in studies of the conjugation of bile acids with taurine [21] or sulfuric acid [22] since it simplifies the assay procedures.

Ethanol and methanol are the most common solvents for extraction of fluids containing proteins, e.g., bile, duodenal contents [1,23–35], and plasma [36–40]. Propanol-2 [41] and methanol/chloroform [42,43] are sometimes used for extraction of duodenal bile, and acetonitrile/ammonium sulfate (giving a two-phase system) has been used in analyses of plasma [44]. Acetone/ethanol, 1:1 (v/v), gives a high yield of steroid mono- and disulfates [45] and should be equally useful for extraction of bile acids. It is not always realized that recoveries are affected by the way in which the extractions are performed and that the yield of added tracers does not necessarily reflect recoveries of endogenous bile acids. The fluid should be added drop by drop to 10–20 volumes of solvent, held in an ultrasonic bath. Using this procedure and acetone/ethanol, plasma proteins form a very fine, white precipitate which is easy to wash and to remove by filtration or centrifugation [45]. Addition of solvent to plasma [37,38,44] can precipitate lumps of protein with inclusion of the bile acids. Extraction with water-immiscible solvents is not recommended.

Some methods combine chemical treatment and extraction. In their methods for quantitative analysis of bile acids in plasma by isotope-dilution GC/MS, Björkhem and associates subject plasma to alkaline hydrolysis in aqueous ethanol followed by extraction with diethyl ether [46–50]. This procedure should only be used in combination with labeled internal standards and has the drawback that the yield of different types of conjugates (amidated, sulfated, glucuronidated) cannot be evaluated. Similar procedures were used by Beppu et al. [51], who did not add ethanol to the hydrolysis mixture (see Section 4.1). Enzymatic hydrolysis has also been used prior to extraction [52].

Javitt and co-workers were the first to describe extraction of serum with 2,2-dimethoxypropane [53,54]. This method has been used in similar forms by others for analysis of bile acids in serum [55] and amniotic fluid [56,57]. Dimethoxypropane reacts with water under acidic conditions to yield methanol and acetone in a molar ratio of 2. When 7 ml of dimethoxypropane is added per milliliter of serum acidified to pH 1 with concentrated hydrochloric acid, proteins precipitate and bile acids are extracted. When a molar excess of dimethoxypropane is added (>9 ml/ml serum) an anhydrous mixture is obtained suitable for solvolysis of sulfates (see Section 4.4) and methylation of carboxyl groups (see Section 4.5). Reaction of vicinal hydroxyl groups to form acetonides is also possible. The quantitative aspects have not been reported for all potential reactions in extractions of plasma.

Ion-pair extraction of bile acids was first described by Hofmann, who used tetraheptylammonium as the counterion dissolved as the chloride in the organic phase [58]. A lower extraction efficiency was reported when the bromide of this counterion was used [59]. Schill and co-workers, who have pioneered in the field of ion-pair extraction, have made a detailed study of factors determining the extraction of taurine-conjugated bile acids as ion pairs with tetrabutyl- and tetrapentylammonium [60]. Colored or ultraviolet-absorbing ion-pairing agents have been used for analytical purposes [61,62] and in HPLC systems [63] (see this volume, Chapter 2). The presence of large amounts of counter-ion in the extract often constitutes a problem in the further purification and analysis. However, it is likely that ion-pair extraction will find increased use in combination with reversed-phase ion-pair HPLC [63–65] (see also Section 2.2.3).

Two-phase solvent extractions can be performed by absorbing the biological fluid in a support, e.g., Kieselguhr in a column (e.g., Extrelut, Merck, Darmstadt, FRG) and eluting with an immiscible solvent. This method has limited applications in bile acid analysis because of the high polarity of many bile acid classes.

2.1.2. Tissues and Feces

Several chapters specifically describe bile acids in these materials, and only general aspects on extraction are discussed here.

Samples containing solid material have to be initially extracted with solvents or be liquefied by chemical or enzymatic means for subsequent solid-phase extraction. The possibility of enzyme digestion was pointed out in the previous review [1], but no systematic studies have been carried out to test this approach in bile acid analysis.

Digestion of tissues in strong alcoholic alkali (10% KOH in 50% ethanol, 12 hr, room temperature) has been used in some studies of tissue bile acids [66–68]. Such conditions may create artifacts depending on the structure of the bile acid and the mode of conjugation (see Section 4.1). The solubilized sample may be extracted by solid-phase methods [68], but application of enzymatic methods for analysis of bile acids in such extracts [68] is not likely to produce accurate estimates. Strong aqueous alkali (3.4 M NaOH, 10 min, 100°C) has also been used for digestion of liver tissue [69]. The resulting solution could then be diluted with buffer and directly treated with cholylglycine hydrolase for subsequent extraction with diethyl ether. Sulfated and glucuronidated bile acids and labile structures would be lost in this procedure. Somewhat milder conditions for solubilization (1.25 M NaOH, 30 min, 80°C) have been used in combination with solid-phase extraction [70,71] (see Section 2.2.1). Recoveries of added common free, conjugated, and sulfated bile acids were in the range 72–96%, but yields of endogenous compounds and possible formation of artifacts have not been studied.

Extraction with ethanol/hydrochloric acid gives high recoveries of ^{14}C after addition of ^{14}C-labeled bile acids [67,68]. However, the yields of endogenous bile acids, sulfates, and glucuronides are not known. Esterification is likely to occur, and the extract has to be subjected to alkaline hydrolysis. Both acid- and alkali-labile bile acids may be lost.

The choice of solvents in solvent extractions depends on the problem to be studied. The polarities of bile acid derivatives range from those of fatty acid or ethyl esters of lithocholic acid to those of glucuronides of polyhydroxycholanoic acids. Generally, it is advisable to use a solvent mixture that will extract compounds of a wide polarity range and then to remove unwanted material. The solvent should also be able to penetrate cell walls, which requires the presence of both polar and nonpolar functional groups. Chloroform/methanol is the classical mixture for such purposes, but the water wash in the Folch extraction procedure cannot be used if all bile acid derivatives are to be recovered in one phase. Also, recoveries of internal standard were low (57%) when this solvent was used in analyses of bile acids in skin [72]. The more recent lipid extraction method utilizing hexane/2-propanol, 3:2 (v/v) [73], only partially extracts conjugated bile acids from liver [74].

Although aqueous ethanol may seem a poor solvent for penetration of lipid layers and bacterial cell walls, it has been found very useful for extraction of bile acids. Refluxing twice in 95% ethanol containing 0.1% ammonium hydroxide extracts the bile acids in homogenized rat liver [75–81]. Hedenborg and co-workers recently evaluated the efficiency of this method using subcutaneous fat, muscle, and skin from cholestatic patients given [24-^{14}C]chenodeoxycholic acid [82]. The recovery of labeled endogenous bile acids was 96–97%. There are many minor variations of the ethanol extraction procedure. Addition of ammonium hydroxide may not be necessary since endogenous cholic acid conjugates labeled by prior injection of [24-^{14}C]deoxycholic acid were quantitatively recovered by extraction with 90% ethanol alone [74]. Acetone was claimed to give quantitative recoveries, but no details were given [80,83].

Bile acids in subcellular fractions can be extracted in the same way as from whole tissue [84–86]. Unconjugated intermediates in bile acid biosynthesis have been extracted with ethyl acetate [87,88].

The presence of lithocholic acid bound in amide linkage to tissue proteins has been reported [76–78]. This finding could not be confirmed in studies of normal and cirrhotic liver [81], nor in muscle, fat, and skin [82]. If present, covalently bound bile acids have to be released by enzymatic or chemical methods prior to extraction [76–78,80].

The presence of nonpolar esters [89,90] and the strong binding to bacterial and dietary residues have to be considered in the extraction of bile acids from feces [91,92]. The recoveries of added compounds and of endogenous bile acids labeled by administration of ^{14}C-labeled cholic and chenodeoxycholic acids seem to be quantitative after sequential refluxing in 90% and 80% aqueous ethanol followed by chloroform/methanol, 1:1 (v/v) [92–94]. Although this method is mild compared to those employing strong alkaline or acidic conditions (see this volume, Chapter 12, and Refs. 1,95–97), the possibility of artifact formation cannot be excluded. Extended Soxhlet extractions with chloroform/methanol [98,99] are more likely to produce artifacts. n-Butanol was used for extraction of acidified meconium, which contains mainly taurine conjugated and sulfated bile acids [100,101]. This procedure would not be applicable to conventional feces or tissues.

2.2. Solid-Phase Extraction

Early solid-phase extractions were performed with a macroreticular poly-styrene/strong anion exchanger, Amberlyst XN1006 [102]. This permitted simultaneous extraction and isolation of an acidic fraction from blood serum. The same or similar ion exchangers, Amberlite A-26 and A-27, have been used for the same purpose after initial enzymatic hydrolysis of conjugated bile acids [103,104]. However, large volumes of ammonium carbonate in aqueous ethanol are required to desorb the bile acids, and inorganic ions are also obtained in the extract. For these reasons, these sorbents are no longer used in bile acid analysis.

2.2.1. Amberlite XAD Resins

These resins were introduced into steroid research by Bradlow [105]. They have been extensively used for extraction of bile acids from plasma [38,106–126], skin interstitial fluid [123], urine [20,118,127–145], bile or duodenal fluid [145–150], and from reaction mixtures, e.g., in the synthesis of bile acid sulfates [151]. They have also been used for extraction of tissues solubilized by aqueous alkali [68,70,71] and for desalting of extracts of urine [20] or feces [152] obtained by previous solvent extraction. From the time of their introduction until now, a number of problems with the use of these resins have been encountered and sometimes reported. If losses in precipitates removed from the samples prior to extraction [143] are excluded, one can distinguish three major reasons for these problems: (1) the chemical properties of the resins have changed; (2) the methods of sorption and desorption have not been appropriate for the type of biological sample studied; (3) new bile acids and bile acid conjugates have become available for recovery studies and found to be unsuitable for extraction with XAD resins.

Amberlite XAD resins are intended to be neutral. Fifteen years ago this may have been the case since sulfated bile acids in urine, endogenously labeled with ^{14}C, were quantitatively recovered using the original method [105] with methanol elution [127–130]. Later studies indicated that both sorption and desorption were dependent on pH [135,150,153] and conditions for washing and storing the resin [153]. Amberlite XAD-4 behaved in a very similar way to XAD-2 [149]. At the same time problems were reported with sorption of polar steroid [154] and bile acid conjugates [151] and desorption of acidic steroid conjugates, particularly sulfates [155]. The latter were solved by washing with aqueous triethylamine sulfate prior to elution with methanol [155]. This was believed to be due to formation of triethylamine salts, but Axelson and Sahlberg showed that a wash with aqueous sodium sulfate was equally effective [156]. Their study shows that desorption is dependent on the ionic strength of the sample applied and that Amberlite XAD-2 behaves as a weak anion exchanger with low capacity and high affinity for sulfates [156]. This would explain the improved yields when ammonium hydroxide [135] or carbonate [153] is added to the alcoholic eluent (cf. elution from the anion exchanger Amberlyst A-26 [102–104]. The same mechanism may explain the improved elution with strong acid [38].

The poor yields of different bile acids under many conditions have resulted in various modifications claimed to be superior to the original procedures. In view of the probable variability of the batches of resins, it is difficult to establish the factors responsible for improved extraction and desorption efficiencies.

When bile acids are to be extracted from plasma, protein binding is an additional problem. In most methods this is overcome by dilution of the sample with about 10 volumes of 0.1 M NaOH (e.g., Refs. 38,112,113,117) as described in the original version of the method [106]. However, dilution with saline and performing the extraction at 64°C, as originally developed for neutral steroids [157,158], may improve the yields [121] (see further in Section 2.2.2).

Batch adsorption on the more polar Amberlite XAD-7 has given much better results than column adsorption on XAD-2, particularly in the extraction of sulfated bile acids [112,113]. However, the yields of the latter are lower than those of nonsulfated bile acids [112,113,117,126]. A difference in the undesired ion-exchanging properties between the two resins has not been excluded and might be one factor explaining the better results with XAD-7.

Irrespective of possible advantages with one resin or the other, it is presently believed that bonded silica of the reversed-phase type is superior to any of the polystyrene polymers so far tested.

2.2.2. *Octadecylsilane-Bonded Silica*

Reversed-phase supports of different types are used extensively in extraction/purification steps in analyses of various endogenous and exogenous lipophilic and amphipatic compounds and their metabolites. Materials of this kind were introduced into steroid and bile acid analysis by Shackleton and Whitney [159,160] and by Goto, Nambara, and co-workers [161]. These groups used Sep-Pak C_{18} cartridges (Waters Assoc., Milford, Massachusetts); alternative sorbents tested are Bond-Elut (Analytichem International, Harbor City, California) [162], Develosil ODS (Nomura Chemical, Seto-Shi, Japan) [163], and Lichrosorb RP-8 (octylsilane-bonded, Merck, Darmstadt, Federal Republic of Germany) [164]. Sep-Pak C_{18} or Bond-Elut is most commonly used. The methods to synthesize these supports differ, as do the particle sizes. This results in different chemical and physical properties. For example, reagents or side products from the synthesis may remain adsorbed to the particles and be eluted with the bile acids. These contaminants from the reaction, e.g., the trimethylsilyl ether and the dimer of dimethyloctadecylsilanol, may be removed by washing the sorbent with methylene chloride in addition to the normal wash with methanol prior to use. The presence of dimer is indicated by a gas chromatographic peak with a retention index of about 3860 on methyl silicone columns.

The stability of the covalently bound substituents may differ, and washing with acidic or basic solutions may result in hydrolytic release of bound material. Whiting reported appearance of contaminant peaks after extraction of an alkaline solution of plasma with Sep-Pak C_{18} [165]. One of the peaks may have been derivatized dimethyloctadecylsilanol. We have obtained large amounts of the dimer and smaller amounts of other contaminants in eluates from Sep-Pak C_{18} following a wash with dilute acid. Similar problems, which may be batch-dependent, were not seen with Bond-Elut [165] or batches of Sepralyte tested by us. The contaminants may not be observed in all cases since they are neutral and less polar than bile acids and may be removed in subsequent purification steps [165]. However, because of their poor solubility in aqueous ethanol, they may also be adsorbed on lipophilic ion exchangers and appear with unconjugated bile acids (Section 3.4.2).

The sorbents can be obtained as ready-made beds. The materials in which these are prepared may also release contaminants, and for high-sensitivity work it is advisable to prepare the beds in glass columns from bulk material (Preparative C18 instead of Sep-Pak, Sepralyte instead of Bond-Elut). This is also necessary when a controlled elevated temperature is needed [156,157]. The particle size (smaller for Bond-Elut) may be of importance when viscous biological fluids are extracted.

The mechanisms of sorption of bile acids on octadecylsilane-bonded silica have not been established. Bile acids with a very wide range of polarities are sorbed, and nonpolar adsorption to the bonded phase is unlikely to be the only mechanism. Polar bile acids, e.g., glucuronides, might be adsorbed to polar groups on the support. Both bile acids and salts are sorbed. If there are multiple mechanisms for sorption, the nature of the silica and the substituents and the degree of coverage with the latter are likely to affect sorption capacity and selectivity. Another factor of importance is the solvent. It is notable that tracer amounts of cholesterol or other nonpolar compounds in water are not efficiently sorbed by octadecylsilane-bonded silica [93,156,162,166]. This might be due to the physical state in which these compounds occur in water, i.e., not in monomolecular solution, or to the immiscibility of water and octadecyl groups at the interface. If methanol is added in sufficient concentrations, at least 50–60%, a conventional reversed-phase system is created, lipid aggregates may be dispersed, and sorption occurs [166–168]. Obviously, most bile acids are not sorbed from this solution and the effluent has to be concentrated to water and again passed through the sorbent bed [166].

Octadecylsilane-bonded silica has been used in different ways for extraction/purification of bile acids. The procedures employed are primarily governed by the nature of the sample: urine, plasma, bile, feces, and tissue extracts.

Bile acids and alcohols in urine are usually extracted according to the original procedure: passage of sample, wash with water, elution with methanol [143,160,164,169–175]. Sample flow rates can be 20 ml/min during application and washing [160,162]. A Sep-Pak cartridge (about 0.34 g sorbent) may carry more than 100 μmoles of taurocholate [160], but a capacity of about 30 μmoles is more realistic in work with urine [173]. Thus, in many analytical applications it is possible to use very small beds of sorbent, which minimizes solvent volumes and contamination. This advantage has not been fully appreciated but is obvious from the studies of Ishii and co-workers [163]. We have found 30 mg of sorbent to be sufficient for many purposes, but even smaller beds in Teflon tubing are useful [74,163]. Some procedures include a wash with 1.5% aqueous ethanol [161] and use of 90% aqueous ethanol for elution [160]. The purity of the extract is obviously influenced by the washing-and-elution sequence. Bile acid 3-glucuronides can be eluted with 50% aqueous ethanol [175]. The concentration of alcohol in the washing solvent and of water in the eluting solvent must be kept low if losses of polar bile acid conjugates and incomplete elution of nonpolar compounds are to be avoided. Another cause of loss of nonpolar bile acids is precipitation in the urine [143], and it is advisable to keep the pH of the sample around 7–8 [161].

Extraction of serum and plasma requires special conditions because of the inhomogeneous nature of the sample with presence of lipoproteins and protein binding of bile acids. Although this was considered in the method using Amberlite XAD-2 [106], it was not noted in the early studies of Sep-Pak C_{18} since taurocholate was the only model compound studied [160]. Lithocholic acid and its conjugates are most firmly bound to proteins, and less than 10% of these compounds are extracted [162] if plasma is only diluted with water [160]. Based on a systematic study, Setchell and Worthington [162] recommended a fourfold dilution with 0.1 M sodium hydroxide prior to extraction with Bond-Elut [161], similar to the procedure with Amberlite XAD-2 [106]. Yields of lithocholic acid and its sulfate were 85–90% with this method [162]. Dilution with different amounts of sodium hydroxide has been used by several authors [165,176–179]. Recoveries of lithocholic acid and its conjugates were lower than those of other bile acids in one study [176], but not in another [177]. A selective washing sequence and elution with 75% aqueous methanol was used in the latter study [177]. In contrast, recoveries of only 65% of common conjugated bile acids have been reported [180]. In this case 0.4 M ammonium bicarbonate gave better yields [180]. The reasons for these variations between studies are not known.

Dilution with phosphate buffer, 10 volumes 0.5 M, pH 7 [181,182], four volumes 0.1 M, pH 8 [183,184], 10 volumes 0.01 M, pH 7–8 [163], and undefined [185,186], has been used by several authors. In most studies internal standards labeled with stable isotope have been added, and only relative recoveries have been calculated [183]. When absolute extraction efficiencies were determined with Sep-Pak C_{18}, only 50% of the sulfates of lithocholic acid and its glycine and taurine conjugates were found to be extracted [182]. Dilution with 19 volumes 0.1 M Tris-HCl buffer pH 8 gave recoveries of 82–96% for the common di- and trihydroxy bile acids, but lithocholic acid and its conjugates were not studied [187].

Combined solvent–sorbent methods have also been used for extraction and desalting of plasma [39,40,44,188,189]. An extract obtained with acetonitrile [44,188] or methanol [39,40] is concentrated, resuspended in water [44,188] or phosphate buffer [39,40], and passed through the sorbent bed. Recoveries with these methods have usually been better than 85% for the bile acids studied [39,44], but taurine conjugates may be partly lost [40], and sulfates and glucuronides have not been studied.

Ruben *et al.* mixed serum with eight volumes of a solution of about 0.1 M sodium hydroxide in 30% methanol and passed this mixture through a Sep-Pak C_{18} cartridge [189,190]. After washing with water and 10% acetone, the bile acids were eluted in the usual way. Recoveries were usually better than 90%, but sulfates and glucuronides were not studied and the latter would probably be lost.

A systematic study of conditions for quantitative extraction of all types of bile acids and their conjugates has yet to be performed. It is probable that none of the methods just reviewed is satisfactory in all cases. In a study of steroids ranging from progesterone to estriol glucuronide and androstenediol disulfate, Axelson and Sahlberg considered the importance of ionic and nonpolar interactions with binding proteins and lipoproteins in plasma and milk as well as the properties of the sorbent [156]. They concluded that dilution with an equal volume of 0.5 M triethylamine sulfate, pH 7, and extraction at 64°C were necessary conditions for nearly quantitative recovery of all the steroids tested. The triethylamine sulfate was needed to decrease ionic interactions, possibly by formation of ion pairs [155], while heating decreased nonpolar interactions with binding proteins [157,158]. Hedenborg and Norman compared various methods for extraction of a wide range of bile acids and conjugates and concluded that dilution with two volumes of triethylamine sulfate followed by extraction at 64°C gave the best recoveries [191]. However, details were not provided. The same method was used for extraction of bile acids in saliva, ultrafiltrate of serum, and isolated serum lipoproteins, and the yields of different added bile acid conjugates were better than 80% [192,193]. It should be noted that the sorbent may retain a small amount of triethylamine even after the wash with water, and this may interfere in subsequent steps of an analysis, e.g., enzymatic hydrolysis [156]. However, it does not affect the group separation by ion exchange [191–193]. Although Sep-Pak C_{18} has been used for extraction of bile acids in amniotic fluid [160], possible protein binding has not been considered.

Bile acids in bile have been extracted with octadecylsilane-bonded silica after dilution with 20–50 volumes of phosphate buffer [39,40,65,161,194], in some cases with heating [150] or addition of mobile phase with sodium hydroxide and methanol [189,190]. In many cases the desalting and purification obtained by the solid-phase extraction is not needed, and it is simpler to add bile to ethanol and use an aliquot for further separation and analysis. Both procedures may be used for extraction of bile acids in duodenal and small-intestinal contents. In this case the heterogeneity of the sample "solution" must be considered, and the conditions have to be adjusted to suit the particular sample and compounds to be analyzed.

Octadecylsilane-bonded silica has been used as a desalting stage in combined solvent/solid-phase extractions of tissues and feces [74,93,166]. The problem in this application is to get the solvent extract into a suitable form in an aqueous medium. Dispersion of the dried extract in acidic water is not satisfactory unless a combination of Lipidex 1000 and Bond-Elut is used for sorption [93]. This method gives incomplete recoveries of less polar compounds when the lipid content of the sample is high (unpublished results). Adequate yields are obtained if the extract is dissolved/dispersed in 50–60% aqueous methanol, passed through Sep-Pak C_{18}, and then evaporated to a water solution which is again passed through the sorbent bed [166]. This emphasizes the importance of avoiding micellar or other aggregates of the molecules to be extracted. An alternative method is to pass the aqueous suspension of the extract through the sorbent bed at 64°C [74]. This has yielded high recoveries of bile acids from ethanol extracts of rat liver [74]. When bile acids are extracted with aqueous detergent solutions, they may be recovered by passing the solution through Sep-Pak C_{18} after dilution to a point below the critical micellar concentration of the detergent [74].

As exemplified in many of the references just cited, extraction with octadecylsilane-bonded silica can be used at various stages of an analytical procedure. It may be used after enzymatic [194–196] or microchemical conversions [177,197] and is superior to solvent extraction when conjugated bile acids are studied. Solid-phase extraction is also superior to solvent extraction in synthetic work involving polar bile acid conjugates [151,198].

Extraction methods based on use of octadecylsilane-bonded silica permit simultaneous purification of the sample. This is achieved by selection of appropriate solvents for washing and subsequent elution of the compounds to be analyzed. Some examples, e.g., washing with aqueous acetone and elution with aqueous instead of pure methanol, are found in the references cited in this section. However, the possibility to use different solvent combinations has not been utilized to the full extent. For example, washing with hexane was found advantageous in extractions of plasma bile acids with Amberlite XAD-2 [106]. DeMark *et al.* have extended this principle with Sep-Pak C_{18} [177]. After sorption of serum bile acids from an alkaline solution, they washed the Sep-Pak cartridge with 0.1 M sodium hydroxide, water, hexane, water, and 40% aqueous methanol and eluted the bile acids with 75% aqueous methanol. The recoveries of common bile acids and their conjugates and sulfates (including those of lithocholic acid) were greater than 90%. Glucuronides would be lost in this procedure. Axelson used different combinations of washing and elution in a reversed-phase mode with aqueous methanol and a straight-phase mode with hexane and chloroform/hexane to purify extracted metabolites of vitamin D_3 [168]. A shift between reversed-phase and straight-phase solvents could be done by intermediate drying of the sorbent with a stream of nitrogen. Similar procedures have been used for other classes of compounds and might be useful also in bile acid analysis.

2.2.3. Lipidex Gels

Lipidex 1000 and 5000 (Packard Instr. Co., Downers Grove, IL) are lipophilic–hydrophobic derivatives of Sephadex LH-20 (Pharmacia Fine Chemicals, Uppsala, Sweden) containing 10% and 50% (by weight), respectively, of hydroxyalkyl groups with an average chain length of 14 carbon atoms (see Ref. 19). Beds of these gels in water behave as sorbents of compounds of low or medium polarity. The mechanism of sorption is different from those with Amberlite XAD resins and octadecylsilane-bonded silicas, and resembles that of a solvent extraction. Thus, polar conjugates and salts of bile acids are not sorbed from an aqueous solution [17,199]. Unconjugated bile acids are efficiently extracted from an acidified water solution [199]. Lipidex 1000 gives an extract with less polar contaminants than Amberlite resins and bonded silica, which makes it a convenient alternative to solvent extraction after alkaline hydrolysis [16,17,178,179,199,200]. The procedure is analogous to that with other sorbents: passage of the aqueous sample, washing with water, and elution with methanol. Systematic studies of solvents for washing and elution, aiming at simultaneous purification, have not been carried out. However, it is possible to shift between reversed-phase and straight-phase elution sequences with intermediate removal of solvent by a stream of nitrogen as used in analyses of other types of compounds [201].

Bile acid conjugates, sulfates, and glucuronides may be extracted from urine and plasma as ion pairs with decyltrimethylammonium [200]. This is the only counterion found to be effective, and it is added as a bromide to give a final concentration of 0.03 M in the sample. The probable mechanism is formation in the aqueous solution of ion pairs which are then extracted by Lipidex 1000 [200]. In addition to the ion-pairing effect, decyltrimethylammonium bromide counteracts the protein binding of nonpolar bile acids, and the yield of lithocholic acid from plasma is better than 90% provided the flow rate is not too high (0.5 ml/cm^2 column cross-section area) [200]. However, since the extraction depends on the polarity, ion pairs of glucuronides are less retained than other conjugates and can be eluted with an extended water wash. A reversed-phase separation of the ion pairs according to polarity can be achieved [74], but the separation efficiencies are obviously low compared to those obtained in HPLC-systems (this volume, Chapter 2). As mentioned in Section 2.1.1, the introduction of an ion-pairing agent is often a disadvantage. Decyltrimethylammonium bromide can be removed by passing an aqueous solution through a small bed of Sepralyte and washing with water, but this adds an additional step to the analysis [74].

Lipidex 1000 is useful for combined solvent/sorption extraction of fecal and tissue bile acids [93]. The solvent extract may be dispersed in water and passed through a Lipidex bed for extraction of sterols, fatty acids and unconjugated bile acids [93]. The effluent can be directly passed through Bond-Elut for sorption of the more polar conjugates. In an alternative procedure, used in steroid analysis, the solvent extract is evaporated *in vacuo* together with Lipidex 1000 [167]. When an appropriate solvent is used, lipid-soluble compounds are sorbed by the gel, which can be slurried in water and transferred to a small column of Lipidex 1000 packed in water on top of a Sep-Pak C_{18} cartridge. The combined beds can then be washed and eluted as required. In this system elution with 85% methanol will leave most of the phospholipids on the Sep-Pak cartridge, while steroids (and bile acids, see Section 2.2.2) will appear in the eluate [167].

3. PURIFICATION AND GROUP SEPARATION

Depending on the problem studied and the analytical method to be used, subfractionation of the extract is usually required, e.g., to decrease the load on sensitive analytical columns or other systems. The subfractionation is usually based on differences in polarity or charge of the compounds present. Extraction, partition, and adsorption chromatographic methods discussed in Volume 1 of this series [1] are still being used, although to a lesser extent. Extractive subfractionation (e.g., Ref. 25) is rarely used. Partition chromatography with two immiscible liquid phases has been replaced by chromatography on bonded phases. Early examples of such phases are the substituted Sephadex gels [202–205], which are useful for group fractionation, while the development of bonded silica phases and HPLC has increased the separation efficiencies for individual bile acids dramatically (Chapter 2). It is not possible to review all possible combinations of purification and group separation procedures used, and only examples are given. The classification of methods is made for convenience; some methods are under the wrong heading, and it is realized that the distinction between adsorption and partition chromatography is floating and that separation mechanisms are usually multiple.

3.1. Adsorption Chromatography

3.1.1. Thin-Layer Chromatography

A variety of solvent systems have been used to separate contaminating lipids from bile acids in biological extracts. The methods aim either at a minimum of separation of individual bile acids or at separation of groups of bile acids from each other and contaminating lipid classes. The former is exemplified by a system for separation of bile acids from triglycerides in plasma extracts using an alkaline solvent system [206]. Several systems have been described for group separation of unconjugated, glycine-, and taurine-conjugated bile acids from each other and from phospholipids, fatty acids, cholesterol, cholesterol esters, and triglycerides in extracts of bile and duodenal contents [207–211]. Unconjugated and glycine-conjugated bile acids overlap in some systems [208], and individual components within each group are more [207,210,212,213] or less [208,209,211] separated in the different systems. Systems for isolation of unconjugated bile acids as a group have also been studied [214].

In early methods for analysis of fecal bile acids Grundy *et al.* used TLC to isolate unconjugated bile acids as a group free of contaminating lipids (see Ref. 1 and this volume, Chapter 12). This principle has been used in many subsequent studies with various modifications, e.g., Refs. 96, 215–217. In some cases separation of isomeric bile acids are needed; one study used Eneroth's solvent mixtures (see Ref. 1) to separate 3α and 3β isomers of lithocholic and deoxycholic acids in feces [218]; another aimed at group separations according to number of oxygen substituent [219]. TLC has also been used for purification of bile acids in extracts of tissues and serum [54,66,67,69,76]. The results of some studies [66,67] indicate that tissue extracts are too crude for direct purification by TLC.

Little attention has been paid to the occurrence of sulfated and glucuronidated bile acids when TLC is used for purification of biological extracts. However, detailed studies of systems for separation of groups and individuals of sulfated bile acids and their glycine and taurine conjugates have been performed by Parmentier and Eyssen [198] and by Hofmann and co-workers [220,221]. The former study [198] is very systematic and covers a large number of isomeric sulfated bile acids. A new reversed-phase system based on octadecylsilane-bonded silica gel is described by the latter authors, who also study a wide range of bile acid sulfates [221]. Complete separation of 3- and 7-monosulfates of cholic and chenodeoxycholic acids and their conjugates were achieved [198,221]. Reversed-phase C_{18} plates have also been evaluated for separation of nonsulfated bile acids, and a two-dimensional development with a nonaqueous solvent (straight-phase) in the first direction followed by an aqueous solvent (reversed-phase) in the second direction gave excellent separations of biliary bile acids [4].

Three more recent TLC techniques have been applied to bile acid analysis. One involves the use of Chromarods (Iatron Laboratories, Tokyo, Japan), i.e., silica gel coated on glass rods [149]. This is an analytical rather than a purification technique, and the resolution is no better than with regular TLC. In contrast, better resolution is obtained with high-performance TLC [3]. After some corrections and modifications [2], this method may be simple and rapid for routine analysis of bile. Its possible use for preparative group separations in the ng–μg scale remains to be studied. Overpressure-layer chromatography (Labor MIM, Budapest, Hungary), which shows analogies with column HPLC, also has a higher separation efficiency than TLC [5,222] and shorter analysis times. Its use for micropreparative purposes has to be evaluated.

Although TLC is frequently used, it cannot be recommended as a general method for group separation and purification of bile acids in biological extracts. Exposure to oxygen and dehydrating conditions on the silica surface can lead to chemical transformations; quantitative extraction of polar, strongly adsorbed compounds without use of solvents that partially dissolves the silica is difficult; and contamination from the silica gels, which adsorb a variety of compounds from the surrounding, is a problem. A second TLC of eluted zones on silica gel-sintered plates has been used to overcome the latter problem in analyses of bile acid methyl esters by infrared spectroscopy [223]. Suspension of the eluted material in buffer and extraction with Sep-Pak C_{18} was used in a method for fecal bile acid analysis [96].

3.1.2.   Silicic Acid and Aluminum Oxide

The use of these adsorbents in different forms in columns (see Ref. 1) has decreased with the development of bonded phases. However, new and modified methods have been developed for group separations based on polarity and charge. Alumina chromatography was used to purify bile acid methyl esters in analysis of serum and tissues [75,84,102,103,108], and modifications were made to include methyl lithocholate [75,84,103]. Alumina continues to be used for frac-tionation of mixtures of methylated bile acid metabolites [127,130]. Protonated alumina has been used extensively by Haslewood and co-workers for group isolation of anionic substances from bile [224,225].

Silicic acid can be used for group separation of unconjugated, glycine-, and taurine-conjugated bile acids [226]. Different systems were tested, mostly con-sisting of mixtures of water-saturated chloroform and ethanol and with added acetic acid for group separation of conjugates. Partial separation within groups was also obtained. The study confirmed that silicic acid chromatography is useful for purification of synthetic sulfated bile acids from reaction mixtures [227]. Although the fraction volumes in this method are large [226], an adaptation using Sep-Pak Sil cartridges (Waters) and chloroform/ethanol/water/acetic acid mixtures of increasing polarity at 4°C gave excellent separation of unconjugated, glycine-, and taurine-conjugated bile acids in small volumes of solvent [44]. This is certainly a valuable alternative to ion exchange separations (see Section 3.4.2.), but it should be realized that the separations are based on polarity and information regarding the charge of compounds in different fractions is not obtained.

Unisil (Clarkson Chemical Company, Williamsport, PA), used for purifi-cation of solvolyzed steroid sulfates from plasma [45], is also useful for group separation of bile acid methyl esters according to polarity [26]. The benzene may be replaced by toluene or hexane/toluene mixtures. Unisil is a very consistent and clean preparation of silicic acid, and group separation of methyl esters is preferable to separation of the free unconjugated acids [98]. Purification of methyl or ethyl esters has also been performed on silica gel in hexane/ethyl acetate and diethyl ether/methanol mixtures [54,70,80]. This was used in analyses of bile acids in serum [54], liver [70], and colon [80].

3.1.3. Other Adsorbents

Amberlite XAD-2 has been used in ethyl acetate for purification of methylated bile acids in extracts of feces [228]. The resin adsorbs pigments, and subsequent purification by TLC gives higher yields of the bile acid esters. The mechanism is unknown, and the authors call Amberlite XAD-2 an ion exchanger, indicating a misunderstanding of the nature of the resin [228].

Javitt and co-workers have reported on glycophase G bonded to controlled pore glass (Pierce Biochemicals) for group separation of mono-, di-, and trihydroxy bile acid methyl esters [229,230]. The solvents were hexane/ethyl acetate with an increasing proportion of the latter. Sulfated bile acid esters required ethyl acetate/methanol (1:1, v/v) for elution.

3.2. Partition Chromatography

3.2.1. Sephadex LH-20 and Lipidex

These are substituted dextran gels, essentially neutral and differing in polarity [19,205].

Sephadex LH-20 (and earlier methylated Sephadex) forms straight-phase systems in mixtures of organic solvents [202–205]. Separations are combinations of partition, adsorption, and molecular sieving, the predominant mechanism depending on the solvent used. Group separations according to polarity can be achieved with hydrocarbon/chloroform/ethanol mixtures [202–205,231]. Acetic acid is added when acids are to be separated [232]. These systems are only suitable for unconjugated bile acids and their esters. They have been used for subfractionation of methyl esters of urinary bile acids [132] and microbial metabolites of lithocholic acid [233–235].

Lipidex 1000 and 5000 are more widely applicable to separation of unconjugated bile acids than Sephadex LH-20 since they can form both straight-phase systems in nonaqueous solvents and reversed-phase systems in aqueous solvents [205,236]. A group fractionation of mono-, di-, tri-, and tetrahydroxy bile acid methyl esters can be performed with Lipidex 5000 in hexane/chloroform in proportions 1:4 (mono-), 3:7 (di-), 1:1 (tri-), and final elution with methanol (tetrahydroxycholanoates) [26,139]. This is a convenient method when the separation within groups should be minimal. However, solvent volumes are large, and modifications of the recent Sep-Pak Sil systems [44] are likely to become superior.

The inertness of Lipidex gels makes them suitable for purification of labile derivatives. Trimethylsilyl ethers and reaction mixtures with trimethysilylimidazole/methoxyamine hydrochloride can be rapidly purified on small columns of Lipidex 5000 in hexane containing low percentages of pyridine, hexamethyldisilazane, and in some cases dimethoxypropane [157,158]. The usefulness of this method has not been adequately appreciated in bile acid analysis.

Lipidex 1000 and 5000 form reversed-phase systems in aqueous methanol with or without addition of 10–20% chloroform or ethylene chloride [205,236]. Such systems also permit group separations according to polarity and have been used for this purpose in studies of biliary [23] and fecal [93] bile acids, for separation of bile alcohols [141] and methyl esters of bile acid glucuronides [140]. The choice of system depends on the problem. Lipidex 1000 in methanol/water/chloroform/n-butanol has to be used for the more polar compounds [140,141]. Lipidex 1000 or 5000 in aqueous methanol with or without addition of chloroform is suitable for removal of sterols from bile acids and hormonal steroids [44,93,157,167].

Sephadex LH-20 has been extensively used for group separation of sulfated from nonsulfated bile acids. Chloroform/methanol is the solvent, containing about 0.01 M NaCl or KCl [203,237]. The mobility of the sulfates is determined by the chloroform/methanol ratio and the nature of the cation. Sulfated steroids and bile acids are strongly retained compared to nonsulfated compounds, more so with increasing concentration of chloroform. The mechanism seems to be a liquid–gel partition of the undissociated salts, potassium salts being more polar and retained than sodium salts [203,237]. Addition of water destroys the system. The need for the inorganic salt in the solvent depends on the presence of a low concentration of anionic sites on the gel [203,237], and the salt is not needed when the sample is large enough to provide an excess of cation. This also means that different salt forms of sulfated steroids can be separated from each other [203,237]. It has been claimed that the method does not work with conjugated lithocholic acid sulfate because of the low polarity of this compound. This is not correct, as is clearly seen from the separations obtained with conjugated monohydroxycholanoic acid sulfates in rat bile [238]. The reasons for the disagreement may be found in one or several of the important variables discussed above.

The use of the method can be exemplified by the original [238] and other [24,26] applications to the analysis of bile, urine [108,118,128–131,135,239–241], serum or plasma [36,38,107,109,117,126,133,241], meconium [100,101], mouse feces [95,242], and in synthetic work [151,198,227]. Sodium salts have been used in most studies, but Norman and Strandvik preferred potassium salts [128–130]. More detailed separations of the conjugate classes can be obtained by elution with larger volumes of chloroform/methanol before changing to methanol [239] or by starting elution with a higher content of chloroform in the solvent and decreasing it stepwise [117,130,135,242]. An overlap between taurine-conjugated bile acids and sulfated unconjugated bile acids [38], not observed in the early studies, can be avoided in this way [117]. Although the separation within groups is small in some systems [135], this is not always the case [38], and even the position of the sulfate group influences the mobility [38]. This indicates that solvent and salt should be carefully selected depending on the problem studied. Furthermore, there is a slight memory effect (1–2%) [26,95,117,133,241], which may give errors in analyses of samples when the amount of nonsulfated bile acids is much higher than that of sulfated ones [26]. However, judging from the detailed studies by Norman and Strandvik [128–130], using urine containing endogenously radiolabeled bile acids, both separations and recoveries of different forms of bile acids can be very satisfactory.

Bile acid monoglucuronides have a similar mobility as the monosulfates [133,240]. This and a number of other factors discussed earlier indicate that chromatography on lipophilic ion exchangers is to be preferred to Sephadex LH-20 chromatography in most analytical applications. In some cases aiming at analysis of selected bile acid sulfates and in synthetic work, the latter method may be simpler and more rapid for group separation. In addition, Haslewood and co-workers found that chromatography on Sephadex LH-20 in chloroform/methanol, 1:1, without salt was a simple and quick method to obtain a purified bile salt fraction from bile of different species [225,243].

3.2.2. Bonded Silica Phases

These are the phases used in HPLC and, in coarser grades, for extraction purposes. HPLC is described in this volume, Chapter 2, and purification in connection with the extraction step was discussed in Section 2.2.2. It can be safely assumed that automated purification procedures for bile acids and related compounds are going to be developed based on these phases and column-switching methods. Development of fully automated procedures in drug [244] and steroid [245] analysis has been more rapid because of the need for greater degrees of purification and larger number of analyses in these areas. Automated sample processors based on the use of neutral and charged sorbents of the Bond-Elut and Sep-Pak types are already commercially available. The micro-HPLC system of Ishii et al. [163] constitutes a beginning to an automated system for bile acids.

3.3. Electrophoresis

Since the different classes of conjugated bile acids differ in acidity, electrophoresis should be an attractive method for group separation. However, there are very few applications of this technique in bile acid analysis. Kaplowitz and

Javitt separated unconjugated, glycine conjugates, taurine conjugates, and the monosulfates of the conjugates into five groups on paper strips in a pyridine/acetic acid buffer, pH 3.7 [6]. The method was applied to the analysis of duodenal fluid, which could be applied directly on the starting line. Beke *et al.* [31] used cellulose acetate strips and a similar buffer, pH 2.2. This system gave bands of glycine and taurine conjugates with separation of dihydroxy and trihydroxy acids within each group [7]. The bands could be readily eluted and the bile acid content determined with a 3α-hydroxysteroid dehydrogenase method. More recently, analysis of biliary bile acids by capillary isotachophoresis in 95% methanol was described [8]. The glycine and taurine conjugates of dihydroxy and trihydroxy bile acids could be determined separately after injection of the methanolic eluate from a Sep-Pak extraction of bile diluted with phosphate buffer.

3.4. Ion Exchange

Separations of bile acids according to mode of conjugation were originally performed on ion exchangers based on cross-linked polystyrene matrices (Dowex 1). This work was reviewed in detail by Kuksis in Volume I of this series [246]. Adsorption of amphiphilic compounds to the matrix and incompatibility with organic solvents are the main problems with these ion exchangers. Large volumes of solvent containing hydrochloric acid are needed to elute the bile acids, and there is considerable tailing of the peaks [246]. One modification of the original system has been published, which does not solve these problems [247]. The authors referred to modified Sephadex LH-20, apparently not knowing that it was an ion exchanger [247].

Anion exchangers based on a macroreticular, rigid polystyrene matrix, more compatible with the use of organic solvents (Amberlite A-26 and A-27), were used to obtain an acidic fraction from serum and tissue extracts [75,84,102–104]. Also, in this case large solvent volumes were needed for elution, and group separations were not attempted.

The synthesis of lipophilic ion exchangers based on a cross-linked, hydroxypropylated dextran matrix (Sephadex LH-20) [19,205] has greatly improved group separations of bile acids. Ion exchange occurs rapidly in aqueous ethanol or methanol (usually about 70% alcohol), and the matrix shows no adsorptive properties toward the amphiphilic compounds in this solvent. The syntheses of ion exchangers are simple, and material sufficient for a year's use can be prepared in 2 days. Both weak and strong cation and anion exchangers can be synthesized, and the polarity of the matrix can be varied depending on the need for additional separation effects [19]. The latter possibility has not been utilized in bile acid separations, but has been found useful, e.g., in steroid analysis [19,158].

3.4.1. Cation Exchangers

In the original use of anion exchange chromatography for group separation of bile acids [135,246] the sample was first taken through a cation exchanger. This had two objectives: (1) to achieve purification and (2) to exchange cations that may form ion pairs with bile acids. Dowex 50 was used in early studies [246] and then Amberlyst A-15, a strong cation-exchanging macroreticular resin [135,248,249], more compatible with the use of organic solvents. Amberlyst A-15 has been extensively used in bile acid analysis [29,30,94,99,121,135–140,142, 145,148,173]. Since losses of steroid conjugates have occasionally been observed with use of Amberlyst A-15 [249], other cation exchangers have been tested. Sulfoethyl and carboxymethyl Sephadex LH-20 were studied in steroid analyses, and only the strong cation exchanger (sulfoethyl) proved satisfactory prior to separation on a weak anion exchanger, DEAP-LH-20 (see Section 3.4.2) [249]. Carboxymethyl Sephadex LH-20 was used prior to separation on PHP-LH-20 (see Section 3.4.2) [144]. A comparative study of Amberlyst A-15, SE-LH-20, and SP-Sephadex C-25 (Pharmacia Fine Chemicals, Uppsala, Sweden) in steroid analysis showed that SP-Sephadex C-25 (in H^+-form) permitted use of the smallest columns, 0.4 g ion exchanger for the equivalent of 20 ml urine extracted with Amberlite XAD-2 [250]. A similar study has not been performed in bile acid analysis, but SP-Sephadex has been used in all recent methods in our laboratories [93,141,169–171]. A lipophilic strong cation exchanger, sulfohydroxypropyl Sephadex LH-20 (SP-LH-20), has been preferred in analysis of fecal bile acids since it permits use of less polar solvents [166].

The early studies of the need for a cation exchanger prior to group separation on DEAP-LH-20 were made with biological samples extracted with Amberlite XAD-2 [135,249]. Hedenborg and Norman have reported that a cation exchanger is not needed in analyses of serum and other fluids [191–193]. This may be due to the use of octadecylsilane-bonded silica rather than Amberlite XAD-2 for extraction. The two sorbents do not give identical extracts, and it has been shown that this can affect recoveries in subsequent purification steps in ultramicroanalyses of estradiol [251]. Thus, the need for a cation-exchanging step in a bile acid analysis is determined both by the nature of the extract and by the anion exchanger used for the subsequent group separation (see Section 3.4.2). Finally, it should be pointed out that the use of a strong cation exchanger in acid form may catalyze formation of artifacts, e.g., esterification, hydrolysis of esters, and other transformations of acid-labile structures.

3.4.2. Anion Exchangers

With few exceptions the group separations are carried out on one of two predominant ion exchangers: diethylaminohydroxypropyl Sephadex LH-20 (DEAP-LH-20 or Lipidex-DEAP, Packard Instr. Co. Downers Grove, Illinois) [135,249] or piperidinohydroxypropyl Sephadex LH-20 (PHP-LH-20) [147,150]. The former is a weak anion exchanger, the latter a stronger one. Both have been used in the acetate form. For unknown reasons, Goto *et al.* were unable to separate unconjugated glycine- and taurine-conjugated bile acids into the three groups on DEAP-LH-20 [135] and therefore synthesized PHP-LH-20 [147].

Solvent composition, buffer salts, pH, and ionic strength will obviously influence the chromatographic behavior of bile acid conjugates. In the original method with DEAP-LH-20, used for analysis of a desalted extract of 25 ml urine, a column bed of about 250 $\times$ 0.4 mm containing 0.6 g and about 1 meq of ion exchanger was used in 72% aqueous ethanol. Following passage of neutral compounds, unconjugated bile acids were eluted with 7.4 ml of 0.1 M acetic acid, glycine conjugates with 16.5 ml of 0.3 M acetic acid/ammonium hydroxide (HAc/Am), apparent pH 5.0, taurine conjugates with 11 ml 0.15 M HAc/Am, apparent pH 6.6, monosulfates with 11 ml 0.34 M HAc/Am, apparent pH 7.6, and di- and trisulfates with 13 ml HAc/Am, apparent pH 9.6, all in 72% ethanol [135]. The column size can often be decreased in analytical applications, and buffer volumes are then decreased proportionally. Two important factors have to be considered: (1) only 15–20% of the ion-exchanging sites are sterically available to bile acid molecules [135] and overloading can occur; (2) the pH of the solution applied should be about 7 to avoid elution of weak acids among neutral compounds and subsequent disturbances of the separations.

When information about conjugation is not required, the biological sample may be hydrolyzed, extracted on Sep-Pak C_{18}, and the unconjugated bile acids isolated by chromatography on DEAP-LH-20 [120].

The original scheme has been used with slight modifications, depending on the needs and possible batch variations, in many studies, e.g., Refs. 29,93,94,99,101,120–123,136–140,173,174,252. Bile acid glucuronides appear partly in the taurine conjugate but mainly in the monosulfate fraction [136,140]. Hedenborg and Norman have used rechromatography with stepwise increase of pH and ionic strength to separate glucuronides and sulfates [173]. In a recent study they describe a new elution scheme which gives a separate glucuronide fraction with very little overlap into the preceding taurine conjugate and subsequent sulfate fractions [191]. The glucuronides are eluted with 0.1 M formic acid in 72% ethanol, as adapted from previous methods for isolation of steroid and bile alcohol glucuronides [141,169–171,249]. This method for isolation of glucuronides was also used by Stiehl and co-workers, who first eluted unconjugated bile acids and glycine and taurine conjugates with 0.1 M sodium acetate in 70% methanol, apparent pH 7 [142,145]. It should be noted that the separations are somewhat different in 70% methanol and 72% ethanol, probably depending on differences in dissociation of buffers and bile acids in the two solvents. Judging from the chromatograms presented, 70% methanol may be the preferred solvent for isolation of glucuronides [142]. This is the solvent used in the separation of unconjugated bile alcohols and their glucuronides and sulfates [141,169–171] in a manner analogous to that used for metabolites of hormonal steroids [249]. In the group separation of bile alcohols and their conjugates, the glucuronide and sulfate fractions will obviously be contaminated by bile acid conjugates. The latter are removed on rechromatography on DEAP-LH-20 after hydrolysis/ solvolysis to produce the neutral bile alcohols [141,169–171].

The yields in group separations on DEAP-LH-20 (Lipidex-DEAP) are usually better than 90–95% as determined by addition of radiolabeled bile acids or in work with samples containing metabolites of administered labeled bile acids [93,94,99,121,135,139]. Samples of the latter type have been carefully studied from a methodological point of view by Hedenborg and Norman [82,173,174, 191–193]. These authors have also made comparative studies of fractions from DEAP-LH-20 using radioimmunoassay, 3α- and 7α-hydroxysteroid dehydrogenases, and gas chromatography as analytical methods [253,254].

The yields of bile acids in group separation and purification steps are also determined by the materials used in column systems and other equipment. We use only glass and Teflon, and glass surfaces should usually be siliconized to prevent adsorptive losses [44,158,191,250]. This is particularly important as the degree of purification increases and when the bile acid concentrations are low. The contaminating compounds act as carriers and prevent the surface adsorption. As they are removed, adsorption, e.g., to glass, increases and recoveries decrease.

PHP-LH-20 has been used for group separation of bile acid and bile alcohol conjugates in an analogous way to DEAP-LH-20 [30,35,39,40,96,144,147, 150,161,164,175,176,181–186,196]. With few exceptions [30,144] the methods do not include a preceding cation exchange. However, most studies were performed with bile or extracts obtained with Sep-Pak C_{18} when a cation exchange step does not seem to be required. The only method for urine extracted with Amberlite XAD-2 utilized a preceding column of carboxymethyl Sephadex LH-20 [144].

Following the original report of the separation of unconjugated, glycine-, and taurine-conjugated bile acids in bile on PHP-LH-20 [147], a detailed method for group separation including sulfates was described [150]. A 20×6 mm column bed of 110 mg PHP-LH-20 in acetate form was used in 90% ethanol. Following passage of neutral compounds, unconjugated bile acids were eluted with 4 ml 0.1 M acetic acid, glycine conjugates with 4 ml 0.2 M formic acid, taurine conjugates with 4 ml 0.3 M acetic acid/potassium acetate, apparent pH 6.3, all in 90% ethanol [147]. Sulfated bile acids were finally eluted with 4 ml 1% ammonium carbonate in 70% ethanol similar to an eluent previously used with DEAP-LH-20 [135]. The system of Goto *et al.* [147,150] has been used in most other methods employing PHP-LH-20. Sodium carbonate has replaced ammonium carbonate in one method [182].

With one exception (analysis of urinary bile acids using Amberlite XAD-

2 extraction [144]) recoveries of bile acid derivatives from PHP-LH-20 have been better than 90–95% [30,39,40,96,147,150,161,176,182].

Bile acid glucuronides are distributed between the glycine-conjugated, taurine-conjugated, and sulfated bile acid fractions [175]. The entire group can be eluted with 0.5% ammonium carbonate following elution of unconjugated bile acids with 0.1 M acetic acid in 90% ethanol [175]. This method is not a comprehensive group separation but is intended only for glucuronides. Some loss of nonglucuronidated bile acids is also likely, since 50% methanol is used for elution in the Sep-Pak extraction [175]. When information about glycine and taurine conjugation of glucuronidated and sulfated bile acids is not required, hydrolysis with cholylglycine hydrolase followed by separation of unconjugated, glucuronidated, and sulfated bile acids on PHP-LH-20 is achieved by sequential elution with 0.1 M acetic acid, 0.4 M formic acid (both in 90% ethanol), and 1% ammonium carbonate in 70% ethanol [183–185].

PHP-LH-20 has also been used for group separation of unconjugated and glycine- and taurine-conjugated bile acids after their conversion into 3-(1-anthroyl) derivatives for fluorescence detection [181]. The same eluting solvents could be used as for the underivatized compounds.

PHP-LH-20 has been used in analogy with DEAP-LH-20 for group separation of bile alcohols and their conjugates [35,164]. After collection of neutral compounds with 90% ethanol, glucuronides are eluted with 0.2 M formic acid in 90% ethanol and sulfates with 1% ammonium carbonate in 70% ethanol. Obviously, glycine-conjugated and glucuronidated bile acids will be present in the glucuronide fraction and taurine-conjugated and sulfated bile acids in the sulfate fraction [35]. This overlap is resolved after hydrolysis/solvolysis and rechromatography on PHP-LH-20, as in the methods using DEAP-LH-20.

A few other ion exchangers have also been used. When DEAP-LH-20 (Lipidex-DEAP) is used in 70% methanol, some tailing of nonpolar lipids, e.g., cholestanones and cholestenones, into the fraction of unconjugated bile acids may occur [255]. This was avoided by use of triethylaminohydroxypropyl Sephadex LH-20 (TEAP-LH-20) [166], which permits use of less polar solvents [19,166]. The ion exchanger was used in bicarbonate form, which retains carboxylic acids but not phenols and is less likely to produce artifacts than the free-base form. After elution of neutral lipids with 90% methanol and chloroform/methanol, unconjugated bile acids were eluted with 0.15 M acetic acid in 95% methanol or with methanol saturated with CO_2. A column bed size of 80 × 4 mm is suitable for most analytical purposes.

DEAE-Sepharose CL-6B in acetate form in 72% ethanol permitted separation of unconjugated, glycine-, and taurine-conjugated bile acids [148]. Unconjugated bile acids appeared mainly in the neutral fraction, which makes this system less useful. Another weak anion exchanger, TSK Gel IEX 540 DEAE (Toyo Soda Manufacturing Co., Tokyo), has been used in HPLC systems for group separation of glycine and taurine conjugates [256]. The method is rapid and has the potential of being developed with column switching into an automated procedure.

Ion exchange chromatography is probably the best method for group separation of bile acids in analytical applications to biological materials. The fractions can be readily desalted using octadecylsilane-bonded silica and can then be analyzed by a variety of methods, e.g., HPLC, TLC, enzyme assays, radioimmunoassays, mass spectrometry. The group separation is needed for a complete analysis by HPLC [174]. Ion exchange is also an important method for purification of hydrolyzed and solvolyzed bile acids [135], as is evident from all applications using gas chromatography in the final analysis. Thus, the hydrolysis/solvolysis of bile acid and bile alcohol conjugates changes the charge or acidity of the compounds, which then appear in another fraction from the ion exchanger. The importance of this purification is readily appreciated in analyses of bile acids in urine of pregnant women where large amounts of steroid glucuronides and sulfates contaminate the conjugated bile acid fractions [135,137]. Other chemical transformations of the moieties containing charged groups can also aid in the purification of conjugate classes, as exemplified by the methylation of glucuronides followed by rechromatography on the anion exchanger [136,140].

3.4.3. Isolation of Ketonic Bile Acids

Lipophilic strong cation exchangers can be used for group separation of ketonic from nonketonic bile acids [166,255]. The method is based on the sorption of unsubstituted oximes of ketonic steroids in methanolic solution by the strong cation exchanger in acid form [257,258]. The ketonic bile acids are converted into oximes with 5 mg hydroxylammonium chloride in 50 μl pyridine, 60°C, 30 min and extracted with ethyl acetate or Sepralyte. They are separated on a 40 × 4 mm column bed of SP-LH-20 in H^+ form in methanol [255]. Nonketonic bile acids appear in the first 4 ml of effluent followed by oximes of 7-oxo and 12-oxo bile acids between 8 and 18 ml. Oximes of 3-oxo bile acids are finally eluted with 4 ml of methanol/pyridine, 20:1. The yields from the column are better than 90–95%, and the separations are the same with free acids, methyl esters, and ketonic neutral steroids. The method is useful for gas chromatographic and mass spectrometric analyses of low concentrations of ketonic bile acids and sterols in the presence of large amounts of nonketonic compounds.

4. HYDROLYSIS AND SOLVOLYSIS

Although many analytical methods can be used for analysis of intact conjugated bile acids, glycine conjugates [188] and glucuronides [136,140,259] are the only conjugates that can be directly analyzed by GLC. For this reason and to permit detailed analyses of complex mixtures by GC/MS, it is often necessary to remove the conjugating moeity. This is a critical step in the analysis, and quantitative cleavage and absence of artifact formation cannot be guaranteed with any available method. A discussion of advantages and limitations of the most common methods seems justified, also because this step can provide additional purification of the sample.

4.1. Alkaline Hydrolysis

Conjugated bile acids cannot be subjected to acid hydrolysis without destruction of the nuclear structure [1,51,260]. Hydrolysis in 15% NaOH in 50% ethanol for 10 hr at 110–120°C was considered to give the best yields of unconjugated from amidated bile acids [1]. Although losses of ketonic bile acids had been observed [1], these were not thought to be as large as found in more recent studies [261–263]. The presence of ethanol results in reduction of an oxo group in the 3-position, but only to a minor extent for those in 7- or 12-positions [262]. It can be assumed that the vigorous conditions in alkaline ethylene glycol [67,95,118,127,242] will create similar artifacts. Other causes of loss also seem likely since the yields of 3-oxo bile acids were good after hydrolysis in aqueous alkali in one study [262] but not in another [263]. Also, the low yields of hydroxy acids in one study using aqueous alkali [261] are not confirmed by the other studies [262,263]. In addition to the reduction of 3-keto bile acids, formation of ethyl esters may occur when an ethanolic hydrolyzate is acidified for extraction of the bile acids [264]. This was more likely when the mixture was acidified to pH < 1 rather than to pH 3. In this case, chloroform/methanol was used for extraction [264], which could also have influenced the results since more ethanol would be present in the organic extract during evaporation than when diethyl ether or ethyl acetate is used.

It seems likely that differences in concentrations of alkali and ethanol, temperatures and times, solvents for extraction, and materials in the vessels used for hydrolysis [1,102] may explain differences in yields between the few systematic studies performed [261–263].

As a result of these studies, alkaline hydrolysis is now usually carried out in a nonalcoholic medium. The conditions vary: sodium or potassium hydroxide 1.25–4 M, temperature 110–120°C, time 3–12 hr, the longer times usually with the weaker alkali [34,36,38,43,46–48,51,54,55,70,71,81,97,108,114,115, 131,182]. The mildest conditions are used by Javitt and co-workers (1.25 M NaOH, 120°C, 3 hr), who also add EDTA to the hydrolysis mixture [54,55]. Similar conditions were used in a method for combined hydrolysis and solvolysis, although in this case the alkaline solution contained 25% methanol [265] (see Section 4.4). However, yields of bile acids may differ with the structures [261,262], and neither the degree of hydrolysis nor the stability of the nuclear or side-chain structures can be predicted for less common bile acids.

The conjugates of C_{27} acids require more drastic conditions for alkaline hydrolysis than those of C_{24} acids, probably because their carboxyl group is more hindered [225,266,267]. These conditions, e.g., 6.25 M aqueous NaOH, 116°C, 16 hr or 6.25 M NaOH in diethylene glycol 215°C, 25 min [267], result in epimerization at C-25 [266,267]. Alkaline hydrolysis also destroys side-chain hydroxylated C_{27} bile acids [225], e.g., varanic acid [118], probably causing dehydration. In contrast, 23-hydroxycholic acid was not epimerized during hydrolysis of its taurine conjugate in 1.8 M KOH at 135°C for 2 hr [268]. The yield of hydrolyzed bile acid was not given.

In addition to the destruction of sensitive structures and incomplete cleavage of conjugated C_{27} acids, alkaline hydrolysis will affect sulfated and glucuronidated bile acids. Sulfates may be stable toward conditions of ester hydrolysis [151,269], but elimination of sulfuric acid with formation of unsaturated bile acids occurs under more vigorous conditions [38,90,95,227,242,270,271]. However, there is not complete agreement on this point, since van Berge-Henegouwen *et al.* claim good recoveries of sulfated bile acids when solvolysis is preceded by a "mild" alkaline hydrolysis [265], similar to that used by Galeazzi and Javitt [271]. The other groups who have reported on formation of unsaturated artifacts have used more drastic hydrolytic conditions [38,95,227,242,270]. Ether glucuronides are stable toward conditions of ester hydrolysis [259], but some hydrolysis and considerable formation of by-products occur under the more drastic conditions needed for cleavage of amide bonds [183,271]. Ester glucuronides are obviously hydrolyzed by alkali [259]. If sulfates and glucuronides are to be analyzed, alkaline hydrolysis should not be the first hydrolytic step. In addition, the decomposition products from sulfates may give peaks with the same retention characteristics as other bile acids in the sample and disturb the analysis of nonsulfated bile acids [51,114].

4.2. Cholylglycine Hydrolase

This enzyme purified from clostridia was introduced in bile acid analysis by Nair and co-workers [272,273] and has been used by them in many studies (e.g., Refs. 76,77). The reaction is carried out in acetate or phosphate buffer, pH 5.6–6.0, in the presence of β-mercaptoethanol and EDTA [272]. The original method or modifications have been widely used (e.g., Refs. 40,44,52,79,81, 93,97,111,114,116,117,119,120,124,132,145,173,177, 178,183,185,197,198, 240,274). The variations are in amount and source of enzyme and in the time of incubation. Plasma can be directly subjected to hydrolysis either with a high concentration of enzyme [52,119] or with a long time of incubation [120]. The reasons why long incubation times (at least 12 hr [116]) are needed in many studies are not clear, but we have also found [93] that the yields of unconjugated bile acids from fractions of biological samples are lower after the shorter incubation times needed for quantitative hydrolysis of pure bile acid conjugates

[261,272,273]. The solubilities of some bile acids may be a problem, and the samples may be dissolved in a small amount of ethanol or methanol before the buffer and enzyme are added [76,81,132]. Another group found the hydrolysis of taurine conjugates to be difficult [40].

Cholylglycine hydrolase from clostridia shows a specificity for glycine- and taurine-conjugated C_{24} bile acids [261,272,275]. The activity toward lithocholyl-ε-L-lysine is controversial [76,81,310]. Following the failure of Yanagisawa *et al.* to achieve hydrolysis of this bile acid [81], Nair *et al.* described appropriate conditions for storage and use of cholylglycine hydrolase and indicated that the particular subspecies of enzyme, which is active on the lysine conjugate, is relatively unstable at temperatures above 0°C and requires storage at below −20°C to maintain its activity [310]. Conjugates of C_{25} and C_{27} acids are hydrolyzed poorly or not at all [267,275,276]. The enzyme is inactive toward neutral esters of the conjugated bile acids [272], and the presence of oxo groups decreases or abolishes activity [263,272,275]. Again, this may be due to solubility problems. Glyco- and tauro-tetrahydroxycholanoates in urine (having an additional hydroxyl group at C-1 or C-6) have been reported to be resistant to hydrolysis [145]; however, this is not confirmed by our own studies in which the glycine- and taurine-conjugated bile acid fractions isolated from the meconium and feces readily released a series of C-1 and C-6 tetrahydroxylated bile acids when hydrolyzed by cholylglycine hydrolase [311]. Sulfated [191,198, 227,270,277] and 3-glucuronidated [183,185,191,240] bile acids appear to be hydrolyzed, but quantitative studies of the latter are lacking. Glycine and taurine conjugates of common bile acids sulfated at C-3, C-7, or C-12 are quantitatively hydrolyzed with an incubation time of 16 hr [227,277].

A comparison of alkaline and enzymatic hydrolysis shows that neither is suitable in all cases. The methods are complementary and the choice depends on the analytical problem. If a screening of all bile acids is attempted, both methods should be used in parallel or in sequence. The most difficult bile acid conjugates are those of C_{27} bile acids, which either are not hydrolyzed or suffer decomposition [118,266,267,275,276]. In this case a rat fecal suspension may be used, as has been done for hydrolysis of tauro-trihydroxycoprostanic acid [267]. However, side reactions are likely to occur, depending on the conditions of incubation. Other bacterial preparations may be found useful, e.g., from *Pasteurella* [278].

4.3. β-Glucuronidase

Glucuronides of bile acids and bile alcohols have been hydrolyzed by different preparations of β-glucuronidase under conditions similar to those used for hydrolysis of glucuronides of neutral steroids [279]. A comparative study of enzymes from different sources has been performed [280]. Important findings were that pH optima and rates of hydrolysis depended both on the enzyme and on the number of hydroxyl groups in the steroid nucleus, but not on the presence or absence of a glycine or taurine moiety. This makes it difficult to obtain optimal conditions for simultaneous hydrolysis of a mixture of the 3-glucuronidated bile acids studied. However, a pH of 4.5 for *Helix pomatia, Patella vulgata,* and beef liver enzymes and of 6.5 for *Escherichia coli* enzyme was suggested. After 6 hr at 37°C, the hydrolysis of the glucuronides studied was almost quantitative with the two former, but only partial with the two latter enzymes. Stiehl and co-workers stated that enzyme preparations by *H. pomatia* and beef liver yielded the same results in analyses of serum, bile, and urine [281]. The incubations were performed with 1000 units of enzyme/ml at pH 4.5 and 37°C for 24 hr [240]. Similar conditions have been used by others [131,133,136,140,142,173,282], the pH varying between 4.0 and 5.0, the enzyme concentration between 1000 and 12000 units/ml, and the incubation time between 20 and 48 hr, sometimes with addition of more enzyme after 24 hr as used in steroid analysis [279]. One group reported that an incubation time of 2 hr was sufficient [183,185]; another used 6 hr [283]. As noted by a number of authors, beef liver preparations contain bile acids [131,133,136,240,281], which make them less suitable for studies in this area. Such preparations also partially hydrolyzed the amide bond in conjugated bile acids [175]. Methyl esters of bile acid glucuronides are not hydrolyzed as readily as the acids and should be subjected to alkaline hydrolysis prior to the incubation with enzyme [136].

Enzyme from *E. coli* has been used by other groups, and yields were reported to be good after 24 hr at pH 6.5–6.8 and 37°C with widely varying concentrations of enzyme (100–10000 units/ml) [175,284–286].

Bile acids may be glucuronidated at different positions, C-3, C-6, or in certain cases at the carboxyl group [140,175,259]. Enzymatic hydrolysis has only been tested with synthetic 3-glucuronides, but judging from analyses of biological samples, 6-glucuronides are also readily hydrolyzed [140]. When in doubt, chemical hydrolysis may be attempted, e.g., periodate oxidation [140], solvolysis [287], or other methods used in steroid analysis [279].

Bile alcohol glucuronides have been hydrolyzed by methods similar to those of other steroid glucuronides using incubation times of 20–48 hr and in most cases *H. pomatia* enzyme [35,141,169–172,288,289].

The formation of bile acid glucosides has recently been demonstrated with microsomal preparations from human liver [290]. The importance of this metabolic pathway *in vivo* is not yet known. The glucosides formed can be hydrolyzed with β-glucosidase from almonds, 100 units/ml, pH 5, 5 hr at 25°C [290]. The conjugate is not hydrolyzed by β-glucuronidase.

Enzyme preparations, particularly from *H. pomatia* and beef liver, contain a variety of low-molecular-weight contaminants which can disturb subsequent analysis. These can be removed if the final enzyme solution is passed through an Amberlite XAD-2, Sep-Pak C_{18}, or Sepralyte bed immediately before use [169,171]. This method is routinely used in our laboratories whenever enzymatic reactions are performed and the enzyme is stable to the treatment.

4.4. Solvolysis

In this context solvolysis means cleavage of sulfates and glucuronides by acids in a nonaqueous medium. The basic principles were established by Burstein and Liebermann [279,287,291], who proposed a reaction mechanism in which an unstable protonated species of the sulfate reacts with solvent molecules to give the free alcohol [291]. The factors of main importance are the solvent and the structure of the sulfate. In principle, the reaction rate increases with decreasing polarity of the solvent, but this effect is counteracted by the poorer solubility of the steroid sulfates in such solvents. For this reason alcohols are often added to the reaction mixture. This creates the problem of esterification of carboxyl groups. The position and orientation of the sulfated hydroxyl group is of great importance, the rate of solvolysis decreasing with increasing hindrance [291].

Several conditions have been used for solvolysis of bile acid sulfates: (1) acidified ethanol(methanol)/acetone, 1:9; (2) acidified ethyl acetate, with or without addition of 10% ethanol; (3) 2,2-dimethoxypropane/methanol/ethyl acetate/hydrochloric acid; (4) the methanol/diethyl ether phase after extraction from an acidified hydrolyzate containing 25% methanol in water; (5) acidified butanone; (6) dioxane/trichloroacetic acid; (7) acidified tetrahydrofuran.

Acidified ethanol/acetone has been widely used following Palmer's studies of lithocholic acid sulfates [270,292]. The conditions have been varied, often for no obvious reason. This may have affected the results and may explain differences between studies. The original procedure, dissolution in ethanol acidified to pH 1 or less with 2 M hydrochloric acid, addition of nine volumes of acetone, and incubation for 1–3 days at room temperature, has been widely used, e.g., Refs. 36,70,108,127,132,134,135,239,270,292. Short incubation times at 37°C have been used in some cases [71,126]. Solvolysis is followed by hydrolysis of the ethyl esters formed in 2–5% methanolic KOH at reflux temperature for 2 hr. Since esters of conjugated bile acids are poor substrates of cholylglycine hydrolase [272,275], omission of this step will lead to poor recoveries in a subsequent enzyme hydrolysis [173]. However, methods have been published where this problem has not been considered.

In a modified procedure [95,151,227,242] methanol/acetone, 1:9, acidified with 2 drops–0.1 ml 6 M hydrochloric acid, is used with incubation for 18 hr at 37°C, e.g., Refs. 38,93,95,115,116,151,178,182,183,185,227,242. Methylation of bile acids was not mentioned [151], but ester hydrolysis was performed by others [93,182,183]. Solvolysis of the 3-sulfate of cholic acid was complete in 3 hr [151,182] and that of the 12- and 7-sulfates in 6 and 18 hr, respectively

[151]. Di- and trisulfates were also solvolyzed [151]. The 7-sulfate of 7α-hydroxy-3,12-dioxo-5β-cholanoic acid gave an additional unsaturated compound [151], and formation of unsaturated by-products from the 7-sulfate of chenodeoxycholic acid was also noted with the same procedure [293]. The water content of the reaction mixture is probably of great importance [291], and Karlaganis and Paumgartner used dry methanolic hydrochloric acid/acetone and shorter incubation times [115,116]. However, a systematic study has not been carried out to determine the rates and yields in solvolysis of different bile acid sulfates. While the 3-sulfate of chenodeoxycholic acid is solvolyzed by both methods [151,270], the yield of chenodeoxycholic acid from its 7-sulfate is 30% with the Palmer [270] and 100% with the Parmentier–Eyssen [151] method [293].

Javitt and co-workers introduced a convenient solvolysis in 2,2-dimethoxypropane (DMP) [53]. When mixed with water and acid, this forms a methanol/acetone mixture. Different conditions have been used: nine volumes DMP per milliliter biological fluid acidified to pH 1 [53], 7.5 ml DMP, 1 ml fluid, and 0.1 ml concentrated hydrochloric acid, 1–3 hr at 20°C [271], and 4 ml DMP, 3 ml methanol, 3 ml ethyl acetate, and 0.1 ml concentrated hydrochloric acid, 3 and 24 hr at 25°C [293]. Only the latter conditions solvolyzed chenodeoxycholic acid 7-sulfate quantitatively [293]. Also, in this case small amounts of an unsaturated by-product are formed. Solvolysis in DMP has been used by several authors [43,55–57,117,124,177]. The treatment with DMP and acid can also result in methylation and formation of acetonides from *cis* α-glycol structures. The optimal conditions differ for the three types of reactions (see Section 4.5). An ester hydrolysis has to be carried out prior to a subsequent hydrolysis with cholylglycine hydrolase.

Solvolysis in ethyl acetate has been used extensively in steroid analysis [45,279,291,294,295]. The original procedure involved extraction with ethyl acetate of the aqueous solution of steroid sulfate acidified to 2 M sulfuric acid, followed by incubation for 48 hr [294]. Addition of sodium chloride aided in the transfer of the sulfate to the organic phase [294]. This method has also been used for solvolysis of bile acid [269] and bile alcohol [35,164,225] sulfates. However, more polar sulfates may remain in the aqueous phase, and addition of ethyl acetate equilibrated with 2–4 M sulfuric acid to 0.1 volume of an ethanol solution of the bile acid sulfates followed by incubation for 16 hr at 39°C has been more commonly used [35,81,135,137,164], in analogy with the application to polar steroid sulfates [295]. Addition of concentrated sulfuric acid may be needed in some cases [137]. Solvolysis in ethyl acetate results in ethylation and acetylation [135,269], and alkaline hydrolysis has to be performed prior to analysis or hydrolysis with cholylglycine hydrolase. For unknown reasons, acetylation is more extensive than when bile alcohols sulfated in the side chain are solvolyzed [225,269]. In analyses of urinary bile acids, a cleaner extract is obtained by solvolysis in ethanol/ethyl acetate than in ethanol/acetone [135].

However, the former is a milder method and 7-sulfates are poorly solvolyzed [293], which is a disadvantage.

A combined alkaline hydrolysis/acid solvolysis has been described [265]. The solvolysis occurs in acidified diethyl ether [279], and the bile acids are obtained in this phase by extraction of the acidified alkaline hydrolyzate with three volumes of ether [51,182,265]. The conditions of alkaline hydrolysis must be of great importance to ascertain complete hydrolysis with no formation of unsaturated bile acid from the sulfates (see Section 4.1). The presence of a defined percentage of methanol in the hydrolyzate is probably very important for the extraction of the sulfates into the ether phase. Thus, little or no solvolysis of different sulfates of deoxycholic, chenodeoxycholic, and cholic acids takes place in a mixture of 1 M hydrochloric acid and ether at 40°C for 5 hr [38,51,182]. Sulfates of the two former acids can be solvolyzed by a modified procedure in which a fraction of the codistillate of diethyl ether and concentrated hydrochloric acid is added to the sample in 0.1 volume of methanol [94,296]. Optimal conditions for solvolysis in ether probably differ with the polarity of the bile acid sulfate, and neither the original [265] nor the modified [296] method has yet been tested with polar sulfates such as those of cholic acid or glycine- and taurine-conjugated di- and trihydroxycholanoic acids. In the absence of such studies, it is advisable to use the more forceful single-phase systems for solvolysis even though these were claimed to give poorer yields [265].

Tetrahydrofuran with sulfuric or perchloric acid will solvolyze both steroid sulfates and glucuronides, depending on the conditions [287,291]. Reaction rates can be varied within wide limits by appropriate selection of water concentration and acid [287,291]. The rate constants for 3-sulfates of 3β-hydroxy-Δ^5 steroids are very high [291], and 15 min at room temperature in tetrahydrofuran/0.1 M sulfuric acid, 200:1, can be used for solvolysis of the 3-sulfates of the acid-labile 3β,7α-dihydroxy- and 3β,7α,12α-trihydroxy-5-cholenoic acid sulfates in fractions from urine [297]. Under these conditions solvolysis of 3β-hydroxy-5-androsten-17-one sulfate and recovery of 3β,7α-dihydroxy-5-cholenoic acid are better than 80%. In contrast, tetrahydrofuran/perchloric acid/water mixtures did not yield recoveries better than 36% of 7α-hydroxycholesterol in meconium [298]. Various conditions for solvolysis of endogenous steroid mono- and di-sulfates in urine have been studied, and tetrahydrofuran/4 M sulfuric acid, 1000:1, at 50°C for 1 hr gave high yields also in the case of difficult C_{19} steroid disulfates [299], but systematic studies for bile acid sulfates are lacking. Solvolysis of bile acid sulfates using tetrahydrofuran/1 M trifluoracetic acid dissolved in dioxane (9:0.1) after dissolution of the extract in one volume of methanol was recently described [312]. The reaction is relatively rapid (2 hr at 45°C) and is terminated by evaporation of the volatile reagents in a stream of nitrogen. Hydrolysis of glycolithocholic-3-sulfate was reported to be quantitative ($> 97\%$) with no ester or artifact formation. When this solvolysis technique was applied to serum samples and compared to the method of Kornel [295], a good agreement was obtained for the concentrations of the principal bile acids.

Other conditions have been less commonly used. Haslewood solvolyzed bile alcohols in acidified butanone and dioxane/trichloroacetic acid [266,276]. Also in these cases hydrolyzable by-products of unknown structure are formed [269].

As can be expected from the studies by Burstein and Lieberman [287,291,294], the conditions used for solvolysis of bile acid sulfates can give partial cleavage of glucuronides. The methods of Palmer [270] and Parmentier and Eyssen [151] gave about 10% and 60% solvolysis of tested bile acid glucuronides, respectively [183].

None of the comparative studies [38,51,93,126,135,182,265] are sufficiently complete or unbiased to permit selection of a "best" solvolysis procedure. The choice must be made depending on the problem. Although a systematic evaluation of the efficiency of the THF/TFA solvolytic method for a broad spectrum of bile acid sulfates has yet to be performed, our own experiences suggest it to be the most appropriate method at present.

Enzymatic hydrolysis provides an alternative to solvolysis. Hedenborg and Norman compared hydrolysis with enzyme preparations from *H. pomatia*. Using 120,000 Roy units/ml at pH 4.6 for 48 hr at 37°C, as much bile acid was released from sulfate fractions from urine as with solvolysis in acetone or ethyl acetate [173]. Bile acids sulfated at C-3 or C-7 were fully hydrolyzed in 15–20 hr, while hydrolysis of disulfates required 25–30 hr [300]. Pure reference compounds were hydrolyzed more rapidly than biological samples for which a 48-hr incubation was used [173,300].

Bacterial sulfatases have been studied by Eyssen's group [301,302]. Enzyme(s) from a *Clostridium* species was specific for an equatorial sulfate group at C-3 and was dependent on the length of the side chain [302]. Thus, bacterial sulfatases are presently of limited analytical use.

4.5.   Esterification

It is outside the scope of this chapter to review derivatization procedures. However, esterification of the carboxyl group is used so frequently and to improve the chromatographic behavior in purification steps that a brief discussion is justified.

Treatment with diazomethane has been a predominant method for methylation of bile acids [303,304]. However, side reactions may occur even with freshly distilled reagent, particularly with ketonic bile acids [303], but also with hydroxyl groups which may become methylated [25,305]. The side reactions are probably catalyzed by Lewis acids present in solvents [305] or the biological sample. For this reason, alternative methods have been studied.

2,2-Dimethoxypropane, previously employed in lipid analysis, was introduced by Javitt and co-workers for methylation of bile acids [53,54]. Original conditions were 2 ml methanol, 1.4 ml DMP, and 10 μl of conc. hydrochloric acid [54], later used in several methods [43,124,177]. A detailed study of conditions and yields has been performed using chenodeoxycholic acid as model compound [305]. Dissolving the sample in 50 μl methanol, 50 μl DMP, and 10 μl conc. hydrochloric acid results in essentially complete methylation in 3–16 hr with no visible formation of side products [305]. Similar yields were obtained of lithocholic and deoxycholic acid methyl esters. The *cis vic*-glycol structures of hyocholic and β-muricholic acids reacted to form acetonides. In our experience, conditions for optimal methylation and formation of acetonides differ, and the yield of acetonides [139] is often low under the conditions described for methylation. The latter have also been tested with 3-, 7- and 12-oxo bile acids and appear not to produce artifacts. However, difficulties have been experienced in obtaining quantitative methylation of 3-oxo-5β-cholanoic acid both with methanol-DMP-HCl and with diazomethane (Axelson and Sjövall, unpublished results).

Methylation and ethylation with the dry alcohol containing 4–5% hydrogen chloride has been extensively used by Miyazaki and co-workers [25,70,71,81,306] and others [36]. Side products are not formed from common bile acids [306]. Simultaneous extraction and methylation by use of methanol/hydrochloric acid has also been described [152]. Possible reactions between components in the biological material were not considered.

Esterification can also be catalyzed by strong cation exchangers in hydrogen form. Quantitative methylation of common bile acids occurred on a column of the lipophilic SP-LH-20 in 30–45 min [166]. However, later studies have shown considerable losses of ketonic bile acids (unpublished), and this method cannot be recommended for general use.

Although the acid-catalyzed esterifications appear to be best for common bile acids, they cannot be used for acid-labile compounds. Examples are 3,7-dihydroxy-Δ^4 and 3,7-dihydroxy-Δ^5 bile acids, whose allylic hydroxyl group will be rapidly methylated, epimerized, and eliminated. The extent of these reactions depends on conditions [307,308].

Permethylation has been used by some groups [252,309]. This results in methylation also of hydroxyl and possibly oxo groups. Conditions for quantitative methylation are demanding, and the reaction has only been studied with common bile acids. At present, it is probably best suited for special applications rather than as a general method.

As indicated by this discussion, no single method can be stated to be superior for esterification of all bile acids. The nature of the sample and the bile acids to be analyzed have to be considered in the choice of method.

5. CONCLUSIONS

Significant progress in techniques for isolation and group separation of bile acids has been made since the previous review was written for Volume 1 of this series [1]. The development has followed the lines suggested in the conclusion of that chapter. Thus, there have been major improvements in column liquid-liquid chromatography; new column-packing materials have been developed with specific groups covalently bound to inert matrices [1]. These have made extraction as well as group separation much simpler, more quantitative, and milder. The next decade will probably see automation of these and improved methods with on-line connection to HPLC, LC/MS, and LC/MS/MS systems. The number of forms in which bile acids occur has increased, and other forms are likely to exist. New bile acid structures continue to be found, and it is necessary to design future methods to include known and unknown structures. The ultimate importance of such detailed analysis of bile acids remains to be seen. The pathophysiological importance of steroidal compounds is not directly correlated to their concentration, as exemplified by the steroid hormones. Unusual structures of bile acids may well have important effects on the cell biology. The definition of such structures by use of improved analytical methods remains an important area of research.

ACKNOWLEDGMENTS.   This chapter would not have been completed without the assistance of Ms. Agneta Sjövall during the preparation of the manuscript. Work in the laboratory of J.S. was supported by the Swedish Medical Research Council (Grant No. 03X-219) and Karolinska Institutet.

REFERENCES

1. P. Eneroth and J. Sjövall, in "The Bile Acids" (P. P. Nair and D. Kritchevsky, eds.), Vol. 1, p. 121, Plenum Press, New York (1971).
2. T. A. Robb and G. P. Davidson, Ann. Clin. Biochem. 21, 137 (1984).
3. R. W. Shepherd, P. S. Bunting, M. Khan, J. G. Hill, S. J. Soldin, and D. G. Gall, Clin. Biochem 11, 106 (1978).
4. L. Lepri, D. Heimler, and P. G. Desideri, J. Chromatogr. 288, 461 (1984).
5. G. Szepesi, K. Dudas, A. Pap, and Z. Vegh, J. Chromatogr. 237, 127 (1982).
6. N. Kaplowitz and N. B. Javitt, J. Lipid Res. 14, 224 (1973).
7. R. Beke, G. A. De Weerdt, J. Parijs, and F. Barbier, Clin. Chim. Acta 71, 27 (1976).
8. J. C. Reijenga, H. J. L. A. Slaats, and F. M. Everaerts, J. Chromatogr. 267, 85 (1983).

9. C. A. Bloch and J. B. Watkins. *J. Lipid Res.* **19**, 510 (1978).

10. B. Gardner and M. S. Chenouda. *J. Lipid Res.* **19**, 985 (1978).

11. M. J. Whiting and A. M. Edwards. *J. Lipid Res.* **20**, 914 (1979).

12. C. A. Barth and K. Wirthensohn, *J. Lipid Res.* **22**, 1025 (1981).

13. A. Roda, S. Girotti, S. Ghini, B. Grigolo, G. Carrea, and R. Bovara, *Clin. Chem.* **30**, 206 (1984).

14. J. Sjövall, *in* "Liver and Bile" (L. Bianchi, W. Gerok, and K. Sickinger, eds.). p. 67, MTP Press, Lancaster, England (1977).

15. J. M. Street, D. J. H. Trafford, and H. L. J., Makin, *J. Lipid Res.* **24**, 491 (1983).

16. K. D. R. Setchell and A. Matsui. *Clin. Chim. Acta.* **127**, 1 (1983).

17. K. D. R. Setchell, *in* "Bile Acids in Gastroenterology" (L. Barbara, R. H. Dowling, A. F. Hofmann, and E. Roda, eds). p. 1, MTP Press, Lancaster (1982).

18. J. Sjövall, *in* "Advances in Bile Acid and Bile Alcohol Research in Japan" (S. Ikawa, H. Kawasaki, and N. Kaibara, eds.). p. 39, Tokyotanabesyoji, Tokyo (1984).

19. J. Sjövall and M. Axelson, *J. Pharm. Biomed. Anal.* **2**, 265 (1984).

20. P. Back, *Clin. Chim. Acta* **44**, 199 (1973).

21. S. Barnes, R. Waldrop, and S. Neighbors, *Anal. Biochem.* **133**, 470 (1983).

22. J. Bremer, *Acta Chem. Scand.* **9**, 683 (1955).

23. T. Cronholm, I. Makino, and J. Sjövall, *Eur. J. Biochem.* **24**, 507 (1972).

24. T. Laatikainen, J. Perheentupa, R. Vihko, I. Makino, and J. Sjövall, *J. Steroid Biochem.* **3**, 715 (1972).

25. H. Miyazaki, M. Ishibashi, M. Inoue, M. Itoh, and T. Kubodera, *J. Chromatogr.* **99**, 553 (1974).

26. S. Matern, J. Sjövall, E. W. Pomare, K. W. Heaton, and T. S. Low-Beer, *Med. Biol.* **53**, 107 (1975).

27. F. Kern Jr., H. Eriksson, T. Curstedt, and J. Sjövall, *J. Lipid Res.* **18**, 623 (1977).

28. C. H. Bolton, T. S. Low-Beer, E. W. Pomare, A. C. B. Wicks, J. Yeates, and K. W. Heaton, *Clin. Chim. Acta.* **83**, 177 (1978).

29. H. Eriksson, W. Taylor, and J. Sjövall, *J. Lipid Res.* **19**, 177 (1978).

30. K. Kuriyama, Y. Ban, T. Nakashima, and T. Murata, *Steroids* **34**, 717 (1979).

31. M. S. Sian and A. J. H. Rains. *Clin. Chim. Acta.* **98**, 243 (1979).

32. K. Maruyama, H. Tanimura, and Y. Hikasa, *Clin. Chim. Acta* **100**, 47 (1980).

33. F. Nakayama and M. Nakagaki, *J. Chromatogr.* **183**, 287 (1980).

34. M. Tohma, Y. Nakata, H. Yamada, T. Kurosawa, I. Makino, and S. Nakagawa, *Chem. Pharm. Bull.* **29**, 137 (1981).

35. S. Kuroki, K. Shimazu, M. Kuwabara, M. Une, K. Kihira, T. Kuramoto, and T. Hoshita, *J. Lipid Res.* **26**, 230 (1985).

36. C. B. Campbell, C. McGuffie, and L. W. Powell, *Clin. Chim. Acta* **63**, 249 (1975).

37. P. E. Ross, C. R. Pennington, and I. A. D. Bouchier, *Anal. Biochem.* **80**, 458 (1977).

38. J. F. Pageaux, B. Duperray, D. Anker, and M. Dubois, *Steroids* **34**, 73 (1979).

39. S. Kamada, M. Maeda, A. Tsuji, Y. Umezawa, and T. Kurahashi, *J. Chromatogr.* **239**, 773 (1982).

40. S. Kamada, M. Maeda, and A. Tsuji, *J. Chromatogr.* **272**, 29 (1983).

41. P. S. Tietz, J. L. Thistle, L. J. Miller, and N. F. LaRusso, *J. Chromatogr.* **336**, 249 (1984).

42. K. Imai, Z. Tamura, F. Mashige, and T. Osuga, *J. Chromatogr.* **120**, 181 (1976).

43. F. Stellaard, P. D. Klein, A. F. Hofmann, and J. M. Lachin, *J. Lab. Clin. Med.* **105**, 504 (1985).

44. J. M. Street, D. J. H. Trafford, and H. L. J. Makin, *J. Chromatogr.* **343**, 259 (1985).

45. O. Jänne, R. Vihko, J. Sjövall, and K. Sjövall, *Clin. Chim. Acta* **23**, 405 (1969).

46. B. Angelin and I. Björkhem, *Gut* **18**, 606 (1977).

47. B. Angelin, I. Björkhem, and K. Einarsson, *J. Lipid Res.* **19**, 527 (1978).
48. B. Angelin, I. Björkhem, K. Einarsson, and S. Ewerth, *J. Clin. Invest.* **70**, 724 (1982).
49. I. Björkhem and O. Falk, *Scand. J. Clin. Lab. Invest.* **43**, 163 (1983).
50. S. Ewerth, B. Angelin, K. Einarrsson, K. Nilsell, and I. Björkhem, *Gastroenterology* **88**, 126 (1985).
51. T. Beppu. Y. Seyama, T. Kasama, and T. Yamakawa, *J. Biochem.* **89**, 1963 (1981).
52. G. P. Van Berge Henegouwen, A. Ruben, and K. H. Brandt, *Clin. Chim. Acta* **54**, 249 (1974).
53. N. B. Javitt, U. Lavy, and E. Kok, *in* "The Liver" (R. Preisig, J. Bircher, and G. Paumgartner, eds.), p. 249, Editio Cantor, Aulendorf (1976).
54. S. S. Ali and N. B. Javitt, *Can. J. Biochem.* **48**, 1054 (1970).
55. P. Bonazzi, C. Calaresu, and R. Galeazzi, *Pharmacol. Res. Commun.* **16**, 549 (1984).
56. G. Délèze, D. Sidiropoulos, and G. Paumgartner, *Pediatrics* **59**, 647 (1977).
57. G. Délèze, G. Paumgartner, G. Karlaganis, W. Giger, M. Reinhard, and D. Sidiropoulos, *Eur. J. Clin. Invest.* **8**, 41 (1978).
58. A. F. Hofmann, *J. Lipid Res.* **8**, 55 (1967).
59. W. E. Kurtin and W. H. Schwesinger, *Anal. Biochem.* **147**, 511 (1985).
60. B. Fransson and G. Schill, *Acta Pharm. Suec.* **12**, 417 (1975).
61. W. H. M. Christie, I. A. Macdonald, and C. N. Williams, *J. Lab. Clin. Med.* **85**, 505 (1975).
62. G. R. Webb, I. A. Macdonald, and C. N. Williams, *Clin. Chem.* **23**, 460 (1977).
63. N. Parris, *Anal. Biochem.* **100**, 260 (1979).
64. H. J. Wildgrube, U. Füssel, H. Lauer, and H. Stockhausen, *J. Chromatogr.* **282**, 603 (1983).
65. W. Swobodnik, U. Klüppelberg, J. G. Wechsler, M. Volz, G. Normandin, and H. Ditschuneit, *J. Chromatogr.* **339**, 263 (1985).
66. J. Dupont, S.-Y. Oh, L. A. O'Deen, and S. Geller, *Lipids* **9**, 294 (1974).
67. S.-Y. Oh and J. Dupont, *Lipids* **10**, 340 (1975).
68. J. Dupont, S.-Y. Oh, and P. Janson, *in* "The Bile Acids" (P. P. Nair and D. Kritchevsky, eds.), Vol. 3, p. 17, Plenum Press, New York (1976).
69. H. Greim, P. Czygan, F. Schaffner, and H. Popper, *Biochem. Med.* **8**, 280 (1973).
70. J. Yanagisawa, M. Itoh, M. Ishibashi, H. Miyazaki, and F. Nakayama, *Anal. Biochem.* **104**, 75 (1980).
71. Y. Akashi, H. Miyazaki, and F. Nakayama, *Clin. Chim. Acta* **133**, 125 (1983).
72. C. N. Ghent, J. R. Bloomer, and G. Klatskin, *Gastroenterology* **73**, 1125 (1977).
73. A. Hara and N. S. Radin, *Anal. Biochem.* **90**, 420 (1978).
74. T. Masui and J. Sjövall (to be published).
75. T. Okishio, P. P. Nair, and M. Gordon, *Biochem. J.* **102**, 654 (1967).
76. P. P. Nair, A. I. Mendeloff, M. Vocci, J. Bankoski, M. Gorelik, G. Herman, and R. Plapinger, *Lipids* **12**, 922 (1977).
77. P. P. Nair, R. Solomon, J. Bankoski, and R. Plapinger, *Lipids* **13**, 966 (1978).
78. N. Turjman and P. P. Nair, *Cancer Res.* **41**, 3761 (1981).
79. H. Greim, D. Trülzsch, J. Roboz, K. Dressler, P. Czygan, F. Hutterer, F. Schaffner, and H. Popper, *Gastroenterology* **63**, 837 (1972).
80. A. M. Gelb, C. K. McSherry, J. R. Sadowsky, and E. H. Mosbach, *Am. J. Gastroenterol.* **77**, 314 (1982).
81. J. Yanagisawa, Y. Akashi, H. Miyazaki, and F. Nakayama, *J. Lipid Res.* **25**, 1263 (1984).
82. G. Hedenborg, A. Norlander, and A. Norman, *Scand. J. Clin. Lab. Invest.* **46**, 539 (1986).
83. U. Leuschner, H. J. Wildgrube, E. Reber, and W. Erb, *Histochemie* **29**, 178 (1972).
84. T. Okishio and P. P. Nair, *Biochemistry* **5**, 3662 (1966).
85. H. J. Wildgrube, U. Leuschner, and A. Al-Fureyh, *Verh. Dtsch. Ges. Inn. Med.* **78**, 1398 (1972).
86. W. Kurtz, U. Leuschner, A. Hellstern, and P. Janka, *Hepato-gastroenterol.* **29**, 227 (1982).

87. H. Oftebro, I. Björkhem, S. Skrede, A. Schreiner, and J. I. Pedersen, *J. Clin. Invest.* **65,** 1418 (1980).
88. I. Björkhem, H. Oftebro, S. Skrede, and J. I. Pedersen, *J. Lipid Res.* **22,** 191 (1981).
89. A. Norman, *Br. J. Nutr.* **18,** 173 (1964).
90. A. Norman and R. H. Palmer, *J. Lab. Clin. Med.* **63,** 986 (1964).
91. B. E. Gustafsson and A. Norman, *Scand. J. Gastroenterol.* **3,** 625 (1968).
92. B. E. Gustafsson and A. Norman, *Br. J. Nutr.* **23,** 429 (1969).
93. K. D. R. Setchell, A. M. Lawson, N. Tanida, and J. Sjövall, *J. Lipid Res.* **24,** 1085 (1983).
94. N. Breuer, P. Dommes, R. Tandon, and H. Goebell, *J. Clin. Chem. Clin. Biochem.* **22,** 623 (1984).
95. H. J. Eyssen, G. G. Parmentier, and J. A. Mertens, *Eur. J. Biochem.* **66,** 507 (1976).
96. L. R. Ferguson, G. W. Rewcastle, J. M. Lello, P. G. Alley, and R. N. Seelye, *Anal. Biochem.* **143,** 325 (1984).
97. O. J. Roseleur and C. M. van Gent, *Clin. Chim. Acta* **82,** 13 (1978).
98. P. Eneroth, K. Hellström, and J. Sjövall, *Acta Chem. Scand.* **22,** 1729 (1968).
99. R. W. Owen, M. H. Thompson, and M. J. Hill, *J. Steroid Biochem.* **21,** 593 (1984).
100. P. Back and K. Ross, *Hoppe-Seyler's Z. Physiol. Chem.* **354,** 83 (1973).
101. P. Back and K. Walter, *Gastroenerology* **78,** 671 (1980).
102. D. H. Sandberg, J. Sjövall, K. Sjövall, and D. A. Turner, *J. Lipid Res.* **6,** 182 (1965).
103. E. D. Pellizzari, F. S. O'Neil, R. W. Farmer, and L. F. Fabre Jr., *Clin. Chem.* **19,** 248 (1973).
104. H. Ngoc-Chau, J. P. Bali, and M. Solere, *Clin. Chim. Acta* **75,** 41 (1977).
105. H. L. Bradlow, *Steroids* **11,** 265 (1968).
106. I. Makino and J. Sjövall, *Anal. Lett.* **5,** 341 (1972).
107. I. Makino, S. Nakagawa, K. Shinozaki, and K. Mashimo, *Lipids* **7,** 750 (1972).
108. I. Makino, K. Shinozaki, S. Nakagawa, and K. Mashimo, *J. Lipid Res.* **15,** 132 (1974).
109. P. Back, J. Sjövall, and K. Sjövall, *Med. Biol.* **52,** 31 (1974).
110. H. P. Schwarz, K. V. Bergmann, and G. Paumgartner, *Clin. Chim. Acta* **50,** 197 (1974).
111. T. Laatikainen and A. Hesso, *Clin. Chim. Acta* **64,** 63 (1975).
112. G. P. Van Berge Henegouwen and A. F. Hofmann, *Clin. Chim. Acta* **73,** 469 (1976).
113. S. Barnes and A. Chitranukroh, *Ann. Clin. Biochem.* **14,** 235 (1977).
114. J.-F. Pageaux, B. Duperray, M. Dubois, Y. Pacheco, H. Pacheco, N. Herne, and D. Hauteville, *Clin. Chim. Acta* **85,** 131 (1978).
115. G. Karlaganis and G. Paumgartner, *J. Lipid Res.* **19,** 771 (1978).
116. G. Karlaganis and G. Paumgartner, *Clin. Chim. Acta* **92,** 19 (1979).
117. A. Cantafora, M. Angelico, A. F. Attili, L. Ercoli, and L. Capocaccia, *Clin. Chim. Acta* **95,** 501 (1979).
118. G. G. Parmentier, G. A. Janssen, E. A. Eggermont, and H. J. Eyssen, *Eur. J. Biochem.* **102,** 173 (1979).
119. P. T. Clayton and D. P. R. Muller, *Clin. Chim. Acta* **105,** 401 (1980).
120. G. Karlaganis, R. P. Schwarzenbach, and G. Paumgartner, *J. Lipid Res.* **21,** 377 (1980).
121. A. Bremmelgaard and B. Almé, *Scand. J. Gastroenterol.* **15,** 593 (1980).
122. K. D. R. Setchell, A. M. Lawson, E. J. Blackstock, and G. M. Murphy, *Gut* **23,** 637 (1982).
123. T. C. Bartholomew, J. A. Summerfield, B. H. Billing, A. M. Lawson, and K. D. R. Setchell, *Clin. Sci.* **63,** 65 (1982).
124. F. Stellaard, R. Schubert, and G. Paumgartner, *Biomed. Mass Spectrom.* **10,** 187 (1983).
125. S. Hasegawa, R. Uenoyama, F. Takeda, J. Chuma, S. Baba, F. Kamiyama, M. Iwakawa, and M. Fushimi, *J. Chromatogr.* **278,** 25 (1983).
126. M. Tohma, H. Wajima, R. Mahara, T. Kurosawa, and I. Makino, *Steroids* **44,** 47 (1984).
127. A. Norman and B. Strandvik, *J. Lab. Clin. Med.* **78,** 181 (1971).

128. A. Norman and B. Strandvik, *Acta Paediatr. Scand.* **62**, 253 (1973).
129. A. Norman, B. Strandvik, and Ö Ojamäe, *Acta Paediatr. Scand.* **63**, 97 (1974).
130. A. Norman and B. Strandvik, *Acta Paediatr. Scand.* **63**, 92 (1974).
131. P. Back, K. Spaczynski, and W. Gerok, *Hoppe-Seyler's Z. Physiol. Chem.* **355**, 749 (1974).
132. J. A. Summerfield, B. H. Billing, and C. H. L. Shackleton, *Biochem. J.* **154**, 507 (1976).
133. P. Back, *Hoppe-Seyler's Z. Physiol. Chem.* **357**, 213 (1976).
134. G. P. Van Berge Henegouwen, K-H. Brandt, H. Eyssen, and G. Parmentier, *Gut* **17**, 861 (1976).
135. B. Almé, A. Bremmelgaard, J. Sjövall, and P. Thomassen, *J. Lipid Res.* **18**, 339 (1977).
136. B. Almé, Å. Nordén, and J. Sjövall, *Clin. Chim. Acta* **86**, 251 (1978).
137. P. A. Thomassen, *Eur. J. Clin. Invest.* **9**, 425 (1979).
138. A. Bremmelgaard and J. Sjövall, *Eur. J. Clin. Invest.* **9**, 341 (1979).
139. A. Bremmelgaard and J. Sjövall, *J. Lipid Res.* **21**, 1072 (1980).
140. B. Almé and J. Sjövall, *J. Steroid Biochem.* **13**, 907 (1980).
141. G. Karlaganis, B. Almé, V. Karlaganis, and J. Sjövall, *J. Steroid Biochem.* **14**, 341 (1981).
142. A. Stiehl, R. Raedsch, G. Rudolph, P. Czygan, and S. Walker, *Clin. Chim. Acta* **123**, 275 (1982).
143. J. Yanagisawa, H. Ichimiya, M. Nagai, and F. Nakayama, *J. Lipid Res.* **25**, 750 (1984).
144. Y. Tazawa, M. Yamada, M. Nakagawa, T. Konno, and K. Tada, *Tohoku J. Exp. Med.* **143**, 361 (1984).
145. A. Stiehl R. Raedsch, G. Rudolph, U. Gundert-Remy, and M. Senn, *Hepatology* **5**, 492 (1985).
146. R. W. Baker, J. Ferrett, and G. M. Murphy, *J. Chromatogr.* **146**, 137 (1978).
147. J. Goto, M. Hasegawa, H. Kato, and T. Nambara, *Clin. Chim. Acta* **87**, 141 (1978).
148. S. Okuyama, *Gastroenterol. Jpn.* **14**, 129 (1979).
149. R. Beke, G. A. De Weerdt, and F. Barbier, *J. Chromatogr.* **193**, 504 (1980).
150. J. Goto, H. Kato, Y. Saruta, and T. Nambara, *J. Chromatogr.* **226**, 13 (1981).
151. G. Parmentier and H. Eyssen, *Steroids* **26**, 721 (1975).
152. A. Van den Ende, C. E. Rädecker, W. M. Mairuhu, and A. P. van Zanten, *Clin. Chim. Acta* **121**, 95 (1982).
153. S. Ikawa, T. Yamamoto, D. Yoshioka, M. Takita, M. Ogura, and Y. Kishimoto, *Yonago Acta Med.* **21**, 76 (1977).
154. M. Matsui, M. Hakozaki, and Y. Kinuyama, *J. Chromatogr.* **115**, 625 (1975).
155. H. L. Bradlow, *Steroids* **30**, 581 (1977).
156. M. Axelson and B.-L. Sahlberg, *Anal. Lett.* **14**, 771 (1981).
157. M. Axelson and J. Sjövall, *J. Steroid Biochem.* **5**, 733 (1974).
158. M. Axelson and J. Sjövall, *J. Steroid Biochem.* **8**, 683 (1977).
159. C. H. L. Shackleton and J. O. Whitney, *Clin. Chim. Acta.* **107**, 231 (1980).
160. J. O. Whitney and M. M. Thaler, *J. Liq. Chromatogr.* **3**, 545 (1980).
161. J. Goto, H. Kato, Y. Saruta, and T. Nambara, *J. Liq. Chromatogr.* **3**, 991 (1980).
162. K. D. R. Setchell and J. Worthington, *Clin. Chim. Acta* **125**, 135 (1982).
163. D. Ishii, S. Murata, and T. Takeuchi, *J. Chromatogr.* **282**, 569 (1983).
164. K. Shimazu, M. Kuwabara, M. Yoshii, K. Kihira, H. Takeuchi, I. Nakano, S. Ozawa, M. Onuki, Y. Hatta, and T. Hoshita, *J. Biochem.* **99**, 477 (1986).
165. M. J. Whiting, *Clin. Chim. Acta* **141**, 261 (1984).
166. M. Axelson and J. Sjövall, in "Enterohepatic Circulation of Bile Acids and Sterol Metabolism" (G. Paumgartner, A. Stiehl, and W. Gerok, eds.), p. 249, MTP Press, Lancaster (1985).
167. S. H. G. Andersson and J. Sjövall, *Anal. Biochem.* **134**, 309 (1983).
168. M. Axelson, *Anal. Lett.* **18**, 1607 (1985).
169. G. Karlaganis, A. Németh, B. Hammarskjöld, B. Strandvik, and J. Sjövall, *Eur. J. Clin. Invest.* **12**, 399 (1982).

170. G. Karlaganis, V. Karlaganis, and J. Sjövall, *in* "Bile Acids and Cholesterol in Health and Disease" (G. Paumgartner, A. Stiehl, and W. Gerok, eds.), p. 119, MTP Press, Lancaster (1983).

171. G. Karlaganis, V. Karlaganis, and J. Sjövall, *J. Lipid Res.* **25**, 693 (1984).

172. B. G. Wolthers, M. Volmer, J. van der Molen, B. J. Koopman, A. E. J. de Jager, and R. J. Waterreus, *Clin. Chim. Acta* **131**, 53 (1983).

173. G. Hedenborg and A. Norman, *Scand. J. Clin. Lab. Invest.* **44**, 725 (1984).

174. G. Hedenborg, A. Norlander, and A. Norman, *Scand. J. Clin. Lab. Invest.* **45**, 157 (1985).

175. J. Goto, K. Suzaki, M. Ebihara, T. Nambara, and A. Masu, *J. Chromatogr.* **345**, 241 (1985).

176. S. Onishi, S. Itoh, and Y. Ishida, *Biochem. J.* **204**, 135 (1982).

177. B. R. DeMark, G. T. Everson, P. D. Klein, R. B. Showalter, and F. Kern, Jr., *J. Lipid Res.* **23**, 204 (1982).

178. F. Stellaard, M. Sackmann, T. Sauerbruch, and G. Paumgartner, *J. Lipid Res.* **25**, 1313 (1984).

179. F. Stellaard and G. Paumgartner, *Biomed. Mass Spectrom.* **12**, 560 (1985).

180. A. Hernanz and R. Codoceo, *Clin. Chim. Acta* **145**, 197 (1985).

181. J. Goto, M. Saito, T. Chikai, N. Goto, and T. Nambara, *J. Chromatogr.* **276**, 289 (1983).

182. N. Murata, T. Beppu, H. Takikawa, H. Otsuka, T. Kasama, and Y. Seyama, *Steroids* **42**, 575 (1983).

183. H. Takikawa, H. Otsuka, T. Beppu, Y. Seyama, and T. Yamakawa, *J. Biochem.* **92**, 985 (1982).

184. H. Takikawa, H. Otsuka, T. Beppu, and Y. Seyama, *Biochem. Med.* **33**, 393 (1985).

185. H. Takikawa, H. Otsuka, T. Beppu, Y. Seyama, and T. Yamakawa, *Digestion* **27**, 189 (1983).

186. Y. Tazawa, M. Yamada, M. Nakagawa, T. Konno, and K. Tada, *J. Pediatr. Gastroenterol. Nutr.* **3**, 394 (1984).

187. K. Linnet, *Scand. J. Clin. Lab. Invest.* **42**, 455 (1982).

188. J. M. Street, D. J. H. Trafford, and H. L. J. Makin, *J. Lipid Res.* **27**, 208 (1986).

189. A. T. Ruben and G. P. Van Berge-Henegouwen, *Clin. Chim. Acta.* **119**, 41 (1982).

190. M. Muraca and Y. Ghoos, *J. Lipid Res.* **26**, 1009 (1985).

191. G. Hedenborg and A. Norman, *Scand. J. Clin. Lab. Invest.* **45**, 151 (1985).

192. G. Hedenborg, A. Norlander, and A. Norman, *Scand. J. Clin. Lab. Invest.* **47**, 83 (1987).

193. G. Hedenborg, A. Norlander. A. Norman, and A. Svensson, *Scand. J. Clin. Lab. Invest.* **46**, 745 (1986).

194. A. D. Reid and P. R. Baker. *J. Chromatogr.* **260**, 115 (1983).

195. A. D. Reid and P. R. Baker. *J. Chromatogr.* **268**, 281 (1983).

196. T. Kawasaki, M. Maeda, and A. Tsuji, *J. Chromatogr.* **272**, 261 (1983).

197. Y. Ghoos, P. Rutgeerts, and G. Vantrappen, *J. Liq. Chromatogr.* **5**, 175 (1982).

198. G. Parmentier and H. Eyssen. *J. Chromatogr.* **152**, 285 (1978).

199. A. Dyfverman and J. Sjövall, *in* "Biological Effects of Bile Acids" (G. Paumgartner, A. Stiehl, and W. Gerok, eds.), p. 281. MTP Press, Lancaster (1979).

200. A. Dyfverman and J. Sjövall. *Anal. Biochem.* **134**, 303 (1983).

201. K. Norén and J. Sjövall, *J. Chromatogr. Biomed. Appl.* **414**, 55 (1987).

202. E. Nyström and J. Sjövall. *Anal. Biochem.* **12**, 235 (1965).

203. J. Sjövall, E. Nyström, and E. Haahti, *in* "Advances in Chromatography" (J. C. Giddings and R. A. Keller, eds.), Vol. 6, p. 119. Dekker, New York (1968).

204. J. Ellingboe, E. Nyström, and J. Sjövall, *Meth. Enzymol.* **14**, 317 (1969).

205. E. Nyström and J. Sjövall, *Meth. Enzymol.* **35**, 378 (1975).

206. Y. Ghoos, G. Vantrappen, and D. Mayer. *J. Chromatogr.* **76**, 425 (1973).

207. S. Ikawa and M. Goto. *J. Chromatogr.* **114**, 237 (1975).

208. R. Beke, G. A. De Weerdt. J. Parijs. W. Huybrechts, and F. Barbier, *Clin. Chim. Acta* **70**, 197 (1976).

209. S. K. Goswami and C. F. Frey, *Biochem. Med.* **17**, 20 (1977).
210. M. Nakagaki and F. Nakayama, *J. Chromatogr.* **177**, 343 (1979).
211. K. Shimada, M. Hasegawa, J. Goto, and T. Nambara, *J. Chromatogr.* **152**, 431 (1978).
212. A. K. Batta, G. Salen, and S. Shefer, *J. Liq. Chromatogr.* **3**, 1865 (1980).
213. A. Van den Ende, C. E. Rädecker, and W. M. Mairuhu, *Anal. Biochem.* **134**, 153 (1983).
214. C. T. L. Huang and B. L. Nichols, *J. Chromatogr.* **109**, 427 (1975).
215. B. I. Cohen, R. F. Raicht, G. Salen, and E. H. Mosbach, *Anal. Biochem.* **64**, 567 (1975).
216. R. Spears, D. Vukusich, S. Mangat, and B. S. Reddy, *J. Chromatogr.* **116**, 184 (1976).
217. W. T. Beher, S. Stradnieks, G. J. Lin, and J. Sanfield, *Steroids* **38**, 281 (1981).
218. M. Takahashi, R. F. Raicht, A. N. Sarwal, E. H. Mosbach, and B. I. Cohen, *Anal. Biochem.* **87**, 594 (1978).
219. C. T. L. Huang, P. A. Szczepanik-Van Leeuwen, and B. L. Nichols, *J. Chromatogr.* **196**, 150 (1980).
220. O. W. Cass, A. E. Cowen, A. F. Hofmann, and S. B. Coffin, *J. Lipid Res.* **16**, 159 (1975).
221. R. Raedsch, A. F. Hofmann, and K. Y. Tserng, *J. Lipid Res.* **20**, 796 (1979).
222. K. Dudas, S. Szepesi, A. Pap, T. Fehér, E. Mincsovics, and E. Tyihak, in "Advances in Steroid Analysis" (S. Görög, ed.), p. 417, Adadémiai Kiado, Budapest (1982).
223. S. Ikawa, *J. Chromatogr.* **117**, 227 (1976).
224. G. A. D. Haslewood, *Biochem. J.* **126**, 27P (1972).
225. I. G. Anderson, G. A. D. Haslewood, R. S. Oldham, B. Amos, and L. Tökés, *Biochem. J.* **141**, 485 (1974).
226. S. Ikawa, *Anal. Biochem.* **85**, 197 (1978).
227. G. Parmentier and H. Eyssen, *Steroids* **30**, 583 (1977).
228. N. Turjman and C. Jacob, *Am. J. Clin. Nutr.* **40**, 957 (1984).
229. P. F. Miskovitz, N. B. Javitt, E. Kok, S. Burstein, and M. Gut, in "Bile Acid and Cholesterol in Health and Disease" (G. Paumgartner, A. Stiehl, and W. Gerok, eds.), p. 99, MTP Press, Lancaster (1983).
230. E. Kok, S. Burstein, N. B. Javitt, M. Gut, and C. Y. Byon, *J. Biol. Chem.* **256**, 6155 (1981).
231. E. Nyström and J. Sjövall, *Ark. Kem.* **29**, 107 (1968).
232. E. Nyström and J. Sjövall, *Acta Chem. Scand.* **21**, 1974 (1967).
233. M. I. Kelsey and R. J. Thompson, *J. Steroid Biochem.* **7**, 117 (1976).
234. M. I. Kelsey and S. A. Sexton, *J. Steroid Biochem.* **7**, 641 (1976).
235. M. I. Kelsey and S. A. Sexton, *J. Chromatogr.* **133**, 327 (1977).
236. J. Ellingboe, E. Nyström, and J. Sjövall, *J. Lipid Res.* **11**, 266 (1970).
237. J. Sjövall and R. Vihko, *Acta Chem. Scand.* **20**, 1419 (1966).
238. T. Cronholm, I. Makino, and J. Sjövall, *Eur. J. Biochem.* **26**, 251 (1972).
239. A. Stiehl, *Eur. J. Clin. Invest.* **4**, 59 (1974).
240. W. Fröhling and A. Stiehl, *Eur. J. Clin. Invest.* **6**, 67 (1976).
241. P. Back, *Klin. Wochenschr.* **60**, 541 (1982).
242. G. Parmentier, J. Mertens, and H. Eyssen, in "Advances in Bile Acid Research" (S. Matern, J. Hackenschmidt, P. Back, and W. Gerok, eds.), p. 139, Schattauer, Stuttgart (1975).
243. E. S. Haslewood and G. A. D. Haslewood, *Biochem. J.* **130**, 89P (1972).
244. W. Roth and K. Beschke, *J. Pharm. Biomed. Anal.* **2**, 289 (1984).
245. M. Schöneshöfer, A. Kage, and B. Weber, *Clin. Chem.* **29**, 1367 (1983).
246. A. Kuksis, in "The Bile Acids" (P. P. Nair and D. Kritchevsky, eds.), Vol. 1, p. 173, Plenum Press, New York (1971).
247. A. Smythe, D. Mangnall, and A. G. Johnson, *Anal. Biochem.* **118**, 65 (1981).
248. J.-Å. Gustafsson and J. Sjövall, *Eur. J. Biochem.* **8**, 467 (1969).
249. K. D. R. Setchell, B. Almé, M. Axelson, and J. Sjövall, *J. Steroid Biochem.* **7**, 615 (1976).
250. M. Tetsuo, M. Axelson, and J. Sjövall, *J. Steroid Biochem.* **13**, 847 (1980).

251. M. Tetsuo, H. Eriksson, and J. Sjövall, *J. Chromatogr.* **239**, 287 (1982).

252. G. A. De Weerdt, R. Beke, H. Verdievel, and F. Barbier, *Biomed. Mass Spectrom.* **7**, 515 (1980).

253. G. Hedenborg, A. Norman, and K. Samuelson, *Scand. J. Clin. Lab. Invest.* **44**, 761 (1984).

254. G. Hedenborg, A. Norman, and K. Samuelson, *Scand. J. Clin. Lab. Invest.* **44**, 765 (1984).

255. R. Tandon, M. Axelson, and J. Sjövall, *J. Chromatogr.* **302**, 1 (1984).

256. S. Miyazaki, H. Tanaka, R. Horikawa, H. Tsuchiya, and K. Imai, *J. Chromatogr.* **181**, 177 (1980).

257. M. Axelson and J. Sjövall, *J. Chromatogr.* **126**, 705 (1976).

258. M. Axelson and J. Sjövall, *J. Chromatogr.* **186**, 725 (1979).

259. A. Radominska-Pyrek, P. Zimniak, M. Chari, E. Golunski, R. Lester, and J. St. Pyrek, *J. Lipid Res.* **27**, 89 (1986).

260. T. Harano, C. Fujita, K. Harano, and K. Yamasaki, *Steroids* **30**, 393 (1977).

261. O. J. Roseleur and C. M. Van Gent, *Clin. Chim. Acta* **66**, 269 (1976).

262. G. Lepage, A. Fontaine, and C. C. Roy, *J. Lipid Res.* **19**, 505 (1978).

263. W. T. Beher, S. Stradnieks, G. R. Beher, and G. J. Lin, *Steroids* **32**, 355 (1978).

264. W. S. Harris, L. Marai, J. J. Myher, and M. T. R. Subbiah, *J. Chromatogr.* **131**, 437 (1977).

265. G. P. van Berge-Henegouwen, R. N. Allan, A. F. Hofmann, and P. Y. S. Yu, *J. Lipid Res.* **18**, 118 (1977).

266. G. A. D. Haslewood, "Bile Salts," Methuen, London (1967), p. 28.

267. A. K. Batta, G. Salen, F. W. Cheng, and S. Shefer, *J. Biol. Chem.* **254**, 11907 (1979).

268. A. Kutner, R. Jaworska, W. Kutner, and A. Grzeszkiewicz, *in* "Advances in Steroid Analysis" (S. Görög, ed.), p. 333, Akadémiai Kiado, Budapest (1982).

269. E. S. Haslewood and G. A. D. Haslewood, *Biochem. J.* **155**, 401 (1976).

270. R. H. Palmer and M. G. Bolt, *J. Lipid Res.* **12**, 671 (1971).

271. R. Galeazzi and N. B. Javitt, *J. Clin. Invest.* **60**, 693 (1977).

272. P. P. Nair, *in* "Bile Salt Metabolism" (L. Schiff, J. B. Carey, Jr., and J. M. Dietschy, eds.), p. 172, Thomas, Springfield, IL (1969).

273. P. P. Nair and C. Garcia, *Anal. Biochem.* **29**, 164 (1969).

274. M. Une, N. Matsumoto, K. Kihira, M. Yasuhara, T. Kuramoto, and T. Hoshita, *J. Lipid Res.* **21**, 269 (1980).

275. A. K. Batta, G. Salen, and S. Shefer, *J. Biol. Chem.* **259**, 15035 (1984).

276. G. A. D. Haslewood, "The Biological Importance of Bile Salts," North-Holland, Amsterdam (1978), p. 32.

277. G. Lepage, C. C. Roy, and A. M. Weber, *J. Lipid Res.* **22**, 705 (1981).

278. E. Sacquet, C. Leprince, M. Riottot, C. Mejean, and P. Raibaud, *Steroids* **32**, 1(1978).

279. H. L. Bradlow, *in* "Chemical and Biological Aspects of Steroid Conjugation" (S. Bernstein and S. Solomon, eds.), p. 131, Springer Verlag, New York (1970).

280. J. Goto, A. Sato, K. Suzaki, and T. Nambara, *Chem. Pharm. Bull.* **29**, 1975 (1981).

281. A. Stiehl, M. Becker, P. Czygan, W. Fröhling, B. Kommerell, H. W. Roffhauwe, and M. Senn, *Eur. J. Clin. Invest.* **10**, 307 (1980).

282. E. Sacquet, M. Parquet, M. Riottot, A. Raizman, P. Jarrige, C. Huguet, and R. Infante, *J. Lipid Res.* **24**, 604 (1983).

283. H. Matern, S. Matern, and W. Gerok, *Anal. Biochem.* **133**, 417 (1983).

284. J. M. Little, J. St. Pyrek, and R. Lester, *J. Clin. Invest.* **71**, 73 (1983).

285. M. Parquet, M. Pessah, E. Sacquet, C. Salvat, A. Raizman, and R. Infante, *FEBS Lett.* **189**, 183 (1985).

286. J. M. Little, M. V. Chari, and R. Lester, *J. Lipid Res.* **26**, 583 (1985).

287. G. M. Jacobsohn and S. Lieberman, *J. Biol. Chem.* **237**, 1469 (1962).

288. T. Hoshita, M. Yasuhara, M. Une, A. Kibe, E. Itoga, S. Kito, and T. Kuramoto, *J. Lipid Res.* **21**, 1015 (1980).

289. A. K. Batta, S. Shefer, M. Batta, and G. Salen, *J. Lipid Res.* **26**, 690 (1985).

290. H. Matern, S. Matern, and W. Gerok, *Proc. Natl. Acad. Sci. USA* **81**, 7036 (1984).

291. S. Burstein and S. Lieberman, *J. Am. Chem. Soc.* **80**, 5235 (1958).

292. R. H. Palmer, *Proc. Natl. Acad. Sci. USA* **58**, 1047 (1967).

293. B. I. Cohen, K. Budai, and N. B. Javitt, *Steroids* **37**, 621 (1981).

294. S. Burstein and S. Lieberman, *J. Biol. Chem.* **233**, 331 (1958).

295. L. Kornel, *Biochemistry* **4**, 444 (1965).

296. P. Dommes, N. F. Breuer, and H. Goebell, *Clin. Chim. Acta* **154**, 237 (1986).

297. P. T. Clayton, J. V. Leonard, R. Dinwiddie, A. M. Lawson, K. D. R. Setchell, S. Andersson, B. Egestad, and J. Sjövall, *J. Clin. Invest.* **79**, 1031 (1987).

298. U. Lavy. S. Burstein, M. Gut, and N. B. Javitt, *J. Lipid Res.* **18**, 232 (1977).

299. M. Axelson, B.-L. Sahlberg, and J. Sjövall, *J. Chromatogr. Biomed. Appl.* **224**, 355 (1981).

300. G. Hedenborg, "Synthesis, Transport and Excretion of Conjugated Bile Acids in Cholestasis" Thesis, Karolinska Institutet, Stockholm (1986).

301. S. M. Huijghebaert, J. A. Mertens, and H. J. Eyssen, *Appl. Environ. Microbiol.* **43**, 185 (1982).

302. S. M. Huijghebaert and H. J. Eyssen, *Appl. Environ. Microbiol.* **44**, 1030 (1982).

303. J. Sjövall, *Meth. Biochem. Anal.* **12**, 97 (1964).

304. P. Eneroth and J. Sjövall, *Meth. Enzymol.* **15**, 237 (1969).

305. R. Shaw and W. H. Elliott, *J. Lipid Res.* **19**, 783 (1978).

306. A. Fukunaga, Y. Hatta, M. Ishibashi, and H. Miyazaki, *J. Chromatogr.* **190**, 339 (1980).

307. T. Harano, K. Harano, and K. Yamasaki, *Steroids* **32**, 73 (1978).

308. M. Ikeda and K. Yamasaki, *Steroids* **32**, 85 (1978).

309. S. Barnes, D. G. Pritchard, R. L. Settine, and M. Geckle, *J. Chromatogr.* **183**, 269 (1980).

310. P. P. Nair, G. Kessie, and V. P. Flanagan, *J. Lipid Res.* **27**, 905 (1986).

311. J. M. Street, W. F. Balistreri, and K. D. R. Setchell, *Gastroenterology* **90**, 347 (1986) (Abstract).

312. Y. Hirano, H. Miyazaki, S. Higashidate, and F. Nakayama, *J. Lipid Res.* **28**, 1524 (1987).

Chapter 2

HIGH-PERFORMANCE LIQUID CHROMATOGRAPHY

Toshio Nambara and Junichi Goto

1. INTRODUCTION

Considerable attention has recently been focused on the biosynthesis and metabolism of bile acids in patients with hepatobiliary diseases, and the development of reliable methods is urgently required for the analysis of profiles of bile acids in biological materials. High-performance liquid chromatography (HPLC) is well recognized as an ideal tool for the analysis of a wide range of compounds, in particular compounds that are polar and unstable.

This method is suitable for the separation and determination of unconjugated and conjugated bile acids without prior hydrolysis and/or solvolysis. This chapter will review the application of HPLC to the analysis of bile acids in biological fluids.

2. CHROMATOGRAPHIC BEHAVIOR OF BILE ACIDS

2.1. Unconjugated, Glycine-, and Taurine-Conjugated Bile Acids

Many different methods are available to separate bile acids by means of HPLC. The chromatographic mode can be divided into two types: separation on straight-phase and reversed-phase columns. In a straight-phase system, an organic acid is usually added to the mobile phase to prevent tailing and/or leading of a peak in the chromatogram [1,2]. Bile acids are eluted in the order of decreasing number of the hydroxyl groups on the steroid nucleus. However, difficulty is often encountered with the resolution of glycine and taurine conjugates of chen-

Toshio Nambara and Junichi Goto   Pharmaceutical Institute, Tohoku University, Sendai, Japan

odeoxycholic and deoxycholic acids. The chemically bonded stationary phase, octadecylsilyl (ODS) reversed-phase column is the most suitable and widely used column for the resolution of bile acids.

Several factors affecting the chromatographic behavior of bile acids on an ODS column have been clarified [3,4]. The capacity ratio (k') of bile acids is markedly influenced by the pH of a mobile phase (Table I). In the pH range 7.0–8.0, unconjugated, glycine-, and taurine-conjugated bile acids exhibit similar k' values to one another. On the other hand, k' values of unconjugated and glycine-conjugated bile acids increase with decreasing pH from 6.5 and 5.0, respectively. It has been demonstrated that the k' value of an organic acid on an ODS column is dependent on the acidity of an eluent and can be estimated from the pK value of the organic acid and pH of the eluent [5]. The differences in the chromatographic behavior between these three groups of bile acids can be similarly explained in terms of pK values: 6.0 for unconjugated, 4.5 for glycine-conjugated, and 1.5 for taurine-conjugated bile acids [6]. Accordingly, with an anion exchange column, bile acids are readily separated into glycine

TABLE I.   Relative Capacity Ratios of Unconjugated, Glycine-, and Taurine-Conjugated Bile Acids[a]

Bile acid	Rk'		
	A[b]	B[c]	C[d]
UDCA	—	—	0.23
CA	—	—	0.27
CDCA	—	—	0.75
DCA	—	—	0.93
LCA	—	—	2.66
GUDCA	1.08	0.53	0.21
GCA	1.19	0.71	0.25
GCDCA	1.89	1.12	0.69
GDCA	2.26	1.26	0.87
GLCA	3.87	2.07	2.51
TUDCA	0.53	0.46	0.24
TCA	0.62	0.60	0.28
TCDCA	0.86	0.90	0.82
TDCA	1.00	1.00	1.00
TLCA	1.41	1.59	2.81

[a] Conditions: (A) fatty-acid analysis column, isopropanol/8.8 mM potassium phosphate buffer (pH 2.5) (160:340), (B) μBondapak C18 column, acetonitrile/methanol/0.03 M phosphate buffer (pH 3.4) (10:60:30), (C) Radial-Pak A column, acetonitrile/0.3% ammonium phosphate buffer (pH 7.7) (8:19).
[b] From Ref. 12.
[c] From Ref. 16.
[d] From Ref. 4.

and taurine conjugates depending on the pK value [7]. This group separation method provides information about the ratio of glycine- to taurine-conjugated bile acids in bile and serum, which may be diagnostically useful in reflecting the clinical state of hepatobiliary diseases.

The relationship between structure and chromatographic behavior on an ODS column has been investigated [8,9] (Tables II, III). Bile acids having an equatorial hydroxyl group at the 6α, 7β, or 12β position exhibit smaller k' values than the corresponding epimers with axial hydroxyl groups. On the other hand, the 3α-hydroxyl group shows a larger k' value than the 3β-hydroxyl group in both 5β- and 5α-cholanoic acids. The effect of a hydroxyl group on the retention

TABLE II.   Relative Capacity Ratios of Unconjugated 5β-Cholanoic Acids[a,b]

| | Rk' | | | |
| | pH 5.0 | | pH 7.5 | |
Substituent	A	B	A	B
3α-OH	1.00	1.00	1.00	1.00
3β-OH	0.93	0.57	0.77	0.60
7α-OH	2.02	1.74	2.11	2.20
7β-OH	1.41	1.32	1.45	1.50
12α-OH	2.06	1.84	2.42	2.82
12β-OH	2.22	2.11	2.48	2.68
3α,7α-(OH)$_2$	0.26	0.46	0.24	0.47
3α,7β-(OH)$_2$	0.11	0.17	0.082	0.16
3α,12α-(OH)$_2$	0.28	0.53	0.27	0.57
3α,12β-(OH)$_2$	0.16	0.27	0.11	0.20
3β,7α-(OH)$_2$	0.17	0.28	0.11	0.22
3β,7β-(OH)$_2$	0.11	0.16	0.066	0.11
3β,12α-(OH)$_2$	0.18	0.28	0.12	0.25
3β, 12β-(OH)$_2$	0.17	0.26	0.10	0.20
7α,12α-(OH)$_2$	0.54	0.73	0.48	0.90
7α,12β-(OH)$_2$	0.27	0.23	0.17	0.26
7β,12α-(OH)$_2$	0.21	0.23	0.15	0.30
7β,12β-(OH)$_2$	0.20	0.24	0.15	0.25
3α,7α,12α-(OH)$_3$	0.079	0.18	0.084	0.24
3α, 7α, 12β-(OH)$_3$	0.021	0.036	0.019	0.034
3α,7β,12α-(OH)$_3$	0.019	0.039	0.023	0.043
3α,7β,12β-(OH)$_3$	0.006	0.007	0.002	0.005
3β,7α,12α-(OH)$_3$	0.003	0.006	0.003	0.008
3β,7β,12α-(OH)$_3$	0.011	0.018	0.008	0.018
3β,7β,12β-(OH)$_3$	0.009	0.012	0.005	0.010

[a] Conditions: column. µBondapak C$_{18}$; mobile phase. (A) acetonitrile/0.3% ammonium phosphate buffer, (B) ethanol/0.3% ammonium phosphate buffer.
[b] From Ref. 9.

TABLE III.   Relative Capacity Ratios of
Unconjugated 5α-Cholanoic Acids[a,b]

Substituent	Rk'
3α-OH	2.13
3β-OH	1.26
6α-OH	2.03
6β-OH	3.30
7α-OH	4.28
7β-OH	4.20
12α-OH	5.35
3 = O	1.62
6 = O	2.65
3α,6α-(OH)$_2$	0.46
3α,6β-(OH)$_2$	0.82
3α,7α-(OH)$_2$	0.92
3α,12α-(OH)$_2$	1.81
7α,12α-(OH)$_2$	0.33
3 = O,7α-OH	0.41
3 = O,12α-OH	0.48
3α,6α,7α-(OH)$_3$	0.32
3α,7α,12α-(OH)$_3$	0.36

[a] Conditions: column, μBondapak C$_{18}$; mobile phase, isopropanol/10 mM potassium
   phosphate buffer (pH 7.0).
[b] From Ref. 8.

value is also dependent on the position and increases with decreasing polarity in the following sequence: 3-OH = 6-OH < 7-OH < 12-OH. There is no significant difference in the mobility between 5β- and 5α-epimers.

For oxygenated bile acids, the retention value is similarly influenced by the position, configuration, and number of oxygen function. Polyoxygenated cholanoic acids having a β-hydroxyl group and/or oxo group exhibit smaller k' values than the corresponding poly-α-hydroxylated ones. These chromatographic behaviors can be explained as follows. The mobility of a compound in reversed-phase chromatography is closely related to its affinity with the mobile and stationary phases. An increase in the polarity of a molecule enhances the solubility in the mobile phase, whereas the availability of a hydrophobic surface favors the affinity with the stationary phase. As for nonvicinal poly-α-hydroxylated bile acids, a constant β-surface area is available for hydrophobic binding with the stationary phase. Therefore, a change in the number of hydroxyl group on the α-face would alter only the polarity of a molecule and, hence, its solubility in the aqueous mobile phase. On the other hand, the β-hydroxyl or oxo group on the cholanoic acid nucleus reduces the hydrophobic surface area and increases the polarity of the molecule, resulting in a decrease in the retention value.

Introduction of an olefinic bond into the steroid nucleus causes a decrease in the retention value, while elongation of the side chain leads to an increase.

The resolution of the two positional isomers, chenodeoxycholic acid and deoxycholic acid, is of particular interest [10]. The pH effect of the mobile phase on the resolution of these two is illustrated in Fig. 1. Deoxycholic acid provides a larger k' value than chenodeoxycholic acid in the higher pH region, whereas both exhibit almost identical values in the lower pH region. On the other hand, glyco- and taurodeoxycholic acids exhibit larger k' values than the corresponding chenodeoxycholic acid in the pH range 4.0–7.5. These chromatographic phenomena will be discussed in Section 2.2.

For the separation and determination of bile acids in biological specimens with an ODS column, two different conditions are employed: an acidic mobile phase [11–18] and a neutral or weakly alkaline mobile phase [3,4]. Under the acidic condition the separation of glycine- and taurine-conjugated bile acids can be readily attained. In this case, however, elution of unconjugated bile acids takes a longer time, and resolution of unconjugated chenodeoxycholic acid and deoxycholic acid is unsatisfactory, as mentioned earlier. This condition may be applicable to bile specimens where unconjugated bile acids are usually present in low concentrations. Under a neutral pH condition, separation of bile acids discriminating the conjugated form is somewhat difficult, unless gradient elution is used. Preliminary fractionation into unconjugated, glycine-, and taurine-conjugated groups, followed by HPLC in isocratic mode, is to be recommended for the analysis of a complex mixture of bile acids [4,19].

Bile acids are excreted in feces principally in the unconjugated form. Therefore, the carboxyl group is available for esterification of these bile acids. Unconjugated bile acids are easily derivatized into the methyl [20], p-nitrobenzyl [21,22], phenacyl [23], p-bromophenacyl [24,25], m-methoxyphenacyl [25], and 4-nitrophthalimidomethyl [25] esters. Esterification with 4-bromomethyl-7-methoxycoumarin [26,27], 9-anthryldiazomethane [25], and 1-bromoacetylpyrene [28] is also employed for fluorescence labeling. For the separation of these esterified bile acids, the combined use of a reversed-phase column and an aqueous organic

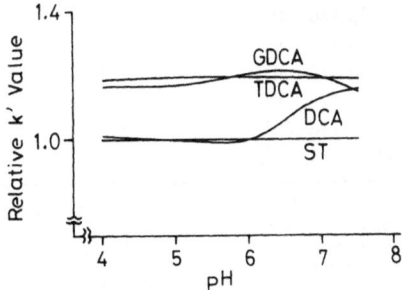

FIGURE 1. Effect of pH of mobile phase on k' values of deoxycholic acids relative to chenodeoxycholic acids. Conditions: column, Radial-Pak A; mobile phase, acetonitrile/0.3% phosphate buffer. ST, unconjugated, glyco- and taurochenodeoxycholic acids. (From Ref. [10].)

Toshio Nambara and Junichi Goto

solvent without acids or inorganic salts is preferable. It is evident from the data in Table IV that there is a close similarity in the relationship between structure and chromatographic behavior for esterified and unesterified bile acids.

Derivatization of bile acids through a 3α-hydroxyl group has also been attempted with the fluorescence labeling reagents 1-anthroyl nitrile and 4-dimethylamino-1-naphthoyl nitrile [29–31]. It is of interest that the 3-aroyl derivatives of unconjugated, glycine-, and taurine-conjugated ursodeoxycholic acids

TABLE IV. Relative Capacity Ratios of the C-24 Esters of 5β-Cholanoic Acids[a,b]

| | Rk' | | | |
Substituent	BP	MP	NPM	AM
3α-OH	2.04	2.17	2.13	2.27
3β-OH	1.94	2.03	2.13	2.79
7α-OH	2.84	2.99	3.28	2.88
7β-OH	2.33	2.47	2.85	2.59
12α-OH	3.02	3.12	3.28	2.79
12β-OH	3.39	3.42	4.13	3.41
3α,6α-(OH)$_2$	0.44	0.45	0.47	0.61
3α,7α-(OH)$_2$	0.90	0.91	0.94	1.00
3β,7α-(OH)$_2$	0.54	0.55	0.54	0.72
3α,7β-(OH)$_2$	0.30	0.32	0.36	0.45
3β,7β-(OH)$_2$	0.34	0.36	0.39	0.56
3α,12α-(OH)$_2$	1.00	1.00	1.00	1.00
3β,12α-(OH)$_2$	0.56	0.55	0.53	0.64
3α,12β-(OH)$_2$	0.46	0.45	0.53	0.62
3β,12β-(OH)$_2$	0.44	0.45	0.56	0.68
7α,12α-(OH)$_2$	0.56	1.12	1.20	1.00
7β,12α-(OH)$_2$	0.29	0.24	0.36	0.36
7α,12β-(OH)$_2$	0.36	0.36	0.41	0.49
7β,12β-(OH)$_2$	0.24	0.19	0.32	0.35
3α,6α,7α-(OH)$_3$	0.50	0.46	0.47	0.52
3α,7α,12α-(OH)$_3$	0.15	0.15	0.14	0.15
3α,7α,12β-(OH)$_3$	0.13	0.10	0.13	0.10
3α,7β,12α-(OH)$_3$	0.26	0.22	0.22	0.20
3β,7α,12α-(OH)$_3$	0.05	0.03	0.04	0.06
3α,7β,12β-(OH)$_3$	0.10	0.07	0.09	0.10
3β,7α,12β-(OH)$_3$	0.08	0.06	0.07	0.08
3β,7β,12α-(OH)$_3$	0.06	0.04	0.06	0.07
3β,7β,12β-(OH)$_3$	0.50	0.46	0.49	0.40

[a] Conditions: column, Zorbax ODS; BP, mobile phase, *p*-bromophenacyl ester, methanol/water (80:20); MP, *m*-methoxyphenacyl ester, methanol/water (80:20); NPM, 4-nitrophthalimidomethyl ester, methanol/water (80:20); AM, 9-anthrylmethyl ester, methanol/water (90:10).
[b] From Ref. 25.

are eluted later than those of the corresponding cholic acids. This result implies that introduction of the bulky group into the 3α-position alters the hydrophobic binding with the stationary phase. These derivatization methods are described in detail in Section 3.3.

Higher bile acids and bile alcohols with an additional hydroxyl group on the side chain may provide a new asymmetrical center. Reversed-phase partition chromatography is effective for the resolution of these stereoisomers. It is evident from the data in Table V that the epimeric pairs of *p*-bromophenacyl esters are efficiently resolved [32].

2.2. Bile Acid Sulfates and Glucuronides

In recent years much interest has been focused on the metabolism of bile acid sulfates and glucuronides. The separation and determination of these conjugates has hitherto been carried out by gas–liquid chromatography with prior hydrolysis and/or solvolysis. A drawback of this approach is the loss of information about the conjugated form and position. Bile acid sulfates and glucuronides are very polar and lacking in volatility and thermostability. For these reasons, HPLC on a reversed-phase column affords a more suitable method for the separation of these conjugates.

TABLE V.   Relative Retention Times of *p*-Bromophenacyl Esters of Higher Bile Acids[a,b]

Bile acid	Relative retention time
3α,7α,12α-Trihydroxy-5β-cholanoic acid	1.00
3α,7α,12α-Trihydroxy-26,27-dinor-5β-cholestan-25-oic acid	1.29
3α,7α,12α-Trihydroxy-27-nor-5β-cholestan-26-oic acid	1.65
(25R)-3α,7α,12α-Trihydroxy-5β-cholestan-26-oic acid	1.92
(25S)-3α,7α,12α-Trihydroxy-5β-cholestan-26-oic acid	1.89
3α,7α,12α-Trihydroxy-5β-cholestane-24-carboxylic acid	2.16
3α,7α,12α-Trihydroxy-5β-cholest-23-en-26-oic acid	1.68
(24E)-3α,7α,12α-Trihydroxy-5β-cholest-24-en-26-oic acid	1.55
(24Z)-3α,7α,12α-Trihydroxy-5β-cholest-24-en-26-oic acid	1.66
3α,7α,12α-Trihydroxy-5β-cholest-22-ene-24-carboxylic acid	1.96
(24S)-3α,7α,12α,24-Tetrahydroxy-27-nor-5β-cholestan-26-oic acid	0.80
(24R)-3α,7α,12α,24-Tetrahydroxy-27-nor-5β-cholestan-26-oic acid	0.83
(24R,25R)-3α,7α,12α,24-Tetrahydroxy-5β-cholestan-26-oic acid	1.00
(24R,25S)-3α,7α,12α,24-Tetrahydroxy-5β-cholestan-26-oic acid	1.08
(24S,25R)-3α,7α,12α,24-Tetrahydroxy-5β-cholestan-26-oic acid	0.90
(24S,25S)-3α,7α,12α,24-Tetrahydroxy-5β-cholestan-26-oic acid	0.98

[a] Conditions: column, TSK GEL LS-410; mobile phase, methanol/water (80:20).
[b] From Ref. 32.

The chromatographic behavior of bile acid sulfates on an ODS column has been examined [10,33]. The capacity ratios of sulfated bile acids relative to taurodeoxycholic acid 3-sulfate are listed in Table VI. The k' value for the 3-sulfate is dependent on the structure of the side chain and the acidity of the mobile phase. The relative k' (Rk') values of the unconjugated bile acid 3-

TABLE VI. Relative Capacity Ratios of Sulfated Bile Acids[a,b]

Sulfate	Rk'		
	pH 3.5	pH 5.0	pH 7.0
3-Sulfate			
CA	2.19	1.72	0.24
UDCA	3.68	2.97	0.12
CDCA	5.95	5.27	0.77
DCA	6.34	5.70	0.82
LCA	—	—	1.31
GCA	0.72	0.36	0.27
GUDCA	0.69	0.19	0.12
GCDCA	2.02	1.05	0.74
GDCA	2.18	1.18	0.85
GLCA	—	—	1.66
TCA	0.35	0.35	0.32
TUDCA	0.20	0.19	0.18
TCDCA	0.90	0.89	0.88
TDCA	1.00	1.00	1.00
TLCA	—	—	2.24
7-Sulfate			
CA	1.50	1.39	0.19
UDCA	5.28	4.31	0.25
CDCA	4.99	4.49	0.73
GCA	0.57	0.28	0.21
GUDCA	1.27	0.40	0.28
GCDCA	1.87	0.94	0.69
TCA	0.30	0.27	0.25
TUDCA	0.43	0.36	0.40
TCDCA	0.90	0.88	0.87
12-Sulfate			
CA	1.50	1.33	0.21
DCA	5.41	5.07	0.85
GCA	0.63	0.31	0.22
GDCA	2.24	1.19	0.88
TCA	0.32	0.30	0.26
TDCA	1.06	1.06	1.05

[a] Conditions: column, μBondapak C_{18}; mobile phase, acetonitrile/0.5% phosphate buffer.
[b] From Refs. 10, 33.

sulfates increase with decreasing pH down to 4.5. As for the glycine-conjugated 3-sulfates, an inflection of the Rk' value is observed at pH 5.0, with a marked increase in the value toward the lower pH region (Fig. 2). Irrespective of the structure of the side chain, the 3-sulfates are eluted earlier with an increasing number of the hydroxyl group on the steroid nucleus, except for 3-sulfated ursodeoxycholic acids, which provide smaller k' values than corresponding cholic acid 3-sulfates. It is of interest that the elution order of 3-sulfated bile acids is the same as that of unsulfated bile acids.

The resolution of 3-sulfated chenodeoxycholic acid and deoxycholic acid is affected by the salt concentration as shown in Fig. 3. The peak shape of chenodeoxycholic acid 3-sulfate is almost constant in the range 0.1–0.7% salt concentrations, whereas deoxycholic acid 3-sulfate shows a minimum value at 0.5%. Thus, the maximum resolution factor for the two bile acid 3-sulfates is observed at 0.5% salt concentration.

The chromatographic behavior of 7- and 12-sulfates is similar to that of 3-sulfates, with the only exception being for deoxycholic acid 12-sulfate. The 12-sulfates of glyco- and taurodeoxycholic acids exhibit larger k' values than the corresponding 3-sulfates over the pH range 3.0–7.5. The elution order of the 12- and 3-sulfates of unconjugated deoxycholic acid is reversed at pH 6.0 (Fig. 4). As illustrated in Fig. 5, other 12-sulfates exhibit similarly unusual chro-

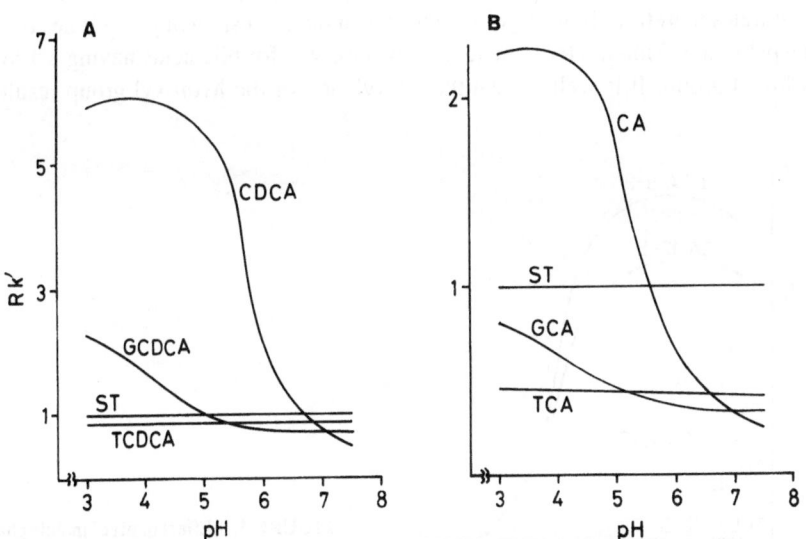

FIGURE 2. Effect of pH of mobile phase on k' values of bile acid 3-sulfates relative to taurodeoxycholic acid 3-sulfate. Conditions: column, μBondapak C_{18}; mobile phase, acetonitrile/0.5% phosphate buffer. (A) chenodeoxycholic acid 3-sulfates; (B) cholic acid 3-sulfates. ST, taurodeoxycholic acid 3-sulfate. (From Ref. [33].)

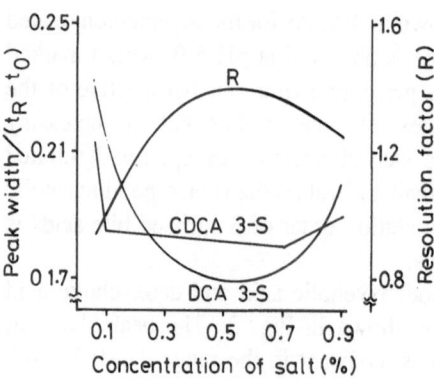

FIGURE 3. Effect of salt concentration on resolution of 3-sulfated chenodeoxycholic acid and deoxycholic acid. Conditions: column, ODS SC-02; mobile phase, acetonitrile/ammonium carbonate solution (8:26). The peak width/(t_R–t_0) value is taken as a parameter to represent the peak shape. (From Ref. [33].)

matographic behaviors. The Rk' value of the 12-sulfate increases significantly with increasing pH above 3.5, and this is particularly marked above 6.0. This pH effect, however, is not observed for bile acid 3- and 7-sulfates. This phenomenon may be explained by the fact that the sulfate at C-12 is sterically close to the carboxylic acid or sulfonic acid residue on the side chain, and hence, steric interaction between these two groups may occur at the higher pH when the carboxyl group becomes dissociated, resulting in larger retention values. Since glycine- and taurine-conjugated bile acids have lower pK values, the dissociated species are dominant in this pH region. Conjugated deoxycholic acid 12-sulfates therefore show larger k' values than the corresponding 3-sulfates over this pH range. Similar steric interaction is observed for bile acids having a 12α-hydroxyl group. It is well known that acetylation of the hydroxyl group results

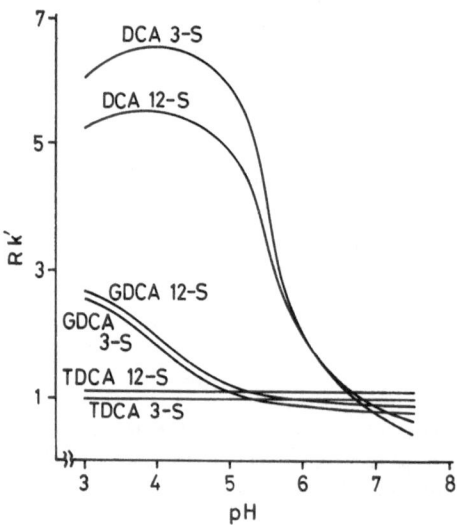

FIGURE 4. Effect of pH of mobile phase on relative k' values of monosulfated deoxycholic acids. Conditions: column μBondapak C_{18}; mobile phase, acetonitrile/0.5% phosphate buffer. (From Ref. [33].)

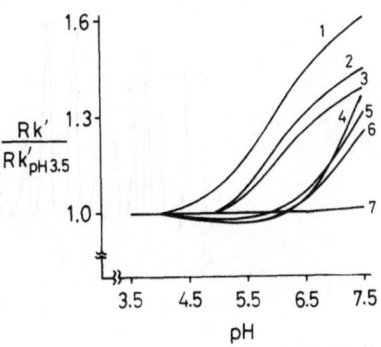

FIGURE 5. Effect of pH of mobile phase on relative k' values of monosulfated bile acids. Conditions: column, ODS SC-02; mobile phase, acetonitrile/0.5% phosphate buffer. 1, deoxycholic acid 12-sulfate; 2, cholic acid 3-acetate 12-sulfate; 3, cholic acid 7-acetate 12-sulfate; 4, 7-dehydrocholic acid 12-sulfate; 5, 7-dehydrocholic acid 3-acetate 12-sulfate; 6, 3-dehydrocholic acid 12-sulfate; 7, 3-dehydrocholic acid 12-acetate 7-sulfate. (From Ref. [10].)

in an increase in the retention value on a reversed-phase column. For 12α-hydroxylated bile acids, the effect of acetylation on the retention value is dependent on pH and decreases in the pH region above 6.0 (Fig. 6). In the previous section, it was demonstrated that unconjugated chenodeoxycholic acid and deoxycholic acids are efficiently resolved under weakly alkaline conditions by an ODS column. The separation of deoxycholic acid and chenodeoxycholic acids would be ascribable to the steric interaction between the hydroxyl group at C-12 and acidic moiety of the side chain. For the simultaneous determination of 3-sulfated bile acids on an ODS column, a weakly alkaline mobile phase should be employed. As illustrated in Fig. 7, 3-sulfated bile acid are efficiently separated on an ODS SC-02 column using acetonitrile/0.5% ammonium carbonate as a mobile phase [34].

The chromatographic behavior of bile acid 3-glucuronides has also been examined and found to resemble those of bile acid 3-sulfates [35]. It is of interest that the elution orders of unconjugated, glycine-, and taurine-conjugated chenodeoxycholic acid 3-glucuronides and the corresponding deoxycholic acid 3-glucuronides are reserved at pH 4 (Table VII). This phenomenon may be explained by steric interaction in the higher pH region, between the 7α-hydroxyl

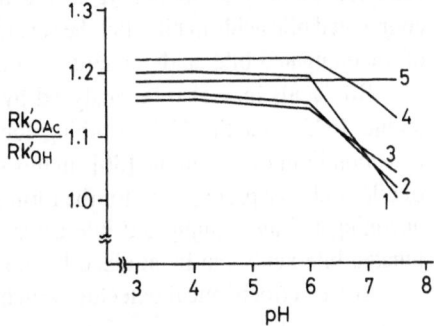

FIGURE 6. Effect of acetylation on relative k' values of sulfated bile acids. Conditions: column, ODS SC-02; mobile phase, acetonitrile/0.5% phosphate buffer. 1, 7-dehydrocholic acid 12-acetate 3-sulfate; 2, cholic acid 12-acetate 3-sulfate; 3, cholic acid 12-acetate 7-sulfate; 4, 3-dehydrocholic acid 12-acetate 7-sulfate; 5, 3-dehydrocholic acid 7-acetate 12-sulfate. (From Ref. [10].)

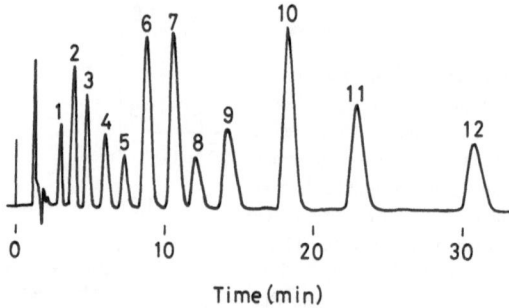

Time (min)

FIGURE 7.   Separation of 3-sulfated bile acids. Conditions: column, ODS SC-02; mobile phase, acetonitrile/0.5% ammonium carbonate solution (8:26), 2.0 ml/min. 1, cholic acid; 2, glycocholic acid; 3, taurocholic acid; 4, chenodeoxycholic acid; 5, deoxycholic acid; 6, glycochenodeoxycholic acid; 7, glycodeoxycholic acid; 8, taurochenodeoxycholic acid; 9, taurodeoxycholic acid; 10, lithocholic acid; 11, glycolithocholic acid; 12, taurolithocholic acid. (From Ref. [33,34].)

and carboxyl groups of the glucuronic acid moiety, which has a pK value of approximately 3.5.

3.   DETECTION METHODS

3.1.   Direct Determination

A variety of methods are presently available for the detection of bile acids separated by HPLC. The differential refractometer is widely used for the detection of compounds that do not exhibit ultraviolet (UV) absorption or fluorescence properties. The sensitivity obtainable by refractive index detection, however, is insufficient for the determination of bile acids in biological fluids [11].

Bile acids are usually monitored by UV detection at wavelength 190–210 nm. The limits of detection monitored at 205 nm for unconjugated, glycine-, and taurine-conjugated bile acids are 1, 0.1, and 0.3 nmole, respectively. This detection method is applicable to the determination of glycine- and taurine-conjugated bile acids in bile, but the sensitivity is insufficient for the measurement of unconjugated bile acids in serum, in particular, from normal subjects.

Bile acids may also be analyzed by HPLC using an ion pair reagent, Hyamine 1622, and the UV-absorbing ion pair is monitored at 254 nm with a detection limit of 0.5 nmole [36]. In this system dissociation of the acidic moiety of bile acids is a prerequisite for the formation of an ion pair. Distinction between unconjugated and conjugated bile acids and between glycine- and taurine-conjugated bile acids can be attained by adjusting the pH of mobile phase.

The electrochemical detector, which is currently used for the trace analysis

TABLE VII.   Relative Capacity Ratios of Bile Acid 3-Glucuronides[a,b]

	Rk'		
Glucuronide	pH 3.5	pH 5.0	pH 7.0
CA	1.16	2.58	0.35
UDCA	1.11	2.38	0.31
CDCA	3.09	6.91	0.92
DCA	3.43	5.90	0.88
LCA	6.35	13.10	1.68
GCA	0.32	0.43	0.39
GUDCA	0.29	0.38	0.32
GCDCA	0.88	1.00	1.01
GDCA	1.00	1.00	1.00
GLCA	1.90	2.11	1.75
TCA	0.16	0.47	0.52
TUDCA	0.15	0.41	0.42
TCDCA	0.31	1.02	1.45
TDCA	0.38	1.05	1.39
TLCA	0.72	1.79	2.15

[a] Conditions: column. ODS SC-02; mobile phase. acetonitrile/0.5% phosphate buffer.
[b] From Ref. 35.

of catecholamines, is also applicable to the direct measurement of unconjugated bile acids, with a detection limit of 1 nmole [37]. Conductometric detection, which provides response to the electrolytes and ionizable compounds, is more sensitive than UV detection and has a sensitivity of 10 pmoles [38]. However, only a weakly alkaline mobile phase may be used in combination with a suppressor column of cation exchanger.

Since the direct-detection methods have a disadvantage in selectivity, determination of bile acids in biological fluids is remarkably influenced by the cleanup procedure employed. For solid-phase extraction of bile acids (see this volume, Chapter 1), Amberlite XAD resin, Sep-Pak C_{18}, and Bond Elut cartridges are currently used. Comparative data on the recovery tests for serum bile acids by use of these three are listed in Table VIII and have been reported [39,40].

Amberlite XAD-2 resin has been widely used for the extraction of conjugated steroids. This method, however, is not always satisfactory with respect to the recovery rate and reproducibility. When 0.4 nmole each of bile acids is applied to a column of Amberlite XAD-2, recoveries are less than 50%. In the case of 2 nmoles, the recoveries are improved but not quantitative, especially for conjugated bile acids. On the other hand, when more than 20 nmoles each

TABLE VIII. Recovery Rates of Bile Acids Added to Human Serum[a]

Bile acid	Bond Elut (0.4 nmol)	Sep-Pak C_{18} (0.4 nmol)	Amberlite XAD-2 (2 nmoles)
		Recovery (%)	
CA	96.0 ± 5.6	91.2 ± 7.4	75.2 ± 31.5
UDCA	98.9 ± 3.0	99.6 ± 2.2	85.0 ± 9.6
CDCA	96.6 ± 5.0	89.1 ± 2.1	73.8 ± 6.7
DCA	100.7 ± 2.9	94.4 ± 2.9	75.5 ± 6.3
LCA	97.6 ± 2.2	76.5 ± 3.0	42.4 ± 0.9
GCA	90.9 ± 5.3	88.3 ± 4.3	44.1 ± 26.1
GUDCA	92.3 ± 3.7	95.6 ± 3.9	65.9 ± 14.1
GCDCA	100.8 ± 6.4	89.5 ± 3.4	50.7 ± 15.8
GDCA	95.8 ± 3.7	84.7 ± 2.2	59.1 ± 9.8
GLCA	92.9 ± 1.4	91.4 ± 3.6	39.1 ± 2.2
TCA	92.7 ± 4.6	85.7 ± 7.9	19.4 ± 34.9
TUDCA	98.9 ± 4.7	83.0 ± 1.5	24.7 ± 12.7
TCDCA	96.1 ± 5.8	97.4 ± 1.5	26.5 ± 16.4
TDCA	90.3 ± 3.5	98.5 ± 1.8	21.2 ± 11.6
TLCA	91.0 ± 1.8	88.1 ± 2.7	25.3 ± 2.8

[a] From Ref. 39.

of bile acids is loaded on this column, satisfactory recoveries are obtained. Moreover, the recovery of bile acid sulfates is influenced by the pH for adsorption and desorption and by the concentration of alcohol used for elution [34]. For this reason, solid-phase extraction with Amberlite XAD resin can be applied to bile, but is unsuitable for serum and urine specimens.

In recent years Sep-Pak C_{18} and Bond Elut cartridges have been developed for use in the cleanup of biological materials. This method is highly efficient for the extraction of bile acids in biological fluids and for removing interfering substances in a complex matrix [4,34,39,40]. The biological sample is first diluted with 2–5 volumes of either phosphate buffer (pH 7.0) or 0.1 M sodium hydroxide solution and then filtered through the cartridge. After being washed with water and 1.5% ethanol to remove coexisting inorganic salts and polar substances, bile acids are quantitatively recovered by elution with 90% ethanol or methanol. In the case of bile acid sulfates, the use of a large volume of water instead of 1.5% ethanol is recommended for washing the cartridge because losses of polar sulfates may occur.

When a neutral or weakly alkaline mobile phase is used for the HPLC separation of bile acids, little difference in the k' value is observed among unconjugated, glycine-, and taurine-conjugated bile acids, though satisfactory separation of chenodeoxycholic acid and deoxycholic acid is obtained.

In addition, a solid-phase extraction procedure is not always suitable for purification of sulfated bile acids. Therefore, a group separation into the unconjugated, glycine-, and taurine-conjugated, and sulfated bile acids prior to HPLC is recommended. This may be achieved on a lipophilic ion exchange gel, such as diethylaminohydroxypropyl Sephadex LH-20 [41], piperidinohydroxypropyl Sephadex LH-20 (PHP-LH-20) [4,19,30,34], or triethylaminohydroxypropyl Sephadex LH-20 [42].

HPLC has been criticized as having limited value in structural elucidation studies; however, utilizing various combinations of stationary and mobile phases and inspection of the chromatographic behavior of bile acids under different conditions is helpful for the characterization of bile acids [34]. The fraction corresponding to a peak on the chromatogram is collected and, after the addition of an internal standard, is subjected to HPLC using different mobile phases of varying pH and salt concentration. Comparison of the Rk' values and the peak area ratios to the internal standard with those of standard bile acid are of great use for unambiguous characterization.

3.2. Coupling with Enzymic Reaction

The sensitivity obtainable by the direct detection method is insufficient for determination of bile acids in serum, and numerous attempts have been made to improve the sensitivity of the detection system. 3α-Hydroxysteroid dehydrogenase (3α-HSD) is widely used for the enzymic determination of bile acids. This enzyme is capable of transforming the 3α-hydroxyl group of 5β-steroids into the 3-oxo group in the presence of NAD^+ under alkaline condition. In this enzymic reaction NAD^+ is converted into NADH exhibiting ultraviolet (UV) absorption and fluorescence, and this enzymic reaction has been developed as a monitoring system for bile acids.

In the initial work [43] the eluate is mixed with the 3α-HSD and NAD^+ solutions, and NADH yielded by enzymic reaction is monitored by fluorescence detection (excitation at 350 nm; emission at 460 nm). Since this method consumes a considerable amount of the enzyme, a modified system was then developed using a column of immobilized enzyme in a continuous flow mode [44–47]. A flow diagram of the analytical system is illustrated in Fig. 8. The enzymic detection system is more selective than the direct determination methods and is capable of detecting 10–20 pmoles of bile acids. This enzymic reaction may be further coupled with electrochemical detection [48]. A solution of phenazine methosulfate is supplied and mixed with the eluate from the enzyme column. In the presence of NADH, phenazine methosulfate is readily converted into its reduced form, which is highly responsive to electrochemical detection. There is

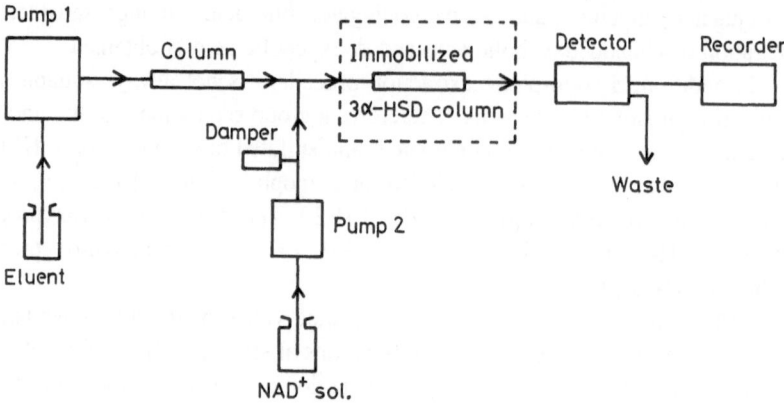

FIGURE 8.   Flow diagram of HPLC coupled with the immobilized 3α-HSD reactor.

no significant difference in the detection limit between electrochemical and fluorescence detection systems, but the electrochemical detector is more favorable with respect to selectivity.

Using a detection system coupled with the enzymic reaction, linear gradient elution is usually employed. The limited use of a weakly alkaline mobile phase, which is suitable for enzymic reaction, is of disadvantage. Also, difficulties are encountered with the resolution of bile acids and the reproducibility and reliability of enzymic reaction. These problems may be overcome by prior separation of bile acids into groups before HPLC analysis.

3.3.   Derivatization of Carboxyl and Hydroxyl Groups

In HPLC, derivatization is widely used for biological substances that have neither significant UV absorption nor fluorescence properties to render them sensitive to detection. Precolumn labeling of bile acids with a chromophore or fluorophore usually utilizes the carboxyl group in the side chain. Various derivatization methods for use in UV detection have been devised [21–25]. For instance, precolumn labeling of unconjugated and glycine-conjugated bile acids with 1-*p*-nitro-benzyl-3-*p*-tolytriazene or O-*p*-nitrobenzyl-*N*, *N*'-diisopropylisourea affords the *p*-nitrobenzyl ester, which exhibits UV absorption properties at 254 nm with the detection limit of 0.2 nmole. This sensitivity, however, is still insufficient for the determination of bile acids at low levels found in normal serum. The fluorescence reagents, 4-bromomethyl-7-methoxycoumarin [26,27] and 1-bromoacetylpyrene [28], are suitable for precolumn labeling. Bile acids are subjected to derivatization with these reagents in the presence of 15-crown-5-ether or 18-crown-6-ether, and the derivatives formed are separated on a reversed-phase HPLC column. Fluorescence detection affords a detection limit for bile acids of 5–10 pmoles. These derivatization methods are of disadvantage in resolving chenodeoxycholic acid and deoxycholic acids, and more important, prior deconjugation by chemical or enzymic means is a prerequisite for determination of the taurine conjugates which do not form derivatives with these reagents.

Introduction of a fluorophore into the C-3 position of bile acids has also been developed. Bile acids are transformed by enzymic reaction with immobilized 3α-HSD into the 3-oxo compounds, which in turn are derivatized with dansylhydrazine under acidic conditions. The resulting dansylhydrazone shows high sensitivity to the fluorescence detector with a detection limit of 0.5 pmole [49,50]. In this method a neutral mobile phase is more effective for the resolution of chenodeoxycholic acid and deoxycholic acid, but not necessarily suitable for removal of interfering peaks in the chromatogram derived from the reagent.

It is sufficiently substantiated that the 3α-hydroxyl group is more reactive toward acylating agents than the axial 7α- and 12α-hydroxyl groups and the equatorial 7β-hydroxyl group [51–54]. The fluorophore can be introduced selectively into the 3α-hydroxyl group to form the ester. Several derivatization reagents have been proposed for use in HPLC of alcoholic hydroxyl compounds. These reagents, however, are not necessarily satisfactory with respect to reactivity, stability, and sensitivity. Combination of carbonyl nitrile with fluorophores yields aroyl nitriles, a new type of fluorescence labeling reagent.

In actuality, 1-anthroyl nitrile reacts with bile acids in the presence of quinuclidine in acetonitrile to provide solely the 3-(1-anthroyl) derivatives [29,30] (Fig. 9). The derivatized bile acids are efficiently separated on an ODS column using methanol/0.3% phosphate buffer (pH 6.0) as a mobile phase and can be monitored by fluorescence detection (excitation at 370 nm; emission at 470 nm), with a limit of detection of 20 fmoles (Fig. 10). It is to be noted that in this condition the more sterically hindered 7α-, 12α-, and 7β-hydroxyl groups undergo no acylation owing to their steric hindrance. This derivatization method is also applicable to characterization and quantification of 7- and 12-sulfated bile acids in human urine [55]. In this instance urine specimen is subjected successively

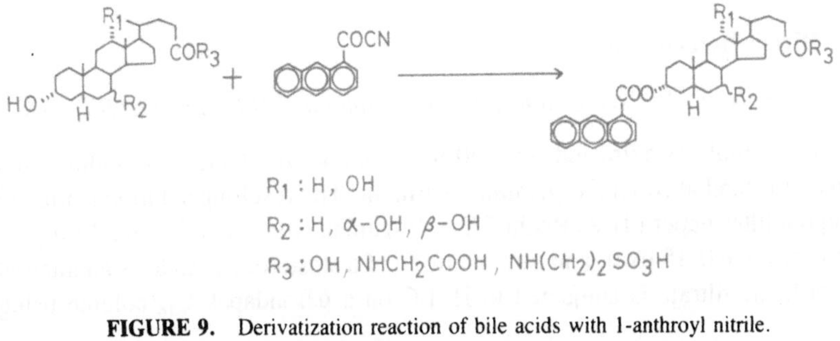

R_1 : H, OH

R_2 : H, α-OH, β-OH

R_3 : OH, NHCH$_2$COOH, NH(CH$_2$)$_2$SO$_3$H

FIGURE 9. Derivatization reaction of bile acids with 1-anthroyl nitrile.

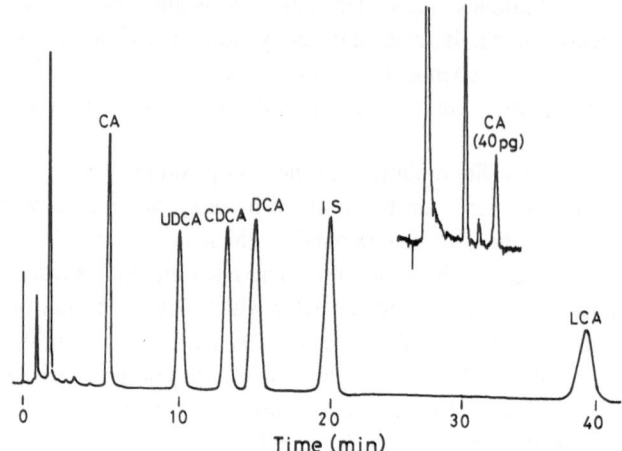

FIGURE 10. Chromatogram of 3-(1-anthroyl) derivatives of unconjugated bile acids. Conditions: column, Cosmosil 5C$_{18}$; mobile phase, methanol/0.3% potassium phosphate buffer (pH 6.0) (1:5), 1.6 ml/min. IS, internal standard, deoxycholic acid 12-propionate. (From Ref. [30].)

to cleanup on a Sep-Pak C$_{18}$ cartridge and group separation on PHP-LH-20. The sulfate fraction isolated is derivatized through the 3α-hydroxyl function with 1-anthroyl nitrile and the resulting 3-(1-anthroyl) derivatives are separated on an ODS column and monitored by fluorescence detection.

It should be stressed that when applying precolumn labeling methods to biological specimens, efficient cleanup procedures are necessary to eliminate interfering substances and excess reagents.

4. APPLICATION OF HPLC TO BIOLOGICAL SPECIMENS

The following section will outline methods that have been applied to the determination of bile acids from biological fluids using HPLC.

4.1 Direct Determination

4.1.1. Analysis of Glycine- and Taurine-Conjugated Bile Acids in Bile [16]

Bile sample is extracted with 20-fold volume of ethanol, by boiling, and allowed to stand at room temperature overnight. The resulting solution is filtered through a filter paper (Toyo Roshi 5A or Whatman No. 43) and then a Millipore filter (pore size 0.45 μm), and after addition of testosterone acetate as an internal standard, the filtrate is subjected to HPLC on a μBondapak C$_{18}$ column using a mobile phase of acetonitrile/methanol/0.3 M phosphate buffer (pH 3.4) (1:6:3). Flow rate is 0.5 ml/min, and bile acids are monitored by UV detection (at 200 nm).

4.1.2. Analysis of Bile Acids and Their Sulfates in Bile [4,34]

A bile sample (10–100 μl) is diluted with 0.5 M phosphate buffer (pH 7.0, 4 ml), heated at 60°C for 1 hr and passed through a Sep-Pak C_{18} cartridge. After being washed with water (12 ml), bile acids and their sulfates are eluted with 90% ethanol (4 ml). The eluate is applied to a column (20 mm × 6 mm i.d.) of PHP-LH-20 acetate form (110 mg). Neutral compounds are removed by washing with 90% ethanol (4 ml), and the bile acids are then separated into unconjugated, glycine-, and taurine-conjugated fractions by stepwise elution with 0.1 M acetic acid in 90% ethanol (4 ml), 0.2 M formic acid in 90% ethanol (4 ml), and 0.3 M acetic acid–potassium acetate in 90% ethanol (pH 6.3, 4 ml). After addition of estriol, as an internal standard, each fraction is subjected to HPLC on a Radial-Pak A column using acetonitrile/0.3% ammonium phosphate buffer (pH 7.7) as mobile phase (8:19 and 8:23). Further elution with 1% ammonium carbonate in 70% ethanol (4 ml) provides the sulfates. Glycocholic acid and 7α-acetoxy-12α-hydroxy-3-oxo-5β-cholan-24-oic acid 12-sulfate are added as internal standards, and the whole is subjected to a sulfoethyl Sephadex LH-20 column (50 mm × 10 mm i.d.) for elimination of inorganic salts. The sulfate fraction obtained is subjected to HPLC on an ODS SC-02 column using acetonitrile/0.5% ammonium carbonate (8:31, 8:26, and 8:23) as mobile phases with a flow rate of 2 ml/min, and the eluate is monitored by UV detection (at 205 nm).

4.2. Derivatization Methods

4.2.1. Fluorescence Labeling via the Hydroxyl Group

4.2.1a. 1-Anthroyl Nitrile [30]. The serum sample (100 μl) is mixed with unconjugated, glycine-, and taurine-conjugated deoxycholate 12-propionates as internal standards, diluted with 0.5 M phosphate buffer (pH 7.0, 1 ml), and applied to a Bond Elut cartridge. After successive washing with water (2 ml) and 1.5% ethanol (1 ml), bile acids are eluted with 90% ethanol (2 ml). An aliquot (1/5) of the effluent is dried and treated with 1-anthroyl nitrile (200 μg) and 0.08% quinuclidine in acetonitrile (200 μl) for 20 min at 60°C. The excess reagent is decomposed by addition of methanol, and the resulting solution evaporated to dryness. After separation into unconjugated, glycine-, and taurine-conjugated fractions, bile acids of each group are resolved by HPLC on a Cosmosil $5C_{18}$ column using methanol/0.3% potassium phosphate buffer (pH 6.0)

(5:1) as a mobile phase with a flow rate of 1.6 ml/min. The eluate is monitored by fluorescence detection (excitation at 370 nm; emission at 470 nm).

4.2.1b. Dansylhydrazine [49]. The serum sample (100 μl) is mixed with methanol (500 μl), sonicated for 15 min, and centrifuged. The supernatant (300 μl) is removed and evaporated. The residue is dissolved in 0.05 M phosphate buffer (pH 7.0, 1 ml), purified by passage through a Sep-Pak C_{18} cartridge followed by group separation on PHP-LH-20. The unconjugated, glycine-, and taurine-conjugated fractions obtained are dissolved in methanol (100 μl) and applied to an immobilized 3α-HSD column (36 mm × 6 mm i.d.). After elution with 0.5 mM NAD$^+$ solution in 0.1 M pyrophosphate buffer (pH 9.0), the eluate is passed through a Sep-Pak C_{18} cartridge. 3-Oxo bile acids thus formed are treated with 0.2% dansylhydrazine in 0.1% trichloroacetic acid/benzene (200 μl) for 10 min at 30°C. Derivatized bile acids are separated by HPLC on a Zorbax ODS column using acetonitrile/methanol/0.03 M phosphate buffer (pH 3.2) (15:68:17) as a mobile phase with a flow rate of 1 ml/min, and the eluate is monitored by fluorescence detection (excitation at 340 nm, emission at 520 nm).

4.2.2. Fluorescence Labeling via the Carboxyl Group

1-Bromoacetylpyrene [28]. Bile acids are extracted from serum (100–200 μl) with a Sep-Pak C_{18} cartridge and then fractionated on a PHP-LH-20 column. After addition of lauric acid as an internal standard, the unconjugated and glycine-conjugated fractions are dissolved in 0.005% methanolic potassium hydroxide solution (100 μl) and dried. To the residue are added 0.1% dicyclohexyl-18-crown-6 in acetonitrile (100 μl) and 25 mM 1-bromoacetylpyrene in acetonitrile (100 μl), and the mixture is heated at 40°C for 30 min. The derivatives are subjected to HPLC on a Radial-Pak A column using acetonitrile/methanol/water (10:5:4 and 10:5:2) as mobile phases in stepwise gradient mode. The taurine-conjugated fraction is desalted on a Sep-Pak C_{18} cartridge followed by hydrolysis with cholylglycine hydrolase. The deconjugated bile acids thus obtained are derivatized with 1-bromoacetylpyrene, as described above.

5. CONCLUSIONS

In this chapter we have reviewed recent advances in HPLC for the separation and determination of bile acids. During the last decade various methods for the analysis of bile acids have been developed, with satisfactory reproducibility, sensitivity, and selectivity. In spite of much effort, however, the analytical procedure still remains unsolved with respect to simplicity and rapidity. The method with high specificity and reliability is often tedious, time-consuming, and, hence, not suitable for routine analysis. This is a particularly important problem in clinical studies.

A general scheme for analysis of bile acids in biological fluids consists of three steps: pretreatment of a biological specimen, separation into individual bile acids, and determination by HPLC with a detection system. Significant progress has recently been made in the cleanup procedure, e.g., solid-phase extraction with an ODS cartridge and group separation on a lipophilic ion exchange gel. In the future an online cleanup system in a continuous flow mode may be devised. In addition, a new type of column capable of separating bile acids as efficiently as the capillary column in gas–liquid chromatography will be developed. Although inspection of the chromatographic behavior can serve for unequivocal characterization of bile acids, the HPLC method must overcome its inherent disadvantage in structural elucidation. A novel system consisting of liquid chromatography combined with mass spectrometry seems to offer the most promising method of solving the problem of the analysis of bile acids in biological fluids.

REFERENCES

1. R. Shaw and W. H. Elliott, *Anal. Biochem.* **74,** 273 (1976).
2. R. W. R. Baker, J. Ferrett, and G. M. Murphy, *J. Chromatogr.* **146,** 137 (1978).
3. K. Shimada, M. Hasegawa, J. Goto, and T. Nambara, *J. Chromatogr.* **152,** 431 (1978).
4. J. Goto, H. Kato, Y. Saruta, and T. Nambara, *J. Liquid Chromatogr.* **3,** 991 (1980).
5. C. Horváth, W. Melander, and I. Molnár, *Anal. Chem.* **49,** 142 (1977).
6. D. M. Small, *in* "The Bile Acids" (P. P. Nair and D. Kritchevsky, ed.), Vol. 1, p. 249, Plenum Press, New York (1971).
7. S. Miyazaki, H. Tanaka, R. Horikawa, H. Tsuchiya, and K. Imai, *J. Chromatogr.* **181,** 177 (1980).
8. R. Shaw, M. Rivetna, and W. H. Elliott, *J. Chromatogr.* **202,** 347 (1980).
9. J. Goto, T. Chikai, T. Nambara, and T. Iida (in preparation).
10. J. Goto, H. Kato, K. Kaneko, and T. Nambara, *J. Liquid Chromatogr.* **4,** 1351 (1981).
11. N. A. Parris, *J. Chromatogr.* **133,** 273 (1977).
12. R. Shaw, J. A. Smith, and W. H. Elliott, *Anal. Biochem.* **86,** 450 (1978).
13. R. Shaw and W. H. Elliott, *Lipids.* **13,** 971 (1978).
14. C. A. Bloch and J. B. Watkins, *J. Lipid Res.* **19,** 510 (1978).
15. M. S. Sian and A. J. H. Rains, *Clin. Chim. Acta* **98,** 243 (1979).
16. F. Nakayama and M. Nakagaki, *J. Chromatogr.* **183,** 287 (1980).
17. A. D. Reid and P. B. Baker, *J. Chromatogr.* **247,** 149 (1982).
18. A. T. Ruben and G. P. Van Berge Henegouwen, *Clin. Chim. Acta* **119,** 41 (1982).
19. J. Goto, M. Hasegawa, H. Kato, and T. Nambara, *Clin. Chim. Acta* **87,** 141 (1978).
20. W. E. Jefferson, Jr. and F. C. Chang, *Anal. Lett.* **9,** 429 (1976).
21. S. Okuyama, D. Uemura, and Y. Hirata, *Chem. Lett.* **1976,** 679.
22. B. Shaikh, N. J. Pontzer, J. E. Molina, and M. I. Kelsey, *Anal. Biochem.* **85,** 47 (1978).
23. F. Stellaard, D. L. Hachey, and P. D. Klein, *Anal. Biochem.* **87,** 359 (1978).
24. G. Mingrove, A. V. Greco, and S. Passi, *J. Chromatogr.* **183,** 277 (1980).

25. T. Iida, Y. Ohnuki, F. C. Chang, J. Goto, and T. Nambara, *Lipids*, **20**, 187 (1985).
26. G. L. Carlson and D. T. E. Belobaba, *in* "The ACS/CSJ Chemical Congress at Hawaii" Abstract (1979).
27. S. Okuyama, D. Uemura, and Y. Hirata, *Chem. Lett.* **1979**, 461.
28. S. Kamada, M. Maeda, and A. Tsuji, *J. Chromatogr.* **272**, 29 (1983).
29. J. Goto, N. Goto, F. Shamsa, M. Saito, S. Komatsu, K. Suzaki, and T. Nambara, *Anal. Chim. Acta* **147**, 397 (1983).
30. J. Goto, M. Saito, T. Chikai, N. Goto, and T. Nambara, *J. Chromatogr.* **276**, 289 (1983).
31. J. Goto, S. Komatsu, N. Goto, and T. Nambara, *Chem. Pharm. Bull.* **29**, 899 (1981).
32. M. Une, F. Nagai, and T. Hoshita, *J. Chromatogr.* **257**, 411 (1983).
33. J. Goto, H. Kato, and T. Nambara, *J. Liquid Chromatogr.* **3**, 645 (1980).
34. J. Goto, H. Kato, Y. Saruta, and T. Nambara, *J. Chromatogr.* **226**, 13 (1981).
35. J. Goto, K. Suzaki, and T. Nambara, *J. Chromatogr.* **348**, 151 (1985).
36. N. Parris, *Anal. Biochem.* **100**, 260 (1979).
37. W. Kemula and W. Kutner, *J. Chromatogr.* **204**, 131 (1981).
38. Y. Hashimoto, Y. Asai, M. Moriyasu, and A. Uji, *Anal. Lett.* **14**, 1483 (1981).
39. J. Goto, T. Chikai, K. Makiuchi, and T. Nambara, *Bunseki Kagaku* (in press).
40. K. D. R. Setchell and J. Worthington, *Clin. Chim. Acta* **125**, 135 (1982).
41. B. Almé, A. Bremmelgaard, J. Sjövall, and P. Thomassen, *J. Lipid Res.* **18**, 339 (1977).
42. M. Axelson and J. Sjövall, *Biochim. Biophys. Acta* **751**, 162 (1983).
43. S. Baba, R. Uenoyama, K. Suminoe, F. Takeda, S. Hasegawa, and Y. Kameno, *Kobe J. Med. Sci.* **26**, 89 (1980).
44. S. Okuyama, N. Kokubun, S. Higashidate, D. Uemura, and Y. Hirata, *Chem. Lett.* **1979**, 1443.
45. K. Arisue, Z. Ogawa, K. Kohda, C. Hayashi, and Y. Ishida, *Jpn. J. Clin. Chem.* **9**, 104 (1980).
46. S. Ohnishi, S. Itoh, and Y. Ishida, *Biochem. J.* **204**, 135 (1982).
47. S. Hasegawa, R. Uenoyama, F. Takeda, J. Chuma, and S. Baba, *J. Chromatogr.* **278**, 25 (1983).
48. S. Kamada, M. Maeda, A. Tsuji, Y. Umezawa, and T. Kurahashi, *J. Chromatogr.* **239**, 773 (1982).
49. T. Kawasaki, M. Maeda, and A. Tsuji, *J. Chromatogr.* **272**, 29 (1983).
50. A. D. Reid and P. R. Baker, *J. Chromatogr.* **260**, 115 (1983).
51. J. Goto, H. Kato, F. Hasegawa, and T. Nambara, *Chem. Pharm. Bull.* **27**, 1402 (1979).
52. J. Goto, H. Kato, K. Kaneko, and T. Nambara, *Chem. Pharm. Bull.* **28**, 3389 (1980).
53. J. Goto, K. Suzaki, and T. Nambara, *Chem. Pharm. Bull.* **28**, 1258 (1980).
54. J. Goto, K. Suzaki, and T. Nambara, *Chem. Pharm. Bull.* **30**, 4422 (1982).
55. J. Goto, T. Chikai, and T. Nambara, *Anal. Sci.* **2**, 175 (1986).

Chapter 3

NUCLEAR MAGNETIC RESONANCE SPECTROSCOPY OF BILE ACIDS

Stephen Barnes and David N. Kirk

---

1. INTRODUCTION: WHY USE NMR FOR BILE ACIDS?

It is perhaps to the detriment of those investigators who are currently working in research areas involving bile acids that so much of the chemistry of this group of substances was worked out in the first half of this century (see Fieser and Fieser [1] for a summary of this body of knowledge). As a result, several of the analytical techniques that have been developed over the past 20 years have not been systematically applied to the study of bile acids.

Nuclear magnetic resonance (NMR) spectroscopy has become the most important spectroscopic technique for the study of inorganic and organic molecules in solution, and it is increasingly being applied, where appropriate, to the solid or gaseous states. It can provide a wealth of information on the molecular structure and physical properties of a compound. When a chemical substance is known to belong to a particular class, like the bile acids, its detailed structure, including the configurations of substituents, is often accessible from a straightforward analysis of its NMR spectrum. Use of more elaborate NMR methods has led to the assignment of molecular formulae to many organic compounds of previously unknown structures.

Another application of NMR which is developing rapidly concerns the size of the molecular complex in solution, its tumbling rate in the solvent medium, and the time-averaged solution structure. NMR methods are also being applied to the study of organic molecules immobilized in cell membranes. The time scale of the physical process underlying the NMR experiment is appropriate in

---

Stephen Barnes Departments of Pharmacology and Biochemistry and Comprehensive Cancer Center, University of Alabama at Birmingham, Birmingham, Alabama **David N. Kirk** Department of Chemistry, Queen Mary College, University of London, London, England

many cases for the observation of two or more chemical states of a system in equilibrium, and for measurement of the kinetics of the interchange between them.

One distinct advantage of NMR over other techniques is that it is nondestructive and allows complete recovery of samples. This is most vividly demonstrated by the recent rapid development of *in vivo* NMR methodology, using both imaging methods and spectroscopy. The latter technique has enabled the study of metabolism in living cells, isolated organs, small animals, and even the human body.

2. BASIC PRINCIPLES OF NMR

2.1. The NMR Phenomenon

NMR spectroscopy, like other spectroscopic techniques, depends on the absorption and emission of quanta of radiation, associated with transitions between energy levels in a molecule or atom. In the case of ultraviolet spectroscopy the transitions are between energy levels of the outermost electrons of atoms or molecules, whereas NMR spectroscopy involves the energy levels of the spinning atomic nucleus in a magnetic field.

The hydrogen nucleus (a proton, ^1H), and certain other nuclei, have magnetic properties associated with their electric charge and spin. They therefore tend to align themselves in an external magnetic field, like a compass needle in the earth's magnetic field. Quantum restrictions limit the number of orientations of the proton to two: its magnetic field may either be aligned with or opposed to the external field, corresponding to low- and high-energy states, respectively. The energy *difference* between these states for an individual proton (unlike a compass needle) is so small that there is only a very slight excess population of the lower over the higher-energy state at thermal equilibrium, given by the equation for the Boltzmann distribution as of the order of 1 part in 10^6. Although small, this difference in populations is critical for the NMR phenomenon.

When an external field is applied to a spinning magnetic nucleus, the axis of spin precesses around the direction of the applied field, just as a spinning top precesses around the vertical direction under the earth's gravitational pull. The precession frequency, ν, is directly proportional to the strength, B_o, of the applied field: $\nu \propto B_o$. Electromagnetic radiation of frequency ν is able to invert the spin orientation, with absorption of energy (*resonance*). This process can occur until the populations of protons in the two spin states are equalized (*saturation*), but the excitation energy may be reradiated or lost by transfer to the surrounding environment (*relaxation*), so that the population difference between energy states tends to be maintained. For the magnetic fields normally employed in NMR spectrometers, in the range 1.4–11.7 tesla (T) (14,000–117,000 gauss), the resonance condition for protons requires radiofrequencies in the range 60–500 megahertz (MHz).

Protons have a spin quantum number $I = 1/2$. All other nuclei with nonzero values of I are also magnetic, although only those with $I = 1/2$ (e.g., ^1H, ^3H, ^{13}C, ^{19}F, ^{31}P) provide NMR spectra free from excessive multiplicity of peaks. The deuterium nucleus (^2H; $I = 1$), and others for which $I > 1/2$, give complex spectra, with highly split peaks. Nuclei with *even* mass number and *even* charge number have zero spin and are accordingly not observable by NMR. Such nuclei include the common isotopes ^{12}C, ^{16}O, and ^{32}S. NMR spectroscopy of carbon nuclei is therefore based on the ^{13}C isotope, which, however, has a natural abundance of only 1.1%. The resonance frequency for ^{13}C in any given field is roughly 1/4 of that for ^1H. ^{13}C NMR is considerably less sensitive than ^1H NMR, although modern instruments provide ^{13}C spectra at natural abundance. In some cases isotopic enrichment can be helpful.

The most important features that can be derived from the NMR spectrum of a chemical compound are *chemical shifts*, peak *integrals*, and the *multiplicities* and widths of split peaks. The following sections explain these terms briefly.

2.2. Chemical Shift

An atomic nucleus that is part of a molecule is to some extent shielded from an applied magnetic field by the electrons and other atomic nuclei that surround it. Since the immediate environments of individual nuclei in a molecule are normally not identical, the shielding they experience will differ slightly. They therefore resonate over a narrow band of frequencies rather than at a single frequency, giving rise to a spectrum of resonance lines, or *signals*.

The position of a signal relative to that of a standard is termed the *chemical shift*. The usual standard, except for aqueous solution, is tetramethysilane [$(CH_3)_4Si$; TMS]. Its ^1H and ^{13}C nuclei are more shielded than in almost all purely organic molecules, so their chemical shifts are arbitrarily assigned zero value. Sodium trimethylsilyl-1,1,2,2-tetradeutero propionate is a suitable standard for aqueous solution. The solvents used in NMR are generally the deutero analogs in order to avoid the otherwise enormous proton resonances from the proteo forms. However, they usually have ~1% of the proton form, which can be used as a reference in the ^1H NMR spectrum, e.g., chloroform, dioxane, methanol.

Signals in the spectra of other compounds are said to be *deshielded*, or *downfield* from the TMS signal, with chemical shifts measured in parts per million of the applied radiofrequency and designated by the symbol δ. Because the chemical shifts are unitless and independent of the applied radiofrequency, simple comparison can be made of data obtained from different NMR spectrometers.

Values of δ normally lie in the range 0–10 ppm for protons. (An older scale, now rare, used the symbol τ, with TMS assigned a chemical shift of 10 ppm. Interconversion is simple: $\delta = 10 - \tau$). Within the range δ 0–10, it is possible to distinguish a number of characteristic regions corresponding to the resonance frequencies of protons in particular environments. The examples in Table I refer particularly to those structural features which occur commonly in bile acids and their derivatives. For further information on chemical shifts and more extensive listings the reader is referred to Williams and Fleming [2] and Kemp [3]. The most important general principle affecting chemical shifts is that proximity of a proton to an electronegative atom or group, which withdraws electronic charge, or attachment to unsaturated carbon tends to cause a shift to lower field (larger value of δ; *deshielding*).

The NMR spectra of all compounds of the steroid class are complicated by the appearance of a "methylene envelope" in the region δ 0.5–2.5, due to overlapping resonance peaks from all those protons around the ring system and in the side chain that are not shifted to lower field by substituents or unsaturation (Fig. 1). Various special methods (see Sections 5.3–5.7), and the use of NMR spectrometers with superconducting magnets (see Section 5.1), now make it possible to resolve the methylene envelope into its component parts.

TABLE I. Characteristic Chemical Shifts for Protons

	Approximate range [δ (ppm)]
Methyl groups (C$\underline{\text{H}}_3$, remote from unsaturation or electronegative substituents)	0.5–1.5
Methylene (C$\underline{\text{H}}_2$) and methine (C$\underline{\text{H}}$) groups, remote from unsaturation or electronegative substituents	1–2
Methyl groups adjacent to carbonyl (—C—C$\underline{\text{H}}_3$ in acetates or ketones) ‖ O	2.0–2.3
Methylene groups adjacent to carbonyl (—C—C$\underline{\text{H}}_2$—) ‖ O	2.1–2.4
Methyl groups attached to unsaturated carbon (C = C or Ar—C$\underline{\text{H}}_3$) ∣ CH$_3$	2–2.5
Methoxy groups (ethers, C$\underline{\text{H}}_3$OR)	3–3.5
Methyl esters of carboxylic acids (C$\underline{\text{H}}_3$OCOR)	3.5–3.8
Protons geminal to hydroxy groups (C$\underline{\text{H}}$OH)	3.5–4.5
Protons geminal to acetoxy groups (C$\underline{\text{H}}$OAc)	4–5
Protons on unsaturated carbon (H$\underline{\text{C}}$=C)	5–6
Protons on an aromatic (benzene) ring (Ar$\underline{\text{H}}$)	6.5–8
Proton of an aldehyde group (C$\underline{\text{H}}$O)	9.5–10

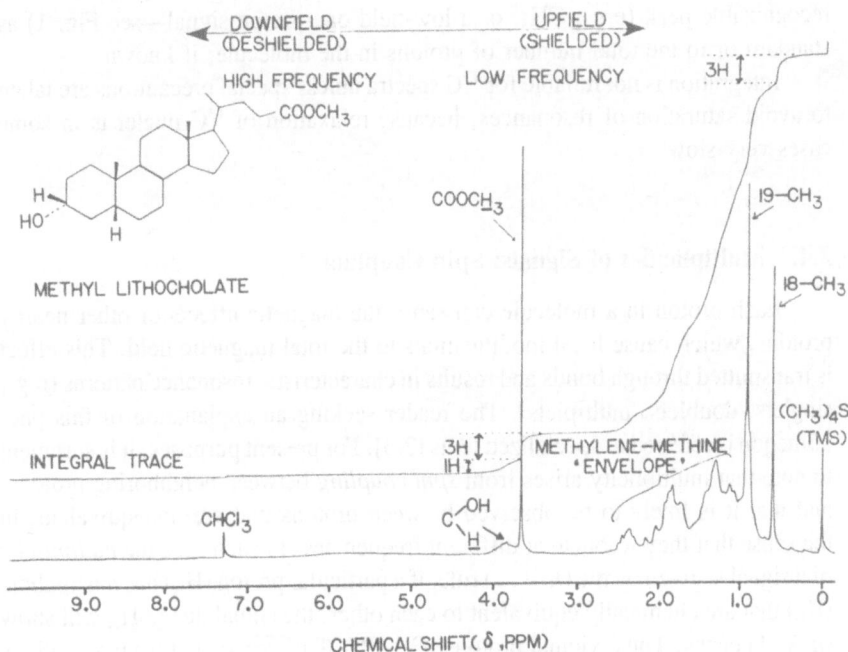

FIGURE 1. Proton NMR spectrum of methyl lithocholate in deuterochloroform. Some of the terms used in the text are illustrated: the internal standard, tetramethylsilane (TMS); the singlet resonances of the methyl groups, 18-CH₃, 19-CH₃, and ester methyl; the methylene–methine envelope, the proton geminal to the 3α-hydroxyl group; the resonance (at 7.25 ppm) from the small trace of protiochloroform in deuterochloroform; the integral trace, which is usually standardized on the 18-CH₃ resonance.

Resonances of ^{13}C nuclei are spread over a much wider range (0–200 ppm). The basic principles that determine the individual chemical shifts within this range are similar to those for protons. As a very rough "rule of thumb," ^{13}C chemical shifts are usually about 20 times larger than those of any attached protons.

2.3. Peak Integrals

Provided the *saturation* condition is not approached (Section 2.1), the area of a peak in the ^1H NMR spectrum is proportional to the number of protons giving rise to the signal. The spectrometer has an integration facility and is able to plot the integral trace on the same chart as the spectrum (Fig. 1). The relative heights of the steps in the integral trace indicate the ratios of peak areas and hence permit "counting" of protons under each peak by reference either to a recognizable peak (e.g., CH_3, or a low-field one-proton signal—see Fig. 1) as standard or to the total number of protons in the molecule, if known.

Integration is not reliable for ^{13}C spectra unless special precautions are taken to avoid saturation of resonances, because relaxation of ^{13}C nuclei is in some cases very slow.

2.4. Multiplicities of Signals: Spin Coupling

Each proton in a molecule can sense the magnetic effects of other nearby protons, which cause local modifications to the total magnetic field. This effect is transmitted through bonds and results in characteristic resonance patterns (e.g., singlets, doublets, multiplets). The reader seeking an explanation of this phenomenon is referred to specialized texts [2,3]. For present purposes, it is sufficient to note that multiplicity arises from *spin coupling* between neighboring protons, and that it is likely to be observed between protons that are nonequivalent, in the sense that they resonate at different frequencies. In such case the *multiplicity* of a signal is given by the $(n + 1)$ rule: if a particular proton (H_A) has n neighbors (H_B) that are chemically equivalent to each other, the signal due to H_A will show $(n + 1)$ peaks. Thus, vicinal protons -CH-CH- that are isolated from any others will each give *doublet* resonances, whereas for a system $-CH_2-CH_3$, the two methylene protons (CH_2) will give a *quartet* signal, while the methyl protons

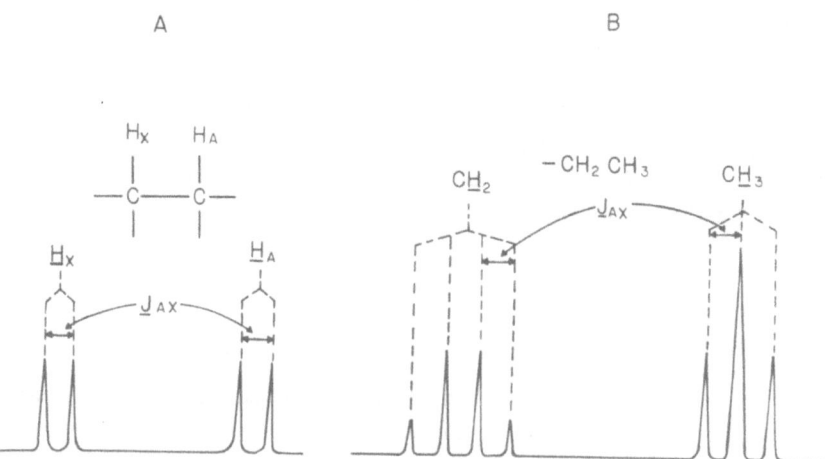

FIGURE 2.   Idealized splitting of NMR signals due to spin coupling. (A) vicinal protons, giving a pair of doublets; (B) ethyl group, giving a triplet and a quartet.

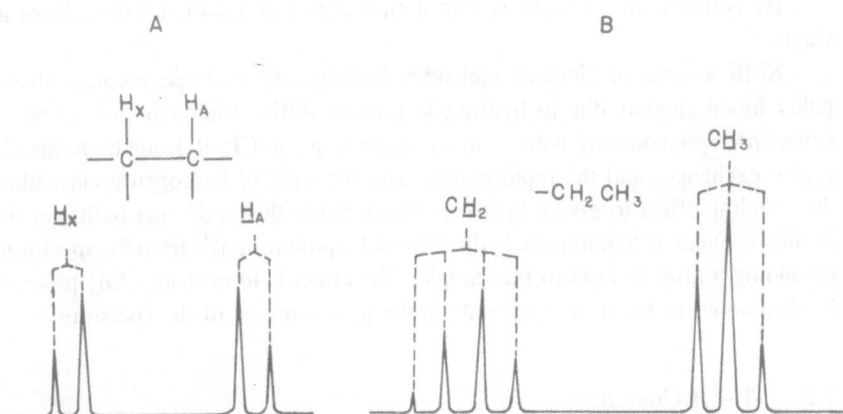

FIGURE 3. Split NMR signals perturbed as they appear in the spectrum if the separation of signals from different protons (the chemical shift difference) is not large compared to the splitting (J values). Values of J_{ax} remain as in Fig. 2.

(CH_3) will resonate as a *triplet*, but integrating for a total of three protons (Fig. 2). The protons of methyl groups attached to quaternary carbon (CH_3-C-), like the angular methyl group at C-18 and C-19 in bile acids, resonate as *singlets* because there are no vicinal protons (see Fig. 1).

The magnitude of spin coupling is expressed by the *coupling constant, J*, with the dimensions of frequency (Hz). Coupling constants are independent of the applied magnetic field, being a property of the molecule itself. Typical values of J range from 0 to 20 Hz, with $J \simeq 5$–7 Hz where free rotation about C-C bonds is possible. The value of J can be an important indicator of conformation in relatively rigid molecules, as we shall see in Section 3.5.

The shapes of multiplets, i.e., the relative heights of their component signals, are important in recognizing and interpreting them. Where the signals from spin-coupled sets of protons are well separated in the spectrum, the component peaks of doublets are in 1:1 ratio; of triplets in 1:2:1 ratio; and of quartets in 1:3:3:1 ratio. [In the general case of coupling to n equivalent protons, relative intensities are given by the coefficients of terms in the expansion of $(1 + x)^n$]. Signals that are relatively close in the spectrum show perturbations that render them unsymmetrical: the innermost peaks of coupled resonances tend to be larger at the expense of the outer peaks (Fig. 3).

Excellent examples of a simple two-spin system are provided by 1-en-3-ones and 1,4-dien-3-ones, where H-1 and H-2 each resonate as doublets. Their chemical shifts differ by only ca. 0.7 ppm, largely owing to their β- and α-relationships respectively to the 3-keto group. The pair of doublets is moderately perturbed (cf. Fig. 3).

By contrast, the isolated H-4 in 4-en-3-ones and 1,4-dien-3-ones gives a singlet.*

NMR spectra of alcohols and other hydrogen-bonded species may show rather broad singlets due to hydroxylic protons (O$\underline{\text{H}}$). Although such protons experience spin coupling with vicinal protons (e.g., in CHOH), the high rate of proton exchange, and the rapid making and breaking of hydrogen bonds, blur the coupling effect to give a broadened peak rather than a distinct multiplet. It is often convenient to eliminate hydroxylic (O$\underline{\text{H}}$) proton signals from the spectrum by adding a drop of D_2O to the sample. The effect is to exchange O$\underline{\text{H}}$ protons by deuterons, which do not resonate in the proton region of the spectrum.

2.5. ^{13}C–^{1}H Coupling

^{13}C-nuclei also interact with the different spin states of each of the protons that are attached to them and give spectra in which signals of all except quaternary and carbonyl carbon atoms are multiplets ($\underline{\text{C}}$H$_3$, quartet; $\underline{\text{C}}$H$_2$, triplet; $\underline{\text{C}}$H; doublet). Interpretation of such spectra would be virtually impossible for complex molecules, but the problem may be avoided by applying *broad-band decoupling* to remove the effects of protons. Since all the protons need to be decoupled, a spread of frequencies in the ^{1}H region is required. This is obtained by modulating an appropriate radiofrequency by audiofrequency noise, hence an alternate name, *noise* decoupling. With the ^{1}H–^{13}C coupling removed, the ^{13}C spectrum consists of singlet resonance lines which are normally well separated in the spectrum (Fig. 4). This improves the sensitivity (signal-to-noise ratio), an important factor for ^{13}C NMR experiments, and permits counting the number of carbon atoms in the molecule. The natural abundance of ^{13}C is so low (1.1%) that the chance of two ^{13}C atoms being linked is minute. ^{13}C–^{13}C coupling is therefore not a problem (see, however, Section 4). In situations where enrichment of ^{13}C has been achieved chemically, the effects of ^{13}C–^{13}C coupling and ^{13}C–^{1}H coupling will be observed. The latter results in additional splitting of the attached protons.

2.6. Preparation of Samples for NMR

As in other types of spectroscopy, high-quality NMR spectra of a compound can only be obtained on a "pure" sample. Pure in this context is at least 95% of the *total* organic material. In addition, it is important to remove any paramagnetic metal ions, which can cause the chemical shifts of certain resonances to move downfield and exhibit band broadening owing to relaxation. This may

---

* Scale-expanded spectra from a high-resolution spectrometer may reveal a very narrow doublet ($J \approx 1$ Hz) as a result of long-range coupling with 6β-H and/or 2α-H (in 4-en-3-ones).

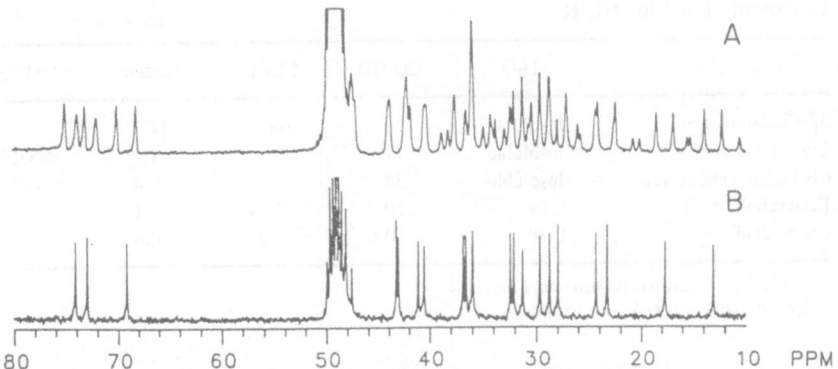

FIGURE 4. Effect of broad-band, noise decoupling of proton–carbon-13 coupling on the ^{13}C NMR spectrum of cholic acid in deuteromethanol. (A) with off-resonance decoupling, showing multiplicities of signals; (B) fully decoupled (broad band). Note the 1 : 3 : 3 : 1 quartet pattern for the 18-C̲H$_3$ and 19-C̲H$_3$ resonances (A). Two resonances, 14-C̲H and 17-C̲H, are hidden by the proton forms of the solvent deuteromethanol-d$_4$.

be readily accomplished by the use of chelating resins. Samples arising from biological sources should be processed through this step.

Since water contains protons, their resonance (at δ 4.75) would swamp the ^{1}H resonances arising from the compound being studied. To avoid this difficulty, samples must be thoroughly dried before being analyzed. The sample is then dissolved in a perdeuterated solvent, i.e., D$_2$O, deuteromethanol-d$_4$ (CD$_3$OD), deuterochloroform (CDCl$_3$), deuterodioxane-d$_8$. The choice of solvent reflects the solubility of the bile acid or its derivative being studied. No single solvent is ideal for all situations. For ^{1}H NMR, deuteromethanol and deuterochloroform can be used for most instances, since only dilute (0.5–10 mM) solution are needed. However, ^{13}C NMR requires somewhat higher concentrations (10–500 mM).

Dimethylsulfoxide-d$_6$ is an excellent solvent for bile acids; however, unlike CD$_3$OD and CDCl$_3$, it is not easily removed by evaporation because of its high boiling point (196°C). A summary of the solubilities of the common bile acids in different solvents is shown in Table II. Perdeuterated solvents are readily available from several major suppliers of organic chemicals.

The sample should be fully in solution, or free of particulate materials, since this will upset the field homogeneity. The clean solution is transferred to long, thin (5- or 10-mm diameter) NMR tubes, which after sealing are placed inside the magnet. They are spun during data acquisition to average out any field inhomogeneity.

TABLE II.   Solubilities[a] (mg/ml) of Bile Acids in Solvents
Commonly Used for NMR

	D$_2$O	CD$_3$OD	CDCl$_3$	Benzene	DMSO
5β-Cholanoic acid	Insoluble	6	66	17	—
Lithocholic acid	Insoluble	11	7	1	>330[b]
Chenodeoxycholic acid	Insoluble	34	3	0.4	—
Deoxycholic acid	0.24	220.7	2.94	0.1	—
Cholic acid	0.28	30.6	5.9	0.4	—

[a] Data taken from Volume 9 of the *Merck Index* and Ref. 4.
[b] Unpublished observation by Vincent Waterhous.

3.   PROTON NMR OF BILE ACIDS

3.1.   Introduction

Until fairly recently only ¹H spectra at up to 100 MHz were routinely
recorded. The main features of these are listed in Tables III–VI. They permitted
recognition of the main structural features and functional groups of bile acids,
although little information could be obtained about the skeletal parts of a molecule, which resonate at high field in the "methylene envelope" region (ca. 1–2.5
ppm).

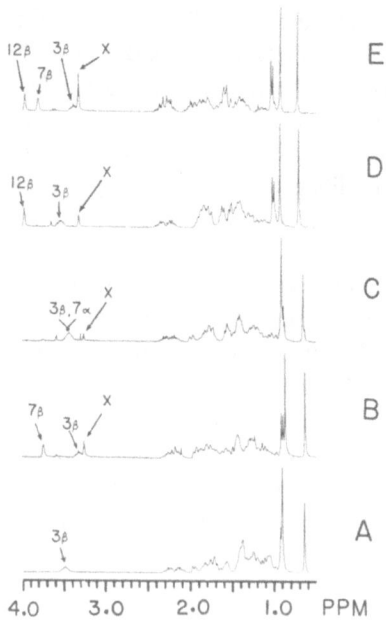

FIGURE 5.   The proton NMR spectra of five
common bile acids recorded at 300 MHz: (A) lithocholic acid, (B) chenodeoxycholic acid, (C) ursodeoxycholic acid, (D) deoxycholic acid, and (E)
cholic acid. The peak marked with an X is from
the small amount of the proton form of CD$_3$OD.
Despite the complexity of the methylene–methine
envelope, some differences can be observed at this
field strength.

TABLE III. ¹H-NMR Data for Unsubstituted and Monooxygenated Bile Acids[a]

Compound/substituent	Solvent	18-CH₃	19-CH₃	21-CH₃	Protons geminal to –OH	Ester OCH₃	Other protons	Reference
C₂₀ bile acid 3α-Hydroxy-5α-etianic acid	CDCl₃	0.65 (s)	0.79 (s)	n.g.	4.04 (3β)	3.65 (s)		5
C₂₁ bile acid Methyl 3β-hydroxy-pregn-5-en-21-oate	CDCl₃	0.62 (s)	1.02 (s)	—	3.50 (3α)	3.66 (s)	2.30 20CH₂ (bd) 5.36 6H (bd, J = 4 Hz)	5
C₂₂ bile acid Pregn-5-en-20α-carboxy-3β-hydroxy, 3-acetate, norethisterone ester	CDCl₃	0.70 (s)	1.00 (s)	n.g.	5.27 (3β)	—		6
A-nor-5β-cholanoates Me 3α-ol	CDCl₃	0.70 (s)	1.00 (s)	n.g.	4.28 (3β)	n.g.		7
3-Acetate	CDCl₃	0.68 (s)	0.98 (s)	n.g.	5.30 (3β)	n.g.		7
Me 3β-ol	CDCl₃	0.68 (s)	1.02 (s)	n.g.	4.23 (3α)	n.g.		7
3-Acetate	CDCl₃	0.65 (s)	0.98 (s)	n.g.	5.17 (3α)	n.g.		7
Me 3α-carboxymethyl-12α-ol	CDCl₃	n.g.	0.97 (s)	n.g.	3.95 (12β)	3.68 (s)		7

(continued)

TABLE III. *(Continued)*

Compound/substituent	Solvent	18-CH₃	19-CH₃	21-CH₃	Protons geminal to -OH	Ester OCH₃	Other protons	Reference
Cholan-24-oic acids								
Δ⁴ and Δ⁵ series								
Me Δ⁵-3β-ol	CDCl₃	0.69 (s)	1.00 (s)	n.g.		3.63 (s)		8
3-Acetate	CDCl₃	0.68 (s)	1.00 (s)	n.g.		3.65 (s)		8
Me Δ⁵,⁷-3β-ol	CDCl₃	0.71 (s)	0.96 (s)	n.g.		3.63 (s)	5.5–5.9 Olefin	8
Et Δ⁵,²²-3β-ol tetrahydropyranyloxy	CDCl₃	0.70 (s)	1.00 (s)	1.07 (d, J = 7 Hz)			6.91 22$\underline{H}$ (dd, J = 6, 16 Hz)	9
Me Δ⁴-3-one	CDCl₃	0.60 (s)	1.06 (s)	0.85		3.56 (s)	5.65 4$\underline{H}$ (s)	10
Me Δ¹,⁴-3-one	CDCl₃	0.79 (s)	1.21 (s)	0.93		3.67 (s)	6.24 2$\underline{H}$ 6.50 4$\underline{H}$ 7.01 1$\underline{H}$	11
Me Δ⁴,⁶-3-one	CDCl₃	0.76 (s)	1.11 (s)	0.93		3.68 (s)	5.67 4$\underline{H}$	12
5β series								
Unsubstituted	CDCl₃	0.67 (s)	0.93 (s)	0.93 (d)		—	2.25 (23-C$\underline{H}_2$)	13
	CD₃OD	0.68 (s)	0.94 (s)	0.95 (d)		—	2.20 (23-C$\underline{H}_2$)	13
3α-ol (Lithocholic acid)	CDCl₃	0.66 (s)	0.93 (s)	0.93 (d)	3.60 (3β)	—	2.20 (23-C$\underline{H}_2$)	13
	CD₃OD	0.68 (s)	0.93 (s)	0.93 (d)	3.50 (3β)	—	2.20 (23-C$\underline{H}_2$)	13
Me 3α-ol	Pyr-d₅	0.64 (s)	0.95 (s)	—	n.g.	3.69 (s)		14
	CDCl₃	0.65 (s)	0.92 (s)	0.91 (d)	n.g.	n.g.		14
3-Acetate	Pyr-d₅	0.65 (s)	0.94 (s)	n.g.	4.74 (br m,3β)	3.69 (s)		14
3-Formate	CDCl₃	0.65 (s)	0.95 (s)	n.g.	4.55 (m,3β)	—		6
Norethisterone ester	CDCl₃	0.67 (s)	0.96 (s)	n.g.	4.64 (3β)	n.g.		15
3-Mesylate 3-Trimethylsilyl ether	CDCl₃	0.64 (s)	0.92 (s)	0.92 (d)				16

Me Δ^{2},Δ^{3}	CCl_4	—	—	—	—	n.g.	5.50 (m, 2H, olefin)	17
Δ^{6}-3α-ol, 3-Acetate	CCl_4	0.82 (s)	0.89 (s)	—	—	—	5.40 6H (s)	17
Δ^{6}-7-Methyl-3α-ol	n.g.	0.77 (s)	1.14 (s)	1.06 (d, $J = 6$ Hz)	3.64 (3β)	—	5.42 6H (d, $J = 4$ Hz); 1.77 (7-CH_3)	18
Δ^{7}-3α-ol	n.g.			1.05 (d, $J = 6$ Hz)	3.70 (3β)	—	4.72 7H; 4.76	18
3-Acetate	CCl_4	—	—	—	—	—	5.15 6H (s)	17
Me 7α-ol	Pyr-d_5	0.65 (s)	0.91 (s)	n.g.	3.86 (br m, 7β)	3.68 (s)		14
7-Acetate	Pyr-d_5	0.64 (s)	0.89 (s)	n.g.	4.86 (br m, 7β)	3.67 (s)		14
7-Mesylate	$CDCl_3$	0.66 (s)	0.91 (s)	n.g.	4.90 (7β)	3.63 (s)		19
Me 3β-chloro-7α-ol	Pyr-d_5	0.67 (s)	0.97 (s)	n.g.	3.87 (br m, 7β)	3.67 (s)	4.50 (br m, 3α)	14
7-Acetate	Pyr-d_5	0.60 (s)	0.97 (s)	n.g.	4.90 (br m, 7β)	3.66 (s)	4.56 (br m, 3α)	14
Me 3α-diethylamino-7α-ol	Pyr-d_5	0.69 (s)	0.94 (s)	n.g.	3.88 (7β) (1H, br m)	3.70 (s)		14
7-Acetate	Pyr-d_5	0.66 (s)	0.93 (s)	n.g.	4.91 (7β) (1H, br m)	3.68 (s)		14
7β-ol	$CDCl_3$	0.68 (s)	0.93 (s)	n.g.	3.58 (7β)	3.68 (s)		19
Me 7β-ol	$CDCl_3$	0.84 (s)	0.99 (s)	n.g.	3.60 (7α)	3.68 (s)		19
Me 12α-ol	Pyr-d_5	0.71 (s)	0.92 (s)	n.g.	4.15 (12β)	3.62 (s)		14
12-Acetate	Pyr-d_5	0.66 (s)	0.89 (s)	n.g.	5.25 (12β) (1H, br m)	3.65 (s)		14
Δ^{3}-12α-ol, 12-Acetate	CCl_4	—	—	—	—	—	5.48 (m, 2H, olefin)	17
12α-ol, 12-Mesylate	$CDCl_3$	0.77 (s)	0.91 (s)	n.g.	5.11 (12β)	—		
3α-Acetoxy-5β-cholan-24-oyl chloride	$CDCl_3$	0.66 (s)	0.92 (s)	0.96 (d)	4.60 (3β)	—	3.57 (23-CH_2)	20
3α-Acetoxy-5β-cholan-24-onitrile	$CDCl_3$	0.66 (s)	0.92 (s)	0.96 (d)	4.59 (3β)	—	2.34 (23-CH_2)	20

ᵃ s, singlet; d, doublet; bd, broad doublet; dd, doublet of doublets; t, triplet; m, multiplet; br m, broad multiplet; J = proton-proton coupling constant; J_{CH} = carbon 13-proton coupling constant; n.g., chemical shift value not given.

TABLE IV. ^{1}H-NMR Data for Dioxygenated Bile Acids[a]

Compound/substituent	Solvent	18-CH₃	19-CH₃	21-CH₃	Protons geminal to hydroxyl groups	Ester OCH₃	Other protons	Reference
23,24-Bisnor-cholan-22-oic acids								
Δ⁴ Series								
Me Δ^4-6α-ol-3-one	CDCl₃	0.82 (s)	1.23 (s)	1.25 (d, J = 6 Hz)	4.54 (6β) (1H, m)	3.72 (s)	6.45 4H̲ (s)	21
Me $\Delta^{1,4}$-6α-ol-3-one	CDCl₃	0.80 (s)	1.24 (s)	1.26 (d, J = 6 Hz)	4.52 (6β) (1H, m)	3.71 (s)	6.29 2H̲ (d, J = 10 Hz); 6.55 4H̲ (s); 7.06 1H̲ (d, J = 10 Hz)	21
Me Δ^4-7α-ol-3-one	CDCl₃	0.79 (s)	1.22 (s)	1.24 (d, J = 6 Hz)	4.05 (7β) (1H, s)	3.69 (s)	6.15 4H̲ (s)	22
Me $\Delta^{1,4}$-7α-ol-3-one	CDCl₃	0.79 (s)	1.19 (s)	1.22 (d, J = 6 Hz)	4.08 (7β) (1H, s)	3.68 (s)	6.06 4H̲ (s); 6.16 2H̲ (d, J = 10 Hz); 7.00 1H̲ (d, J = 10 Hz)	22
Me Δ^4-12α-ol-3-one	CDCl₃	0.75 (s)	1.17 (s)	1.24 (d)	3.99 (12β) (1H, m)	3.64 (s)	5.73 4H̲ (s)	23
Me $\Delta^{1,4}$-12α-ol-3-one	CDCl₃	0.78 (s)	1.22 (s)	1.24 (d, J = 6 Hz)	3.96 (12β) (1H, t, J = 3Hz)	3.62 (s)	6.03 4H̲ (s); 6.17 2H̲ (d, J = 10 Hz); 6.97 1H̲ (d, J = 10 Hz)	24
	CDCl₃	0.76 (s)	1.20 (s)	1.26 (d)	3.98 (12β) (1H, m)	3.63 (s)	6.08 4H̲ (br s); 6.20 2H̲ (dd, J = 2,10 Hz); 7.01 1H̲ (d, J = 10 Hz)	23
Me $\Delta^{1,4,17(20)}$-12α-ol-3-one	CDCl₃	1.03 (s)	1.24 (s)	n.g.	4.54 (12β) (1H, m)	3.71 (s)	6.07 4H̲ (br, s); 6.21 2H̲ (dd, J = 2,10 Hz); 7.00 1H̲ (d, J = 10 Hz)	23
Me Δ^4-3,6-dione	CDCl₃	0.75 (s)	1.17 (s)	1.24 (d)		3.66 (s)	6.18 6H̲ (s)	5
5β-Series								
Me 3α,12α-diol	CDCl₃	0.69 (s)	0.92 (s)	1.23 (d)	3.64 (3β) (1H, br m); 3.93 (12β) (1H, m)	3.64 (s)		25

Compound	Solvent	18-CH₃	19-CH₃	21-CH₃	3-H	6-/12-H	Other	OCH₃	Ref
Me 12α-ol-3-one	CDCl₃	0.74 (s)	1.02 (s)	1.25 (d)		3.97 (12β) (1H, m)		3.65 (s)	25
A-nor-5β-cholanoic acids									
Me 3α,12α-diol	CDCl₃	0.77 (s)	1.00 (s)	n.g.	4.27 (3β) (s)	3.95 (12β) (s)		n.g.	7
Diacetate	CDCl₃	0.77 (s)	0.98 (s)	n.g.	5.28 (3β) (s)	5.16 (12β) (s)		n.g.	7
Me 3β,12β-diol	CDCl₃	0.77 (s)	1.08 (s)	n.g.	4.32 (3β) (1H, br m)			n.g.	7
Me 3α-ol-12-one	CDCl₃	1.02 (s)	1.02 (s)	n.g.				n.g.	7
Me 12α-ol-3-one	CDCl₃	0.75 (s)	1.17 (s)	n.g.		3.98 (12β) (s)		n.g.	7
3-acetate	CDCl₃	0.78 (s)	1.00 (s)	n.g.				n.g.	7
Me 3,12-dione	CDCl₃	1.03 (s)	1.23 (s)	n.g.				n.g.	7
24-Nor-5β-cholan-23-oic acids									
Me Δ⁴-7α,12α-diol	CDCl₃	0.68 (s)	1.16 (s)	1.03 (d, J = 4 Hz)			5.80 4H	3.66 (s)	26
Cholan-24-oic acids									
5α; Δ⁴ & Δ⁵ series									
Diols									
3α,6β-Diol	CDCl₃	0.69 (s)	1.02 (s)	0.92 (d, J = 6 Hz)	4.18 (3β) (w½ = 8 Hz) (1H, m)	3.76 (6α) (w½ = 7 Hz) (1H, t)		—	b
Me 5α-bromo-3β,6β-diol, 3-acetate	CDCl₃	0.68 (s)	1.32 (s)	n.g.	5.47 (3α) (1H, m)	4.16 (6α) (1H, t)		3.64 (s)	27
Me 5α-bromo-6β,19-epoxy 3β-ol, 3-acetate	CDCl₃	0.68	—	—	5.18 (3α) (1H, m)	4.04 (6α) (1H, m)	3.68 (19-CH₂) (d, J = 7 Hz), 3.92 (19-CH₂) (d, J = 7 Hz)	3.64 (s)	27
Me Δ⁵-3β,7α-diol	CDCl₃	0.68 (s)	0.98 (s)	n.g.			5.59 6H (d, J = 6 Hz)	3.67 (s)	8
7-Methyl ether	CDCl₃	0.66 (s)	0.97 (s)	n.g.			5.74 6H (d, J = 5 Hz), 3.33 (7β-OCH₃) (s)	3.63 (s)	8
3-Acetate, 7-methyl ether	CDCl₃	0.69 (s)	0.98 (s)	n.g.			5.77 6H (d, J = 5 Hz), 3.33 (7β-OCH₃) (s)	3.64 (s)	8

(continued)

TABLE IV. (*Continued*)

Compound/substituent	Solvent	18-CH₃	19-CH₃	21-CH₃	Protons geminal to hydroxyl groups	Ester OCH₃	Other protons	Reference
Me Δ⁵-3β,7β-diol	CDCl₃	0.70 (s)	1.05 (s)	n.g.		3.68 (s)	5.35 6H̲ (s)	8
3-Acetate	CDCl₃	0.68 (s)	1.03 (s)	n.g.		3.64 (s)	5.47 6H̲ (s) 3.24 (7β-OCH₃) (s)	8
7-Methyl ether	CDCl₃	0.69 (s)	1.03 (s)	n.g.		3.64 (s)	5.46 6H̲ (s) 3.26 (7β-OCH₃) (s)	8
Me Δ⁵-3β,19-diol, 3-acetate	CDCl₃	0.72 (s)	—	n.g.	4.64 (3α) (1H, br m)	3.65 (s)	5.76 6H̲ (t) 3.64 19-CH̲₂ (d.J = 10 Hz) 3.84 19-CH̲₂ (d.J = 10 Hz)	27
Me Δ⁴-7α,12α-diol	CDCl₃	0.72 (s)	0.97 (s)	n.g.	3.78 (7β) 3.99 (12β) (1H, m) (1H, m)	3.67 (s)	5.60 4H̲ (2H, br m)	19
ol-ones								
3α-ol-6-one	CDCl₃	0.67 (s)	0.73 (s)	0.95	4.16 (3β) (w½ = 12 Hz)	—		b
Me 3α-ol-6-one	CDCl₃	0.68 (s)	0.74 (s)	0.94	4.15 (3β) (w½ = 12 Hz)	3.67 (s)		b
3β-ol-6-one	CDCl₃	0.68 (s)	0.76 (s)	0.98	3.59 (3α) (w½ = 12 Hz)	—		b
Me 3β-ol-6-one	CDCl₃	0.66 (s)	0.75	0.94	3.56 (3α) (w½ = 12 Hz)	3.66 (s)		b
Me Δ⁵-3β-ol-19-one, 3-Acetate	CDCl₃	0.62 (s)	—	n.g.	4.57 (3α) (1H, br m)	3.64 (s)	5.88 6H̲ (m)	27
Me Δ⁴-6α-ol-3-one	CDCl₃	0.69 (s)	1.18 (s)	0.92 (d.J = 6Hz)	4.35 (6β) (1H, m)	3.67 (s)	6.16 4H̲ (s)	28

Compound	Solvent								Ref.
Me Δ^{1,4}-6α-ol-3-one	CDCl$_3$	0.72 (s)	1.20 (s)	0.90 (d,J = 6Hz)		4.46 (6β) (1H, m)	3.66 (s)	6.23 2<u>H</u> (d,J = 10 Hz) 6.46 4<u>H</u> (s) 6.98 1<u>H</u> (d,J = 10 Hz)	28
<u>Δ^4</u>-12α-ol-3-one	CDCl$_3$	0.70 (s)	1.15 (s)	0.97 (d)	3.69 (12β)		—	5.64 4<u>H</u> (s)	29
12-Formate	CDCl$_3$	0.81 (s)	1.17 (s)	0.84 (d)	5.27 (12β) (1H, br s)		—	5.73 4<u>H</u> (s)	29
Me <u>Δ^4</u>-12α-ol-3-one	CDCl$_3$	0.79 (s)	1.05 (s)	1.30 (d,J = 6Hz)	4.11 (12β) (1H,t,J = 3Hz)		3.72 (s)	6.77 4<u>H</u> (s)	28
Δ^4-12α-ol-3-one	CDCl$_3$	0.75 (s)	1.18 (s)	0.99 (d)	4.05 (12β) (1H, m)		3.67 (s)	5.73 4<u>H</u> (s)	23
Me <u>Δ^{1,4}</u>-12α-ol-3-one	CDCl$_3$	0.78 (s)	1.04 (s)	1.28 (d,J = 6Hz)	4.00 (12β) (1H,t,J = 3Hz)		3.72 (s)	6.02 4<u>H</u> (s) 6.15 1<u>H</u> (d,J = 10 Hz) 6.96 2<u>H</u> (d,J = 10 Hz)	28
	CDCl$_3$	0.77 (s)	1.23 (s)	0.99 (d)	4.06 (12β) (1H, m)		3.67 (s)	6.06 4<u>H</u> (br s) 6.20 2<u>H</u> (dd,J = 2,10 Hz) 7.00 1<u>H</u> (d,J = 10 Hz)	23
<u>Δ^{4,6}</u>-12α-ol-3-one	CDCl$_3$	0.74 (s)	1.08 (s)	0.98 (d)	3.96 (12β)		—	5.56 4<u>H</u> (s) 6.10 6<u>H</u> (s) 6.10 6<u>H</u> (s)	29
12-Formate	CDCl$_3$	0.84 (s)	1.05 (s)	0.84 (d)	5.26 (12β) (1H, br s)		—	5.65 4<u>H</u> (s) 6.09 6<u>H</u> (s) 6.09 7<u>H</u> (s)	29
Me <u>Δ^{4,6}</u>-12α-ol-3-one	CDCl$_3$	0.79 (s)	1.10 (s)	0.99 (d,J = 6 Hz)	5.08 (12β) (1H,t,J = 3 Hz)		3.68 (s)	5.68 4<u>H</u> (s) 6.12 6<u>H</u> (s) 6.12 7<u>H</u> (s)	30
Me <u>Δ^{1,4,22E}</u>-12α-ol-3-one	CDCl$_3$	0.82 (s)	1.23 (s)	1.15 (d)	4.06 (12β) (1H, m)		3.73 (s)	5.74 23<u>H</u> (d,J = 16 Hz) 6.06 4<u>H</u> (br s) 6.02 2<u>H</u> (dd,J = 2,10 Hz) 6.85 22<u>H</u> (dd,J = 9,16 Hz) 7.00 1<u>H</u> (d,J = 10 Hz)	9

(continued)

TABLE IV. *(Continued)*

Compound/substituent	Solvent	18-CH_3	19-CH_3	21-CH_3	Protons geminal to hydroxyl groups		Ester OCH_3	Other protons	Reference
Diones									
Me 3,6-dione	$CDCl_3$	0.70 (s)	0.96 (s)	0.94 (d,J = 6 Hz)			3.67 (s)		b
Me $\Delta^{4,6}$-3,12-dione	$CDCl_3$	0.88 (s)	1.20 (s)	1.13 (s)			3.68 (s)	5.73 4$\underline{\text{H}}$ (s) 6.14 6$\underline{\text{H}}$ (s) 6.14 7$\underline{\text{H}}$ (s)	30
5β-Series									
Me 1α,3α-diol	n.g.	n.g.	n.g.	n.g.	3.3 (1β) (d)	3.15 (3β) (d)	3.60 (s)		31
3,7-Diols									
3α,7α-diol (Chenodeoxycholic acid)	$CDCl_3$; DMSO	0.64 (s)	0.88 (s)	0.94 (d)	3.38 (3β) (1H, br m)	3.78 (7β) (1H, m)			32
	CD_3OD	0.67 (s)	0.90 (s)	0.94 (d)	3.36 (3β)	3.78 (7β)			32
	CD_3OD	0.67 (s)	0.90 (s)	0.92 (d)	3.50 (3β) (1H, br m)	3.77 (7β) (1H, m)		2.15 23-$\underline{\text{CH}}_2$ (2H,d)	13
	$CDCl_3$	0.69 (s)	0.92 (s)	0.92 (d)	3.50 (3β)	3.90 (7β)		2.20 23-$\underline{\text{CH}}_2$ (2H,d)	13
	D_2O	0.68 (s)	0.92 (s)	0.92 (d)	3.50 (3β) (1H, br m)	3.88 (7β) (1H, m)		2.10 23-$\underline{\text{CH}}_2$ (2H,d)	13
7β-Methyl	n.g.	0.79 (s)	0.97 (s)	1.05 (d,J = 6Hz)	3.56 (3β) (1H, m)			1.32 (7β-$\underline{\text{CH}}_3$)	18
7-Formate	$CDCl_3$; DMSO-d_6	0.67 (s)	0.93 (s)	n.g.	3.46 (3β) (1H, br m)	5.03 (7β) (1H, br m)	—		33
Diformate, norethisterone ester	n.g.	0.75 (s)	0.92 (s)	n.g.	4.55 (3β)	5.20 (7β)	—		6
Me 3α,7α-diol	$CDCl_3$	0.68(s)	0.92 (s)	n.g.	3.42 (3β)	3.85 (7β)	—		34
	$CDCl_3$	0.66 (s)	0.90 (s)	0.92 (d)	n.g.	n.g	—		16
[2α,2β,3β,4α,4β-$^2\text{H}_5$]	$CDCl_3$	0.67 (s)	0.90 (s)	0.93 (d,J = 6Hz)		3.77 (7β) (1H, m)	3.60 (s)		35

Compound	Solvent					3β	7β		Ref
[6α,6β,7β,8β-²H₄]	CDCl₃	0.67 (s)	0.90 (s)	0.93 (dJ = 6Hz)		3.43 (3β) (1H, br m)		3.62 (s)	35
[23,24-¹³C₂-22ε,23ε-²H₂]	CDCl₃	0.68 (s)	0.94 (s)	0.93 (dJ = 6Hz)		3.55 (3β,7β) (2H, br m)		3.65 (dJ$_{CH}$ = 3 Hz)	9
3-Acetate	Pyr-d₅	0.66 (s)	0.91 (s)	n.g.		4.65 (3β) (1H, br m)	3.86 (7β) (1H, br m)	3.68 (s)	14
7-Acetate	CDCl₃	0.67 (s)	0.93 (s)	0.93 (dJ = 6Hz)		3.43 (3β) (1H, br m)	4.80 (7β) (1H, m)	3.63 (s)	35
Diacetate	CDCl₃	0.64 (s)	0.92 (s)	n.g.		4.56 (3β) (1H, br m)	4.88 (7β) (1H, br m)	3.63 (s)	36
	Pyr-d₅	0.67 (s)	0.93 (s)	n.g.		4.56 (3β) (1H, br m)	4.93 (7β) (1H, br m)	3.70 (s)	14
3-Acetate, 7-mesylate	CDCl₃	0.71 (s)	1.00 (s)	n.g.		4.61 (3β) (1H, br m)	4.91 (7β) (1H, m)	3.64 (s)	15
3-Cathylate	CDCl₃	0.67 (s)	0.91 (s)	n.g.		4.44 (3β) (1H, br m)	3.82 (7β) (1H, m)	3.64 (s)	37
3-Cathylate, 7-mesylate	CDCl₃	0.67 (s)	0.95 (s)	n.g.		4.42 (3β) (1H, br m)	4.89 (7β) (1H, m)	3.64 (s)	37
Dimesylate	CDCl₃	0.64 (s)	0.93 (s)	n.g.		4.49 (3β) (1H, br m)	4.89 (7β) (1H, m)	3.62 (s)	15
3-Mesylate	Pyr-d₅	0.67 (s)	0.93 (s)	n.g.		4.51 (3β) (1H, br m)	3.84 (7β) (1H, m)	3.69 (s)	14
3-Mesylate, 7-acetate	Pyr-d₅	0.70 (s)	0.96 (s)	n.g.		4.52 (3β) (1H, br m)	4.92 (7β) (1H, br m)	3.70 (s)	14
7β-Methyl	n.g.	0.79 (s)	0.97 (s)	1.05 (dJ = 6Hz)	1.32 (7β-C$\underline{H}_3$)	3.56 (3β) (1H, m)		—	18
3-Methyl ether	CDCl₃	0.68 (s)	0.92 (s)	n.g.		3.04 (3β)	3.86 (7β)	—	34
7-Methyl ether	CDCl₃	0.64 (s)	0.91 (s)	n.g.		3.38 (3β)	3.22 (7β)	—	34
3-Succinate	Pyr-d₅	0.67 (s)	0.92 (s)	n.g.		4.60 (3β) (1H, br m)	3.88 (7β) (1H, br m)	3.69 (s)	14
3-Tosylate	Pyr-d₄	0.64 (s)	0.87 (s)	n.g.		4.33 (3β)	3.83 (7β)	3.67 (s)	14

(continued)

TABLE IV. (Continued)

Compound/substituent	Solvent	18-CH$_3$	19-CH$_3$	21-CH$_3$	Protons geminal to hydroxyl groups		Ester OCH$_3$	Other protons	Reference
3-Tosylate, 7-acetate	Pyr-d$_5$	0.63 (s)	0.87 (s)	n.g.	4.33 (3β)	4.82 (7β)	3.66 (s)		14
3-Tosylate, 7-mesylate	CDCl$_3$	0.64 (s)	0.90 (s)	n.g.	4.33 (3β) (1H, m)	4.89 (7β) (1H, m)	3.64 (s)		15
Bis trimethylsilyl ether	CDCl$_3$	0.63 (s)	0.88 (s)	0.93 (d)	n.g.	n.g.	—		16
Et 3α,7α-diol [23,24-^{13}C-22ε,23ε-^{2}H$_2$]	CDCl$_3$	0.66 (s)	0.91 (s)	0.93 (d,J = 6 Hz)	3.41 (3β) (1H, br m)	3.82 (7β) (1H, m)	—		9
Me Δ^{11}-3α,7α-diol, diacetate	n.g.	—	—	—	—	—	—	5.46 (d,J = 10 Hz) 6.18 (d,J = 10 Hz)	17
Et Δ^{22}-3α,7α-diol [23,24-^{13}C$_2$]	CDCl$_3$	0.71 (s)	0.92 (s)	1.10 (d,J = 6 Hz)	3.49 (3β) (1H, br m)	3.87 (7β) (1H, m)	—	5.67 22H̲ (ddd,J = 16 Hz, J_{CH} = 160 Hz) 6.94 22H̲ (dd,J = 7,16 Hz)	9
Dietetrahydropyranyloxy [23,24-^{13}C$_2$]	CDCl$_3$	0.67 (s)	0.90 (s)	1.09 (d,J = 7 Hz)			—	5.67 22H̲ (ddd,J = 16 Hz, J_{CH} = 160 Hz) 6.94 22H̲ (dd,J = 7,16 Hz)	9
3α,7β-Diol (ursodeoxycholic acid)	CDCl$_3$; DMSO-d$_6$	0.69 (s)	0.96 (s)	n.g.	3.62 (3β,7α) (2H, br m)		—		37
3-Mesylate	CDCl$_3$; DMSO-d$_6$	0.66 (s)	0.95 (s)	0.92 (d)	3.45 (3β) (1H, br m)	3.45 (7α) (1H, br m)	—		32
3-Mesylate	CDCl$_3$	0.70 (s)	0.97 (s)	n.g.	4.58 (3β) (1H, br m)	3.56 (7α) (1H, br m)			15
Me 3α,7β-diol	CDCl$_3$	0.69 (s)	0.94 (s)	n.g.	3.55 (3β,7α) (2H, br m)		3.65 (s)		37
[2α,2β,3β,4α,4β-^{2}H$_5$]	CDCl$_3$	0.68 (s)	0.93 (s)	0.94 (d,J = 6 Hz)		3.55 (7α) (1H, br m)	3.65 (s)		35
[6α,6β,7α,8β-^{2}H$_4$]	CDCl$_3$	0.67 (s)	0.93 (s)	0.92 (d,J = 6 Hz)	3.63 (3β) (1H, m)		3.67 (s)		35

Compound	Solvent								Ref
7-Acetate	CDCl₃	0.70 (s)	0.98 (s)	0.98 (d J = 6 Hz)		3.50 (3β) (1H, br m)	4.73 (7α) (1H, br m)	3.63 (s)	35
Diacetate	CDCl₃	0.70 (s)	0.98 (s)	0.97 (d J = 6 Hz)		4.67 (3β,7α) (2H, br m)		n.g.	35
7α-Methyl	n.g.	0.77 (s)	1.09 (s)	1.06 (d J = 6 Hz)	1.40 (7α-C$\underline{H}_3$)	3.62 (3β) (1H, m)		n.g.	38
3β,7α-Diol	n.g.	0.67 (s)	0.94 (s)	n.g.		4.07 (3α) (1H, m)	3.87 (7β) (1H, m)	—	15
Me 3β,7α-diol	CDCl₃	0.68 (s)	0.96 (s)	n.g.		4.06 (3α) (1H, m)	3.87 (7β) (1H, m)	—	15
3-Acetate	CDCl₃	0.69 (s)	0.98 (s)	n.g.		5.02 (3α) (1H, m)	3.88 (7α) (1H, m)	—	15
3-Acetate, 7-mesylate	CDCl₃	0.69 (s)	0.99 (s)	n.g.		5.06 (3α) (1H, m)	4.93 (7α) (1H, m)	—	15
3β,7β-Diol	CDCl₃	0.71 (s)	1.00 (s)	n.g.		4.07 (3α) (1H, m)	3.65 (7α) (1H, br m)		39
Me 3β,7β-diol	CDCl₃	0.71 (s)	1.00 (s)	n.g.		4.04 (3α) (1H, m)	3.55 (7α) (1H, br m)	3.65 (s)	39
3,12-diols									
3α,12α-Diol (deoxycholic acid)	CD₃OD	0.70 (s)	0.92 (s)	1.00 (d)	2.20 23-C$\underline{H}_2$ (1H, m)	3.50 (3β) (1H, br m)		3.95 (12β) (s)	13
	D₂O	0.70 (s)	0.92 (s)	0.92	2.20 23-C$\underline{H}_2$ (1H, m)	3.65 (3β) (1H, br m)		4.05 (12β) (s)	13
12-Formate	CDCl₃	0.75 (s)	0.88 (s)	n.g.		3.62 (3β) (1H, br m)		5.27 (12β) (1H, m)	33
Me 3α,12α-diol	CDCl₃	0.68 (s)	0.91 (s)	0.97 (d)			—	—	16
Bis trimethylsilyl ether	CDCl₃	0.65 (s)	0.89 (s)	0.94 (d)			—	—	16
3,12-Diol-enes									
Δ⁶-3α,12α-Diol, diacetate	n.g.	n.g.	n.g.	n.g.	5.50 (br s)			—	17
Δ⁷-3α,12α-Diol, diacetate	n.g.	n.g.	n.g.	n.g.	5.10 (br s)			—	17
Δ⁹⁽¹¹⁾-3α,12α-Diol, diacetate	n.g.	n.g.	n.g.	n.g.	5.65 (d J = 6 Hz)			—	17

(continued)

TABLE IV. *(Continued)*

Compound/substituent	Solvent	18-CH₃	19-CH₃	21-CH₃	Protons geminal to hydroxyl groups	Ester OCH₃	Other protons	Reference
3,12-Diol-dienes								
Me Δ^7,9(11),3α,12α-diol, 3-acetate	CDCl₃	0.47 (s)	1.00 (s)	n.g.	4.62 (3β) (1H, br m) / 4.00 (12β) (d J = 6 Hz)	3.63 (s)	5.77 11H	40
Diacetate	CDCl₃	0.55 (s)	1.02 (s)	n.g.	4.70 (3β) (1H, br m) / 5.08 (12β) (d J = 6 Hz)	3.65 (s)	5.40 7H (d J = 6 Hz) / 5.42 7H	40
3-Acetate, 12-methyl ether	CDCl₃	0.48 (s)	1.02 (s)	n.g.	4.68 (3β) (1H, br m) / 5.37 (12β) (d J = 6 Hz)	3.65 (s)	5.80 11H (d J = 6 Hz) / 5.37 7H	40
Me Δ^7,9(11),3α,12β-diol, 3-acetate	CDCl₃	0.53 (s)	1.03 (s)	n.g.	4.68 (3β) (1H, br m) / 4.22 (12α) (1H, br m)	3.63 (s)	5.87 11H (d J = 6 Hz) / 5.30 11H / 5.23 7H	40
Me Δ^7,9(11),3α,12β-diol, diacetate	CDCl₃	0.61 (s)	1.02 (s)	n.g.	4.63 (3β) (1H, br m) / 5.13 (12α) (1H, br m)	3.63 (s)	5.39 11H / 5.23 7H	40
Me Δ^7,9(11),3α,12α-diol, 3-acetate, 12-methyl ether	CDCl₃	0.50 (s)	1.03 (s)	n.g.	4.28 (3β) (1H, br m)	3.65 (s)	5.65 11H / 5.28 7H	40
7α,12α-Diols								
Me 7α,12α-diol	CDCl₃	0.68 (s)	0.89 (s)	n.g.	3.84 (7β) (1H, m) / 3.96 (12β) (1H, m)	3.63 (s)		19
Δ^2,3-7α,12α-diol, diacetate	n.g.	n.g.	n.g.	n.g.	n.g.	—	5.35 (m)	17
Me 7α,12α-diol, dimesylate	CDCl₃	0.78 (s)	0.92 (s)	n.g.	4.91 (7β) (1H, m) / 5.12 (12β) (1H, m)	3.66 (s)		19
7α,12β-Diol	CDCl₃: DMSO-d₆	0.72 (s)	0.91 (s)	n.g.	3.82 (7β) (1H, m) / 3.51 (12α) (1H, br m)	—		19
Me 7α,12β-diol	CDCl₃	0.72 (s)	0.91 (s)	n.g.	3.84 (7β) (1H, m) / 3.47 (12α) (1H, m)	3.64 (s)		19
7β,12α-Diol	CDCl₃: DMSO-d₆	0.67 (s)	0.90 (s)	n.g.	3.49 (7β) (1H, br m) / 3.88 (12β) (1H, m)	—		19
Me 7β,12α-diol	CDCl₃	0.70 (s)	0.93 (s)	n.g.	3.57 (7α) (1H, br m) / 3.98 (12β) (1H, m)	3.63 (s)		19

	Solvent					Ketal		Ref.
7β,12β-Diol	CDCl₃	0.72 (s)	0.93 (s)	n.g.	3.41 (7α,12α) (2H, br m)		—	19
Me 7β,12β-diol	CDCl₃	0.76 (s)	0.96 (s)	n.g.	3.41 (7α,12α) (2H, br m)		—	19
ol-ones								
Me 1α-ol-3-one	n.g.	n.g.	n.g.	n.g.	3.42 (1β)		3.60 (s)	31
3α-ol-6-one	CDCl₃	0.68 (s)	0.76 (s)	0.95 (d,J = 6Hz)	3.58 (3β) (w = 30 Hz)		—	b
Me 3α-ol-7-one	CDCl₃	0.65 (s)	1.20 (s)	0.95 (d,J = 6Hz)	3.50 (3β) (1H, br m)		3.62 (s)	35
Me 7α-ol-3-one	Pyr-d₅	0.71 (s)	1.02 (s)	n.g.	3.92 (7β) (br m)		3.68 (s)	14
7-Acetate	CDCl₃	0.68 (s)	1.03 (s)	0.92 (d,J = 6Hz)	4.90 (7β) (1H, m)		3.60 (s)	35
Me 7β-ol-3-one, 7-acetate	CDCl₃	0.75 (s)	1.08 (s)	0.93 (d,J = 6Hz)	4.73 (7α) (1H, br m)		3.62 (s)	35
Me 3,3-ethylenedioxy-7α-ol	Pyr-d₅	0.66 (s)	0.99 (s)	n.g.	4.02 (7β) (1H, br s)	3.88 ketal (s)	3.65 (s)	14
7α-Acetate	Pyr-d₅	0.60 (s)	0.99 (s)	n.g.	5.05 (7β) (1H, br s)	3.87 ketal (s)		14
7α-ol-12-one	CDCl₃: DMSO-d₆	1.00 (s)	1.02 (s)	n.g.	3.84 (7β) (1H, m)			19
Me 7α-ol-12-one	CDCl₃	0.99 (s)	1.01 (s)	n.g.	3.92 (7β) (1H, m)		3.63 (s)	19
7-Acetate	CDCl₃	1.01 (s)	1.01 (s)	n.g.	4.97 (7β) (1H, m)		3.64 (s)	19
7-Mesylate	CDCl₃	1.01 (s)	1.01 (s)	n.g.	4.96 (7β) (1H, m)		3.64 (s)	19
7β-ol-12-one	CDCl₃	1.04 (s)	1.04 (s)	n.g.	3.59 (7α) (1H, br m)		—	19

(continued)

TABLE IV. *(Continued)*

Compound/substituent	Solvent	18-CH$_3$	19-CH$_3$	21-CH$_3$	Protons geminal to hydroxyl	Ester OCH$_3$	Other protons	Reference
Me 7β-ol-12-one	CDCl$_3$	1.04 (s)	1.04 (s)	n.g.	3.56 (7α) (1H, br m)	3.63 (s)		19
12α-ol-3-one	CDCl$_3$	0.72 (s)	1.00 (s)	1.00 (d)	4.03 (12β) (1H, br s)	—		29
12-Formate	CDCl$_3$	0.79 (s)	1.01 (s)	0.84 (d)	5.27 (12β) (1H, br s)	—		29
Me 12α-ol-3-one	CDCl$_3$	0.73 (s)	1.02 (s)	1.00 (d)	4.04 (12β) (1H, m)	3.67 (s)		25
4β-Bromo-12α-ol-3-one, 12-formate	CDCl$_3$	0.77 (s)	1.05 (s)	0.82 (d)	5.25 (12β) (1H, br s)	—	4.90 (4α) (1H, d, J = 12 Hz)	29
Diones								
Me 3,6-dione	CDCl$_3$	0.70 (s)	0.96 (s)	0.94 (d, J = 6 Hz)		3.67 (s)		b
Et 3,7-dione [23,24-^{13}C-22ε,23e-^{2}H$_2$]	CDCl$_3$	0.67 (s)	1.20 (s)	0.93 (d, J = 6 Hz)		—		9
Me 3,12-dione	CDCl$_3$	1.07 (s)	1.12 (s)	n.g.		n.g.		7
27-Nor-5β-cholestan-26-oic acids								
3α,7α-diol	CD$_3$OD	0.69 (s)	0.92 (s)	0.94 (d, J = 7.5 Hz)		—		41
24R-methyl	CD$_3$OD	0.69 (s)	0.93 (s)	0.94 (d, J = 6.6 Hz)		—	0.93 24R CH$_3$ (d, J = 6.5 Hz)	41
24S-methyl	CD$_3$OD	0.69 (s)	0.93 (s)	0.94 (d, J = 6.6 Hz)		—	0.94 24S CH$_3$ (d, J = 6.6 Hz)	41

[a] See Table III for list of abbreviations
[b] D. N. Kirk and M. Nadim, unpublished data.

TABLE V. ^{1}H-NMR Data for Trioxygenated Bile Acids[a]

Compound/substituent	Solvent	18-CH_3	19-CH_3	21-CH_3	Protons geminal to hydroxyls	Methyl ester OCH_3	Other protons	Reference
23,24-Bisnorcholan-22-oic acids								
Me Δ⁵-7α,12α-diol-3-one	$CDCl_3$	0.80 (s)	1.24 (s)	1.26 (d,J = 6 Hz)	4.11 (7β) (s) 3.95 (12β) (t,J = 3 Hz)	3.70 (s)	6.17 4H (s)	42
Me Δ¹,⁴-7α,12α-diol-3-one	$CDCl_3$	0.81 (s)	1.25 (s)	1.27 (d,J = 6 Hz)	4.10 (7β) (s) 3.96 (12β) (t,J = 3 Hz)	3.71 (s)	6.20 4H (s) 6.30 2H (d,J = 10 Hz) 7.06 1H (d,J = 10 Hz)	42
24-Norcholan-23-oic acids 5α-Series								
3α,7α,12α-triol	CD_3OD	0.77 (s)	0.93 (s)	n.g.	3.33 (3β) (1H, m) 3.90 (7β) (1H, m) 4.33 (12β) (1H, m)	—		26
Me 3α,7α,12α-triol	$CDCl_3$	0.72 (s)	0.76 (s)	1.02 (d,J = 4 Hz)	3.50-4.16 (3β,7β,12β) (3H, br m)	3.66 (s)		26
	$CDCl_3$	0.72 (s)	0.88 (s)	1.03 (d,J = 4 Hz)		3.66 (s)		26
Me 3β,7α,12α-triol	$CDCl_3$	0.73 (s)	0.80 (s)	1.02 (d,J = 4 Hz)	3.50-4.00 (3α,7β,12β) (3H, br m)	3.66 (s)		26
Me 7α,12α-diol-3-one	$CDCl_3$	0.68 (s)	1.01 (s)	1.06 (d,J = 4 Hz)	3.92 (7β) (1H, m) 4.09 (12β) (1H, m)	3.70 (s)		26
5β-Series								
3α,7α,12α-Triol	$CDCl_3$	0.77 (s)	0.93 (s)	n.g.	3.33 (3β) (1H, m) 3.90 (7β) (1H, m) 4.33 (12β) (1H, m)			26
Triacetate	Pyr-d_5	0.70 (s)	0.85 (s)	0.94 (d,J = 6 Hz)	4.72 (3β) (1H, m) 5.08 (7β) (1H, m) 5.28 (12β) (1H, m)	—		43

(continued)

TABLE V. (Continued)

Compound/substituent	Solvent	18-CH₃	19-CH₃	21-CH₃	Protons geminal to hydroxyls	Methyl ester OCH₃	Other protons	Reference
Me 3β,7α,12α-triol	CDCl₃	0.68 (s)	0.87 (s)	n.g.		3.66 (s)		26
Me 7α,12α-diol-3-one	CDCl₃	0.77 (s)	1.00 (s)	1.08 (d J = 4 Hz)		3.68 (s)		26
Cholan-24-oic acids **5α- & Δ⁴ Series**								
Me 3α,7α,12α-triol	CDCl₃	0.68 (s)	0.76 (s)	n.g.		n.g.		44
Me Δ⁴-4,12α-diol-3-one, diacetate	CDCl₃	0.79 (s)	1.22 (s)	n.g.	5.09 (12β) (1H, m)	3.67 (s)		45
Me 2α-bromo-7α,12α-diol-3-one	CDCl₃	0.74 (s)	1.08 (s)	n.g.		3.67(s)		46
Me Δ⁴-7α,12α-diol-3-one	CDCl₃	0.75 (s)	1.18 (s)	1.01 (d J = 4 Hz)	4.03 (7β,12β) (2H, m)	3.70 (s)	5.88 4H̲	26
Δ⁴-7α,12α-diol-3-one	CDCl₃	0.70 (s)	1.16 (s)	0.97 (d)	3.90 (7β,12β) (2H, br m)	—	5.68 4H̲ (s)	29
Diformate	CDCl₃	0.81 (s)	1.19 (s)	0.82 (d)	5.13 (7β) (1H, br s); 5.23 (12β) (1H, br s)	—	5.68 4H̲ (s)	29
5β-Series								
Me 1β,3α,12α-triol	CDCl₃	0.68 (s)	1.04 (s)	n.g.	3.94 (1α) (1H, m); 4.08 (3β) (1H, br m); 3.87 (12β) (1H, m)	3.65 (s)		47
3,7,11-Triols								
Me 12α-bromo-3α,7α,11β-triol, diacetate	CDCl₃	1.06 (s)	1.15 (s)	n.g.	5.00 (3β) (1H, m); 4.47 (7β) (1H, m)	3.65 (s)	4.47 (12β) (1H, m)	48
Me 12α-iodo-3α,7α,11β-triol, diacetate	CDCl₃	0.72 (s)	1.09 (s)	n.g.	4.68 (3β) (1H, m); 4.86 (7β) (1H, m); 4.06 (11α) (1H, m)	3.67 (s)	4.50 (12β) (1H, m)	48

3.7.12-Triols

Compound	Solvent	18-, 19-CH₃	21-CH₃	3β-H	7β-H	12β-H	OCH₃	Other	Ref.
3α,7α,12α-Triol (cholic acid)	CD₃OD	0.71 (s) 0.92 (s)	1.01(d)	3.45 (3β) (1H, br m)	3.80 (7β) (1H, m)	3.96 (12β) (1H, m)	—	2.20 23-CH₂	13
Na salt	CD₃OD	0.72 (s) 0.93 (s)	1.02(d)	3.49 (3β)	3.91 (7β)	4.07 (12β)	—		[b]
	D₂O	0.73 (s) 0.93 (s)	0.98 (d, $J = 6$ Hz)	3.51 (3β) (1H, m)	3.91 (7β) (1H, m)	4.08 (12β) (1H, m)	—		49
	D₂O	0.70 (s) 0.92 (s)	0.95(d)	3.50 (3β) (1H, br m)	3.85 (7β) (1H, m)	4.00 (12β) (1H, m)	—	2.20 23-CH₂	13
7,12-Diformate	CDCl₃	0.77 (s) 0.93 (s)	n.g.	3.53 (3β) (1H, br m)	5.10 (7β) (1H, br s)	5.32 (12β) (1H, m)	—		33
Triformate, norethisterone conjugate	CDCl₃	0.65 (s) 0.95 (s)	n.g.	4.55 (3β) (1H, m)	5.28 (7β) (1H, m)	5.08 (12β) (1H, m)	—		31
3-Tosylate	CDCl₃	0.66 (s) 0.84 (s)	0.96 (s, br)	4.20 (3β) (1H, br m)	3.77 (7β) (1H, s)	3.90 (12β) (1H, s)	—	2.25 23-CH₂ (2H, m)	50
3,7-Diacetate	CDCl₃	0.68 (s) 0.96 (s)	1.00 (d, $J = 4$ Hz)	4.59 (3β) (1H, m)	4.85 (7β) (1H, m)	3.95 (12β) (1H, m)	—	2.25 23-CH₂ (2H, m)	50
Me 3α,7α,12α-triol	CDCl₃	0.70 (s) 0.90 (s)	n.g.	3.35 (3β) (1H, br m)	3.79 (7b) (1H, m)	3.89 (12β) (1H, m)	n.g.		44
	CDCl₃	0.68 (s) 0.88 (s)	1.00 (d, $J = 4$ Hz)	3.45 (3β) (1H, br m)	4.05 (7β) (1H, m)	4.18 (12β) (1H, m)	n.g.	2.28 23-CH₂ (2H, m)	50
	Pyr-d₅	0.78 (s) 0.98 (s)	1.14 (d, $J = 5.5$ Hz)	3.45 (3β) (1H, br m)	4.05 (7β) (1H, br m)	4.18 (12β) (1H, m)	n.g.		46
	CDCl₃	0.70 (s) 0.90 (s)	0.99 (d, $J = 6$ Hz)				n.g.		16
[2α,2β,3β,4α,4β-²H₅]	CD₃OD	0.71 (s) 0.91 (s)	1.00 (d, $J = 6$ Hz)	3.37 (3β)	3.79 (7β) (1H, m)	3.94 (12β) (1H, m)	3.64 (s)		51
	CDCl₃	0.67 (s) 0.91 (s)	0.97 (d, $J = 7$ Hz)		3.86 (7β) (1H, m)	3.98 (12β) (1H, m)	3.67 (s)		35
3-Acetate	Pyr-d₅	0.69 (s) 0.90 (s)	n.g.	4.56 (3β) (1H, br m)	3.86 (7β) (1H, br m)	3.96 (12β) (1H, br m)	3.68 (s)		14
3-Acetate, 12-methyl ether	Pyr-d₅	0.70 (s) 0.93 (s)	n.g.	4.57 (3β) (1H, br m)	3.85 (7β) (1H, br m)		3.69 (s)	3.32 (s) 12α-OCH₃	14

(continued)

TABLE V. (Continued)

Compound/substituent	Solvent	18-CH₃	19-CH₃	21-CH₃	Protons geminal to hydroxyls	Methyl ester OCH₃	Other protons	Reference
3-Cathylate. 12-mesylate	CDCl₃	0.74 (s)	0.88 (s)	n.g.	3.85 (7β)(1H, m); 5.11 (12β)(1H, m)	3.62 (s)		39
3.7-Diacetate	Pyr-d₅	0.68 (s)	0.92 (s)	n.g.	4.56 (3β)(1H, m); 4.90 (7β)(1H, m); 4.00 (12β)(1H, br m)	3.67 (s)		14
	CDCl₃	0.69 (s)	0.96 (s)	n.g.	4.53 (3β)(1H, m); 4.88 (7β)(1H, m); 4.00 (12β)(1H, m)	3.63 (s)		48
3.7-Diacetate-12-methyl ether	Pyr-d₅	0.70 (s)	0.92 (s)	n.g.	4.57 (3β)(1H, br m); 4.90 (7β)(1H, br m)	3.67 (s)	3.33 (s) 12α-OCH₃	14
	CDCl₃	0.74 (s)	0.92 (s)	0.82 (d.J = 6 Hz)	4.60 (3β)(t.t.J = 31 Hz); 4.92 (7β) (J = 3Hz); 5.09 (12β) (J = 6Hz)	3.66 (s)		52
12-Methyl ether	Pyr-d₅	0.70 (s)	0.92 (s)	n.g.	n.g.; 3.87 (7β)(1H, br m)	n.g.		14
3-Tosylate	CDCl₃	0.64 (s)	0.83 (s)	n.g.	4.29 (3β)(1H, br m); 3.82 (7β)(1H, m); 3.92 (12β)(1H, m)	3.63 (s)		19
3-Tosylate, 7-acetate	CDCl₃	0.67 (s)	0.88 (s)	n.g.	4.56 (3β)(1H, br m); 4.90 (7β)(1H, br m); 5.07 (12β)(1H, br m)	3.65 (s)		53
Triacetate	Pyr-d₅	0.69 (s)	0.87 (s)	n.g.		3.65 (s)		14
Et 3α.7α.12α-triol [23.24-¹³C₂-22ε.23ε-²H₂]	CDCl₃	0.68 (s)	0.89 (s)	1.02 (d.J = 6 Hz)	3.45 (3β)(1H, br m); 3.86 (7β)(1H, br m); 3.98 (12β)(1H, m)			9
Mc Δ⁹⁽¹¹⁾-3α.7α. 12α-triol. 3.7-diacetate	CDCl₃	0.57 (s)	1.12 (s)	n.g.	4.57 (3β); 5.07 (7β); 3.82 (12β) (d.J = 6 Hz)	3.65 (s)	5.75 11H (d.J = 6 Hz)	40
Triacetate	CDCl₃	0.63 (s)	1.10 (s)	n.g.	4.63 (3β); 5.12 (7β); 5.00 (12β) (d.J = 6 Hz)	3.65 (s)	5.85 11H (d.J = 6 Hz)	40

Et Δ^{22}-3α,7α,12α-triol [23,24-$^{13}C_2$]	CDCl$_3$	0.72 (s) 0.90 (s)	1.17 (d,J = 6Hz)	3.50 (3β) (1H, br m)	3.86 (7β) (1H, m)	3.97 (12β) (1H, m)		5.67 22 $\underline{H}$ (ddd,J = 16 Hz) 6.94 22 $\underline{H}$ (dd,J = 16, 7 Hz)	9
Tri(tetrahydropyranyloxy) [23,24-$^{13}C_2$]	CDCl$_3$	0.70 (s) 0.90 (s)	1.07 (d,J = 6Hz)					5.67 22 $\underline{H}$ (ddd,J = 16 Hz) 6.94 22 $\underline{H}$ (dd,J = 16,7 Hz)	9
3α,7α,12β-Triol	Acetone-d$_6$: D$_2$O	0.71 (s) 0.91 (s)	n.g.	3.40 (3α,12α) (2H, br m)			—		53
Me 3α,7α,12β-triol	CDCl$_3$	0.72 (s) 0.90 (s)	n.g.	3.40 (3α,12α) (2H, br m)	3.86 (7β)		3.54 (s)		53
Me $\Delta^{9(11)}$-3α,7α,12β-triol, 3,7-diacetate	CDCl$_3$	0.63 (s) 1.10 (s)	n.g.	4.55 (3β) (1H, br m)	5.03 (7β) (1H, br m)	4.08 (12α) (1H, br m)	3.64 (s)	5.30 11$\underline{H}$ (br)	40
Triacetate	CDCl$_3$	0.73 (s) 1.12 (s)	n.g.	4.60 (3β) (1H, br m)	5.08 (7β) (1H, br m)	5.17 (12α) (1H, br m)	3.66 (s)	5.35 11$\underline{H}$ (br)	40
Me 11β,12β-epoxy-3α, 7α-diol, diacetate	CDCl$_3$	0.75 (s) 1.09 (s)	n.g.	4.62 (3β) (1H, m)	4.90 (7β) (1H, br m)	3.08 (11.12) (1H, m)	3.63 (s)		48
3α,7β,12α-Triol (ursocholic acid)	CDCl$_3$: DMSO:D$_2$O	0.69 (s) 0.91 (s)	n.g.	3.51 (3β.7α) (2H, br m)		3.91 (12β) (1H, m)	—		39
Me 3α,7β,12α-triol	CDCl$_3$	0.71 (s) 0.93 (s)	n.g.	3.59 (3β.7α) (2H, br m)		3.98 (12β) (1H, m)	3.64 (s)		39
3α,7β,12β-Triol	CDCl$_3$: DMSO:D$_2$O	0.71 (s) 0.92 (s)	n.g.	3.44 (3β.7α.12α) (3H, br m)			—		39
Me 3α,7β,12β-triol	CDCl$_3$	0.76 (s) 0.96 (s)	n.g.	3.52 (3β.7α.12α) (3H, br m)			3.64 (s)		39
Me 3β,7α,12α-triol	CDCl$_3$	0.68 (s) 0.91 (s)	n.g.	4.01 (3α,12β) (2H, m)	3.86 (7β) (1H, m)		3.64 (s)		53
3-Formate, 7-acetate	CDCl$_3$	0.68 (s) 0.96 (s)	n.g.	5.16 (3α) (1H, m)	4.95 (7β) (1H, m)	4.00 (12β) (1H, m)	3.66 (s)		53
3-Formate	CDCl$_3$	0.68 (s) 0.93 (s)	n.g.	5.14 (3α) (1H, m)	3.85 (7β) (1H, m)	3.96 (12β) (1H, m)	n.g.		53

(continued)

TABLE V. *(Continued)*

Compound/substituent	Solvent	18-CH₃	19-CH₃	21-CH₃	Protons geminal to hydroxyls			Methyl ester OCH₃	Other protons	Reference
3β,7α,12β-Triol	Acetone-d₆:D₂O	0.73 (s)	0.95 (s)	n.g.	3.97 (3α) (1H, m)	3.82 (7β) (1H, m)	3.43 (12α) (1H, br m)	—		53
Me 3β,7α,12β-triol	CDCl₃	0.73 (s)	0.95 (s)	n.g.	4.04 (3α) (1H, m)	3.88 (7β) (1H, m)	3.45 (12α) (1H, br qrt)	3.66 (s)		53
3β,7β,12α-Triol	CDCl₃; DMSO:D₂O	0.70 (s)	0.94 (s)	n.g.	3.96 (3α,12β) (2H, m)	3.56 (7α) (1H, br m)		—		39
Me 3β,7β,12α-triol	CDCl₃	0.71 (s)	0.96 (s)	n.g.	4.06 (3α) (1H, m)	3.58 (7α) (1H, br m)	3.99 (12β) (1H, m)	3.64 (s)		39
3β,7β,12β-Triol	CDCl₃; DMSO:D₂O	0.70 (s)	0.94 (s)	n.g.	3.93 (3α) (1H, m)	3.40 (7α,12α) (2H, br m)		—		39
Me 3β,7β,12β-triol	CDCl₃	0.76 (s)	0.98 (s)	n.g.	4.07 (3α) (1H, m)	3.42 (7α,12α) (2H, br m)		3.63 (s)		39
3,7,22-Triols										
Me 3α,7α,22R-triol	Pyr-d₅	0.69 (s)	0.97 (s)	1.17 (d,J = 6 Hz)	3.4–4.4 (3β,7β,22S) (3H, br m)			—		54
Me 3α,7α,22S-triol	Pyr-d₅	0.74 (s)	0.96 (s)	1.16 (d,J = 6 Hz)	3.4–4.4 (3β,7β,22R) (3H, br m)			—		54
3,12,15-Triols										
Me 3α,12α,15β-triol	CDCl₃	0.94 (s)	0.94 (s)	n.g.	4.18 (15α) (t,J = 7.5 Hz)			n.g.		47
Diols-ones										
Me Δ¹-2α,12α-diol-3-one, diacetate	CDCl₃	0.78 (s)	1.22 (s)	n.g.	5.07 (12β) (1H, m)			3.68 (s)	6.31 1H (s)	45

Compound	Solvent							Ref.
Me 3α,7α-diol 11-one, diacetate	CDCl₃	0.62 (s) 1.15 (s)	n.g.	4.53 (3β) (1H, m)	5.00 (7β) (1H, m)		3.65 (s)	48
Me 12α-bromo-3α,7α-diol, 11-one, diacetate	CDCl₃	0.80 (s) 1.13 (s)	n.g.	4.56 (3β) (1H, m)	4.90 (7β) (1H, m)	4.24 (12β) (1H, m)	3.65 (s)	48
3α,7α-Diol-12-one	DMSO-d₆	0.94 (s) 0.96 (s)	0.76 (d,J = 5 Hz)	3.25 (3β) (1H, br m)	3.70 (7β) (1H, m)	2.10 23-CH₂ (2H, m)	—	50
3-Tosylate	CDCl₃	1.02 (s) 1.00 (s)	0.88 (d,J = 4 Hz)	4.30 (3β) (1H, br m)	3.89 (7β) (1H, m)	2.37 23-CH₂ (2H, m)	—	50
Diacetate	CDCl₃	1.05 (s) 1.05 (s)	0.87 (d,J = 5 Hz)	4.50 (3β) (1H, br m)	4.92 (7β) (1H, m)	2.30 23-CH₂ (2H, m)	—	50
11ε-Azido-3α,7α-diol-12-one	DMSO-d₆; CDCl₃	1.08 (s) 1.00 (s)	0.80 (d,J = 5 Hz)	3.30 (3β) (1H, m)	3.69 (7β) (1H, m)	2.13 23-CH₂ (2H, m)	—	50
11ε-Azido-3α,7α-diol-12-one, diacetate	DMSO-d₆; CDCl₃	1.13 (s) 0.98 (s)	0.78 (d,J = 5 Hz)	4.40 (3β) (1H, m)	4.79 (7β) (1H, m)	2.12 23-CH₂ (2H, m)	—	50
11ε-Bromo-3α,7α-diol-12-one, diacetate	DMSO-d₆; CDCl₃	1.20 (s) 0.97 (s)	0.76 (d,J = 5 Hz)	4.45 (3β) (1H, br m)	4.57 (7β) (1H, m)	2.15 23-CH₂ (2H, m)	—	50
Me 3α,7α-diol-12-one	CDCl₃	1.03 (s) 1.03 (s)	n.g.	3.50 (3β) (1H, br m)	3.93 (7β) (1H, m)		3.70 (s)	53
3-Acetate, 7-mesylate	CDCl₃	1.04 (s) 1.04 (s)	n.g.	4.56 (3β) (1H, m)	4.98 (7β) (1H, m)		3.64 (s)	39
3-Cathylate	CDCl₃	1.09 (s) 1.09 (s)	n.g.	4.49 (3β) (1H, m)	3.98 (7β) (1H, m)		3.70 (s)	39
3-Cathylate, 7-mesylate	CDCl₃	1.08 (s) 1.08 (s)	n.g.	4.50 (3β) (1H, m)	5.01 (7β) (1H, m)		3.67 (s)	39
Diacetate	CDCl₃	1.04 (s) 1.04 (s)	n.g.	4.59 (3β) (1H, m)	5.03 (7β) (1H, m)		3.65 (s)	48
	Pyr-d₅	1.03 (s) 1.03 (s)	0.82 (d)	4.60 (3β) (1H, br m)	5.04 (7β) (1H, s)		3.68 (s)	14
3-Tosylate	CDCl₃	0.94 (s) 0.99 (s)	n.g.	4.30 (3β) (1H, m)	3.92 (7β) (1H, m)		3.63 (s)	39

(continued)

TABLE V. *(Continued)*

Compound/substituent	Solvent	18-CH₃	19-CH₃	21-CH₃	Protons geminal to hydroxyls		Methyl ester OCH₃	Other protons	Reference
3-Tosylate, 7-acetate	CDCl₃	0.98 (s)	0.98 (s)	n.g.	4.30 (3β) (1H, br m)	4.97 (7β) (1H, m)	3.64 (s)		53
3-Tosylate, 7-mesylate	CDCl₃	1.00 (s)	1.00 (s)	n.g.	4.40 (3β) (1H, m)	4.96 (7β) (1H, m)	3.64 (s)		39
12-Tosylhydrazone	CDCl₃; DMSO-d₆	0.78 (s)	0.93 (s)	n.g.	3.33 (3β) (1H, br m)	4.21 (7β) (1H, m)	—		15
12-Tosylhydrazone	CHCl₃	0.78 (s)	0.97 (s)	n.g.	3.33 (3β) (1H, br m)	3.84 (7β) (1H, m)	3.64 (s)		15
12-Tosylhydrazone-7-acetate	n.g.	0.81 (s)	0.98 (s)	n.g.	3.40 (3β) (1H, br m)	3.90 (7β) (1H, m)	n.g.		15
12-Tosylhydrazone diacetate	CDCl₃	0.81 (s)	0.96 (s)	n.g.	4.51 (3β) (1H, br m)	4.91 (7β) (1H, m)	3.65 (s)		15
3α,7β-Diol-12-one	CDCl₃; DMSO-d₆	1.08 (s)	1.08 (s)	n.g.	3.58 (3β,7α) (2H, br m)		—		39
Me 3α,7β-diol-12-one	CDCl₃	1.04 (s)	1.04 (s)	n.g.	3.56 (3β,7α) (2H, br m)		3.64 (s)		39
3-Tosylate	CDCl₃	1.02 (s)	1.02 (s)	n.g.	4.40 (3β) (1H, m)	3.50 (7α) (1H, br m)	n.g.		39
12-Tosylhydrazone	CDCl₃; DMSO-d₆	0.79 (s)	0.99 (s)	n.g.	3.50 (3β,7α) (2H, br m)		3.64 (s)		15
3β,7α-Diol-12-one	Acetone-d₆	1.07 (s)	1.07 (s)	n.g.	3.91 (3α) (4H, m)	3.50 (7β) (1H, br m)	—		53
Me 3β,7α-diol-12-one	CDCl₃	1.03 (s)	1.05 (s)	n.g.	4.05 (3α) (1H, m)	3.95 (7β) (1H, m)	3.65 (s)		53

Compound	Solvent	18-CH₃	19-CH₃	21-CH₃	3-H	7-H	12-H	Other	OCH₃	Ref.
7-Acetate	CDCl₃	1.03 (s)	1.05 (s)	n.g.	4.08 (3α) (1H, m)	5.00 (7β) (1H, m)			3.66 (s)	15
3-Formate	CDCl₃	1.04 (s)	1.04 (s)	n.g.	5.13 (3α) (1H, m)	3.96 (7β) (1H, m)			3.64 (s)	39
3-Formate, 7-acetate	CDCl₃	1.03 (s)	1.06 (s)	n.g.	5.16 (3α) (1H, m)	5.00 (7β) (1H, m)			3.64 (s)	53
3-Formate, 7-mesylate	CDCl₃	1.04 (s)	1.07 (s)	n.g.	5.18 (3α) (1H, br m)	4.98 (7β) (1H, m)			3.65 (s)	39
3β,7β-Diol-12-one	CDCl₃; DMSO:D₂O	1.05 (s)	1.05 (s)	n.g.	4.00 (3α) (1H, m)	3.49 (7α) (1H, br m)			—	39
Me 3β,7β-diol-12-one	CDCl₃	1.04 (s)	1.07 (s)	n.g.	4.04 (3α) (1H, m)	3.58 (7α) (1H, br m)			3.64 (s)	39
3α,12α-Diol-7-one	DMSO-d₆	0.61 (s)	0.93 (s)	0.93 (d)	3.30 (3β) (1H, m)		3.74 (12β) (1H, m)	2.11 23-CH₂ (2H, m)	—	50
3β,12α-Diol-7-one	n.g.	0.69 (s)	1.21 (s)	n.g.	4.01 (3α,12β) (2H, m)				—	39
Me 3β,12α-diol-7-one	CDCl₃	0.69 (s)	1.22 (s)	n.g.	4.02 (3α,12β) (2H, m)				3.66 (s)	39
7α,12α-Diol-3-one	CDCl₃	0.72 (s)	0.98 (s)	0.98 (d)		3.90 (7β) (1H, br s)	4.01 (12β) (1H, br m)		—	29
Diformate	CDCl₃	0.79 (s)	1.03 (s)	0.85 (d)		5.13 (7β) (1H, br s)	5.28 (12β) (1H, br s)		—	29
Me 7α,12α-diol-3-one	CDCl₃	0.67 (s)	0.96 (s)	1.21 (d)		4.05 (7β) (1H, br s)	4.00 (12β) (1H, t)		3.66 (s)	28
4β-Bromo-7α,12α-diol-3-one, diformate	Pyr-d₅	0.76 (s)	1.04 (s)	n.g.		4.04 (7β,12β)			3.71 (s)	14
	CDCl₃	0.77 (s)	1.07 (s)	0.82 (d)		5.15 (7β)	5.25 (12β)	5.26 4α (d, J = 12 Hz)	—	29

(continued)

TABLE V. *(Continued)*

Compound/substituent	Solvent	18-CH$_3$	19-CH$_3$	21-CH$_3$	Protons geminal to hydroxyls			Methyl ester OCH$_3$	Other protons	Reference
ol-Diones										
Me 3β-ol-7,12-dione	CDCl$_3$	1.04 (s)	1.33 (s)	n.g.	4.07 (3α) (1H, m)			3.65 (s)		39
Cholestan-26-oic acids										
5α-Series										
3β,7α,12α-Triol	CDCl$_3$	0.80 (s)	0.90 (s)	1.23 (d,J = 6 Hz)	3.80 (3α) (1H, m)	4.09 (7β) (1H, m)	4.23 (12β) (1H, m)	—	1.32 27-C$\underline{H}_3$ (d,J = 6 Hz)	55
7α,12α-Diol-3-one	CDCl$_3$	0.75 (s)	1.05 (s)	0.81 (d,J = 6 Hz)		4.98 (7β) (1H, m)	5.07 (12β) (1H, m)	—	1.23 27-C$\underline{H}_3$ (d,J = 6 Hz)	55
5β-Series										
Me 3α,7α,12α-triol	CDCl$_3$	0.68 (s)	0.89 (s)	0.95 (d,J = 6.5 Hz)	3.45 (3β) (t,t,J = 31 Hz)	3.85 (7β) (q,J = 9 Hz) (t,J = 6 Hz)	3.98 (12β) (t,J = 6 Hz)	3.67 (s)		52
Triacetate	CDCl$_3$	0.72 (s)	0.92 (s)	0.79 (d,J = 6 Hz)	4.60 (3β) (t,t,J = 31 Hz)	4.92 (7β) (q,J = 9 Hz) (t,J = 6 Hz)	5.10 (12β) (t,J = 6 Hz)	3.67 (s)		52

									Ref.
Δ24E-3α,7α,12α-Triol	Pyr-d5	0.78 (s) 0.97 (s)	1.19 (d, J = 6 Hz)	3.65 (3β) (1H, m)	4.05 (7β) (1H, m)	4.20 (12β) (1H, m)	—	1.80 27-CH$_3$ (s) 7.13 24H (t, J = 6 Hz)	38
Δ24Z-3α,7α,12α-triol	Pyr-d5	0.78 (s) 0.97 (s)	1.19 (d, J = 6 Hz)	3.65 (3β) (1H, m)	4.05 (7β) (1H, m)	4.20 (12β) (1H, m)	—	1.80 27-CH$_3$ (s) 5.85 24H (t, J = 6 Hz)	38
Me 3α,7α,22ε-triol	Pyr-d5	0.83 (s) 0.99 (s)	1.38 (d, J = 6.5 Hz)	3.90 (3β) (1H, br m)		4.30 (12β) (1H, m) 4.01 (22) (1H, br m)	3.65 (s)	1.22 27-CH$_3$ (d, J = 7 Hz)	56
27a-Dihomo-27b-carboxymethyl 5β-cholestan-26, 27-dioic acid, Me 3α,7α,12α-triol, triacetate		0.72 (s) 0.92 (s)	0.79 (d, J = 6.5 Hz)	4.60 (3β) (t.t, J = 31Hz)	4.92 (7β) (q, J = 9.5 Hz)(t, J = 6 Hz)	5.10 (12β)	3.65 (s)		52

[a] See Table III for abbreviations.

[b] V. Waterhous, S. Barnes, and D. Muccio, unpublished data.

TABLE VI. ¹H-NMR Data for Tetraoxygenated Bile Acids[a]

Compound	Solvent	18-CH₃	19-CH₃	21-CH₃	Protons geminal to hydroxyl				Ester OCH₃	Other protons	Reference
Cholan-24-oic acids											
5α-Series											
2β,3α,7α,12α-Tetrol	Pyr-d₅	0.79 (s)	1.40 (s)	1.20	4.27 (3β)		4.09 (7β) (1H, m)	4.16 (12β) (1H, m)	—		46
Me 2β,3α,7α,12α-tetrol, 2,3,7-Triacetate	CDCl₃	0.68 (s)	0.92 (s)	0.97 (d,J = 5.5 Hz)	4.91 (2α,3β,7β) (3H, m)			3.96 (12α) (1H, t)	n.g.		46
	C₆D₆	0.48 (s)	0.73 (s)	0.84 (d,J = 5.5 Hz)	4.95-5.13 (2α,3β,7β) (3H, m)			3.64 (12β) (1H, m)	n.g.		46
Tetraacetate	CDCl₃	0.71 (s)	0.90 (s)	0.79 (d,J = 5.5 Hz)	4.89 (2α,3β,7β) (3H, m)			5.03 (12β) (1H, t)	n.g.		46
	C₆D₆	0.49 (s)	0.70 (s)	0.79 (d,J = 5 Hz)	4.95-5.15 (2α,3β,7β,12β) (4H, m)				n.g.		46
5β-Series											
Me 1β,3α,7α,12α-etrol	CD₃OD	0.71 (s)	1.01 (s)	0.99 (d)	3.82 (1α)	3.84 (3β)	3.79 (7β)	3.92 (12β)	3.64 (s)		51
2β,3α,7α,12α-Tetrol	Pyr-d₅	0.76 (s)	1.01 (s)	1.19 (d,J = 5 Hz)	4.20 (2α) (1H, br m)	3.50 (3β) (1H, br m)	4.05 (7β) (1H, m)	4.19 (12β) (1H, m)	—		46
Me 2β,3α,7α,12α-tetrol, 2,3,7-triacetate	CDCl₃	0.67 (s)	0.95 (s)	0.97 (d)			4.85 (7β) (1H, m)	4.02 (12β) (1H, m)	n.g.		46
	C₆D₆	0.45 (s)	0.60 (s)	0.81 (d)			4.90 (7β) (1H, m)	3.65 (12β) (1H, m)	n.g.		46

Compound	Solvent	18-CH₃	19-CH₃	21-CH₃	27-CH₃	OCH₃	2β/3β/6β	7β	12β	24	Ref.
Tetraacetate	CDCl₃	0.72 (s)	0.94 (s)	0.81 (d, J = 5.5Hz)		n.g.	4.5-5.2 (3β) (1H, br m)	4.89 (7β) (1H, m)	5.10 (12β) (1H, m)		46
	C₆D₆	0.46 (s)	0.53 (s)	0.82 (d, J = 5Hz)		n.g.	4.7-5.4 (3β) (1H, br m)	4.87 (7β) (1H, m)	5.22 (12β) (t, $w_{1/2}$ = 6Hz)		46
Me 2β,7α,12α-triol-3-one	CDCl₃	0.76 (s)	1.03 (s)	n.g.		3.65 (s)					46
Me 3α,6α,7α,12α-tetrol	CD₃OD	0.71 (s)	0.91 (s)	1.00 (d)		3.64 (s)	3.33 (3β); 3.77 (6β)	3.77 (7β)	3.94 (12β)		51
5β-Cholestan-26-oic acids											
Me 2α,3α,7α,12α-tetrol	Pyr-d₅	0.78 (s)	1.03 (s)	1.17 (d, J = 5.5Hz)	1.29 27-CH₃ (d, J = 6.5 Hz)	n.g.	4.0-4.4 (2β) (1H, br m); 3.5-3.9 (3β) (1H, br m)	4.08 (7β) (1H, m)	4.20 (12β) (1H, m)		46
Tetraacetate	CDCl₃	0.72 (s)	0.94 (s)	0.80 (d, J = 5 Hz)	1.13 27-CH₃ (d, J = 6.5 Hz)	n.g.					46
Me 3α,7α,12α,22ξ-tetrol	Pyr-d₅	0.90 (s)	1.03 (s)	1.40 (d, J = 6 Hz)	1.22 27-CH₃ (d, J = 7 Hz)	3.63 (s)	3.72 (3β) (1H, br m)	4.11 (7β) (1H, m)	4.35 (12β) (1H, m)		56
Et 3α,7α,12α,22ξ-tetrol	Pyr-d₅	0.91 (s)	1.04 (s)	1.42 (d, J = 6 Hz)	1.23 27-CH₃ (d, J = 7 Hz)	n.g.	3.74 (3β) (1H, br m)	4.10 (7β) (1H, m)	4.35 (12β) (1H, m)		56
24S,25R-3α,7α,12α,24-Tetrol	Pyr-d₅	0.83 (s)	1.00 (s)	1.24 (d, J = 6 Hz)	1.24 27-CH₃ (s)	—	3.70 (3β) (1H, m)	4.10 (7β) (1H, m)	4.27 (12β) (1H, m)	4.35 (24) (1H, m)	38
24S,25S-3α,7α,12α,24-Tetrol	Pyr-d₅	0.82 (s)	1.00 (s)	1.25 (d, J = 6 Hz)	1.25 27-CH₃ (s)	—	3.70 (3β) (1H, m)	4.09 (7β) (1H, m)	4.25 (12β) (1H, m)	4.29 (24) (1H, m)	38
24R,25R-3α,7α,12α,24-Tetrol	Pyr-d₅	0.81 (s)	0.98 (s)	1.25 (d, J = 6 Hz)	1.25 27-CH₃ (s)	—	3.70 (3β) (1H, m)	4.07 (7β) (1H, m)	4.25 (12β) (1H, m)	4.30 (24) (1H, m)	38
24R,25S-3α,7α,12α,24-Tetrol	Pyr-d₅	0.81 (s)	0.98 (s)	1.25 (d, J = 6 Hz)	1.25 27-CH₃ (s)	—	3.70 (3β) (1H, m)	4.08 (7β) (1H, m)	4.25 (12β) (1H, m)	4.30 (24) (1H, m)	38

ᵃ See Table III for abbreviations

^{1}H NMR spectra of bile acids are typified by the spectra in Fig. 5 for 5β-cholanic acid, lithocholic acid, chenodeoxycholic acid, ursodeoxycholic acid, deoxycholic acid, and cholic acid. In each case the 18-methyl, 19-methyl, and 21-methyl group resonances are the most prominent features of the spectrum at high field. The 18-methyl and 19-methyl protons give singlets (see Section 2.4). The 21-methyl signal is a doublet (J = 5–6 Hz), being coupled to 20-H. Resonances of protons geminal to hydroxy (or acetoxy) groups appear at lower field (δ 3–5) as broad signals or multiplets, split by spin coupling with any vicinal protons. The significance of widths and multiplicities is discussed in Sections 3.4 and 3.5. It is to be regretted that many authors have omitted to report the coupling constants for these easily observed proton resonances (Tables III–VI).

The remainder of the CH_2 and CH protons in the bile acid molecule give overlapping resonances in the range 1.0–2.35 ppm. Even at high fields (e.g., 400 MHz spectra), and with resolution enhancement, these resonances cannot be fully separated in the spectrum as normally recorded. Special techniques that take advantage of the power of computer-controlled FT–NMR spectrometers have had to be employed for the complete analysis of such spectra (see Sections 5.1 and 5.2).

The following sections detail the main features of bile acid spectra in relation to molecular structure.

3.2.   Methyl Proton Resonances

Zurcher [57] showed that the chemical shifts of proton signals from the two angular methyl groups (C-18 and C-19) are affected by the pattern of substitution in a steroid molecule and can be calculated by summation of empirically determined increments for the particular steroid skeleton and for each of the substituents, including any side chain at C-17. Zurcher's original list of group increments, and a subsequent one compiled by steroid chemists at Oxford University [58], were derived from study (in $CDCl_3$) of several hundreds of steroid derivatives, with widely varying patterns of substitution involving all the ring positions. Calculated chemical shifts for the methyl protons usually agree with experimental values (to ± ca. 0.02 ppm), unless substituents are close enough to interact strongly with each other or profoundly alter the molecular configuration.

Selected group increments appropriate to bile acids are presented in Table VII. The isomeric 5α- and 5β-cholan-24-oic acids are treated here as "parent" compounds. The available group increments for hydroxy and oxo functions, and for those olefinic bonds likely to be found in bile acids, permit calculation or prediction of the angular methyl proton chemical shifts for bile acids with reasonable confidence.

As an example, cholic acid (3α,7α,12α-trihydroxy-5β-cholan-24-oic acid) and its methyl ester or sodium salt are reported (Table V) to exhibit two methyl

TABLE VII. Chemical Shift Increments for Protons at C-18 and C-19[a]

	ppm from TMS	
	18-H	19-H
5α-, Δ⁴, and Δ⁵-Series		
5α-Cholan-24-oic acid	0.65	0.78
1-Oxo	0.02	0.38
1α-OH	0.00*	0.01*
1β-OH	0.00*	0.05
2-oxo	0.01	−0.03
2α-OH	0.00*	0.02*
2β-OH	0.01	0.25
3-Oxo	0.04	0.24
3α-OH	0.01	0.00
3β-OH	0.01	0.03
4-Oxo	0.02	−0.03
4α-OH	0.00*	0.02*
4β-OH	0.00*	0.24*
5α-OH	0.00*	0.18*
6β-OH	0.04	0.23
Δ¹-3-Oxo	0.05	0.25
Δ⁴-3-Oxo	0.08	0.42
Δ¹,⁴-3-Oxo	0.10	0.46
Δ⁴,⁶-3-Oxo	0.11	0.33
Δ⁴(without 3-Oxo)	0.04	0.25
Δ5	0.03	0.23
6-Oxo	0.02	−0.05
5β-Series		
5β-Cholan-24-oic acid	0.65	0.92
1-Oxo	−0.02*	0.20*
1α-OH	n/a[b]	n/a
1β-OH	0.00*	0.12*
2-Oxo, 2α-OH, or 2β-OH	n/a	n/a
3-Oxo	0.04	0.12
3α-OH	0.01	0.01
3β-OH	0.01	0.05
4-Oxo	0.00	0.20
4α-OH	0.01	0.01
4β-OH	n/a	n/a
5β-OH	n/a	n/a
Ring B, C, D substituents (applicable to all types)		
6-Oxo	0.03	−0.09
6α-OH	0.00*	0.03*
6β-OH(5α)	0.04	0.23
6β-OH(5β or Δ⁴)	0.04	0.19

(continued)

TABLE VII.   *(Continued)*

	ppm from TMS	
	18-H	19-H
Δ^6	0.05	0.03
7-Oxo	0.01	0.28
7α-OH	0.01	−0.01
7β-OH	0.03	0.03
Δ^7	−0.12	−0.01
$\Delta^9(11)$	−0.07	0.14
11-Oxo	−0.03	0.22
11α-OH	0.03	0.12
11β-OH	0.24	0.26
12-Oxo	0.38	0.10
12α-OH	0.04	−0.01
12β-OH	0.07	0.01
15-Oxo	0.08	0.01
15α-OH	0.03	0.01
15β-OH	0.28*	0.02*
16-Oxo	0.17*	0.04*
16α-OH	0.01*	−0.01*
16β-OH	0.25*	0.02

[a] Derived from Zurcher [57,59,60] and Ref. [58] (indicated by an asterisk).
[b] n/a, not available.

proton singlets at about δ 0.7 and δ 0.9. The calculated chemical shifts are obtained by the following summations:

	18-H	19-H
5β-cholan-24-oic acid	0.65	0.92
3α-OH	0.01	0.01
7α-OH	0.01	−0.01
12α-OH	0.04	−0.01
Calculated totals for cholic acid	0.71 ppm	0.91 ppm

The individual assignments of the observed methyl proton signals are obvious. Slight variations, affecting only the second decimal place, are attributable to solvent effects, or the formation of salts or esters of the carboxyl function.

Inspection of the group increments in Table VII shows that in general each methyl group is affected appreciably only by substituents in the immediately adjoining rings. For a more detailed account of the nature and origins of these substituent effects, the reader should consult either Zurcher's original papers [59,60] (in German), or a summary by Bhacca and Williams [61]. The Zurcher method is of greatest value for steroids (or bile acids) with unusual patterns of substitution.

Protons on the C-21 methyl group normally give a doublet signal ($J \simeq 4$–6 Hz), being split by spin coupling with 20-H. The doublet is usually, but not invariably, seen at slightly lower field than the 18- and 19-methyl singlets (ca. δ 0.9–1.2). The precise value of its chemical shift does not have great diagnostic value.

The methyl proton singlet for methyl esters of bile acids is a distinctive feature of these derivatives at δ 3.6–3.7.

3.3. Solvent Effects

Zurcher's chemical shift data refer to spectra in $CDCl_3$. The limited range of examples of spectra determined in aqueous systems or in CD_3OD indicates little if any variation in these solvents (that is, except for possible conformational effects—see Table XIII). Deuteriopyridine (C_5D_5N) [62] and deuteriobenzene (C_6D_6) [63,64], however, cause appreciable solvent shifts, owing to their association with polar substituents in the bile acid and the anisotropy of magnetic susceptibility of the aromatic π-system. Shifts to lower field (deshielding) are generally observed in deuteriopyridine (C_5D_5N). The shifts are largest for solutes with free OH groups, which associate with pyridine by hydrogen bonding (e.g., by ± 0.08 ppm for methy cholate—see Table V), whereas the shift is upfield in C_6D_6. Deuteriopyridine is particularly useful as a solvent for the more highly oxygenated bile acids, which have limited solubilities in $CDCl_3$ (Table II). Unfortunately, the range of data available for solutions in C_5D_5N is too small at present for a reliable set of group increments to be deduced to parallel those for $CDCl_3$ in Table VII.

3.4. Derivatives of Bile Acids

Esterification of the carboxy group has little, if any, effect on the C-18 and C-19 methyl chemical shifts for cholan-24-oic acids, or those with longer side chains, so that Zurcher's increments can be applied without regard to possible esterification. In the 24-nor bile acids, the carboxy group seems to be sufficiently close to 18-CH_3 to produce small esterification effects (up to 0.04 ppm), although the data are too limited for a precise value to be derived.

Derivatives of bile acids at their hydroxyl groups add further proton resonances to the spectrum and also cause downfield shifts of the proton geminal to the hydroxyl group. The proton resonances of these substituents, e.g., acetate, formate, mesylate, tosylate, are given in Table VIII.

TABLE VIII. Typical ^{1}H Chemical Shifts for Acyl and Sulfonyl Groups in Esters

Group	^{1}H chemical shift (δ)
Formate (—OC$\underline{H}$) ‖ O	8.0–8.1 (s)
Acetate (—OCC$\underline{H}_3$) ‖ O	2.0–2.2 (s)
Propionate (—OCC$\underline{H}_2$C$\underline{H}_3$) ‖ O	2.3–2.4 (C$\underline{H}_2$; q) 1.1–1.2 (C$\underline{H}_3$; t)
Hemisuccinate (OCC$\underline{H}_2$C$\underline{H}_2$CO$_2$H) ‖ O	2.5–2.6 (s, or d,d; C$\underline{H}_2$C$\underline{H}_2$)
Benzoate (OC—C$_6\underline{H}_5$) ‖ O	7.2–8.1 (complex; C$_6\underline{H}_5$)
Methanesulfonate (mesylate) O ‖ (—OSC$\underline{H}_3$) ‖ O	2.9–3.1 (s)
Toluene-p-sulfonate (tosylate) O ‖ (—OS—⟨◯⟩—CH$_3$) ‖ O	7.4 (d) and 7.8 (d) (C$_6\underline{H}_4$) 2.4–2.6 (C$\underline{H}_3$)

3.5. Spin Coupling: Splitting of Proton Signals

The splitting of the 21-methyl proton signal into a doublet has already been mentioned in Section 3.2.

More important for structural studies is the multiplicity and width of the signal due to a "methine" proton geminal to a hydroxy function, as a result of coupling to as many as *four* vicinal protons on the ring system.

According to a principle established by Karplus [65], the numerical value of the coupling constant J between two vicinal protons in a molecule of rigid conformation depends on the torsion angle (ω) between the two C-H bonds, as

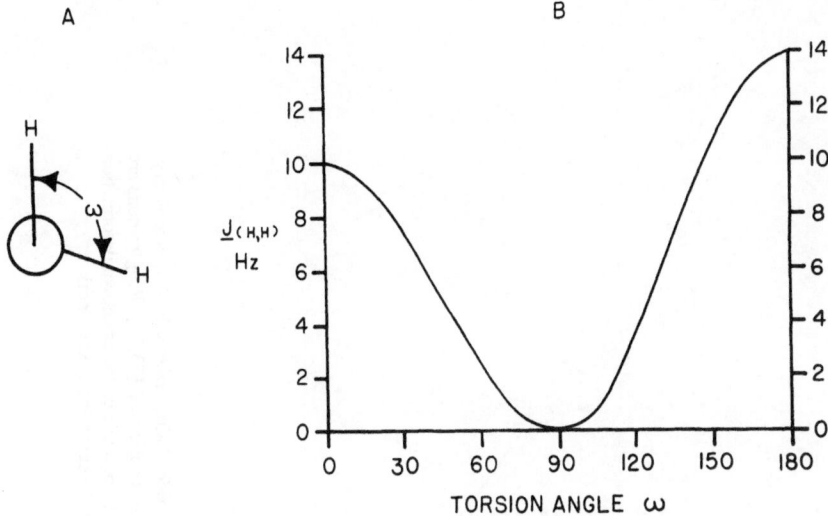

FIGURE 6. Dependence of $J_{(H,H)}$ for vicinal protons on HCCH torsion angle ω. (A) Newman projection; (B) graphical representation of the Karplus equation (see text).

viewed in projection along the C-C axis (Fig. 6). The value of J varies approximately as the square of the cosine of ω (Fig. 6B): that is, $J = k \cos^2\omega$. (A more complex expression, derived mathematically, is no more reliable in practice than this simple form of the "Karplus equation.")

The proportionality "constant" k is not a true constant, as its value depends on circumstances. For ω in the range 0–90°, k typically has a value of 8–10, whereas for ω in the range 90–180°, k is likely to be 12–16. The precise values for k within these ranges depend on the electronegativities of nearby substituents: electron withdrawal by strongly electronegative groups tends to reduce the value of k. For the present discussion we shall adopt the equations

$$J = 10 \cos^2\omega \qquad 0 < \omega < 90°$$
$$J = 14 \cos^2\omega \qquad 90° < \omega < 180°$$

while recognizing that the proportionality constants are only approximate.

We shall also treat each six-membered ring of the bile acid (steroid) framework as having the ideal chair conformation (Fig. 7).* The torsion angles between

---

* The true conformation of each ring is a slightly flattened chair, depending on the detailed structure. Actual values of ω and J therefore vary slightly from the ideal assumed here.

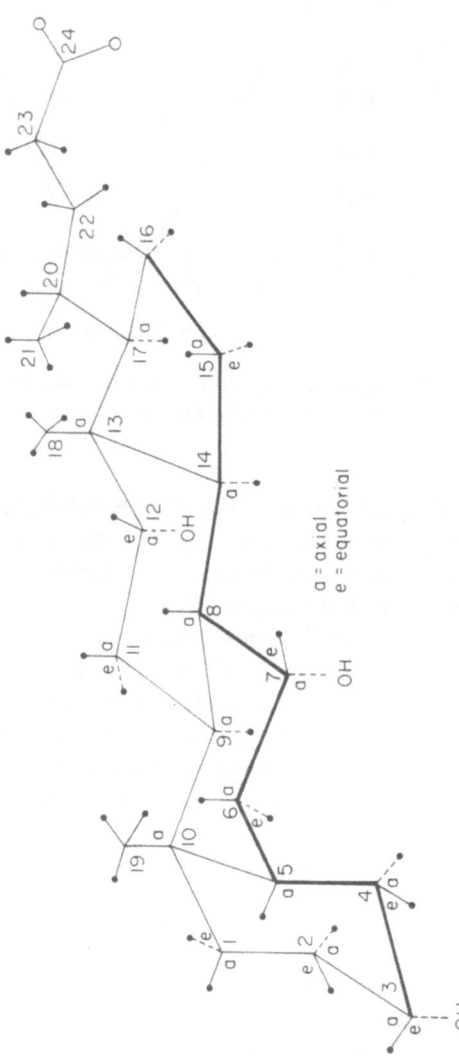

FIGURE 7. Perspective drawing of the cholate molecule. The orientation of the side chain, carbons 20–24, is arbitrary, although a similar orientation is observed in crystals of calcium and sodium salts of cholic acid [73,74]. Ring protons are equatorial (e) or axial (a); no designation is given for the two protons on carbon 16, because of the unknown deformation of the five-membered ring (From Ref. [49]; reproduced with permission of Lipid Research Inc., New York.)

vicinal pairs of C-H bonds then depend on their respective conformations, as follows (ax = axial; eq = equatorial):

$$\omega_{ax/ax} \simeq 180°, \text{ whence } J_{ax/ax} \simeq 14 \text{ Hz}$$

$$\omega_{ax/eq} \simeq \omega_{eq/eq} \simeq 60°, \text{ whence } J_{ax/eq} \simeq J_{eq/eq} \simeq 2.5 \text{ Hz}$$

The total width of the signal due to a methine proton with several vicinal neighbors is the sum of the relevant J values. Consider the signal splitting experienced by the 3β-proton in a 3α-hydroxy bile acid of the common 5β-series (Fig. 8A). The 3α-OH group is equatorial, leaving 3β-H axial. It is evident from Fig. 7 that there will be axial/axial couplings between the 3β-proton and each of the 2α- and 4α-protons, and also axial/equatorial couplings between the 3β-proton and each of the 2β- and 4β-protons. The total signal width is therefore estimated to be $14 + 14 + 2.5 + 2.5 = 33$ Hz. Values in the range 30–35 Hz are commonly observed.

In a similar way, considering the epimeric 3β-hydroxy-5β-bile acid, the 3α-proton is equatorial (Fig. 7). It therefore experiences *two* equatorial/equatorial and *two* equatorial/axial couplings to vicinal protons, giving a total signal width of $4 \times 2.5 = 10$ Hz (Fig. 8B). This is a typical value for an equatorial methine proton.

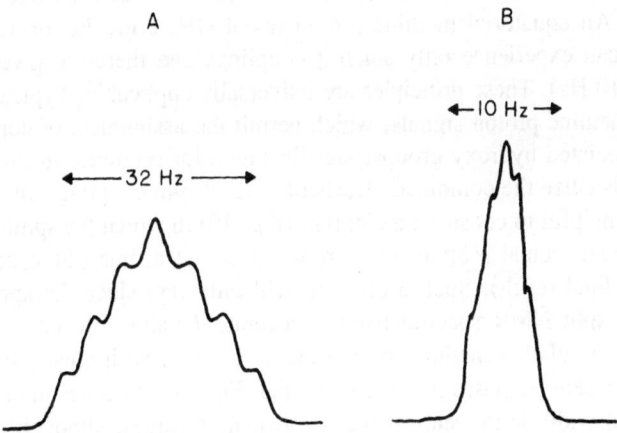

FIGURE 8. Typical resonance profiles for protons at C-3 in 3-hydroxy bile acids: (A) 3β-H (axial) in a 3α-hydroxy 5β-bile acid, or 3α-H (axial) in a 3β-hydroxy 5α-bile acid; (B) 3α-H (equatorial) in a 3β-hydroxy 5β-bile acid, or 3β-H (equatorial) in a 3α-hydroxy 5α-bile acid. The fine structure may be more or less pronounced than illustrated here.

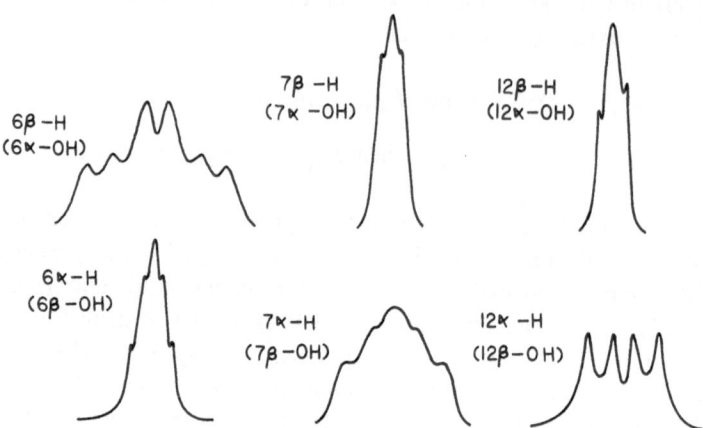

FIGURE 9. Typical profiles for methine proton resonance signals with geminal hydroxyl (axial H, broad signal; equatorial H, narrow signal).

If the hydroxy group is adjacent to a ring junction, there will be fewer vicinal protons to couple with the methine proton. For 7-OH there are *three* (6α, 6β, 8β), whereas for 12-OH there are only *two* (11α, 11β). The signals for the methine protons are correspondingly narrower, but can still be clearly distinguished on the basis that an *axial* methine proton (equatorial OH, e.g., 7β- or 12β-OH) will exhibit a broad signal (≥ 16 Hz), with at least one large $J_{ax/ax}$ included. An equatorial methine proton (axial OH, e.g., 7α- or 12α-OH), in contrast, can experience only small J couplings and therefore gives a narrow signal (≤ 10 Hz). These principles are universally applicable. Typical band profiles for methine proton signals, which permit the assignment of configurations to the associated hydroxy groups, are illustrated for reference in Fig. 9.

To visualize the combined effects of several spin couplings on one proton, it is often helpful to construct a diagram (Fig. 10) in which the splittings caused by individual vicinal protons are introduced one at a time (the order does not affect the final result). Such a diagram will only reproduce the appearance of the actual split NMR spectral band if accurate J values are known and used (computer simulation methods permit this to be done with great precision).

For present purposes it is sufficient that Fig. 10 illustrates the fact that the total signal width is the sum of the component J values, although the number of peaks that actually appear in the split signal will depend on whether the separate components overlap, as often occurs. Broad signals like that for an axial methine proton at C-3 may appear as lumpy multiplets of uncertain structure, or in "good" cases can be seen to have as many as seven constituent peaks with relative intensities in the ratio 1 : 2 : 3 : 4 : 3 : 2 : 1. This is a consequence of

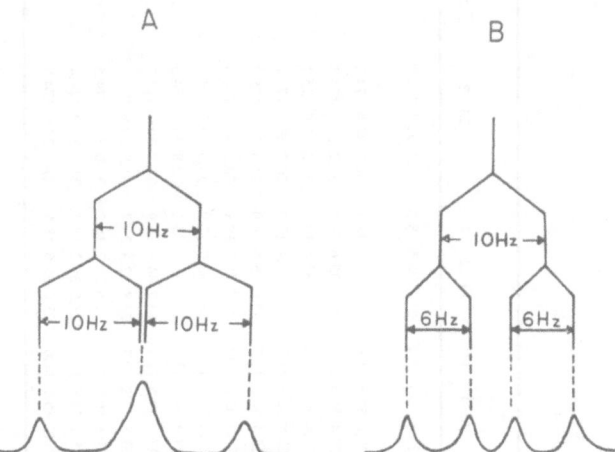

FIGURE 10. Construction of splitting diagrams to illustrate the effects of spin coupling to (A) two vicinal protons with identical J values (triplet) and (B) two vicinal protons with different J values (double doublet).

overlapping among the $2^4 = 16$ individual components which result from splitting by four vicinal protons.

Derivative formation (e.g., acetylation), although it causes a downfield shift of the signal from the methine proton, does not appreciably alter its profile.

4. ^{13}C NMR OF BILE ACIDS

The last few years have seen the rapid development of ^{13}C NMR spectroscopy of bile acids, to obtain direct information about the molecular framework. The lower intrinsic receptivity for the ^{13}C nucleus compared with protons and its low natural abundance (1.1%) make ^{13}C NMR inherently less sensitive than for protons by a factor of about 6000. This problem is overcome by using higher concentrations and longer periods of data collection. However, the chemical shift range observed for ^{13}C nuclei is so large (about 200 ppm) that adequate spectral resolution can usually be obtained even at low field strengths (2 T). This has enabled assignments of all the ^{13}C resonances to be made for quite a large number of bile acids [66–69], particularly in recent studies by Iida and Chang (Tables IX–XI).

Recognition of a large proportion of the singlet peaks in a proton-decoupled ^{13}C spectrum can usually be achieved by inspection and comparison with tab-

TABLE IX. ^{13}C-NMR Chemical Shift Data for Monoxygenated C_{24} 5β-Bile Acids

Compound/ substituent	Solvent	1	2	3	4	5	6	7	8	9	10	11	12	13	14	15	16	17	18	19	20	21	22	23	24	Reference
Unsubstituted methyl ester	CD$_2$Cl$_2$	37.9	21.6	27.4	27.8	44.1	27.5	26.9	36.2	40.8	35.6	21.1	40.6	43.0	56.9	24.4	28.4	56.4	12.1	24.2	35.6	18.3	31.1	31.1	175.0	66
3α-ol	CDCl$_3$	37.5	21.2	26.9	27.4	43.6	27.1	26.4	35.7	40.4	35.2	20.7	40.2	42.6	56.5	24.1	28.0	55.8	11.9	24.1	35.2	18.1	30.8	30.8	174.2	67
	CD$_3$OD	35.6	30.4	71.8	36.5	42.5	27.5	26.7	36.1	40.6	34.8	21.1	40.5	43.0	56.7	24.4	28.4	56.5	11.8	23.2	35.6	18.1	31.2	31.1	174.5	66
Me 3α-ol	CD$_2$Cl$_2$	35.6	30.9	71.9	36.8	42.5	27.5	26.7	36.1	40.7	34.8	21.1	40.5	43.0	56.8	24.4	28.4	56.3	11.0	23.4	35.6	18.3	31.3	31.0	175.1	66
	CDCl$_3$	35.0	30.1	71.0	36.0	41.8	26.9	26.2	35.5	40.1	34.2	20.5	39.9	42.4	56.2	23.9	27.8	55.6	11.7	23.1	35.1	17.9	30.7	30.7	174.2	67
3-Acetate	Dioxane-d$_8$	35.1	26.4	74.4	32.2	42.0	27.3	26.7	36.1	40.6	34.8	21.1	40.5	43.0	56.7	24.4	28.4	56.5	11.7	23.0	35.6	18.0	31.3	31.1	175.2	66
Me 3β-13	CDCl$_3$	29.8	27.8	66.7	33.4	36.3	26.5	26.1	35.5	39.6	34.9	20.9	40.2	42.6	56.4	24.0	28.0	55.8	11.9	23.7	35.2	18.1	30.8	30.8	174.4	67
Me 7α-ol	CDCl$_3$	37.4	21.1	27.5	30.2	43.0	35.5	68.1	39.2	32.6	35.0	20.3	39.4	42.3	50.1	23.3	27.8	55.5	11.4	23.3	35.0	17.9	30.6	30.6	174.1	67
Me 7β-ol	CDCl$_3$	37.4	21.0	26.7	28.4	44.0	37.1	71.2	43.6	39.1	34.7	20.9	40.1	43.6	55.7	26.9	27.9	54.8	12.0	24.1	35.1	18.2	30.8	30.8	174.3	
Me 12α-ol	CDCl$_3$	37.2	21.0	26.7	27.2	43.5	26.9	26.0	35.8	33.4	34.6	28.5	72.8	46.2	48.0	23.5	27.2	49.6	12.5	23.7	34.9	16.9	30.7	30.7	174.2	67
Me 12β-ol	CDCl$_3$	37.2	21.0	26.8	27.2	43.1	26.9	25.9	34.4	39.1	35.0	29.3	79.1	47.6	54.4	23.4	23.8	57.2	7.7	23.8	32.4	20.7	31.9	30.9	174.3	67
Me 3-one	CDCl$_3$	37.1	36.9	212.9	42.3	44.3	25.7	26.6	35.6	40.9	34.9	21.1	40.1	42.8	56.5	24.1	28.1	56.0	12.0	22.6	35.2	18.2	31.0	31.0	174.5	68
Me 7-one	CDCl$_3$	36.4	21.6	26.5	28.1	47.6	45.8	212.2	49.5	42.8	35.8	20.6	39.0	42.8	49.0	24.8	28.1	54.8	12.0	23.8	35.2	18.3	31.0	31.0	174.4	68
Me 12-one	CDCl$_3$	37.5	21.0	26.9	27.4	43.1	26.9	26.2	35.8	44.3	36.2	38.1	214.9	57.5	58.7	24.3	27.4	46.5	11.6	23.6	35.6	18.5	31.3	30.5	174.5	68

TABLE X. ¹³C-NMR Chemical Shift Data for Dioxygenated C_{24} 5β-Bile Acids

Compound/substituent	Solvent	1	2	3	4	5	6	7	8	9	10	11	12	13	14	15	16	17	18	19	20	21	22	23	24	Reference
Me 3α,6α-diol	C₂DCl₂	36.3	29.7	71.8	35.3	48.8	68.1	30.6	35.3	40.2	36.1	21.0	40.2	43.1	56.3	24.5	28.4	56.3	12.1	23.6	35.8	18.3	31.3	31.1	175.1	66
3α,7α-Diol	CD₃OD	35.8	31.0	71.4	39.7	42.1	35.4	67.8	39.9	33.3	35.5	21.1	39.9	43.0	50.9	24.0	28.5	56.3	11.9	23.0	35.8	18.3	31.2	31.1	174.6	66
	CD₃OD	36.5	31.2	72.7	40.3	43.1	36.1	68.9	40.7	33.9	35.8	21.7	41.0	43.6	51.4	24.6	29.1	57.2	12.2	23.4	36.6	18.8	32.2	31.9	175.1	69
Me 3α,7α-diol	CD₂Cl₂	35.8	31.3	72.1	39.9	42.1	35.5	68.6	39.9	33.3	35.5	21.1	39.9	43.0	50.9	24.0	28.5	56.3	12.1	23.2	35.8	18.4	31.3	31.1	175.2	66
	CD₃OD	35.1	30.7	71.6	39.2	41.3	34.4	68.2	39.2	32.6	34.9	20.3	39.2	42.4	50.1	23.4	27.9	55.6	11.5	22.6	35.1	18.0	30.7	30.3	174.7	69
	CDCl₃	35.2	30.5	71.7	39.6	41.5	34.7	68.2	39.3	32.7	35.0	20.5	39.6	42.5	50.3	23.5	28.0	55.7	11.7	22.7	35.2	18.2	30.9	30.9	174.5	67
3,7-Diacetate	Dioxane-d₈	35.3	31.6	74.6	35.2	41.2	31.2	71.2	38.1	34.8	35.8	21.1	39.9	43.0	51.2	23.9	28.3	56.3	12.1	23.0	35.7	17.3	31.2	31.0	175.3	66
Me 3β,7α-diol	CDCl₃	29.8	27.7	66.7	36.6	35.9	35.5	68.5	39.3	32.0	34.2	20.8	39.6	42.6	50.4	23.6	28.0	55.7	11.7	23.1	35.3	18.2	30.9	30.9	174.5	67
3α,7β-Diol	CD₃OD	36.1	30.9	71.7	38.4	44.3	37.9	71.9	43.9	40.5	35.0	22.3	41.4	44.6	56.4	27.8	29.6	57.3	12.7	24.0	36.5	19.0	32.2	31.9	174.7	69
Me 3α,7β-diol	CDCl₃	34.8	30.1	70.9	37.2	42.4	37.0	70.9	43.4	39.2	33.9	21.1	40.1	43.6	55.8	26.8	28.4	54.9	12.0	23.3	35.1	18.2	30.9	30.9	174.5	67
	CD₃OD	34.8	30.4	70.9	40.0	43.4	36.9	70.9	42.3	36.9	33.9	21.0	39.1	43.5	54.8	26.7	28.4	55.6	11.9	23.2	35.1	18.2	30.4	29.9	174.7	69
Me 3β,7β-diol	CDCl₃	29.3	27.2	65.7	34.1	36.7	36.5	70.7	43.0	38.2	34.1	21.1	39.8	43.2	55.6	26.5	28.2	54.5	11.7	23.5	34.8	18.0	30.6	30.6	174.2	67
Me 3,7-dione	CDCl₃	35.4	36.6	210.7	42.9	47.7	44.9	209.7	49.5	42.9	35.4	22.1	38.9	42.9	48.9	24.7	28.1	54.8	12.0	22.3	35.1	18.3	31.0	31.0	174.3	68
3α,12α-Diol	CD₃OD	35.7	30.5	71.7	36.1	42.7	27.6	26.6	36.4	33.9	34.5	29.0	72.8	46.8	48.4	24.2	26.5	47.4	12.9	23.4	35.7	17.5	31.3	31.2	177.2	66
Me 3α,12α-diol	Dioxane-d₈	35.7	30.7	71.8	36.1	42.7	27.6	26.6	36.4	33.9	34.5	29.0	73.3	46.8	48.4	24.2	26.5	47.7	12.9	23.4	35.7	17.5	31.4	31.2	175.2	67
	CDCl₃	35.1	30.2	71.4	36.2	42.0	27.1	26.0	35.9	33.3	33.9	28.5	72.8	46.3	47.9	23.6	27.4	47.0	12.5	22.9	35.1	17.1	31.0	30.8	174.5	67
3,12-Diacetate	Dioxane-d₈	35.5	26.7	74.2	32.2	43.1	27.6	26.6	34.9	36.4	34.5	27.3	75.8	44.7	50.1	23.8	26.5	48.4	12.8	23.5	35.7	17.5	31.4	31.2	175.2	66
Me 3β,12α-diol	CDCl₃	29.7	27.6	66.9	33.3	36.4	26.5	25.9	35.7	32.7	34.5	28.8	75.0	46.4	48.3	23.5	27.4	47.2	12.6	23.5	35.0	17.2	31.0	30.8	174.5	67
Me 3α,12β-diol	CDCl₃	35.2	30.3	71.1	35.9	41.4	26.9	25.8	34.4	39.3	34.2	29.5	79.1	47.8	54.8	23.5	24.0	57.3	7.8	23.0	32.5	20.7	32.0	31.2	174.5	67
Me 3β,12β-diol	CDCl₃	29.5	27.3	66.1	33.0	35.7	26.3	25.5	33.9	38.3	34.5	29.1	78.9	47.5	54.2	23.3	23.6	57.0	7.5	23.3	32.1	20.5	31.8	31.0	174.4	67
Me 3,12-dione	CDCl₃	36.8	36.8	211.6	42.0	43.7	25.5	26.6	35.4	44.3	35.6	38.3	213.7	57.5	58.5	24.2	27.4	46.5	11.7	22.1	35.6	18.5	31.3	30.5	174.4	68
Me 7β,12α-diol	CDCl₃	37.2	20.7	26.8	27.8	43.9	36.8	71.1	43.7	31.9	34.1	28.7	72.2	47.1	47.2	26.2	27.6	45.7	12.6	23.8	34.8	17.1	30.8	30.8	174.4	67
Me 7α,12β-diol	CDCl₃	37.4	21.2	27.5	30.2	42.7	35.1	68.0	37.9	31.9	35.6	29.2	78.8	47.4	48.4	22.9	23.7	56.9	7.5	23.2	32.4	20.8	32.0	30.8	174.5	67
Me 7β,12β-diol	CDCl₃	37.5	21.1	26.9	28.0	43.8	37.1	71.2	42.3	37.8	34.7	29.0	79.4	48.6	53.9	26.2	23.6	56.5	8.0	24.0	32.3	21.1	32.3	31.2	174.6	67
Me 7,12-dione	CDCl₃	36.4	20.4	26.2	28.2	46.7	45.8	209.8	49.0	45.6	36.6	38.4	212.8	56.9	52.0	25.3	27.7	45.6	11.8	23.3	35.6	18.7	31.3	30.5	174.5	68

TABLE XI.   **¹³C-NMR Data for Tri- and Tetraoxygenated Bile Acids**

Compound/substituent	Solvent	1	2	3	4	5	6	7	8	9	10	11
5β-Cholan-24-oic acid												
3α,7α,12α-triol	CD₃OD	35.7	30.6	71.6	39.6	42.0	35.1	68.0	39.9	26.7	35.1	28.5
Me 3α,7α,12α-triol	CD₃OD	36.5	31.1	72.8	40.4	43.1	35.8	69.0	41.0	27.8	35.9	29.5
	CD₂Cl₂	35.7	30.9	72.2	39.9	42.0	35.1	68.9	39.9	26.7	35.1	28.5
Triacetate	CDCl₃	35.3	30.1	71.7	39.4	41.4	34.7	68.3	39.4	26.2	34.7	28.0
	Dioxane-d₈	35.4	27.4	74.3	35.5	41.5	31.6	71.2	38.0	29.6	35.1	27.1
	CDCl₃	—	—	74.3	—	41.2	—	71.0	—	—	—	—
Me 3α,7α,12β-triol	CDCl₃	34.8	30.6	71.4	39.3	41.2	35.2	67.7	38.0	32.0	34.8	29.3
Me 3α,7β,12α-triol	CDCl₃	34.2	29.1	71.0	36.9	42.6	36.6	71.0	43.5	31.2	33.9	27.7
Me 3β,7α,12α-triol	CDCl₃	29.8	27.7	66.9	36.6	35.9	35.2	68.5	39.5	25.9	34.3	28.6
Me 3α,7β,12β-triol	CDCl₃	34.8	30.1	70.6	36.9	42.1	36.9	70.6	42.1	37.9	33.8	29.1
Me 3β,7α,12β-triol	CDCl₃	29.8	27.7	66.6	36.5	35.7	35.4	68.2	38.0	31.3	34.4	29.3
Me 3β,7β,12α-triol	CDCl₃	29.3	27.7	66.2	34.3	36.9	36.6	71.0	43.5	31.2	33.9	29.2
Me 3β,7β,12β-triol	CDCl₃	29.0	27.6	66.1	34.4	36.9	36.7	71.0	42.1	37.2	34.4	29.5
Me 3,7,12-trione	CDCl₃	35.2	36.4	208.6	42.7	46.7	44.9	208.4	49.0	45.7	36.0	38.6
Me 3α,7α,22R-triol	CD₃OD	36.5	31.4	72.9	40.6	43.1	35.9	69.1	40.8	34.0	36.1	21.7
Me 3α,7α,22S-triol	CD₃OD	36.6	31.5	73.0	40.6	43.2	36.0	69.2	40.9	34.1	36.2	21.8
Me 1β,3α,7α,12α-tetrol	CD₃OD	74.1	38.0	67.3	40.0	37.0	35.5	68.8	41.3	29.5	40.3	29.6
Me 3α,6α,7α,12α-tetrol	CD₃OD	36.7	31.2	72.8	33.2	49.5	70.7	72.8	40.3	27.7	36.7	29.7
5β-Cholestan-26-oic acid												
Me 3α,7α,12α-triol triacetate	CDCl₃	—	—	74.4	—	41.3	—	71.1	—	—	—	—
27α-dihomo-27b-carboxymethyl 5β-Cholanoate-26,27-dioate												
Me 3α,7α,12α-triol triacetate	CDCl₃	—	—	74.7	—	41.3	—	71.1	—	—	—	—

ulated data. Any problems of assignment for new compounds may be tackled by a suitable combination of the methods originally used. These include various double-resonance and multipulse techniques which provide distinctions between ¹³C resonances from methyl, methylene, methine, and quaternary carbon atoms (see Section 5.7). In difficult cases it may be helpful to resort to lanthanide shift reagents (see Section 5.6), measurements of spin-lattice relaxation times (T_1 values) (see Section 5.5), or even isotopic labeling. With computer-controlled high-field spectrometers it is now possible to obtain two-dimensional spectra that display both the ¹H and the ¹³C spectral features and provide a ready means of full spectral assignments for both nuclei (¹H/¹³C-heteronuclear correlated 2D-NMR) (Section 5.7).

Figure 11 shows a typical set of ¹³C chemical shifts for a bile acid. Methyl carbon atoms resonate at the highest field (ca. 7–24 ppm), followed by methylene (ca. 23–40 ppm) and ring-junction carbons (ca. 34–55 ppm).

12	13	14	15	16	17	18	19	20	21	22	23	24	25	26	27	Reference
72.5	46.8	42.0	23.5	28.0	47.4	12.8	22.8	35.7	17.6	31.3	31.2	174.5				66
73.9	47.4	42.9	24.2	28.6	47.9	13.0	23.2	36.7	17.6	32.2	31.8	176.3				51
73.4	46.8	42.0	23.5	28.0	47.4	12.8	22.8	35.7	17.6	31.3	31.2	175.3				66
73.0	46.3	41.4	23.1	27.4	46.8	12.3	22.3	35.3	17.4	31.0	31.0	174.7				67
75.6	45.7	43.9	23.5	28.0	48.4	12.8	23.1	35.6	18.5	31.4	31.2	175.4				66
75.6	45.3	43.6	23.0	—	47.6	12.4	22.7	—	17.7	—	—	174.8				52
78.9	47.5	48.5	22.9	23.7	57.0	7.6	22.5	32.5	20.9	32.1	30.8	174.7				67
72.2	47.5	47.2	26.1	27.4	45.7	12.6	23.4	34.8	17.2	30.9	30.9	174.6				67
72.8	46.6	41.9	23.2	27.4	47.2	12.5	22.9	35.2	17.4	31.1	30.9	174.7				67
79.1	48.6	54.2	26.2	23.8	56.5	8.0	23.1	32.3	21.0	32.1	31.2	174.8				67
79.0	47.6	48.6	23.1	23.8	57.1	7.7	23.0	32.6	21.0	32.1	31.2	174.7				67
72.2	47.5	47.2	26.1	27.4	45.8	12.6	23.4	34.8	17.2	30.9	30.9	174.5				67
79.3	48.6	53.9	26.2	23.6	56.4	8.0	23.6	32.3	21.1	32.3	31.4	174.7				67
211.6	56.9	51.8	25.1	27.5	45.6	11.7	21.8	35.4	18.5	31.3	30.4	174.3				68
40.8	43.9	51.2	24.7	28.4	54.6	12.1	23.4	43.1	12.8	71.2	36.5	—				54
41.2	43.5	51.4	24.7	28.7	53.8	12.1	23.4	42.0	12.4	70.9	41.5	—				54
73.8	47.3	42.9	24.3	28.5	48.0	13.0	17.9	36.8	17.6	32.2	31.8	176.4				51
73.7	47.5	42.9	24.1	28.6	47.9	13.0	23.5	36.8	17.6	32.2	31.8	176.4				51
															17.2(R)	
75.8	45.4	43.7	23.1	—	48.0	12.5	22.8	—	18.1	—	24.0	—	—	177.7	17.7(S)	52
75.8	45.4	43.7	23.1	—	48.0	12.5	22.8	—	18.2	—	23.5	—	45.0	176.4	173.7	52

Electron withdrawal by hydroxy groups or their esters results in significant deshielding of the carbon atoms that bear them (to ca. 65–80 ppm), whereas the ^{13}C atom of a carbonyl group is shifted strongly downfield (to ca. 175 ppm for carboxy or >200 ppm for keto groups). The C-19 chemical shift provides a useful indicator of the steroid configuration at C-5, being typically around 12 ppm for a 5α-saturated skeleton, but shifted some 10–12 ppm further downfield in a similar compound of the 5β-configuration. However, no ^{13}C spectra have so far been reported for 5α-bile acids.

Several attempts have been made to devise "rules" for the prediction of ^{13}C chemical shifts in molecules of the steroid type [70,71]. In general, it may be said that, for any purely hydrocarbon framework, the chemical shift of a carbon atom tends to be larger the more carbon atoms are attached to it. However, the presence of more remote carbon atoms in a chain or ring (carbons are commonly designated α-, β-, γ-, etc., from the atom in question) exerts a considerable

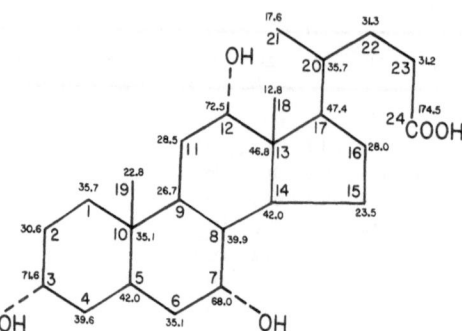

FIGURE 11. ^{13}C chemical shifts for cholic acid.

influence: remote effects of polar substituents and unsaturation must also be taken into account. Chang and Iida, from their series of assignments of ^{13}C resonances of isomeric 3-, 7-, and 12-hydroxylated bile acids, have calculated the α, β, and γ effects of substituent hydroxyl [67] and oxo-groups [68] (Table XII).

A procedure at present in its infancy, but holding great promise for the future, involves determination of carbon–carbon connectivities in any molecular framework, making use of extremely weak NMR signals from the very rare isotopic species in which pairs of ^{13}C atoms are directly bonded (see Section 5.7, Pulse Sequences). The practical problems can now only be overcome by using large samples, but future generations of instruments with improved sensitivity may make this a widely used device.

5. RECENT ADVANCES IN NMR SPECTROMETRY

5.1. Superconducting Magnets

Since chemical shifts and coupling constants are independent of the field strength, it follows that at higher fields the absolute separations (in Hz) between signals from individual nuclei will increase in proportion to the field strength. This results in better resolution of the spectrum with increasing field strength.

Prior to the mid 1970s, NMR was carried out using large iron core magnets. These had an upper limit of 2 T (i.e., protons resonating at 100 MHz). Heat dissipation was a problem, as was field inhomogeneity. Because of the latter, the sample had to be spun to reduce inhomogeneity.

A significant breakthrough in NMR technology occurred as a result of pioneering work on superconducting magnets in the Department of Biochemistry at the University of Oxford. In these magnets field strengths up to 12 T (protons resonating at 600 MHz) have been achieved. The coils of the magnet are cooled to the temperature of liquid helium, −269°C, whereby electrical resistivity falls

TABLE XII. Oxygen (Hydroxy- and Oxo-) Substituent Effects[a] on ^{13}C-Resonances for Methyl 5β-Cholan-24-oic Acids

Carbon	3α-OH	3β-OH	3-Keto	7α-OH	7β-OH	7-Keto	12α-OH	12β-OH	12-Keto
1	−2.5 (γ)	−7.7 (γ)	−0.4 (γ)	−0.1	−0.1	−1.1	−0.3	−0.3	0
2	8.9 (β)	6.6 (β)	15.7 (β)	−0.1	−0.2	0.4	−0.2	−0.2	−0.2
3	44.1 (α)	39.8 (α)	186.0 (α)	0.6	−0.2	−0.4	−0.2	−0.1	0
4	8.6 (β)	6.0 (β)	14.9 (β)	2.8 (δ)	1.0 (δ)	0.7 (δ)	−0.2	−0.2	0
5	−1.8 (γ)	−7.3 (γ)	0.7 (γ)	−0.6 (γ)	0.4 (γ)	4.0 (γ)	−0.1	−0.5	−0.5
6	−0.2 (δ)	−0.6 (δ)	−1.4 (δ)	8.4 (β)	10.0 (β)	18.7 (β)	−0.2	−0.2	−0.2
7	−0.2	−0.3	0.2	41.7 (α)	44.8 (α)	185.8 (α)	−0.4	−0.5	−0.2
8	−0.2	−0.2	−0.1	3.5 (β)	7.9 (β)	13.9 (β)	0.1 (δ)	−1.3 (δ)	+0.1 (δ)
9	−0.3	−0.8	0.5	−7.8 (γ)	−1.3 (γ)	1.9 (γ)	−7.0 (γ)	−1.3 (γ)	3.9 (γ)
10	−1.0 (δ)	−0.3 (δ)	−0.3 (δ)	−0.2 (δ)	−0.5 (δ)	0.9 (δ)	−0.6 (δ)	−0.2 (δ)	1.0 (γ)
11	−0.2	0.2	0.4	−0.4 (δ)	0.2 (δ)	−0.5 (δ)	7.8 (β)	8.6 (β)	17.4 (β)
12	−0.3	0	−0.1	0.8	−0.1	−1.1	32.6 (α)	38.9 (α)	174.7 (α)
13	−0.2	0	0.2	−0.3	1.0		3.6 (β)	5.0 (β)	14.9 (β)
14	−0.3	−0.1	0	−6.4 (γ)	−0.8 (γ)	−7.5 (γ)	−8.5 (γ)	−2.1 (γ)	2.2 (γ)
15	−0.2	−0.1	0	−0.8 (δ)	2.8 (δ)	0.7 (δ)	−0.6 (δ)	−0.7 (δ)	0.2 (δ)
16	−0.2	0	0.1	−0.2	−0.1	0.1	−0.8 (δ)	−4.2 (δ)	−0.6 (δ)
17	−0.3	0	0.2	−0.3	−1.0	−1.0	−8.9 (γ)	1.4 (γ)	−9.3 (γ)
18	−0.2	0	0.1	−0.5	0.1	0.1	0.6 (γ)	−4.2 (γ)	−0.3 (γ)
19	−1.0	−0.4	−1.5	−0.8	0	−0.3	−0.4	−0.3	−0.5
20	−0.1	0	0	−0.2	−0.1		−0.3 (δ)	−2.8 (δ)	+0.4 (δ)
21	−0.2	0	0	−0.2	0.1	0.2	−1.2	2.6	+0.4
22	−0.1	0	0.2	−0.2	0	0	−0.1	1.1	+0.5
23	−0.1	0	0.2	−0.2	0	0	−0.1	0.1	−0.3
24	0	0.2		−0.1	0.1		0	0.1	
25	−0.1	0.1		−0.1	0.1		0	0	

[a] These values were determined from data in Table IX. A positive value indicates a downfield shift of the resonance (relative to its position in methyl 5β-cholan-24-oate), whereas negative values imply upfield shifts.

to almost zero. This allows the use of higher current flow through the coils and hence greater field strength. In addition, there is a marked improvement of field homogeneity, which has resulted in spectra being obtained from living biological systems, e.g., brain, heart, liver, leg.

5.2. Fourier-Transform NMR

Until the mid-1960s NMR experiments were performed using continuous wave operation. This is analogous to obtaining absorption spectra with an ultraviolet or visible wavelength scanning spectrophotometer. The absorption of energy to sustain resonance is measured using a narrow-band pass.

In contrast, in *Fourier-transform* NMR (FT-NMR) a short, broad-band pulse (1–50 μsec duration) is applied at right angles to the magnetic field. This excites all the proton resonances, which then begin decaying according to their own resonance frequencies. The composite signal observed is known as the *free induction decay* (FID). Fourier transformation of the FID transfers this information from the time domain to the frequency domain whereby the usual NMR spectrum is obtained, yielding chemical shift and coupling information. Since the data needed to produce the entire spectrum can be obtained with one pulse, FT-NMR is 2–3 orders of magnitude more efficient than continuous-wave NMR, and consequently is more sensitive. The mathematical treatment of the FID is tedious and necessitates use of a minicomputer. The declining cost of computers in general has been a boon to NMR.

For ^1H NMR studies good signal-to-noise ratios can be obtained without repetitive scanning, down to a 10 mM concentration. Below this concentration multipulse experiments are performed. The FIDs obtained for each pulse are added to each other—Fourier transformation is only carried out on the accumulated FID. With this approach a suitable spectrum at a concentration of 0.5 mM requires ca. 1000 pulses. To maintain the signal-to-noise ratios, each twofold reduction in concentration requires a fourfold increase in the number of pulses. The low natural abundance of ^{13}C and its weaker nuclear receptivity make a large number of pulses necessary (1000–4000) even for a 20-mM solution.

A further factor in the multipulse approach is the time taken for reestablishment of the initial distribution of energy states. A suitable delay between pulses is essential—leaving out the delay causes saturation of the resonances and hence loss of sensitivity. For NMR experiments on bile salts in D_2O a 2- to 4-sec delay is used.

5.3. Spin Decoupling

In Section 2.5. it was shown that the use of broad-band (noise) decoupling of the interaction between protons and carbon-13 led to the simplification of the ^{13}C NMR spectrum. In ^1H NMR experiments it is possible to decouple proton spins selectively. This is done by applying, at right angles to the field of the magnet, a second, narrow-band radiofrequency signal at a frequency corresponding to the resonance of the proton to be decoupled [72]. By applying the appropriate power so that the product of the proton magnetogyric ratio and the field strength greatly exceeds 2 Π J, protons previously coupled to the irradiated protons cease to be so. This results in elimination of one component of the observed splitting. In a simple two-proton A-X system, decoupling of proton A causes the doublet pattern of the X proton to be replaced by a single resonance, centered between the two previous resonances, with an increased signal intensity. In systems where all resonances are well separated, interpretation of decoupling experiments can be carried out by simple inspection. However, this is not appropriate for the high-field region of bile acid spectra because of the extensive overlap of proton resonances. This difficulty can be overcome by carrying out an additional experiment in which the decoupler has the same power, but the frequency is adjusted so as to be off-resonance, e.g., at 2.75 ppm for most bile acids. A difference spectrum between the on- and off-resonance spectra then reveals the position of the decoupled proton(s). The resonance pattern thus

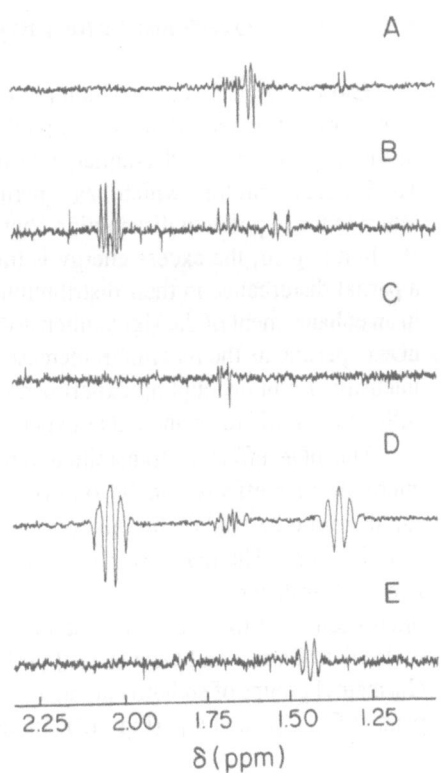

FIGURE 12. [^1]H NMR decoupled difference spectra of aqueous sodium cholate. Each spectrum is the difference between on- and off-resonance decoupled spectra. In A, on-resonance decoupling was at 12β-H (δ 4.08). The peak patterns are due to the overlapping 11α-H and 11β-H protons. In B and C, on-resonance decoupling was at 7β-H (δ 3.91). In B, the nonrelaxed spectrum contains the 6α-H (δ 2.01), 8β-H (δ 1.68), and 6β-H (δ 1.52) protons. In C, decoupling and partial relaxation have been combined to produce a methine [^1]H spectrum containing only the 8β-H resonance. In D, on-resonance decoupling was at 3β-H (δ 3.51). The peak patterns are due to 4α-H (δ 2.04), 2β-H and 4β-H (δ 1.60), and 2α-H (δ 1.36). In E, on-resonance decoupling was at 21-H (δ 0.98). As for C, decoupling and partial relaxation were combined to produce a methine [^1]H spectrum showing only 20-H. (From Ref. [49]; reproduced with permission of Lipid Research Inc., New York.)

obtained is necessarily complex since it will contain both negative (eliminated splitting) and positive (simplified splitting) components.

This approach has been used for proton resonance assignments for bile acids in some recent studies [49,51]. In each case, the investigators started by decoupling the protons geminal to the hydroxyl groups, since these resonances are readily assignable and separated from other resonances. Thus, for cholic acid, saturating the 3β-proton resonance decouples, and hence simplifies, the resonance patterns for the 2α-, 2β-, 4α-, and 4β-protons (Fig. 12D). Saturating the 7β-proton resonance decouples the 6α- and 6β-protons (methylene protons) and the 8β-proton (a methine proton) (Fig. 12B). Assigning the resonance patterns in the difference spectrum still remains a problem, although knowledge of expected coupling constants in a molecule of known conformation can be used. Three other NMR techniques, discussed elsewhere (the nuclear Overhauser effect, Section 5.4, the use of partial relaxation combined with decoupling, Section 5.5, and two-dimensional homonuclear correlated NMR spectroscopy, Section 5.7), have been used to overcome this difficulty.

5.4. Nuclear Overhauser Effect Experiments

Closely related to the technique of homonuclear spin decoupling is another double-irradiation procedure, which produces the *nuclear Overhauser effect* (nOe). In this experiment the decoupler is turned *off* during the acquisition of signal. The irradiated proton, which has a perturbed distribution of energy levels, will return to the population distribution appropriate for the prevailing magnetic field, B_o. In doing so, the excess energy is transferred to neighboring nuclei, causing a partial disturbance in their distribution of energy levels. This generally results in an enhancement of the signal intensity of these nuclei. For homonuclear proton nOe experiments the maximum increase in signal intensity is 50% of the original intensity. As in decoupling experiments, a difference spectrum can be obtained following an off-resonance nOe experiment [73].

The nOe effect is transmitted through space. Therefore both intra- and intermolecular effects can be observed. Concentration-dependent aggregation can therefore be studied by this technique, e.g., the aggregation of bile salts to form micelles. The nOe effect is inversely dependent on the sixth power of the internuclear distance and can be measured up to ca. 4 Å. This phenomenon has often been used to obtain information on configurations and conformations of molecules. Barnes and Geckle [49] performed nOe experiments on the C_{18} and C_{19} methyl groups of sodium cholate to distinguish between the methylene proton pairs at C_6, C_{11}, and C_{15} (Fig. 13). In each case the proton in the β orientation

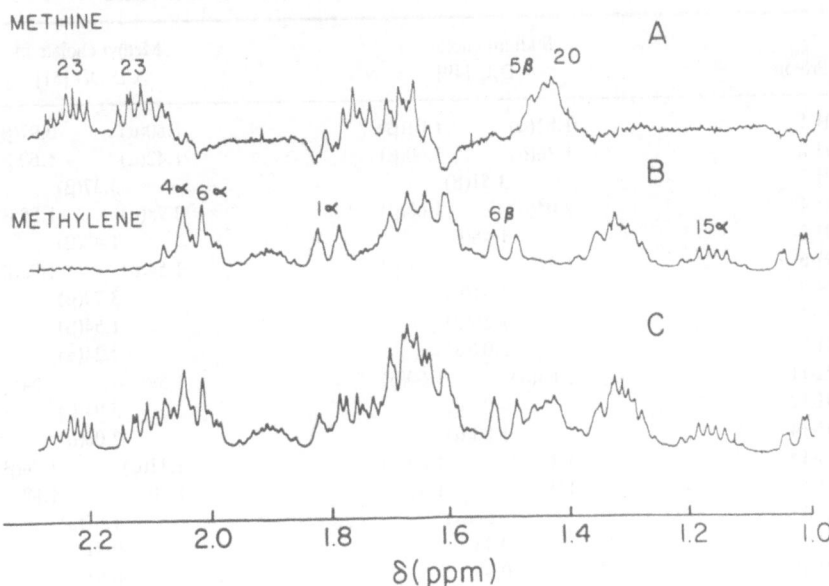

FIGURE 13. Methine and methylene ¹H NMR spectra of aqueous sodium cholate. The methine spectrum (A) was obtained with a 180°–τ–90° pulse sequence with a τ value of 155 msec and the methylene spectrum (B) with a τ value of 300 msec. The normal spectrum (C) is also shown. (From Ref. [49]; reproduced with permission of Lipid Research Inc., New York.)

was at a distance of 2.5 Å, or less, from the irradiated protons and therefore showed a nOe effect, whereas the α-proton did not. Similarly, an enhancement was observed for the 12β-proton in a nOe experiment on the C_{21} methyl group. This suggested that the C_{21} methyl group has a restricted rotation about the C_{17}-C_{20}-C_{22} chain, causing it to lie in the approximate plane of the B, C, and D rings. This conformation has also been observed in crystals of several salts of cholic acid [74–76].

Despite the extensive overlap of the ¹H resonances of protons on the steroid nucleus, the assignments of all the ¹H resonances of bile acids have been reported [49,51] (Table XIII). Most interestingly, there were marked differences in chemical shifts for ¹H resonances of methyl cholate compared to sodium cholate. We have recently shown that these differences are due to the solvents used (deuteromethanol and D_2O, respectively) [77]. The solvent effect was not the same for each proton; the resonances of protons on the hydrophilic face move downfield in deuteromethanol, whereas those of protons on the hydrophobic face move upfield (see Table XIII).

TABLE XIII. ¹H Chemical Shifts for Individual Protons of Bile Acids

Proton	Sodium cholate in D$_2$O [49]		Methyl cholate in CD$_3$OD [51]	
H-1	1.81(α)	1.01(β)	1.80(α)	0.97(β)
H-2	1.36(α)	1.60(β)	1.42(α)	1.60(β)
H-3		3.51(β)		3.37(β)
H-4	2.04(α)	1.60(β)	2.28(α)	1.55(β)
H-5		1.48(β)		1.40(β)
H-6	1.52(α)	2.01(β)	1.54(α)	1.95(β)
H-7		3.91(β)		3.79(β)
H-8		1.68(β)		1.54(β)
H-9		2.07(α)		2.24(α)
H-11	1.64(α)	1.60(β)	1.59[a]	1.54[a]
H-12		4.08(β)		3.94(β)
H-14		1.80(α)		2.00(α)
H-15	1.17(α)	1.69(β)	1.11(α)	1.74(β)
H-16	1.32[a]	1.92[a]	1.30[a]	1.88[a]
H-17		1.66(α)		1.87(α)
H-18		0.73		0.71
H-19		0.93		0.91
H-20		1.45		1.40
H-21		0.98		1.00
H-22	1.32	1.74	1.35	1.85
H-23	2.12	2.25	2.26	2.37

[a] The authors were unable to distinguish between this pair of resonances.

5.5. Relaxation

The detailed theory of relaxation phenomena is beyond the scope of this chapter, but some understanding of their basic principles is useful. The alignment of nuclear magnetic dipoles, or more strictly their axes of precession, in an external magnetic field has been noted in Section 2.1 as the basis of the NMR phenomenon. In Fig. 14 the resulting magnetization of the sample is shown as directed along the Z axis, although the precessions of individual nuclei are not in phase. When a radiofrequency pulse is applied in the Y direction, in the NMR experiment, the precessions of nuclear dipoles are brought into phase, with the result that the magnetization acquires a component rotating at the precessional frequency in the XY plane. At the same time the pulse disturbs the populations of nuclear dipoles aligned with and against the field. Depending on the pulse, the net magnetization in the Z direction may be either reduced, nullified (a so-called "90° pulse"), or even reversed ("180° pulse"). Relaxation processes then gradually restore the system to its equilibrium state, over a time scale of the order of 1 sec.

The relaxation process has two separable components. Return toward the equilibrium magnetization (populations) in the Z direction is termed *"longitudinal"* or *"spin-lattice"* relaxation, characterized by a relaxation time T_1. Energy transfer takes place both intermolecularly, to the surrounding medium, and intramolecularly (hence "lattice"), by dipole–dipole interactions which depend on relative motions, including tumbling of the molecule in solution.

The rotating magnetization vector in the XY plane also decays, as the phase coherence of the precessing nuclear dipoles is lost through a "fanning out" of magnetic vectors due to minute differences in precession frequencies. The resulting loss of XY magnetization is characterized by the *"transverse"* or *"spin–spin"* relaxation time T_2. The values of T_1 and T_2 are generally similar in solution, but T_1 is lengthened in solids and very viscous solvents by the slowing of molecular motions, which reduce opportunities for resonant interactions with the lattice.

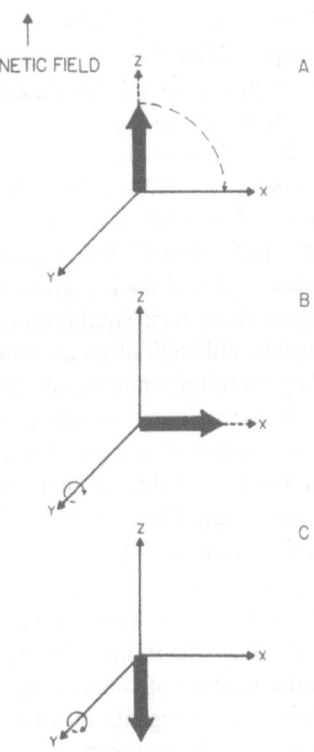

MAGNETIC FIELD

FIGURE 14. In A the nuclear spins precess about the Z axis (the direction of the external magnetic field) giving rise to a component in the Z direction. When a pulse is applied in the Y direction, it causes the spins to rotate toward the X axis. In B a 90° pulse in the Y direction has tipped the spins into the transverse XY plane. Measurement of the NMR signal is carried out by induction of an a.c. signal in a receiver coil situated in the XY plane. If a second 90° Y pulse is applied immediately after the first 90° Y pulse (a net 180° Y pulse), the spins are fully inverted in the $-Z$ direction with no component in the XY plane (C). Spin-lattice relaxation times (T_1's) are determined in the inversion recovery experiment (180°–τ–90°) by applying a 180° Y pulse, thus inverting the spins, allowing a certain time (τ) for the spins to reorientate in the direction of the applied magnetic field, and then applying a 90° Y "read" pulse, which produces a component in the XY plane. Variation of the value of τ yields different degrees of relaxation. When $\tau = T_1 \ln 2$, the magnetization in the XY plane, following the 180°–τ_{null}–90° pulse sequence, is zero. This phenomenon was used in Figs. 12 and 13 to obtain partially relaxed spectra.

Both relaxation processes occur exponentially. Spin-lattice relaxation toward the equilibrium magnetization $M_z(\infty)$ proceeds according to the expression

$$M_z(t) = M_z(\infty)(1 - e^{-t/T_1})$$

where $M_z(t)$ is the z-polarized magnetization at time t.

Spin–spin transverse relaxation proceeds toward zero according to

$$M_{x,y}(t) = M_{x,y}(o)e^{-t/T_2}$$

where $M_{x,y}(t)$ is the residual transverse magnetization at time t.

The spin-lattice relaxation time T_1 is the more useful for chemical purposes. Measured values of T_1, or of its reciprocal, the *relaxation rate* R_1 [70], for individual nuclei, have been shown to have some value as an aid to structural assignments, particularly on the basis that relaxation is assisted by any neighboring nuclei (through space). Relaxation of sterically crowded nuclei therefore occurs more rapidly.

Measurements of T_1 are best made by the *inversion-recovery* experiment. A 180° pulse is applied in the Y direction to invert the net magnetization of the sample. After allowing relaxation to proceed for a time interval τ, a 90° pulse is applied to tip the residual Z magnetization vector into the XY plane, allowing acquisition of the spectrum of the partly relaxed sample in the usual way. Since individual nuclei have relaxed longitudinally (Z axis) to different extents, in accordance with their different T_1 values, some may still have excess populations in the -Z direction and will produce NMR signals of negative sign. Those which are "half relaxed," with equalized populations, will give no signal. The most rapidly relaxed nuclei, given a sufficient value of the time interval τ, will be approaching their equilibrium populations and will give normal positive NMR signals, although of reduced intensity unless relaxation is essentially complete. By performing a number of "180°–τ–90°" experiments with varying pulse delays τ, for example, over the range 0.05–1 sec, a series of "partially relaxed" spectra can be obtained, with the different relaxation rates of individual signals revealed by the time scales of their change from negative through the "null point" to positive sign. For each signal, $T_1 = \tau_n/ln2$, where τ_n is the delay corresponding to the null point for that particular signal [Alternatively [78], $R_1 = ln2/\tau_n$.]

In general, T_1 values for ring methylene protons in molecules of steroids or bile acids are closely grouped, as are those for methine resonances, the latter being twice the former [49,78]. It is possible to use this difference to advantage in the analyses of complex ^1H spectra, particularly in the "methylene envelope" region where signals overlap. Spectra obtained by the 180°–τ–90° pulse sequence, with appropriate choice of τ, will show only those signals not nullified by partial relaxation. Examples are illustrated in Fig. 13, where the methine and methylene parts of the ^1H spectrum of sodium cholate are displayed separately by choice of appropriate values of τ [49]. When combined with the "decoupling difference" technique, the partial relaxation method becomes even more powerful (Fig. 13C).

5.6. Paramagnetic Shift Reagents

Shift reagents is a term used to describe complexes of the rare earth elements the lanthanides. These metals have unpaired electrons in the 4f valence shell and affect both chemical shifts and relaxation rates of adjacent nuclei. Lanthanides contribute not only the usual *diamagnetic* effect on the resonances of nuclei, but also a *paramagnetic* effect. The lanthanides at the beginning (lanthanum) and at the end of the series (lutetium) have filled electron shells and, therefore, contribute only diamagnetic effects. Gadolinium, the central member of the series, has a large relaxation effect several orders larger than for the other lanthanides. However, it has no effect on chemical shift.

The spectral changes that result from their coordination with organic molecules have found popular use in the assignment of resonances [79]. The reagents most commonly used are organic solvent-soluble complexes of europium or praesodymium with ligands such as 2,2,6,6-tetramethylheptane-3,5-dionato or 1,1,1,2,2,3,3-heptafluoro-7,7-dimethyl-4,6-octadionato (fod). Such complex [e.g., $Eu(fod)_3$] bind reversibly to polar groups (such as -OH and $-NH_2$) on the substrate. The exchange between the free substrate and the paramagnetic complex is fast on the NMR time scale, so that an averaged spectrum is observed. Signals for protons close to the coordination site undergo large lanthanide-induced changes in chemical shift (as much as 30 ppm downfield). To a reasonable approximation the induced shifts fall off as the inverse cube of the distance between the lanthanide atom and the affected nucleus, although stereochemical orientation of the lanthanide complex with respect to the substrate molecule must be considered in a more precise analysis. In practice the lanthanide complex is added in increasing amounts until satisfactory separation of the overlapping resonances of the substrate is obtained (see Fig. 15). A graphical treatment of induced-shifts versus lanthanide concentration is often helpful, giving linear plots for individual nuclei for moderate lanthanide concentrations. This approach is limited to substances with a suitable functional group. Compounds with a large number of sites for complex binding can cause major problems in the interpretation of the resulting spectra.

In aqueous systems the lanthanide metal ions can be used instead of the lanthanide complexes [80]. These ions interact with anionic groups such as carboxylate in organic acids (fatty acids, bile acids, and amino acids), as well as with the -OH, $-NH_2$, and carbonyl groups. If the lanthanide interacts with a single binding site of a substrate, the paramagnetically induced chemical shift

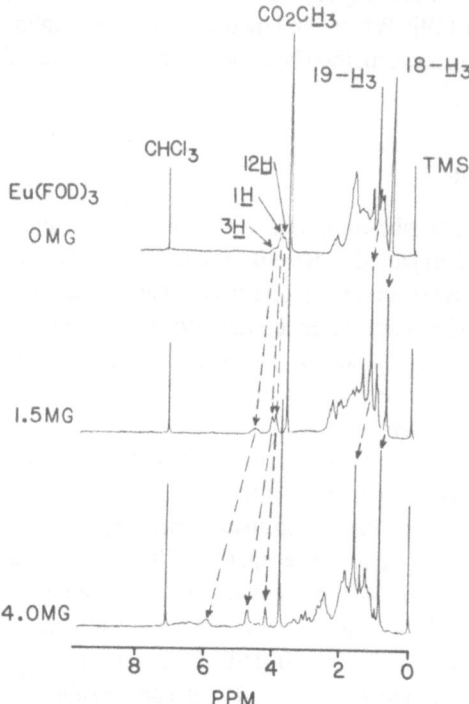

FIGURE 15. ^{1}H NMR spectra of methyl $1\beta,3\alpha,12\alpha$-trihydroxy-5-β-cholan-24-oate, showing the effects of added Eu(fod)$_3$. Note particularly the progressive downfield shifts and separation of the 3α-H, 1β-H, and 12β-H signals.

and relaxation rate change with respect to the concentration of the lanthanide. The changes can be interpreted to yield distances (and vectors in detailed experiments) from the lanthanide to the individually observed nuclei [81]. In doing so, a three-dimensional reconstruction of the average solution structure of the lanthanide–substrate complex can be obtained. Since lanthanides are isomorphous with calcium, this method has been used to map calcium-binding sites on proteins [82]. A similar approach has recently been used to determine the aqueous solution structure of a nonmicellar complex of glycocholate with dysprosium [83]. Details of the paramagnetic lanthanide NMR mapping technique can be found in Refs. 80, 84–86.

5.7. Pulse Sequences

Computer-controlled pulse sequences permit the unraveling of important information regarding the structure of molecules. We have seen how a $180°$-τ-$90°$ sequence permits the calculation of the spin-lattice relaxation time, T_1 (see Section 5.5). This was developed further to find values of τ that led to the suppression of methylene or methine proton resonances.

In recent years pulse sequences have been developed that allow signals from different carbon atoms (methyl, methylene, methine, and quaternary) to be distinguished. These have a variety of euphonic acronyms: APT (attached proton test) [87], INEPT (insensitive nuclei enhanced by polarization transfer) [88], and DEPT (distortionless enhancement by polarization transfer) [89]. The important aspect of these methods is the evolution (delay) times between pulses; these delays are dependent on the reciprocal of the J_{CH} coupling constant (120–150 Hz, i.e., ≈ 8 msec). The delay times are adjusted so that methine carbon signals (and methyl carbon signals) are fully in the negative direction in the spectrum; the methylene and quaternary carbon signals are fully positive. By adding this spectrum to a regular ^{13}C spectrum, the methine and methyl carbon signals are eliminated. By subtracting the two spectra the methylene and quaternary carbon signals are eliminated (Fig. 16). These methods for distinguishing individual carbon resonances have recently been applied to the study of bile acids [51,77].

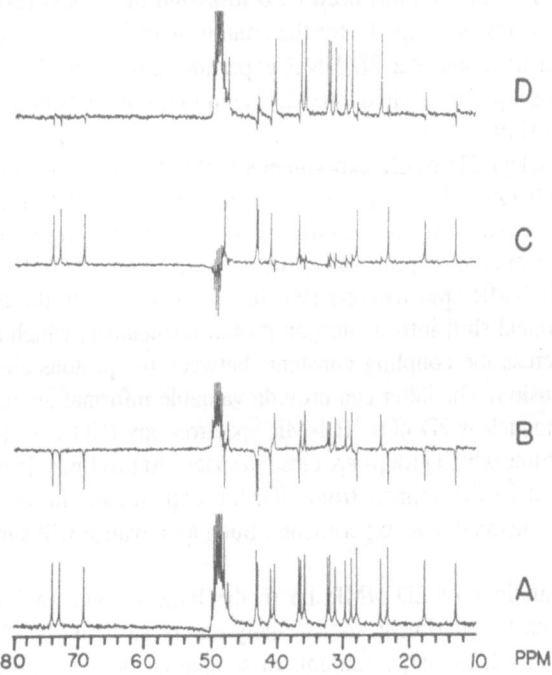

FIGURE 16. The APT experiment on cholic acid in deuteromethanol. The standard ^{13}C spectrum is shown in A. In B the proton evolution time has been adjusted so that the methyl and methine ^{13}C resonances are fully negative, whereas the methylene and quaternary ^{13}C resonances are fully positive. Spectrum C, containing only the methyl and methine ^{13}C resonances, was obtained by subtracting B from A; spectrum D, showing methylene and quaternary ^{13}C resonances, was obtained by adding B to A.

An important new area of pulse sequences, which will have much application to the study of bile acids, is *two-dimensional NMR* (2D NMR). It also uses modulation of a delay in a pulse sequence. Many FIDs are recorded with known increments of the delay. These FIDs are functions of both the variable delay and the collection time. If each FID is fourier-transformed, the resulting spectra are a function of frequency and the variable delay. Therefore, for each time increment there will be an associated signal intensity in the spectrum. This new array of intensities—one for each time increment—may also be fourier-transformed to yield an additional spectrum. This enables the creation of a two-dimensional plot, as either a stacked or a contour plot.

Several types of 2D NMR have been described. In heteronuclear $^{13}C/^{1}H$ 2D-correlated NMR spectroscopy [90,91], the resulting plot is of ^{13}C chemical shifts versus ^{1}H chemical shifts. This method allows assignment of the proton resonances for each proton attached to a carbon atom. If the ^{13}C resonances are known (this is often the case), then the proton resonances can be directly read off the plot. This method has been used to assign the proton resonances in six common bile acids (see Fig. 17 for the contour map for chenodeoxycholic acid) [77]. Another heteronuclear 2D NMR experiment, heteronuclear 2D J-resolved NMR spectroscopy [92], yields the C,H coupling constants correlated with the ^{13}C chemical shifts.

Homonuclear 2D NMR experiments yield structural information directly. In correlated (COSY) NMR spectroscopy, proton–proton [93] or carbon–carbon [94,95] connectivities are determined (this is equivalent to an extensive series of one-dimensional decoupling experiments—see Section 5.3). Homonuclear 2D J-resolved ^{1}H NMR spectroscopy [96] has no acronym. In the first dimension it yields chemical shift information on proton resonances, which are reduced to singlets, whereas the coupling constants between the protons are shown in the second dimension. The latter can provide valuable information on the structure. Finally, homonuclear 2D nOe ^{1}H NMR spectroscopy (NOESY) [97,98] can be used to determine which groups are close in space. At this time, distances between protons cannot be determined from NOESY experiments in the way they can from one-dimensional nOe experiments. Such an advance will surely take place soon.

The limitations of 2D NMR lie in the large amount of instrument time required for each experiment and in many cases, particularly when ^{13}C information is involved, the large amounts of sample needed. For instance, to carry out a heteronuclear ($^{13}C/^{1}H$) correlated 2D NMR experiment on a bile acid, at least 50–100 mg of sample is needed and data collection takes 20–40 hr. The limiting factor is the low abundance of ^{13}C. Enrichment of the compound being studied with ^{13}C can reduce the total acquisition time. Future improvements in detector sensitivity will also help in 2D NMR, as they will in all other aspects of NMR.

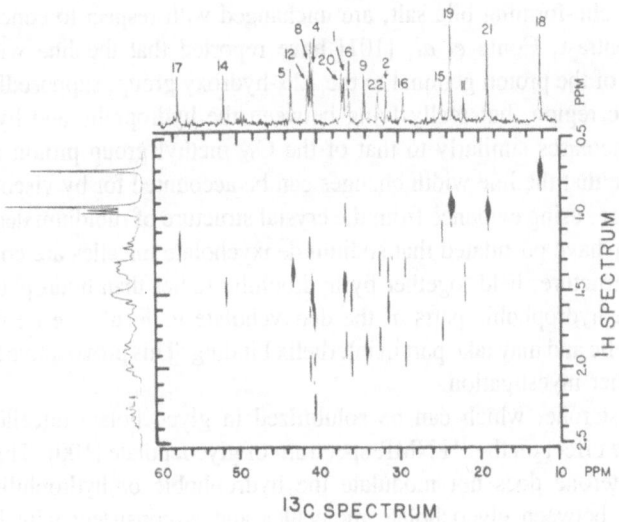

FIGURE 17. $^{13}C/^1H$-Heteronuclear-correlated 2D NMR contour plot of chenodeoxycholic acid in deuteromethanol. The ^{13}C spectrum on the horizontal axis is a projection of the peaks in the contour plot and does not contain the quaternary ^{13}C resonances. The ^1H spectrum along the vertical axis is part of a standard one-dimensional spectrum shown in Fig. 5B. Proton resonances associated with peaks on the contour plot are assigned by identification of the carbon resonances from their chemical shifts (horizontal axis), location of the center of the peaks, and measurement of the proton chemical shifts (vertical axis). Resonances of the hydroxylated carbons, C_3 and C_7, are off scale and not shown. The carboxylate carbon C_{24} and the quaternary carbons C_{10} and C_{13} do not have attached protons and do not give rise to peaks in this experiment.

6. APPLICATION OF NMR TO THE PHYSICAL CHEMISTRY OF BILE ACIDS

Despite the great potential of NMR in the study of simple bile salt micelles and mixed micelles with other organic compounds, as yet only a few studies have been reported.

Several groups [4,49,99–101] have reported line broadening of the C_{18} and C_{19} methyl group ^1H resonances as the concentration of bile salt is raised above the critical micellar concentration. On the other hand, resonances from protons in the hydrophilic regions of the molecule are mostly unaffected [4,100]. Although part of the line broadening may be attributable to changes in viscosity [49,100], the data have been interpreted to show that the hydrocarbon (the methyl groups) parts of the bile salt molecule interact during the formation of micelles, with the hydrophilic parts of the bile salt molecule being on the outside of the micelle in direct contact with the aqueous phase. Interestingly, Paul et al. [99]

found that the line widths of the methyl group ^1H resonances for dehydrocholate, a poor micelle-forming bile salt, are unchanged with respect to concentration.

In contrast, Conte et al. [101] have reported that the line width of the resonance of the proton geminal to the 12α-hydroxy group, supposedly from the hydrophilic region, but really lying between the hydrophilic and hydrophobic regions, increases similarly to that of the C_{18} methyl group proton resonance. They argue that the line width changes can be accounted for by viscosity alone. Furthermore, using evidence from the crystal structure of rubidium deoxycholate [102], they have postulated that sodium deoxycholate micelles are composed of a helical structure, held together by hydrophilic rather than hydrophobic forces [101]. The hydrophobic parts of the deoxycholate molecule are on the outside of the micelle and may take part in interhelix binding. This provocative hypothesis needs further investigation.

Testosterone, which can be solubilized in glycocholate micelles, has no measurable effect on the ^1H NMR spectrum of glycocholate [100]. This suggests that testosterone does not modulate the hydrophobic or hydrophilic micellar interaction between glycocholate molecules and is consistent with binding of testosterone to the surface, rather than the interior, of the micelles.

Fung and Peden [103] have applied deuterium (^2H) NMR to the study of the aggregation of deoxycholate, using $[7,7'-^2H_2]$-deoxycholate. Deuterium spin-lattice relaxation times indicated that deoxycholate forms a tetramer in aqueous solution above the critical micellar concentration. Addition of NaCl causes the formation of larger aggregates, which cannot be observed directly because of their very short deuterium T_1s (less than 1 msec); however, a reduction in FID signal intensity permits estimation of the proportion of the remaining tetramers. Smith and Barnard [104] used a similar approach to study cholate aggregation by ^2H NMR with $[7-^2H]$-cholic acid. They concluded from T_1 measurements that a tetramer was formed at high concentration ($>0.1\ M$). In addition, they also showed that temperature-related changes in T_1 values are a function of changing solution viscosity rather than changing aggregation number.

Elgavish and Barnes (unpublished data) found that the T_1 values of ^1H resonances of sodium cholate in D_2O are essentially unchanged as the concentration is increased from 3.5 mmole. This contradicts the line-broadening data mentioned earlier. However, it is noteworthy that Lindman et al. [105] observed that cholate molecules have considerable motion while part of a micellar aggregate, thus mitigating an affect on T_1.

The formation of reversed micelles of bile acid methyl esters in hydrophobic solvents (deuterochloroform and carbon tetrachloride) has been reported. Smith [106], using ^{13}C NMR, showed that ^{13}C T_1s fell as a function of the number of hydroxyl groups and the solute concentration. Robeson et al. [107], on the basis of concentration-dependent changes in hydroxyl proton chemical shifts, concluded that association of bile acid methyl esters is due to hydrogen binding by the hydroxyl groups.

The motion of molecules solubilized within bile salt micelles has been studied by NMR [66,108,109]. Using deuterium-labeled benzene and naphthalene, Fung and Thomas [108] found that both these materials have librational motion in deoxycholate and cholate micelles; in addition, benzene, but not naphthalene, rotates about its C_6 axis. Motion is greater in deoxycholate micelles than in cholate micelles. Similarly, Leibfritz and Roberts [66], using ^{13}C NMR, also showed that the fused two-ring system of naphthalene has a restricted motion in deoxycholate micelles, whereas p-xylene (cf. benzene) does not. Menger et al. [109] observed upfield changes in the chemical shift of C_{18} and C_{19} methyl group ^1H resonances of glycodeoxycholate (0.2 M) when naphthalene was added. The larger effect on the C_{19} methyl ^1H resonance led them to conclude that binding of naphthalene occurs close to the A/B ring junction. This effect was also observed for deoxycholate and chenodeoxycholate, but to a lesser extent for ursodeoxycholate [110]. A similar effect of p-xylene, phenanthrene, and styrene on the C_{19} methyl group ^1H resonance of deoxycholate was reported by Conte et al. [101].

A few studies have been reported on the interaction of bile salts and phospholipids. Ulmius et al. [111] used ^2H NMR to show that incorporation of sodium cholate into lecithin bilayers causes disordering of the fatty acids, the effect being much greater for the distal end of the fatty-acid chain. Stark et al. [112] also used ^2H NMR with palmitoyl-2-oleoylphosphatidyl choline labeled with deuterium at specific sites on the palmitoyl chain. They found a gradient of mobility along the fatty-acid chain as evidenced by the large increase in deuterium T_1s. Saito et al. [113] used [6,6,7,7,8-$^2H_5$]-deoxycholate and [11,11,12,12-$^2H_4$]-chenodeoxycholate to study the incorporation of these bile salts into lecithin bilayers. They observed that at low bile salt : phospholipid ratios the bile salts are incorporated so that the B, C, and D rings lie parallel to the surface of the bilayer. A similar orientation has been reported for cholate in lecithin bilayers [111]. Stark and Roberts [114] used 500 MHz ^1H NMR to show that both lecithin and taurocholate in mixed micelles retain substantial motional freedom, which is only lost as the phase limit (2 : 1 mole ratio of lecithin and taurocholate) is approached.

Hunt and Jawaharlal [115] examined the effect of bile salts on the permeability of phosphatidylcholine vesicles to Pr^{3+}. The ^1H NMR resonance of the choline methyl head group, $N^+ (CH_3)_3$, sited on the exterior of the vesicle, is shifted downfield because of the presence of Pr^{3+}, whereas the interior one is not. Thus two signals are observed. Penetration of the lipid bilayer by Pr^{3+} eliminates the difference. Bile salts increase the rate of penetration of Pr^{3+}, possibly as a complex of praesodymium and an everted bile salt micelle. Similar results were reported by Castellino and Violand [116], who used ^{31}P NMR to study the permeability of phosphatidylcholine vesicles to another lanthanide, Eu^{3+}.

Brouillette *et al.* [117] used ^{1}H NMR to examine the effect of deoxycholate on the signals from the choline *N*-methyl head groups of phosphatidylcholine in unilamellar vesicles. In the absence of deoxycholate two signals are observed owing to the head groups in the inner monolayer of the vesicle being packed together more closely than those in the outer monolayer. The difference in the signals increases as the vesicle radius of curvature decreases. As the mole ratio of deoxycholate is raised, the upfield resonance from the inner monolayer broadens and moves downfield, becoming a single broad peak at a deoxycholate : lecithin ratio of 1 : 3. When the proportion of deoxycholate is raised further, the line width of the peak becomes less and the peak moves downfield. Those data were interpreted as indicating, first, the formation of flattened vesicles (sic, bilayers) and then their intercalation by deoxycholate, eventually to form deoxycholate : lecithin mixed micelles.

7. NMR IN THE BIOSYNTHESIS AND METABOLISM OF BILE ACIDS

Two studies have been reported on the use of NMR to determine the labeling positions in bile salts originating from the use of [2,2,2-^{2}H]-ethanol and [^{13}C-2,2-^{2}H]-ethanol [118,119]. Rats with a biliary cannula to collect bile were treated with the labeled ethanols. The ^{13}C-label was found at carbon atoms 2, 4, 6, 8, 10, 11, 12, 14, 16, 20, and 23. Enrichment (over the natural abundance of ^{13}C) ranged from 13-fold to 26-fold [118]. Studies with deuterium-labeled ethanol revealed that 60% of the acetyl CoA which served as a precursor for cholesterol and bile acids originates from ethanol (note: in bile fistula rats given 100 mg ethanol per hour) [119].

The potential for studies on bile acid metabolism *in vivo,* or *in vitro* using isolated organs or cells, is certainly a possibility, although the use of ^{13}C-enriched bile acids would be necessary to make most use of the large chemical shift range of this nucleus. Although ^{1}H NMR has greater sensitivity without the need for enrichment, it suffers from the extensive overlap of resonance patterns and hence poor discriminating power between metabolites. However, metabolic steps involving olefinic bonds may be studied by NMR.

8. CONCLUDING REMARKS

The improvements in NMR instrumentation (magnets, detectors, and computer software) in recent years have made NMR an important tool in the study of the structural chemistry and kinetics of many biomolecules. Such advances are equally applicable to bile acids and their derivatives. It is our hope in compiling the chemical shift data on bile acids and by describing the terminology of NMR experiments that more investigators will be encouraged to use this technique in the field of bile acid research.

ACKNOWLEDGMENTS.   S.B. is supported by a grant (AM-25511) from the National Institutes of Arthritis, Metabolism, Digestive Disease, and Kidney Disease, and D.N.K. by a grant from the U.K. Medical Research Council for the Steroid Reference Collection. We warmly appreciate the typing skills and patience of Ms. Beverly Callens. Mr. Clyde L. Greene prepared most of the drawings.

REFERENCES

1. L. F. Fieser and M. Fieser, in "Steroids," Reinhold, New York (1959).
2. D. H. Williams and I. Fleming, "Spectroscopic Methods in Organic Chemistry," 3rd ed., McGraw-Hill, London, New York (1980).
3. W. Kemp, "Organic Spectroscopy," Macmillan, London (1975).
4. D. M. Small, in "The Bile Acids. Chemistry, Metabolism and Physiology." Vol. I. "Metabolism (D. Kritchevsky and P. P. Nair, eds.), Plenum Press, New York (1971).
5. J. St. Pyrek, R. Sterzycki, R. Lester, and A. Adcock, Lipids 17, 241 (1982).
6. J. E. Herz and J. Sandoval, Steroids 41, 327 (1983).
7. H. R. Nace and E. M. Holt, J. Org. Chem. 34, 2692 (1969).
8. T. Harano, K. Harano, and K. Yamasaki, Steroids 32, 73 (1978).
9. C. K. Lai, C. Y. Byon, M. Gut, D. Mostowicz, and W. G. Anderson, J. Labld. Cpds. Radiopharm. 21, 627 (1984).
10. R. W. Owen, A. N. Mason, and R. F. Bilton, J. Lipid Res. 24, 1500 (1983).
11. R. W. Owen and R. F. Bilton, J. Ster. Biochem. 19, 1355 (1983).
12. R. W. Owen, M. J. Hill, and R. F. Bilton, J. Lipid Res. 24, 1109 (1983).
13. D. M. Small, S. A. Penkett, and D. Chapman, Biochem. Biophys. Acta 176, 178 (1969).
14. J. F. Baker and R. T. Blickenstaff, J. Org. Chem. 40, 1579 (1975).
15. T. Iida and F. C. Chang, J. Org. Chem. 47, 2966 (1982).
16. Y. Shalon and W. H. Elliott, FEBS Lett. 44, 223 (1974).
17. P. Child, A. Kuksis, and L. Marai, Can. J. Biochem. 57, 216 (1979).
18. M. Une, B. I. Cohen, and E. H. Mosbach, J. Lipid Res. 25, 407 (1984).
19. T. Iida and F. C. Chang, J. Org. Chem. 48, 1194 (1983).
20. D. L. Hachey, P. A. Szczepanik, O. W. Berngruber, and P. D. Klein, J. Label. Cmpds. 9, 703 (1973).
21. M. R. Tenneson, J. D. Baty, R. F. Bilton, and A. N. Mason, J. Ster. Biochem. 11, 1227 (1979).
22. M. R. Tenneson, J. D. Baty, R. F. Bilton, and A. N. Mason, J. Ster. Biochem. 10, 311 (1979).
23. R. A. Leppik, Biochem. J. 210, 829 (1983).
24. P. J. Barnes, J. D. Baty, R. F. Bilton, and A. N. Mason, Tetrahedron 32, 89 (1976).
25. R. A. Leppik, Biochem. J. 202, 747 (1982).

26. Y. Shalon and W. H. Elliott, *Steroids* **28**, 655 (1976).
27. S. Yamauchi, M. Kojima, and F. Nakayama, *Steroids* **41**, 155 (1983).
28. R. W. Owen and R. F. Bilton, *Biochem. J.* **216**, 641 (1983).
29. R. A. Leppik, *Steroids* **41**, 475 (1983).
30. R. F. Bilton, A. N. Mason, and M. E. Tenneson, *Tetrahedron* **37**, 2509 (1981).
31. J. E. Herz and R. Ocampo, *Steroids* **40**, 661 (1982).
32. P. K. Bhattacharyya and Y. G. Bankawala, *Anal. Chem.* **50**, 1462 (1978).
33. K-Y. Tserng and P. D. Klein, *Steroids* **29**, 635 (1977).
34. R. Shaw and W. H. Elliott, *J. Lipid Res.* **19**, 783 (1978).
35. C. K. Lai, C. Y. Byon, and M. Gut, *J. Labld. Cpds. Radiopharm.* **21**, 615 (1984).
36. T. Iida and F. C. Chang, *J. Org. Chem.* **46**, 2786 (1981).
37. T. Iida, H. R. Taneja, and F. C. Chang, *Lipids* **16**, 863 (1981).
38. M. Une, F. Nagai, K. Kihira, T. Kuramoto, and T. Hoshita, *J. Lipid Res.* **24**, 924 (1983).
39. T. Iida and F. C. Chang, *J. Org. Chem.* **47**, 2972 (1982).
40. T. Dahl, Y-H. Kim, D. Levy, and R. Stevenson, *J. Chem. Soc.*, 2723 (1969).
41. K. Kihira, A. K. Batta, E. H. Mosbach, and G. Salen, *J. Lipid Res.* **20**, 421 (1979).
42. M. E. Tenneson, J. D. Baty, R. F. Bilton, and A. N. Mason, *Biochem. J.* **184**, 613 (1979).
43. K. Kihira, A. Kubota, and T. Hoshita, *J. Lipid Res.* **25**, 871 (1984).
44. R. Shaw and W. H. Elliott, *Lipids* **15**, 805 (1980).
45. J. R. Dias and B. Nassum, *Steroids* **35**, 405 (1980).
46. G. A. D. Haslewood and L. Tokes, *Biochem. J.* **126**, 1161 (1972).
47. K. Carlstrom, D. N. Kirk, and J. Sjovall, *J. Lipid Res.* **22**, 1225 (1981).
48. G. Halperin, *Steroids* **34**, 295 (1979).
49. S. Barnes and J. M. Geckle, *J. Lipid Res.* **23**, 161 (1982).
50. W. Kramer and G. Kurz, *J. Lipid Res.* **24**, 910 (1983).
51. P. Back, H. Fritz, and C. Populoh, *Hoppe-Seyler's Z. Physiol. Chem.* **365**, 479 (1984).
52. G. Janssen, S. Toppet, and G. Parmentier, *J. Lipid Res.* **23**, 456 (1982).
53. F. C. Chang, *J. Org. Chem.* **44**, 4567 (1979).
54. K. Kihira, Y. Morioka, and T. Hoshita, *J. Lipid Res.* **22**, 1181 (1981).
55. K. Kihira, Y. Akashi, S., Kuroki, J. Yanagisawa, F. Nakayama, and T. Hoshita, *J. Lipid Res.* **25**, 1330 (1984).
56. G. A. D. Haslewood, S. Ikawa, L. Tokes, and D. Wong, *Biochem. J.* **171**, 409 (1978).
57. R. F. Zurcher, *in* "Progress in Nuclear Magnetic Resonance Spectroscopy" (J. W. Ensley and L. H. Sutcliffe, eds.), Vol. 2, p. 205, Pergamon Press, New York (1967).
58. J. E. Bridgeman, P. C. Cherry, A. S. Clegg, J. M. Evans, Sir E. R. H. Jones, A. Kasal, V. Kumar, G. D. Meakins, Y. Morisawa, E. E. Richardo, and P. D. Woodgate, *J. Chem. Soc.* (C), 250 (1970).
59. R. F. Zurcher, *Helv. Chim. Acta* **44**, 1755 (1961).
60. R. F. Zurcher, *Helv. Chim. Acta* **46**, 2054 (1963).
61. N. S. Bhacca and D. H. Williams, *in* "Applications of NMR Spectroscopy in Organic Chemistry, Illustrations from the Steroid Field," Holden Day, San Francisco (1964).
62. S. Ricca, B. Rindone, and C. Scolastico, *Gazzetta* **99**, 1284 (1969).
63. P. V. Demarco and L. A. Spangle, *J. Org. Chem.* **34**, 3205 (1969).
64. R. G. Wilson, D. E. A. Rivette, and D. H. Williams, *Chem. Ind. (London)*, 109 (1969).
65. M. Karplus, *J. Chem. Phys.* **30**, 11 (1959).
66. D. Liebfritz and J. D. Roberts, *J. Am. Chem. Soc.* **95**, 4996 (1973).
67. T. Iida, T. Tamura, T. Matsumoto, and F. C. Chang, *Org. Magn. Reson.* **21**, 305 (1983).
68. T. Iida, T. Tamura, T. Matsumoto, and F. C. Chang, *J. Jpn. Oil Chem.* **32**, 46 (1983).
69. A. Baillet-Guffroy, D. Baylocq, A. Rabaron, and F. Pellerin, *J. Pharm. Sci.* **73**, 847 (1984).
70. H. Eggert, C. L. VanAntwerp, N. S. Bhacca, and C. Djerassi, *J. Org. Chem.* **41**, 71 (1976).
71. J. W. Blunt and J. B. Stothers, *Org. Magn. Reson.* **9**, 439 (1977).

72. W. A. Gibbons, C. F. Beyer, J. Dadok, F. R. Sprecher, and H. R. Wyssbrod, *Biochemistry* **14**, 420 (1974).
73. J. H. Noggle and R. E. Schirmer, "The Nuclear Overhauser Effect," Academic Press, New York (1971).
74. V. M. Coiro, E. Giglio, S. Morosetti, and A. Palleschi, *Acta Crystallogr.* **B36**, 1478 (1980).
75. R. E. Cobbledick and F. W. B. Einstein, *Acta Crystallogr.* **B36**, 287 (1980).
76. A. Hogan, S. E. Ealick, C. E. Bugg, and S. Barnes, *J. Lipid Res.* **25**, 791 (1984).
77. D. V. Waterhous, S. Barnes, and D. D. Muccio, *J. Lipid Res.* **26** 1068 (1985).
78. L. D. Colebrook and L. D. Hall, *Org. Magn. Reson.* **21**, 532 (1983).
79. C. C. Hinckley, *J. Am. Chem. Soc.* **91**, 5160 (1969).
80. J. Reuben, *Prog. Nucl. Magn. Res. Spectr.* **9**, 1 (1973).
81. J. Reuben and G. A. Elgavish, *in* "Handbook on the Physics and Chemistry of Rare Earths" (K. A. Gschneider, Jr., and L. Eyring, eds.), Vol. 4, p. 483, North-Holland, Amsterdam (1979).
82. L. Lee and B. D. Sykes, *Biochemistry* **22**, 4336 (1983).
83. E. Mukidjam, G. A. Elgavish, and S. Barnes, *Biochemistry* **26**, 6785 (1987).
84. W. D. E. Horrocks, Jr., and D. R. Sudnick, *J. Am. Chem. Soc.* **101**, 334 (1979).
85. G. A. Elgavish and J. Reuben, *J. Magn. Reson.* **42**, 242 (1981).
86. F. Inagaki and T. Miyazawa, *Prog. Nucl. Magn. Reson. Spectr.* **14**, 67 (1981).
87. S. L. Patt and J. N. Schoolery, *J. Magn. Reson.* **46**, 535 (1982).
88. G. A. Morris and R. Freeman, *J. Am Chem. Soc.* **101**, 760 (1979).
89. D. M. Dodrell, D. T. Pregg, and M. R. Bendall, *J. Magn. Reson.* **48**, 323 (1982).
90. A. A. Maudsley, L. Muller, and R. R. Ernst, *J. Magn. Reson.* **28**, 463 (1977).
91. A. Bax, *in* "Two Dimensional Nuclear Magnetic Resonance in Liquids," Delft University Press, Delft, Holland (1982).
92. M. H. Levitt and R. Freeman, *J. Magn. Reson.* **34**, 675 (1979).
93. A. Bax and R. Freeman, *J. Magn. Reson.* **44**, 542 (1981).
94. R. Freeman, J. A. Frenkiel, and M. B. Rubin, *J. Am. Chem. Soc.* **104**, 5545 (1982).
95. G. Lukacs and A. Neszmelyi, *J. Chem. Soc. Chem. Commun.* 1275 (1981).
96. A. Bax, R. Freeman, and J. A. Frenkiel, *J. Am. Chem. Soc.* **103**, 2102 (1981).
97. G. Wider, R. Baumann, K. Nagayama, R. R. Ernst, and K. Wuthrich, *J. Magn. Reson.* **42**, 73 (1981).
98. G. Bodenhauser and R. R. Ernst, *J. Am. Chem. Soc.* **104**, 1304 (1982).
99. R. Paul, M. K. Mathew, R. Narayanan, and P. Balaram, *Chem. Phys. Lipids* **25**, 345 (1979).
100. L. Martis, N. A. Hall, and A. L. Thakkar, *J. Pharm. Sci.* **61**, 1757 (1972).
101. G. Conte, R. Di Blasi, E. Giglio, A. Paretta, and N. V. Pavel, *J. Phys. Chem.* **88**, 5720 (1984).
102. A. R. Campanelli, S. Candeloro De Sanctis, E. Giglio, and S. Petriconi, *Acta Crystallogr.* **C40**, 631 (1984).
103. B. M. Fung and M. C. Peden, *Biochem. Biophys. Acta* **437**, 237 (1976).
104. W. B. Smith and G. D. Barnard, *Can. J. Chem.* **59**, 1602 (1981).
105. B. Lindman, N. Kamenka, H. Fabre, J. Ulmius, and T. Wieloch, *J. Colloid. Interface Sci.* **73**, 556 (1980).
106. W. B. Smith, *J. Phys. Chem.* **82**, 234 (1978).
107. J. Robeson, B. W. Foster, S. N. Rosenthal, E. T. Adams, Jr., and E. J. Fendler, *J. Phys. Chem.* **85**, 1254 (1981).
108. B. M. Fung and L. Thomas, Jr., *Chem. Phys. Lipids* **25**, 141 (1979).
109. F. M. Menger, J-U. Rhee, and L. Mandell, *J. Chem. Soc. Commun.* 918 (1973).
110. J. C. Montet, C. Merienne, and G. Bram, *Tetrahedron* **38**, 1159 (1982).
111. J. Ulmius, G. Lindblom, H. Wennerstrom, L. B-A. Johansson, K. Fontell, O. Sodermann, and G. Arvidson, *Biochemistry* **21**, 1553 (1982).

112. R. E. Stark, J. L. Manstein, W. Curatolo, and B. Sears, *Biochemistry* **22**, 2486 (1983).
113. H. Saito, Y. Sugimoto, R. Tabeta, S. Suzuki, G. Izumi, M. Kodama, S. Toyoshima, and C. Nagata, *J. Biochem.* **94**, 1877 (1983).
114. R. E. Stark and M. F. Roberts, *Biochim. Biophys. Acta* **770**, 115 (1984).
115. G. R. A. Hunt and H. Jawaharlal, *Biochim. Biophys. Acta* **601**, 678 (1980).
116. F. J. Castellino and B. N. Violand, *Arch. Biochem. Biophys.* **193**, 543 (1979).
117. G. C. Brouillette, J. P. Segrest, T. C. Ng, and J. L. Jones, *Biochemistry* **21**, 4579 (1982).
118. D. M. Wilson, A. L. Burlinghame, T. Cronholm, and J. Sjövall, *Biochem. Biophys. Res. Commun.* **56**, 828 (1974).
119. T. Cronholm, J. Sjövall, D. M. Wilson, and A. L. Burlinghame, *Biochim. Biophys. Acta* **575**, 193 (1979).

Chapter 4

X-RAY CRYSTALLOGRAPHY

S. E. Ealick and C. E. Bugg

1. INTRODUCTION

Single-crystal X-ray diffraction provides a powerful technique for precisely determining the positions of atoms in a crystal (Refs. 1,2, which are excellent texts on crystallography). Since the first crystal structure was worked out in 1913 by W. H. Bragg, tens of thousands of crystals have been analyzed using this experimental technique. Today crystal structure analysis by X-ray diffraction is usually routine for molecules containing less than 50 or so nonhydrogen atoms. The results of these structure determinations have provided answers to key questions in nearly every area of chemistry, biochemistry, and molecular biology.

In addition to providing the chemical structure of a molecule (i.e., the connectivity and identity of individual atoms), various conformational parameters, such as bond distances, bond angles, and torsion angles, can be derived. X-ray crystallography can also be used to determine the "hand" of a molecule if a single enantiomer is present in the crystal. Because a crystal is an aggregate of molecules, it may be possible to infer information about preferred or possible modes of intermolecular interaction.

The disadvantages of single-crystal X-ray diffraction lie primarily in the nature of the sample itself. Since the molecule is part of a crystalline array, it is possible that the molecule assumes a different conformation in the solid state than in solution or *in vivo*. Consequently, one must be aware of the assumptions being made when extrapolating crystallographic data to biological systems. A second disadvantage is that a single crystal of suitable size (usually a few tenths of a millimeter) must be prepared. The factors that affect crystal preparation will

S. E. Ealick and C. E. Bugg   Departments of Pharmacology and Biochemistry, Comprehensive Cancer Center and Center for Macromolecular Crystallography, University of Alabama at Birmingham, Birmingham, Alabama

be discussed in Section 3. Another disadvantage of a single-crystal X-ray diffraction study is the time scale. Since the actual experiment may take several days, and the measurements are averaged over millions of molecules, the resulting structure will be a time average. Dynamic structural information is lost, although recent advances using intense synchrotron radiation sources provide some hope that dynamic structural information in the millisecond range can be obtained from single crystals [3].

Although the theory of single-crystal X-ray diffraction is well established, most crystal structure analyses are carried out by a practicing crystallographer, primarily because of the requirement of specialized equipment and computer software packages. In addition, perhaps as many as 1 in 10 structure analyses present a problem which requires modifications of apparatus or data collection techniques, or computer software development in order to achieve a correct molecular structure. Consequently, the purpose of this chapter is not to teach the reader how to determine crystal structures, but rather to make the reader aware of the utility, the requirements, and the limitations of the technique.

2. X-RAY DIFFRACTION BY SINGLE CRYSTALS

Single crystals are characterized by having a regular three-dimensional repeating lattice. The smallest repeat unit is called a unit cell and contains the structural motif from which the crystal is composed (Fig. 1). Each unit cell contains one or more asymmetric units, depending on the crystal symmetry. Ordinarily an asymmetric unit is equivalent to one molecule; however, it can contain more than one molecule, in which case the molecules will be nonidentical,

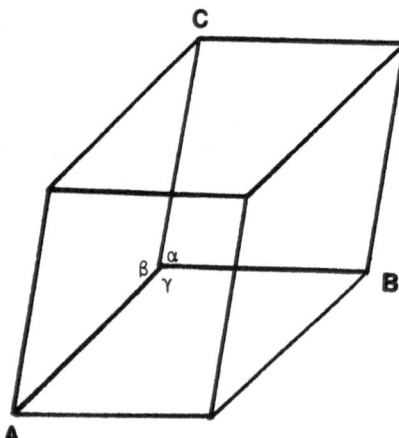

FIGURE 1. The crystallographic unit cell. The lengths of the cell edges are designated by A, B, and C and the angles by α, β, and γ.

or less than one molecule if the molecule possesses internal symmetry. The positions of all atoms in the asymmetric unit must be determined in a crystal structure analysis. Although the crystal lattice was proposed as early as the seventeeth century from morphological studies of crystals, it was not until 1912 that M. V. Laue used X-rays to show the existence of a crystal lattice.

When an X-ray beam is directed on a crystal, the regular array of atoms acts as a diffraction grating. Most of the X-ray beam passes through the crystal, some of the X-ray beam is absorbed, and some of the X-ray beam is diffracted. The diffraction pattern and the intensities of the diffracted beams are dependent on the arrangement of atoms within the crystal and result from the constructive interference of X-rays scattered from individual atoms. X-ray diffraction theory allows the experimenter to use the diffraction pattern to determine the arrangement of atoms from which it was created.

Although M. V. Laue described diffraction of X-rays by crystals in terms of a three-dimensional diffraction grating, W. H. Bragg represented X-ray diffraction by reflection from planes through the crystal lattice. He proposed the following equation, which is known as Bragg's law:

$$n\lambda = 2d \sin\theta$$

In this equation λ is the wavelength of radiation, d is the spacing between planes of atoms, θ is the angle that both the incident and diffracted beam make with respect to the crystal planes, and n is the order of diffraction. Because the angle of diffraction is inversely related to the actual distances that define the lattice, it is convenient to describe the diffraction pattern in terms of a reciprocal lattice. The diffraction pattern can then be indexed with respect to this reciprocal lattice using the Miller indices h, k, and l (Fig. 2).

In an X-ray diffraction experiment the intensity corresponding to each unique reciprocal lattice point, I_{hkl}, is measured. From I_{hkl}, one can derive the structure factor magnitude, $|F_{hkl}|$, but not its direction in phase space. Because the set of F_{hkl}s is the Fourier transform of the electron density, the electron density distribution can be derived by taking the Fourier transform of the derived $|F_{hkl}|$s if the associated phase angle, α_{hkl}, can be determined. The following equation summarizes the calculation:

$$\rho(x,y,z) = \frac{1}{V_{hkl}} \sum_h \sum_k \sum_l |F_{hkl}| \cos(2\pi(hx + ky + lz) - \alpha_{hkl})$$

In this equation $\rho(x,y,z)$ is the electron density at any point x,y,z; V is the volume of the unit cell; $|F_{hkl}|$ is the derived structure factor magnitude as determined from the X-ray diffraction experiment; h, k, and l are the Miller indices; and α_{hkl} is the phase angle. The triple summation is over all reciprocal lattice points

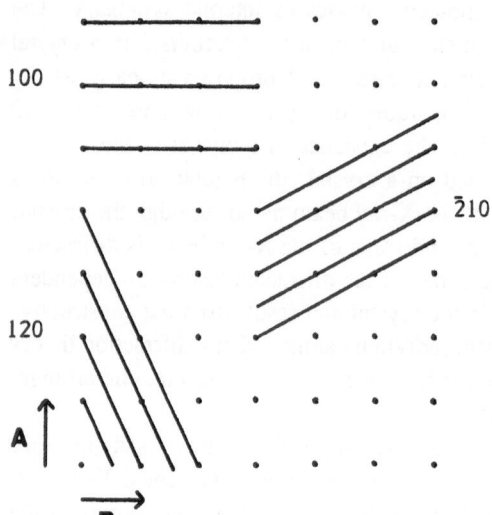

FIGURE 2. A two-dimensional lattice illustrating the Miller indices for certain planes.

for which measurements have been made. Since the set of α_{hkl}s is unknown, the crux of the problem is their determination, which is commonly referred to as the "phase problem."

If the positions of all the atoms in the unit cell are known, the structure factors, F_{hkl}, can be calculated as follows:

$$F_{hkl} = \sum_j f_j e^{2\pi i(hx_j + ky_j + lz_j)}$$

In this equation, f_j is the atomic scattering factor for the jth atom, and x_j, y_j, z_j are the coordinates of the jth atomic position. In addition, Debye showed that thermal motion causes a decrease in intensity. This can be accounted for by multiplying the atomic scattering factor, f_j, by the term $e^{-\beta \sin^2\theta/\lambda^2}$, where u^2 and u^2 is the mean square displacement normal to the diffraction plane. This decrease in intensity as a function of d spacing is the primary factor limiting the resolution of a crystal structure analysis. The smallest separation, r, which can be resolved by X-ray diffraction $r = 0.71\ d_{min}$, where d_{min} is the minimum d spacing for which intensities were measured or, from Bragg's law, $r = 0.71\ \lambda/(2\sin\theta_{max})$ where θ_{max} is the corresponding maximum angle of diffraction.

3. SAMPLE PREPARATION

In order to use the technique of single-crystal X-ray diffraction, one must obtain a well-formed single crystal that can be uniformly bathed in the X-ray beam of the experimental apparatus. The size of the crystal required is usually a few tenths of a millimeter in average dimension. The intensity of a diffracted X-ray beam is proportional to the size of the crystal. However, absorption also increases with the path length that X-rays travel through the crystal, as given by $I = I_o{}^{-ut}$, where I_o is the absorption-free intensity, I is the actual observed intensity, u is the linear absorption coefficient, and t is the path length. One can therefore show that the optimum crystal size is approximately $2/u$. In the case of organic crystals containing no atoms heavier than oxygen, u will be approximately 10 cm^{-1} for copper radiation, so one might use crystals as large as 2.0 mm. In fact, the cross-sectional size of the X-ray beam usually limits the size to 0.7–0.8 mm. In the case of a molecule containing a heavy atom such as iodine, then u $\approx$ 200 cm^{-1}, and consequently the observed intensity would decrease as a crystal became larger than 0.1 mm.

When examined under a binoccular microscope at 10–40X magnification, the crystal should show clear, shiny faces and sharp edges. The crystal should not be cracked or appear to be a conglomerate. If the crystal appears opaque, it may have lost loosely bound solvent of crystallization such as methanol or water and no longer diffract X-rays.

A more critical assessment of crystal quality can be made using crossed polarizers on a polarizing microscope. It a crystal is placed between crossed polarizers, it will appear either uniformly dark or bright. If the crystal appears bright, it will completely extinguish over a small angular range when rotated about an axis normal to the polarizing plane. If the crystal does not follow this behavior, it may be a conglomerate, contain imperfections, or be poorly ordered, in which case the diffraction pattern and subsequent structure analysis will be of poor quality if not impossible.

Suitable single crystals of organic materials are usually prepared by slowly bringing a solution containing the material to saturation or beyond. This can be accomplished on a scale from microgram to gram quantities. The most common ways of bringing a solution to saturation are (1) equilibration with a miscible solvent in which the material is insoluble, (2) evaporation of the solution, and (3) cooling the solution. Figure 3 shows an example of a vapor equilibration experiment in which the inner vial contains a solution of the compound to be crystallized and the outer jar contains a liquid in which the compound is insoluble. The entire system is sealed and allowed to equilibrate. If the original solution is near saturation, crystals usually form within a few days.

Crystallization is also affected by such factors as pH, temperature, type of buffer, ionic strength, and others. When a sample is reluctant to crystallize, one should try many combinations of conditions. If many small crystals form, a possible cause could be too many nucleation sites, in which case filtration of all solutions may be helpful in increasing the size of individual crystals. As a final consideration, it should be noted that although recrystallization is a classic tech-

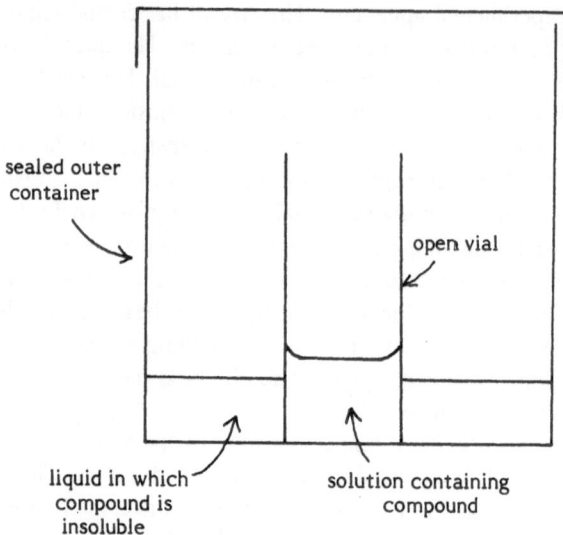

sealed outer
container

open vial

liquid in which
compound is
insoluble

solution containing
compound

FIGURE 3. Experiment for crystallization using vapor diffusion in a closed system.

nique for purification of materials, crystals suitable for X-ray diffraction can sometimes only be obtained from highly purified samples.

Once a batch of potential crystals has been obtained, the next objective is to mount the specimen for insertion into the X-ray beam. Nearly all single crystal X-ray diffraction equipment utilizes a standard sample holder called a goniometer head. This device, shown in Fig. 4, provides two arcs for orienting a sample and three translations for centering the crystal in the X-ray beam.

In general, crystals can be divided into two classes: those which are stable in air for the duration of the experiment and those which are not. If a crystal is stable in air, mounting is much simpler. The mount consists of a brass pin that is 1/8 inch in diameter and 1/2 inch long drilled lengthwise with a hole approximately 1/64 inch. A glass fiber that is a few hundredths of a millimeter in diameter is glued into the brass pin and cut so as to extend about 5 or 6 mm beyond the end of the brass pin. A crystal is selected and glued to the fiber using an adhesive such as fast-drying model cement, epoxy, or nail polish. The final step is usually carried out under a microscope. The final product is illustrated in Fig. 5A.

If the crystal is air sensitive, it must be sealed in a glass or quartz capillary. The crystal is usually sucked into the capillary directly from the crystallization solution using a syringe. The excess liquid is removed and the capillary is sealed at both ends with wax or glue. Sometimes a plug of liquid is left in one end of the capillary to provide a stable atmosphere. Such a mount is shown in Fig. 5B.

FIGURE 4. Standard goniometer head used for orienting a crystalline sample.

Another technique used to stabilize moderately air-sensitive crystals is to mount as in the air-stable case and apply a thin coating of epoxy over all surfaces of the crystal. If the X-ray diffraction apparatus is equipped with a low-temperature device, a sensitive crystal can sometimes be stabilized by cooling to temperatures below room temperature (Ref. 4, which is a comprehensive text on low-temperature X-ray diffraction).

4. EXPERIMENTAL MEASUREMENTS

To begin this section, the authors would like to make a statement about safety. Direct exposure to X-rays is dangerous. The energy of the radiation used

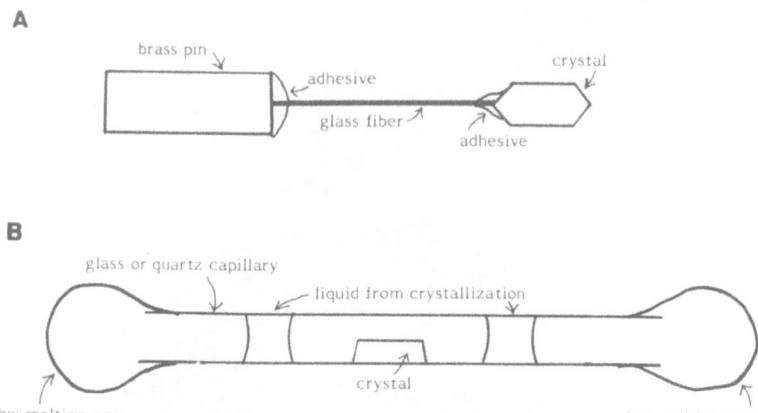

FIGURE 5. Two methods for mounting single crystals: (A) stable crystal on a glass fiber; (B) air-sensitive crystal sealed in a capillary.

is much lower and less penetrating than medical X-rays. Therefore, when any human tissue is exposed, the X-rays are absorbed by surface tissues and serious injury probably will result. If the X-ray tube is properly shielded, the beam properly collimated, and a beam stop is in place, the experimenter is not at risk. Most states have laws requiring radiation enclosures and regular measurement of background radiation levels around experimental apparatus. In any event, the experimenter must always be aware of whether the X-rays are on and whether the X-ray shutter is open.

The actual X-ray diffraction experiment can be divided into two phases. First is characterization of the crystal, which includes space group assignment, determination of unit cell dimensions, and general assessment of crystal quality and suitability. The second step is to measure all unique X-ray diffraction intensities to the highest possible resolution.

All single-crystal X-ray diffraction systems have three major components: (1) an X-ray source, (2) a crystal orienter, (3) an X-ray detection device. The X-ray source is usually an X-ray tube in which accelerated electrons strike an anode material such as copper or molybdenum followed by emission of X-ray photons of characteristic wavelength. The choice of anode material is usually dictated by the nature of the sample. Copper radiation is most commonly used because it provides high intensity. Molybdenum radiation is used if the crystal contains elements that heavily absorb copper radiation. The X-ray beams may be initially reflected off a crystal such as graphite to provide a more monochromatic beam. The use of a monochromator is more common when molybdenum radiation is used. The radiation is usually collimated before striking the crystal. The diameter of the collimator should be chosen to fully bathe the crystal, but it should not be too large since unused beam increases the background of the experimental measurement through air scattering.

The preliminary examination of a crystal is best carried out on either a Weissenberg camera or a precession camera (Fig. 6). In either case, the crystal undergoes a repetitive motion while centered in the X-ray beam, and a two-dimensional X-ray diffraction pattern is recorded on X-ray film. From the X-ray films, the lattice symmetry is determined, and the space group is either uniquely defined as one of the possible 230 or narrowed down to one of two or three related space groups. Unit cell dimensions can be measured from the films, although more accurate values can be obtained through least-squares analysis of the fit to setting angles measured on an X-ray diffractometer.

Each type of camera has its advantages; however, we rely almost exclusively on the Weissenberg camera. The crystal motion for a Weissenberg camera is an oscillation about an axis normal to the X-ray beam, and the resulting X-ray diffraction pattern is recorded on an X-ray film mounted in a cylindrical cassette. As a result the X-ray pattern is systematically distorted. The advantage is that a larger area of reciprocal space can be recorded and the spot shape on the film can be directly related to crystal mosaicity (i.e., the angular range over which a given reflection remains in the diffraction condition).

In the case of the precession camera, both the camera and X-ray film undergo a precession about an axis colinear with the incident X-ray beam. The resulting diffraction pattern is undistorted but records much less of the diffraction pattern than the corresponding Weissenberg exposure.

The films from either camera can be used to obtain individual X-ray intensities either by visual estimation or by microdensitometry. However, the most common method today is the use of a computer-controlled single-crystal X-ray diffractometer (Fig. 7). The X-ray diffractometer measures one intensity at a time using an X-ray-sensitive scintillation crystal (thallium doped sodium iodide) and a photomultiplier tube. The most important advantage is that resulting X-ray intensities are more accurately determined. Not only is the electronic detection of X-ray photons inherently more accurate, but errors by the experimenter are removed. The most important disadvantage is the initial cost of the equipment, which at the time of this writing is in the $150,000–200,000 range for a complete system.

The crystal orienter of a single-crystal X-ray diffractometer is more complicated than that of an X-ray camera. In order to measure one intensity at a time from a set of intensities distributed in three-dimensional space, at least three independent angular motions are required. The alignment of the instrument is critical, and the positioning of angles should be better than 0.01°. These design considerations, along with the need for photon-counting electronics and computer interfacing, are the primary factors affecting the cost of the instrument.

The manufacturers of these instruments have developed "user friendly"

FIGURE 6. Two types of X-ray cameras: (A) Weissenberg camera; (B) Buerger precession camera.

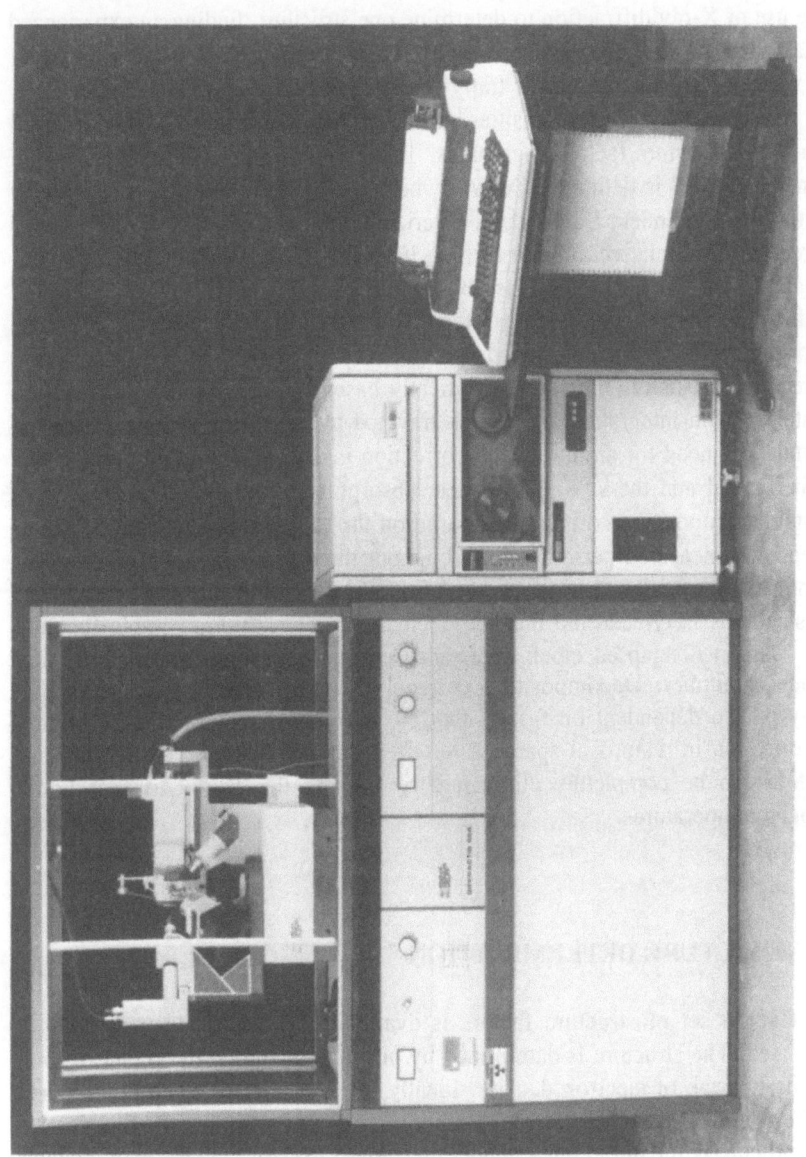

FIGURE 7. Computer controlled single-crystal X-ray diffractometer.

software packages for data collection. It is possible with a modest amount of training for the uninitiated scientist to measure a set of X-ray diffraction intensities without assistance. If one intends to use this technique extensively, the effort should be made to become independent. If the needs are limited, such as one-time use of X-ray diffraction to determine one structure, finding an experienced collaborator might be advised. At any rate, one should be constantly aware of the variety of "crystallographer traps" which will be discussed later.

Once a set of unique intensities has been obtained, the data must be reduced to a set of structure factor magnitudes, $|F|$s, for structure determination and refinement. $|F|$ is defined as $\sqrt{I/Lp}$ where I is the net intensity, and L and p are the $\sin\theta$-dependent Lorentz and polarization corrections. The Lorentz factor corrects for the relative amount of time it takes for a reciprocal lattice point to sweep through the diffraction position, and polarization is a result of the change of direction of the X-ray wave after diffraction from a crystal. In addition, absorption and decomposition corrections may be necessary. The absorption correction is usually computed analytically based on the shape and size of the crystal and the integrated path length traveled by the X-ray beam through the crystal. The need for an absorption correction is determined by the dimensions of the crystal and the size of the linear absorption coefficient. The need for a crystal decomposition correction is based on the monitoring of several standard reflections, measured periodically throughout the data collection. If significant decomposition occurs, intensities are corrected assuming that the behavior of the standards is typical and that the decrease in intensity is a simple function (e.g., linear) of elapsed clock time, accumulated exposure time, or reflection sequence number. Decomposition corrections are usually incomplete in that decay is also dependent on factors that are not considered, such as resolution and direction in reciprocal space. It has frequently been observed that decomposition can be completely eliminated by cooling the crystal to near liquid nitrogen temperatures.

5. STRUCTURE DETERMINATION

Once a set of structure factors is available, an electron density map is calculated. The structure is determined by placing an atom at the center of each significant peak of electron density. Ideally, the type of atom can be assigned by the magnitude of the electron density. However, both the structure factors and particularly the initial phase angles are subject to error, and the experimenter may need to rely on prior knowledge of the structural formula to obtain the initial trial structure. Subsequent improvements in the set of phase angles will in most cases lead to an unambiguous interpretation of the electron density. If the data are of good quality and sufficient resolution, hydrogen atom positions are normally assigned.

Determination of an initial set of phase angles for all or part of the X-ray diffraction data is the essential problem of crystallography. Consequently, a great deal of effort has been invested in developing methods for deriving phase angles. The development of more powerful techniques continues to be the major thrust of a large number of research groups worldwide. The three methods most applicable to structure determination of bile acids and related compounds are (1) the heavy-atom method, (2) direct methods, and (3) molecular replacement. Most crystallographic software systems contain the necessary computer programs to apply these methods.

The heavy-atom method relies, as the name implies, on the presence of a heavy atom. The position of the heavy atom is first located from a Patterson map and then used to calculate phase angles, which in turn are used to calculate an electron density map. The Patterson map is in effect a vector map and is calculated by performing a three-dimensional Fourier summation in which the Fourier coefficients are F^2. Alternatively, the Patterson map can be thought of as the self-product of the electron density map and thus has positive values only if the two points from the electron density map are both positive. The important aspects are that a Patterson map can be calculated without knowing the phase angles and that the heavy-atom vectors contained therein can be identified by their relative magnitudes. That is, the largest values in the Patterson map will correspond to those vectors connecting two heavy-atom positions. From the knowledge of the space group symmetry, the values of the positional coordinates can be solved for directly.

The most important question is how heavy a heavy atom has to be in order for this method to be used successfully. The answer depends on several factors, but in general the technique will be highly effective if the heavy atom represents about 50% of the total scattering power, where scattering power is defined as the sum of the squares of the number of electrons for each atom. The total scattering power is the sum over all atoms of the asymmetric unit. With some patience, the heavy-atom method can be used when the percent scattering due to a heavy atom is as low as 15%; however, the initial phases will contain larger errors and the electron density map may reveal only part of the structure and contain spurious electron density peaks as well.

The second technique is known as "direct methods" because the phases are determined directly from the intensity distribution using statistical considerations. In this case, phases are generated for a subset of data consisting of the largest normalized structure factors, E, and a subsequent Fourier map is calculated using E as the coefficient. The E value is generated by normalizing the structure factor with respect to resolution; thus, a weak intensity at high resolution can be more important than a strong intensity at low resolution if the high-resolution intensity is much larger than other reflections at high resolution. During the 1950s and 1960s, Hauptman and Karle worked out the key phase probability distribution for both the centrosymmetrical (i.e., phases restricted to 0° or 180°) and the noncentrosymmetrical (phase can be any value between 0° and 360°) cases. In general, they showed that if the Miller indices (h, k, l) of certain groups of reflections obeyed certain rules, then the sum of the phase angles of the group tended to be 0. The probability that the sum is 0 is a function of the product of the absolute values of the E's for those reflections. The most useful relationship is as follows: given three reflections, if $h_1 + h_2 + h_3 = 0$ and $k_1 + k_2 + k_3 = 0$, and $l_1 + l_2 + l_3 = 0$, then $\alpha_1 + \alpha_2 + \alpha_3$ is a constant value. The larger the value $| E_1E_2E_3 |$, the more likely this is to be true. The usefulness is that if the values of any two phases are known, the third can be estimated. By examining a large number of such relationships, enough phases can be generated to calculate an interpretable Fourier map.

Direct methods have become more and more powerful as the phase probability distributions have become more sophisticated. Several automated computer programs have been written with which direct methods can be applied to crystal structure determination. The most popular computer programs are MULTAN [5] and SHELX [6], although many in-house computer programs have been effectively used by crystallographic research groups.

The success of direct methods has made structure determination of molecules containing 30–50 nonhydrogen atoms nearly routine, and reports of structures containing as many as 200 nonhydrogen atoms have been made. Unfortunately, bile acids present a special problem. One of the underlying assumptions in the derivation of phase probability distributions is that the atoms are essentially randomly distributed throughout the unit cell. In the case of bile acids, the framework contains a repeating fused-ring arrangement. This gives rise to a nonrandom distribution of atoms and a biased intensity distribution. The resulting E values and probability estimations are less reliable and structure determination more difficult. One of the more recent advances in direct methods has been the use of known structural information, which would be particularly relevant to bile acid structure determination.

The final method to be discussed here will be molecular replacement, which also uses known structural information (Ref. 7, which is a comprehensive text on applications of molecular replacement techniques). To apply this technique, the experimenter must be able to propose a model of the structure to be determined. The model can contain all or part of the atoms in the asymmetric unit; however, the better the model, the greater the likelihood for success. The model is then rotated through all possible orientations and translated through all possible locations to find the best fit to the X-ray intensity data. The search can be carried out in either Patterson space (vector space) or reciprocal space. Once a possible orientation and location of the model structure are obtained, phase angles can be calculated followed by electron density calculations which will reveal the changes and additions with respect to the original search molecule. This technique or variations of this technique have been used successfully in cases where other methods have failed to yield a trial structure.

In all structure determinations, the initial phase angles contain significant errors. Improvement of the phase angles is accomplished by refining the initial trial model and by making additions to the trial model. Refinement of the model is done by least-squares analysis in which the atomic positional parameters are adjusted to minimize the quantity $\Sigma w(\Delta F)^2$ where ΔF is the difference between the observed and calculated structure factors after appropriate scaling and $w = 1/\sigma_F^2$ where σ_F is the standard deviation of the observed structure factor. The least-squares problem is nonlinear and is made linear through Taylor series expansion. Consequently, the process is iterative and requires multiple cycles for convergence.

Additions to the initial trial model are often made by examination of a difference Fourier map. In this case, the coefficients for the Fourier summation are ΔF, and as a result, the map contains only the electron density that has not been previously accounted for in the model. This technique usually reveals the positions of molecules of solvation, hydrogen atoms, or weakly scattering atoms such as disordered or highly mobile atoms of side chains.

Improvements to the model can be made by refining temperature factors. The temperature factor represents the distribution of positions the atom occupies as a result of thermal vibration. The temperature factor can either be isotropic (spherical distribution) or anisotropic (ellipsoidal distribution). The isotropic representation requires one additional parameter while the anisotropic representation requires six additional ones.

At the completion of refinement, the quality of the model can be judged in several ways. The most common is by the crystallographic R value, defined as $R = \Sigma \mid \Delta F \mid /\Sigma F$. Although the R value has no expected value and is calculated in a variety of ways (e.g., omitting weak reflections), one can usually assume that a structure with an R value less than 5% has been determined very well. If the R value is between 5 and 10%, the structure is probably correct but some factor is mildly affecting the quality of model or data. If the R value is above 10%, it is possible that some aspect of the model is incorrect or poorly defined. A more meaningful criterion is the goodness-of-fit, defined as $\sqrt{\Sigma w \Delta F^2/(m - n)}$ where m is the number of observations and n is the number of variables. If the model is correct and the weighting scheme is unbiased, the value will be 1. In practice, the value for a well-determined structure will be less than 3.0. A final way of judging the quality of a structure is to examine the final difference Fourier map. If the model has accounted for all the electron density, the difference Fourier map will be featureless, containing only random fluctuations indicative of the error level of the electron density. For a light-atom structure (e.g., only, C, H, N, and O atoms), the values of a final difference Fourier map will usually fall between $\pm 0.3e/\text{Å}^3$. If the structure contains a heavy atom (e.g., Rb, I, Cu), peaks as large as $1.0e/\text{Å}^3$ or higher may be found near the heavy atom.

6. INTERPRETATION OF RESULTS

Several types of useful data can be derived from an X-ray diffraction study. If the structural formula of the molecule was previously unknown, the X-ray diffraction study will reveal the chemical bonding pattern and the type of each atom. This is particularly useful for a newly isolated product or to verify the results of the chemical modification of a known molecule. It should also be noted that a three-dimensional X-ray structure can unambiguously show the stereochemistry of chiral atoms.

Frequently, the experimenter knows the structural formula of the molecule of interest but would also like to know something about its conformation. The three-dimensional structure will allow the calculation of torsion angles, least-squares planes, dihedral angles, coordination geometry, and other factors. The goal of determining the conformation of a bile acid or related compound is to gain insight as to how it functions biologically. This point has been responsible for generating many lively discussions regarding the significance of molecular conformation obtained from a crystal.

On this subject, we would like to make the following points. One can imagine a molecule existing in a variety of environments, such as (1) in a vacuum, (2) in a gas, (3) in solution, (4) in a crystal lattice, and (5) *in vivo*. Although it is possible that the molecular conformation is the same under any condition, it is likely that the conformation varies as a result of the additional interactions with its surroundings. The most interesting conformation is usually that of the molecule when functioning biologically (e.g., bound to receptor, forming a micelle). In that case, the molecule will be experiencing intermolecular contacts, such as van der Waals interactions and hydrogen bonds, just as it is in the solid state. Although it is unlikely that all the solid-state interactions are identical to the *in vivo* contacts, the solid-state environment may not be any less typical of the *in vivo* situation than a dilute solution, a gas, or a vacuum. In any event, the X-ray structure determination will reveal at least one of the possible low-energy conformations of the molecule.

In some instances, additional useful information can be obtained by examination of the relative thermal parameters. Since the thermal parameters are related to the r.m.s. displacement of the atoms from their mean positions over the duration of the experiment, the atoms with the largest thermal parameters are most likely to occur in the region of greatest conformational flexibility. It should be pointed out, however, that thermal parameters may be affected by other factors including systematic experimental error.

Another advantage of single-crystal X-ray diffraction is that standard deviations for atomic positions can be estimated from the least-squares analysis. As a result, standard deviations can be calculated for bond distances, bond angles, torsion angles, deviations from least-squares planes, and other factors. The standard deviations can in turn be used for statistical testing to determine whether any derived quantity differs significantly from normal values. For example, if a bond distance is significantly longer than normal, then that region of the molecule could be highly stressed. This in turn may correlate with chemical data such as point of bond cleavage in a chemical reaction.

Finally, a crystal structure may reveal information about the way in which molecules aggregate. If a molecule contains both hydrophobic and hydrophilic regions, then the crystal will form such that hydrophobic regions of different molecules come together and likewise hydrophilic regions come together. Frequently this results in alternating hydrophobic and hydrophilic layers. The intermolecular interactions observed in a crystal may be similar to interactions found in dimers or higher aggregates.

In summary, X-ray crystallography provides precise structural data about both intramolecular and intermolecular interactions. This information can be useful in both explaining and predicting the chemical behavior of the molecule under investigation. The most serious potential limitation is the crystalline nature of the sample and the resulting interactions which contribute to the overall energy of the solid-state system.

7. PITFALLS

When single-crystal X-ray diffraction works, it usually works well. In addition, least-squares analysis provides an estimation of the reliability of the quantities determined in the experiment. There are, however, instances in which the technique fails or the results are misleading as a result of oversignt on the part of the experimenter.

The primary assumption in the experimental analysis is that the sample is a single crystal. The two most common problems are: (1) the crystal is a conglomerate, or (2) the crystal is twinned. The first situation occurs most often for crystals with platelike morphology in which the apparent single crystal is in fact several thinner plates slightly misaligned by a rotation about the normal to the plate face. This results in multiple diffraction spots and systematic error in the measured intensities. To avoid the problem, the investigator should select a crystal that does not display this phenomenon or obtain a different crystal form. As a last resort, the data can be measured with a large aperture in front of the dector, but at the expense of reduced signal-to-noise.

A much more difficult problem is twinning. Macroscopic twinning can be detected as an interpenetrating growth of two or more crystals and should be avoided. Microscopic twinning is not apparent through visual examination of the crystal and occurs in one of two ways. First, the reciprocal lattice points of the two twins overlap only partially. In this case, each lattice can be indexed separately and corrections can be applied for reciprocal lattice points which overlap. In the second case, the reciprocal lattice points overlap exactly but come from two different orientations of the lattice. In many instances, this type of twinning becomes apparent only after repeated attempts to determine the structure have failed. One should be suspicious of twinning if the reciprocal lattice has unusual patterns of absences. Although twinning can be dealt with, it would be best to find a different crystal form. This can often be accomplished by recrystallization from a different solvent system.

The advent of automatic diffractometers has greatly increased the productivity of crystallographers but has also created some new problems. Most software packages contain computer programs for space group determination by automatic search and indexing. The convenience of this technique has led some investigators to bypass the X-ray photography, which in some cases has led to incorrect space group determination [8]. This is particularly a problem if the crystal has weak zones of intensities which the search routine fails to detect. Without knowledge of such reciprocal lattice points, the unit cell volume will be incorrectly determined. In other cases, the automatic routines have not detected all symmetry elements, and as a result the space group is reported as a subgroup of the correct space group. The important point to remember is that X-ray photography has a role that can be only partially replaced by an automatic diffractometer. The detection of pseudosymmetry or twinning, assessment of crystal quality, and certainty of determination of space group are best accomplished by film techniques.

Problems during data collection can also arise from several sources. The optical alignment of the diffractometer must be true. Generally, X-ray tube position, monochromator angle and collimator tilt can be adjusted and should be checked every month or 2 to ensure that the alignment has not changed. The electronic counting circuitry should also be checked occasionally. In particular, the pulse-height analyzer, which determines whether a pulse is likely due to an X-ray photon or electronic noise, should be calibrated. The experimenter who does not know about these things should consult with somebody who does.

During data reduction, corrections for decomposition and absorption are sometimes critical. Decomposition can be nonlinear and anisotropic and is usually more severe at higher resolution. Several standard intensities should be recorded every few hours to determine changes in the crystal during the data collection. In most cases, decomposition can be completely eliminated by collecting data near liquid nitrogen temperatures, but it is necessary to beware of phase changes. If the absorption correction is large (i.e., > 50 cm^{-1}), the morphology of the crystal must be carefully determined for input into an absorption correction algorithm. Absorption effects can be minimized by using shorter wavelength radiation (e.g., Mo instead of Cu). The form of the correction can be simplified by grinding the crystal into a sphere prior to data collection.

Once a trial structure has been determined, the least-squares refinement process becomes the most sensitive indicator of problems. Abnormal thermal parameters can indicate incorrect atom type assignment or disorder. If all thermal parameters are affected, then a general systematic error in the data probably exists, such as incorrect decomposition or decay corrections or instrument or sample misalignment.

Disorder frequently occurs in crystals. An atomic position is considered to be disordered if the atom is distributed over two or more discrete positions in the crystal lattice. Disorder is usually detected in difference Fourier maps as significant electron density peaks which provide an alternative and geometrically reasonable interpretation of the molecular structure. Usually the fraction of atoms occupying each position (occupancy) can be determined, and all atoms can be included in the least-squares analysis.

The most important statistical parameter is the goodness-of-fit, as defined in Section 5. If its value is greater than 3 or 4, then either the model or the weighting scheme is inadequate. Large values can result from systematic errors in only a few structure factors; therefore, a list of individual residuals, $w\Delta F^2$, should be examined for outliers. In addition, the average value of $w\Delta F^2$ should be independent of resolution, structure factor magnitude, or any special function such as Miller indices $(h, k,$ or $l)$.

8. EXAMPLES

Altogether several hundred steroid structures have appeared in the literature. Of these about two dozen are analogs of cholanic acid and will be discussed here. A complete cataloguing of all steroid structures determined prior to 1975 has been compiled by Duax and Norton [9] and a second volume has just been completed [10]. The earliest report of a crystallographic study of a cholanic acid was in the 1930s by Katky, Giacomello, and co-workers (for a summary of this work see Ref. 11). They examined crystals of deoxycholic acid and a variety of its choleic acid derivatives and determined that crystals of deoxycholic acid must contain accessible channels based on the fact that only small changes in unit cell parameters were observed upon addition of aliphatic groups.

The first report of a cholanic acid structure at atomic resolution was not until 1972, when the structure of deoxycholic acid *p*-bromoanilide was determined by Schaefer and Reed [12]. This work confirmed the stereochemistry of the chiral atoms and showed the expected *cis* juncture for the A:B rings and *trans* junctures for the B:C and C:D rings. The six-membered rings were in an approximate chair conformation, while the D ring was in an approximate envelope conformation, with C(14), C(15), C(16), and C(17) making the best plane.

The cholanic acid crystal structures that have been determined to date are listed in Table I. These structures can be divided into three groups: (1) mixed crystals, (2) cholanic acid analogs, and (3) salts of cholanic acids. Before considering each of these groups, we shall give a general description of the cholanic acid analogs and define the conformational parameters necessary to characterize a structure. Figure 8 shows a stereoview of a molecule of cholic acid. The

TABLE I.   Summary of Crystal Structures of Cholanic Acids

Compound	Abbreviation	Reference
Mixed crystals		
Cholic acid-ethanol	CAETH	13
Deoxycholic acid-*p*-diiodobenzene 2:1	DCAIB	14
Deoxycholic acid-phenanthrene 3:1	DCAPHE	14
Deoxycholic acid-acetic acid	DCAAC	15
Deoxycholic acid-di-*t*-butyldiperoxy carbonate 4:1	DCABPC	16
Deoxycholic acid-acetone 2:1	DCAACE	17
Apocholic acid-acetone	ACAACE	17
Deoxycholic acid-ethanol-water 3:2:1	DOAEW:HEX	18
Deoxycholic acid-ethanol-water 2:1:1	DCAEW:TET	19
Deoxycholic acid-dimethyl sulfoxide-water 2:1:1	DCADSW	20
Deoxycholic acid-palmitic acid-ethanol 8:1:1	DCAPAL	21
Deoxycholic acid-norbornadiene	DCANOR	22
Deoxycholic acid-camphor	DCACAM	23
Deoxycholic acid-*p*-fluoroacetophenone	DCAFAP	24
Deoxycholic acid-pinacolone 2:1	DCAPIN	25
Cholanic acids		
Deoxycholic acid *p*-bromoanilide	DCABA	12
3α,6α-dihydroxy-5β-cholan-24-oic acid	HDCA	26
Lithocholic acid	LCA	27
Chenodeoxycholic acid	CDCA	28
Cholic acid	CA	29
Salts		
Sodium cholate	NACA	30
Rubidium deoxycholate	RBDCA	31
Calcium cholate	CACA	32
Sodium deoxycholate	NADCA	33

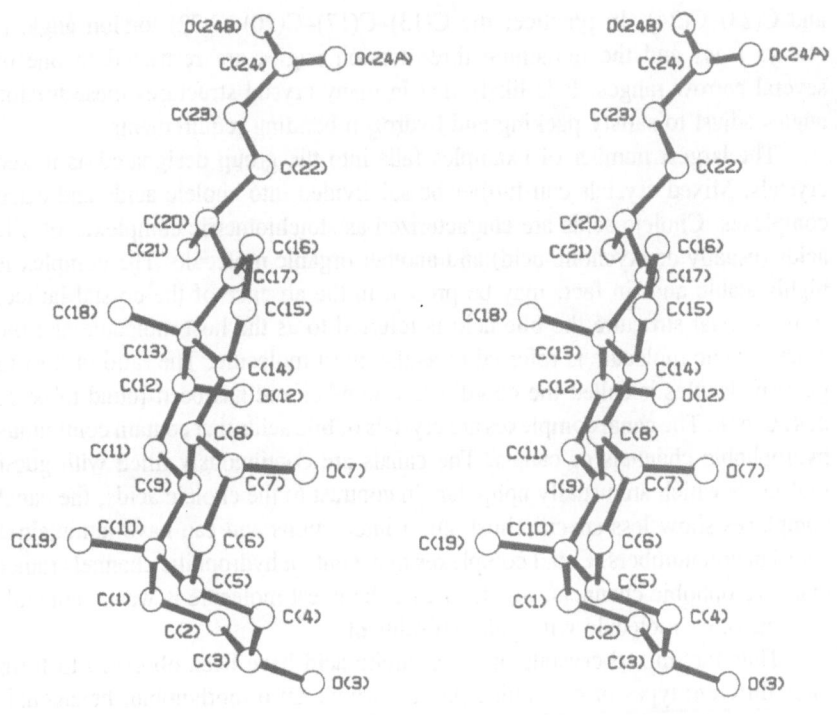

FIGURE 8. Stereoview of a hypothetical cholanic acid structure. The hydrogen atoms have been omitted for clarity.

molecule consists of a rigid steroid nucleus and a flexible carboxylic acid tail extending out from C(17) of the D ring. The fused-ring system that makes up the steroid nucleus always has substitutions at C(10), C(13), and C(17) and possible $-OH$ substitutions at C(3), C(6), C(7), and C(12). These substituents, along with the ring junctures, provide torsional barriers that make deviations from the conformation shown in Fig. 8 very costly in terms of energy. In addition, the steroid nucleus possesses an amphipathic character (i.e., one side of the molecule is hydrophilic and the opposite side is hydrophobic). Consequently, it is likely that when the molecules aggregate or pack in a crystal lattice, entropic considerations will also favor the steroid structure shown in Fig. 8. The lowest-energy structure for the D ring is intermediate between the envelope and half-chain conformations. This is primarily because C(15) and C(16) are unsubstituted, thus allowing a torsion angle about the C(15)–C(16) bond approaching 0°.

The carboxylic acid tail has, in principle, four degrees of freedom corresponding to the torsion angles about C(17)–C(20), C(20)–C(22), C(22)–C(23), and C(23)–C(24). In practice, the C(13)–C(17)–C(20)–C(22) torsion angle is always *trans* and the remaining three torsion angles are restricted to one of several narrow ranges. It is likely that in many crystal structures these torsion angles adjust to satisfy packing and hydrogen bonding requirements.

The largest number of examples falls into the group designated as mixed crystals. Mixed crystals can further be subdivided into choleic acids and canal complexes. Choleic acids are characterized as stoichiometric complexes of bile acids (usually deoxycholic acid) and another organic molecule. The complex is highly stable and, in fact, may be present in the absence of the crystal lattice. In the crystal structure the bile acid is referred to as the host molecule and the other organic molecule is referred to as the guest molecule. The ratio of host to guest molecules is called the coordination number and has been found to be 1, 2, 4, 6, or 8. The canal complexes are crystals of bile acids that contain continuous hydrophobic channels or canals. The canals are continuously filled with guest molecules which are usually nonpolar. In contrast to the choleic acids, the canal complexes show less specific host–guest interactions and can have nonintegral coordination numbers. Canal complexes may contain hydrophilic channels rather than hydrophobic channels, in which case the guest molecule is water, ethanol, or some other molecule with polar substituents.

Thus far, mixed crystals of deoxycholic acid have been observed to form three different types of crystalline phases, which are orthorhombic, hexagonal, or tetragonal. In the orthorhombic phase (Fig. 9), the bile acids are associated

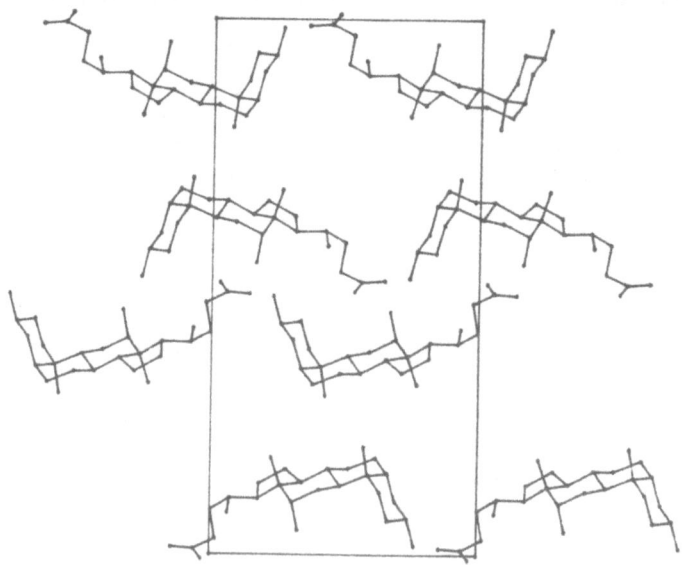

FIGURE 9. Orthorhombic phase of deoxycholic acid projected down the crystallographic *c* axis.

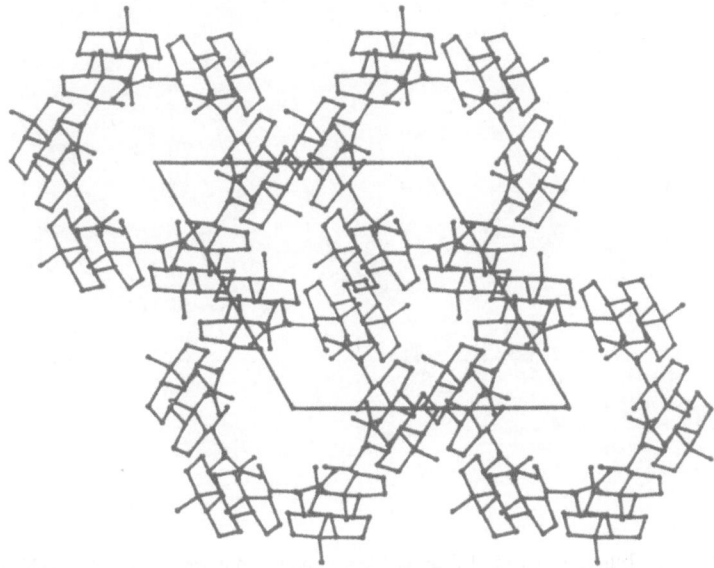

FIGURE 10. Hexagonal phase of deoxycholic acid projected down the crystallographic c axis.

through hydrogen bonding into pleated bilayers with individual molecules arranged in head-to-tail fashion. The region between the bilayers is hydrophobic, and its size and shape may adjust to fit the guest molecule. The space group in each case is $P2_12_12_1$, with unit cell dimensions of approximtely $a = 26$ Å, $b = 13.5$ Å, $c = n$ (7.2 Å), where $n = 1, 2$, or 3. The canals run primarily in the c directions. The doubling or tripling of the c axis occurs where the guest molecule spans two or more unit cells. The interactions between the host and guest molecules are primarily hydrophobic.

In the hexagonal phase (Fig. 10), deoxycholic acid molecules form helices about the sixfold symmetry axis. The interior of the helix is hydrophilic, and the guest molecule participates in hydrogen bonding with deoxycholic acid. The exterior of the helix is mostly hydrophobic and packs against the hydrophobic exterior of other deoxycholic acid helices. In the two cases reported, the space group was $P6_5$ and the guest molecules are water plus either ethanol or dimethyl sulfoxide.

In the tetragonal phase (Fig. 11), the formation of bilayers is once again observed. The exterior of the bilayers is hydrophobic, while the interior is hydrophilic and contains the guest molecules, ethanol and water. Adjacent bilayers are related by the crystallographic fourfold axis and run in perpendicular directions. Within the bilayer, the head-to-tail arrangement of deoxycholic acid molecules is similar to that observed in the orthorhombic phase.

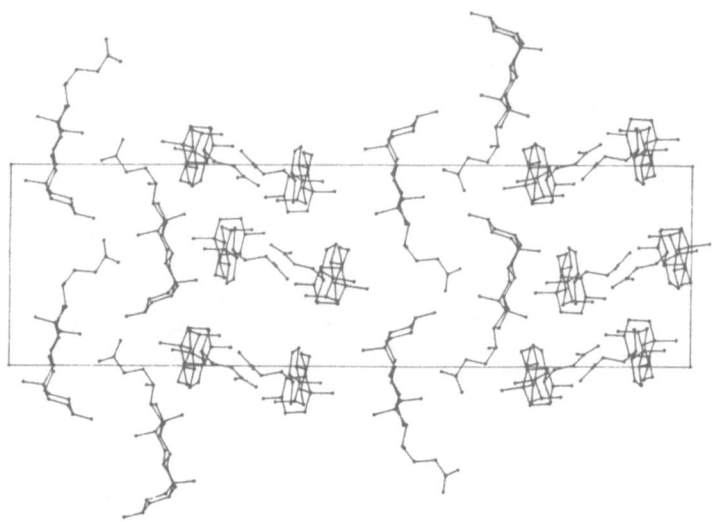

FIGURE 11.   Tetragonal phase of deoxycholic acid projected down the crystallographic c axis.

Only a few examples of additional compounds where the bile acid is something other than deoxycholic acid have been reported. Crystals of cholic acid–ethanol are orthorhombic, space group $P2_12_12_1$, but the packing is generally different than the orthorhombic phase described for deoxycholic acid. The head-to-tail arrangement of bile acid molecules is not observed, and instead a hydrogen bond is formed between $O(3)$ and the ethanol guest molecule.

A handful of bile acid structures have been determined in which no other molecule is present in the crystal lattice. These molecules show a variety of crystal-packing arrangements and hydrogen-bonding schemes. Together these crystal structures provide an opportunity to examine the flexible regions of the molecule in different environments. As in other bile acid structures, the D ring is in a conformation intermediate between β-envelope and half-chair. Quantitatively, the conformation of the five-membered ring is best described using the pseudorotation parameters Δ and ϕ_m where Δ is the phase angle and ϕ_m is the maximum torsion angle obtainable in that conformation [34]. The value of Δ is $+36°$ for a β-envelope, $0°$ for a half-chair, and $-36°$ for an α-envelope. The values of Δ and ϕ_m for a number of molecules are given in Table II. It is interesting to note that in the case of chenodeoxycholic acid, where two independent molecules exist in the asymmetrical unit, one D ring is primarily β-envelope while the other is primarily half-chair.

The carboxylic acid tail is the other region of conformational flexibility. The values for the four main torsion angles for several bile acid molecules are

**TABLE II.   Internal Torsion Angles and Pseudorotation Parameters for the
D Rings of Selected Cholanic Acid Compounds**

Compound	C(13)–C(14)	C(14)–C(15)	C(15)–C(16)	C(16)–C(17)	C(17)–C(13)	Δ	ϕm
DCABA	47	−36	10	20	−40	12	48
DAETH	48	−33	5	24	−43	24	53
DCAPHE	50	−39	12	19	−42	7	50
DCAEW:HEX	45	−31	6	22	−41	22	48
DCAEW:TET	44	−33	9	18	−37	11	45
	50	−38	12	16	−39	4	50
DCADSW	48	−33	5	25	−45	24	53
DCAPAL	46	−34	8	20	−39	14	48
DCANOR	44	−34	10	17	−37	9	45
	47	−36	12	16	−39	5	47
DCACAM	47	−37	13	16	−38	4	47
DCAPIN	47	−36	11	18	−39	8	48
	48	−39	14	15	−39	1	48
HDCA	44	−33	8	19	−38	15	46
LCA	48	−33	5	25	−44	24	53
CDCA	48	−34	7	23	−42	19	51
	47	−36	11	17	−39	7	47
NACA	46	−34	8	20	−40	15	48
RBDCA	48	−36	10	20	−41	12	50
CACA	46	−33	8	20	−39	15	47

given in Table III. Although it is clear that the C(13)–C(17)–C(20)–C(22) torsion angle is near 180° for all cases, the remaining torsion angles show a distribution over several possible ranges. The C(17)–C(20)–C(22)–C(23) torsion angle is near either 60° or 180°. A value near −60° would bring C(23) into close contact with C(16) and has never been observed. It has been suggested that if the carboxylate is involved in hydrogen bonding, this torsion angle will be near 60°, and if not, the value will be near 180°. The structure of chenodeoxycholic acid provides at least one counter example to this hypothesis. The value of the torsion angle C(20)–C(22)–C(23)–C(24) shows a strong preference for the *trans* conformation. The largest variation is seen in the orientation of the carboxyl plane with respect to rotation about the C(23)–C(24). In principle, the C(22)–C(23)–C(24)–O(2) [where O(2) is the protonated oxygen atom] torsion angle can fall near ±60°, ±120°. In fact, a wide range of values has been observed, although it should be pointed out that X-ray diffraction data do not always clearly differentiate between the protonated and unprotonated oxygen atoms.

Only four examples of salts of bile acids have appeared in the literature. They are the sodium and calcium salts of cholic acid and the rubidium and

TABLE III. Side-Chain Torsion Angles for Selected Cholanic Acid Compounds

Compound	C(13)–C(17)–C(20)–C(22)	C(17)–C(20)–C(22)–C(23)	C(20)–C(22)–C(23)–C(24)	C(22)–C(23)–C(24)–X[a]
DCABA	176	–171	173	132
CAETH	179	–173	–175	168
DCAPHE	175	62	179	124
DCAEW:HEX	180	–164	–161	–155
DCAEW:TET	177	66	178	119
	170	50	178	160
DCADSW	179	–168	165	–153
DCAPAL	176	59	173	121
DCANOR	174	66	178	121
	176	67	173	150
DCACAM	172	64	–179	124
DCAPIN	173	63	177	124
	177	64	174	148
HDCA	–179	–168	–98	164
LCA	172	64	176	–165
CDCA	173	–176	65	170
	–176	180	179	–179
NACA	–179	–176	180	160
RBDCA	172	65	–174	–100
CACA	–170	–166	–172	122
	171	69	–179	–103

[a] X is chosen as the atom that gives the torsion angle closer to 180°.

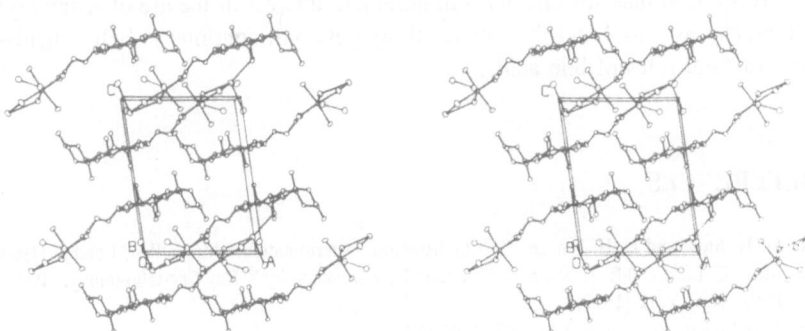

FIGURE 12. Crystal packing in the calcium salt of cholic acid.

sodium salts of deoxycholic acid. With the exception of sodium deoxycholate, the other three crystallize in the monoclinic space group $P2_1$ and show remarkable similar packing arrangements. The crystal structure consists of bilayers in which the hydrophobic surfaces of the bile acid face inward and the hydrophilic surfaces face outward. The bile acids are arranged in essentially a head-to-tail pattern, although no hydrogen bonds are observed between the hydroxyl group at C(3) and the carboxylate. The cations are sandwiched between the bilayers and interact through hydroxyl groups, carboxylate groups, or water molecules (Fig. 12). The size and shape of the hydrophilic channel shows variation in the three structures, while the hydrophobic contacts in the interior of the bilayer are almost invariant. All three molecules show different conformations for the carboxylate tail. This is almost certainly due to the fact that the carboxylate group must adjust to serve as a bidentate ligand for the cation.

In summary, the cholanic acid analogs that have been examined crystallographically confirm the expected conformation and stereochemistry for the six-membered A, B, and C rings of the steroid nucleus. The flexible regions of the molecule are the five-membered D ring and the carboxylic acid tail attached at C(17). Although the D ring is restricted to a narrow range of conformations, the side chain assumes a wide variety of conformations associated with a wide variety of functions. Examination of intermolecular interactions and crystal packing reveals a tendency for these compounds to form bilayers with a hydrophobic core and hydrophilic exterior. The core of the bilayer may accommodate other nonpolar molecules as in the mixed crystals. When bilayers are formed, the arrangement of individual molecules is head-to-tail. The kinds of hydrophobic interactions observed in the different types of bilayers also show variation with function.

It is hoped that this chapter will stimulate interest in the use of X-ray crystallography as a tool and that the resulting data will continue to help elucidate the structural roles of bile acids.

REFERENCES

1. G. H. Stout and L. H. Jensen, "X-Ray Structure Determination," Macmillan, London (1968).
2. M. F. C. Ladd and R. A. Palmer, "Structure Determination by X-Ray Crystallography," Plenum Press, New York (1977).
3. J. Helliwell, *Rep. Prog. Phys.* **47,** 1403 (1984).
4. R. Rudman, "Low-Temperature X-Ray Diffraction," Plenum Press, New York (1976).
5. P. Main, S. J. Fiske, S. E. Hull, J. Lessinger, G. Germain, J-P. Declercq, and M. M. Woolfson, "MULTAN80. A System of Computer Programs for the Automatic Solution of Crystal Structures from X-ray Diffraction Data," University of York, England (1980).
6. G. M. Sheldrick, "SHELX76. Program for Crystal Structure Determination," University of Cambridge, England (1976).
7. M. G. Rossmann, ed., "The Molecular Replacement Method," Gordon and Breach, New York (1972).
8. F. H. Herbstein and R. E. Marsh, *Acta Cryst.* **B38,** 1051 (1982).
9. W. L. Duax and D. A. Norton, eds., "Atlas of Steroid Structure," Vol. 1, Plenum Press, New York (1975).
10. W. L. Duax, ed., "Atlas of Steroid Structure," Vol. 2, Plenum Press, New York (1983).
11. H. Sobotka, "Chemistry of Steroids," Williams & Wilkins, Baltimore (1938).
12. J. P. Schaefer and L. L. Reed, *Acta Cryst.* **B38,** 1743 (1972).
13. P. L. Johnson and J. P. Schaefer, *Acta Cryst.* **B28,** 3083 (1972).
14. S. C. De Sanctis, E. Giglio, V. Pavel, and C. Quagliata, *Acta Cryst.* **B28,** 3656 (1972).
15. B. M. Craven and G. T. DeTitta, *Chem. Commun.* **9,** 530 (1972).
16. N. Friedman, M. Lahav. L. Leisevowitz, R. Popovitz-Biro, C-P. Tang, and Z. Zaretskii, *J. C. S. Chem. Commun.* 864 (1975).
17. M. Lahav. L. Leiserowitz, R. Popovitz-Biro, and C-P. Tang. *J. Am. Chem. Soc.* **100,** 2542 (1978).
18. S. C. DeSanctis, V. M. Coiro, E. Giglio, S. Paglivca, N. V. Pavel, and C. Quagliata, *Acta Cryst.* **B34,** 1928 (1978).
19. V. M. Coiro, A. D'Andrea, and E. Giglio, *Acta Cryst.* **B35,** 2941 (1979).
20. S. C. DeSanctis, E. Giglio, F. Petri, and C. Quagliata, *Acta Cryst.* **B35,** 226 (1979).
21. V. M. Coiro, A. D'Andrea, and E. Gaglio, *Acta Cryst.* **B36,** 848 (1980).
22. A. D'Andrea, W. Fedeli, E. Giglio, F. Mazza, and N. V. Pavel, *Acta Cryst.* **B37,** 368 (1981).
23. J. G. Jones, S. Schawzbaum, L. Lessinger, and B. W. Low, *Acta Cryst.* **B38,** 1207 (1982).
24. H. C. Chang, R. Popovitz-Biro, M. Lahav, and L. Leiserowitz, *J. Am. Chem. Soc.* **104,** 614 (1982).
25. V. M. Coiro, F. Mazza, G. Pochetti, E. Giglio, and N. V. Pavel, *Acta Cryst.* **C41,** 229 (1985).
26. S. R. Hall, E. N. Maslen, and A. Cooper, *Acta Cryst.* **B30,** 1441 (1974).
27. S. K. Arora, G. Germain, and J. P. Declercq, *Acta Cryst.* **B32,** 415 (1976).
28. P. F. Lindley, M. M. Mahmoud, F. E. Watson, and W. A. Jones, *Acta Cryst.* **B36,** 1893 (1980).
29. L. Lessinger, *Cryst. Struct. Commun.* **11,** 1787 (1982).

30. R. E. Cobbledick and F. W. B. Einstein, *Acta Cryst.* **B36,** 287 (1980).
31. V. M. Coiro, E. Giglio, S. Morosetti,and A. Palleschi, *Acta Cryst.* **B36,** 1478 (1980).
32. A. Hogan, S. E. Ealick, C. E. Bugg, and S. Barnes, *J. Lipid Res.* **25,** 791 (1984).
33. G. Conte, R. DiBlasi, E. Giglio, A. Peretta, and N. V. Pavel, *J. Phys. Chem.* **88,** 5720 (1984).
34. C. Altona, H. J. Geise, and C. Romers, *Tetrahedron* **24,** 13 (1968).

Chapter 5

MASS SPECTROMETRY OF BILE ACIDS

A. M. Lawson and K. D. R. Setchell

1. INTRODUCTION

The continued use over many years of mass spectrometry (MS) for bile acid analysis can be attributed to its value as a method for providing definitive qualitative and quantitative information. Its combination with gas chromatography (GC–MS) has been particularly important for investigating the stereochemical variety of the structure of bile acids in relation to their biosynthesis, transport, and metabolism.

The mass spectra of bile acids were presented in Volume 1 of this series by Sjövall, Eneroth, and Ryhage [1], and this remains a valuable reference. This chapter provides information on MS development that has taken place since and extends discussion on features of fragmentation behavior that may assist in compound identification. For this purpose a limited but hopefully useful compendium of mass spectra is included. The library has been restricted to the spectra of methyl ester-trimethylsilyl (Me-TMS) ether derivatives of bile acids for reasons explained later, but this should not be taken as minimizing the past and future value of other derivatives. The quantitative evaluation of bile acids by MS has received considerable attention in recent years as the sensitivity, selectivity, and use of stable isotopically labeled standards and tracers have opened up new areas of research.

In line with the objectives of the book, the application of qualitative and quantitative mass spectrometric methodology to the measurement of bile acids in biological fluids has been reviewed.

2. PRINCIPLES AND INSTRUMENTATION

There have been many changes to mass spectrometric instrumentation in the last 25 years since it was first used in the structural analysis of bile acids.

A. M. Lawson   Section of Clinical Mass Spectrometry, Clinical Research Centre, Harrow, Middlesex, England   K. D. R Setchell   Department of Pediatric Gastroenterology and Nutrition, Clinical Mass Spectrometry Laboratories, Children's Hospital Medical Center, Cincinnati, Ohio

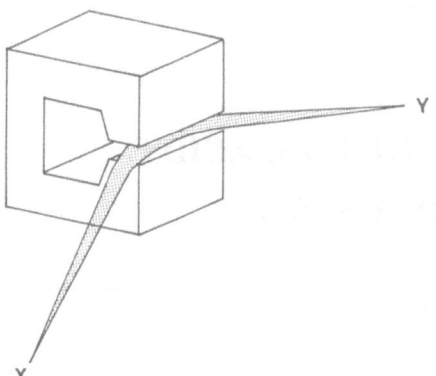

FIGURE 1. Schematic of the magnetic deflection principle for the separation of charged molecules or fragments.

While the underlying principles of the technique remain unaltered, new concepts and methods have been introduced. In order to place these developments in context, the basis of mass spectrometry should be considered by those new to the field.

A mass spectrum consists of a record of the abundances and mass/charge ratios (m/z) of the ions formed from a molecule during an ionization process. The relative abundance of the different ions is characteristic of a particular compound and is reproducible for a given set of instrumental conditions. In addition, the total number of ions formed from a sample is directly proportional to the sample concentration. Thus, both qualitative and quantitative measurements are possible by MS.

The steps in producing a mass spectrum include sample introduction, sample ionization, ion separation, ion detection, and ion recording. Although all aspects are important, the mode and energy of the ionization process are the primary influences on the appearance of the ion spectrum, and thus the principal ionization methods of relevance to bile acid analysis merit separate discussion later. Methods of sample introduction that are suited to bile acids are also given separate coverage, leaving the more general aspects of ion separation, detection, and recording to be described here.

2.1. Ion Separation

Charged molecules or molecular fragments of individual mass can be separated from each other by several methods, but most commonly by passing them through magnetic or electrostatic fields. Ions projected into a magnetic field are deflected to a degree dependent on their momenta and the field strength (B). For an ion produced at X (Fig. 1) and accelerated by voltage V to reach point Y, the centripetal force resulting from magnetic deflection is balanced by the

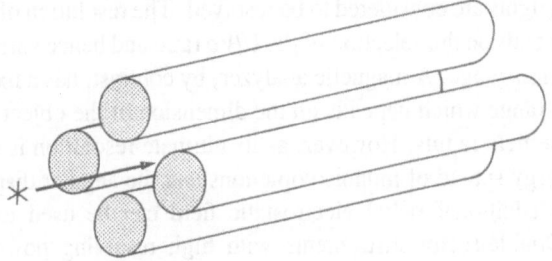

FIGURE 2. Separation of ions in the quadrupole mass spectrometer relies on their stability in the fringing hyperbolic electrostatic/radiofrequency field created in the square array of rods shown.

centrifugal force due to its mass and velocity. Hence, for an ion of particular mass and charge (unity in most cases) to reach a detector placed at Y, the following simple equation holds $\dfrac{m}{z} = \dfrac{B^2 r^2}{2V}$ (where r is the magnet radius).

Directly from this equation follows the means for ion separation by magnetic deflection. By maintaining a constant acceleration voltage (V) and sweeping the field strength from a low to a high value, ions of increasing mass but with equivalent charge will pass across the detection point Y. The reason for referring to m/z, rather than mass alone, is also apparent as, for example, a doubly charged ion will be focused with singly charged ions of half its mass.

At the end of the 1960s the first commercial quadrupole filter instruments became available which separated ions of varying m/z on the basis of their stability in electrostatic and radio frequency fields. The ions follow oscillating paths under the influence of an electric field created in a square array of four electrodes (Fig. 2) by particular combinations of radio frequency (U) and DC voltage (Vo) applied to diagonal pairs of the electrodes. A small voltage is used to introduce ions into the quadrupole along the axis of the rods. The ratio U/Vo is held constant but increased in amplitude to allow increasingly larger ions to survive passage through the filter. An advantage of quadrupole analysis in GC–MS applications is the ease and speed with which U and Vo can be manipulated.

Magnetic and quadrupole analyzers can cope with either positive or negative ions. Although negative ion mass spectrometry is useful for some aspects of bile acid analysis (e.g., fast atom bombardment ionization), almost exclusive use has been made of the positive ion mode.

An important point of comparison between some magnetic and quadrupole systems is in their ability to resolve ions from one another throughout the mass range. Mass resolution is expressed as $\dfrac{M}{\Delta M}$ where ΔM is the mass difference between mass M and the next higher mass. Arbitrarily peaks overlapping by 5%

of their peak heights are considered to be resolved. The resolution of a quadrupole instrument depends on the selection of the U/Vo ratio and hence varies throughout a scan of the mass range. A magnetic analyzer, by contrast, has a fixed resolution throughout its range which depends on the dimension of the object and collector slits and on the field radius. However, as its ultimate resolution is dependent on the kinetic energy spread of monoisotopic ions and the angular dispersion of the ion beam, an additional radial electrostatic field can be used to give higher resolutions. Double-sector instruments with high resolving power permit the separation of ions differing by a fraction of a mass unit and allow the mass of each ion to be accurately determined (third or fourth decimal place) and hence their atomic compositions to be computed from exact elemental mass values.

Metastable ions are those ions which decompose after leaving the ion source but before reaching the detector. The products can only be recorded with a magnetic deflection instrument, and usually appear in the normal mass spectrum as low-intensity broad peaks at mass values $m*$ which are related to the parent (m_1) and product (m_2) ions by the equation $m* = \dfrac{m_2^2}{m_1}$. Knowledge of the genesis of ions from metastable transitions can greatly assist in interpretation of a spectrum.

In magnetic double-focusing instruments the accelerating voltage (V), the magnetic sector field (B), and the electric sector field (E) can be scanned separately or in one of several combined modes to detect the metastable transitions occurring in its field free regions (i.e., regions other than V, B, or E). If V is scanned, for example, and fields E and B are fixed, then all precursor ions leading to a selected product can be recorded.

A more recent innovation, which uses linked scanning procedures, is the technique of collision-induced dissociation (CID) (alternatively called collision-activated dissociation), in which secondary ionization is encouraged by passing the ion from the initial ionization process through a gas cell containing helium or other gases. Detection of the secondary ions formed gives direct parent–product ion information and may also provide unique spectra differentiating isomeric molecules that have almost identical primary spectra. An example of this in bile acid analysis is the mass-analyzed ion kinetic energy (MIKE) mass spectra of ions produced by CID from isomeric bile acid conjugates [2]. MIKE spectra are obtained in a reverse geometry instrument (B sector before E sector) [3], and in this case with V constant an ion is selected by the magnetic sector to be collisionally dissociated. The various product ions are then recorded by scanning the electric voltage (E). Sodium glycodeoxycholate and glycochenodeoxycholate give identical fast atom bombardment ionization spectra (Section 3.4), but the structural differences are reflected in the middle mass region of the CID–MIKE spectra of their $[M + Na]^+$ ions (Fig. 3).

CID can be carried out in quadrupole instruments in which three quadrupoles are placed in series. Ions separated in the first quadrupole are passed through

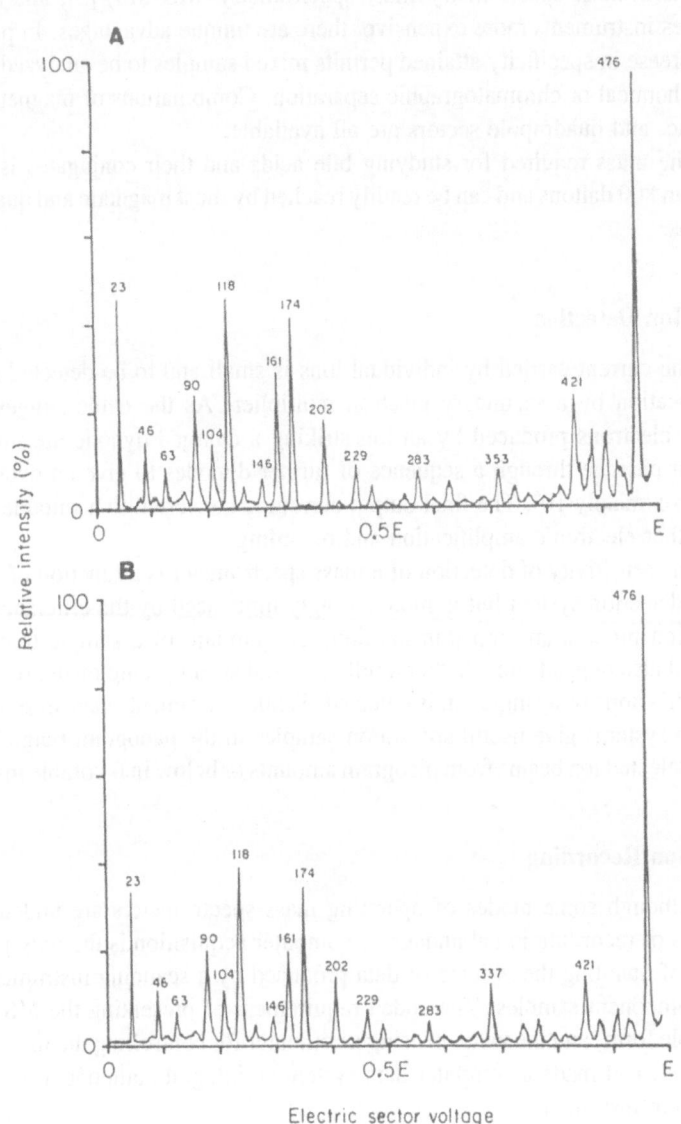

FIGURE 3.   The CID–MIKE spectra of the [M + Na]$^+$ ion (*m/z* 494) from the (A) FAB ionization of sodium glycodeoxycholate and (B) sodium glycochenodeoxycholate acids. (Reproduced from Ref. [2], with permission).

the second, which acts as a gas cell, and the third sector separates the secondary ions produced [4]. The idea of placing analysis sectors in tandem has given rise to the term mass spectrometry–mass spectrometry (MS–MS) [5], and although it makes instruments more expensive, there are unique advantages. In principle, the increase in specificity attained permits mixed samples to be analyzed without prior chemical or chromatographic separation. Combinations of magnetic, electrostatic, and quadrupole sectors are all available.

The mass required for studying bile acids and their conjugates is usually less than 800 daltons and can be readily reached by most magnetic and quadrupole systems.

2.2. Ion Detection

The current carried by individual ions is small and to be detected requires amplification by a secondary electron multiplier. As the name suggests, secondary electrons produced by an ion striking a charged dynode are multiplied by their passage through a sequence of further dynodes to give an overall gain of approximately 10^6. The final output current is converted to a suitable voltage for further electronic amplification and recording.

The sensitivity of detection of a mass spectrometer is a function of the gain of the detection system but is more strongly influenced by the efficiency of the ionization process and ion transmission. The amount of a sample that can be detected also depends on whether a full or partial scan is being made, or whether only a few ions or a single ion is detected. While instruments vary in sensitivity, modern systems give useful spectra on samples in the nanogram range but will detect selected ion beams from picogram amounts or below in favorable instances.

2.3. Ion Recording

Although some modes of operating mass spectrometers are undemanding in terms of recording ion abundances, computer acquisition is the only practical means of handling the volume of data produced by a scanning instrument from multicomponent samples. The added requirement of presenting the MS data in a suitable form, and the value in using a computer for controlling the instrument's operation, had made a computer data system an integral component of present-generation instruments.

Detailed discussion of computer systems can be found in the literature (e.g., Ref. 6), but some aspects are relevant to the recording of ion signals in the two principal modes of MS operation, mass scanning and selected ion monitoring.

2.3.1. Repetitive Scanning Technique

During a mass scan, signals from the detector pass through a filter and are converted to digital form in the interface to the computer. In the scan mode the digitization rate is set to record sufficient samples across each peak to permit accurate determination of its centroid. Individual ion mass signals are reduced to a time and intensity value, and by use of a previously generated reference table of the times of known masses, a mass value can be assigned to each peak. Some magnetic instruments employ the calibrated field strength as a basis for determining mass. Quadrupole systems generate a linear mass scale and as a result are more simply manipulated than magnet machines.

The most usual method of scanning in GC–MS is to make repeated scans every few seconds or even several times a second during the elution of the sample from the GC. This ensures maximum data collection, and hence, for complex mixtures of compounds, mass spectra are available for each component for subsequent interpretation (e.g., Refs. 7–9). Some of the principal aspects of repetitive scanning are apparent from Fig. 4. By summing the ion intensities in

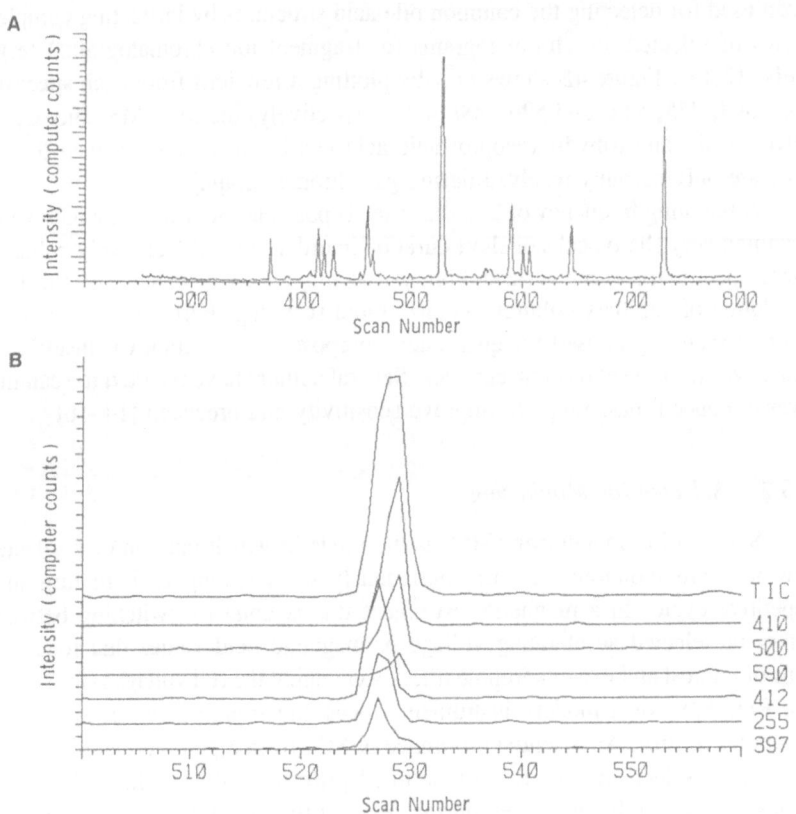

FIGURE 4. (A) The TIC chromatogram of a Me-TMS ether derivatized bile acid extract from a patient's plasma and (B) the selected ion chromatograms of the ions m/z 412, 255, 397, and m/z 590, 500, 410, which indicate the presence of 3α,7α-dihydroxy-5β-cholestan-26-oic and 3α,7α,12α-trihydroxy-5β-cholestan-26-oic, respectively, in the major peak.

each scan, the total ionization chromatogram (TIC) is obtained, and this is closely comparable for bile acid samples with the gas chromatogram using flame ionization detection. The TIC output can be inspected manually, and mass spectra of peaks of interest can be requested. Alternatively, computer programs are available to screen all spectra to detect significant scans in which ions are found to be maximizing and hence represent component compounds [10]. After subtraction of background ions these spectra may be plotted or automatically compared with a reference library file of spectra to identify individual compounds. Retention indices of component peaks can also be assigned automatically by recognition of coinjected hydrocarbon standards [11], and this information can be used in conjunction with the mass spectrum to make an identification.

The detection of known compounds in complex mixtures can be made by looking for fragment ions characteristic of these compounds in the data file acquired from repetitive scans of the sample eluted. Computer programs have been used for detecting the common bile acid structures by inspecting complete series of selected ion chromatograms (or fragment ion chromatograms) (e.g., Refs. 12,13). Figure 4B shows how by plotting a few ions from each spectrum (m/z 397, 255, 412, and 590, 500, 410, respectively) the Me-TMS ether derivatives of di- and trihydroxycoprostanic acid can be distinguished even though they are only partially resolved during gas chromatography.

A scanning frequency of 2–3 scans/sec is possible, but necessary only when scanning very sharp peaks of short duration in order to obtain relatively unbiased mass spectra for computerized library matching. Full advantage of the high resolution of capillary columns requires rapid scanning. Furthermore, when repetitive scanning is used for quantitative purposes (see Section 6), insufficient scans will lead to serious inaccuracies. Several authors have resorted to scanning over a reduced mass range to improve sensitivity and precision [14–16].

2.3.2. Selected Ion Monitoring

Selected ion monitoring (SIM) is the mode in which one ion or a selected few ions are monitored almost continuously by detecting each in turn in a repetitive cycle. In a magnetic instrument this is done by switching between either preselected accelerating voltage or magnetic field values that focus the ions of interest and in a quadrupole filter by changing the rod voltages (for review see Ref. 17). Most modern instruments focus and switch between ions under computer control. An essential difference of SIM from repetitive scanning is the need to preselect the ions to be monitored prior to analysis, although several series of ions can be monitored during selected time periods [18,19] of the GC elution profile.

SIM developed from the need to detect or quantify compounds with much greater sensitivity than was possible by normal scanning. This is achieved from the much longer signal integration times that are the essence of the technique, although the large dynamic range of signals that often results requires a computer interface with the appropriate digitization facilities. Sensitivity is influenced by a variety of factors in addition to normal instrumental sensitivity, such as the number of ions being monitored, interference from background ions, adsorption of analytes in the inlet system, and fragmentation characteristics. For these reasons, the search for volatile derivatives that give a restricted number of intense high-mass ions is often justified (see Section 6).

3. IONIZATION METHODS

Bile acids and their derivatives can be ionized by a variety of methods, and the selection of the most appropriate method depends largely on the type of information and sensitivity required. In the following sections a selection of ionization techniques are described in relation to their value for bile acid analysis.

3.1. Electron Impact

Electron impact ionization (EI) spectra of bile acids give more comprehensive structural information than any other method and hence are more commonly used and recorded in the literature. The ion spectrum is produced by the initial impact of high-energy electrons (20–70 eV) on sample molecules in the gas phase. The resulting loss of an electron from the molecule and the energy transferred leads to fragmentation of this positive radical ion to give a series of ions.

Selection of electron energy should be given consideration. At low potentials (20 eV) less fragmentation is produced, and this may give more diagnostically useful information, as has been reported for Me-TMS ether derivatives of bile acids [12]. As the energy is increased (70 eV), more fragmentation usually, but not necessarily, occurs. The spectra in the present compilation (Section 9) were acquired at 70 eV, and although some information may be less obvious at this voltage, it is important with any library file to compare spectra obtained at a uniform energy. The most suitable electron voltage for producing ions for quantitative measurement should be experimentally determined, as sensitivity and selectivity may have to be optimized. It is a common practice to protect hydroxyl functions by derivatization to improve both stability and volatility of the molecule at the ion source temperature necessary to avoid adsorption.

3.2. Chemical Ionization

Chemical ionization (CI) spectra of bile acids are recorded as derivatives, e.g., methyl esters [20], methyl ester acetates [21–24], Me-TMS ethers [25,26], and only infrequently as free compounds [27].

The principal advantage of CI is the much lower energy involved in the ionization process and the reduced fragmentation that results. The ion source pressure is adjusted to about 1.33×10^{-2} kilopascals with a reagent gas, and the ions formed from the gas by an electron impact process are used to ionize the sample molecules. Choice of the reagent gas gives a range of ionizing conditions and hence control of the information available. Strongly acidic ionic species, such as those produced from methane gas, react by proton transfer to give a protonated molecular ion.

i.e. Ionization of methane by EI

$$CH_4 \rightarrow CH_4^{+\bullet}, CH_3^{+}$$

$$CH_4^{+\bullet} + CH_4 \rightarrow CH_5^{+} + CH_3^{\bullet}$$

$$CH_3^{+} + CH_4 \rightarrow C_2H_5^{+} + H_2$$

Then $\qquad M + CH_5^{+}$ (or $C_2H_5^{+}$) $\rightarrow MH^{+} + CH_4^{+}$ (or C_2H_4)

Also $\qquad M + CH_5^{+} \rightarrow (M + CH_5)^{+}$

The extent of fragmentation is greater the more exothermic the proton transfer reaction, while the abundance of molecular ion species (quasimolecular ion) increases with decreasing exothermicity.

Methane, for example, gives spectra of the methyl esters of lithocholate, deoxycholate, chenodeoxycholate, and cholate that favor dehydration to production of the quasimolecular ions [20]. Similarly, Me-acetate derivatives lose acetic acid [21], and Me-TMS ether derivatives lose trimethylsilanol [25] under isobutane CI conditions. However, the reduced exothermicity of N_2/NO reagent gas gives a molecular ion for the chenodeoxycholate Me-TMS ether derivative [25,26] (Fig. 5). The second reagent gas, in this case 6% NO, is chosen such that the ions produced from it by the major reactant gas will be of lower reactivity and hence create less fragmentation on reaction with the sample. This process is known as charge exchange.

The value of reducing fragmentation, in addition to giving molecular weight information, is the increase in sensitivity that often accrues from the ionization being limited to fewer ions. This is the case for bile acids Me-acetates [23] and for cholic acid [27] when ammonia is used as reagent gas. DeMark and Klein [23] reported the quasimolecular ions of 28 bile acids to be the base peaks with little or no fragmentation and only slight loss of acetic acid. In these cases attachment ions were the principal quasimolecular ions.

There are also reagent gases [28] that extract a proton from the molecule to give negative quasimolecular ions $[M - H]^{-}$.

Although molecular weight and sensitivity advantages favor CI, it is a less useful ionization method for distinguishing isomeric structures than EI unless used in conjunction with additional chromatographic data. Techniques of EI and CI give complementary information and should be used accordingly.

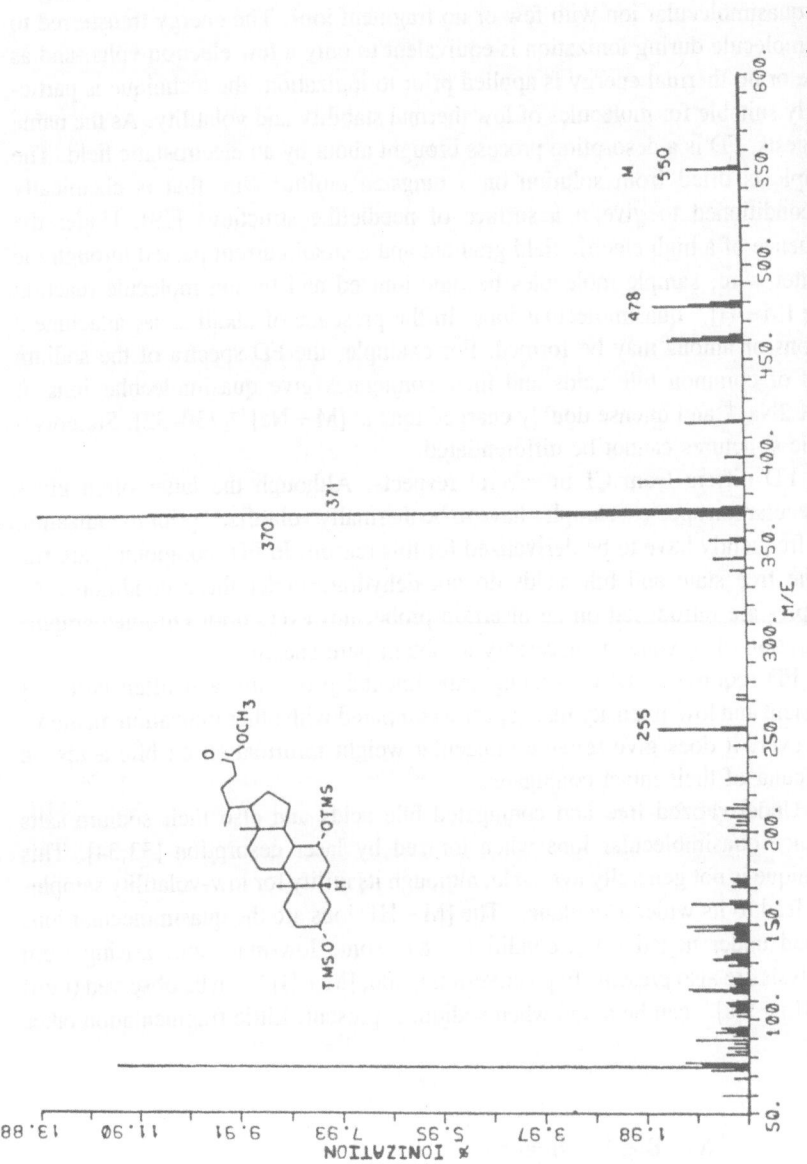

FIGURE 5. The charge exchange mass spectrum of chenodeoxycholate (Me-TMS ether) with N₂/NO reagent gas mixture, showing an abundant molecular ion. (Reproduced from Ref. [261], with permission).

3.3. Field and Laser Desorption

Field desorption ionization (FD) generally gives mass spectra consisting of the quasimolecular ion with few or no fragment ions. The energy transferred to the molecule during ionization is equivalent to only a few electron volts, and as little or no thermal energy is applied prior to ionization, the technique is particularly suitable for molecules of low thermal stability and volatility. As the name suggests, FD is a desorption process brought about by an electrostatic field. The sample is dried from solution on a tungsten emitter wire that is chemically preconditioned to give it a surface of needlelike structures [29]. Under the influence of a high electric field gradient and a small current passed through the emitter wire, sample molecules become ionized and by ion molecule reaction give $[M+H]^+$ quasimolecular ions. In the presence of alkali salts, attachment cations or anions may be formed. For example, the FD spectra of the sodium salts of common bile acids and their conjugates give quasimolecular ions of $[M+2Na]^+$ and intense doubly charged ions at $[M+Na]^{2+}$ [30–32]. Stereoisomeric structures cannot be differentiated.

FD differs from CI in several respects. Although the latter often gives molecular ion species, samples have to be thermally volatilized prior to ionization and frequently have to be derivatized for this reason. In FD, compounds are run in the free state and bile acids do not dehydrate under these conditions. As samples are introduced on an insertion probe, however, prior chromatographic separation of mixtures is necessary to obtain pure spectra.

FD requires a more exacting experimental procedure and often produces transient and low-intensity mass spectra compared with other ionization methods. However, it does give reliable molecular weight information on bile acids, in particular of their intact conjugates.

Underivatized free and conjugated bile acids and also their sodium salts produce quasimolecular ions when ionized by laser desorption [33,34]. This technique is not generally available, although its utility for low-volatility samples may lead to its wider acceptance. The $[M-H]^-$ ions are the quasimolecular ions formed under negative-ion conditions, and some low-mass ions arising from pyrolysis are also present. In positive-ion mode, $[M+H]^+$ can be observed (rare) and $[M+Na]^+$ can be noted when sodium is present. Little fragmentation takes place.

3.4. Fast Atom Bombardment

One of the most important developments in MS in the last few years has been the introduction of fast atom bombardment (FAB) ionization. Its importance stems from the ability to ionize many of the polar, charged, biologically relevant molecules of molecular weight up to 20,000 daltons and perhaps beyond. Despite the use of a high-energy ionizing atom beam, quasimolecular ions are almost

TABLE I. Intensities of Major Ions from Representative Mono-, Di-, and Trihydroxylated Free Bile Acids Analyzed by FAB-MS[a]

Bile acid	M_r	+ve FAB		−ve FAB	
		$[M+H]^+$	$[M-nH_2O+H]^+$	$[M-H]^-$	$[M-nH_2O-H]^-$
Lithocholic	376	377 (2)	359 (100)	375 (100)	357 (3)
Chenodeoxycholic	392	393 (8)	375 (9)	391 (100)	373 (5)
			357 (100)		
Cholic	408	409 (7)	391 (8)	407 (100)	389 (6)
			373 (44)		
			355 (100)		

[a] Relative intensity values in parentheses.

always produced, although sufficient energy is generally transferred to the ionized molecules to produce fragmentation.

FAB ionization was a development of secondary ion mass spectrometry (SIMS), in which secondary ions are sputtered from a surface layer of sample molecules when struck by ions of high translational energy [35]. By dissolving the sample in a liquid matrix (usually glycerol) and using a beam of neutral atoms, it was found that may organic molecules, which were destroyed under SIMS conditions, ionize to give stable ion beams of long duration (many minutes) [36]. Subsequently it was appreciated that either a neutral or a charged atom source was equally suitable as the ionizing beam, and the technique is more correctly known as liquid SIMS. The liquid matrix provides a surface layer of sample molecules that can be continuously replenished from within the matrix.

Both positive and negative ions, such as $[M+H]^+$ and $[M-H]^-$, can be produced by FAB together with cationized species, e.g., $[M+Na]^+$, and also attachment ions formed with the matrix, e.g., $[MH+glycerol]^+$. Depending on the type of compound, the proportions of positive to negative ions produced may differ, and by recording both spectra, complementary information may be obtained.

Unconjugated bile acids are readily ionized by FAB, and their mass spectral characteristics have been summarized [37,38]. Table I lists the intensities of the major ions from representative mono-, di-, and trihydroxylated free bile acids. As can be seen, positive ion spectra have intense ions owing to loss of water, whereas negative ion spectra do not. However, when unconjugated bile acids are run as their sodium salts, then intense quasimolecular ions $[M+H]^+$ and $[M+Na]^+$ are dominant, in contrast to the negative ions produced, which are almost all in the form of $[M-Na]^-$ (i.e., same mass value as the $[M-H]^-$ ions of free compounds). Glycerol attachment ions m/z 469, 485, and 501 are seen in the positive spectra (467, 483, and 499 in the negative spectra) of the mono-, di-, and trihydroxy-

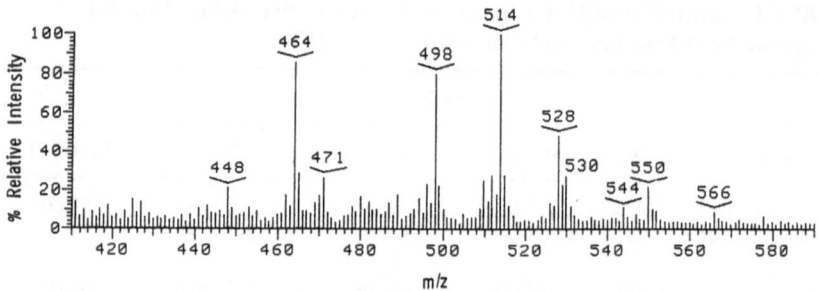

FIGURE 6. The negative-ion FAB mass spectrum of a crude urine extract from a cholestatic patient. The ions marked indicate the presence of [M-1] ions from the following conjugates: *m/z* 448, glycodihydroxycholanoate; *m/z* 464, glycotrihydroxycholanoate; *m/z* 471, dihydroxycholanoate sulfate; *m/z* 498, taurodihydroxycholanoate; *m/z* 514, taurotrihydroxycholanoate; *m/z* 528, glyco-dihydroxycholanoate sulfate, and its Na adduct ion *m/z* 530, taurotetrahydroxycholanoate; *m/z* 544, glycotrihydroxycholanoate sulfate and its Na adduct ion *m/z* 566.

lated compounds, respectively. The intact molecules also associate as cluster ions, and isomers at *m/z* values of $[2M+1]^+$ and $[2M-1]^-$ are evident and show increasing intensity with increasing oxygen content (i.e., trihydroxy > dihydroxy > monohydroxy).

In bile acid analysis FAB has had the greatest impact for analyzing conjugated bile acids. It is a practically simpler and much more sensitive (10–1000 times) [39] technique than FD and will ionize all bile acid conjugates, including sulfates, that are not amenable to other ionization methods.

FAB mass spectra of conjugated bile acids have been studied by Whitney *et al.* [32,37–41] and others [42–47]. The positive ion spectra of the sodium salts of taurine- and glycine-conjugated bile acids contain both $[M+H]^+$ and $[M+Na]^+$, with $[M+Na]^+$ more intense than the $[M+H]^+$ in taurine-conjugated spectra. This is also true for sulfates, which have an ion containing further sodium $[M+2Na]^+$ (e.g., Refs. 43,45–47).

The presence of sodium also affects the molecular ion region of negative FAB spectra of conjugated bile acids, giving sodium- and non-sodium-containing quasimolecular ions, although the latter are always more intense. Conversion of the salt to its protium form or conversion to a triethylammonium salt [45,46] simplifies the ionization to almost single ion spectra of $[M-H]^-$ and makes them more suitable for the simultaneous analysis of mixtures of conjugated bile acids. Figure 6, for example, shows the negative quasimolecular ions of free bile acid conjugates from a simple single-step extract of urine from a patient.

Differentiation of isomeric bile acid structures from their FAB spectra is not possible, and additional chromatographic data have to be used. The technique of MS–MS, described briefly in Section 2, may prove a solution in some cases, and CID spectra have already been used to differentiate the $[M-H]^-$ ion of 3β-hydroxy-chol-5-enoic acid from the $[M-H-2]^-$ ion of lithocholic acid [48] and of bile salts [2].

4. METHODS OF SAMPLE INTRODUCTION

Introduction of the sample into the ion source is dependent on the chemical nature of the sample compound and the ionization method required to give the necessary analytical data [47]. The specialized inlet probes for FD [29], FAB [36], and LD [33] are described elsewhere, but as the vast majority of bile acid spectra in the literature have been recorded under either EI or CI conditions, the inlet systems suitable for their use will be mentioned here.

4.1. Direct Insertion Probe

This sample inlet method was initially designed to obtain spectra from single compounds that were relatively involatile or were thermally unstable if volatilized at normal pressure. The sample is placed in a small heatable crucible at the tip of a probe, which is passed through a vacuum lock and the tip positioned close to the ionization region. The temperature of the crucible is raised until the sample is volatilized. Rapid heating may help to minimize thermal dissociation of unstable molecules.

Temperatures for volatilization need to be determined experimentally, but free bile acids require temperatures up to 120°C, whereas their glycine conjugates volatilize up to 180°C and the taurine conjugates between 225° and 250°C [31]. Methyl ester and Me-acetate derivatives of bile acids as well as the free compounds can be introduced by direct insertion probe [49,50].

The obvious disadvantage of the direct probe is the need to separate mixed samples prior to analysis if the difficulty of interpreting mixed spectra is to be avoided.

A variation of the direct insertion probe for use with CI is one having a wire tip that is coated with a solution of the sample of interest [51,52]. When this is placed in close proximity to the ionizing plasma and heated rapidly, the sample molecules are desorbed and ionized almost immediately before suffering collisions with a surface. The intention is to reduce thermal decomposition effects that can occur with the direct insertion probe, and although it has not been applied to bile acid samples, it may have future application.

4.2. Gas Chromatographic Inlet

Following the introduction of GC–MS in the early 1960s and some early work on bile acid derivatives (e.g., Refs. 53–55), the GC inlet has contributed most to the elucidation of bile acid structures in biological fluids by providing the means for separating the complex mixtures of acids encountered in such samples. As a stand-alone system GC has found wide utility for their identification and quantification, and the conditions for these analyses have been incorporated in the conditions for GC–MS (for reviews see Refs. 1,47,56,57).

The preparation of volatile derivatives of bile acids has been the subject of many studies, and references to these have been summarized recently [47]. The most commonly applied derivatives are the acetates, trifluoroacetates, and trimethylsilyl ethers of the methyl cholanoates, each having advantages, although latterly the Me-TMS ethers have been the most widely adopted for qualitative purposes. In quantitative studies, where sensitivity and selectivity are the priority, the ammonia CI spectra of Me-acetates [21,22] and the EI spectra of methyl ester alkyldimethylsilyl ethers [58–60] may be a more appropriate choice.

Introduction of capillary columns (15–50 m length, 0.25–0.5 mm internal diameter) and subsequent development to their present level have eliminated many of the problems attendant on the use of packed columns and high gas flow rates. The need for a carrier gas separator as interface between GC and MS is now avoided with the column passing without interruption directly into the ion source. This has removed the potential absorption and thermal effects of the carrier gas separator, and the ion source pumping system can cope easily with the 0.5–2 ml min^{-1} flow of carrier gas. Chemically bonded liquid phases in capillary columns have much greater thermal stability and can be cleaned by solvents. This has virtually eliminated earlier difficulties with columns whose phase would bleed into the ion source to produce a high background of ions. Mass spectrometers have had to increase sensitivity and scan speed to take full advantage of the higher separation efficiencies of capillary columns.

4.3. Liquid Chromatographic Inlet

High-performance liquid chromatography (LC) is now widely used for the separation of many compound types without resorting to chemical pretreatment, and this has greatly encouraged efforts to produce a satisfactory combination LC–MS instrument (see Ref. 61). There are numerous potential advantages to be gained by coupling these two analytical systems.

LC–MS is a reality, but several limitations to the present systems remain. There are two principal methods in use. In one the effluent from the LC column is spotted on a circulating belt that passes through evacuated chambers to remove solvent before passing into the ion source. The solute is flash-evaporated and analyzed by either EI or CI. Laser desorption and FAB can also be used with a belt system inlet [62]. The maximum volume of effluent from the column that can be loaded onto the belt is fixed at about 100 μl min^{-1}, and hence as only a portion of the eluant is applied, this limits the sensitivity. Microbore and microcapillary columns may provide an answer to this problem but are restricted in sample loading. Mechanical transfer works well for many compounds, but the system has some disadvantages, including uneven spreading on the belt, which gives fluctuations in ion beam, and inability to cope with thermally labile compounds, nonvolatile buffers, and other involatile compounds. Spray deposition of the sample on the belt has helped to reduce this problem.

The second system is known as direct liquid introduction in which the LC effluent passes via a capillary and desolvation region into the ion source (e.g., Ref. 63). Chemical ionization is induced in a plasma of solvent-derived ions when the appropriate solvents are employed in the LC separation or added postcolumn. A development of direct liquid introduction is thermospray ionization in which electrolytic eluants are used. The liquid stream from the column is passed through a heated metal capillary [64,65] and, in the resulting vaporized jet, primary ions are produced from the rapidly desolvating charged droplets by surface fields. These fields encourage the formation of ions from the liquid phase. However, optimum temperature conditions of the vaporizer have to be set for different compounds to avoid pyrolysis. Ammonium acetate is the most widely used electrolyte because of its good volatility and its gas phase NH_4^+ proton donor and $CH_3CO_2^-$ proton acceptor ions.

Despite the advances in LC–MS, combination LC and MS are more often employed in discontinuous mode in which chromatographic fractions are individually subjected to mass spectrometry [66,67]. The application of LC on its own is used for bile acid analysis, e.g., glycine and taurine conjugates [68], 5α- and 5β-cholanoates [66], conjugated bile acids [69–71], and sulfates [72,73]. The sensitivity achieved by ultraviolet detection is a limitation for analyzing low concentrations of bile acids. Both column reaction and other detection methods, once fully developed, may improve the situation, but these are unlikely to match the sensitivity and specificity of MS analysis. CI with ammonia as reagent gas and a mechanical belt interface was used for LC–MS operation for separation of lithocholic, cholic, chenodeoxycholic, taurocholic, and glycocholic acids [74]. Most gave the expected spectra, but taurocholic acid eliminated water and showed no quasimolecular ion. Problems of this nature may be avoided by using FAB ionization or the thermospray principle.

5. MASS SPECTROMETRIC FRAGMENTATION OF BILE ACIDS

Much of the important early work on the MS analysis of bile acids involved the methylcholanoates, and aspects of their spectra have been comprehensively reviewed [1,56]. However, the need to protect bile acids from thermal decomposition during analysis and to employ GC separation to study biological samples has concentrated effort on the mass spectra of their derivative forms. The ensuing discussion will deal only with the fragmentation observed for derivatized bile acids in general, and with the Me-TMS ether derivatives in particular. If LC–MS is more widely used in the future, there may be renewed interest in underivatized bile acid spectra.

The Me-TMS ether derivatives give more informative spectra than other derivatives, and their ease of preparation and volatility makes them a good choice for identification studies [1,47,56,57]. Nevertheless, there is considerable experience of the MS fragmentation and GC behavior of acetates and trifluoroacetates of methylcholanoates (e.g., Refs. 1,21,75–79) which supports the con-

TABLE II. Molecular Weights of the Main Groups of Hydroxylated Bile Acids

Structure	Molecular weight	
	Me ester	Me-TMS ether
Methyl cholanoate	374	—
Methyl monohydroxycholanoate	400	472
Methyl dihydroxycholanoate	416	550
Methyl trihydroxycholanoate	432	638
Methyl tetrahydroxycholanoate	448	726

tinuing use of these derivatives. The general literature would be simplified if a single derivative type could be adopted.

The Me-TMS ether EI spectra of bile acids are arranged into groups according to their functionality. The initial step in evaluating a bile acid spectrum is to determine into which group the structure falls by establishing its molecular weight from the general fragmentation pattern. The molecular ion is often small or absent, and other ions have to be used for this purpose. Table II lists the molecular weights of the main groups of hydroxylated bile acids. An additional 14 daltons have to be added for each oxo group present and 2 daltons subtracted for each unsaturated bond. Cholestanoic acids have an eight-carbon side chain and are 42 daltons greater in mass than the corresponding cholanoic analogs. Bile alcohols (C_{27}) have molecular weights two mass units less than bile acids (C_{24}) with equivalent functionality and should not be confused with unsaturated bile acids.

Among the general fragmentations of hydroxylated bile acids are loss of an angular methyl group or loss of a methyl of the trimethysilanol group and the single or multiple rearrangement loss of trimethysilanol (TMSOH). The configuration and position of the TMS ether groups determine whether they are readily lost and influence the intensity of the product ions. The electron energy may also affect the abundance of these ions and should be borne in mind. Cleavage loss of the complete side chain (i.e., -115 daltons) and, depending on substitution, fission across, A, B, C, or D rings are also general fragmentations and occur in combination with the elimination losses of TMSOH and methyl groups.

The composition and origin of many ions in the spectra of methylcholanoates have been postulated from the combined study of low-resolution spectra of a large number of bile acids and the theory of mass spectral fragmentation. The structures of many fragment ions have been confirmed by data from high-resolution measurements (e.g., Ref. 80). Extrapolation of this information to the spectra of the Me-TMS ether derivatives and reference to the wider MS literature of fragmentation processes of steroids [81–83] permit confident interpretation of these spectra despite the genesis of a number of ions remaining uncertain. Most difficult is identification of bile acid spectra for which no reference standards are available, and although chromatographic retention data and the spectra may suggest a particular structure, final proof can only be obtained by comparison with the synthesized compound.

The accepted carbon atom numbering system and ring notation for the bile acid molecule is shown in Fig. 7, and in the following general discussion of the fragmentation of bile acid derivatives, the reader is directed to Section 9, which gives a comprehensive compendium of EI spectra obtained under uniform mass spectrometric conditions.

5.1. Monohydroxy Bile Acids

The varying stability and influence of the TMS ethers in the 3-, 7-, and 12-monohydroxy methylcholanoates are reflected in the intensity of the molecular ions (m/z 462), M-90 (m/z 372), M-90-side chain (m/z 257), and m/z 215 (resulting from D-ring cleavage and charge retention on the ABC ring fragment). The elimination of TMSOH is comparable with water loss in free compounds; this has been summarized [1,56,57]. Low-intensity molecular ions are observed in the Me-TMS ether spectra of the 3α-, 7α-, and 12α-ols of the 5β series and the 7α-, 7β-, and 12α-ols of the 5α series, whereas the 3β- and 7β-ols of the 5β series and the 3α-ol of the 5α series are more intense.

M/z 215 and 372 are intense ions in the spectrum of lithocholate (Me-TMS ether), but m/z 257 is of low abundance. In comparison, m/z 257 is the base peak in 12α-hydroxy-5β-cholanoate, reflecting the rapid elimination of TMSOH followed by the side chain. The 7α-hydroxy analog exhibits m/z 372 as its base peak and m/z 215 and 257 of much lower intensity. The alternative ion to m/z 215 from D-ring fission is the ion m/z 230 (arising by fission of bonds C-13,17 and C-15,16) and consists of C-16, C-17, and the side chain. It is produced from 3- and 7-hydroxylated bile acids but is suppressed in the 12-hydroxy analog.

There are several other diagnostically useful ions in the Me-TMS ether spectra of monohydroxylated bile acids. 12-Hydroxylated bile acids have an ion at m/z 208 which arises by cleavage across the C ring and contains C-12, the D ring, and side chain after loss of TMSOH [1]. A 7-hydroxy group initiates cleavage of the B ring in a similar way to give m/z 249 in which the A and B

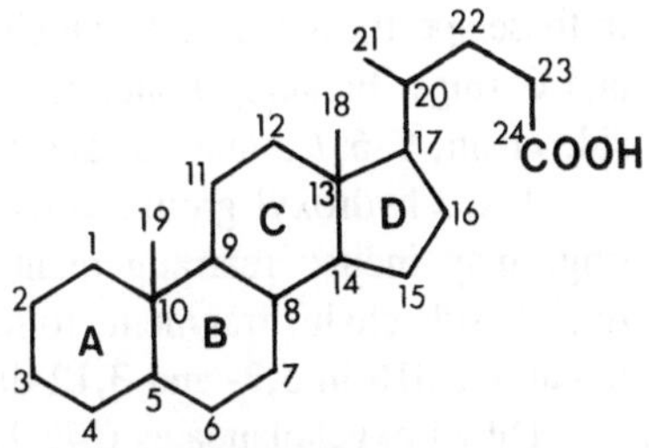

FIGURE 7. The basic bile acid nucleus (cholanoic acid) with systematic numbering of the carbon atoms.

rings have been eliminated. Cleavage of the C-6,7 alpha bond with the C-9,10 bond and loss of TMSOH yield m/z 262.

5.2. Dihydroxy Bile Acids

The molecular ions of most dihydroxy bile acids (Me-TMS ethers) are small or missing, but (M-15) is usually present. Chenodeoxycholic and its $3\beta,7\alpha$ analog also lack the (M-15) ion, but the molecular weight is readily confirmed by the presence of m/z 460, 370, and 255, the latter being equivalent to m/z 257 in monohydroxy bile acids in which both TMS ether and the side chain are lost.

The most commonly encountered dihydroxy bile acids, 3,6-, 3,7-, and 3,12-diols and their epimers, can all be distinguished directly from their Me-TMS ether spectra, although in some cases (e.g., $3\alpha,12\alpha$, and $3\beta,12\alpha$-diols of the 5α series) additional GC retention data are required. Single or double elimination of TMSOH followed by facile cleavage of the side chain is observed in 3,12-dihydroxycholanoates to give m/z 255 as the base peak. The C-12 OTMS is lost before the C-3 OTMS, and as the latter is less readily eliminated in the 5α series, the ratio of m/z 255 to 345 is greater in isomers of the 5β configuration. In the 5β series m/z 370 is more intense than m/z 345. The 3,6- and 3,7-dihydroxy-5β-cholanoates do not readily eliminate the side chain from (M-90), and hence m/z 345 is of very low intensity in comparison to m/z 370 although the $3\beta,12\beta$ analog is an exception. The cleavage occurs more readily from m/z 370 to give m/z 255 of moderate intensity [1].

Other features of the spectra of 3,6- and 3,7-diols clearly differentiate them from 3,12-diols. The B-ring substituents initiate cleavage of the B ring, giving m/z 249 (CD rings + side chain) and m/z 262 (C-7 + CD rings + side chain) (cf. 7-monohydroxy bile acid) [1,57,80,84]. This ion is at m/z 263 in the 3,6-diols. In contrast, 3,12-diols contain a prominent m/z 208 ion, the same ion as in 12-monohydroxy-5β-cholanoate, arising by C-ring cleavage. Rupture across the A and B rings in 3,7-dihydroxycholanoates produces an ion at m/z 243 that contains C-3 to C-7 and their -OTMS groups [1,57]. An ion of low intensity at this mass is usually present in 3,6-diols.

A diagnostic ion that confirms 3,6-dihydroxy substitution involves loss of C-1 to C-4 of the A ring together with the C-3 OTMS group. This ion has a mass of 405 and is more intense for 6β than 6α isomers [85]. Also prominent in these spectra is m/z 323, which is thought to arise by cleavage across both B and C rings, breaking bonds C-5,6, C-8,9, and C-12,13 to retain the C ring, side chain, C-6,7,8 and 18, and the C-6 OTMS group [85].

The 3-hydroxyl group, when eliminated to produce unsaturation in the A ring, may induce rearrangement loss of C-1 to C-4 [1,57,81]. This (BCD ring + side chain) fragment, following TMSOH loss, produces a low-intensity ion at m/z 316 in 3,7- and 3,12-dihydroxycholanoate Me-TMS ether spectra.

Dihydroxycholanoates (Me-TMS ether) give the analogous fragmentations of the D ring to monohydroxy bile acids, in which the side chain with C-15 to C-17 or C-16 and C-17 are lost to produce m/z 228 and 213, respectively, after TMSOH elimination.

5.3. Trihydroxy Bile Acids

An ion representing the molecular weight of Me-TMS ether derivatives of trihydroxy bile acids (638 daltons) is almost always absent. However, many have an (M-15) (m/z 623), although this may also be missing or be of low intensity, particularly in the 5β series of 3,6,7-triols. Nevertheless, the ion sequence of TMSOH eliminations, m/z 548, 458, and 368, coupled with the m/z 253 ion (due to subsequent loss of the side chain) will confirm a trihydroxylated C-24 bile acid. The possibility of unsaturation in the molecule should be remembered when the M and (M-15) ions are absent, as the (M-90) ion of a triol has the same molecular weight as an unsaturated diol (548 daltons).

Many of the trihydroxycholanoates can be differentiated by their spectra, although with the increased number of stereoisomeric possibilities for substituents this becomes more difficult. As TMSOH is eliminated from 3,7,12-triols in the sequence, 12-OTMS before 7-OTMS before 3-OTMS, differences in the stereochemistry of these centers and of C-5 have an effect on ion intensity [1,57,82–85]. The more facile loss of a 7α- than a 7β-OTMS in the 5β series is reflected in the greater intensity of the alternative side-chain loss [i.e., (M-90-115)] and reduced abundance of m/z 253 in the 5α when compared to the 5β series is indicative of the reduced elimination of the 3α-OTMS. The influence of the configuration at C-5 is also evident in a comparison of Me-TMS ether spectra of cholate and *allo*-cholate. The 5α-ring junction encourages B-ring cleavage to give m/z 261 (CD rings + side chain + C-7) following the loss of two TMSOH groups [1,56].

The C-ring cleavage fragment m/z 208, already mentioned for mono- and dihydroxy spectra with a 12-hydroxy group [1,86,87], is of low intensity in both cholate and *allo*-cholate, possibly owing to the competitive process of elimination of the C-7 OTMS group. However, it is a prominent ion in the spectrum of the 3α,7β,12α-trihydroxy-5β-cholanoate isomer [1].

The configurations of the hydroxyl groups in 3,6,7-trihydroxycholanoates have dramatic influence on fragmentation of their Me-TMS ether derivatives. Most striking is the B-ring fission (C-9,10 and C-6,7 bonds broken), which produces an (A ring + C-6) ion at m/z 285 and, by TMSOH loss, m/z 195 which are very intense peaks in the spectra of 7β-isomers of both the 5α and 5β series [1,56,57,88]. These ions are of much lower significance in the spectra of 7α-epimers, which have either the m/z 458 (M-90-90) or m/z 369 (M-90-90-89) as their base peaks. The vicinal arrangement of the 6- and 7-hydroxyl groups accounts for the 89 loss (-OTMS) instead of 90 as the third sequential elimination of the OTMS ether group [1,56,57]. Vicinal hydroxyl groups with a *cis* configuration can be identified by preparing the acetonide, a derivative that gives useful mass spectra and GC retention characteristics [89–91]. Alternatively a cyclic boronate ester can be formed with both *cis-* and *trans-*diols [92].

Some mention should be made of the spectra of less commonly observed trihydroxylated bile acids. A 1-hydroxylated bile acid has been reported from several sources [13,93–95], and the spectrum of the 1β,3α,12α-triol of the 5β series has been interpreted [96]. The salient distinguishing feature is a base peak at m/z 217 consisting of the C-1 to C-3 atoms and their OTMS ether groups. Most other expected ions are present, but of low intensity, in the spectrum of 1β,3α,12α-trihydroxy-5β-cholanoate (Me-TMS ether). The ions m/z 142 and 143 may be due to C-1 to C-4 with a 1-OTMS group and are apparent in the spectrum of a 1,3,7-triol [13,95]. The m/z 182 and 195 ions are also present in the latter spectrum and are likely to arise from B-ring cleavages.

Fragmentation of the A ring in which C-1 to C-4 are lost with charge retention on the [BCD ring (after TMSOH loss) + side chain] fragment gives direct information on the number of hydroxyl groups in the A ring. In the case of 1,3-substitution in a trihydroxy bile acid, the ion is at m/z 316 and not m/z 314, as it would be if two of the three OTMS ethers were present somewhere in the B, C, and D rings.

3,6,12-Trihydroxy bile acids give Me-TMS ether spectra not unlike the 3,7,12 analogs [97] but usually contain a significant m/z 208 ion and also an m/z 261 ion.

The spectrum of 3α,12α,15β-trihydroxy-5β-cholanoate (Me-TMS ether) has been published [96]. Fragmentation induced by the 15-OTMS substituent gives m/z 243 due to the side chain + C-15 to C-17 and C-15 OTMS group and is not to be confused with the m/z 243 produced in 3,7-diols. An ion at m/z 393 due to loss of 245 mass units from the molecular weight confirms this C-ring substitution.

5.4. Tetrahydroxy Bile Acids

Several tetrahydroxylated bile acids have been found in biological samples, and some have been identified from the mass spectra of their Me-TMS ether derivatives [13,91,93,98,99]. Molecular ions are very small or absent, but the [M-15] ion at m/z 711 in most cases confirms the presence of four hydroxyl groups for which m/z 251 (loss of side chain and all OTMS ethers) shows they are located in the nuclear rings. Loss of the side chain can occur before all the OTMS ether groups are eliminated to give a series of ions m/z 521, 431, 341 in order of decreasing intensity. If one of the hydroxyls is in the side chain (e.g., 3α,7α,12α,23-tetrahydroxy-5β-cholanoic) [13], then the m/z 251 is shifted to m/z 253 and the side-chain cleavage m/z 456 → 253 indicates the additional OTMS (i.e., loss of 203 and not 115).

Four sequential losses of TMSOH from the molecular ion give rise to m/z 636, 546, 456, and 366 of varying intensity ratios which reflect the position and configuration of the OTMS substituents in tetrahydroxylated bile acids. In the case of vicinal TMS ethers, losses of 89 (-OTMS) rather than 90 (TMSOH) daltons can occur. In the 70-eV spectrum of $3\alpha,6\alpha,7\alpha,12\alpha$-tetrahydroxy-5β-cholanoate (Me-TMS ether), for example, 89- and 90-dalton losses occur from m/z 546 to give m/z 457 and 456 and the ion 367 almost completely replaces m/z 366.

As described earlier for trihydroxy bile acids, the presence of 1-OTMS group in addition to the 3-OTMS gives rise to an intense m/z 217 ion. This is the base peak in the spectrum of $1\beta,3\alpha,7\alpha,12\alpha$-tetrahydroxy-5β-cholanoate (Me-TMS ether) [13]. Also of note in the latter bile acids are the A-ring ions, probably formed by cleavage of bonds C-9,10 with either C-5,6 or C-6,7 at m/z 182 and 195, respectively. The ion m/z 314 is of diagnostic value to establish the disubstituted A ring as its mass indicates only two OTMS groups are present in the B, C, D rings [1,13].

The same deduction can be made from the spectrum of the $2\beta,3\alpha,7\alpha,12\alpha$-tetrol using m/z 314 in combination with the ions m/z 182 and 195 [91–99]. Of interest in this spectrum is a substantial fragment ion m/z 243 which is present but of lower intensity in 3,7-diols. This is likely to be the same ion, but of increased abundance owing to activation of the C-2,3 bond.

Although only a few of the stereoisomeric forms of 3,6,7,12-tetrols are available [13], the configuration of the 7-OTMS group appears to make a substantial difference to fragmentation, as was discussed for the 3,6,7-triols. The base peak in the spectrum of the $3\alpha,6\alpha,7\beta,12\alpha$-isomer at m/z 285 and large m/z 195 signals the presence of the 7β configuration.

A 2,3,6,7-tetrol has been suggested to be a normal constituent in the urine of human newborns [93]. This structure is proposed on the basis of its Me-TMS ether mass spectrum, which has a series of ions m/z 507, 417, and 327 that may indicate loss of a C-3 fragment from the A ring (with two OTMS groups) and then two respective losses of TMSOH. Other potentially diagnostic ions at m/z 271 and 181 may be A-ring ions resulting from cleavage of C-9,10 and C-5,6 bonds, while m/z 314 confirms a dihydroxylated A ring. 4-Hydroxylated bile acids have not yet been found in nature and, although less likely than 2-hydroxylation, cannot be excluded as a possible explanation of these fragmentations.

It is likely that a number of other tetrahydroxylated and even pentahydroxylated bile acids will be identified in the future [98,99], and although the absolute configuration of each hydroxy substituent may not be substantiated, it may be possible to determine their positions in the bile acid molecule from the ions discussed earlier.

5.5.   Oxo Bile Acids

The MS analysis of methyl oxo-cholanoates has been reviewed in the past [1,56,57] and again recently [47]. The understanding of MS fragmentation mechanisms and the influence of different derivatization methods has been important for identifying unknown hydroxy bile acids following their oxidation. In this section only the mass spectral characteristics will be considered of some of the oxohydroxy bile acids that occur naturally in physiological fluids and of some related compounds for comparison.

Specific derivatization of the oxo group to an oxime, most frequently the methyl oxime, can be helpful for identification purposes [85,96,97,100,101]. In contrast to the EI mass spectra of oxo bile acids, where the oxo group generally directs the major fragmentation [1], the OTMS groups in the Me-TMS ether derivatives of hydroxylated oxo acids dominate the initiation of bond cleavage. However, the individual mass values of fragments containing the oxo function obviously differ from the pure hydroxylated analogs and are readily recognized.

The position of hydroxylation of hydroxy-3-oxo-5β-cholanoates can be determined from differences in the intensity of ions in the Me-TMS ether spectra and from characteristic cleavages. The molecular ions (m/z 476) are present in 6α-hydroxy-3-oxo and 7α-hydroxy-3-oxo compounds but are absent in the 12α-hydroxy-3-oxo isomer. All have (M-15) ions at m/z 461, (M-90) ions at m/z 386, and (M-90-side chain) ions at m/z 271. Their intensities differ for each isomer, the base peak being m/z 386 for the 7α-OTMS bile acid but m/z 271 in the other two spectra. The 6- and 12-ols are readily differentiated, as the 12-OTMS group gives rise to m/z 208 as described previously, while the 6α-hydroxy-3-oxo has a prominent ion at m/z 323, which is considered to be an ABC-ring fragment arising from cleavage of the C-13,17 and C-14,15 bonds and loss of hydrogen.

The 3-oxo group can be eliminated under electron impact as water, and this occurs from several ions, including m/z 371 (M-15-90) and m/z 271, giving rise to m/z 353 and m/z 253, respectively. Rearrangement loss of the A ring and 3-oxo group (-70 daltons) also takes place from several ions (i.e., from m/z 271 → 201 for all isomers and from m/z 386 → 316, most noticeably in the 7α-hydroxy-3-oxo where m/z 386 is a base peak) [1,56,81,102].

Cleavage through the D ring to lose the side chain and C-15 to C-17 is common in 3,6- and 3,7-hydroxylated bile acids producing m/z 213, and this shifts to m/z 229 in the 3-oxo-6α-OTMS and 3-oxo-7α-OTMS spectra [1].

The spectrum of 3α-hydroxy-6-oxo-5β-cholanoate (Me-TMS ether) is distinctive from other monohydroxy spectra having m/z 461 ([M-15]) as the base peak. Most of the remaining ions are of low intensity (i.e., m/z 386, 271, and 253), reflecting a reduction in the elimination of the 3-OTMS group. The D-ring cleavage ion m/z 229 and analog m/z 319 are present.

The 7-oxo group can be lost more readily as water than from the other oxo positions and both [M-18] and [M-90-18] are prominent ions. The 7-oxo group promotes fragmentation of the B ring (cleavage of bonds C-5,6 and C-9,10), giving rise to m/z 292 and possibly m/z 177 when the side chain is also eliminated.

A number of 12-oxo bile acids have been observed in nature and can be recognized from several characteristic fragmentations, in particular, the D-ring cleavage ion M-155 (m/z 321) in monohydroxy-monooxo bile acid spectra [1,56,57]. Loss of TMSOH from this ion produces the base peak in the 3α-hydroxy-12-oxo compounds of both the 5α and 5β series. The intensity of ions m/z 476, 386, 371, 361, 321, 271, and 253 allows these spectra to be differentiated.

The mass spectra of dihydroxy-monooxo bile acids (Me-TMS ether) may be of interest for future identification purposes, and some have been included in the compilation of spectra at the end of the chapter.

5.6. Unsaturated Bile Acids

As was pointed out earlier, some care should be taken when interpreting as unsaturation in the molecule a 2-dalton difference of the molecular ion (see Table II), because this may be due to a fragment ion loss of TMSOH from a higher mass. Several workers have studied the spectra of unsaturated bile acids (Refs. 103,104) and they have been reviewed [47,57].

A 2,3-double bond can be distinguished from a 3,4-double bond as the latter does not facilitate the retro-Diels-Alder rearrangement of the A ring to give [M-54] [81,82,103]. Similarly, a 4,5-double bond will not undergo this cleavage but can be detected in 3β,7α-dihydroxy- and 3β,7α,12α-trihydroxy-substituted bile acids by ions m/z 196 and 209. These ions involve B-ring cleavage by fission of bonds C-9,10 with C-6,7 or C-7,8, respectively, and transfer of a proton following elimination of the 7-OTMS group [105]. The unsaturation in these two bile acids strongly influences the spectra, both having very small [M] and [M-15] ions and a very intense [M-90] ion (ions m/z 458 and 546, respectively) owing to the stability of the resulting conjugated diene formed by the elimination process.

3β-Hydroxy-5-cholenoic acid is one of the more common unsaturated bile acids in certain biological samples, and its Me-TMS ether spectrum has been reported (e.g., Ref. 106). Its characteristic ions at m/z 404 [M-56], m/z 331 [M-129], and m/z 129 are all induced by the 5,6-double bond. The 129-dalton fragment consists of C-1 to C-3 and 3-OTMS, while the 56-dalton loss involves the same carbon atoms but lacking the C-3 OTMS [107,108]. The m/z 249 ion seen in the 6-and 7-hydroxylated bile acids from B-ring cleavage (see earlier) is also present, as it is in the 6,7-double bond compound (e.g., Refs. 1,103).

The presence of unsaturation at C-5,6 in 7-hydroxylated bile acids leads to instability under acid conditions and ready dehydration. In the mass spectrum of 3β,7α-dihydroxy- and 3β,7α,12α-trihydroxy-5-cholenoate (Me-TMS ether) the [M-90] ions are of high intensity, and little further fragmentation takes place. However, ions m/z 129 and 209 provide an indication of the position of unsaturation [105].

A bile acid with a 7,8-double bond has been found in fish [109], but spectra of Me-TMS ether derivatives have not been reported. Me-acetate-trifluoroacetate mass spectra indicate that the B-ring fragmentation which gives m/z 249 (or m/z 247 for 12-hydroxylated compounds) does not occur and thus provides a means of distinguishing C-6,7 from C-7,8 double bonds [1,88,105].

As might be expected, the presence of unsaturation of the C-11,12 bond induces similar fragmentation to a 12-OTMS group, which is mainly the predominant loss of the side chain [1,105,110].

5.7. Modified Side-Chain Bile Acids

Compounds with modifications of the normal five-carbon side chain can be distinguished by mass spectrometry. These modifications include, in addition to hydroxylation or unsaturation, a shorter chain in C_{20}, C_{21}, and C_{22} steroidal acids and in norcholic acid (C_{23}) or a longer side chain as in C_{25} and C_{27}, due to defective metabolism of the cholesterol side chain. The mass spectral characterization of many of these modified side-chain bile acids has been referenced and summarized [47].

The length of the side chain can usually be determined from the mass difference between ions that eliminate the side chain and their product ions. In the case of the Me-TMS ether spectrum of $3\alpha,7\alpha,12\alpha$-trihydroxy-5β-norcholan-23-oate, the molecular ion (m/z 634) sustains three losses of TMSOH (i.e., m/z 534, 444, and 354) [13,111]. Subsequent cleavage of the side chain from m/z 354 yields m/z 253, indicating the side chain (101 daltons) to be 14 daltons (i.e., CH_2) less than the normal five-carbon side chain. Similarly, $3\alpha,7\alpha,12\alpha$-trihydroxy-5β-homocholan-25-oate (Me-TMS ether) has a molecular ion (652 daltons) giving the series m/z 562, 472, and 282 and the consequent deduction of a 129-dalton side chain (similarly for $3\alpha,7\alpha$-dihydroxy-5β-homocholanoate Me-TMS ether [112]).

A high proportion of the reported bile acids with side chains differing from the pentanoic side chain of cholic acid involve acids isolated from animals [113], and their MS behavior provides useful fragmentation information which will aid identification of analogous compounds from human sources should they arise [114–126]. For the purposes of the present chapter, however, only a selection of bile acids with unusual side chains which are found in humans will be considered. The most commonly encountered acids of this type are the C-27 cholestanoic acids and their hydroxylated derivatives.

5.7.1. Bile Acids with Longer Side Chain

The unsubstituted side chain of $3\alpha,7\alpha$-dihydroxy-5β-cholestanoic (Me-TMS ether) and its $3\alpha,7\alpha,12\alpha$-triol isomer are evident from the 157-dalton loss from the respective ABCD-ring fragment ions m/z 412 and 410, respectively. Many of the features of these spectra are similar to those of their C-24 analogs. MS data on other derivatives and on free compounds and their isomers have been more extensively reported (e.g., Refs. 116,127–131).

Varanic acid, $3\alpha,7\alpha,12\alpha,24$-tetrahydroxy-$5\beta$-cholestanoic, an intermediate in the biosynthesis of C-24 bile acids, has been found in humans [116] in addition to its earlier recognition in certain animals [121,124,130,132]. Comparison of the mass spectrum of its Me-TMS ether derivative and of its stereoisomer indicate some variation in intensity with the MS instrumentation used [121,124]. The [M-15] ion confirms the molecular weight at 768, and four sequential losses at 90 daltons indicate the four OTMS groups. The presence of one OTMS being in the side chain is evident from m/z 253 ([ABCD ring − side chain − 3 × TMSOH]), as well as from the 245-dalton loss of the side chain from m/z 498 or, alternatively, of 155 daltons from m/z 408 when the side chain has already lost TMSOH.

There are also ions arising from fission of the side chain itself. The elimination of the 24-OTMS group to give a 24,25-double bond activates the allylic C-20,22 bond and may be responsible for an ion at m/z 281, an analog of the ABCD-ring fragment m/z 253 containing the C-20 and C-21 atoms. The expected side-chain cleavage alpha to the 24-OTMS group was commented on by Une et al. [121] although not by others [124]. Fragmentation of C-24,25 with subsequent losses of TMSOH is responsible for m/z 321 (also m/z 411).

Hydroxylation at C-25 is recognized by a loss of the carboxymethyl group as 59 daltons from the molecular weight to give, in the spectrum of $3\alpha,7\alpha,12\alpha,25$-tetrahydroxy-$5\beta$-cholestanoate (Me-TMS ether), m/z 709 and then a series of ions due to TMSOH loss, m/z 619, 529, 439, and 349 [98,99]. These serve to differentiate the 25- and 26-tetrols, the latter showing no useful ions to locate the position of OTMS group in the side chain. The spectra have a number of ions in common, including m/z 281, 226 and three related peaks m/z 211, 301, and 391 of unknown origin but which may arise from a three-carbon loss from a A or D ring, cleavage of the side chain, and TMSOH elimination.

5.7.2. Shortened Side Chain

Acidic steroids of C_{20}, C_{21}, and C_{22} carbon atoms have been detected in human meconium [133–135] and some of these also in serum [136]. The mass spectra of the methyl ester, Me-acetate, and Me-TMS ether derivatives of these short-chain "bile acids" have been used for identification and the Me-TMS ether derivatives for SIM detection [136].

The 3-hydroxylated saturated etianic (i.e., androstane-17β-carboxylic) acids observed give Me-TMS ether spectra with molecular ions and the confirmatory [M-15] (m/z 391), [M-90] (m/z 316), and [M-90-15] (m/z 301) ions whose intensities vary with the stereochemistry of the 3-hydroxyl group and the ring junction. The carboxylic acid side chain, however, does not readily leave as in the C_{24} bile acids to produce the ABCD ring fragment m/z 257, but is lost with C-15 to C-17 and TMSOH to yield m/z 215.

As expected, 3-hydroxy-bis-norcholan-22-oic (C_{22}) Me-TMS ether derivatives undergo comparable fragmentation to their C_{20} analogs, although low-intensity ions at m/z 255 and 257 may indicate side chain elimination.

Unsaturation of the C-5,6 bond in 3-hydroxylated etianic acids leads to the Me-TMS ether spectra [136] of very similar appearance to that of 3β-hydroxy-chol-5-enoate [106]. A-ring cleavage ions [M-129], m/z 275 and [M-56], m/z 348 are evident, and m/z 129 is the base peak in the spectra. The corresponding C_{21} and C_{22} C-5,6 unsaturated steroidal acids give equally characteristic mass spectra [134,136].

5.8. Conjugated Bile Acids

Analyses of bile acid conjugates by MS require different strategies from their unconjugated analogs. It may be sufficient to hydrolyze the conjugates resulting from conjugate group separation and determine the free bile acids by GC-MS following derivatization [13,137]. This depends on reliable group separation and hydrolysis methods and is not straightforward for mixed conjugates. Alternatively, conjugated bile acids can be isolated and submitted to one of the appropriate procedures listed in Table III.

Soft ionization methods such as FD and FAB are particularly suited to the involatile and thermally labile nature of polar bile acid conjugates (see Sections 3.3 and 3.4) but at present are limited to providing molecular weight information and do not indicate the position of conjugate substitution and the location of hydroxyl groups or their stereochemistry. However, FAB does ionize sulfated bile acids and sulfated mixed conjugates, which has not been possible by other ionization methods.

Attempts to form volatile derivatives of intact conjugates for GC-MS are unsuccessful for taurine conjugates but can be achieved for glycine [146,147] and glucuronide conjugates [97,142]. The glucuronyl moiety plays a dominant role in fragmentation of bile acids conjugated with glucuronic acid. In the Me-TMS ether derivatives, for example, the base peak at m/z 217, the intense peaks at m/z 204 and 317, and the ion at m/z 407 all indicate the sugar moiety. Although there is no evidence for the site of conjugation, this can be deduced from the oxo bile acids produced by a series of reactions in which the conjugate is first oxidized by periodate to a formate ester followed by oxidation of the hydroxyl groups and final hydrolysis of the formate [97]. Elimination of the glucuronyl group with its linking oxygen gives rise to m/z 459, which then loses TMSOH to m/z 369 in trihydroxylated bile acid glucuronides. The presence of ions m/z 371 and 373 indicates di- and monohydroxy analogs respectively. The glucuronide of 3β-hydroxy-chol-5-enoate (Me-TMS ether) also produces a m/z 371 ion. CI of the Me-acetate derivative was successfully used to identify the glucuronide of chenodeoxycholic acid [139,140].

Elliott has reviewed the mass spectra of conjugated bile acids [57] and paid particular attention to the glycine and taurine conjugates [31]. Glycine conju-

TABLE III.   MS Analysis Methods for Conjugated Bile Acids

Conjugate form	MS method	Derivative	Reference
Glycine	Direct probe/EI	Free	31
		Me ester	31,66,138
		Me-acetate	139,140
		Me-trifluoroacetate	141
	FD–MS	Free	31,32
	FAB–MS	Free	31,32,37–39,41–48
Taurine	Direct probe/EI	Free	31
		Trifluoroacetate	141
	FD–MS	Free	31,32
	FAB–MS	Free	31,32,37–42,44–48
Glucuronide	Direct probe/CI	Me-acetate	139,140
	GC–MS/EI	Me-TMS ether	99,142
Sulfate	FAB–MS	Free	2,40,43,45–48
Miscellaneous			
-Arginine, ornithine, and histidine	Direct probe/EI	Free	138,143
		Me ester	
		Me-acetate	
-N-lysine	Direct probe/EI	Me-acetate	144,145
-Leucine, alanine proline, and γ-butyric acid	Direct probe/EI	Me-acetate	44
	FAB–MS	Free	44

gates, unlike their taurine analogs, can be methylated to give derivatives with similar mass spectra to the free compounds. Molecular ions are apparent in the bile acids studied [31], as are ions arising by loss of water. The side chain is prone to rearrangement with elimination of 89, 131, or 172 daltons (in methyl esters) in combination with water loss. The fragment of 131 daltons $[CH_2=C(OH)NHCH_2CO_2CH_3]$ also retains a charge to give an intense ion m/z 131 in most spectra, usually the base peak. Methyl glycocholate and glyco-*allo*-cholate are exceptions to this where m/z 271, $[M–H_2O–H_2O–172]$, is the base peak. These compounds, however, can be differentiated from their underivatized spectra by the ratio of m/z 271 to m/z 253, which is greater for the *allo* isomer. Following loss of the complete side chain, fragmentation of the nucleus is the same as for unconjugated free bile acids [1,56].

Taurine-conjugated bile acid spectra show loss of water, methyl groups, side chain, and rings A and D in agreement with the fragmentation of free bile acids. Rearrangement also occurs with losses of 108 $[CH_2=CHSO_3H]$ and 167 $[CH_2=C(OH)NHCH_2CH_2SO_3H]$ daltons. The origin of most ions present in the spectrum of taurocholic acid has been proposed and may be extrapolated to other taurine conjugates [31].

6. QUANTITATIVE ANALYSIS OF BILE ACIDS

The GC–MS procedures used in quantification, namely, of repetitive scanning and SIM, are discussed in Section 2.3 and later, and many of the applications to bile acids are included in Sections 7 and 8. The experimental details of SIM methods for determining bile acids in biological materials have been summarized (e.g., Refs. 47,147). Despite the apparent advantages of the repetitive scanning techniques and its extensive use by Sjövall and co-workers [13,95,148–151], it has been investigated by only a few others for quantification of bile acids (e.g., Refs. 152,153). The principal strength of acquiring full-mass scans continuously during sample elution from the GC is the completeness of the resulting data file. There need be no prior selection of compounds to be measured in the sample, and any component present can be evaluated by assessing the intensity of its representative ions from the file at the correct retention index. The full spectrum can be used to confirm identity and the existence of interference from coeluting components.

If necessary, the complete range of known bile acids in the biological sample can be automatically determined from a repetitive scan run [13,95,150,151]. Series of ions can be used to screen for the common C_{24} bile acid structures in GC–MS repetitive scan data of Me-TMS ether derivatives [13,47], although only a limited number are necessary for quantitative calculations. The original literature should be consulted for detailed information on ion intensity calculations, background subtraction, component purity checks, and calibration [148,154]. The accuracy and precision associated with quantification of complex bile acid mixtures is dependent on detecting and excluding interference, in addition to factors such as the initial sample treatment and extraction methods, concentration levels, and calibration procedures. Coefficients of variation for standard mixtures of bile acids (5.3–8.6%) and for bile acids from urine (6.1–14.0%) indicate the influence of some of these factors [154].

The recognized limitation of repetitive scanning for quantitative analysis is its inadequate sensitivity for low concentrations of bile acids. As it is dependent on instrumental sensitivity and the GC and MS characteristics of the derivative being used, it is unlikely that concentrations below the low nanogram level can be reached. This is a restriction for measuring normal levels of bile acids in serum and tissues but only in part responsible for the limited use made of repetitive scanning in quantitative studies. Frequently, only a few bile acids need to be quantified, and then SIM is more convenient and less demanding in terms of data-handling facilities than repetitive scanning. The merits of both techniques should be assessed for a particular application [148,155].

If high sensitivity is important, SIM is advantageous, but as a very limited number of ions are usually monitored, care must be taken to identify interference from other sources, particularly from the spectra of bile acids that contain the same ions. Sensitivity can be maximized by selecting derivatives of bile acids that have intense high-mass ions in their spectra. Methyl ester alkyldimethylsilyl ether derivatives, for example, give intense ions at $[M\text{-alkyl}]^+$ [156–158] (see Fig. 8) that have great potential for SIM measurement as long as adequate GC resolution is available to separate isomeric structures. Similarly, CI may be used to increase sensitivity. DeMark and Klein [23] demonstrated a 130–270% increase in the relative sensitivities of measuring the Me-acetate derivatives of lithocholate, deoxycholate, chenodeoxycholate, and cholate using the $[M + NH_4^+]$ ions from the ammonia CI spectra compared to isobutane CI spectra.

7. BILE ACIDS LABELED WITH STABLE ISOTOPES

The mass spectra in Fig. 9 demonstrate the shift in mass value of ions containing isotopically labeled atoms. The shift in mass indicates the number of heavy atoms present, and ion intensity provides abundance data. In addition, information on the location of the label can be obtained from the fragmentation. For example, in the spectrum of deuterated ursodeoxycholate (Me-TMS ether) shown in Fig. 9, m/z 257 indicates the deuterium atoms must be in the nucleus and not in the side chain, m/z 215 and 230 prove they are not in C-15, 16, 17, m/z 243 excludes the substitution on C-3,4,5,6, or 7 and B-ring cleavage ions further pinpoint the location of the label to the C ring. Data of this type are much more difficult to obtain by conventional radioactive isotope techniques with which there is the added radiation hazard for *in vivo* human studies. As a result, stable labeled bile acids have considerable potential value for kinetic studies on metabolism, pool size, turnover rate, absorption, and excretion (e.g., Refs. 159–166), and a number of these investigations are discussed in Section 8. Furthermore, the *in vivo* metabolic conversion of administered compounds that have been isotopically labeled can be followed. The incorporation of [13]C- and [2]H-labeled ethanol and ^{18}O-labeled oxygen into bile acids in rats has been investigated [16,167–173], giving data on pools and turnover of cholesterol and bile acids.

The most common use of labeled bile acids has been their application as standards for quantitative assay (e.g., Refs. 174–184). By direct addition of the stable isotope to serum (or other biological fluid) compensation is made for any losses occurring during the extraction, isolation, purification, and derivatization steps, and if used in excess, the standard may act as a carrier to limit adsorption losses in the GC–MS instrument. In the selection of an internal standard consideration should be given to (1) the chemical stability of the stable isotopic atoms incorporated in the molecule, (2) the degree of isotopic enrichment and the number of atoms incorporated, which should be high to minimize interference from naturally occurring isotopes, and vice versa, and (3) isotope effects, which are greater for ^{2}H atoms than for ^{13}C or [18]O atoms.

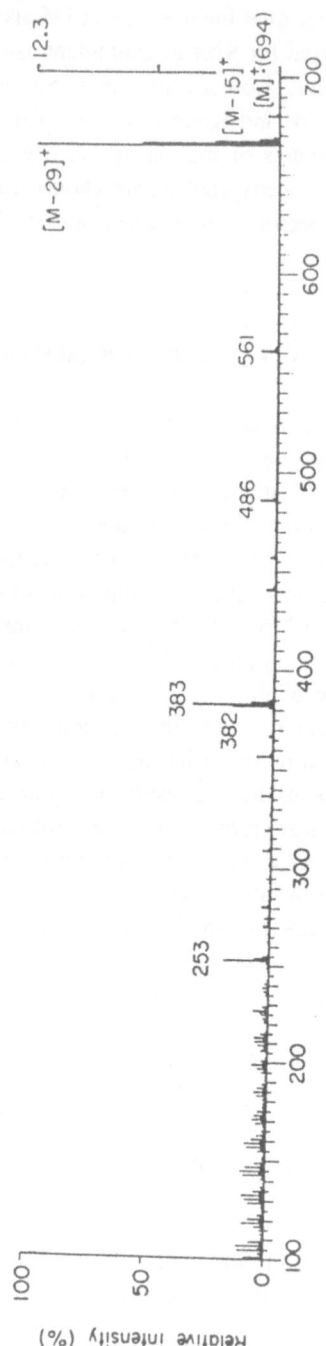

FIGURE 8. The EI (70 ev) mass spectrum of the dimethylethylsilyl ether derivative of ethyl cholate. (Reproduced from Ref. [58], with permission).

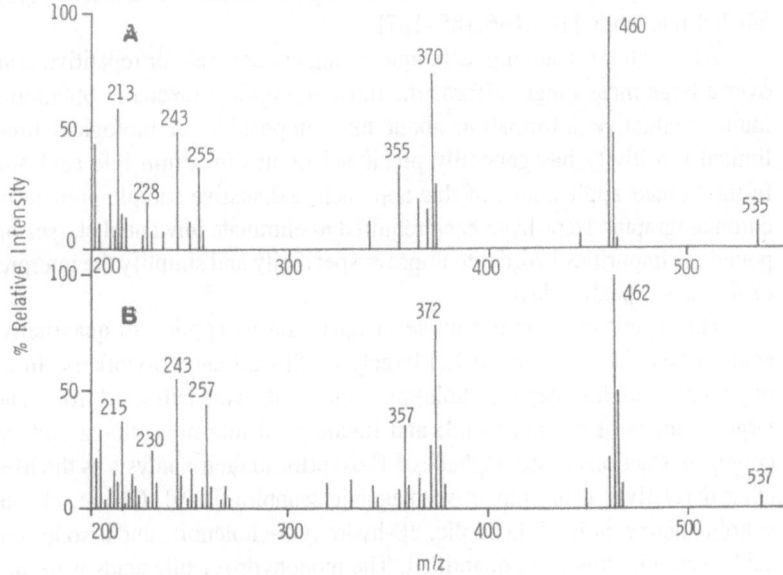

FIGURE 9. The mass spectra of Me-TMS ether derivatives of (A) ursodeoxycholic acid (3α,7β-dihydroxy-5β-cholan-24-oic) and (B) 11,12-[²H₂]ursodeoxycholic acid. (Reproduced from Ref. [158], with permission).

8. APPLICATION OF MASS SPECTROMETRY TO THE DETERMINATION OF BILE ACIDS IN BIOLOGICAL FLUIDS

In this section the applications of MS to the qualitative and quantitative analysis of biological fluids are reviewed. The literature in this area is vast, and for this reason only those applications relating to MS studies of bile acid synthesis and metabolism in humans in health and disease will be discussed.

8.1. Serum

Relatively low concentrations of bile acids are found in the peripheral circulation of healthy subjects, and consequently MS analysis of serum bile acids has largely been restricted to specific measurements of the principal individual primary and secondary bile acids, utilizing the high sensitivity and specificity of stable isotope dilution and/or selected ion monitoring (e.g., Ref. 147).

When repetitive scanning techniques have been employed, this has generally been for characterization of individual compounds in a GC profile. Intact bile acid conjugates can be analyzed directly by FAB ionization MS, but the limited sensitivity of the technique has restricted its application to biliary bile acids [37–41] and serum from cholestatic subjects [45]. One of the most recent and clinically useful developments in the application of MS is determination of bile acid kinetics from the analysis of serum samples following administration of stable isotopically labeled bile acids [165,166,185–187].

Although the scanning technique, using either single or repetitive scanning, over a large mass range, affords the most appropriate means of obtaining maximum qualitative information about the composition of biological fluids, its limited sensitivity has generally precluded its use in serum bile acid analysis. In the limited applications of this approach, exhaustive sample preparative and chromatographic steps have been required to eliminate unwanted classes of compounds or impurities in order to improve specificity and simplify the interpretation of the mass spectral data.

The repetitive scanning mode of operation as applied to quantitative bile acid analysis has been pioneered largely by Sjövall and co-workers. In a study of patients with intrahepatic cholestasis, bile acids were extracted from relatively large volumes of plasma (5 ml) and fractionated into nonsulfated and sulfated conjugate fractions using Sephadex LH-20 prior to their analysis as the Me-TMS ether derivatives using repetitive magnetic scanning [149]. Cholic, chenodeoxycholic, deoxycholic, lithocholic, 3β-hydroxy-5-cholenoic, and ursodeoxycholic acids were identified and quantified. The monohydroxy bile acids were predominantly found as sulfate conjugates in the serum, an observation later corroborated when selected ion monitoring was employed for determination of serum bile acids in patients with cholestasis, with and without pruritic liver disease [188].

In a later study of patients with cholestatic liver disease [151], bile acids were first separated into conjugate classes using Lipidex-DEAP and measured by repetitive magnetic scanning of the Me-TMS ether derivatives. Although only a small number of samples were analyzed, serum bile acid concentrations ranged from 14 to 252 μmole/liter. Cholic acid comprised 44–89% of the total bile acids identified and exceeded the concentration of chenodeoxycholic acid in all patients with primary biliary cirrhosis. Several monohydroxy bile acids were identified, of which 3β-hydroxy-5-cholenoic acid accounted for approximately 13% of the total serum bile acids. Lesser amounts of trihydroxy and tetrahydroxy bile acids, several tentatively identified as 6-hydroxy isomers, were present. Of the two healthy subjects studied, chenodeoxycholic acid was the predominant serum bile acid, and unconjugated bile acids accounted for 30% and 53%, respectively, of the total serum bile acids. This is higher than the values reported in an early GC study by Sandberg et al. [189] but, in accord with the observations of Makino et al. [190] and with the recent GC–MS studies by Setchell et al. [191,192], highlights the importance of using group fractionation techniques in detailed metabolic studies.

The definitive identification of the 6-hydroxy bile acid, hyocholic acid (3α,6α,7α-trihydroxy-5β-cholanoic acid), in the serum from patients with cholestasis was described following the extraction, hydrolysis, and fractionation of bile acids into groups based on polarity, from a 300-ml pool of plasma [193]. In this study considerable interference by neutral steroids in the GC analysis of bile acids points to the drawback of isolation procedures using partition chromatography, which do not distinguish between bile acids and sterols of similar polarity. Ion exchange chromatography using the lipophilic gel Lipidex-DEAP affords a complete separation of neutral sterols from bile acids and in these circumstances simplifies identification of trace levels of bile acids. In a later investigation of serum bile acids in patients with cholestatic liver disease using selected ion monitoring, hyocholic acid was found to be present almost exclusively as a taurine conjugate [188]. In a study of two siblings with familial progressive intrahepatic cholestasis [194], hyocholic acid was reported to be present in trace amounts in the serum; however, the mass spectrum of the Me-TMS ether derivative of hyocholic published by these authors is not consistent with the reference compound of this bile acid derivative (see Section 9 for comparison) but is identical to the partially silylated methyl ester derivative of cholic acid, as is the retention time on SE-30 liquid phase. When samples contain excessively large concentrations of cholic acid, as was the case in this study, some on-column degradation of the fully silylated derivative can occur. Under these conditions the partial derivative of cholic acid (i.e., methyl ester-*bis*-trimethylsilyl ether) is retained on the column longer than the methyl ester-*tris*-trimethysilyl ether. The molecular ion and corresponding fragment ions in the mass spectrum are thus 72 daltons lower than for the fully silylated derivative.

In most instances and because of limited sensitivity, the scanning mode of operation when applied to serum bile acid analysis has been used in conjunction with selected ion-monitoring methods, generally to confirm the identity of individual peaks in SIM chromatograms. In a study of the serum bile acids in premature and newborn infants using capillary column GC–MS, quantification of selected primary bile acids was achieved by SIM, while the scanning mode was used to identify chenodeoxycholic and cholic acids [195]. These studies established a marked increase in the concentrations of serum bile acids in the first week immediately after birth in both premature and term infants. Concentrations of chenodeoxycholate exceeded those of cholate in all cases, but to a greater extent in preterm infants. Similar observations have been made using radioimmunoassay procedures, which has led to the description of a normal physiological cholestasis in early life [196]. In this GC–MS study, bile acids were analyzed as their Me-TMS ether derivatives, and in addition to monitoring general ions for the principal conventional bile acids, Barnes *et al.* [195] also monitored m/z 331 to detect 3β-hydroxy-5-cholenoic acid, m/z 410 typical of C_{27}-trihydroxy, and m/z 412 for C_{27}-dihydroxy bile acids. Trace amounts of 3β-hydroxy-5-cholenoic acid were detected in only one preterm infant, despite its occurrence in relatively large quantities in meconium [95], and traces of a C_{27} bile acid, suggested to be a dihydroxy bile acid, were found but at levels too low to enable a full mass spectrum to be obtained. Hyocholic acid was also tentatively reported, but these authors failed to obtain evidence for the presence of C_{20} or C_{21} bile acids, which have recently been identified in meconium [133–135] and serum from cholestatic patients [136].

MS has been invaluable in the characterization of unusual bile acids and in elucidating abnormal pathways of hepatic bile acid synthesis. In the serum of three children with cholestasis (one of whom probably had Zellweger syndrome), high levels of coprostanic acids, and in particular an unusual C_{29}-dicarboxylic bile acid, were identified in addition to elevated concentrations of cholic and chenodeoxycholic acids [197].

This unusual bile acid, with a very long GC retention index relative to the C_{24} bile acid derivatives, accounted for 27.6–41.3% of the total serum bile acids, while only traces of this compound were found in bile and feces. The authors failed to find this bile acid in over 100 serum samples from children with cholestasis in the absence of coprostanic acidemia. Subsequent careful examination by MS and nuclear magnetic resonance spectroscopy confirmed the structure of this C_{29}-dicarboxylic acid as $3\alpha,7\alpha,12\alpha$-trihydroxy-27a,27b-dihomo-5β-cholestan-26,27b-dioic acid [198,199].

The occurrence of this bile acid with an elongated side chain raised the question of its origins. It has been suggested that β-sitosterol may act as a substrate, although there is no evidence to support this, while very recently there has been some indication that trihydroxycoprostanic acid might be the substrate [200]. In the same patients, small amounts of 3α-hydroxy-5β-cholestan-26-oic, 3β-hydroxy-5-cholesten-26-oic, $3\alpha,7\alpha$-dihydroxy-5β-cholestan-26-oic (DHCA), and $3\alpha,7\alpha,12\alpha$-trihydroxy-5β-cholestan-26-oic (THCA) acids were also identified [197]. THCA and DHCA are intermediates in the conversion of cholesterol to cholic and chenodeoxycholic acids, respectively, and were previously found in substantial amounts in the serum, bile, and urine of two siblings with cholestasis [116,131,201] and the urine of patients with Zellweger syndrome. Zellweger syndrome, or cerebrohepatorenal syndrome, is an autosomal recessive disorder characterized by a virtual absence of peroxisomes in the liver and kidney. The metabolic block in bile acid synthesis is localized to the final side-chain cleavage requiring peroxisomal β-oxidation activity and appears not to involve mitochondrial 26-hydroxylation [200]. In a study of 25 patients clinically diagnosed as having Zellweger syndrome, GC–MS was used to identify and measure the proportions of DHCA, THCA, and the C_{29}-dicarboxylic bile acid in the serum, bile, and urine [202]. Only in 10 patients were these bile acid intermediates detected, and in all instances the increased proportion of the C_{29}-dicarboxylic acid was accompanied by elevated amounts of the coprostanic acids. A detailed longitudinal study of a single patient with Zellweger syndrome using GC–MS was performed by Kase et al. [203]. Serum bile acids were measured on four separate occasions from week 4 to week 16 of life, when the patient died. Selected ion monitoring of the ABCD ring ions m/z 253 and m/z 255 was used to quantify the amounts of the Me-TMS ethers of trihydroxy- and dihydroxycholanoic and -cholestanoic acids and the C_{29}-dicarboxylic bile acid using [$^2H_5$]cholic and [$^2H_5$]chenodeoxycholic acids as internal standards. No evidence for deoxycholic was obtained, and the concentrations of THCA (range 2.1–11.5 μmole/liter) exceeded that of any of the conventional bile acids while the C_{29}-dicarboxylic acid was present at levels ranging from 3.2 to 5.5 μmole/liter.

Small amounts of $3\alpha,7\alpha,24$-trihydroxy-5β-cholestanoic were found. Confirmation of individual bile acids was not possible from full scans because of the small volumes of serum analyzed, but was made on the basis of multiple selected ion current recordings for each compound. By this criterion, evidence was obtained for the presence of 24-hydroxy-THCA (varanic acid) and small concentrations of a tetrahydroxy-cholestanoic acid, suggested to be a 3,6,7,12-tetrahydroxy-cholestanoic isomer. This study highlights the important role of MS in the diagnosis of this condition. Tetrahydroxylated cholestanoic acids detected by capillary GC–MS, in patients with Zellweger syndrome and infantile Refsum's disease, included the C_{24}-, C_{25}-, and C_{26}-hydroxylated analogs of THCA [303]. A 2ξ, 6ξ- dihydroxy-DHCA was tentatively identified.

Refsum's disease is a condition exhibiting closely similar biochemical characteristics to Zellweger syndrome [204]. In three patients with Refsum's disease, THCA was identified from its methyl ester-trifluoroacetate derivative and found to comprise 21–31% of the total serum bile acids, which exceeded 6 μmole/ liter [205].

Following the finding of a series of short-chain bile acids (etianic and etienic) in meconium [133–135], repetitive scanning GC–MS of the Me-TMS ether derivatives of the monohydroxy bile acid fraction from the serum of cholestatic patients confirmed their presence in serum [136]. These short-chain bile acids were present in amounts varying between 0 and 10% of the total monohydroxy bile acids, and reconstructed mass chromatograms of ions specific for C_{20} and C_{22} bile acid derivatives were reputed to reveal the presence of 15 major and at least 50 minor components, the majority of which were unidentified. Short-chain bile acids that were reported to have identical mass spectra to authentic compounds included etienic (3β-5-ene), etianic (3β,5α), and the bisnor bile acids, 3β-hydroxybisnorchol-5-enoic and 3β-hydroxy-5α-bisnorcholanoic. No attempt, however, was made to quantify the amounts of these bile acids in serum. Although it is possible that the initial exhaustive borohydride reduction of the serum extract might have generated these compounds from 3-oxo analogs, this would not explain the presence of the 3β-hydroxy-5-ene isomers of C_{20} and C_{22} bile acids. Particularly interesting is the fact that the monohydroxyetianic acids exhibit marked choleretic effects [206], but the relevance of these observations awaits further information on the origins of these compounds and their physiological role in humans.

In a unique study, GC–MS was employed for the qualitative analysis of bile acids (as the Me-acetate derivatives) from the serum and bile of patients with myotonic muscular dystrophy and 22 controls [207]. A number of unidentified bile acids were found in the sera from these patients, and a significant increase in deoxycholic acid (2.1 μmole/liter compared with 0.7 μmole/liter for the control group) was reported, while a marked decrease in biliary ursodeoxycholic was found. This led the authors to speculate on the implications of these findings in relation to the effects of bile acids on cellular membrane fluidity [207].

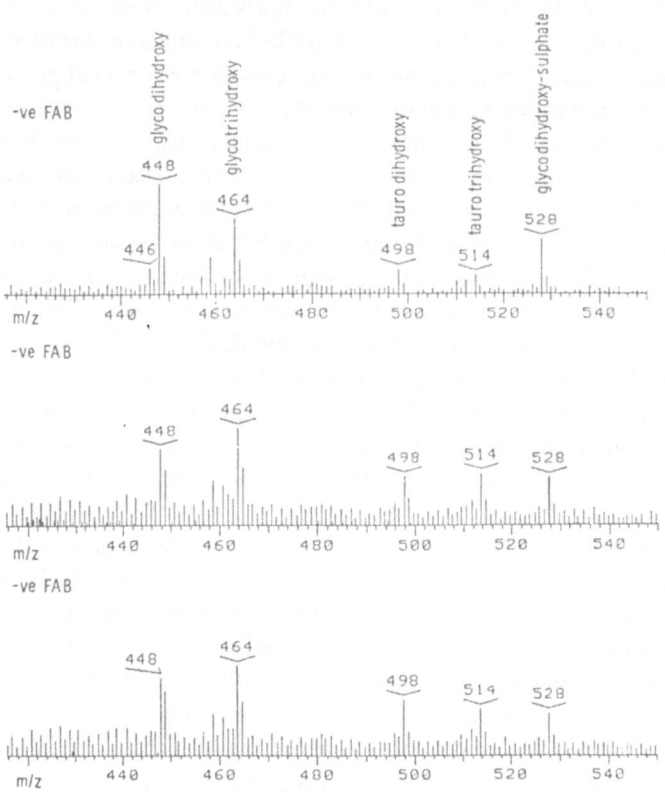

FIGURE 10. Molecular ion region of the negative-ion mass spectra obtained by fast atom bombardment of three different serum samples (20µl) from patients with drug-induced cholestasis. Bile acids were converted to triethylamine salts by ion exchange chromatography. The [M-H]⁻ ions characteristic of glycine, taurine, and sulfate conjugates of dihydroxy and trihydroxy bile salts are indicated.

The direct analysis of intact bile acid conjugates is possible by the FAB ionization technique [36]. However, with the exception of one report [45], the limited sensitivity of this technique has made it difficult for the direct characterization of bile acid conjugates in serum samples. The characteristics of the spectra generated by the FAB ionization mode of operation have been described earlier and by others [37–41,46,47]. Figure 10 indicates the type of partial mass spectra obtained for bile acid extracts of serum (1 ml) from three patients with cholestatic liver disease. The bile acids were first extracted using liquid-solid extraction on Bond–Elut cartridges [208] and bile acid conjugates converted to their triethylamine salts using a cation exchanger [46]. The extract was mixed into a glycerol matrix, introduced directly into the mass spectrometer, and bombarded with fast xenon atoms. Negative-ion FAB spectra indicate the $[M-H]^-$ ions for the presence of the glycine and taurine conjugates of dihydroxy and trihydroxy bile acids. The ion m/z 528 corresponds to the $[M - H]^-$ of a glyco-dihydroxy bile acid sulfate and is consistent with the presence of significant amounts of bile acid sulfates in the serum of patients with cholestatic liver disease. A semiquantitative assessment of the amounts of isomeric bile acid conjugates can be determined from the relative intensities of the various m/z values. The direct analysis of crude bile samples by FAB–MS have been reported [37–41], and the present example illustrates that when serum bile acid concentrations are elevated, FAB–MS may provide a useful rapid tool for screening the relative proportions of isomeric bile acid conjugates with minimal sample pretreatment and without the need to hydrolyze the conjugate moiety [46].

Prior to the advent of the FAB ionization technique, identification of the glucuronide conjugate of chenodeoxycholic in the plasma (8 ml) of a patient with chronic intrahepatic cholestasis was demonstrated using isobutane CI–MS [139]. The conjugate was first isolated by Sephadex LH-20 chromatography and thin-layer chromatography (TLC) of the Me-peracetate and the purified conjugate introduced by direct probe. The mass spectrum revealed fragment ions typical of the 3α-O-β-D-glucuronide and was comparable to an authentic sample of chenodeoxycholic acid 3α-glucuronide. The concentrations of chenodeoxycholic acid glucuronide and 3β-hydroxy-5-cholenoic acid glucuronide in the plasma were reported as 0.4 and 0.8 μg/ml, respectively. These studies provided direct chemical evidence for the conjugation of glucuronic acid to bile acids in humans. More accurate measurements of bile acid glucuronides were later obtained using selected ion monitoring with stable isotopically labeled glucuronide conjugates as internal standards [180,209–211].

Selected ion-monitoring GC–MS affords a considerable increase (100 to 1000-fold) in sensitivity over the full-scanning mode of operation, and for this reason most GC–MS methods for serum bile acids have adopted this approach [47,147]. The high sensitivity of the technique (1–10 pg) allows relatively small volumes of serum containing even low concentrations of bile acids to be measured with a high degree of accuracy and precision.

The requirement of making prior selection of the ions to be monitored introduces a significant bias to the analysis, and since a limited number of masses are generally only recorded, detection of bile acids other than those preselected is not possible. The SIM mode is used primarily to quantify specific bile acids, and therefore the specificity of the assay relies heavily on the GC retention data.

Quantification is performed by using an internal standard and measuring the ratio of the peak height or area response for the bile acid compared with the peak height or area response of an ion monitored for an internal standard. Using a series of calibration standards containing known concentrations of bile acids, it is possible to determine the amount of any unknown from the calibration curve. A variety of internal standards have been employed in selected ion monitoring assays of bile acids [47,147].

In a comprehensive study of the composition of bile acids in the serum and interstitial fluid of patients with cholestatic liver disease, with and without associated pruritus [188], ions were selected which were characteristic of Me-TMS ether derivatives of mono-, di-, and trihydroxy bile acids (*m/z* 368, 370, 372), and the identification of individual isomeric bile acids was based on their GC retention index.

Separation of serum bile acids into groups based on their mode of conjugation was achieved using the lipophilic ion exchanger Lipidex-DEAP [13,137], and after the necessary hydrolysis and deconjugation steps, coprostanol was added as an internal standard prior to derivatization. This comprehensive qualitative study of serum bile acids in pruritus revealed no correlation between serum or interstitial fluid, total bile acid, or individual bile acid concentrations and pruritus, and no correlation with the extent of conjugation of individual bile acids [188]. The qualitative bile acid profiles of interstitial fluid, described for the first time in this study, and serum were similar, and the total bile acid concentrations were linearly and positively correlated. These findings did not support a causative role for bile acids in the pruritus associated with cholestatic liver disease [188], and the etiology of this condition continues to be unknown.

Diurnal changes in serum unconjugated bile acids of healthy humans were studied using the same SIM GC–MS technique [191,192] after prior isolation of this fraction of the serum bile acids by Lipidex 1000 chromatography [137]. This study again highlights the usefulness of group fractionation techniques prior to MS analysis and demonstrated that unconjugated bile acids constituted 30–40% of the total bile acids in serum, in accordance with the early reports of Makino *et al.* [190] using GC alone. The presence in the serum of an array of unusual "secondary" bile acids including 3β,7α-dihydroxy-5β-cholanoic and 3β,7β-dihydroxy-5β-cholanoic acids seems to be indicative of colonic absorption, since the bacterial reactions responsible for the formation of these bile acids are largely confined to the colon. 3β,7β-Hydroxy-5β-cholanoic acid was later identified by GC–MS in the serum of patients treated with ursodeoxycholic acid [212]. More recent studies in which capillary GC was employed for the quantification of serum unconjugated bile acids and GC–MS to confirm the identity of individual peaks in chromatograms have indicated a potential clinical value in the determination of this fraction of bile acids, particularly for patients with conditions of intestinal bacterial overgrowth [192].

As a sequel to early studies of individual bile acid conjugates in serum, GC–MS was employed to examine the fasting and postprandial concentrations of serum bile acids in six healthy individuals [213]. The lipophilic ion exchange system used to group fractionate bile acids prior to their hydrolysis, derivatization, and analysis [13,137,151] was modified to afford the separate isolation of glucuronides. Unlike the amidated or unconjugated bile acids, no postprandial increase in serum bile acid glucuronides or sulfates was observed following meal stimulus [213].

Although several different homologs have been successfully employed as internal standards to quantify by SIM, the most appropriate candidates for internal standards are bile acid analogs labeled with heavy atoms.

The early stable isotope dilution SIM assays for bile acids utilized deuterium-labeled analogs of the principal primary and secondary bile acids [174–177], but these took no account of losses during hydrolytic steps, and in most cases, bile acid sulfates were not determined by the procedure. More recently stable labeled bile acid conjugates have been introduced that overcome some of these problems [179,180,209]. Tables summarizing SIM techniques for bile acids have been published elsewhere [47,147].

Examples of the use of stable isotopically labeled bile acid conjugates include the determination of bile acid sulfates and glucuronides in serum [174,180,209–211]. A highly specific SIM method for 3β-hydroxy-5-cholenoic and its sulfate in human serum was developed following the synthesis of the glycine and sulfate conjugates of this bile acid with deuterium atoms incorporated at carbon atom positions, 2,2,4,4, and 23 [214]. After addition of both penta-deuterated analogs to the serum, bile acids were extracted using Amberlite XAD-7 resin [215] and separated into sulfated and nonsulfated fractions using Sephadex LH-20. After hydrolytic cleavage of the conjugate group, SIM of the ions m/z 370 and 375 afforded a method of quantifying the amounts of 3β-hydroxy-5-cholenoic acid in each fraction, thereby taking into account the significant losses of the sulfates that have been reported to occur in the extraction step using XAD-7 resin [13,216,217]. This bile acid was reported to occur only in the sulfate fraction and at a concentration of 0.482 μg/ml [214].

Takikawa et al. [180] earlier described a SIM technique for the determination of bile acid glucuronides and sulfates in serum. The internal standards [11,11,12-^{2}H$_3$]deoxycholic acid 3-glucuronide and the corresponding 3-sulfate and [2,2,4,4-^{2}H$_4$]lithocholic acid 3-glucuronide were added to serum, and following enzymatic hydrolysis of the glycine and taurine conjugates, the glucuronide and sulfate conjugates were separated on the ion exchanger piperidino-hydroxypropyl Sephadex LH-20 [218]. The glucuronides were hydrolyzed with a β-glucuronidase enzyme and sulfates solvolyzed and the resulting unconjugated bile acids measured as their hexafluoroisopropyl-trifluoroacetate derivatives by the selection of suitable ions of high mass. Bile acid glucuronides and sulfates were reported to comprise 8.7% and 11.2%, respectively, of the total bile acids in the fasting serum of healthy subjects [180].

In a later report using the same procedures, the authors described the composition of bile acid glucuronides and sulfates in the serum of 36 patients with various hepatobiliary diseases, subdivided according to type of liver disease. Although the concentrations of bile acid glucuronides and sulfates increased in hepatobiliary disease, the percent distribution relative to the total serum bile acid concentrations was not very different from healthy subjects and comprised 7–8% for most patients. Chenodeoxycholic was the predominant glucuronide conjugate, while cholic acid was mainly unconjugated. In the patient group with obstructive jaundice, bile acid glucuronides accounted for approximately 5% of the total serum bile acids, while cirrhosis was characterized by decreased proportions of bile acid glucuronides when compared to healthy controls [183].

More recent studies by the same group using identical methodology described the serum concentrations of bile acid glucuronides and sulfates in patients with acute hepatitis [211] and children with cholestasis [210]. In acute hepatitis a wide variation in serum glucuronides and sulfates was reported, where they accounted for as much as 33% and 49%, respectively, of the total bile acids [211]. In healthy children bile acid glucuronides and sulfates comprised 2.6 ± 0.5 and 17 ± 3.1%, respectively, of the total serum bile acid concentration [210].

The earliest examples of the utilization of stable labeled isotopes in SIM GC–MS methods employed pentadeuterated cholic acid and trideuterated deoxycholic acid, and only the major bile acids in serum were quantified [175–177]. In some cases a deuterium-labeled standard was used to quantify more than one bile acid, which is not ideal since corrections are necessary to account for the isotopic contributions of each bile acid to the response obtained from the stable labeled internal standard. Values reported for bile acid concentrations are likely to be underestimated, since bile acid sulfates that are present in the serum of healthy subjects [151] and in patients with cholestatic liver disease (e.g., Ref. 188) were not determined. Nevertheless, a series of articles from Björkhem and co-workers [175–178], on the fasting and meal stimulus postprandial serum bile acid concentrations of healthy subjects and patients with hyperlipidemias, indicate how the mass spectrometer is routinely used as a specific and sensitive GC detector.

The lipid storage disease cerebrotendinous xanthomatosis (CTX) is biochemically characterized by a metabolic defect in side-chain cleavage of C_{27}-bile alcohols and principally involves a deficiency of the enzyme responsible for hepatic C-26 hydroxylation. This results in a marked reduction in bile acid synthesis and a decreased bile acid pool size. Peripheral serum bile acid concentrations are consequently very low, which makes their measurement by more conventional techniques difficult. With SIM GC–MS, however, Beppu et al. [219] determined bile acid profiles of 10 patients with CTX disease and showed cholic acid to comprise 69.5% of the total bile acids at a mean concentration of 0.34 ± 0.29 µg/ml. Chenodeoxycholic concentrations were much lower (0.11 ± 0.13 µg/ml), indicating greater impairment in synthesis of this bile acid, and only trace levels of lithocholic, deoxycholic, and ursodeoxycholic acids were found.

The accurate analysis of oxo bile acids is particularly challenging because of their susceptibility to degradation during workup procedures and their low concentrations relative to hydroxy bile acids. Björkhem et al. [182], however, attempted to circumvent these problems by using sodium borodeuteride to chemically reduce oxo bile acids to their hydroxy analogs. This stable labeled reducing agent leads to the incorporation of a [^{2}H] atom at the site of reduction, and the hydroxylated bile acid products can then be distinguished from the endogenous hydroxy bile acids by their mass difference of one. Prior determination of the concentrations of hydroxy bile acids was performed by their previously described procedures and, following borodeuteride reduction, the ratio of the labeled hydroxylated bile acids to the unlabeled hydroxy bile acids (i.e., [M + 1]/[M]) afforded a semiquantitative assessment of the amounts of monooxo bile acids present [182]. The drawbacks of the technique are the need for two separate determinations in each sample and the inability to accurately determine the exact structure of the oxo bile acids. This information however, could be better obtained using capillary column GC–MS by SIM of specific ions for the oxo bile acid derivatives, as was shown for fecal bile acids [220], and utilization of the GC retention indices [217].

Oxo bile acids were found to constitute 9 ± 1% and 8 ± 3%, respectively, of the nonoxidized bile acids in fasting serum and portal venous blood from six patients with cholelithiasis [182]. Concentrations of the individual monooxo bile acids in peripheral serum were less than 0.08 μmole/liter and much less than in portal venous blood (less than 0.37 μmole/liter).

The potential of stable labeled isotopes for studying metabolic pathways and determining pool sizes, turnover rates, and synthesis rates has been realized for some time. Although bile acids labeled with [^{2}H] or [^{13}C] atoms have been synthesized for use in kinetic studies [110,221–223], their high cost and limited commercial availability have restricted widespread application. The early kinetic studies, like those using radioactive tracers [224], required the collection of bile samples to obtain adequate concentrations for GC–MS bile acid analysis [159,160,162,164,225,226]. Improvements in instrumentation and in methods for isolating bile acids have permitted comparable kinetic data to be obtained from the analysis of serum bile acids [161,163–166,185–187], thus obviating the need for duodenal intubation. In practice the stable labeled bile acid, most commonly a [^{13}C] or [11,12-^{2}H$_2$] analog, is administered, and the relative abundances of the labeled and unlabeled molecules are determined in single serum samples taken over several days. From the plots of atoms % excess versus time, information on pool size and fractional turnover rate is obtained. To obtain adequate sensitivity and precision, SIM is necessary for determining the natural abundances, especially when serum is used.

Using ammonia chemical ionization, DeMark et al. [165] analyzed the Me-acetate derivatives of chenodeoxycholic and cholic acids and their [^{13}C] analogs by SIM of the ammonium adduct ions at m/z 509/508 (for chenodeoxycholic) and m/z 567/566 (cholic). The mean natural abundance of molecules with one [^{13}C] atom was 33.24% and 35.25%, respectively, for the pure compounds, and these values were similar to those observed for the bile acids isolated from serum samples. These measurements were achieved with a precision of between 0.36 and 0.57% (CV), which means that the determination of a 1% excess of labeled bile acid will be possible with a coefficient of variation of 10–20%. Following the oral administration of [24-^{13}C]cholic acid (25 mg), the fractional turnover rates, pool sizes, and synthesis rates determined from serum and bile samples in an adult were in close agreement [201]. Similar agreements between serum and biliary bile acid kinetics were reported after oral [11,12-^{2}H$_2$]chenodeoxycholic acid administration [161].

Taking advantage of the higher resolution of capillary column GC–MS, Stellaard et al. [166] were able to determine [24-¹³C]chenodeoxycholic acid kinetics from the equivalent of 40 μl of serum (initially 2 ml serum was taken for analysis) injected on-column. Bile acids were analyzed as their Me-TMS ether derivatives using electron impact ionization and SIM over the mass ranges m/z 368–373 and m/z 460–461. Measurements of chenodeoxycholic acid pool sizes, fractional turnovers, and synthesis rates in one healthy adult, determined on three separate occasions over a 4-month period, were relatively constant (CV = 20%), and the pool size, 22.4 ± 4.6 μmole/kg per day, compared well with published values using radioactive tracers and biliary bile acid analysis [227]. By this approach the rate of metabolic conversion of chenodeoxycholic to lithocholic and ursodeoxycholic acids was also possible by monitoring the appropriate ions of these derivatives [166]. In a subsequent study [185], the kinetics of [¹³C]chenodeoxycholic and [¹³C]cholic were simultaneously determined in the serum and bile of three healthy subjects and five patients with hepatobiliary disease (three cirrhotic, one cholecystectomy/sphicterotomy, and one sphicterotomy) following oral administration of 20 to 50-mg doses. The Me-TMS ether derivatives were separated on a capillary column, and using electron impact ionization, SIM of the pairs of ions m/z 370, 371 and m/z 458, 459 afforded simultaneous detection of chenodeoxycholic and cholic acid with their [¹³C] analogs, respectively. To avoid interferences between ions in the mass chromatograms, baseline separation of cholic and chenodeoxycholic acid derivatives is essential (Fig. 11). Under the GC and MS scanning conditions at least 20 scans were obtained over each peak, and these were integrated to determine the ¹³C/¹²C isotope ratios, which were expressed as atom % excess according to the equation [228]

$$\text{Atom \% excess} = \frac{R - R_o}{1 + (R - R_o)} \times 100\%$$

where R = isotope ratio and R_o = natural abundance.

A plot of ln atom % excess versus time (Fig. 12) allows calculation of the pool size and fractional turnover rate, from which the synthesis rate is determined. Pool sizes for chenodeoxycholic acid and cholic acid were 32.9 ± 9.9 and 31.8 ± 16.0 μmole/kg, and fractional turnover rates of 0.24 ± 0.13 day⁻¹ and 0.48 ± 0.22 day⁻¹, respectively, were obtained for healthy subjects, which correlated well (r = 0.99 and 0.97) with values obtained from biliary bile acid analysis [185] and were in close agreement with the literature values using radioisotopes [227]. Similar information can also be obtained for deoxycholic acid kinetics and other bile acids for which stable labeled analogs are available [186,187] provided the appropriate ions are monitored. These studies demonstrate the validity of this noninvasive technique for studying bile acid kinetics. This approach permits investigations to be carried out in pregnant women and children where the use of radioisotopes to obtain these data would otherwise be precluded.

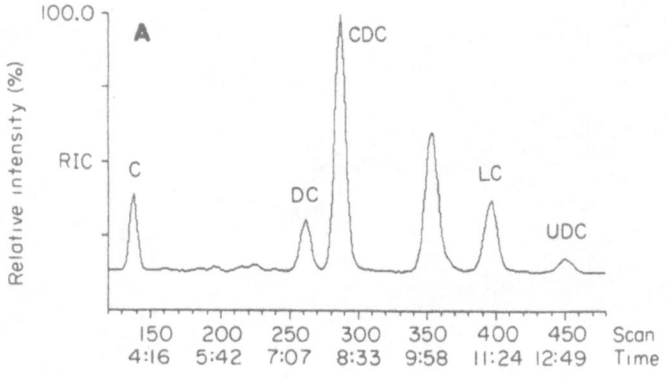

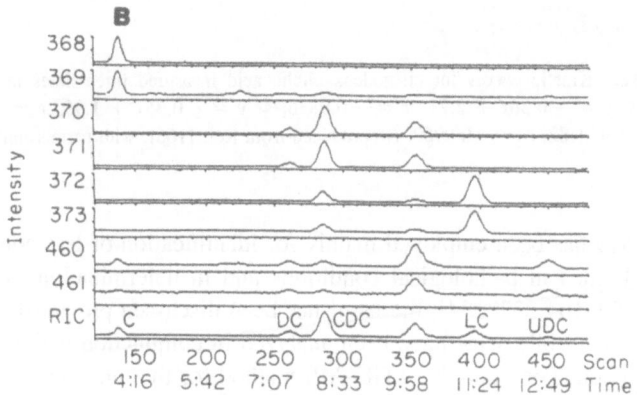

FIGURE 11. Reconstructed ion current chromatogram (A) and selective mass chromatograms (B) after selected ion monitoring of serum bile acids as their Me-TMS ether derivatives. (Reproduced from Ref. [166], with permission).

8.2. Bile

The analysis of biliary bile acids is significantly less challenging than for other body fluids primarily because of their presence in relatively high (mmole) concentrations. Although sampling of bile involves invasive techniques, this has nevertheless not impeded the many investigations which have been carried out by a variety of techniques. The major bile acids in the bile of healthy human adults are cholic, chenodeoxycholic, and deoxycholic acid, while lesser amounts of lithocholic, ursodeoxycholic, and *allo*-cholic are present. Biliary bile acids are almost exclusively conjugated with glycine and taurine in the ratio of 3 : 1 for healthy humans [229–231]. Marked species differences, however, exist in biliary bile acid composition, and conjugation patterns (e.g., Ref. 113) and differences between the adult and newborn infant are also apparent.

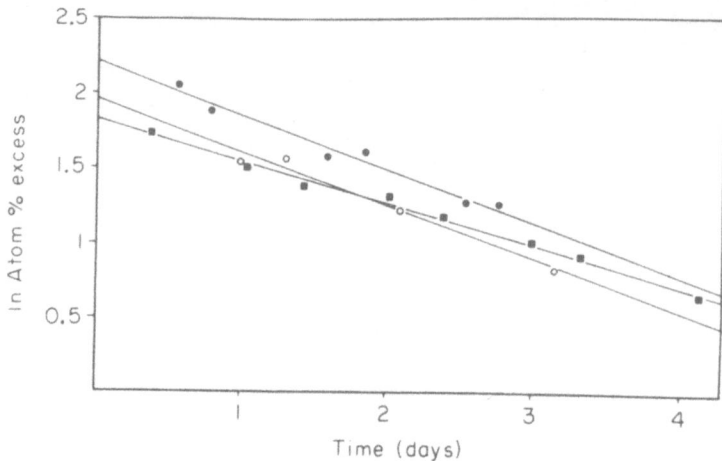

FIGURE 12. Kinetic curves for chenodeoxycholic acid measured three times in one healthy volunteer. ● $y = -0.36x + 2.21$, $r = -0.9836$; ○ $y = -0.35x + 1.95$, $r = -0.9871$; ■ $y = -0.28x + 1.82$, $r = -0.9960$. (Reproduced from Ref. [166], with permission).

GC–MS has been employed mainly for identification of less common bile acid metabolites in pathological conditions and in determination of bile acid kinetics [159–164,232–234]; the latter has been discussed previously. In many instances biliary bile acids have been analyzed in conjunction with other fluids for comparison. For example, a GC–MS method for the microdetermination of bile acids in liver tissue was applied to biliary bile acids [179] and comparisons were made. Deuterium-labeled internal standards of glycine conjugates of lithocholic, deoxycholic, chenodeoxycholic, ursodeoxycholic, and cholic acids were utilized and bile acids determined by capillary column GC–MS with SIM of ions characteristically found in the ethyl-dimethylethylsilyl ether derivatives. A good correlation between bile acid levels in liver tissue and bile was found. However, the ratios of cholic:chenodeoxycholic and primary:secondary bile acids were higher for liver tissue [179].

With the use of repetitive scanning GC–MS the kinetics of [^{2}H$_2$]chenodeoxycholic acid was determined in patients with Crohn's disease [164]. [11,12-^{2}H$_2$]Chenodeoxycholic acid (50 mg) was administered orally to patients and control subjects and bile obtained on days 1–4. The extracted bile acids were analyzed by GC–MS as their Me-propionate derivatives and 12-oxo-chenodeoxycholic acid was added as an internal standard. The incorporation of isotope into the bile acid pool was determined from the ratio of the ions m/z 372/370. A decreased pool size and half-life of chenodeoxycholic were found in patients with Crohn's disease, indicating an impaired absorption of this bile acid in this condition [164].

By a similar methodological approach, the fate of orally administered [^{2}H$_2$]

ursodeoxycholic was determined from its appearance in bile [184]. SIM GC–MS of a range of ions for the Me-TMS ether derivatives afforded a means of quantifying the amount of deuterated ursodeoxycholic acid and its metabolites. The glycine conjugate of 5β-chol-3-enoic acid was added to bile as an internal standard [184] and linearity demonstrated over the range 0.1–0.5 μg. This study indicated the rapid appearance in bile of [^{2}H$_2$] ursodeoxycholic acid, where 88% of the administered dose was recovered in 2 hr and 6 hr, respectively [184].

Comparisons between hepatic and renal excretion of individual bile acids have been made using GC–MS. In a study to observe the changes in biliary, serum, and urinary bile acids following percutaneous transhepatic biliary drainage and bile refeeding for extrahepatic cholestasis, bile acids were analyzed in groups according to their mode of conjugation. Glycine and taurine conjugates of cholic and chenodeoxycholic acid decreased rapidly, but the clearance of sulfated chenodeoxycholic acid was slow, and there was a concomitant gradual increase in biliary bile acid sulfates. Bile refeeding was marked by a restoration in the appearance of the secondary bile acids lithocholate and deoxycholate [235]. A recent study from the same group examined the fate of [^{14}C]chenodeoxycholic acid in similar patients, but used HPLC and GC–MS techniques [236]. A comparison of biliary and urinary excretion of bile acid sulfates and glucuronides in patients with cirrhosis has also been made [237]. An extensive MS study of biliary bile acids from patients with gallstones before and during treatment with chenodeoxycholic acid was recently described [238]. In this study, 255 samples of duodenal bile were examined. Cholic, chenodeoxycholic, deoxycholic, lithocholic, and ursodeoxycholic acids accounted for > 97% of the biliary bile acids. Small amounts varying between 1 and 11% each of 3α-hydroxy-12-oxo-5β-cholanoic, 3α-hydroxy-7-oxo-5β-cholanoic, 3α,12α-dihydroxy-7-oxo-5β-cholanoic, 3α,7β,12α-trihydroxy-5β-cholanoic, and traces of 3β,7β-dihydroxy-5β-cholanoic acids and the 3-oxo derivatives of chenodeoxycholic and deoxycholic were found in the bile of gallstone patients. Following treatment with chenodeoxycholic acid, bile became enriched in this bile acid and its bacterial metabolites at the expense of cholic and deoxycholic, which decreased proportionally.

Bile acid metabolism in the newborn infant and in children in health and disease is markedly different from that of the adult (e.g., see reviews in Refs. 239,240). In a study of the metabolic profiles of bile acids in the bile of the human fetus during early gestation a wide spectrum of bile acids was identified using capillary column GC–MS [241]. Chenodeoxycholic and cholic acids were quantitatively the predominant bile acids in bile, confirming the early observations using less advanced methodology [242]. Significant amounts of polar tetrahydroxy bile acids were present, characterized by hydroxylation at positions C-1, C-2, and C-6, while traces of oxo bile acids, lithocholic, and deoxycholic were evident. These latter "secondary" bile acids suggest a degree of placental transfer from the maternal compartment. Of interest was the finding of a very little 3β-hydroxy-5-cholenoic in fetal bile [241] since this is a major monohydroxy bile acid in meconium [94,243]. In the cited study of fetal bile, bile acids were quantified by capillary GC and correlated with gestational age of the fetus. Biliary bile acid concentrations of chenodeoxycholic acid, cholic acid were < 0.05 mmole/liter prior to week 17 of gestation, but a surge in bile acid synthesis occurred between weeks 17 and 20. The greater concentrations of chenodeoxycholic acid relative to cholic acid indicated a relative immaturity of hepatic 12α-hydroxylase in early life [241], which was confirmed from recent *in vitro* experiments [244].

In a study of microsomal preparations of human fetal liver obtained following legal abortions at weeks 13–24 of gestation, GC–MS was employed to confirm 1β- and 7α-hydroxylation of taurodeoxycholic acid [245]. The Me-TMS ether derivatives showed that between 52 and 80% of the products formed were $1\beta,3\alpha,12\alpha$-trihydroxy-5β-cholanoic acid. These findings explain the appearance of 1β-hydroxylated metabolites in the fetal bile [241] and meconium [94] of healthy newborns.

Limited studies of biliary bile acids of the human newborn have been reported [98,99,242,246–248]. In an attempt to identify bile acids that might reflect primary products of synthesis by the newborn liver, GC–MS studies were made of the gastric aspirate of infants with high intestinal blockage. In this condition the normal enterohepatic circulation of bile acids is impaired, and bile secreted into the duodenum refluxes into the stomach. The condition is thus analogous to a bile fistula since bile is diverted away from the intestine. The biliary bile acid composition should therefore reflect hepatic synthesis with minimal, if any, bacterial degradation [98,99]. The presence of intermediates in the pathway of bile acid synthesis from cholesterol indicated that side-chain oxidation can precede the nuclear changes. For example, $7\alpha,12\alpha$-dihydroxy-3-oxo-4-cholen-24-oic and $7\alpha,12\alpha$-dihydroxy-5β-cholanoic were identified. The 3-oxo- and 3β-epimers of chenodeoxycholic and cholic acids found may arise from the action of hepatic 3α-hydroxysteroid dehydrogenase. The presence of tetrahydroxylated C_{24} bile acids with hydroxy groups at positions C-1, C-2, and C-6 was again indicative of the capacity of the immature liver to hydroxylate the steroid nucleus at positions other than C-3, C-7, and C-12. A series of cholestanoic acids with side-chain hydroxyls at positions C-24, C-25, and C-26 was identified, primarily in the unconjugated bile acid fraction [98,99]. These investigations indicate that the features of bile acid metabolism of the fetus [241] are expressed for some time after birth.

The advantages of FAB ionization MS for direct analysis of bile acid conjugates have been discussed previously. Several papers describe the use of FAB–MS for rapid analysis of biliary bile acid content of raw bile aspirated from normal children and patients with cystic fibrosis and celiac disease [32,37–40]. Additionally, the technique has been used to confirm peaks in HPLC analysis of bile samples after collection of individual fractions. While there are distinct practical advantages to this approach, the inability to distinguish stereo- and positional isomeric bile acid structures may be a disadvantage in some studies.

Direct thermospray LC-MS [64,65] could be a more attractive approach for these cases in future. Nevertheless, FAB–MS of raw bile obtained from nasogastric intubation, or by use of a novel string device, has potential clinical use for rapid screening for bile acid abnormalities.

8.3.   Feces

MS has mainly been employed for the qualitative analysis of bile acids in feces. Eneroth *et al.* [249–251] in an excellent series of papers described the detailed composition of bile acids in feces. Fecal bile acids were isolated and fractionated by silicic acid chromatography and the Me-TMS ether and Me-trifluoroacetate derivatives analyzed by GC–MS. Confirmation of several isomers was achieved after oxidation and examination of the products. Stereochemical assignments relied heavily on GC retention data using columns with different types of liquid phases.

Deoxycholic and lithocholic acids are the principal bile acids found in human feces and arise from bacterial 7α-dehydroxylation of the primary bile acids, cholic and chenodeoxycholic acids, respectively. In feces, significant amounts of oxo bile acids are found, of which the 12-oxo isomers are quantitatively the most important [251,252]. The determination of oxo bile acids is intrinsically difficult because of their lower concentrations relative to hydroxy bile acids and their susceptability to degradation during isolation, particularly under conditions of alkaline hydrolysis (this volume, Chapter 1, and references therein).

Conversion of oxo bile acids to their oxime derivatives permits their isolation on a cation exchanger and separate determination from the hydroxylated bile acids [220]. Repetitive scanning MS analysis of the combined oxime–Me-TMS ether derivatives with construction of mass chromatograms of appropriate ions for these derivatives allowed their semiquantitative assessment in feces. Apparent was the chromatographic resolution of the *syn* and *anti* forms of the C-3 oxime. The 3-oxo bile acids constituted 1–20% of the amounts of lithocholic acid in the feces from eight healthy adults. Concentrations of 3-oxo-5β-cholanoic, 12α-hydroxy-3-oxo-5β-cholanoic, and 3,12-dioxo-5β-cholanoic acids were 27 ± 16, 37 ± 30, and 2–10 mg/g wet weight of stool respectively. Concentrations of 3-oxo-4-cholenoic and 3-oxo-5α-cholanoic were too low for reliable quantification [220].

In health, fecal bile acids are predominantly excreted in the unconjugated state, but in pathophysiological conditions, or following the administration of antibiotics or drugs, the proportions of conjugated bile acids may increase significantly. In all the early methods for fecal bile acids, an initial hydrolysis step was included, and therefore the measurement of conjugated bile acids and bile acid esters was impossible. More recently described methods, involving subfractionation of the organic extract of feces on a lipophilic anion exchanger, afford the separate determination of the various conjugated species of bile acids [252,253,254], and using repetitive scanning GC–MS and capillary column GC, the complexity of fecal bile acid composition is evident. Over 30 different bile acids have been identified from human and animal feces by these techniques [252,254–261], and for the first time an accurate assessment of the relative proportions of conjugated bile acids was possible.

In particular, bile acid sulfates, which have been the focus of recent attention [262,263], were shown to constitute < 10% of the total bile acids in feces of healthy subjects [252,254,256–259,261] and in patients with adenomatous polyps and colonic carcinomas [255,259].

Quantitative analysis of fecal bile acids, by repetitive magnetic scanning (100–555 daltons) of the permethyl derivatives, has been described,and polydeuterated bile acids were employed as internal standards for quantification [253,258].

Esterified bile acids were first observed in feces by Norman and Palmer [264], and more recently MS confirmed the identity of 5β-cholan-24-oyl-3β-palmitate as a major bile acid ester [265] while smaller amounts of palmitoleyl, stearyl, and oleyl esters were described. Ethyl esters of bile acids have also been characterized by mass spectrometry [266], but these should be viewed with some caution since transesterification may result from prolonged refluxing of fecal-homogenates in ethanolic solvents.

Unsaturated bile acids have been reported to occur naturally, but may also derive from harsh hydrolytic or solvolytic procedures, or by thermal degradation of labile hydroxy bile acids in the GC–MS system [1,103,104]. As a group, they have commanded attention because of the postulated relationship between bacteria and fecal bile acids in the etiology of colonic carcinogenesis (reviewed in this volume, Chapter 12). Analysis by repetitive scanning GC–MS of the feces from patients with colonic carcinogenesis confirmed the presence of 5β-chol-3-enoic acid, and this bile acid was excreted in amounts ranging from traces to 11.2 μmoles/day. Confirmation of its structure was established from the complete mass spectrum of the methyl ester derivative and by comparison with the authentic compound [257]. It is probably formed from bacterial degradation of lithocholic acid 3-sulfate [267] and interestingly, it was found with increased frequency in patients with colon cancer [257].

3α-Hydroxy-5β-chol-6-enoic acid has been postulated as an intermediate in the metabolic conversion of chenodeoxycholic acid to ursodeoxycholic acid by gut bacteria, yet only recently has this been confirmed [268]. Following incubation of [24-14C]chenodeoxycholic acid, 3α-hydroxy-5β-chol-6-enoic acid was isolated by TLC and its structure confirmed by GC–MS of the Me-acetate derivative and comparison with the mass spectrum of the authentic standard. Distinguishing features of the mass spectra of the 5-ene, 6-ene, and 7-ene structures were discussed [267]. The proportion of chenodeoxycholic acid biotransformed *in vitro* to the 6-ene structure was 5.5–14.0%.

The unsaturated and potent carcinogen 3-methylcholanthrene, which can be formed *in vitro* from deoxycholic acid and its intermediates, has been implicated in the etiology of colonic carcinogenesis. However, there have been no reports identifying the presence of this compound in feces. A SIM GC–MS technique was developed for specific determination of 3-methylcholanthrene in stool samples [269]. After preparation of the [^{14}C] analog, this metabolite was extracted from feces and isolated by high-performance liquid chromatography (HPLC). Under isothermal (290°C) conditions on a 3% OV-17 packed GC column, quantitative MS analysis was carried out by selected ion monitoring of the molecular ion at m/z 268 and the corresponding ion at m/z 270 for [^{14}C]methylcholanthrene. Despite a limit of detection of < 35ng/g of stool, 3-methylcholanthrene was not detected in the stools of five colon cancer patients and two healthy controls [269].

8.4. Meconium

Surprisingly, in view of the marked differences in bile acid metabolism between the newborn infant and adult, there have been few studies on the characterization of bile acid profiles of meconium. One of the earliest studies of the excretion of bile acids in meconium used TLC and GC only to identify the presence of cholic, chenodeoxycholic, deoxycholic, and lithocholic acids, and the study confirmed maternal–fetal transfer of secondary bile acids [242]. Following evidence for the presence of 3β-hydroxy-5-cholenoic acid in urine of infants with extrahepatic biliary atresia [106] and in pregnant women with cholestasis [270], Back and Ross [271] identified this bile acid in the meconium, from the characteristic mass spectrum of the Me-TMS ether derivative, which revealed a base peak at m/z 129 and the diagnostically significant fragment ion at m/z 331 (M-129). As well as those bile acids reported earlier [242], ursodeoxycholic acid and 3α,7β,12α-trihydroxy-5β-cholanoic acid were also identified.

Total bile acid concentrations in meconium of 13 mature neonates (233 ± 125.9 μg/g wet weight) were not significantly different from those of 11 premature neonates (208 ± 95.4 μg/g wet weight). However, the relative proportion(> 10% of total) of 3β-hydroxy-5-cholenoic acid was greater in the latter group. Furthermore, after Sephadex LH-20 fractionation of a single sample of meconium, 69% of the 3β-hydroxy-5-cholenoic acid and 25% of the lithocholic acid was present in the sulfate fraction. Similar observations using capillary column GC–MS were later reported by Laatikainen et al. [272], who compared amniotic fluid and serum concentrations. In the meconium from five healthy newborns the proportion of bile acids present as sulfate conjugates was 53–84% (total concentrations 1.18–3.97 μmole/liter), but unlike the earlier reports, cholic, chenodeoxycholic, and deoxycholic acids were also found in the sulfate fractions. In three meconium samples collected from babies born to mothers with cholestasis, greater concentrations of total bile acids were found [272].

In a relatively comprehensive analysis of meconium from six healthy newborns, bile acids were first separated into nonsulfated and sulfated forms by Sephadex LH-20 and into conjugate groups on DEAP Sephadex LH-20 and then analyzed by packed column GC–MS [94]. Bile acids were found mainly as taurine conjugates (56%) and sulfates (36%), while unconjugated bile acids and glycine conjugates accounted for only 4.8% and 2.9%, respectively, of the total. Unlike previous studies, hyocholic acid was identified as one of the major bile acids in the meconium and provided evidence for 6α-hydroxylation in early life. This was confirmed from studies of fetal bile [241]. This bile acid was found predominantly as a taurine conjugate, consistent with its occurrence in the serum of patients with cholestatic liver disease [188]. Deoxycholic acid was found as a sulfate conjugate in all but one sample. For the first time evidence for several tetrahydroxy bile acids was obtained from the mass spectral data [94]. In one of the compounds a base peak at m/z 217 and molecular ion at m/z 726 provided evidence for a 1,3,7,12-tetrahydroxycholanoate (Me-TMS ether) isomer. However, authentic standards of this and several other isomers were unavailable to enable definitive identification. Identification was based on relative retention indices and characteristic ion current chromatograms. A bile acid with a hydroxyl group at C-23 was tentatively identified, as were norcholic acid and several *allo*-bile acids, but the low levels of these compounds made MS confirmation difficult. The tentative finding of several epimers of cholic, chenodeoxycholic, and deoxycholic acids suggests these may be of fetal origin, or they may arise from maternal–fetal transfer of secondary bile acids. The findings of Back and Walter [94] were recently corroborated, and in addition, a 2-hydroxy bile acid and several oxo bile acids were found in meconium [135]. The oxo bile acids, because of the use of Hi-eff 8-BP GC phase, were not detected in the previous study [94].

In a series of papers by Lester and co-workers [133,134,136,240] a number of short-chain bile acids were found in human meconium samples. Four bile acids with a C_{20} nucleus and a 17β-carboxylic acid group were identified from comparisons of the mass spectra of the Me-acetate derivatives to authentic standards. Further confirmation of the structures was obtained following reduction of the monohydroxy bile acid fraction with $LiA1H_4$ and $LiA1^2H_4$ and examination of the mass spectra of the resulting alcohols. The short-chain bile acids identified were 3α-hydroxy-5α-, 3α-hydroxy-5β-, and 3β-hydroxy-5α-androstane-17β-carboxylic acid (etianic acids) and the unsaturated etienic acid 3β-hydroxy-5-androstene-17β-carboxylic acid [133]. The major isomer, 3α-hydroxy-5α-androstane-17β-carboxylic acid, was found at a concentration of 0.2 nmole/g meconium, which was reported to be 5–10 times greater than the concentrations of lithocholic and 3β-hydroxy-5-cholenoic acids. Direct attempts to establish the state of conjugation were not carried out. However, their recovery following hydrolysis with β-glucuronidase and sulfatase enzyme preparations and chemical solvolysis suggested their occurrence as glucuronide and/or sulfate conjugates When rats were administered radiolabeled etianic acid (3α-hydroxy-5β isomer), this short-chain bile acid was efficiently cleared from plasma, conjugated to glucuronic acid, and rapidly excreted into bile [206].

In addition to C_{20} bile acids, St. Pyrek *et al.* [134] also identified (20S)-

3α-hydroxy-23,24-bisnor-5β-cholan-22-oic, (20S) and (20R)-3β-hydroxy-23,24-bisnor-5β-cholan-22-oic, and 3β-hydroxy-5-pregnen-21-oic acids from the mass spectra of their Me-TMS ether derivatives.

The origin of these short-chain bile acids remains to be definitively established. However, it has been suggested that they may be produced in the maternal intestine and transported across the placenta to the fetus [134]. The possibility that they may be artificially formed by oxidative cleavage of 20-oxocorticosteroids during the analytical procedures should not be ignored. While this may occur to a minimal extent, the extraction and isolation procedures employed included an initial exhaustive sodium borohydride reduction step to minimize this potential side reaction.

More recent studies using extraction and isolation procedures that avoided adverse pH conditions and employed enzymatic and mild solvolytic methods for deconjugation support the natural occurrence of etianic acids [135]. Chromatography on the anion exchanger DEAP-LH-20 provided further evidence for etianic acids as glucuronide and/or sulfate conjugates. However, the 3α-hydroxy-5α-androstane-17β-carboxylic acid isomer was exclusively found as a sulfate conjugate [135]. Quantitatively the etianic acids comprised 2–3% of the total bile acids of meconium [135].

The relatively short retention indices of these etianic acids compared with the conventional 5β-cholanoic acids probably explains why they were not identified previously by others. The advantages of using capillary column GC–MS with repetitive scanning of all peaks eluting from the column is highlighted by these studies [133–136].

The presence of atypical short-chain bile acids poses the question as to their physiological function. The monohydroxy-etianic acids for example are not dissimilar in structure to lithocholic acid, yet these bile acids are markedly choleretic in their action compared to the cholestatic action of the latter. GC–MS will continue to play a role in elucidating the origins and potential physiological function of this group of compounds.

8.5. Amniotic Fluid

Bile acid patterns of the amniotic fluid from 29 patients were described by Délèze et al. [273] using repetitive scanning GC–MS, but only over the restricted mass range m/z 100–470 daltons. Only the principal bile acids normally found in adult bile and serum were measured, and no indication was given as to whether any of the atypical or polyhydroxylated bile acids normally found in meconium were present. However, 3β-hydroxy-5-cholenoic acid was identified from the characteristic mass spectrum of the Me-TMS ether derivative. It comprised 39.8 ± 4.5% (mean ± SEM) of the total bile acids identified in the amniotic fluid between weeks 32 and 37 of gestation, but at term accounted for 20.2 ± 2.9% of the total. Cholic acid concentrations greatly exceeded those of chenodeoxycholic acid, and traces of deoxycholic and lithocholic acids were found. Total amniotic fluid bile acid concentrations were 1.3 ± 0.13 μmole/liter during weeks 32–37 of gestation but increased at term to levels of 2.06 ± 0.34 μmole/liter [272]. The presence of 3β-hydroxy-5-cholenoic acid in relatively large proportions and its absence from the serum of pregnant women indicates a fetal origin while proportions of the primary and secondary bile acids may arise from the maternal circulation by placental transfer.

8.6. Urine

In health, renal clearance and excretion of bile acids in urine is a relatively minor pathway for bile acid elimination. Daily excretion for total bile acids is generally < 50 μmole (Ref. 13 and this volume, Chapter 11). These low levels and the great diversity in bile acids and their conjugates make a comprehensive analysis difficult. Indeed, their satisfactory analysis can only be achieved following some form of prior group separation. GC–MS affords the most suitable technique at present, although recently the HPLC analysis of urinary bile acid glucuronides following group separation was reported (this volume, Chapter 11, and Ref. 274).

The most well defined GC–MS study of urinary bile acid excretion by healthy subjects was made by Sjövall et al. [13]. Unconjugated taurine and glycine conjugates of 24 bile acids were quantified in six healthy subjects using repetitive scanning GC–MS. This study exemplifies the power of the repetitive scanning GC–MS approach where, for example, several bile acid structures were identified for the first time. These included 1-hydroxylated bile acids. Daily excretion of bile acids in health was between 6.4 and 11.0 μmole ($n = 5$), and the urinary bile acid composition was markedly different from that of serum or bile, particularly with respect to types and conjugation state. In general, all the monohydroxy and the majority of disubstituted bile acids were found in the sulfate fraction while the trihydroxy bile acids were mainly nonsulfated. The absence of bile acids with substituents at C-7 or C-12 in the sulfate fraction indicated sulfation at the C-3 position, and chromatographic mobilites confirmed the monosulfate to be amidated with glycine or taurine. The extent of sulfation appeared to depend on the number of hydroxyl groups and the structural features of the bile acid, and the suggestion that this may be explained by renal sulfation [13] was later confirmed by others [275,276]. Very few bile acids were present in the unconjugated fraction, and those found were all trihydroxy structures [13].

Bile acid glucuronides have been reported to occur in urine [277], but in the detailed study by Sjövall's group [13] this fraction of bile acids was not examined. Analysis of bile acid and steroid glucuronides necessitates a modification of the solvent systems used in the group fractionation [278,279].

Alternatively, conversion of bile acid glucuronides to methyl esters and chromatography on DEAP-LH-20 affords a method for isolating purified bile acid glucuronide methyl esters [97]. This method was extended to allow the site of glucuronidation to be determined and involved the further oxidation with periodate, which leads to the formation of the formate ester, followed by chromate oxidation and preparation of the O-methyloxime-TMS ether derivatives [97]. In three healthy subjects excretion of bile acid glucuronides ranged from 0.83 to 16.22 μmole/day, and these represented 12–36% of the total urinary bile acids. The principal urinary glucuronides identified were 6-hydroxylated, and these included hyocholic, hyodeoxycholic, and the first report of $3\alpha,6\alpha,12\alpha$-trihydroxy-5β-cholanoic acid, which were all conjugated at the C-6 position. This contrasts with chenodeoxycholic and cholic acids, which were conjugated at the C-3 position. No evidence for coupling of glucuronic acid to the C-24 carboxylic acid was obtained, despite a recent report that these conjugates can be formed in vitro by rat liver microsomes [280].

A number of bile acids not found in the urine of adults have been identified by GC–MS in normal infant urine. These include significant amounts of a series of tetrahydroxylated bile acids substituted at carbon positions 1,3,7,12 and 3,6,7,12 and the tentative identification of a 2,3,6,7-tetrahydroxy-cholanoate [93].

A vast literature exists on urinary bile acid excretion in hepatobiliary diseases, and this is reviewed in this volume, Chapter 11. In conditions of cholestasis urinary bile acid excretion is elevated. Computerized GC–MS analysis of urinary bile acid profiles in conditions of intra- and extrahepatic cholestasis have been described by Sjövall's group [13,91,95,106,150]. In addition to the bile acids normally found in urine, 3β-hydroxy-5-cholenoic, ursodeoxycholic, hyodeoxycholic, hyocholic, and $3\alpha,7\alpha,12\alpha$-trihydroxy-5β-cholestanoic acids have also been found in significant amounts [281–283]. As in health, sulfation is the predominant conjugation state, although glucuronidation of bile acids may be extensive [279,280]. It is probable that FAB–MS may be a valuable tool for rapid screening of bile acids in urine from cholestatic patients [284].

GC–MS studies have been made of urinary bile acids in patients with recurrent cholestasis of pregnancy [13,91,95,150]. In this intermittent intrahepatic cholestasis of unknown etiology, urinary bile acid excretion increases during the last trimester of pregnancy to levels approaching 100 μmole/day. This increase was mainly in the sulfate fraction where cholic, chenodeoxycholic, and deoxycholic acids were the predominant bile acids. Taurine conjugates were quantitatively more important than glycine conjugates.

Of interest was the presence of the 1- and 6-hydroxylation products of cholic acid, mainly 1,3,7,12- and 3,6,7,12-tetrahydroxy-cholanoic acids, which as they increase in concentration change from being mainly glycine to mainly taurine conjugates. The equivalent hydroxylation products of deoxycholic acid, 1β,3α,

12α-, 3α,6β,12α-, and 3α,6α,12α-trihydroxy-5β-cholanoic acids, were also identified [91].

The comprehensive nature of a quantitative GC–MS approach to the determination of a complex mixture of bile acids is illustrated from these studies, where it is possible to establish the concentrations and conjugation states of the majority of the components without prior knowledge of the constituents.

Several studies have compared serum and urinary bile acid excretion in cholestatic conditions using GC–MS mainly to identify peaks in gas chromatograms [193,276,285]. In a study of the conjugation, metabolism, and excretion of [24-^{14}C]chenodeoxycholic acid in patients with extrahepatic cholestasis, urinary bile acids were separated into groups using DEAP-LH-20, analyzed by HPLC, and the identification of peaks confirmed by GC–MS [236]. Up to 21% of the administered isotope was biotransformed, and the major excretory products were glycine, taurine, and sulfate conjugates. The main urinary metabolites of chenodeoxycholic acid were the 3-sulfates of glycochenodeoxycholic and taurochenodeoxycholic acids. The effects of biliary drainage were observed [236].

In cirrhosis the extent of alteration in bile acid metabolism is largely dependent on the degree of hepatocellular involvement and the presence or absence of portasystemic shunting. In primary biliary cirrhosis, chenodeoxycholic acid is the major urinary bile acid [95,237,286,287]. GC–MS has been used to determine qualitative and quantitative profiles of urinary bile acids in cirrhosis [95,237,285,288]. Features of urinary bile acid excretion are the presence of C-1 and C-6 hydroxylated bile acids, including tetrahydroxycholanoic structures with substituents at the 1,3,7,12- and 3,6,7,12-carbon atom positions. Two C-1 hydroxylated trihydroxy bile acids, the 1,3,7-trihydroxy and 1,3,12-trihydroxy cholanoic acids, were also present [95]. In a study of trihydroxy bile acids in the urine of 20 patients with liver cirrhosis [288], norcholic acid was the major bile acid identified, and a range of other unusual bile acids were present, including 3α,7β,12α-trihydroxy-5β-cholanoic, hyocholic, 1β,3α,12α-trihydroxy-5β-cholanoic, and *allo*-cholic acids. These were identified by GC–MS of the Me-trifluoroacetate and Me-TMS ether derivatives [288]. Hyocholic acid and hyodeoxycholic acid were previously detected by GC-MS in the urine of two patients with PBC and severe cholestasis [193].

A recent study examined the biliary and urinary excretion of sulfates, glucuronides, and tetrahydroxylated bile acids in patients with cirrhosis [237]. GC–MS analysis involved the use of both EI and ammonium CI to verify the structures of the tetrahydroxy bile acids that were not found in bile. Urinary bile acid excretion averaged 28.8 μmoles/day of which > 70% were sulfates or glucuronides. The two tetrahydroxylated bile acids, 3,6,7,12-tetrahydroxy- and 1,3,7,12-tetrahydroxy-5β-cholanoic acids, were only detected in the glycine and taurine fractions [237].

In hepatitis, an increased urinary excretion of bile acids, particularly sulfates, accompanies the increase in serum [283]. Chenodeoxycholic acid is the predominant metabolite in viral and toxic hepatitis, and 3β-hydroxy-12-oxo-5β-cholanoic and 3β-hydroxy-5-cholenoic acids are also excreted [95]. Bile acids hydroxylated at positions C-1 and C-6 were identified from computerized GC–MS analysis of the urine [95].

In extrahepatic bile duct atresia Makino *et al.* [106] used GC–MS to confirm the occurrence of 3β-hydroxy-5-cholenoic and 3α-hydroxy-5α-cholanoic (*allo*-lithocholic) acids in the urine of four infants. The structures were confirmed after oxidation of these compounds to the 3-oxo-5α- and 3-oxo-5β-cholanoates. The amounts of 3β-hydroxy-5-cholenoic acid excreted were comparable to the amounts of chenodeoxycholic acid in urine, and *allo*-lithocholic was the major saturated monohydroxy bile acid. Administration of [14C]cholesterol resulted in < 5% of the radioactivity being recovered in the monohydroxy bile acid fraction, suggesting that this was not the direct precursor for these bile acids [106].

MS has proved invaluable to the understanding of several conditions involving defects in bile acid synthesis. The most extensively studied disorders include Zellweger syndrome (see Section 8.1) and cerebrotendinous xanthomatosis (CTX) disease. Studies of the latter have assisted greatly the elucidation of normal pathways of bile acid synthesis, It is beyond the scope of this review to discuss the numerous studies of CTX disease, where much of the work has centered on the analysis of bile alcohols in biological fluids rather than bile acids (see Ref. 289) or the use of stable labeled intermediates for *in vitro* experiments to delineate the basic metabolic defect (e.g., Ref. 290).

However, it is worthwhile pointing out that FAB–MS analysis of the urine from patients with CTX disease provides the most appropriate definitive method for diagnosing this condition [284,291]. In CTX, diminished urinary bile acid excretion with concomitant increased excretion of bile alcohol glucuronides is apparent from a FAB–MS spectrum (Fig. 13). Ions at m/z 611, 627, 643, and 659 represent the negatively charged molecular ions of glucuronides of bile alcohols containing 27 carbon atoms and four, five, six, and seven hydroxy functions, respectively. These are normally seen at low concentrations in the urine of healthy subjects.

Features of urinary bile acid excretion in Zellweger syndrome reflect that seen for serum. Large amounts of THCA and DHCA and the frequent presence of a C_{29}-dicarboxylic acid are distinguishing characteristics of the urinary bile acid profiles [116,197,200,202,292]. MS has been employed in studies of the metabolic fate of bile acid precursors in this condition [116,200,203], and FAB–MS has been used to detect the presence of taurine-conjugated tetrahydroxy-cholestanoic acids in urine from patients with Zellweger syndrome and other related peroxisomal disorders [293]. Similarly, the sulfate and Glycosulfate conjugates of 3β,7α-dihydroxy and 3β,7α,12α-trihydroxy-5-cholenoic acids associated with familial giant cell hepatitis have been found in urine [105,294].

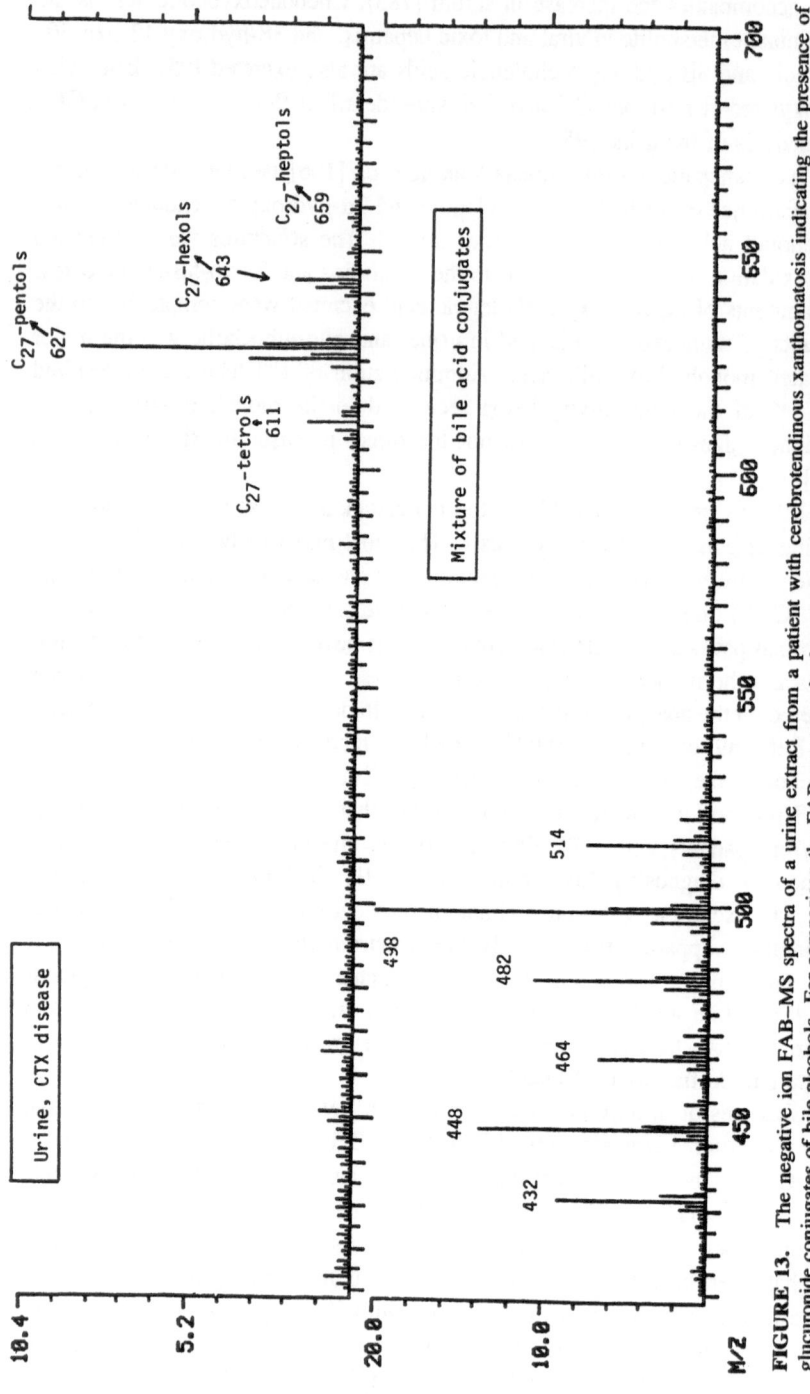

FIGURE 13. The negative ion FAB–MS spectra of a urine extract from a patient with cerebrotendinous xanthomatosis indicating the presence of glucuronide conjugates of bile alcohols. For comparison the FAB spectrum is shown of a mixture of bile acids containing glycolithocholate (*m/z* 432), glycochenodeoxycholate (*m/z* 448), glycocholate (*m/z* 464), taurolithocholate (*m/z* 482), taurochenodeoxycholate (*m/z* 498), and taurocholate (*m/z* 514). (From Ref. [284]).

The presence of significant amounts of monohydroxy bile acids in urinary precipitates was confirmed with the use of MS, and this important finding should be considered a potential source of error in urinary bile acid determination [295].

8.7. Tissue

Evidence indicates that bile acids are hepatoxic and will induce cholestasis in animal models. Furthermore, they may also be carcinogenic or cocarcinogenic and thus an important factor in the development of hepatic and colonic tumors. In particular, the monohydroxy bile acids are markedly cholestatic in their action, and for this reason there has been interest for some time in determining tissue levels of bile acids [296].

Nair et al. [144,297] first reported the existence of a tissue-bound fraction of bile acids, and lithocholic-σ-L-lysine was identified by MS as the predominant bile acid. This tissue-bound species of lithocholic acid was not extractable into a 0.1% solution of ammonia in 95% ethanol [298] and could only be recovered after enzymatic hydrolysis with cholylglycine hydrolase. Furthermore, a trypsin digestion increased the yield of tissue-bound lithocholate. Concentrations of tissue-bound lithocholic acid in 12 human liver biopsy samples were reported to be 4–116 μg/g tissue and in half of the subjects studied represented > 50% of the total lithocholic acid in tissue [144,297].

Using these methods, chenodeoxycholic acid was the predominant bile acid detected in colonic polyp and carcinoma tissues, but no lithocholic acid was found in either the soluble or tissue-bound fraction. No significant difference was found in the tissue bile acid composition of polyp or carcinoma tissue and that of normal bowel tissue which was resected adjacent to the site of the carcinoma [299].

SIM MS has been used to determine bile acid concentrations in liver tissue [300], and comparisons with biliary bile acids have been made [179]. Total bile acid concentrations (mean ± SD) reported by Yanagisawa et al. were 29.56 ± 8.62 μg/g liver and comprised mainly chenodeoxycholic (38%) and cholic (41%) acids.

Following detailed analysis of human liver tissue using GC–MS and SIM, the existence of the tissue-bound fraction of lithocholic acid reported by Nair et al. [144,297] was questioned [145]. Lithocholyl-σ-L-lysine was chemically synthesized and its characteristics checked by TLC in two solvent systems, by its reaction with ninhydrin, and using field desorption (FD) mass spectrometry. The FD-mass spectrum showed the presence of a $[M + H]^+$ ion at m/z 505 (base peak) accompanied by its Na^+ adduct ion at m/z 527 and an intense ion at m/z 461, considered to arise from the loss of the carboxylic acid. The EI spectrum, on the other hand, had a molecular ion at m/z 504, and the synthesized compound was identical in characteristics to the lithocholyl-σ-L-lysine described earlier by Nair et al. [144]. Rigorous and prolonged alkaline hydrolysis of this bile acid released lithocholic acid, which was confirmed from the mass spectrum of the ethyl ester–dimethylethylsilyl ether derivative. However, when lithocholyl-σ-L-lysine was subjected to mild aqueous alkaline hydrolysis and enzymatic hydrolysis, using three different cholylglycine hydrolase preparations and varying incubation conditions, contrary to earlier claims [144], these conditions failed to liberate free lithocholate and the lysine conjugate was recovered unchanged [145]. Using an isotope dilution technique and GC–MS with SIM, soluble and tissue-bound lithocholic acid concentrations were measured in 10 samples of human liver tissue. [²H₄]Lithocholic acid was employed as an internal standard, and the pairs of ions m/z 386 and 461 and m/z 390 and 465 were monitored for the Me-dimethylethylsilyl ether derivatives of lithocholic and [²H₄]lithocholic acids, respectively [145]. Soluble lithocholic acid concentrations were reported as 0.28–12.1 pmole/mg liver tissue (which compared with 29–493 pmole/mg liver reported by Nair *et al.* [144]), while bound lithocholic acid levels were < 0.2 pmole/mg liver and close to the limit of detection of the method. The authors suggested that the lithocholic acid found in the residue reflected traces that remained following the initial organic extraction [145]. However, in response to the findings of Yanagisawa *et al.* [145], Nair *et al.* suggested their inability to hydrolyze the lysine conjugate may have been due to the instability of the enzyme preparation. The cholyglycine hydrolase active on N-ε-lysyllithocholate is relatively unstable at temperatures above 0°C and requires storage at below − 20°C to preserve its activity [301]. This recent study describes the appropriate conditions for storage and use of clostridial cholanoylaminoacid hydrolase from *C. perfringens* and reaffirmed the validity of earlier observations for tissue-bound lithocholic acid [144,302].

9. COMPENDIUM OF MASS SPECTRA OF BILE ACIDS

A large number of mass spectra of individual bile acids and their derivatives are scattered throughout the literature, and the largest single collections of spectra previously published are to be found in Volume 1 of this series [1] and in separate reviews [56,57]. For the analyst, problems arise in comparing unknown spectra with those of published structures, because invariably the MS conditions differ or the derivatives are not comparable. The following compendium of mass spectra of 112 reference bile acids was analyzed under uniform GC–MS conditions using electron impact (70 eV) ionization of the Me-TMS ether derivatives. The spectra are listed according to the number of substituent groups in the cholanoic nucleus.

Note an Appendix (pp. 571–575) has been added to this text that gives a list of retention indices of bile acids chromatographed on nonpolar, nonselective liquid phases as their methyl ester–trimethylsilyl ether derivatives. These values have been compiled from literature sources and elsewhere and a description of the GLC conditions is given in the footnotes. It should be emphasized that no single GLC column, including the high-resolving capillary columns, will permit the complete resolution of the many positional and stereoisomeric forms of bile acids, and definitive identification requires the use of mass spectrometry. Furthermore, some caution should be taken when comparing published retention indices for bile acids, where columns from different sources can give different values for the same bile acid derivative and, in some instances, even the order of elution of isomers may be different. This is evident from these data shown. To alleviate this problem in structural elucidation studies, we recommend that standards be simultaneously analyzed with unknown compounds under identical GLC conditions.

Library List of Bile Acid Standards[a] Analyzed by Mass Spectrometry as the Methyl Ester-Trimethylsilyl (Me-TMS) Ether Derivatives

Spec no.	Systematic name (trivial name)	Mol. wt. (free)	Mol. wt. (Me-TMS)	Base[b] peak
	No hydroxyls			
1.	5α-Cholanoic	360	374	217
2.	5β-Cholanoic	360	374	217
3.	5β-Chol-3-enoic	358	372	215
4.	Chol-3,5,7-trienoic	354	368	135
	Monohydroxy			
5.	3α-Hydroxy-5β-cholanoic (lithocholic)	376	462	215
6.	3β-Hydroxy-5α-cholanoic	376	462	215
7.	3β-Hydroxy-5β-cholanoic (*iso*-lithocholic)	376	462	372
8.	3β-Hydroxy-5-cholenoic	374	460	331
9.	3α-Hydroxy-5β-chol-6-enoic	374	460	145
10.	3β-Hydroxy-7α-methoxy-5-cholenoic	404	490	400
11.	3β-Methoxy-7α-hydroxy-4-cholenoic	404	490	400
12.	7α-Hydroxy-5β-cholanoic	376	462	372
13.	7β-Hydroxy-5β-cholanoic	376	462	372
14.	11β-Hydroxy-5β-cholanoic	376	462	143
15.	12α-Hydroxy-5β-cholanoic	376	462	257
16.	12α-Hydroxy-3,5,7-choltrienoic	370	456	251
	Dihydroxy			
17.	3α,6α-Dihydroxy-5β-cholanoic (hyodeoxycholic)	392	550	370
18.	3α,6β-Dihydroxy-5β-cholanoic	392	550	405
19.	3β,6α-Dihydroxy-5β-cholanoic	392	550	370
20.	3α,7α-Dihydroxy-5β-cholanoic (chenodeoxycholic)	392	550	370
21.	3α,7β-Dihydroxy-5β-cholanoic (ursodeoxycholic)	392	550	460
22.	3β,7α-Dihydroxy-5β-cholanoic	392	550	370
23.	3β,7β-Dihydroxy-5β-cholanoic	392	550	370
24.	3β,7α-Dihydroxy-4-cholenoic	390	548	458
25.	3β,7α-Dihydroxy-5-cholenoic	390	548	458
26.	3α,12α-Dihydroxy-5α-cholanoic (*allo*-deoxycholic)	392	550	255
27.	3α,12α-Dihydroxy-5β-cholanoic (deoxycholic)	392	550	255

(continued)

Library List of Bile Acid Standards[a] Analyzed by Mass Spectrometry as the Methyl Ester-Trimethylsilyl (Me-TMS) Ether Derivatives

Spec no.	Systematic name (trivial name)	Mol. wt. (free)	Mol. wt. (Me-TMS)	Base[b] peak
28.	3α,12β-Dihydroxy-5α-cholanoic	392	550	255
29.	3α,12β-Dihydroxy-5β-cholanoic	392	550	255
30.	3β,12α-Dihydroxy-5α-cholanoic	392	550	255
31.	3β,12α-Dihydroxy-5β-cholanoic	392	550	255
32.	3β,12β-Dihydroxy-5α-cholanoic	392	550	345
33.	3β,12β-Dihydroxy-5β-cholanoic	392	550	255
34.	3α,12α-Dihydroxy-5β-cholen-8(14)-oic	390	548	226
35.	7α,12α-Dihydroxy-5β-cholanoic	392	550	255
	Trihydroxy			
36.	3α,6α,7α-Trihydroxy-5β-cholanoic (hyocholic)	408	638	458
37.	3α,6α,7β-Trihydroxy-5α-cholanoic	408	638	195
38.	3α,6α,7β-Trihydroxy-5β-cholanoic (ω-muricholic)	408	638	195
39.	3α,6β,7α-Trihydroxy-5β-cholanoic (α-muricholic)	408	638	458
40.	3α,6β,7β-Trihydroxy-5β-cholanoic (β-muricholic)	408	638	195
41.	3α,6α,12α-Trihydroxy-5β-cholanoic	408	638	253
42.	3α,7α,12α-Trihydroxy-5α-cholanoic (*allo*-cholic)	408	638	458
43.	3α,7α,12α-Trihydroxy-5β-cholanoic (cholic)	408	638	253
44.	3β,7α,12α-Trihydroxy-5β-cholanoic	408	638	253
45.	3α,7β,12α-Trihydroxy-5β-cholanoic	408	638	253
46.	3α,7α,12β-Trihydroxy-5β-cholanoic	408	638	253
47.	3α,7β,12β-Trihydroxy-5β-cholanoic	408	638	253
48.	3β,7α,12α-Trihydroxy-4-cholenoic	406	636	546
49.	3β,7α,12α-Trihydroxy-5-cholenoic	406	636	546
50.	3α,12α,16α-Trihydroxy-5β-cholanoic	408	638	73
51.	3α,12α,23-Trihydroxy-5β-cholanoic	408	638	255
	Tetrahydroxy			
52.	1β,3α,7α,12α-Tetrahydroxy-5β-cholanoic	424	724	217
53.	2β,3α,7α,12α-Tetrahydroxy-5β-cholanoic	424	724	546
54.	2β,3β,7α,12α-Tetrahydroxy-5β-cholanoic	424	724	546
55.	3α,6α,7α,12α-Tetrahydroxy-5β-cholanoic	424	724	546
56.	3α,6α,7β,12α-Tetrahydroxy-5β-cholanoic	424	724	285
	Oxo			
57.	3-Oxo-4-cholenoic	372	386	124
58.	3-Oxo-5β-cholanoic	374	388	273
59.	6α-Hydroxy-3-oxo-5β-cholanoic	390	476	271
60.	7α-Hydroxy-3-oxo-5β-cholanoic	390	476	386
61.	7α-Hydroxy-3-oxo-4-cholenoic	388	474	384
62.	12α-Hydroxy-3-oxo-5β-cholanoic	390	476	271
63.	12α-Hydroxy-3-oxo-4,6-choldienoic	386	472	267
64.	6α,7α-Dihydroxy-3-oxo-5β-cholanoic	406	564	474
65.	7α,12α-Dihydroxy-3-oxo-5β-cholanoic	406	564	269
66.	7α,12α-Dihydroxy-3-oxo-4-cholenoic	404	562	267

Library List of Bile Acid Standards[a] Analyzed by Mass Spectrometry as the Methyl Ester-Trimethylsilyl (Me-TMS) Ether Derivatives (*Continued*)

Spec no.	Systematic name (trivial name)	Mol. wt. (free)	Mol. wt. (Me-TMS)	Base[b] peak
67.	6α,7α,12α-Trihydroxy-3-oxo-5β-cholanoic	422	652	383
68.	3α-Hydroxy-6-oxo-5α-cholanoic	390	476	447
69.	3α-Hydroxy-6-oxo-5β-cholanoic	390	476	461
70.	3α,12α-Dihydroxy-7-oxo-5β-cholanoic	406	564	341
71.	3α-Hydroxy-7-oxo-5β-cholanoic	390	476	386
72.	12-Oxo-5β-cholanoic	374	388	233
73.	3α-Hydroxy-12-oxo-5α-cholanoic	390	476	121
74.	3α-Hydroxy-12-oxo-5β-cholanoic	390	476	231
75.	3β-Hydroxy-12-oxo-5β-9(11)-cholenoic	388	474	229
76.	3α,7α-Dihydroxy-12-oxo-5β-cholanoic	406	564	229
	Dioxo			
77.	3,6-Dioxo-5α-cholanoic	388	402	107
78.	3,6-Dioxo-5β-cholanoic	388	402	329
79.	3,7-Dioxo-5β-cholanoic	388	402	287
80.	3,12-Dioxo-5β-cholanoic	388	402	247
81.	7α-Hydroxy-3,12-dioxo-5β-cholanoic	404	490	121
82.	3α-Hydroxy-7,12-dioxo-5β-cholanoic	404	490	245
	Trioxo			
83.	3,7,12-Trioxo-5β-cholanoic	402	416	261
	C$_{20}$ bile acids			
84.	3α-Hydroxy-5β-androstan-17β-carboxylic acid	320	406	316
85.	3β-Hydroxy-5α-androstan-17β-carboxylic acid	320	406	391
86.	3β-Hydroxy-5β-androstan-17β-carboxylic acid	320	406	316
87.	3β-Hydroxy-5-androsten-17β-carboxylic acid	318	404	129
	Bisnor-bile acids			
88.	3α,7α-Dihydroxy-5β-bisnorcholan-22-oic	364	522	342
89.	3α,7β-Dihydroxy-5β-bisnorcholan-22-oic	364	522	432
90.	3α,7α,12α-Trihydroxy-5β-bisnorcholan-22-oic	380	610	253
	Nor-bile acids			
91.	5β-Norcholan-23-oic	346	360	217
92.	3α-Hydroxy-5β-norcholan-23-oic (nor-lithocholic)	362	448	358
93.	3α,6α-Dihydroxy-5β-norcholan-23-oic	378	536	356
94.	3α,7α-Dihydroxy-5β-norcholan-23-oic (nor-chenodeoxycholic)	378	536	356
95.	3α,7β-Dihydroxy-5β-norcholan-23-oic (nor-ursodeoxycholic)	378	536	446
96.	3α,12α-Dihydroxy-5β-norcholan-23-oic (nor-deoxycholic)	378	536	255
97.	3α,7α,12α-Trihydroxy-5β-norcholan-23-oic (nor-cholic)	394	624	253
98.	3α,12α-Dihydroxy-7-oxo-5β-norcholan-23-oic	392	622	251
	C$_{25}$ bile acids			
99.	3α,7α,12α-Trihydroxy-5β-homocholanoic	422	652	253
	C$_{27}$ bile acids			
100.	(25R)-3α,7α-Dihydroxy-5β-cholestan-26-oic (dihydroxycoprostanoic)	434	592	412

(*continued*)

Library List of Bile Acid Standards[a] Analyzed by Mass Spectrometry as the Methyl Ester-Trimethylsilyl (Me-TMS) Ether Derivatives

Spec no.	Systematic name (trivial name)	Mol. wt. (free)	Mol. wt. (Me-TMS)	Base[b] peak
101.	(25R)-3α,7α,12α-Trihydroxy-5β-cholestan-26-oic (trihydroxycoprostanoic)	450	680	253
102.	(24ζ)-3α,7α,12α-Trihydroxy-5β-cholest-24-en-26-oic	448	678	253
103.	(25ζ)-3α,7α,12α,24R-Tetrahydroxy-cholestan-26-oic	466	768	253
104.	(25ζ)-3α,7α,12α,24S-Tetrahydroxy-cholestan-26-oic	466	768	253
105.	(25ζ)-3α,7α,12α,25-Tetrahydroxy-5β-cholestan-26-oic	466	768	253
106.	(25ζ)-3α,7α,12α,26-Tetrahydroxy-5β-cholestan-27-oic	466	768	253
	C29 bile acids			
107.	3α,7α,12α-Trihydroxy-27a,27b-dihomo-5β-cholestan-26,27b-dioic	508	752	253
	Glycine conjugates			
108.	3α-Hydroxy-5β-cholanoylglycine (glycolithocholic)	433	519	131
109.	3α,12α-Dihydroxy-5β-cholanoylglycine (glycodeoxycholic)	449	607	131
110.	3α,7α-Dihydroxy-5β-cholanoylglycine (glycochenodeoxycholic)	449	607	131
111.	3α,7α,12α-Trihydroxy-5β-cholanoylglycine (glycocholic)	465	695	253
112.	3α,7β,12α-Trihydroxy-5β-cholanoylglycine	465	695	253

[a] Reference compounds were obtained form the following sources (1) Steralodis Inc., Wildon, NH, (2) Sigma, St. Louis, MO, (3) MRC Steroid Reference Collection, curator Professor D. N. Kirk, Chemistry Dept., Queen Mary College, London, England, (4) The Haselwood Collection, c/o Dr. A. R. Tammer, Division of Biochemistry, Guy's Hospital, University of London, London, England. We gratefully thank and acknowledge Dr. Iida, Nihon University College of Engineering, Koriyama, Fukushima-ken, Japan, who supplied samples of 3,12-dihydroxy-5α-cholanoic isomers; Professor Aldo Roda, Universita Degli Studi Di Bologna, Italy, for samples of 3β,7β-dihydroxy-5β-cholanoic; Prof. Hans Fromm for the sample of 3α-hydroxy-6-cholenoic acid and Professor Alan Hofmann for the samples of nor and bisnor bile acids.

[b] The base peak is defined by the mass/charge (m/z) ratio fo the most abundant ion in the spectrum. The values quoted here (except 9spec. no. 50) exclude the ubiquitous ions present in the Me-TMS ether derivatives, which are of no relevant diagnostic value.

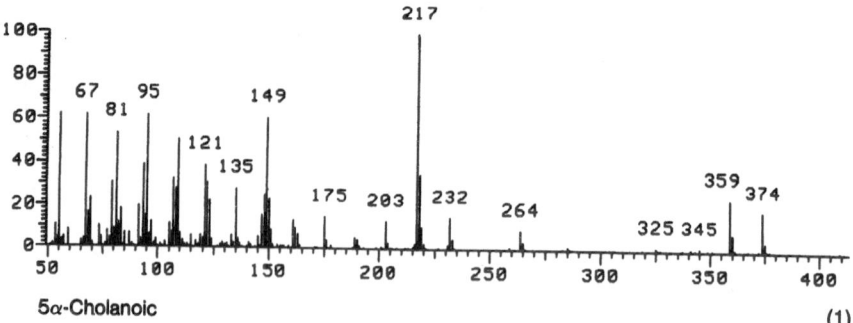

5α-Cholanoic

(1)

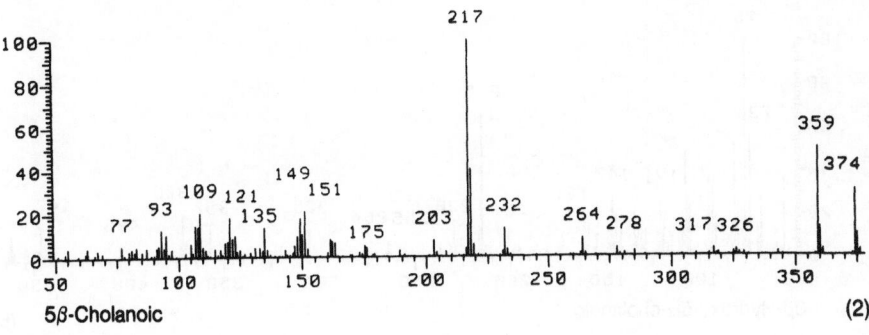

5β-Cholanoic

(2)

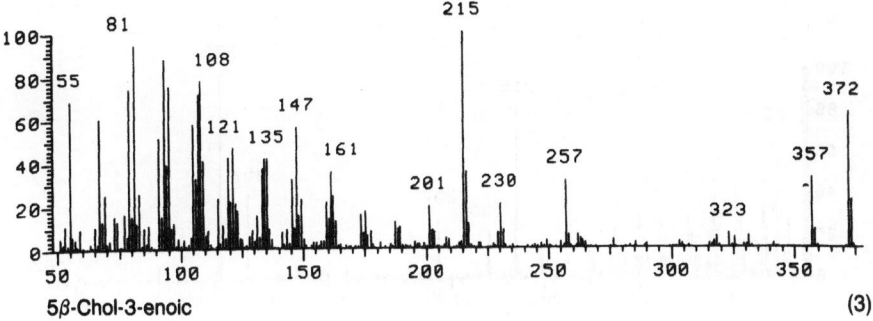

5β-Chol-3-enoic

(3)

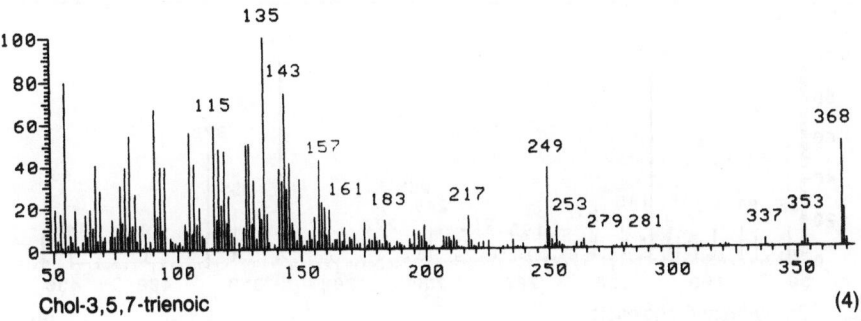

Chol-3,5,7-trienoic

(4)

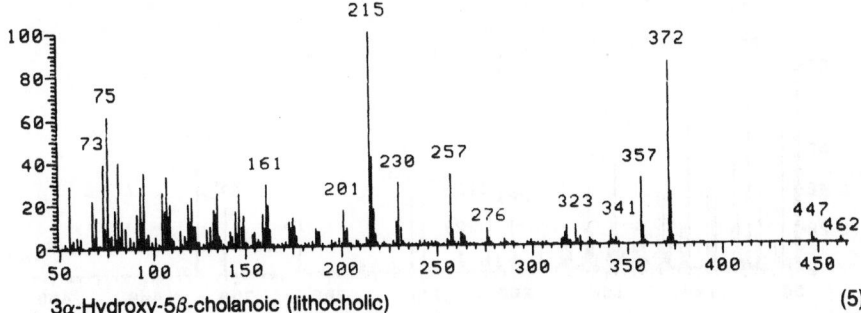

3α-Hydroxy-5β-cholanoic (lithocholic)

(5)

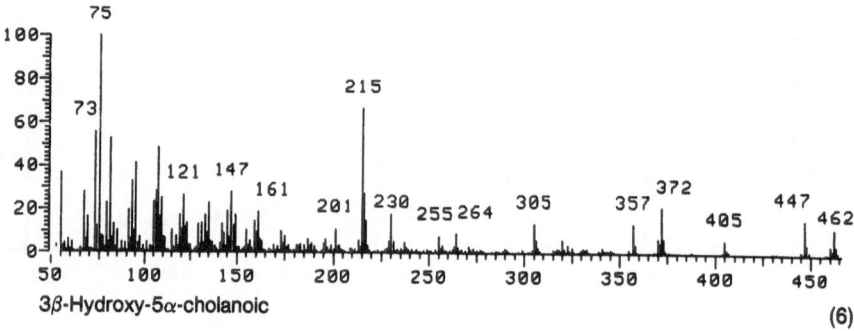

3β-Hydroxy-5α-cholanoic   (6)

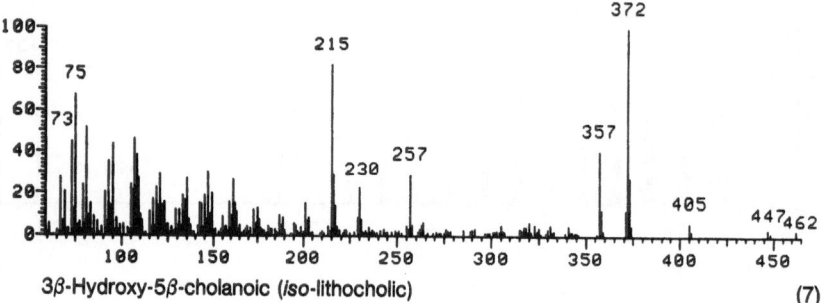

3β-Hydroxy-5β-cholanoic (*iso*-lithocholic)   (7)

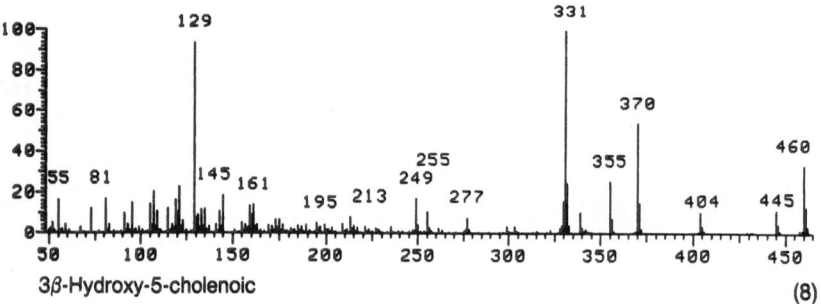

3β-Hydroxy-5-cholenoic   (8)

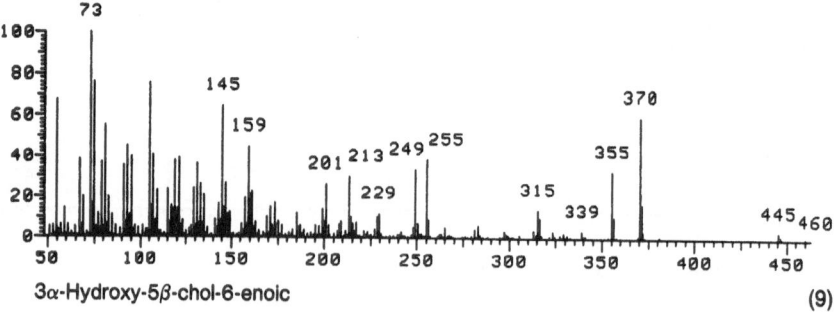

3α-Hydroxy-5β-chol-6-enoic   (9)

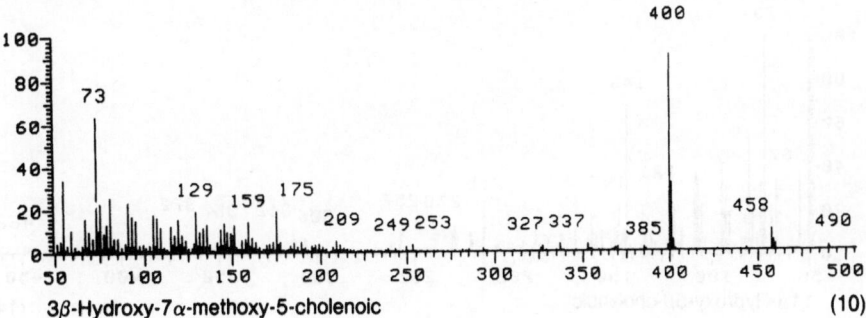

3β-Hydroxy-7α-methoxy-5-cholenoic   (10)

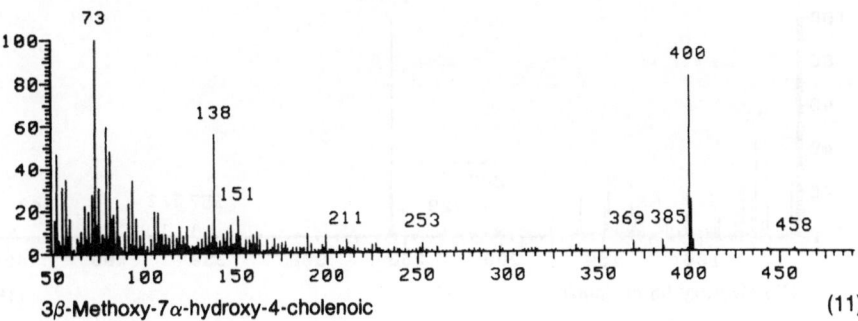

3β-Methoxy-7α-hydroxy-4-cholenoic   (11)

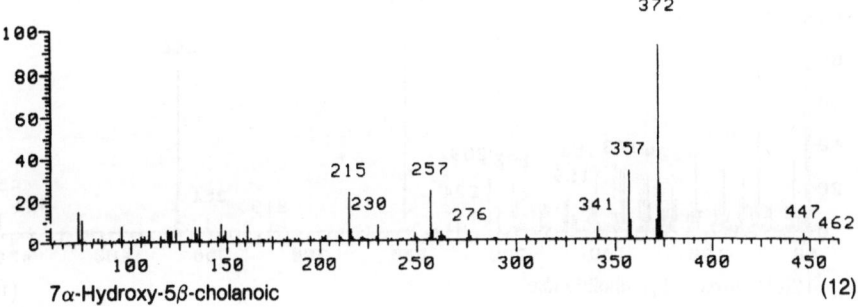

7α-Hydroxy-5β-cholanoic   (12)

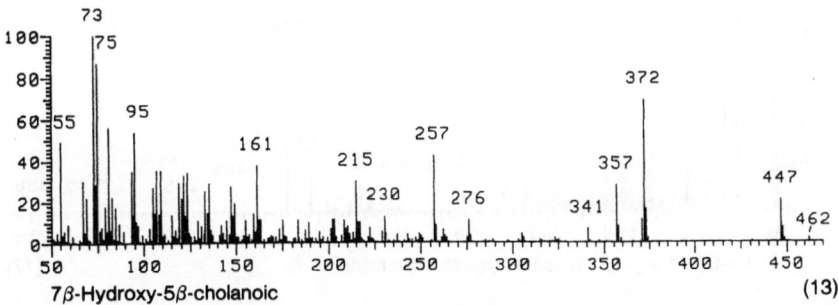

7β-Hydroxy-5β-cholanoic   (13)

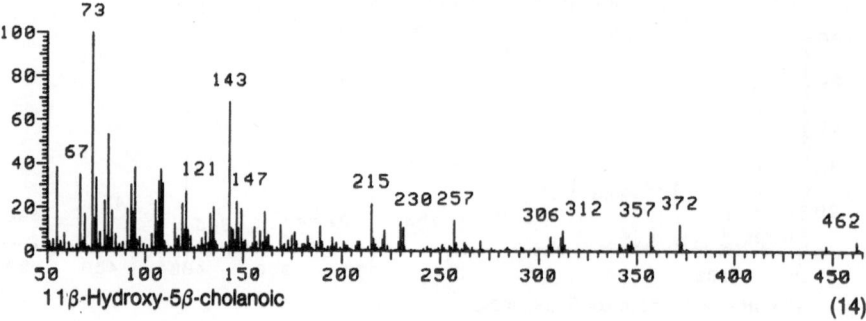

11β-Hydroxy-5β-cholanoic   (14)

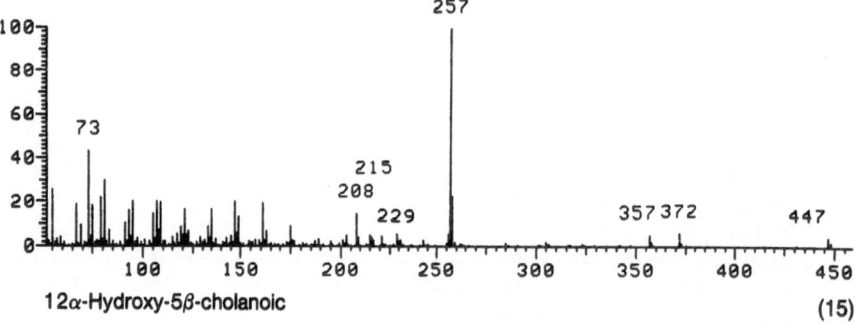

12α-Hydroxy-5β-cholanoic   (15)

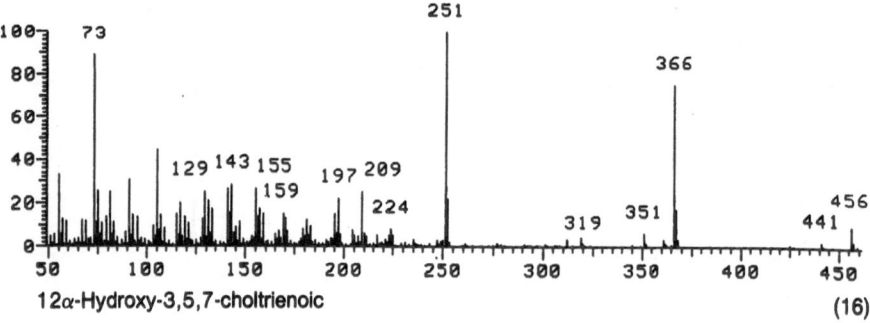

12α-Hydroxy-3,5,7-choltrienoic   (16)

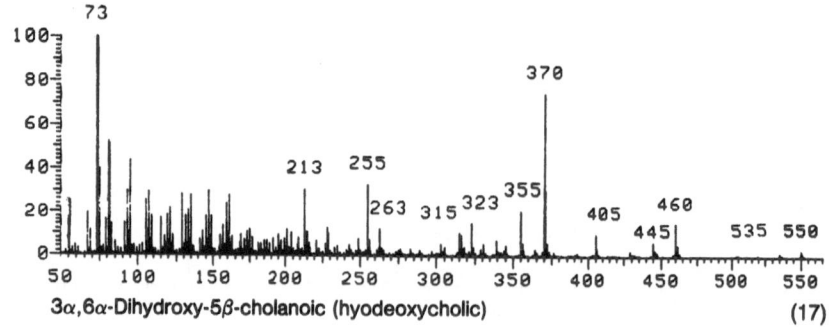

3α,6α-Dihydroxy-5β-cholanoic (hyodeoxycholic)   (17)

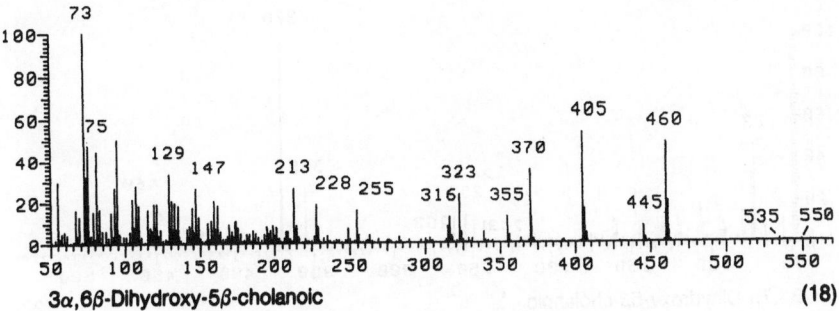

3α,6β-Dihydroxy-5β-cholanoic (18)

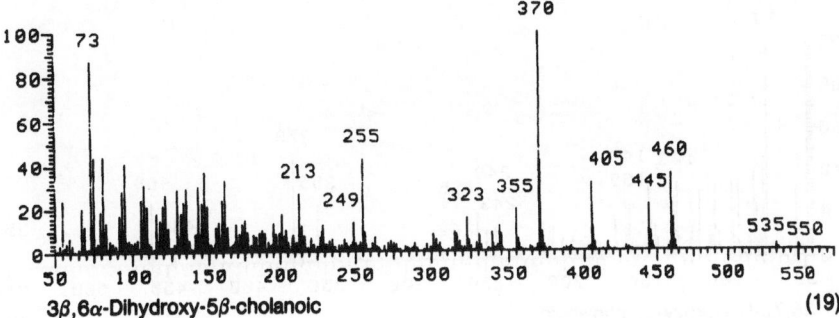

3β,6α-Dihydroxy-5β-cholanoic (19)

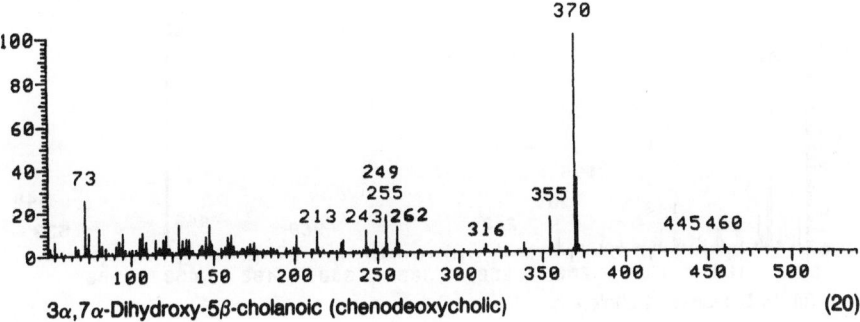

3α,7α-Dihydroxy-5β-cholanoic (chenodeoxycholic) (20)

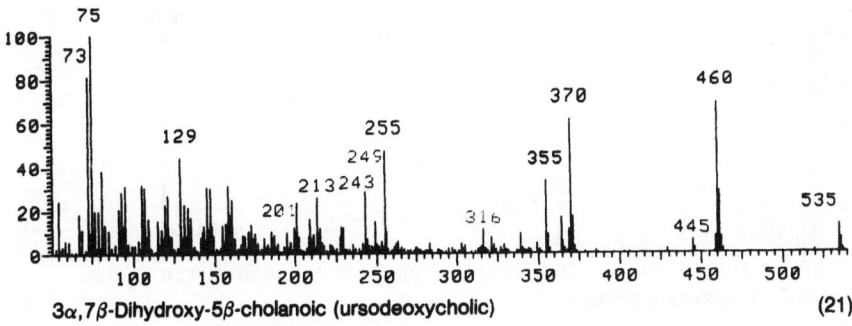

3α,7β-Dihydroxy-5β-cholanoic (ursodeoxycholic) (21)

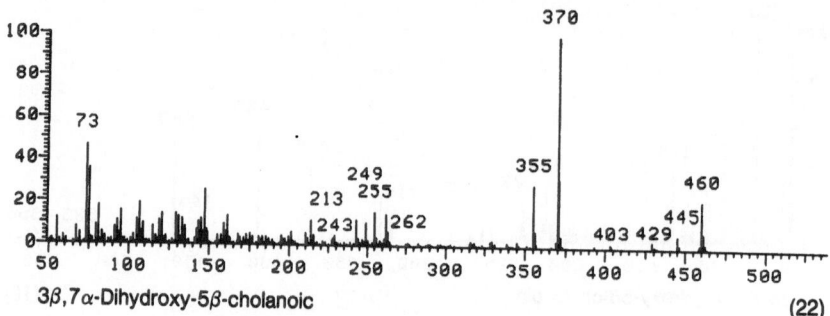

3β,7α-Dihydroxy-5β-cholanoic   (22)

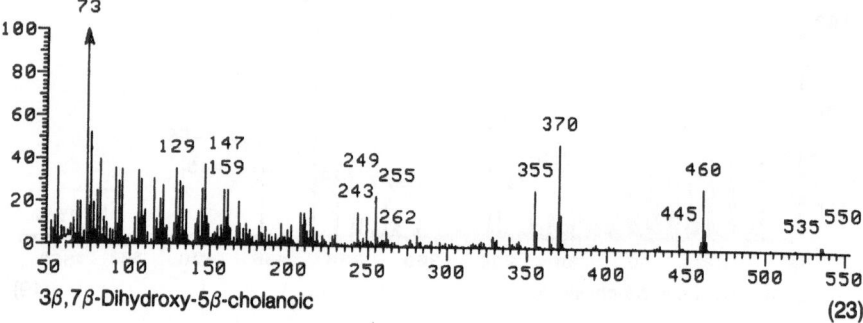

3β,7β-Dihydroxy-5β-cholanoic   (23)

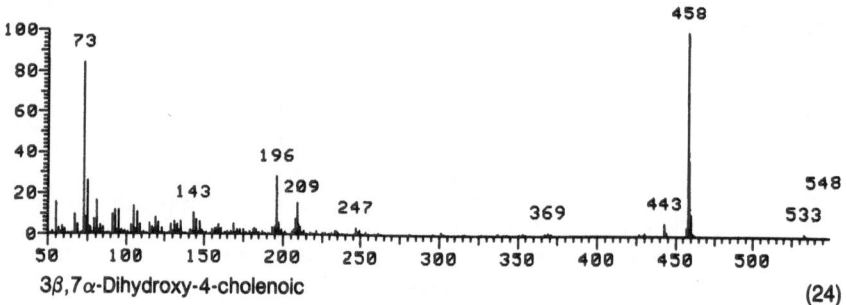

3β,7α-Dihydroxy-4-cholenoic   (24)

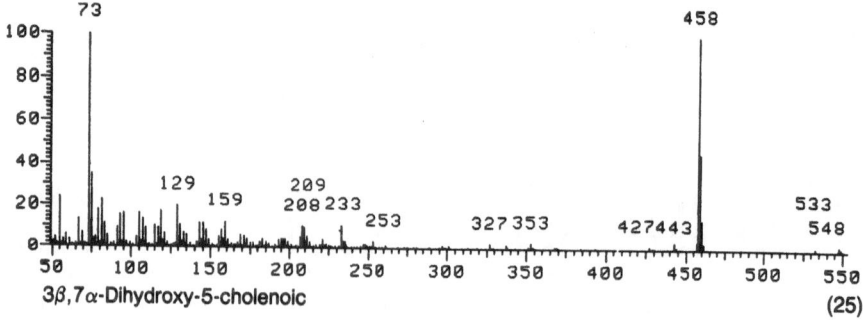

3β,7α-Dihydroxy-5-cholenoic   (25)

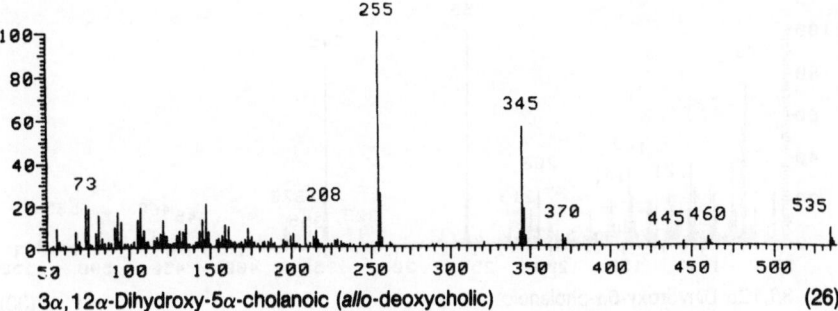

3α,12α-Dihydroxy-5α-cholanoic (*allo*-deoxycholic) (26)

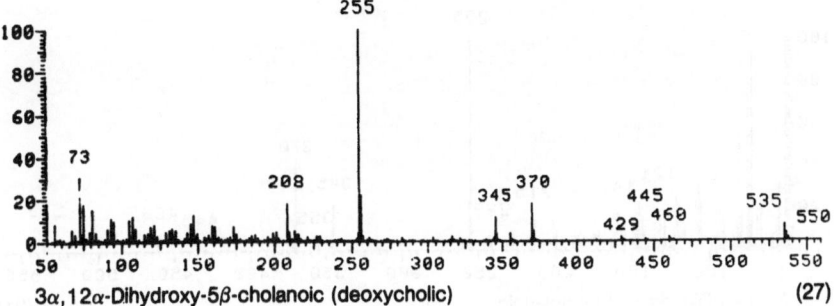

3α,12α-Dihydroxy-5β-cholanoic (deoxycholic) (27)

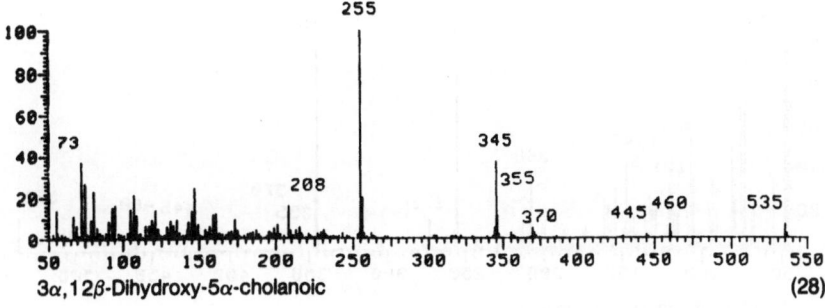

3α,12β-Dihydroxy-5α-cholanoic (28)

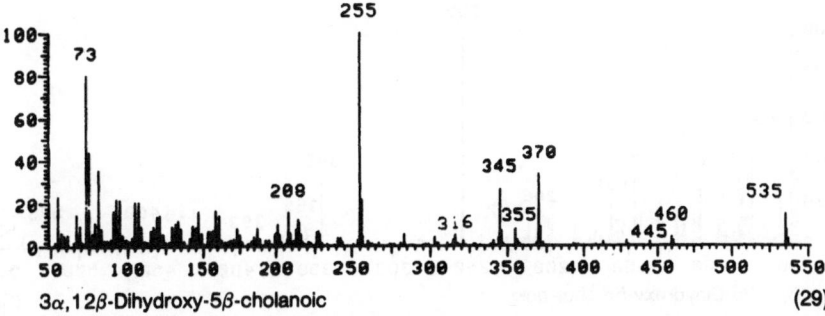

3α,12β-Dihydroxy-5β-cholanoic (29)

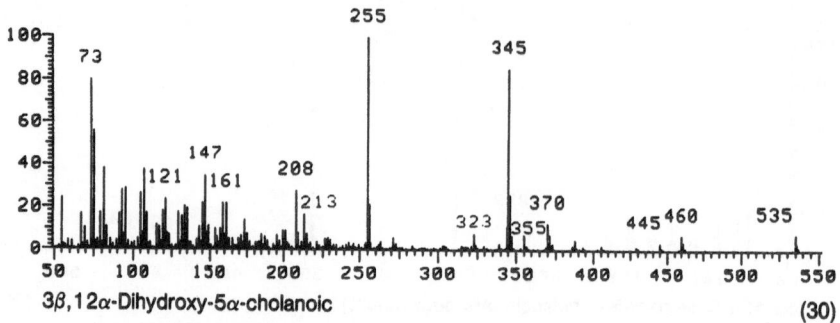

3β,12α-Dihydroxy-5α-cholanoic

(30)

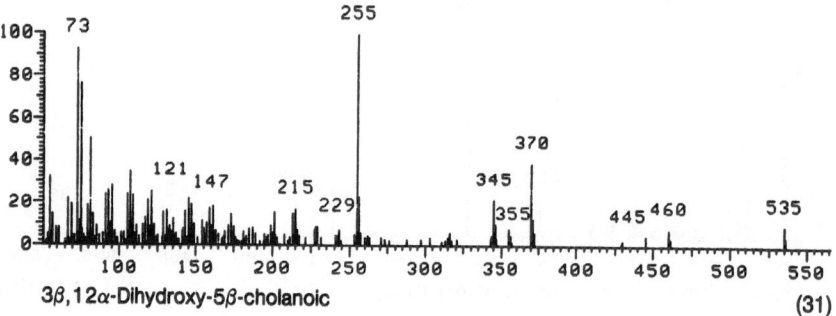

3β,12α-Dihydroxy-5β-cholanoic

(31)

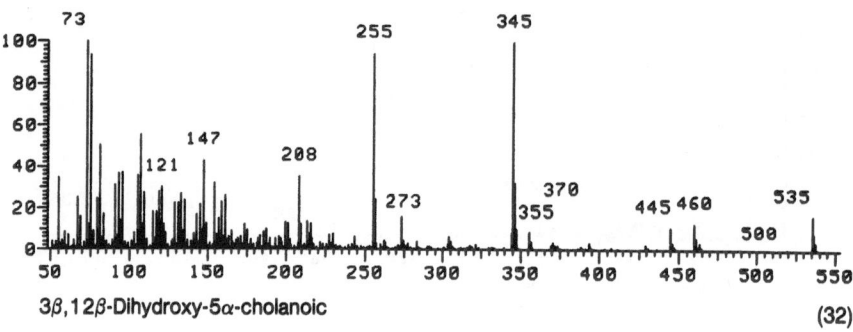

3β,12β-Dihydroxy-5α-cholanoic

(32)

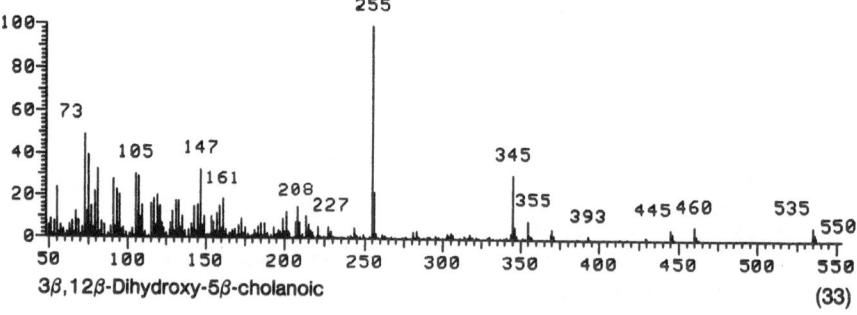

3β,12β-Dihydroxy-5β-cholanoic

(33)

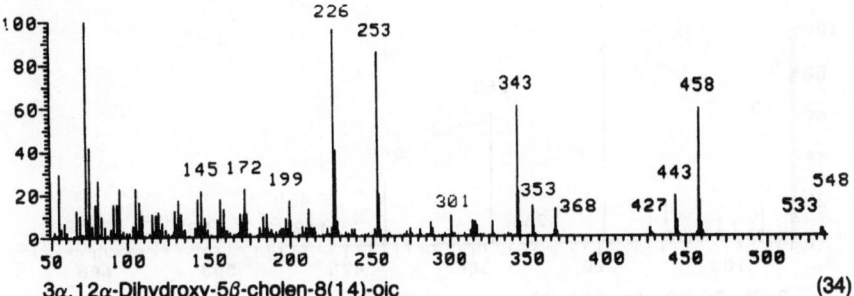

3α,12α-Dihydroxy-5β-cholen-8(14)-oic (34)

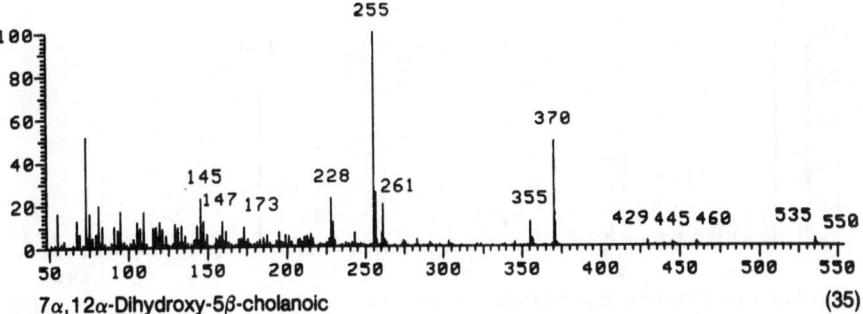

7α,12α-Dihydroxy-5β-cholanoic (35)

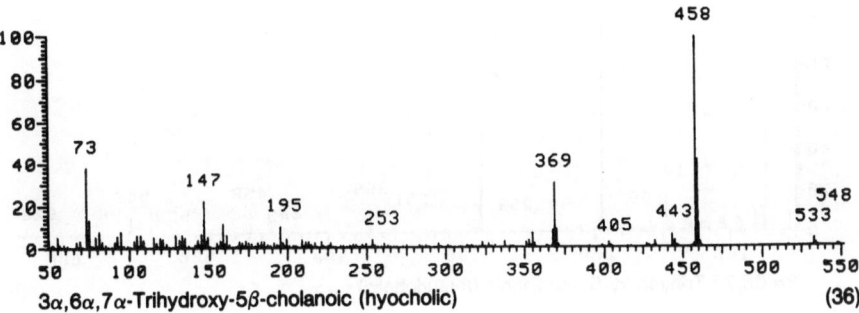

3α,6α,7α-Trihydroxy-5β-cholanoic (hyocholic) (36)

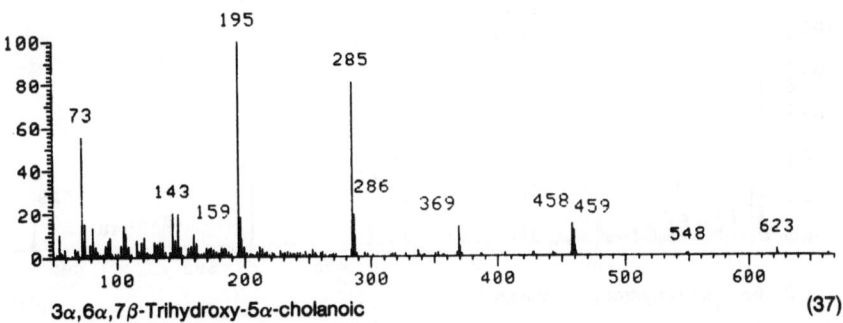

3α,6α,7β-Trihydroxy-5α-cholanoic (37)

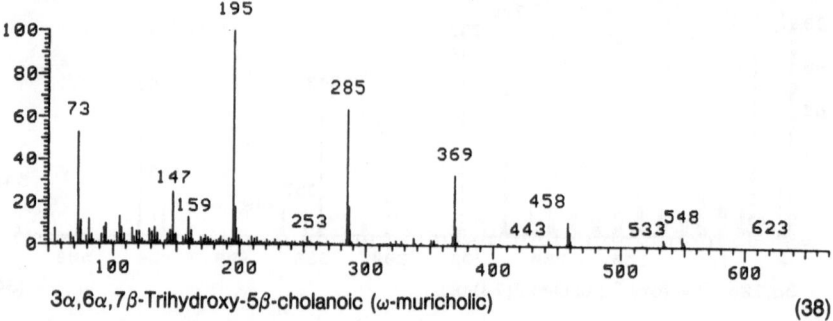

3α,6α,7β-Trihydroxy-5β-cholanoic (ω-muricholic)   (38)

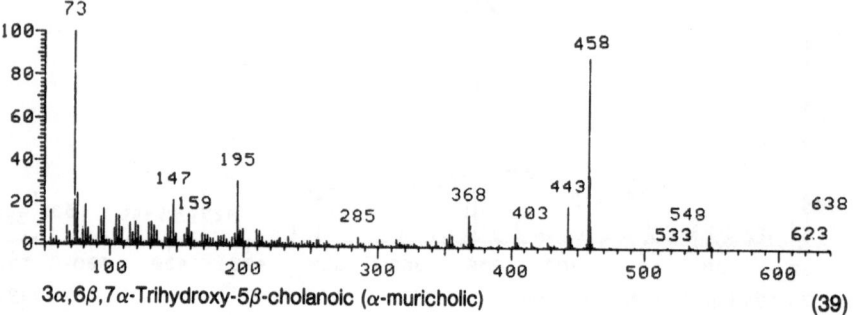

3α,6β,7α-Trihydroxy-5β-cholanoic (α-muricholic)   (39)

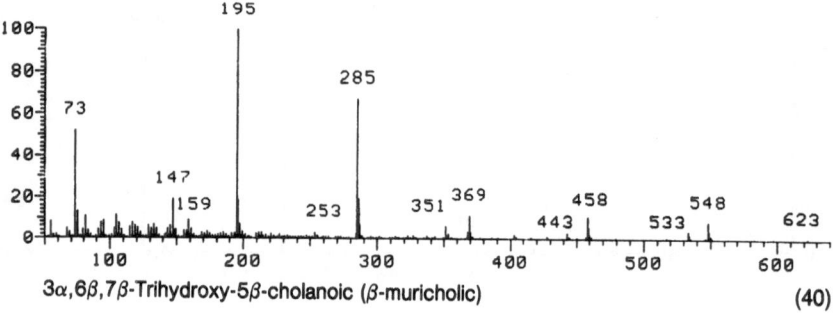

3α,6β,7β-Trihydroxy-5β-cholanoic (β-muricholic)   (40)

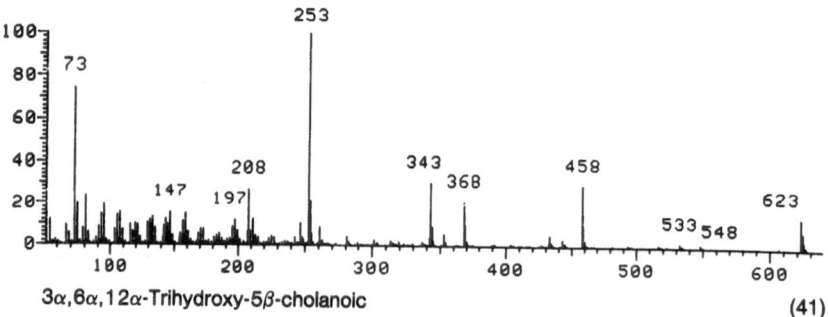

3α,6α,12α-Trihydroxy-5β-cholanoic   (41)

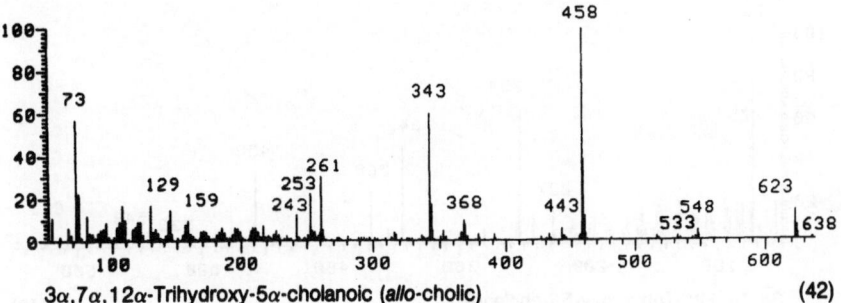

3α,7α,12α-Trihydroxy-5α-cholanoic (*allo*-cholic) (42)

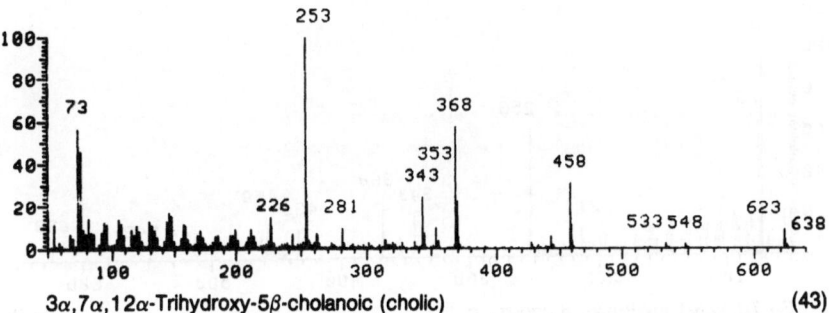

3α,7α,12α-Trihydroxy-5β-cholanoic (cholic) (43)

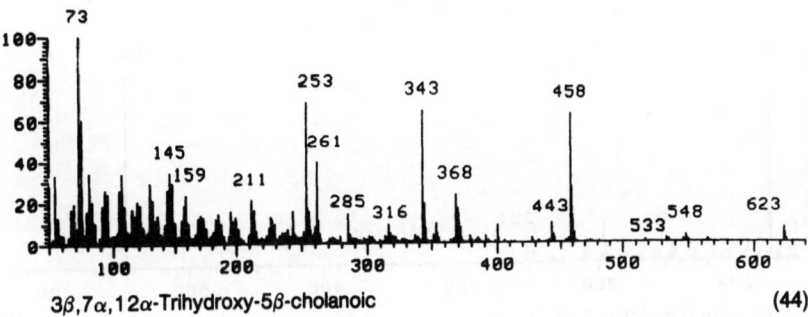

3β,7α,12α-Trihydroxy-5β-cholanoic (44)

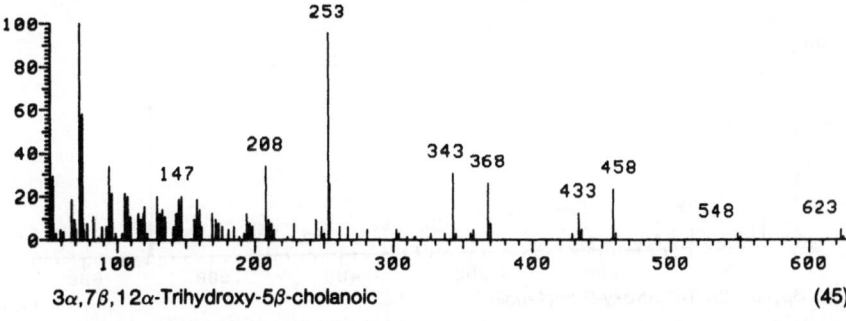

3α,7β,12α-Trihydroxy-5β-cholanoic (45)

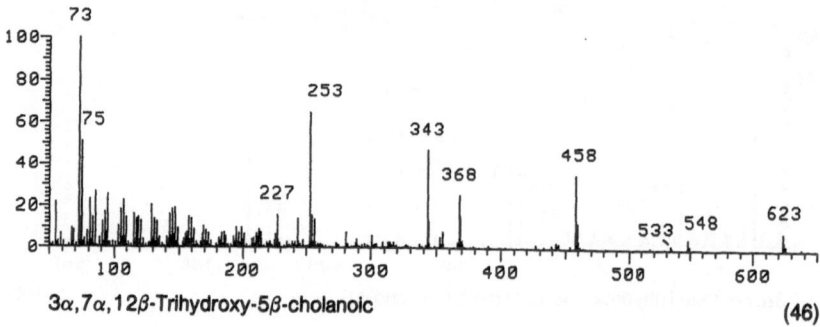

3α,7α,12β-Trihydroxy-5β-cholanoic   (46)

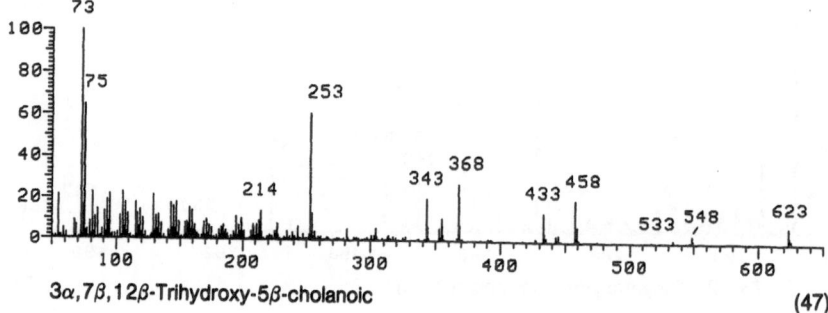

3α,7β,12β-Trihydroxy-5β-cholanoic   (47)

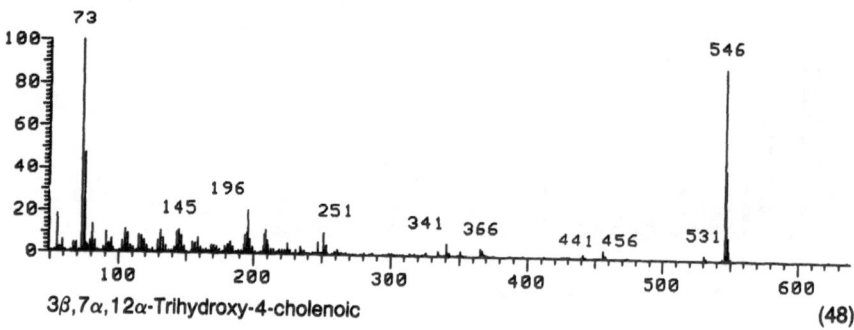

3β,7α,12α-Trihydroxy-4-cholenoic   (48)

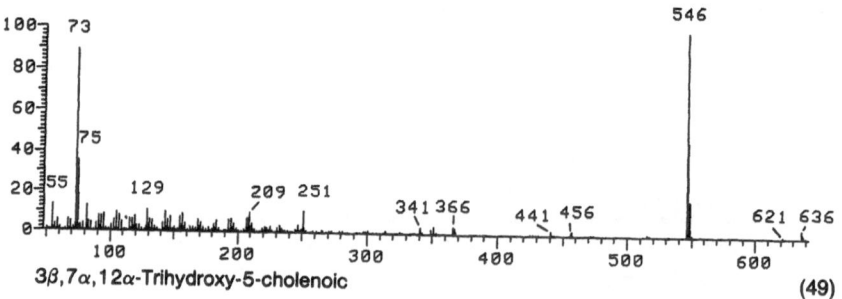

3β,7α,12α-Trihydroxy-5-cholenoic   (49)

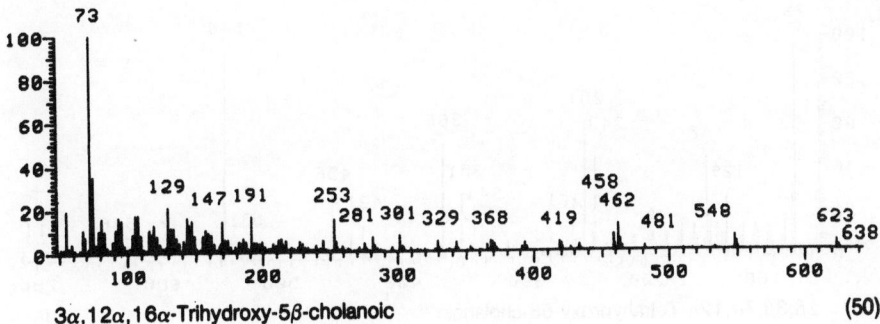

3α,12α,16α-Trihydroxy-5β-cholanoic (50)

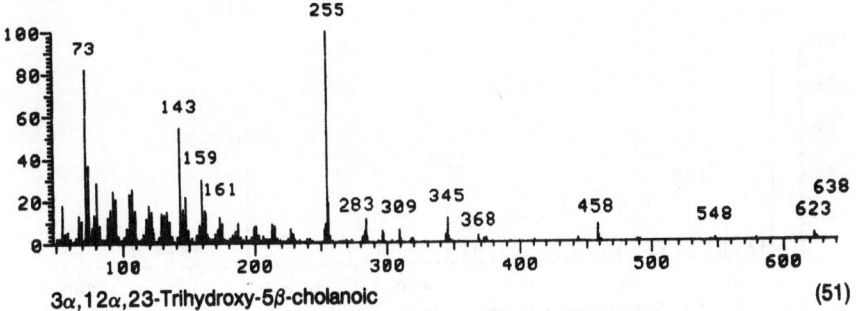

3α,12α,23-Trihydroxy-5β-cholanoic (51)

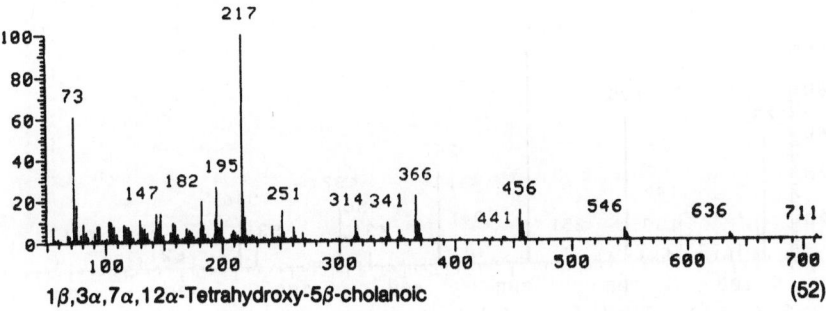

1β,3α,7α,12α-Tetrahydroxy-5β-cholanoic (52)

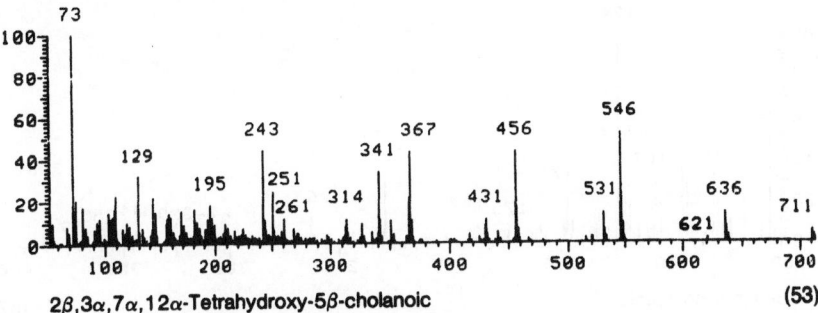

2β,3α,7α,12α-Tetrahydroxy-5β-cholanoic (53)

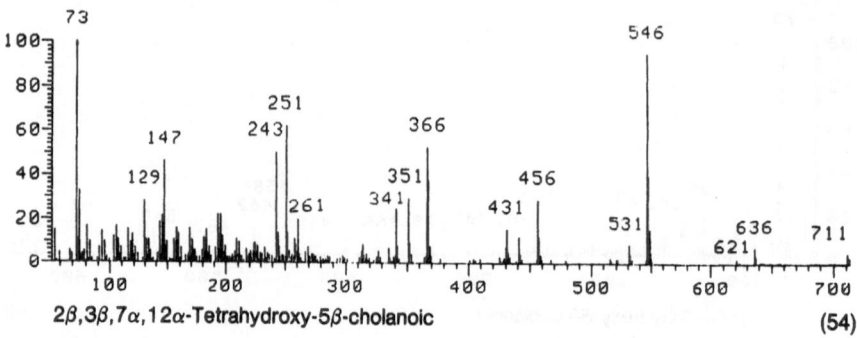

2β,3β,7α,12α-Tetrahydroxy-5β-cholanoic   (54)

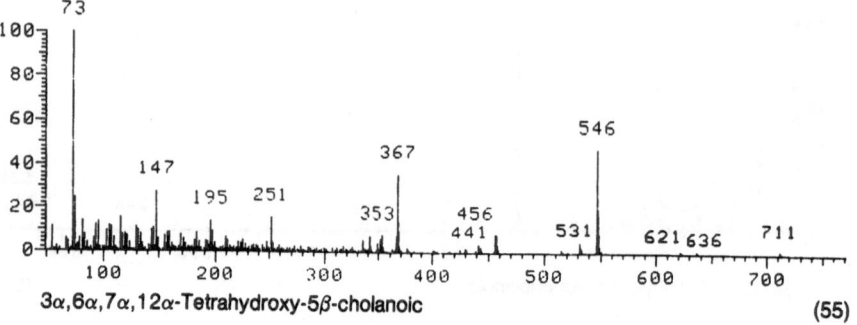

3α,6α,7α,12α-Tetrahydroxy-5β-cholanoic   (55)

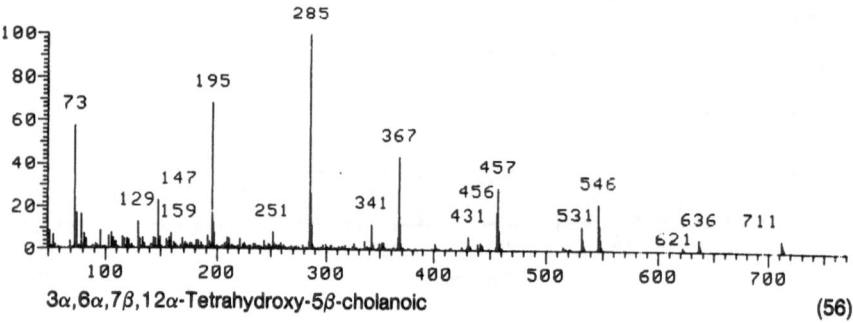

3α,6α,7β,12α-Tetrahydroxy-5β-cholanoic   (56)

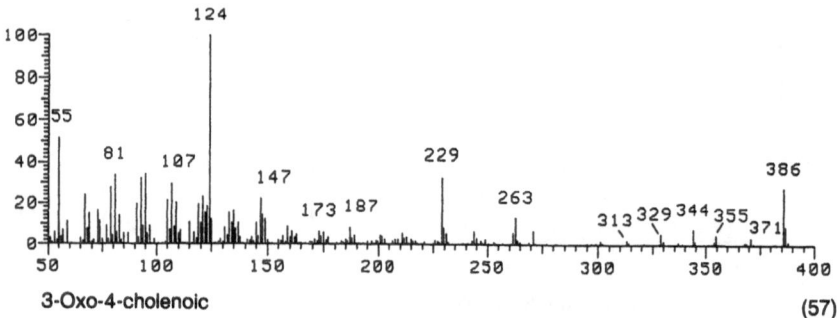

3-Oxo-4-cholenoic   (57)

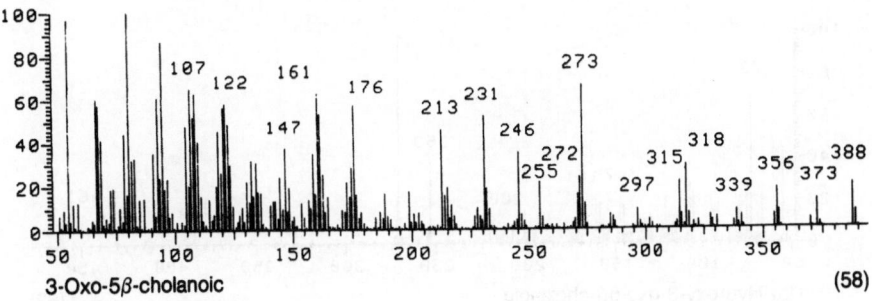

3-Oxo-5β-cholanoic (58)

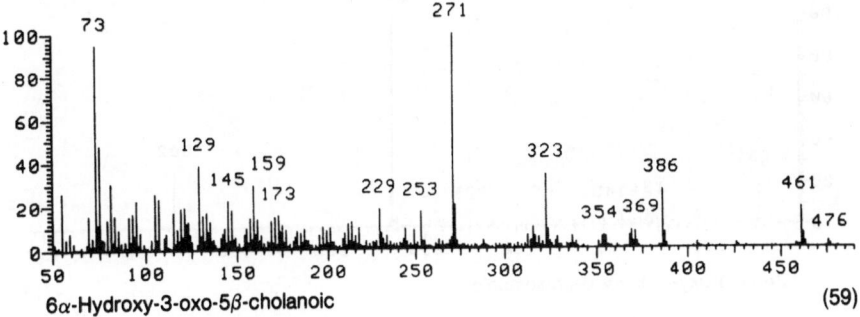

6α-Hydroxy-3-oxo-5β-cholanoic (59)

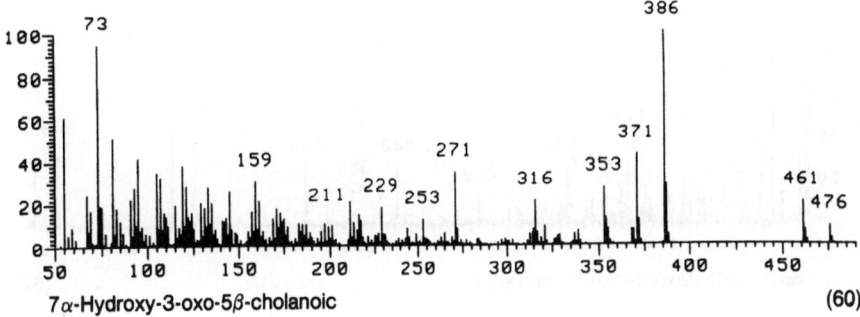

7α-Hydroxy-3-oxo-5β-cholanoic (60)

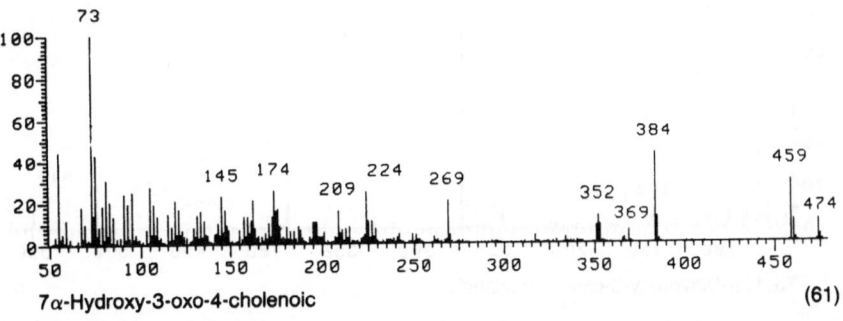

7α-Hydroxy-3-oxo-4-cholenoic (61)

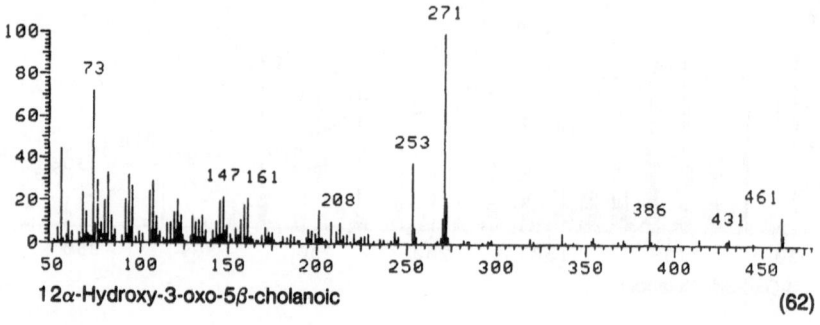

12α-Hydroxy-3-oxo-5β-cholanoic   (62)

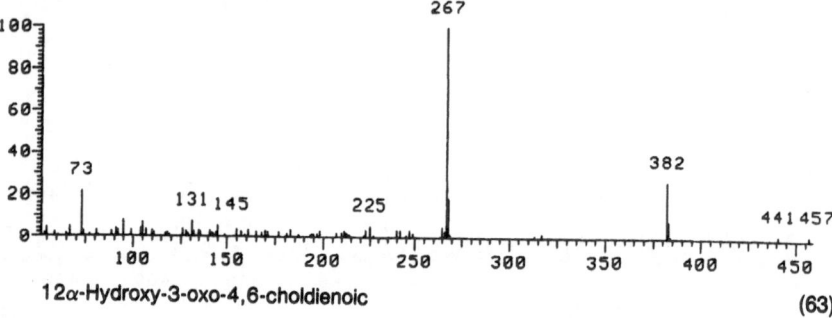

12α-Hydroxy-3-oxo-4,6-choldienoic   (63)

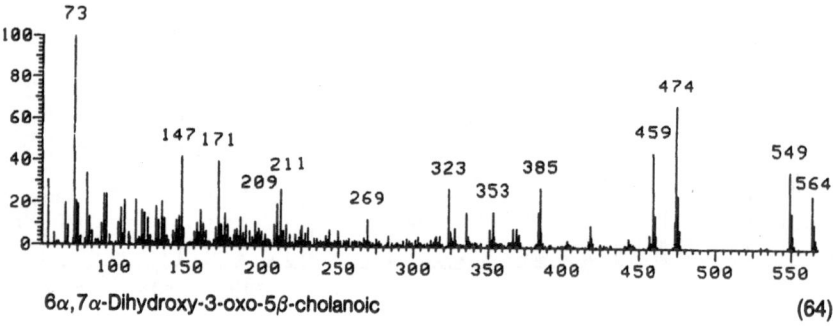

6α,7α-Dihydroxy-3-oxo-5β-cholanoic   (64)

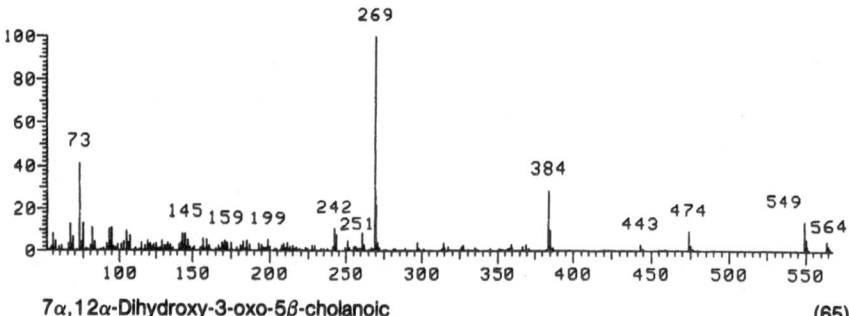

7α,12α-Dihydroxy-3-oxo-5β-cholanoic   (65)

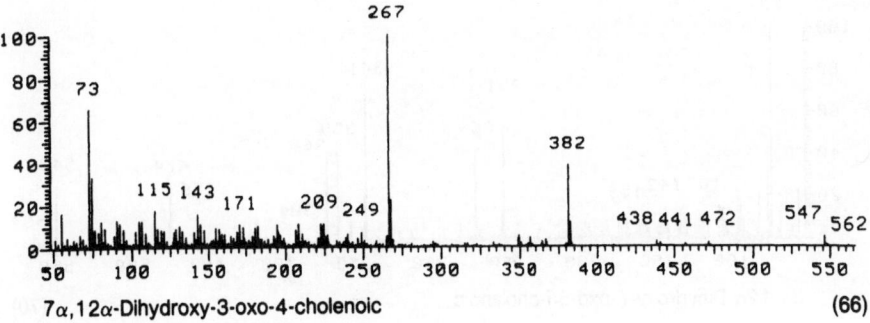

7α,12α-Dihydroxy-3-oxo-4-cholenoic   (66)

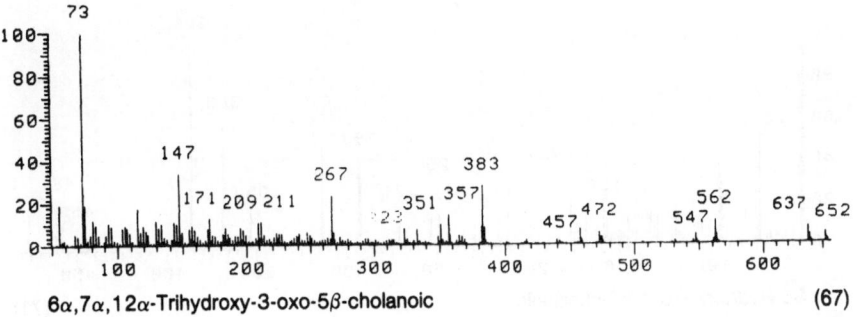

6α,7α,12α-Trihydroxy-3-oxo-5β-cholanoic   (67)

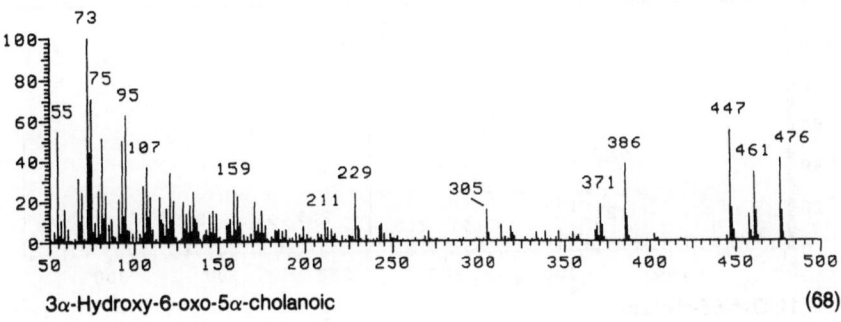

3α-Hydroxy-6-oxo-5α-cholanoic   (68)

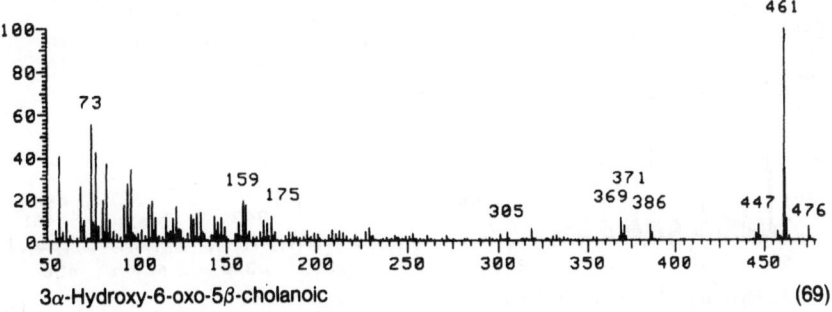

3α-Hydroxy-6-oxo-5β-cholanoic   (69)

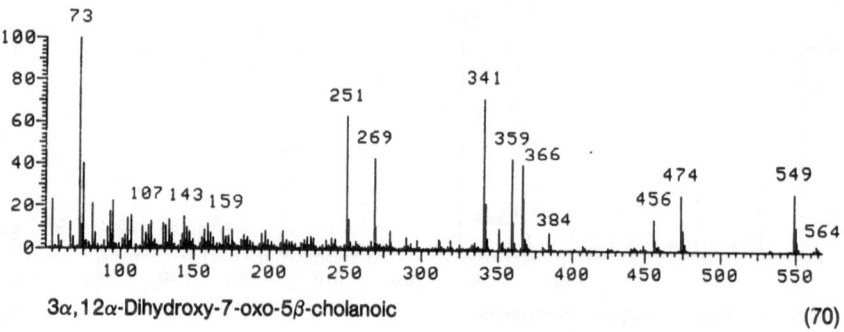

3α,12α-Dihydroxy-7-oxo-5β-cholanoic   (70)

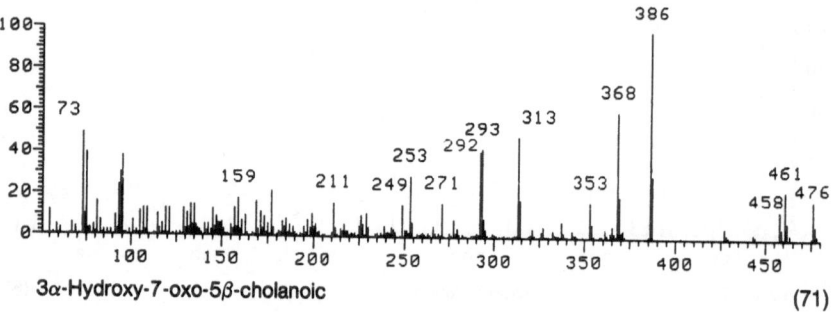

3α-Hydroxy-7-oxo-5β-cholanoic   (71)

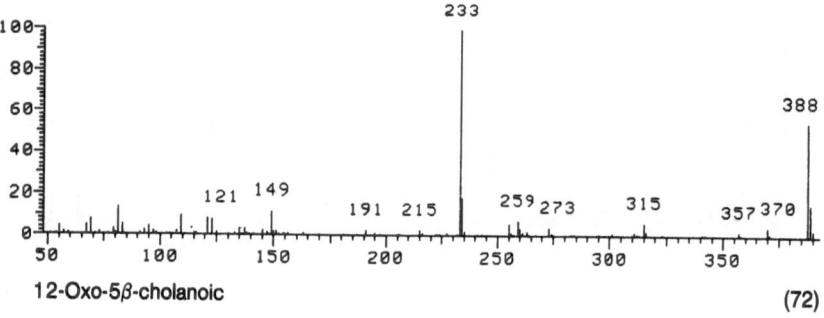

12-Oxo-5β-cholanoic   (72)

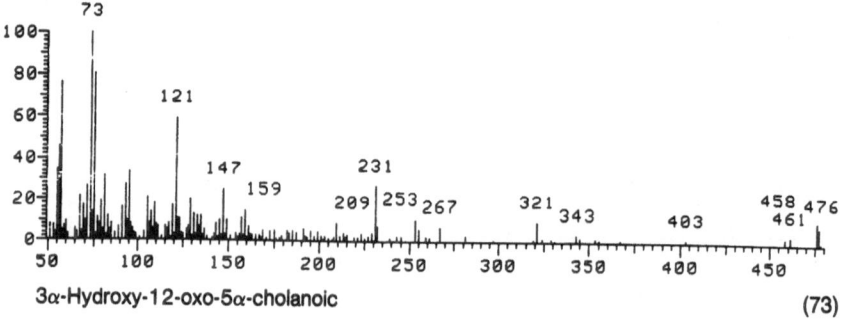

3α-Hydroxy-12-oxo-5α-cholanoic   (73)

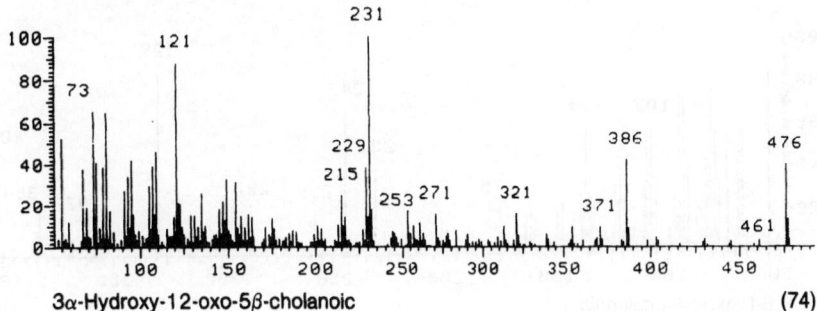

3α-Hydroxy-12-oxo-5β-cholanoic   **(74)**

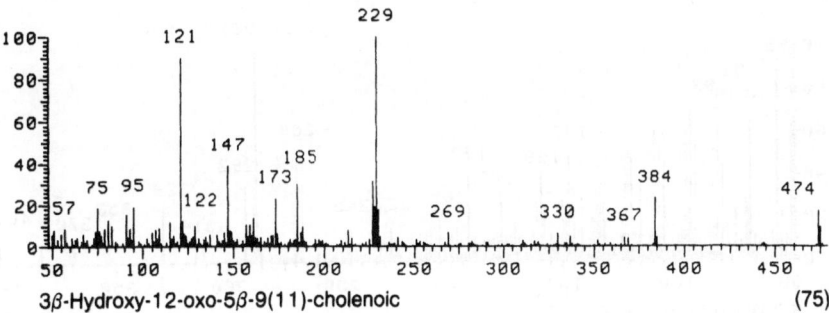

3β-Hydroxy-12-oxo-5β-9(11)-cholenoic   **(75)**

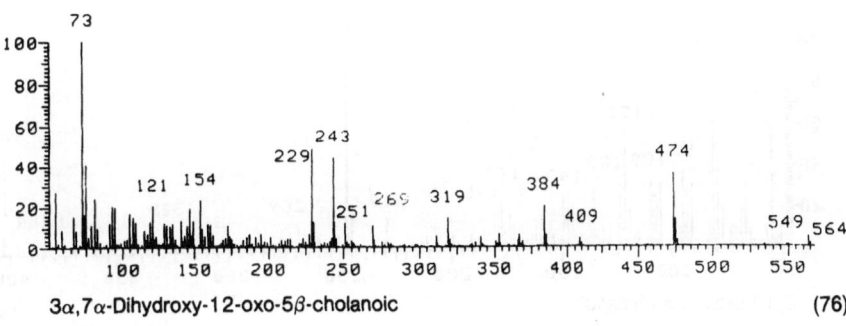

3α,7α-Dihydroxy-12-oxo-5β-cholanoic   **(76)**

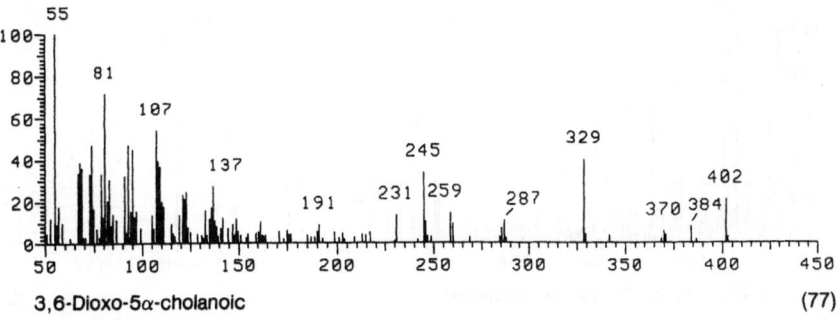

3,6-Dioxo-5α-cholanoic   **(77)**

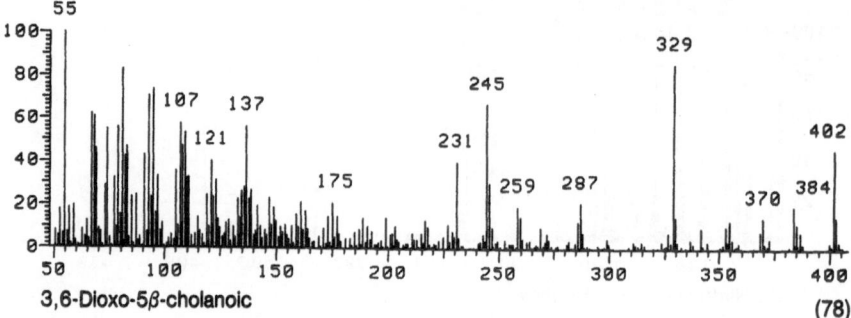

3,6-Dioxo-5β-cholanoic

(78)

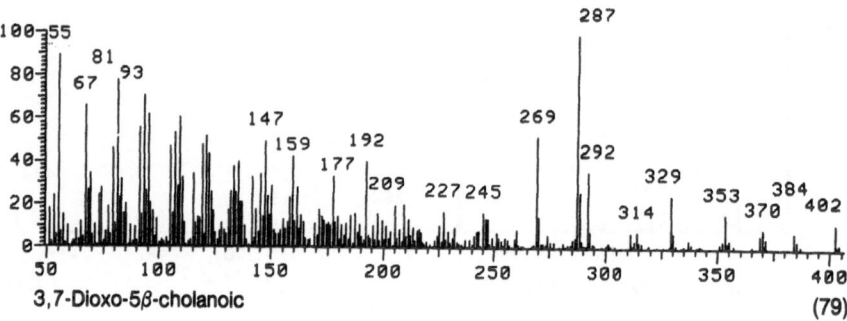

3,7-Dioxo-5β-cholanoic

(79)

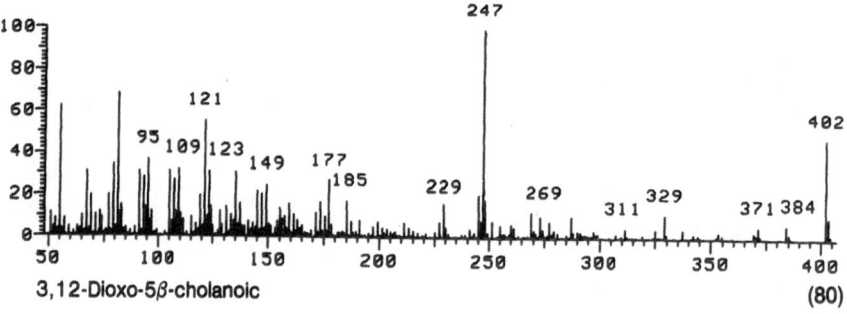

3,12-Dioxo-5β-cholanoic

(80)

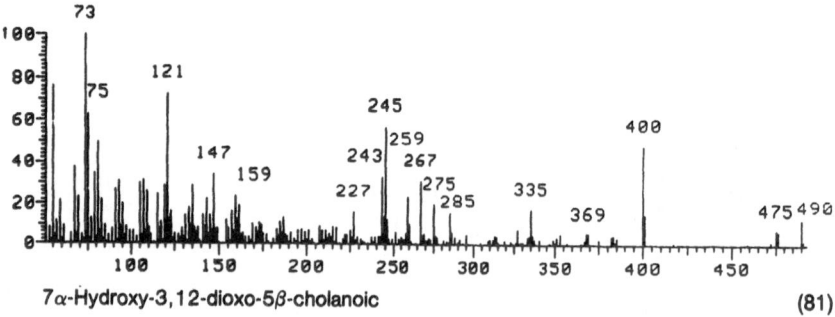

7α-Hydroxy-3,12-dioxo-5β-cholanoic

(81)

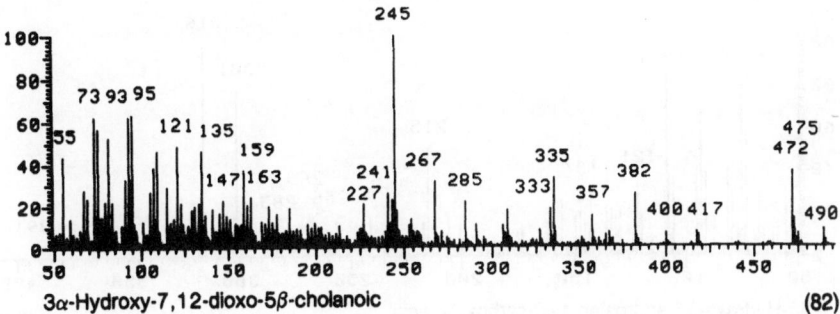

3α-Hydroxy-7,12-dioxo-5β-cholanoic (82)

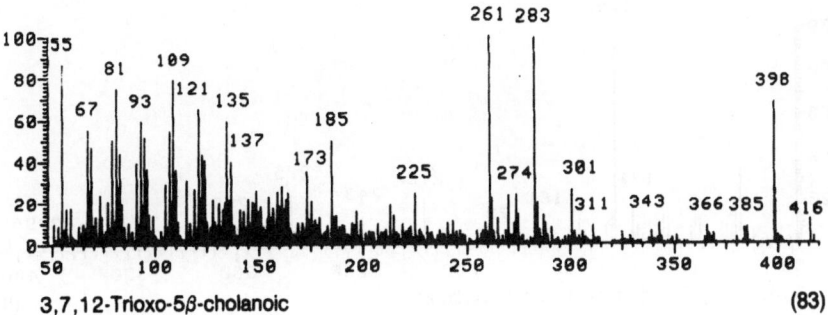

3,7,12-Trioxo-5β-cholanoic (83)

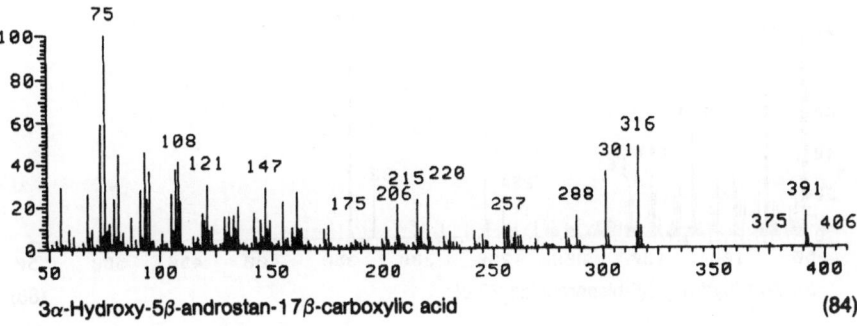

3α-Hydroxy-5β-androstan-17β-carboxylic acid (84)

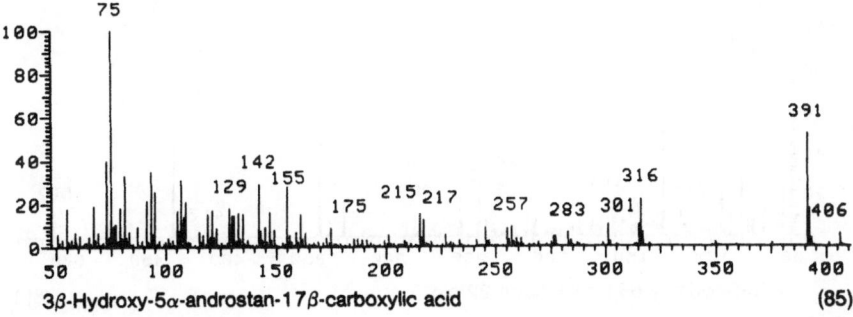

3β-Hydroxy-5α-androstan-17β-carboxylic acid (85)

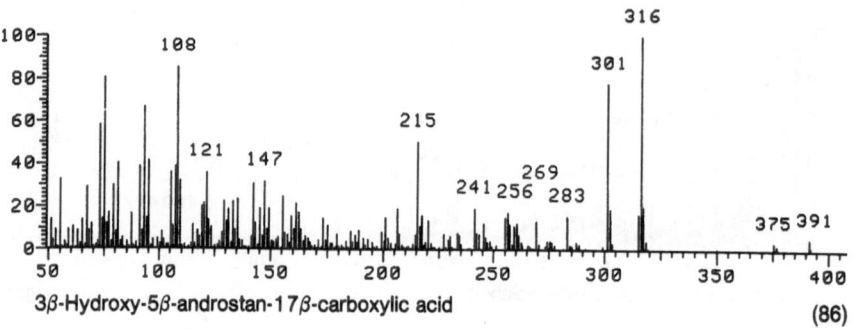

3β-Hydroxy-5β-androstan-17β-carboxylic acid

(86)

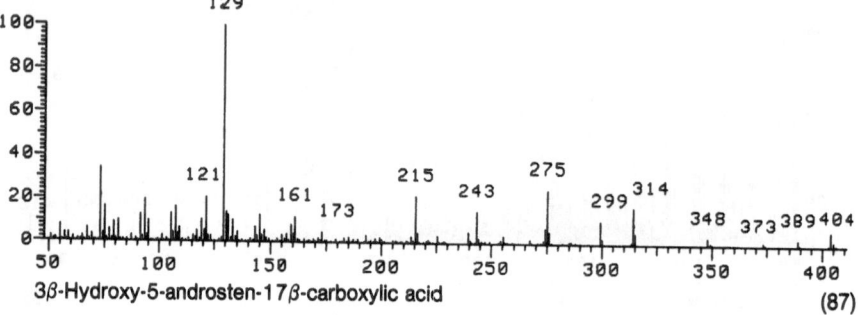

3β-Hydroxy-5-androsten-17β-carboxylic acid

(87)

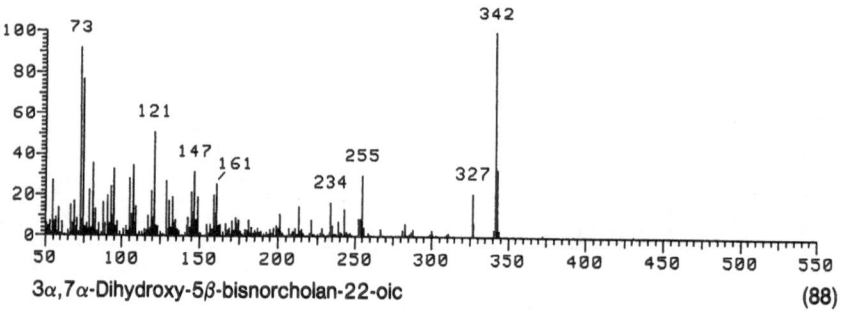

3α,7α-Dihydroxy-5β-bisnorcholan-22-oic

(88)

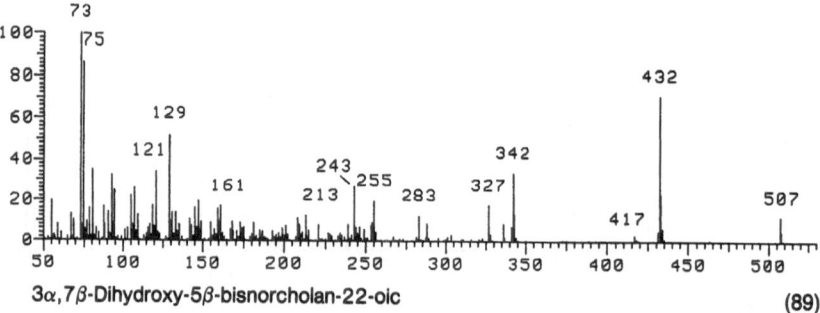

3α,7β-Dihydroxy-5β-bisnorcholan-22-oic

(89)

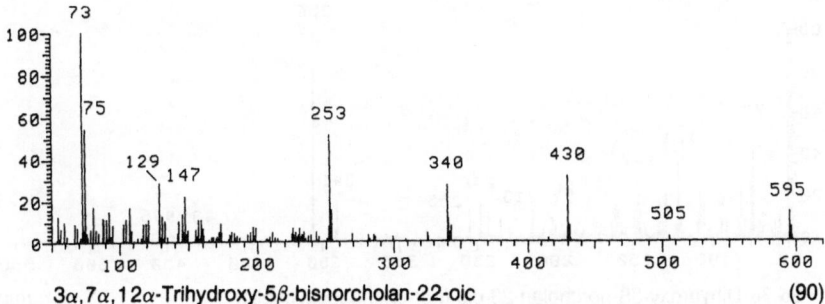

3α,7α,12α-Trihydroxy-5β-bisnorcholan-22-oic (90)

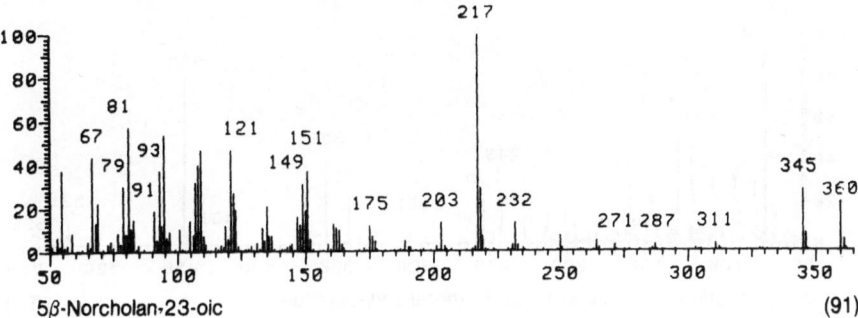

5β-Norcholan-23-oic (91)

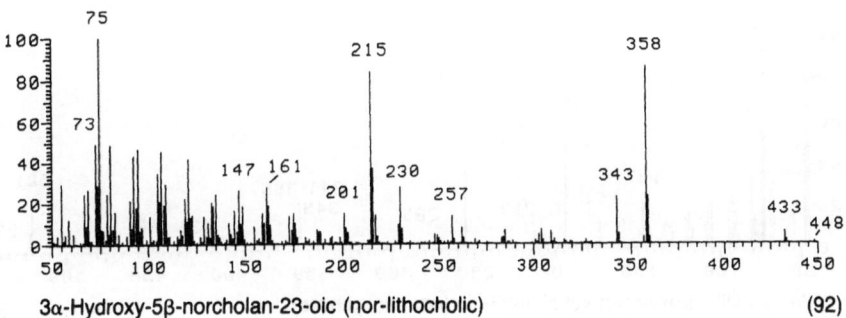

3α-Hydroxy-5β-norcholan-23-oic (nor-lithocholic) (92)

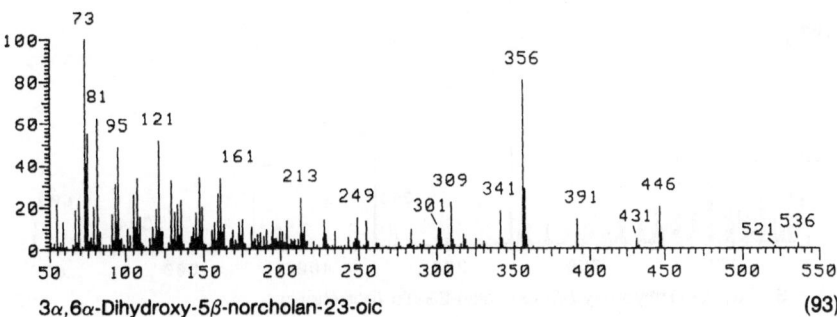

3α,6α-Dihydroxy-5β-norcholan-23-oic (93)

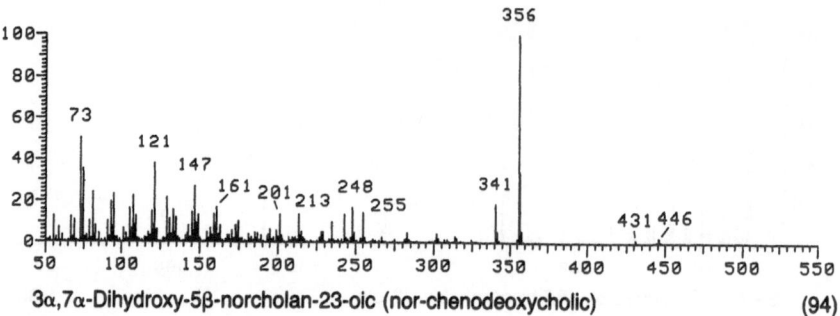

3α,7α-Dihydroxy-5β-norcholan-23-oic (nor-chenodeoxycholic)   (94)

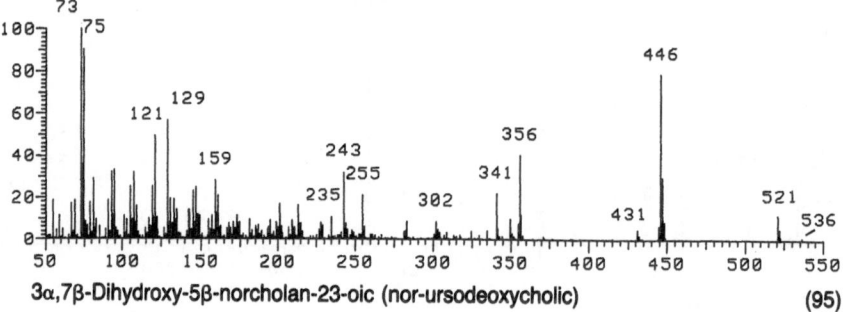

3α,7β-Dihydroxy-5β-norcholan-23-oic (nor-ursodeoxycholic)   (95)

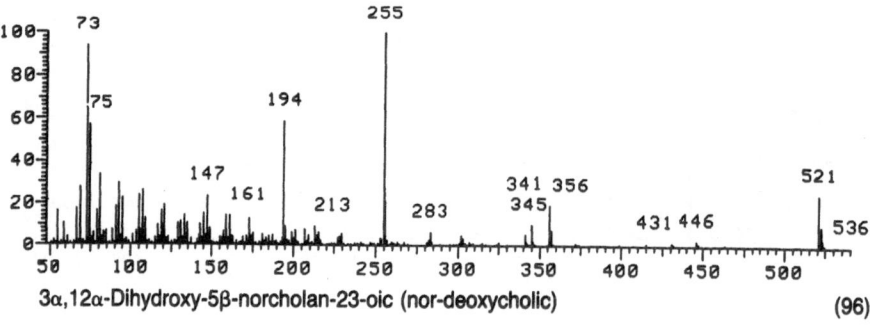

3α,12α-Dihydroxy-5β-norcholan-23-oic (nor-deoxycholic)   (96)

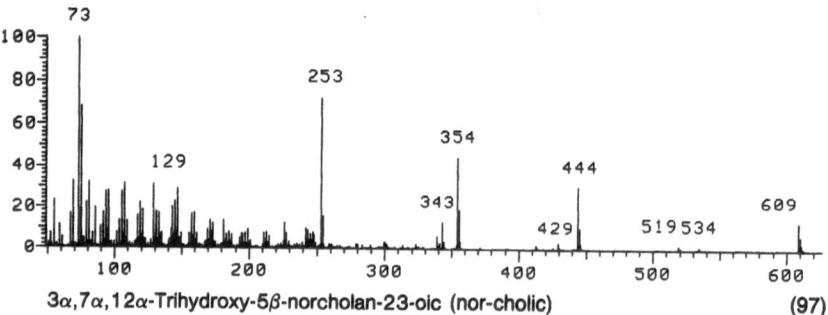

3α,7α,12α-Trihydroxy-5β-norcholan-23-oic (nor-cholic)   (97)

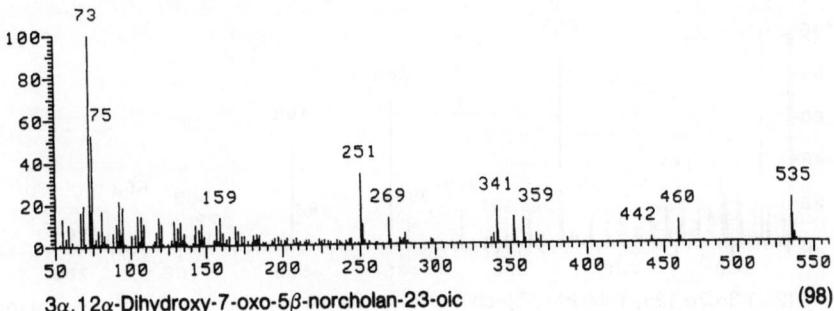

3α,12α-Dihydroxy-7-oxo-5β-norcholan-23-oic   (98)

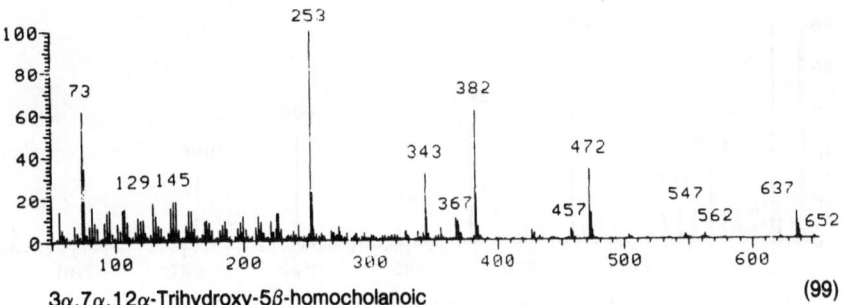

3α,7α,12α-Trihydroxy-5β-homocholanoic   (99)

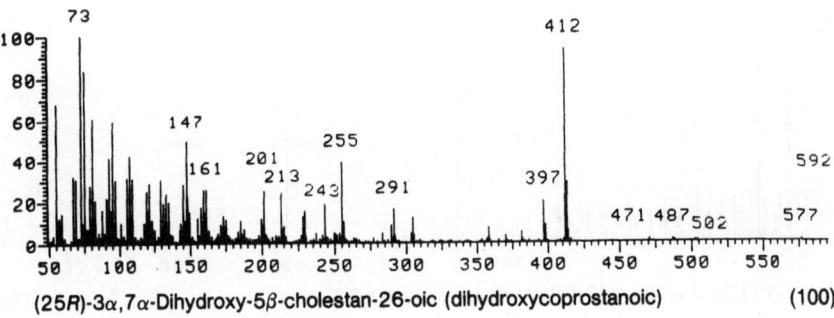

(25R)-3α,7α-Dihydroxy-5β-cholestan-26-oic (dihydroxycoprostanoic)   (100)

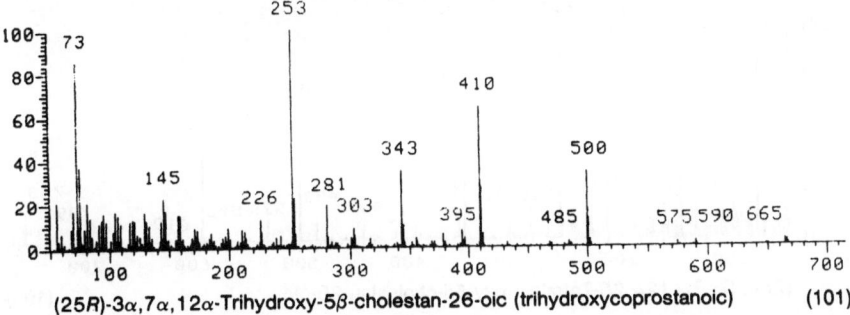

(25R)-3α,7α,12α-Trihydroxy-5β-cholestan-26-oic (trihydroxycoprostanoic) (101)

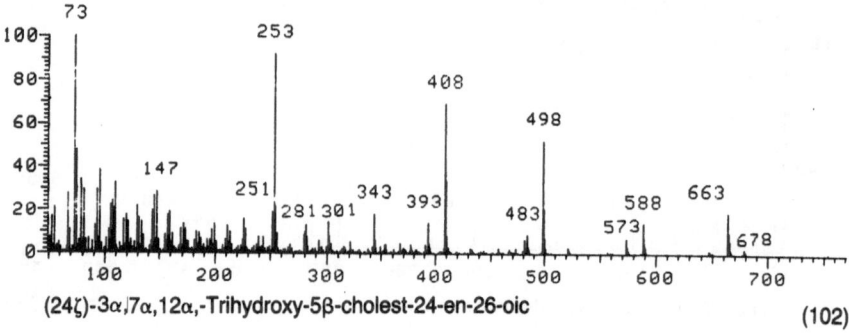

(24ζ)-3α,7α,12α,-Trihydroxy-5β-cholest-24-en-26-oic   (102)

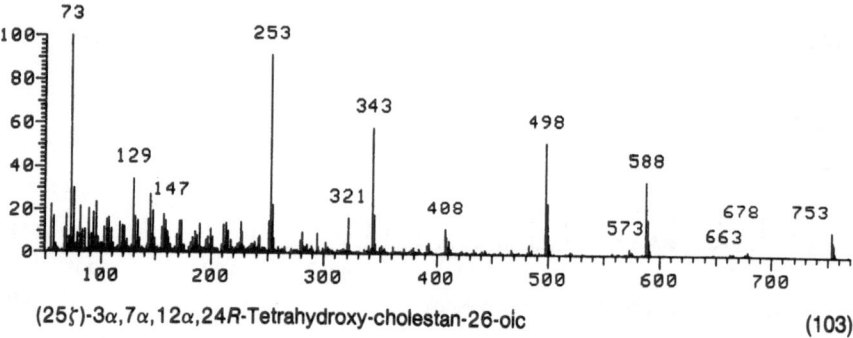

(25ζ)-3α,7α,12α,24R-Tetrahydroxy-cholestan-26-oic   (103)

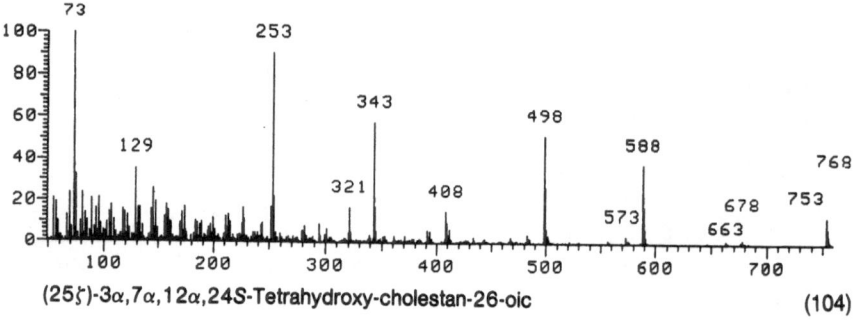

(25ζ)-3α,7α,12α,24S-Tetrahydroxy-cholestan-26-oic   (104)

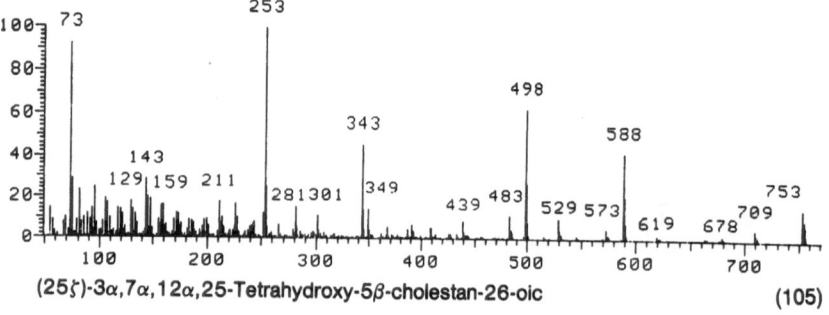

(25ζ)-3α,7α,12α,25-Tetrahydroxy-5β-cholestan-26-oic   (105)

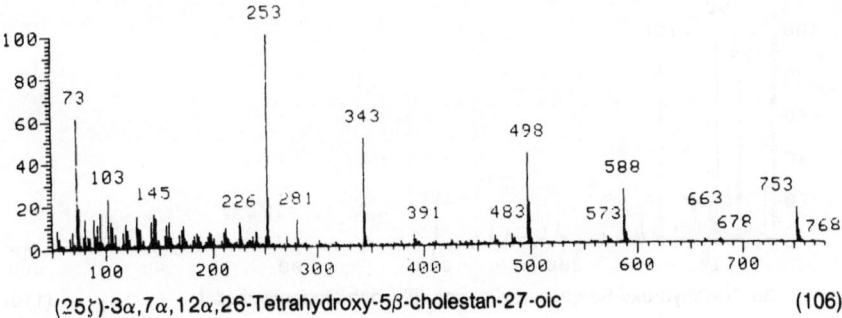

(25ζ)-3α,7α,12α,26-Tetrahydroxy-5β-cholestan-27-oic   (106)

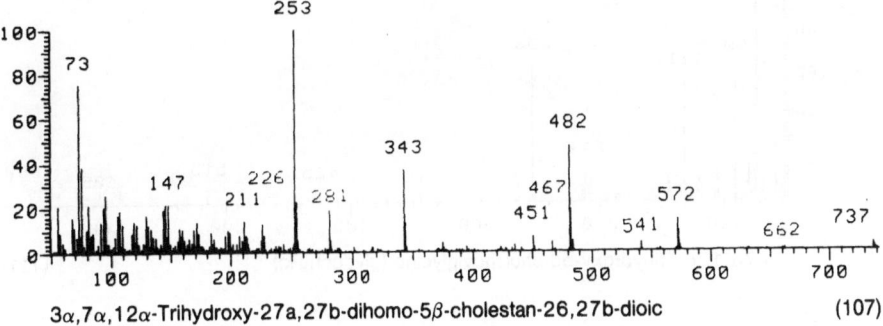

3α,7α,12α-Trihydroxy-27a,27b-dihomo-5β-cholestan-26,27b-dioic   (107)

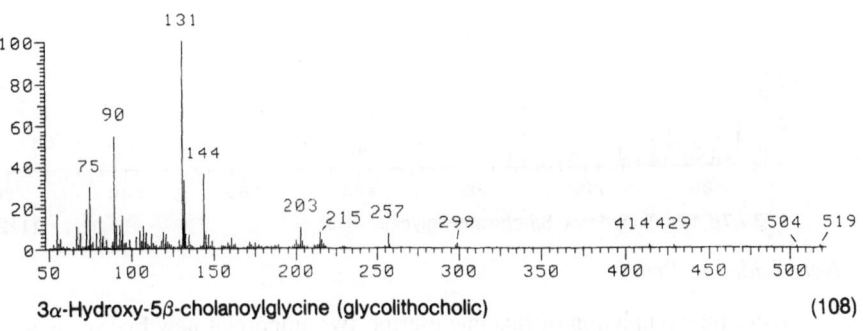

3α-Hydroxy-5β-cholanoylglycine (glycolithocholic)   (108)

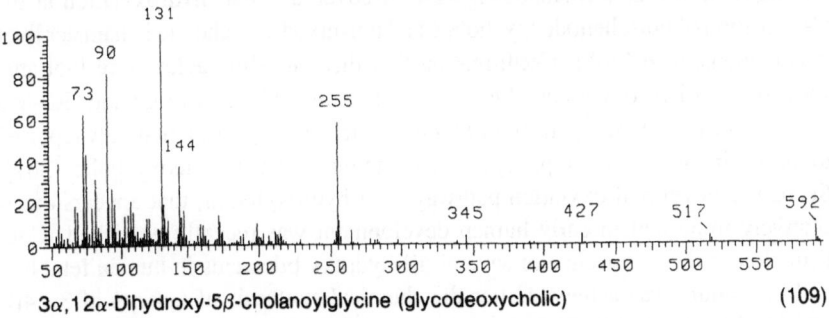

3α,12α-Dihydroxy-5β-cholanoylglycine (glycodeoxycholic)   (109)

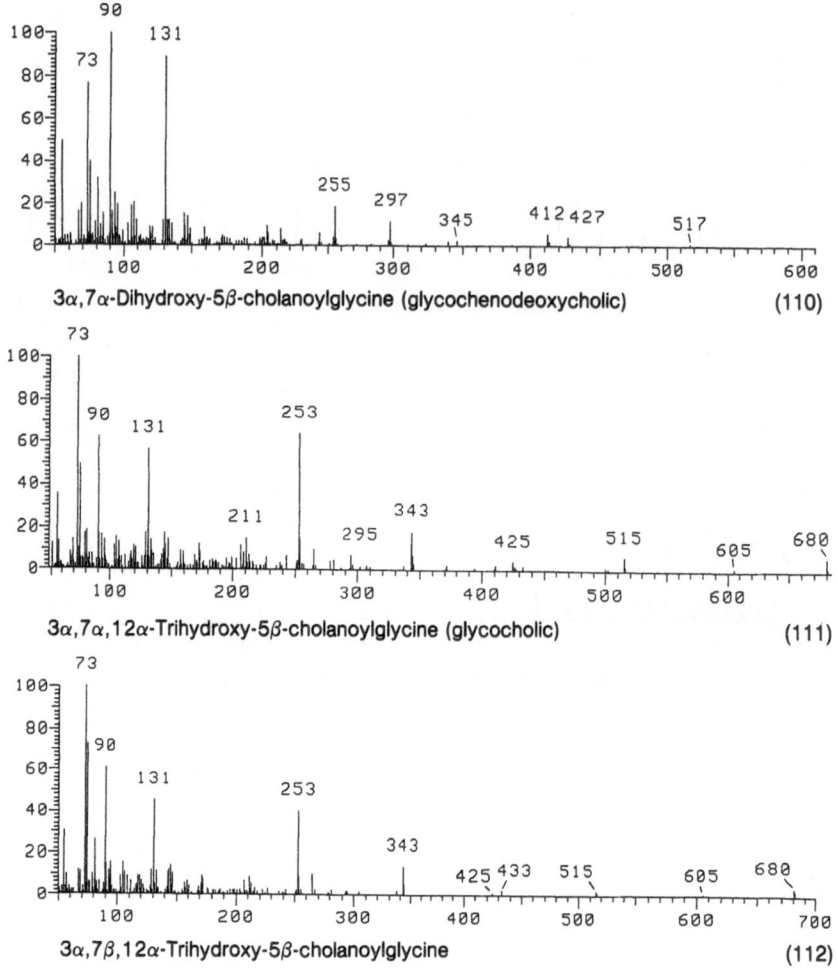

$3\alpha,7\alpha$-Dihydroxy-5β-cholanoylglycine (glycochenodeoxycholic)   (110)

$3\alpha,7\alpha,12\alpha$-Trihydroxy-5β-cholanoylglycine (glycocholic)   (111)

$3\alpha,7\beta,12\alpha$-Trihydroxy-5β-cholanoylglycine   (112)

Note Added in Proof

After the completion of this manuscript, two important new hydroxylation pathways in bile acid metabolism were discovered. First, hydroxylation at the C-5 position of nor-chenodeoxycholate and nor-ursodeoxycholate in hamster liver was demonstrated [305]. Confirmation that these nor-bile acids were biotransformed to 5β-hydroxy metabolites was made from ^{13}C-NMR [305] and GC–MS studies [306]. A 3ξ,5ξ,7β-trihydroxy-cholanoic acid was also tentatively reported to occur in the urine of patients orally administered ursodeoxycholic [307]. Second, a novel hydroxylation pathway, C-4 hydroxylation, that appears quantitatively important in early human development was recently described. Confirmation of the occurrence of several 4β-hydroxy bile acids in human fetal bile and meconium was achieved after the chemical synthesis of a series of 3α,4β- and 3β,4β-hydroxylated bile acids and analysis by GC–MS. The finding that

3α,4β,7α-trihydroxy-5β-cholanoic accounted for 5–15% of the total biliary bile acids of the human fetus in early gestation suggests the liver as the site of 4β-hydroxylation in man [308, 309].

Continued application of FAB–MS to urinary bile acid analysis has led to the recent discovery of a new inborn error in bile acid synthesis in twins with familial giant-cell hepatitis [310, 311]. A defect in Δ^4-3-oxosteroid 5β-reductase was characterized by the accumulation in the plasma and the urinary excretion of relatively large amounts of Δ^4-3-oxo and allo(5α)-bile acids [310, 311].

ACKNOWLEDGMENTS. The help of Mary Jo McCarthy and Mr. R. A. Carruthers in the preparation of the library file of mass spectra is gratefully acknowledged.

REFERENCES

1. J. Sjövall, P. Eneroth, and R. Ryhage, in "The Bile Acids; Chemistry, Physiology and Metabolism" (P. P. Nair and D. Kritchevsky, eds.), pp. 209–248, Plenum Press, New York (1971).
2. J. G. Liehr, E. E. Kingston, and J. H. Beynon, Biomed. Mass Spectrom. 12, 95 (1985).
3. J. H. Beynon, R. G. Cooks, J. W. Amy, W. E. Baitinger, and T. Ridley, Anal. Chem. 45, 1023A (1973).
4. D. F. Hunt, J. Shabanowitz, and A. B. Gyiordani, Anal. Chem. 52, 386 (1980).
5. F. W. McLafferty, "Tandem Mass Spectrometry," New York (1983).
6. J. R. Chapman, "Computers in Mass Spectrometry," Academic Press, London-New York-San Francisco (1978).
7. R. A. Hites, and K. Biemann, Anal. Chem. 42, 855 (1970).
8. C. C. Sweeley, N. D. Young, J. F. Holland, and S. C. Gates, J. Chromatogr. 99, 507 (1974).
9. B. J. Kimble, R. E. Cox, R. V. McPherron, R. W. Olsen, E. Roitman, F. C. Walls, and A. L. Burlingame, J. Chromatogr. Sci. 12, 647 (1974).
10. J. E. Biller, and K. Biemann, Anal. Lett. 7, 515 (1974).
11. H. Nau, and K. Biemann, Anal. Chem. 46, 426 (1974).
12. R. Reimendal, and J. Sjövall, Anal. Chem. 45, 1083 (1973).
13. B. Almé, A. Bremmelgaard, J. Sjövall, and P. Thomassen, J. Lipid Res. 18, 339 (1977).
14. L. Baczynskyji, D. J. Duchamp, J. F. Ziesel, and U. Axen, Anal. Chem. 45, 479 (1973).
15. R. C. Murphy, S. E. Hattox, and H. R. Helbig, Biomed. Mass Spectrom. 5, 444 (1978).
16. Z. R. Vlahcevic, T. Cronholm, T. Curstedt, and J. Sjövall, Biochim. Biophys. Acta 618, 369 (1980).
17. F. C. Falkner, B. J. Sweetman, and J. T. Watson, Appl. Spectrosc. Rev. 10, 51 (1975).
18. S. C. Gates, M. J. Smisko, C. L. Ashendel, N. D. Young, J. F. Holland, and C. C. Sweeley, Anal. Chem. 50, 433 (1978).
19. S. C. Gates, N. Dendramis, and C. C. Sweeley, Clin. Chem. 24, 1674 (1978).
20. G. M. Muschik, L. H. Wright, and J. A. Schroer, Biomed. Mass Spectrom. 6, 166 (1979).
21. P. A. Szczepanik, D. L. Hachey, and P. D. Klein, J. Lipid Res. 17, 314 (1976).
22. K. Kuriyama, Y. Ban, T. Nakashima, and T. Murata, Steroids 34, 717 (1979).
23. B. R. DeMark, and P. D. Klein, J. Lipid Res. 22, 166 (1981).
24. T. Murata, S. Takahashi, S. Ohnishi, K. Hosoi, T. Nakashima, Y. Ban, and K. Kuriyama, J. Chromatogr. 239, 571 (1982).
25. B. L. Jelus, B. Munson, and C. Fenselau, Biomed. Mass Spectrom. 1, 96 (1974).
26. B. Jelus, B. Munson, and C. Fenselau, Anal. Chem. 46, 729 (1974).

27. A. K. Bose, H. Fujiwara, B. N. Pramanik, E. Lazaro, and C. R. Spillert, *Anal. Biochem.* **89,** 284 (1978).
28. A. G. Harrison, "Chemical Ionization Mass Spectrometry," CRC Press, Boca Raton, FL (1983).
29. H. D. Beckey, "Principles of Field Ionization and Field Desorption Mass Spectrometry," Pergamon, Oxford, England (1977).
30. D. E. Games, M. P. Games, A. H. Jackson, A. H. Olavesen, M. Rossiter, and P. J. Winterburn, *Tetrahedron. Lett.* 27, 2377 (1974).
31. R. Shaw and W. H. Elliott, *Biomed. Mass Spectrom.* **5,** 433 (1978).
32. J. O. Whitney, S. Lewis, K. M. Straub, M. M. Thaler, and A. L. Burlingame, (Proc. Jap. Soc. Med. Mass Spectrom.) *Koenshu-Iyo Masu Kenkyukai* **6,** 33 (1981).
33. R. J. Cotter, *Anal. Chem.* **53,** 719 (1981).
34. R. J. Day, J. Zimmerman, and D. M. Hercules, *Spectroscopy Lett.* **14,** 773 (1981).
35. A. Benninghoven, D. Jaspers, and W. K. Sichtermann, *Appl. Physiol.* **11,** 35 (1976).
36. M. Barber, R. S. Bordoli, R. D. Sedgwick, and A. N. Tyler, *Nature* **293,** 270 (1981).
37. J. O. Whitney and A. L. Burlingame, *Koenshu-Iyo Masu Kenkyukai* **7,** 3 (1982).
38. J. O. Whitney, M. M. Thaler, and N. Blanckert, *in* "Bile Acids and Cholesterol in Health and Disease" (G. Paumgartner, A. Stiehl, and W. Gerok, eds.) p. 213, MTP Press Ltd., Lancaster, England (1983).
39. J. O. Whitney, S. Lewis, K. M. Straub, F. C. Walls, A. L. Burlingame, and M. M. Thaler, *in* "Proceedings 29th Annual Conference Mass Spectrometry," Allied Topics, American Society for Mass Spectrometry, Minneapolis, p. 729 (1981).
40. J. O. Whitney, V. Ling, D. Grunberger, M. M. Thaler, and A. L. Burlingame, *Proc. Jap. Soc. Med. Mass Spectrom.* **8,** 47 (1983).
41. J. O. Whitney, D. Grunberger, E. Roitman, V. Ling, Y-M. Yang, P. Rosenthal, and A. L. Burlingame, *in* "Proceedings 32nd Annual Conference Mass Spectrometry," Allied Topics, American Society of Mass Spectrometry, San Antonio, p. 133 (1984).
42. J. G. Liehr, C. F. Beckner, A. M. Ballatore, and R. M. Caprioli, *in* "Proceedings 30th Annual Conference Mass Spectrometry," Allied Topics, American Society of Mass Spectrometry, Honolulu, p. 234 (1982).
43. Y. Itagaki, T. Higuchi, Y. Naito, J. Goto, and T. Nambara, *in* "Proceedings 30th Annual Conference Mass Spectrometry," Allied Topics, American Society of Mass Spectrometry, Honolulu, p. 236 (1982).
44. A. M. Ballatore, C. F. Beckner, R. M. Caprioli, N. E. Hoffman, and J. G. Liehr, *Steroids,* **41,** 198 (1983).
45. A. M. Lawson, C. Bayliss, and K. D. R. Setchell, "Presented at VIII International Bile Acid Meeting, Enterohepatic Circulation of Bile Acids and Sterol Metabolism," Berne, p. 260 (Abstract) (1984).
46. A. M. Lawson, K. D. R. Setchell, and C. Bayliss, *in* "Proceedings British Mass Spectrometry Society," British Mass Spectrometry Society, Edinburgh, p. 119 (1984).
47. J. Sjövall, A. M. Lawson, and K. D. R. Setchell, *in* "Methods in Enzymology" (J. H. Law and H. C. Rilling, eds.), Volume 111, p. 63 Academic Press, London, (1985).
48. K. B. Tomer, N. J. Jensen, and J. O. Whitney, *in* "Proceedings 33rd Annual Conference Mass Spectrometry," Allied Topics, American Society of Mass Spectrometry, San Diego, p. 703 (1985); K. B. Tomer, N. J. Jensen, M. L. Gross, and J. Whitney, *Biomed. Mass Spectrom.* **12,** 265 (1985).
49. P. D. G. Dean and R. T. Aplin, *Steroids* **8,** 565 (1966).
50. H. Egger, *Monatsh. Chem.* **99,** 1163 (1968).
51. M. A. Baldwin and F. W. McLafferty, *Org. Mass Spectrom.* **7,** 1353 (1973).
52. R. J. Cotter, *Anal. Chem.* **52,** 1589A (1980).
53. S. Bergström, R. Ryhage, and E. Stenhagen, *Acta Chem. Scand.* 12, 1349 (1958).
54. R. Ryhage and E. Stenhagen, *J. Lipid Res.* **1,** 361 (1960).

55. S. Bergström, R. Ryhage, and E. Stenhagen, *Svensk. Kem. Tidskr.* **73**, 566 (1961).
56. W. H. Elliott, *in "Biochemical Applications of Mass Spectrometry—Mass Spectra of Bile Acids" (G. R. Waller, ed.), Wiley Interscience, New York, London, p. 291 (1972).*
57. W. H. Elliott, *in "Biochemical Applications of Mass Spectrometry—Mass Spectra of Bile Acids,"* First Supplementary Volume (G. R. Waller and O. C. Dermer, eds.), Wiley-Interscience, New York, London, p. 229 (1980).
58. H. Miyazaki, M. Ishibashi, and K. Yamashita, *Biomed. Mass Spectrom.* **5**, 469 (1978).
59. A. Fukunaga, Y. Hatta, M. Ishibashi, and H. Miyazaki, *J. Chromatogr.* **190**, 339 (1980).
60. S. H. G. Andersson and J. Sjövall, *J. Chromatogr.* **289**, 195 (1984).
61. C. G. Edmonds, J. A. McCloskey, and V. A. Edmonds, *Biomed. Mass Spectrom.* **10**, 237 (1983).
62. T. P. Fan, E. D. Hardin, and M. L. Vestal, *Anal. Chem.* **56**, 1870 (1984).
63. M. Dedieu, C. Juin, P. J. Arpino, J. P. Bounine, and G. Guiochon, *J. Chromatogr.* **251**, 203 (1982).
64. C. R. Blakley, J. J. Carmody, and M. L. Vestal, *J. Am. Chem. Soc.* **102**, 5931 (1980).
65. C. R. Blakley and M. L. Vestal, *Anal. Chem.* **55**, 750 (1983).
66. R. Shaw and W. H. Elliott, *Lipids* **13**, 971 (1978).
67. G. Mingrone, A. V. Greco, L. Boniforti, and S. Passi, *Lipids* **18**, 90 (1983).
68. S. Miyazaki, H. Tanaka, R. Horikawa, H. Tsuchiya, and K. Imai, *J. Chromatogr.* **181**, 177 (1980).
69. S. Okuyama, D. Uemura, and Y. Hirata, *Chem. Lett.* **461** (1979).
70. F. Stellaard, D. L. Hachey, and P. D. Klein, *Anal. Biochem.* **87**, 359 (1978).
71. N. Parris, *Anal. Biochem.* **100**, 260 (1979).
72. J. Goto, H. Kato, and T. Nambara, *Lipids* **13**, 908 (1978).
73. J. Goto, H. Kato, and T. Nambara, *J. Liq. Chromatogr.* **3**, 645 (1980).
74. D. E. Games, C. Eckers, J. L. Gower, P. Hirter, M. E. Knight, E. Lewis, K. R. N. Rao, and N. C. Weerasinghe, *in* "Current Developments in the Clinical Applications of HPLC, GC and MS" (A. M. Lawson, C. K. Lim, and W. Richmond, eds.), Academic Press, London (1980).
75. A. Kuksis, *in* "Lipid Chromatographic Analysis. Gas Chromatography of Bile Acids," Volume 2 (G. V. Marinetti, ed.), Dekker, New York, p. 479 (1976).
76. R. Galeazzi, E. Kok, and N. Javitt, *J. Lipid Res.* **17**, 280 (1976).
77. I. M. Yousef, M. M. Fisher, J. J. Myher, and A. Kuksis, *Anal. Biochem.* **75**, 538 (1976).
78. P. A. Szczepanik, D. L. Hachey, and P. D. Klein, *J. Lipid Res.* **18**, 280 (1978).
79. T. Nakashima, Y. Ban, K. Kuriyama, and T. Takino, *Jpn. J. Pharmacol.* **29**, 667 (1979).
80. T. Iida, F. C. Chang, T. Matsumoto, and T. Tamura, *Biomed. Mass Spectrom.* **9**, 473 (1982).
81. H. Budzikiewicz, *in* "Biomedical Applications of Mass Spectrometry" (G. R. Waller, ed.), p. 251, Wiley, New York (1972).
82. Z. C. Zaretskii, *in* "Mass Spectrometry of Steroids," Wiley, New York (1976).
83. H. Budzikiewicz, *in* "Biomedical Applications of Mass Spectrometry," First Supplementary Volume (G. R. Waller, and O. C. Dermer, eds.), p. 211, Wiley, New York, 1980.
84. T. Cronholm, I. Makino, and J. Sjövall, *Eur. J. Biochem.* **24**, 507 (1972).
85. T. Harano, K. Harano, and K. Yamasaki, *Steroids* **32**, 73 (1978).
86. H. Eriksson, W. Taylor, and J. Sjövall, *J. Lipid Res.* **19**, 177 (1978).
87. L. Aringer, and L. Nordström, *Biomed. Mass Spectrom.* **8**, 183 (1981).
88. M. I. Kelsey, M. M. Mau, and W. H. Elliott, *Steroids* **18**, 261 (1971).
89. T. A. Baillie, H. Eriksson, J. E. Herz, and J. Sjövall, *Eur. J. Biochem.* **55**, 157 (1975).
90. F. Kern, Jr., H. Eriksson, T. Curstedt, and J. Sjövall, *J. Lipid Res.* **18**, 623 (1977).
91. A. Bremmelgaard and J. Sjövall, *J. Lipid Res.* **21**, 1072 (1980).
92. C. J. W. Brooks, W. J. Cole, H. B. McIntyre, and A. G. Smith, *Lipids* **15**, 745 (1980).
93. B. Strandvik, and S-Å. Wikström, *Eur. J. Clin. Invest.* **12**, 301 (1982).

94. P. Back and K. Walter, *Gastroenterology* **78**, 671 (1980).
95. A. Bremmelgaard and J. Sjövall, *Eur. J. Clin. Invest.* **9**, 341 (1979).
96. K. Carlström, D. N. Kirk, and J. Sjövall, *J. Lipid Res.* **22**, 1225 (1981).
97. B. Almé and J. Sjövall, *J. Steroid Biochem.* **13**, 907 (1980).
98. P. T. Clayton, D. P. R. Muller, and A. M. Lawson, *Biochem. J.* **206**, 489 (1982).
99. A. M. Lawson, K. D. R. Setchell, P. T. Clayton, and D. P. R. Muller, *in* Proceedings 31st Annual Conference Mass Spectrometry, Allied Topics, American Society of Mass Spectrometry, Boston, p. 510 (1983).
100. J. G. Allen, G. H. Thomas, C. J. W. Brooks, and B. A. Knights, *Steroids* **13**, 133 (1969).
101. M. Kimura, M. Kawata, M. Tohma, A. Fujino, K. Yamazaki, and T. Sawya, *Chem. Pharm. Bull.* **20**, 1883 (1972).
102. S. V. Hiremath and W. H. Elliott, *Steroids* **38**, 465 (1981).
103. P. Child, A. Kuksis, and L. Marai, *Can. J. Biochem.* **57**, 216 (1979).
104. A. Kuksis and P. Child, *Lipids* **15**, 770 (1980).
105. P. T. Clayton, J. V. Leonard, A. M. Lawson, K. D. R. Setchell, S. Andersson, B. Egestadt, and J. Sjövall, *J. Clin. Invest.* **79**, 1031 (1987).
106. I. Makino, J. Sjövall, A. Norman, and B. Strandvik, *FEBS Lett.* **15**, 161 (1971).
107. C. J. W. Brooks, D. J. Harvey, B. S. Middleditch, and P. Vouros, *Org. Mass Spectrom.* **7**, 925 (1973).
108. I. Björkhem, J-A. Gustafsson, and J. Sjövall, *Org. Mass Spectrom.* **7**, 277 (1973).
109. A. Kallner, *Acta Chem. Scand.* **22**, 2353 (1968).
110. A. E. Cowen, A. F. Hofmann, D. L. Hachey, P. J. Thomas, D. T. E. Belobaba, P. D. Klein, and L. Tokes, *J. Lipid Res.* **17**, 231 (1976).
111. Y. Shalen and W. H. Elliott, *Steroids* **28**, 655 (1976).
112. B. I. Cohen, G. S. Tint, T. Kuramoto, and E. H. Mosback, *Steroids* **25**, 365 (1975).
113. G. A. D. Haslewood, "The Biological Importance of Bile Salts," North-Holland, Amsterdam (1978).
114. S. Y. Karmat and W. H. Elliott, *Steroids* **20**, 279 (1972).
115. I. G. Anderson, G. A. D. Haslewood, R. S. Oldham, B. Amos, and L. Tokes, *Biochem. J.* **141**, 485 (1974).
116. R. F. Hanson, J. N. Isenberg, G. C. Williams, D. Hachey, P. Szczepanik, P. D. Klein, and H. L. Sharp, *J. Clin. Invest.* **56**, 577 (1975).
117. G. A. D. Haslewood, S. Ikawa, L. Tokes, and D. Wong, *Biochem. J.* **171**, 409 (1978).
118. I. G. Anderson, T. Briggs, G. A. D. Haslewood, R. S. Oldham. H. Schuren, and L. Tokes, *Biochem. J.* **183**, 507 (1979).
119. M. Une, N. Matsumoto, K. Kihira, M. Yasuhara, T. Kuramoto, and T. Hoshita, *J. Lipid Res.* **21**, 269 (1980).
120. Y. Noma, M. Une, K. Kihira, M. Yashuda, T. Kuramoto, and T. Hoshita, *J. Lipid Res.* **21**, 339 (1980).
121. M. Une, T. Kuramoto, and T. Hoshita, *J. Lipid Res.* **24**, 1468 (1983).
122. G. S. Tint, B. Dayal, A. K. Batta, S. Shefer, T. Joanen, L. McNease, and G. Salen, *J. Lipid Res.* **21**, 110 (1980).
123. G. S. Tint, B. Dayal, A. K. Batta, S. Shefer, T. Joanen, K. McNease, and G. Salen, *Gastroenterology* **80**, 114 (1981).
124. S. S. Ali, E. Stephenson, and W. H. Elliott, *J. Lipid Res.* **23**, 947 (1982).
125. K. Kihira, Y. Morioka, and H. Takahiko, *J. Lipid Res.* **22**, 1181 (1981).
126. B. Amos, I. G. Anderson, G. A. D. Haslewood, and L. Tokes, *Biochem. J.* **161**, 201 (1977).
127. R. F. Hanson and G. Williams, *Biochem. J.* **121**, 863 (1971).
128. S. Y. Kamat and W. H. Elliott, *Steroids* **20**, 279 (1972).
129. M. M. Mui, S. Y. Kamat, and W. H. Elliott, *Steroids* **24**, 239 (1974).

130. T. Kuramoto, H. Kikuchi, H. Sanemori, and T. Hoshita, *Chem. Pharm Bull.* (*Tokyo*) **21**, 952 (1973).

131. H. Eyssen, G. Parmentier, F. Compernolle, J. Boon, and E. Eggermont, *Biochim. Biophys. Acta* **273**, 212 (1972).

132. J. Gustafsson, *J. Biol. Chem.* **250**, 8243 (1975).

133. J. St. Pyrek, R. Lester, E. W. Adcock, and A. Sanghvi, *J. Steroid Biochem.* **18**, 341 (1983).

134. J. St. Pyrek, R. Sterzycki, R. Lester, and E. W. Adcock, *Lipids* **17**, 241 (1982).

135. J. M. Street, W. F. Balistreri, and K. D. R. Setchell, *Gastroenterology* **90**, Abstract (1986).

136. J. St. Pyrek, J. M. Little, and R. Lester, *J. Lipid Res.* **25**, 1324 (1984).

137. K. D. R. Setchell and A. Matsui, *Clin. Chim. Acta.* **127**, 1 (1983).

138. J. J. Myer, L. Marai, A. Kuksis, I. M. Yousef, and M. M. Fisher, *Can. J. Biochem.* **53**, 583 (1975).

139. P. Back, *Hoppe-Seyler's Z. Physiol. Chem.* **357**, 213 (1976).

140. P. Back and D. V. Bowen, *Hoppe-Seyler's Z. Physiol. Chem.* **357**, 219 (1976).

141. G. Mingrone, A. V. Greco, L. Boniforti, and S. Passi, *Lipids* **18**, 90 (1983).

142. B. Almé, A. Norden, and J. Sjövall, *Clin. Chim. Acta* **86**, 251 (1978).

143. I. M. Yousef and M. M. Fisher, *Can. J. Physiol. Pharmacol.* **53**, 880 (1975).

144. P. P. Nair, A. I. Mendeloff, M. Vocci, J. Bankowski, M. Gorelik, G. Herman, and R. Plapinger, *Lipids* **12**, 922 (1977).

145. J. Yanagisawa, Y. Akashi, H. Miyazaki, and F. Nakayama, *J. Lipid Res.* **25**, 1263 (1984).

146. J. M. Street, D. J. H. Trafford, and H. L. J. Makin, *J. Lipid Res.* **27**, 208 (1986).

147. K. D. R. Setchell, *in* "Proceedings of the 1st International Symposium on Bile Acids in Hepatobiliary Disease," p. 63, IRL Press, Oxford, (1984).

148. R. Reimendal and J. Sjövall, *Anal. Chem.* **44**, 21 (1972).

149. P. Back, J. Sjövall, and K. Sjövall, *Med. Biol.* **52**, 31 (1974).

150. P. A. Thomassen, *Eur. J. Clin. Invest.* **9**, 425 (1979).

151. A. Bremmelgaard and B. Almé, *Scand. J. Gastroenterol.* **15**, 593 (1980).

152. H. Miyazaki, M. Ishibashi, M. Inone, M. Itoh, and T. Kubodera, *J. Chromatogr.* **99**, 553 (1974).

153. G. A. DeWeerdt, R. Beke, H. Verdievel, and F. Barbier, *Biomed. Mass Spectrom.* **7**, 515 (1980).

154. M. Axelson, T. Cronholm, T. Curstedt, R. Reimendal, and J. Sjövall, *Chromatographia* **7**, 502 (1974).

155. B. S. Middleditch and D. M. Desiderio, *Anal. Chem.* **45**, 806 (1973).

156. H. Miyazaki, M. Ishibashi, and K. Yamashita, *Biomed. Mass Spectrom.* **5**, 469 (1978).

157. S. H. G. Andersson and J. Sjövall, *J. Chromatogr.* **289**, 195 (1984).

158. Y. Nishikawa, K. Yamashita, M. Ishibashi, and H. Miyazaki, *Chem. Pharm. Bull.* **26**, 2922 (1978).

159. J. B. Watkins, D. Ingall, P. Szczepanik, P. D. Klein, and R. Lester, *N. Engl. J. Med.* **288**, 431 (1973).

160. W. F. Balistreri, A. E. Cowen, A. F. Hofmann, P. A. Szczepanik, and P. D. Klein, *Pediatr. Res.* **9**, 757 (1975).

161. T. Tateyama, Y. Nezu, M. Shino, K. Sakaguchi, K. Katayama, and J. Tsutsumi, *Koenshulyo Masu Kenkyukai* **3**, 207 (1978).

162. F. Kern, Jr., G. T. Everson, B. DeMark, C. McKinley, R. Showalter, W. Erfling, D. Z. Braverman, P. Sczcepanik-Van Leeuwen, and P. D. Klein, *J. Clin. Invest.* **68**, 1229 (1981).

163. G. Everson, B. DeMark, P. D. Klein, R. Showalter, C. McKinley, and F. Kern, Jr., *Gastroenterology* **80**, 1114 (1981).

164. T. Nishida, H. Miwa, M. Yamamoto, T. Koga, and T. Yao, *Gut* **23**, 751 (1982).

165. B. R. DeMark, G. T. Everson, P. D. Klein, R. B. Showalter, and F. Kern, Jr., *J. Lipid Res.* **23**, 204 (1982).

166. F. Stellaard, R. Schubert, and G. Paumgartner, *Biomed. Mass Spectrom.* **10**, 187 (1983).
167. T. Cronholm, A. L. Burlingame, and J. Sjövall, *Eur. J. Biochem.* **49**, 497 (1974).
168. T. Cronholm, I. Makino, and J. Sjövall, *Eur. J. Biochem.* **26**, 251 (1972).
169. T. Cronholm, H. Eriksson, S. Matern, and J. Sjövall, *Eur. J. Biochem* **53**, 405 (1975).
170. D. M. Wilson, A. L. Burlingame, S. Evans, T. Cronholm, and J. Sjövall, *in* "Stable Isotopes. Applications in Pharmacology, Toxicology and Clinical Research" (T. A. Baillie, ed.), p. 205, Macmillan, London (1978).
171. D. M. Wilson, A. L. Burlingame, D. Hazelby, S Evans, T. Cronholm, and J. Sjövall, Proceedings 25th Annual Conference Mass Spectrometry, Allied Topics, American Society of Mass Spectrometry, p. 357 (1977).
172. T. Cronholm, J. Sjövall, D. M. Wilson, and A. L. Burlingame, *Biochim. Biophys. Acta* **575**, 193 (1979).
173. I. Björkhem and A. Lewenhaupt, *J. Biol. Chem.* **254**, 5252 (1979).
174. T. Beppu, Y. Seyama, T. Kasama, and T. Yamakawa, *J. Biochem. (Tokyo)* **89**, 1963 (1981).
175. B. Angelin and I. Björkhem, *Gut* **18**, 606 (1977).
176. B. Angelin, I. Björkhem, and K. Einarsson, *J. Lipid Res.* **19**, 527 (1978).
177. J. Ahlberg, B. Angelin, I. Björkhem, and K. Einarsson, *Gastroenterology* **73**, 1377 (1977).
178. I. Björkhem and O. Falk, *Scand. J. Clin. Lab. Invest.* **43**, 163 (1983).
179. Y. Akashi, H. Miyazaki, and F. Nakayama, *Clin. Chim. Acta* **133**, 125 (1983).
180. H. Takikawa, H. Otsuka, T. Beppu, Y. Seyama, and T. Yamakawa, *J. Biochem. (Tokyo)* **92**, 985 (1982).
181. M. Shino, Y. Nezu, T. Tateyama, K. Sakaguchi, K. Katayama, J. Tsutsumi, and K. Kawabe, *Yakugaku Zasshi* **99**, 421 (1979).
182. I. Björkhem, B. Angelin, K. Einarsson, and S. Ewerth, *J. Lipid Res.* **23**, 1020 (1982).
183. H. Takikawa, T. Beppu, Y. Seyama, and T. Wada, *Gastroenterol. Jpn.* **18**, 246 (1983).
184. M. Tohma, Y. Nakata, H. Yamada, T. Kurosawa, I. Makino, and S. Nakagawa, *Chem. Pharm. Bull.* **29**, 137 (1981).
185. F. Stellaard, M. Sackmann, T. Sauerbruch, and G. Paumgartner, *J. Lipid Res.* **25**, 1313 (1984).
186. F. Stellaard and G. Paumgartner, *Biomed. Mass Spectrom.* **12**, 560 (1985).
187. F. Stellaard, G. Paumgartner, G. P. VanBerge Henegouwen, and S. D. J. Van der Werf, *J. Lipid Res.* **27**, 1222 (1986).
188. T. C. Bartholomew, J. A. Summerfield, B. H. Billing, A. M. Lawson, and K. D. R. Setchell, *Clin. Sci.* **63**, 65 (1982).
189. D. H. Sandberg, J. Sjövall, and D. A. Turner, *J. Lipid Res.* **6**, 182 (1965).
190. I. Makino, S. Nakagawa, and K. Mashino, *Gastroenterology* **56**, 1033 (1969).
191. K. D. R. Setchell, A. M. Lawson, E. J. Blackstock, and G. M. Murphy, *Gut* **23**, 637 (1982).
192. K. D. R. Setchell, D. L. Harrison, J. M. Gilbert, and G. M. Murphy, *Clin. Chim. Acta* **152**, 297 (1985).
193. J. A. Summerfield, B. H. Billing, and C. H. L. Shackleton, *J. Biochem.* **154**, 507 (1976).
194. Y. Tazawa, M. Yamada, M. Nakagawa, T. Konno, and K. Tada, *J. Pediatr. Gastroenterol. Nutr.* **4**, 32 (1985).
195. S. Barnes, G. Berkowitz, I. Horschowitz, and D. Wirtschafter, *J. Clin. Invest.* **68**, 775 (1981).
196. F. J. Suchy, W. F. Balistreri, J. E. Heubi, J. E. Searcy, and R. S. Levin, *Gastroenterology* **80**, 1037 (1981).
197. G. G. Parmentier, G. A. Janssen, E. A. Eggermont, and H. J. Eyssen, *Eur. J. Biochem.* **102**, 173 (1979).
198. G. Janssen and G. Parmentier, *Steroids*, **37**, 81 (1981).
199. G. Janssen, S. Toppet, and G. Parmentier, *J. Lipid Res.* **23**, 453 (1982).
200. B. F. Kase, J. I. Pedersen, B. Strandvik, and I. Björkhem, *J. Clin. Invest.* **76**, 2393 (1985).

201. R. F. Hansson, P. Szczepanik-Van Leeuwen, G. C. Williams, G. Grabowski, and H. L. Sharp, *Science* **203**, 1107 (1979).
202. H. Eyssen, E. Eggermont, J. Van Eldere, J. Jaeken, G. Parmentier, and G. Janssen, *Acta Pediatr. Scand.* **74**, 539 (1985).
203. B. F. Kase, I. Björkhem, P. Haga, and J. I. Pedersen, *J. Clin. Invest.* **75**, 427 (1985).
204. A. Poulos, P. Sharp, A. J. Fellenberg, and D. M. Danks, *Hum. Genet.* **70**, 172 (1985).
205. A. Poulos and M. J. Whiting, *J. Inher. Metab. Dis.* **8**, 13 (1985).
206. J. M. Little, J. St. Pyrek, and R. Lester, *J. Clin. Invest.* **71**, 73 (1983).
207. K. Tanaka, K. Takeshita, and M. Takita, *Clin. Sci.* **62**, 627 (1982).
208. K. D. R. Setchell and J. Worthington, *Clin. Chim. Acta* **125**, 135 (1982).
209. H. Takikawa, H. Otsuka, T. Beppu, Y. Seyama, and T. Yamakawa, *Digestion* **27**, 189 (1983).
210. H. Takikawa, T. Beppu, Y. Seyama, K. Obinata, and H. Nittono, *Biochem. Med.* **33**, 381 (1985).
211. H. Takikawa, T. Beppu, Y. Seyama, and T. Sugimoto, *Dig. Dis Sci.* **31**, (1986).
212. M. Maeda, H. Ohama, H. Takeda, M. Yabe, M. Nambu, and T. Namihisa, *J. Lipid Res.* **25**, 14 (1984).
213. G. Hedenborg and A. Norman, *Scand. J. Clin. Lab Invest.* **45**, 151 (1985).
214. M. Tohma, H. Wajima, R. Mahara, T. Kurosawa, and I. Makino, *Steroids* **44**, 47 (1984).
215. S. Barnes and A. Chitranukroh, *Ann. Clin. Biochem.* **14**, 235 (1977).
216. J. F. Pageaux, B. Duperray, D. Anker, and M. Dubois, *Steroids* **34**, 73 (1979).
217. K. D. R. Setchell, *in* "Bile Acids in Gastroenterology" (L. Barbara, R. H. Dowling, A. F. Hofmann, and E. Roda, eds.), p. 1, MTP Press, Lancaster, England (1982).
218. J. Goto, M. Hasegawa, H. Kato, and T. Nambara, *Clin. Chim. Acta* **70**, 141 (1978).
219. T. Beppu, Y. Seyama, T. Kasama, S. Serizawa, and T. Yamakawa, *Clin. Chim. Acta* **70**, 141 (1978).
220. R. Tandon, M. Axelson, and J. Sjövall, *J. Chromatogr.* **302**, 1 (1984).
221. D. L. Hachey, P. A. Szczepanik, O. W. Berngruber, and P. D. Klein, *J. Labelled Compd.* **9**, 703 (1973).
222. K-Y. Tserng and P. D. Klein, *J. Lipid Res.* **18**, 400 (1977).
223. T. A. Baillie, M. Karls, and J. Sjövall, *J. Labelled Compd. Radiopharm.* **14**, 849 (1978).
224. S. Lindstedt, *Acta Chem. Scand.* **11**, 417 (1957).
225. P. D. Klein, J. R. Haumann, and W. J. Eisler, *Clin Chim.* **17**, 735 (1971).
226. P. D. Klein, J. R. Haumann, and W. J. Eisler, *Clin. Chim.* **21**, 1253 (1975).
227. A. F. Hofmann and S. A. Cummings, *in* "Bile Acids in Gastroenterology" (L. Barbara, R. H. Dowling, A. F. Hofmann, and E. Roda, eds.), MTP Press, Boston, p. 75 (1982).
228. I. M. Campbell, *Bioorg. Chem.* **3**, 386 (1974).
229. J. Sjövall, *Clin. Chim. Acta* **4**, 652 (1959).
230. J. Sjövall, *Proc. Soc. Exp. Biol. Med.* **100**, 676 (1959).
231. J. Sjövall, *Clin. Chim. Acta* **5**, 33 (1960).
232. R. G. Danzinger, A. F. Hofmann, L. J. Schoenfield, O. W. Berngruber, P. A. Szczepanik, and P. D. Klein, *Gastroenterology* **60**, 192 (1971).
233. P. D. Klein, J. R. Haumann, and D. L. Hachey, *J. Lipid Res.* **22**, 166 (1981).
234. E. J. Norman, J. E. Heubi, M. J. Gelfand, P. Dan, and D. W. Matthews, *Biomed. Mass Spectrom.* **11**, 269 (1984).
235. G. Hedenborg, A. Norlander, and A. Norman, *Scand. J. Clin. Lab. Invest.* **42**, 551 (1982).
236. G. Hedenborg, A. Norlander, and A. Norman, *Scand. J. Clin. Lab. Invest.* **45**, 157 (1985).
237. A. Stiehl, R. Raedsch, G. Rudolph, V. Gundert-Remy, and M. Senn, *Hepatology* **5**, 492 (1985).
238. F. Stellaard, P. D. Klein, A. F. Hofmann, and J. M. Lachlin, *J. Lab. Clin. Med.* **105**, 504 (1985).

239. G. M. Murphy and E. Singer, *Gut* **15**, 151 (1974).
240. R. Lester, J. St. Pyrek, J. M. Little, and E. W. Adcock, *J. Pediatr. Gastroenterol. Nutr.* **2**, 355 (1983).
241. C. Colombo, G. Zuliani, M. Ronchi, J. Breidenstein, and K. D. R. Setchell, *Pediatr. Res.* **21**, 197 (1987).
242. H. L. Sharpe, J. Peller, J. B. Carey Jr., and W. Krivitt, *Pediatr. Res.* **5**, 279 (1971).
243. P. Back and K. Ross, *Hoppe-Seyler's Z. Physiol. Chem.* **354**, 83 (1973).
244. J. Gustafsson, *J. Clin. Invest.* **75**, 604 (1985).
245. J. Gustafsson, *Biol. Neonate* **47**, 26 (1985).
246. A. M. Bongiovanni, *J. Clin. Endocr.* **25**, 678 (1965).
247. J. R. Poley, J. C. Dower, C. A. Owen, and G. B. Stickler, *J. Lab. Clin. Med.* **63**, 838 (1964).
248. J. C. Encrantz and J. Sjövall, *Clin. Chim. Acta* **4**, 793 (1959).
249. P. Eneroth, B. Gordon, R. Ryhage, and J. Sjövall, *J. Lipid Res.* **7**, 511 (1966).
250. P. Eneroth, B. Gordon, and J. Sjövall, *J. Lipid Res.* **7**, 524 (1966).
251. P. Eneroth, K. Hellström, and J. Sjövall, *Acta Chem. Scand.* **22**, 1729 (1968).
252. K. D. R. Setchell, A. M. Lawson, N. Tanida, and J. Sjövall, *J. Lipid Res.* **24**, 1085 (1983).
253. G. A. DeWeerdt, R. Beke, and F. Barbier, *Biomed. Mass Spectrom.* **7**, 515 (1980).
254. N. Tanida, Y. Hikasa, M. Hosomi, M. Satomi, I. Oohama, and T Shimoyama, *Gastroenterol. Jpn.* **16**, 363 (1981).
255. N. Tanida, Y. Hikasa, T. Shimoyama, and K. D. R. Setchell, *Gut* **25**, 824 (1982).
256. K. D. R. Setchell, J. M. Gilbert, and A. M. Lawson, *Br. Med. J.* **286**, 1750 (1983).
257. Y. Hikasa, N. Tanida. T. Ohno, and T. Shimoyama, *Gut* **25**, 833 (1984).
258. G. A. DeWeerdt, H. Verdievel, R. Beke, and F. Barbier, *in* "Proceedings of the 1st International Symposium on Bile Acids in Hepatobiliary and Gastrointestinal Disease," p. 185, IRL Press, Oxford (1984).
259. N. Tanida, Y. Hikasa, T. Shimoyama, and K. D. R. Setchell, *Jpn. J. Cancer Res. (Gann)* **76**, 104 (1985).
260. J. M. Gilbert, K. D. R. Setchell, A. M. Lawson, and P. Royston, *Eur. J. Surg. Oncol.* **12**, 359 (1986).
261. K. D. R. Setchell, J. A. Ives, G. C. Cashmore, and A. M. Lawson, *Clin. Chim. Acta* **162**, 257 (1987).
262. M. T. Podesta, G. M. Murphy, G. E Sladen, N. F. Breuer, and R. H. Dowling, *in* "Biological Effects of Bile Acids" (G. Paumgartner, A. Stiehl, and W. Gerok, eds.), p. 245, MTP Press, Lancaster, England (1978).
263. M. T. Podesta, G. M. Murphy, and R. H. Dowling, *Clin. Chim. Acta* **182**, 243 (1980).
264. A. Norman and R. H. Palmer, *J. Lab Clin. Med.* **63**, 986 (1964).
265. M. I. Kelsey and S. A. Sexton, *J. Chromatogr.* **133**, 327 (1977).
266. M. I. Kelsey and S. A. Sexton, *J. Steroid Biochem.* **7**, 641 (1976).
267. M. I. Kelsey, J. E. Molina, S-K. S. Huang, and K-K. Huang, *J. Lipid Res.* **21**, 751 (1980).
268. M. Malavolti, H. Fromm, B. Cohen, and S. Ceryak, *J. Biol. Chem.* **260**, 11011 (1985).
269. W. C. Duane, J. C. Behrens, S. G. Kelly, and A. S. Levine, *J. Lipid Res.* **25**, 523 (1984).
270. P. Back, J. Sjövall, and K. Sjövall, *Scand. J. Clin. Lab. Invest.* **29**, (Suppl. 126) (1972).
271. P. Back and K. Ross, *Hoppe-Seyler's Z. Physiol. Chem.* **354**, 83 (1973).
272. T. J. Laatikainen, P. J. Lehtonen, and A. E. Hesso, *J. Lab. Clin. Med.* **92**, 185 (1978).
273. G. Déléze, G. Paumgartner, G. Karlaganis, W. Giger, M. Reinhard, and D. Sidiropoulos, *Eur. J. Clin. Invest.* **8**, 41 (1978).
274. J. Goto, K. Suzaki, M. Ebihara, and T. Nambara, *Clin. Chim. Acta* **345**, 241 (1985).
275. J. A. Summerfield, J. Gollan, and B. H. Billing, *Biochem. J.* **156**, 339 (1976).
276. J. A. Summerfield, J. Cullen, S. Barnes, and B. H. Billing, *Clin. Sci. Mol. Med.* **52**, 51 (1977).
277. W. Frohling and A. Stiehl, *Eur. J. Clin. Invest.* **9**, 67 (1976).
278. K. D. R. Setchell, B. Almé, M. Axelson, and J. Sjövall, *J. Steroid Biochem.* **7**, 615 (1976).

279. A. Stiehl, R. Raedsch, G. Rudolf, P. Czygan, and S. Walker, *Clin. Chim. Acta* **123**, 275 (1982).

280. A. Radominska-Pyrek, P. Zimniak, M. Chari, E. Golunski, R. Lester, and J. St. Pyrek, *J. Lipid Res.* **27**, 89 (1986).

281. P. Back, *Clin. Chim. Acta* **44**, 199 (1973).

282. P. Back, K. Spaczynski, and W. Gerok, *Hoppe-Seyler's Z. Physiol. Chem.* **355**, 749 (1974).

283. A. Stiehl, *in* "Clinics in Gastroenterology" (G. Paumgartner, ed.), p. 45, Saunders. London, Philadelphia, Toronto (1977).

284. K. D. R. Setchell and J. M. Street, *Semin. Liver Dis.* **7**, 85 (1987).

285. G. P. VanBerge-Henegouwen, K-H. Brandt, H. Eyssen, and G. Parmentier, *Gut* **17**, 861 (1976).

286. B. Almé, A. Bremmelgaard, J. Sjövall, and P. Thomassen, *in* "Advances in Bile Acid Research" (S. Matern, J. Hackenschmidt, P. Back, and W. Gerok, eds.), p. 145, F. K. Schattanen Verlag, Stuttgart (1974).

287. Y. Amuro, E. Hayashi, T. Endo, K. Higashino, K. Uchida, and Y. Yamamura, *Gastroenterol. Jpn.* **16**, 506 (1981).

288. Y. Amuro, E. Hayashi, T. Endo, K. Higashino, and S. Kishimoto, *Clin. Chim. Acta* **127**, 61 (1983).

289. G. Karlaganis, V. Karlaganis, and J. Sjövall, *J. Lipid Res.* **25**, 693 (1984).

290. I. Björkhem, H. Oftebro, S. Skrede, and J. Pedersen, *J. Lipid Res.* **22**, 191 (1981).

291. B. Egestad, P. Pettersson, S. Skrede, and J. Sjövall, *Scand. J. Clin. Invest.* **45**, 443 (1985).

292. J. Gustafsson, K-H. Gustavson, G. Karlaganis, and J. Sjövall, *Clin. Genet.* **24**, 313 (1983).

293. A. M. Lawson, M. J. Madigan, D. Shortland, and P. T. Clayton, *Clin. Chim. Acta* **161**, 221 (1986).

294. P. T. Clayton, A. M. Lawson, K. D. R. Setchell, S. Anderson, B. Egestad, and J. Sjövall, *in* "Bile Acids and the Liver," (G. Paumgartner, A. Stiehl, and W. Gerok, eds.), p. 259, MTP Press, Lancaster, England (1987).

295. J. Yanagisawa, H. Ichimiya, M. Nagai, and F. Nakayama, *J. Lipid Res.* **25**, 750 (1984).

296. J. Dupont, *in* "The Bile Acids," Volume 1 (P. P. Nair and D. Kritchevsky, eds.), Plenum Press, New York.

297. P. P. Nair, R. Solomon, J. Bankowski, and R. Plapinger, *Lipids* **12**, 922 (1978).

298. T. Okishio, P. P. Nair, and M. Gordon, *Biochem. J.* **102**, 654 (1967).

299. A. M. Gelb, C. K. McSherry, J. R. Sadowsky, and E. H. Mosbach, *Am. J Gastroenterol.* **77**, 314 (1982).

300. J. Yanagisawa, M. Itoh, M. Ishibashi, H. Miyazaki, and F. Nakayama, *Anal. Biochem.* **107**, 75 (1980).

301. P. P. Nair, G. Kessie, and V. P. Flanagan, *J. Lipid Res.* **27**, 905 (1986).

302. N. Turjman and P. P. Nair, *Cancer Res.* **41**, 3761 (1981).

303. P. T. Clayton, B. D. Lake, N. A. Hall, D. M. Shortland, R. A. Carruthers, and A. M. Lawson, *Eur. J. Pedriatr.* **146**, 166 (1987).

304. A. M. Lawson, M. J. Madigan, and P. T. Clayton, *in* "Bile Acids and the Liver," (G. Paumgartner, A. Stiehl, and W. Gerok, eds.), p. 269, MTP Press, Lancaster, England (1987).

305. C. D. Schteingart, L. R. Hagey, and A. F. Hofmann, *Hepatology*, **7**, 1108 (1987).

306. C. D. Schteingart, L. R. Hagey. A. F. Hofmann, and K. D. R. Setchell, *in "Proceedings of the 10th Bile Acid Meeting,"* Freiburg, W. Germany (June 9–11, 1988) (submitted).

307. B. J. Koopman, B. G. Wolthers, J. C. van der Molen, G. T. Nagel, and W. Kruizinga, *Biochim. Biophys. Acta* **917**, 238 (1987).

308. K. D. R. Setchell, and R. Dumaswala, in *"Proceedings of the 10th Bile Acid Meeting,"* Freiburg, W. Germany (June 9–11, 1988) (submitted).

309. K. D. R. Setchell, R. Dumaswala. C. Colombo, and M. Ronchi, *J. Biol. Chem.* (1988) (submitted).

310. K. D. R. Setchell, F. J. Suchy, M. B. Welsh, L. Zimmer-Nechemias, W. F. Balistreri, *Gastroenterology* (1988) (in press).

311. K. D. R. Setchell, F. J. Suchy, M. B. Welsh, L. Zimmer-Nechemias, J. Heubi, W. F. Balistreri, *J. Clin. Invest.* (1988) (submittted).

Chapter 6

IMMUNOLOGICAL METHODS FOR SERUM BILE ACID ANALYSIS

A. Roda, E. Roda, D. Festi, and C. Colombo

1. INTRODUCTION

It is known that bile acid metabolism is abnormal in patients with liver disease, and elevated blood levels have been extensively reported [1,2]. Many studies using various analytical methods have been carried out with the aim of attributing a diagnostic role to serum bile acid determination in different liver diseases [3–5].

In the 15 years since the first sensitive measurement and application of serum bile acids there has been an enormous growth in background information on bile acid metabolism, and the determinants of serum bile acid concentration have been better defined, although many details remain unknown [6].

The momentary balance between intestinal and hepatic uptake of bile acids has been identified as the most important determinant of serum bile acid levels [7]. An increase in serum bile acid levels postprandially or in the fasting state is considered a specific indicator of liver disease reflecting hepatic uptake or systemic shunting of portal venous blood [8,9]. Conversely, decreased levels, especially after meals, are an indicator of bile acid malabsorption that normally reflects ileal dysfunction.

Metabolism of the different bile acids present in the enterohepatic circulation varies, and therefore, blood concentrations and variations related to disease differ accordingly [10,11].

As a result, the qualitative pattern of serum bile acids is highly complex

A. Roda, E. Roda, and D. Festi Institute of Chemical Sciences and Departments of Internal Medicine and Gastrocenterology, University of Bologna, Bologna, Italy   **C. Colombo** Department of Pediatrics and Obstetrics, University of Milan, Milan, Italy

as the serum is enriched by more than 16 major bile acids present at relatively low concentration (micromolar) since the fractional hepatic clearance is very high (50–90%). Several analytical methods have been developed in an attempt to define serum bile acid levels. Gas–liquid chromatography (GLC) combined with mass spectrometry is the reference method, since it is sensitive enough to provide information on a wide range of different bile acids. Analysis of bile acids using high-performance liquid chromatography (HPLC) has been applied to serum, but despite a good resolution of bile acids, the common detector [ultraviolet (UV), refractive index] lacks sensitivity. Enzymatic methods have been widely used for total bile acids, [3α-hydroxysteroid dehydrogenase (3α-HSD)] or particular groups of bile acids [7α-hydroxysteroid dehydrogenase (7α-HSD), 12α-hydroxysteroid dehydrogenase (12α-HSD)].

Conventional enzymatic-colorimetric or fluorimetric techniques have poor sensitivity. Only recently has the combination of bioluminescence detection of the reduced form of nicotinamide-adenine dinucleotide (NADH) produced by the first specific dehydrogenase rendered this enzymatic technique more sensitive and adequate to measure serum bile acid in various liver diseases (see Chapter 7).

In this chapter we shall discuss the method widely applied since the first report in 1973 by Simmonds and co-workers [12], i.e., radioimmunoassay (RIA) (Table I). We will also review enzyme immunoassay (EIA), which in the last few years has partially replaced RIA, because of its safety and practicability in clinical chemistry laboratories. Applications of such methodologies in physiology and pathophysiology will be also discussed with emphasis on the pediatric field and the potential use of this assay as a routine test for patients with hepatobiliary diseases.

2. PHYSICOCHEMICAL STATE OF BILE ACIDS IN SERUM

Bile acids are present in serum as monomers partially bound to proteins. The major carrier protein is albumin. We recently reported [13] some microcalorimetric studies showing that the interaction is essentially hydrophobic and is greatly influenced by the nuclear substituents, while side-chain modifications, such as conjugation with glycine or taurine, play a minor role.

The affinity constant values of bile acids versus albumin [14] are different: lithocholic acid and its conjugates are strongly albumin bound ($K_{aff} = 20 \cdot 10^4$ liters/mole), while trihydroxy bile acids are to a lesser extent ($K_{aff} = 0.3 \cdot 10^4$ liters/mole). Accordingly, in physiological conditions the unbound fraction differs: more than 40% of cholic acid is free, while lithocholic, chenodeoxycholic, and other dihydroxy bile acids are more than 85% albumin bound.

Typical experimental data for the blinding of cholic acid and its conjugates

TABLE I. Characteristics of RIA Methods[a]

Reference	Bile acid measured	Act directly on serum	Labeled antigen	Separation of B/F	Sensitivity (pmole/tube)	Normal values (± SD) (μmole/liter)
12	CCA	Yes	^3H	PEG	5	0.54 ± 0.04
19	CCA	No	^3H	$(NH_4)SO_4$		0.55 ± 1.8
20	CCA	Yes	^3H	$(NH_4)SO_4$	10	0.27 ± 0.03
20	CCDCA	Yes	^3H	$(NH_4)SO_4$	10	0.70 ± 0.03
20	SLCA	Yes	^3H	$(NH_4)SO_4$	10	0.06 ± 0.01
20	DCA	Yes	^3H	$(NH_4)SO_4$	10	0.06 ± 0.01
22	CCA	Yes	^3H	Solid phase		1.4 ± 0.3
23	CCA	Yes	^3H	$(NH_4)SO_4$	5	0.45 ± 0.12
24	CCDCA	Yes	^3H	$(NH_4)SO_4$	2	0.3 ± 0.35
25	CCA+CCDCA	Yes	^{125}I	Charcoal	0.5	3.47 ± 2.16
26	CCA	Yes	^3H	PEG		0.62 ± 0.4
27	DCA	Yes	^3H	PEG	7.5	0.18 ± 0.92
28	LCA	Yes	^3H	PEG	20	0.25 ± 0.016
29	SLCA	Yes	^3H	$(NH_4)SO_4$	10	1.56 ± 0.11
30	CLCA	Yes	^3H	$(NH_4)SO_4$	5	0.085 ± 0.04
31	UDCA	Yes	^3H	PEG	10	0.15 ± 0.11
32	CCA + F	No	^{125}I	PEG	2	0.43 ± 0.17
32	CCDCA + F	No	^{125}I	PEG	0.5	0.47 ± 0.23
32	DCA + F	No	^{125}I	Charcoal	2	0.33 ± 0.11
33	CCA	Yes	^{125}I	Charcoal	9.4	0.4 ± 1.9
34	CCA	Yes	^{125}I	Charcoal		
35	3β Cholenic acid	No	^{125}I	$(NH_4)SO_4$	0.6	0.08 ± 0.45
36	CCA	Yes	^3H	$(NH_4)SO_4$	5	0.49 ± 1.32
36	CCDCA	Yes	^3H	$(NH_4)SO_4$	2	0.55 ± 2.02
37	CCDCA	Yes	^{125}I	$(NH_4)SO_4$	1	1.0 ± 0.6

[a] CCA, Conjugated cholic acid; CCDCA, conjugated chenodeoxycholic acid; DCA, conjugated deoxycholic acid; CLCA, conjugated lithocholic acid; SLCA, sulfolithocholic acid; UDCA, ursodeoxycholic acid; F, unconjugated form.

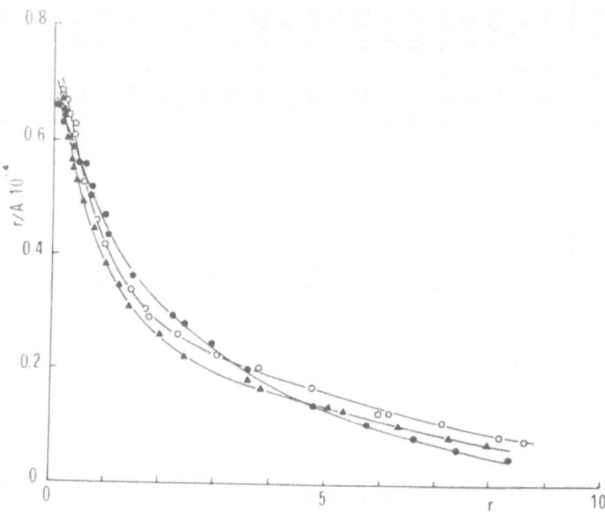

FIGURE 1.   Scatchard plot of data obtained by equilibrium dialyses of ●, glycocholic acid; ▲, taurocholic acid; and ○, cholic acid.

are reported in Fig. 1. The results analyzed by Scatchard procedure indicate that more than one binding class must be implicated, since the relationship is markedly curvilinear.

Variations in albumin or bile acid concentrations, often observed *in vivo*, can modify the proportion of these bile acids in free form and consequently influence their liver uptake, e.g., protein-unbound bile acids can be removed from peripheral blood faster and more efficiently [15].

Studies have shown that other proteins such as high-density lipoproteins (HDL) can bind bile acids, and depending on bile acid structure, 15–40% of the bile acids can be bound to HDL [16].

3.   IMMUNOASSAY

As mentioned in Section 1, bile acids are present in serum at very low concentrations, and the most suitable analytical tools that can be applied routinely in clinical chemistry to measure bile acids are biological methods, including enzymatic and immunometric techniques. Here we describe the characteristics and potential usefulness of the immunometric techniques for serum bile acid analysis.

These methods combine the specificity of antibody detection with the highly sensitive measurement of the analyte, i.e., radioactivity or enzymatic activity, thus offering a unique combination of specificity, sensitivity, precision, and practicability for the determination of analytes at picomolar levels in unfractionated mixture. During the past decade, their application has revolutionized clinical chemistry and other fields.

The major development has been the extension of immunological methods to the field of hapten-specific assays. In this variant, nonpeptide substances of low molecular weight, i.e., bile acids, are not themselves immunogenic, and must first be conjugated with albumin to become antigens.

Antisera to these conjugated preparations contain a mixed population of antibodies, and only those which are directed toward the original low-molecular-weight compound (i.e., hapten bile acid) will be involved in the reaction with the free hapten.

In the assay tube, a competition will take place between a fixed amount of radioactive hapten ($*H$), isotopically diluted in an increased quantity of analyte (H) to be analyzed, and a limited and fixed number of antibody sites. Once equilibrium has been reached, the analyte–antibody complex is separated (bound/free) from the free labeled analyte and the radioactivity of the bound fraction measured. The radioactivity of this fraction will be inversely proportional to the amount of analyte present in the sample:

$$Ab + {*H} + H \rightleftarrows Ab - {*H} + AB - H$$

Free fraction   Bound fraction

This type of assay is called "competitive immunoassay" because in the tube a competition takes place between the "tracer" and the analyte for the antibody. All the assays for bile acids are based on this principle, which is schematically shown in Fig. 2. Other possible immunological procedures, based on others' principles, have never been applied to bile acid analysis.

4. PREPARATION OF ANTIGENIC BILE ACIDS

As previously reported, bile acids are haptens, i.e., molecules that react specifically with antibodies but are not *per se* able to produce antibodies.

In order to render these molecules "antigens," it is necessary to bind them with a macromolecule such as albumin or other proteins. Since 1964 it has been shown that bile acids can produce antibodies when covalently linked to bovine serum albumin (BSA) [17]. In that first study, antibodies against cholan-24-oic acid were produced in rabbit.

One of the most common methods used to chemically bind a protein to a small molecule is to utilize the formation of an amide bond between a terminal-

COMPETITIVE IMMUNOASSAY

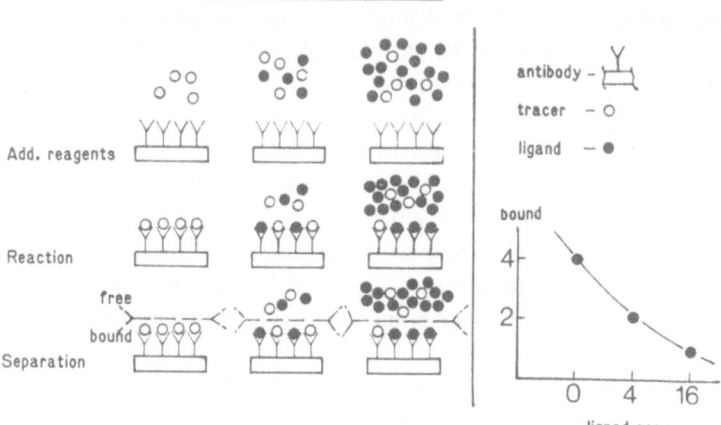

FIGURE 2.   Schematic representation of the competitive immunoassay for bile acids.

NH_2 group of the protein with a carboxyl group on the hapten. In the past, this method has been extensively used for other steroids via a formation of carboxymethyloximes or hemisuccinate derivatives [18].

Bile acids are acidic steroids, and a direct reaction of the C-24 carboxyl with the protein has been the most widely used coupling method. The strategy to obtain a specific antibody is to bind the carrier protein a long way from the more characteristic functional groups of the steroid molecule, such as the hydroxy group and the side chain (Fig. 3). Unfortunately, up to now few attempts have been devoted to that purpose, and the antibodies described lack specificity.

The coupling technique widely used to bind proteins to bile acids involves the carbodiimide method and that of mixed anhydride (Table II). The starting bile acid is usually conjugated with glycine or with a free bile acid [12,19–38].

The molar-bile acid-to-protein ratio is in the order of 15–20; this ratio must be as high as possible in order to achieve maximum antigenicity of the immunogen. Although in most instances the same immunogen has been used and the immunization carried out in the same species, i.e., rabbit, the specificity of the antisera varies greatly (Table III).

Minder *et al.* [33] first developed a different procedure for the production of antibodies for cholic acid conjugates. Cholic acid was used as a hapten with the aim of producing antisera specific for both taurine-, glycine-conjugated, and unconjugated bile acids. Thyroglobulin instead of BSA was used as a carrier protein of the immunogen, as it had previously been reported to increase immune response (higher molecular weight than BSA). With their procedure, the molar bile acid/thyroglobulin ratio was much higher than with BSA: 250 molecules of

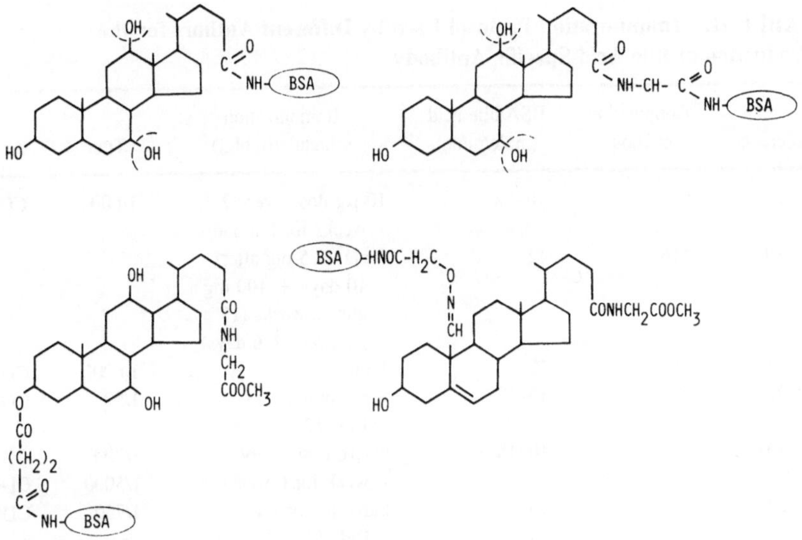

FIGURE 3. Different bile acid antigens used to produce specific antibodies. BSA, bovine serum albumin.

cholic acid were incorporated per molecule of thyroglobulin. With this modification, the antibodies produced were more specific but not entirely satisfactory.

More recently, Orbàn et al. [39] reported the production of an extremely monospecific antibody for glycocholic acid. They adopted a completely new approach to the problem and synthesized a hemisuccinate of cholic acid so that the C-24 carboxy group was protected. In this way both the hydroxy group and the side chain were far from the carrier protein and the antibody produced could probably recognize both functions as they were not masked (Fig. 3).

A similar approach has recently been described for the analysis of glycolithocholic acid [40]. A 3-(O-carbomethyl)-oxime derivative of 3-dehydrolithocholylglycine was synthesized, and its BSA conjugate was used to produce antibodies in rabbit. The specificity of this antiserum was for lithocholylglycine (100%) and sulfolithocholylglycine (100%); other bile acids did not cross-react appreciably. In this case, the specificity of antiserum was from the side chain, but the 3-position in the nucleus was completely masked.

More recently the same authors developed the first monospecific antibody described in the literature [41]. They prepared a 3-hydroxy-5-cholen-24-oyl-glycine-BSA conjugate, in which the hapten was linked to the carrier protein through an O-carboxymethyl oxime bridge at the C-19 position of the steroid (Fig. 3). In this case the protein is far from both the 3-hydroxy group and the side chain. The antibody produced was extremely specific for 3-β-hydroxy-5-

TABLE II.   Immunization Protocol Used by Different Authors for the Production of Bile Acid-Specific Antibody

Reference	Conjugation method[a]	BSA/bile acid (mole/mole)	Immunization schedule (rabbit)	Titer[b]	Antigen[c]
12	C	10–12	10-μg dose every 2 weeks for 6 months	1/6000	CG
19	MA	12	5 mg + 5 mg after 10 days + 100 mg after 2 weeks (15 μg every 2–6 days)		
21	MA	12	1 mg	1/6000	CG
22	C	10–17	Same protocol as Ref. 12	1/50	CG
23	C	10–15	50-μg dose every week for 6 weeks	1/3000 1/5000	CG CDCG
24	C	10–15	Same protocol as Ref. 12	1/750	CDCG
25	C	N[d]	50 μg every 2 weeks	1/80000	CG
27	MA	12	Same protocol as Ref. 12	1/100	DCG
28	C	25–30	100 μg every 2 weeks (6 doses)	1/400	LCG
29	C	25–30	100 μg every 2 weeks (6 doses)	1/10000	SLCG
30	C	20	400-μg dose every week for 4 weeks	1/600	LCG
33	C	250 (TBG)	94 μg monthly (*guinea pig*)	1/10000	CG

[a] C, Carbodiimide; MA, mixed anhydride.
[b] Final dilution at 50% bound.
[c] CG, cholylglycine; CDCG, chenodeoxycholylglycine; DCG, deoxycholylglycine; LCG, lithocholylglycine; SLCG, sulfocholylglycine.
[d] N, not reported.

cholen-24-oyl-glycine, exhibiting no significant cross-reaction with other bile acids, in particular for the 3α analog or taurine-conjugated form.

5.   ANTIBODIES

As mentioned, specific antibodies have commonly been produced in rabbits (New Zealand white). The immunization protocol used by different authors is reported in Table II. Usually the antigen dose varied from 30 μg to 1 mg over a 6- to 12-month period. The resulting antibody titer is always low, and this is a limitation in developing reliable and accurate RIA methods.

TABLE III. Specificity of the Antisera[a] as Reported by the Cited Authors[b]

Reference	a	b	c	d	e	f	g	h	i	j	k	l	m	n	o	p	q	r	s
12	10	10	100	100	1	1	—	—	—	—	1	4	0.2	—	—	0	—	—	—
19	10	10	100	100	5	5	—	—	—	—	2.5	10	2.5	0.1	—	0.1	—	—	—
20	4	0.2	100	18	1	0.1	3	0.1	—	—	4	8	0.1	0.1	—	—	1	—	—
20	100	8	—	0.1	0.1	3	0.3	0.3	—	—	1	—	0.1	0.1	—	—	—	—	—
20	—	—	—	—	100	—	0.1	—	—	—	0.1	—	16	0.1	—	—	0.1	—	—
20	0.4	—	0.1	—	0.1	—	6	—	—	—	—	—	—	—	—	—	100	—	5
21	13	—	100	100	3	3	1	1	—	—	5	56	1	1	—	—	—	—	1
22	10	10	100	100	10	10	—	—	—	—	5	5	5	—	—	—	—	—	—
23	0	0	0	0	1	0.1	0	0	0	0	2	6	0	0	0	0	—	—	—
23	100	100	1	1	0	0	0	0	2	3	4	0	0	—	1	0	—	—	—
24	100	100	1	—	1	—	1	9	9	—	12.5	1	1	1	1	0	—	—	—
25	26	3.5	33	100	0.5	100	0.1	—	—	—	1.3	1.7	0.5	0.5	—	—	—	—	—
27	0.1	—	2	100	100	0.01	100	—	—	—	0.1	0.1	30	0.1	—	—	—	—	—
28	5	0.01	0.01	3	2	0.01	0.5	—	—	—	5	0.01	0.01	65	—	—	0.01	—	—
29	0.01	0.01	0.01	0.01	0.01	0.01	100	—	—	—	0.01	0.01	0.01	0.5	—	—	100	100	—
30	1	1	1	0.5	0.5	0	100	—	—	—	0.5	0	0	0	25	—	—	2	—
31	0.1	0.1	0	0	0	0	0.2	180	180	180	0.1	0	0.2	0.2	100	0	—	—	—
34	14.5	32	100	16.5	2.8	1.8	0.4	—	—	—	3.3	7.8	1	0.1	—	—	—	—	—
32	4.2	1.3	100	130	1.2	3.7	1.4	—	—	—	1.5	50	0.9	0.1	—	—	—	—	—
32	100	100	0.8	0.4	0	0	8	—	—	—	100	0.1	0	0.6	30	—	—	—	—
32	0.5	1.1	0.1	1.3	100	100	10	—	—	—	—	—	0.2	100	—	—	—	—	—
33	0.5	0.6	100	100	0.7	—	0.07	—	—	—	—	7.8	—	—	—	—	—	—	—
	1.4	2.8	—	—	2.1	—	0.3	—	—	—	—	1.8	—	—	—	—	—	—	—
39	0.1	0.1	100	0.5	0.8	0.0	0.0	—	—	—	0.1	0.1	0.1	0.1	—	—	—	—	—
40	1.8		0.01		0.1		100	0.2					0.09						

[a] Antisera: a. chenodeoxycholyl-glycine; b. chenodeoxycholyl-taurine; c. cholylglycine; d. cholyltaurine; e. deoxycholylglycine; f. deoxycholyltaurine; g. lithocholylglycine; h. ursodeoxycholylglycine; i. ursodeoxycholyl-taurine; j. chenodeoxycholic acid; k. cholic acid; l. deoxycholic acid; m. lithocholic acid; n. ursodeoxycholic acid; o. cholesterol; p. sulfolithocholylglycine; q. sulfodeoxycholic acid; r. sulfochenodeoxycholic-glycine; s. sulfocholyltaurine; t. lithocholyltaurine.

[b] Values expressed as relative cross-reactivity (%) at 50% displacement.

We developed an original procedure to produce antibody in rabbit with higher titers and in a shorter time based on preliminary pretreatment of the animal with an antitubercular vaccine (VDS) and using microquantities of antigen (50 μg). The data were compared with those obtained without pretreatment with VDS. Despite the range of variations, the pretreatment induces faster antibody production and higher titers. A typical response versus time for glycochenodeoxycholic acid is shown in Fig. 4.

Minder *et al.* [33] used guinea pigs instead of rabbits and prolonged the immunization period. In this case the titer increased, reaching optimal values. A possible explanation is that rabbits can metabolize bile acids faster than other animals, with a consequent disruption of the administered antigen.

Also, antibody titer is a function of the specific activity of the tracer. In the case of bile acids, the commercially available labeled bile acids show a low specific activity (under 10 Ci/mmole) that limits the titer of antisera and in turn the sensitivity of the assay.

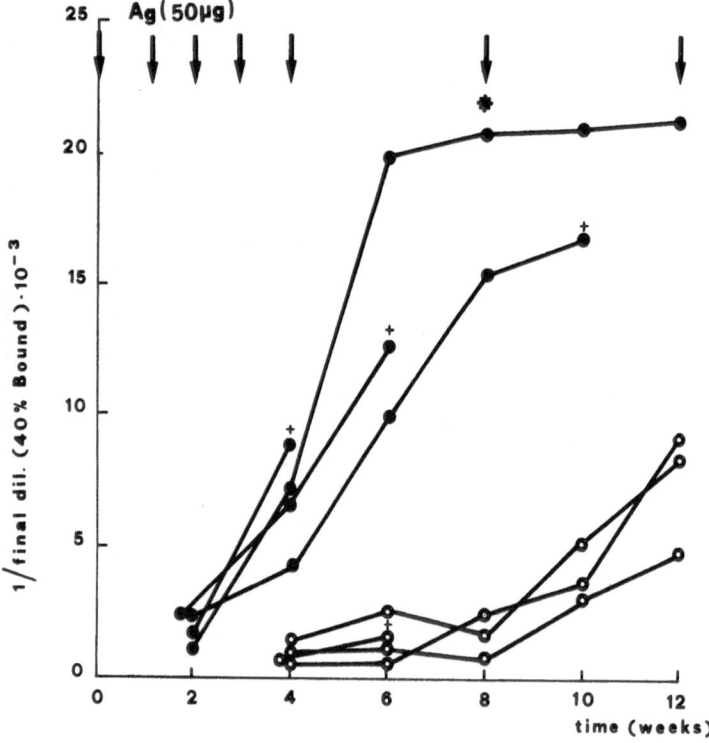

FIGURE 4. Antibody production in rabbit expressed as the reciprocal of the greatest antibody dilution capable of binding 40% of a tracer dose of [^{3}H]glycochenodeoxycholic acid plotted against the time. The open circles refer to control rabbits and the closed circles to sera from rabbits prevaccinated with VDS. Ag, glycochenodeoxycholic acid-BSA.

5.1. Antibody Characterization: Titer and Affinity

As noted earlier, antibody affinity can vary over a broad range, resulting in marked variations in assay sensitivity. The importance of antibody-binding affinity can be revised by a simple calculation in the law of mass action:

$$K_a = \frac{[Ab - H]}{[Ab] \cdot [H]} \quad (K_a = \text{affinity constant})$$

where $[Ab]$; $[H]$ and $[Ab\text{-}H]$ are the equilibrium concentrations of the antibody, hapten, and antibody–hapten complex under conditions in which 50% of the total hapten is antibody bound $[AB - H] = [H]$, and the equation reduces to

$$K_a = \frac{1}{[Ab]}$$

The concentration of free antibody is inversely proportional to K_a.

Thus, if the K_a is 1.10^8 liters/mole, as reported by different authors (Table II), the concentration of antibody required to obtain 50% binding will be about 1.10^{-8} M (approximately 80 mg antibody protein/ml), and the sensitivity of the assay will be in the normal range. But if the K_a is over 1.10^{10} liters/mole, the sensitivity can extend into picomolar levels.

The affinity constant of the antiserum is usually calculated by Scatchard plot and by Michaelis-Menton hyperbole. The former method is directly applied to the RIA computerized program for the standard curve data [30]. By the Michaelis–Menton hyperbole, a direct measurement of the antiserum K_a is obtained by adding increasing amounts of tracer to the same amount of antiserum, until saturation of the antibody-binding site is achieved (Fig. 5).

The affinity constant (K_a) of the antisera obtained by the standard curves of the assay (see Section 7) for all the commercially available assays for bile acid and for those developed by us is reported in Table IV.

The K_a values obtained by different authors are in the range 4×10^6– 1.4×10^8, allowing for the development of direct methods with a sensitivity ranging from 0.7 to 0.06 μmole.

5.2. Specificity

An accurate immunomethod is contingent on antibody specificity, i.e., the capacity to recognize the analyte in the presence of other structural analogs. We defined specificity using "cross-reactivity," i.e., the percent ratio between the mass of specific bile acid over the mass of interferent bile acids at 50% of binding.

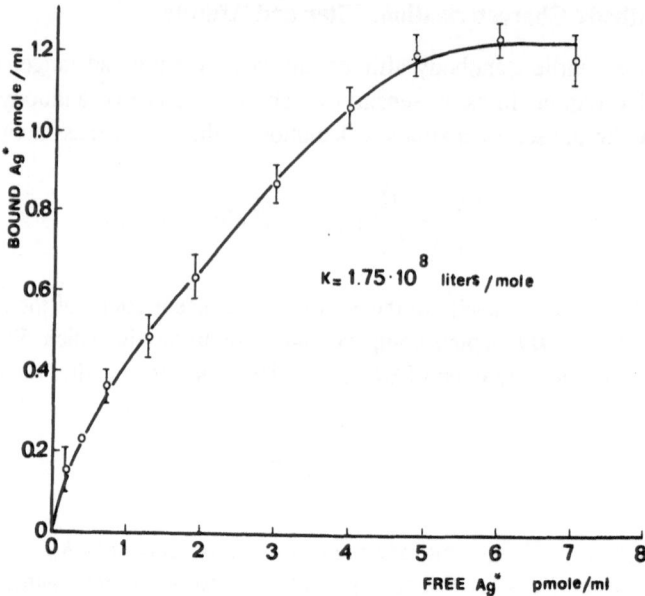

FIGURE 5. Saturation of the antibody-binding sites with increasing amounts of [³H]glycochenodeoxycholic acid (specific activity 17 Ci/mmole) (Michaelis-Menton hyperbole).

Table III illustrates the specificity of various antisera as reported by different authors; cross-reactivity varies widely for different bile acids. Few of the antisera can recognize a single bile acid with high specificity, and antibody specificity is usually towards the glycine- and taurine-conjugated bile acids. This parameter has not always been carefully investigated, and many bile acids present in serum which could interfere in the developed assay have not been included in cross-reactivity studies.

6.   LABELED BILE ACIDS

A fundamental prerequisite for the development of a sensitive assay is radiopurity and a highly specific "tracer" activity. The four major bile acids, i.e., cholic acid, chenodeoxycholic acid, deoxycholic acid, and lithocholic acid, either free or conjugated to glycine, are commercially available as tritium-labeled compounds with a specific activity ranging from 1.0 to 25 Ci/mmole. In the assay with this specific activity, the mass of the labeled compound is not less than 1 pmole and consequently the sensitivity is below these values. The commercially available glycine-conjugated tritium-labeled bile acids contain only two ³H atoms per molecule (Fig. 6). In order to increase the specific activity,

TABLE IV. Affinity Constant (K_{aff}), Concentration (Ab_0) of the Antisera, and Logit Parameters of the Standard Curves

Method	A[a] (%)	B	C (pmole/tube)	D (%)	K_{aff} (liters/ mole)	Ab_0 (mole/ liter)
³H-CCA	91	1.40	13	3.6	1.4×10^8	4×10^{-9}
¹²⁵I-Nordic Lab cholic acid	94	0.89	75	0	7.3×10^6	5×10^{-8}
¹²⁵I-Abbott cholylglycine	100	0.88	135	1.2	4.2×10^6	2×10^{-7}
³H-CCDCA	91	1.40	13.4	3.6	1.4×10^8	6×10^{-9}
¹²⁵I-Nordic Lab chenodeoxycholic acid	98	0.83	35	0	1.7×10^7	6×10^{-8}
¹²⁵I-Becton Dickinson primary bile acids	97	0.95	63	0	1.5×10^7	2×10^{-9}

[a] A, percent response for zero dose; B, slope factor; C, midrange; D, percent response for infinite dose.

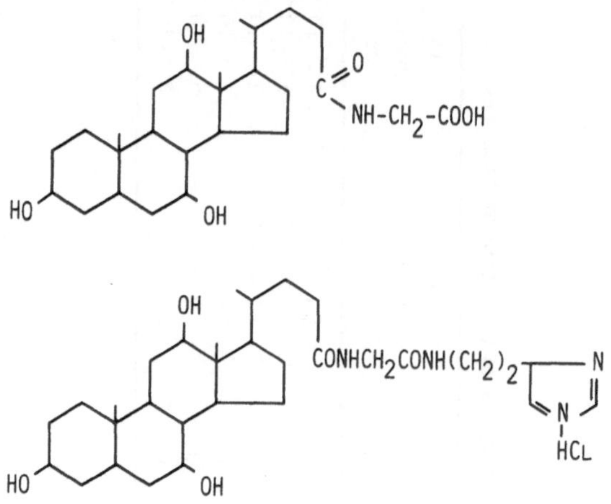

FIGURE 6. Bile acid derivatives used for preparation of the radioactive tracers for the various RIA.

we have recently synthesized a taurine-conjugated bile acid, starting from free chenodeoxycholic acid labeled with four ^{3}H which has a higher specific activity. The low specific activity of ^{3}H in the tracer together with the use of a liquid scintillation technique allowed us to prepare ^{125}I-labeled bile acid.

The most widely used methods involve the synthesis of a histamine derivative followed by iodination of the histamine moiety, while other authors have reported the use of tyrosine [32,33,37]. Despite the use of a tracer that is potentially more sensitive, the developed assays showed a sensitivity similar to that of the tritium-labeled assay.

7. ASSAY PROCEDURE

As mentioned earlier, all the radioimmunoassays described for bile acids are based on a competitive principle. The stepwise procedure is as follows:

- Add the sample, standard, labeled bile acid and antibody in appropriate conditions (pH, ionic strength, protein content).
- Once equilibrium is reached (usually 1–4 hr at 20–25°C), the antibody–antigen complex is separated and radioactivity measured. Most of the developed RIA are performed directly on the serum samples (usually 10–25 µl). Some methods still require preliminary extraction of bile acid from the serum matrix, generally carried out using ethanol or methanol. The antibody is usually in solution, and the bound/free separation is performed using PEG-6000 solution or by salt precipitation using $(NH_4)_2SO_4$.

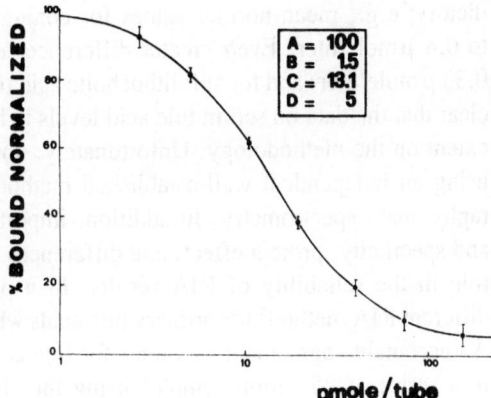

FIGURE 7. A typical dose–response curve for the radioimmunoassay of bile acid. Each point represents the mean ± SD. A, response for zero dose; B, slope; C, midrange; D, response extrapolated on infinite dose.

A solid-phase system has recently been developed in which the antibody is immobilized on a solid bead [22]. The specific antibody is generally immobilized on a sepharose 6B-CNBr activated bead or on a polystyrene tube. The latter has the advantage that it obviates a centrifugation step but can only be applied if ^{125}I-labeled bile acids are used.

The antibody-bound fraction was expressed as a percentage of B/B_o versus log dose. The dose–response curve has been described by a four-parameter logit function:

$$y = \frac{A - D}{1 + \left(\dfrac{x}{C}\right)^B} + D$$

where $y = B/B_o$; A = response for zero dose; D = response for infinite dose; C = midrange/effective dose for 50% dose–response (pmole/tube); and B = slope factor.

The best sigmoidal curve was computed by the least-squares method, and the amount of bile acid was interpolated automatically by a suitable computer program. Figure 7 shows a typical standard curve for conjugated chenodeoxycholic and cholic acid assays developed in our laboratory.

8. COMPARISON OF METHODS

Normal values using RIA techniques for individual classes of bile acids, as reported by different authors, are listed in Table II. Data are often contradictory; e.g., mean normal values for conjugated cholic acid range from 0.18 to 0.4 μmole/liter. Even greater differences exist for deoxycholic acid (0.06–0.33 μmole/liter) and for sulfolithocholic acid (0.06–1.56 μmole/liter). It appears clear that the data on serum bile acid levels in health and disease depend to some extent on the methodology. Unfortunately, not all methods have been validated using an independent well-established method, such as gas–liquid chromatography–mass spectrometry. In addition, important factors such as antibody titer and specificity, protein effect, and differences in "tracer" may play an important role in the reliability of RIA results. In a previous study, we compared six different RIA methods for primary bile acids which were commercially available. An acceptable agreement of values for bile acids was observed when measured in a series of 25 serum samples using the six different kits. Figures 8 and 9 show reasonably close agreement at both low and high bile acid concentration when results for conjugated cholic acid and conjugated chenodeoxycholic acid were intercompared. Figure 10 shows the distribution of normal values. In ad-

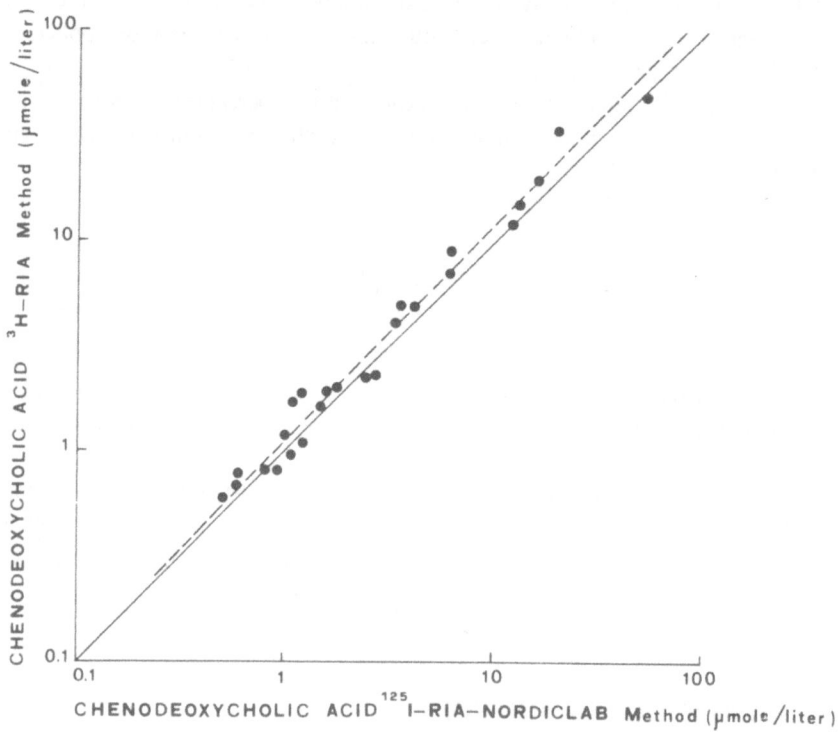

FIGURE 8. Relationship between chenodeoxycholic acid conjugated in serum as measured by a ^{3}H-RIA and ^{125}I-RIA from Nordic Lab. $y = 1.05x + 0.03$; $r = 0.98$; $n = 25$.

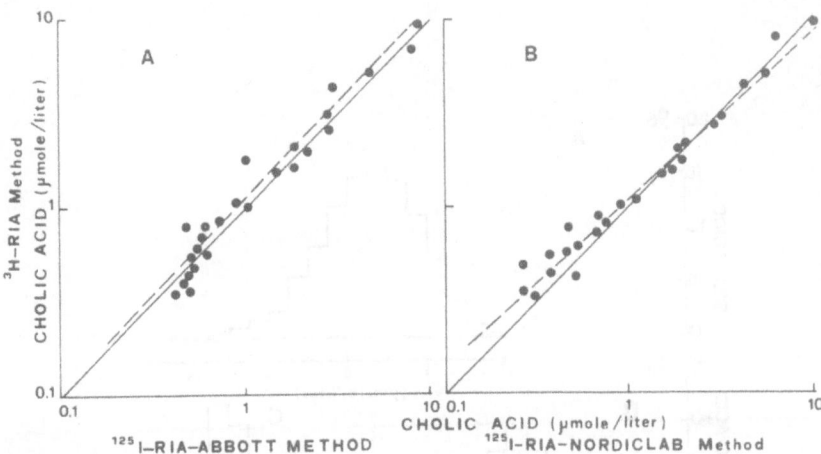

FIGURE 9. Relationship between cholic acid concentration in serum as measured by (A) ³H-RIA and ¹²⁵I-Abbott ($y = 1.03x + 0.02$; $r = 0.98$; $n = 25$) and (B) ³H-RIA and ¹²⁵I-Nordic Lab methods $y = 0.91x + 0.07$; $r = 0.98$; $n = 25$.

dition, normal values using these kits were similar. We concluded that for analysis of the two primary bile acids, the current commercially available methods offer reliable results and, despite differences in antibody specificity, results are comparable. We also evaluated same parameters of precision and sensitivity of these kits, and data are reported in Table V. All the procedures fulfill the standard requisite of precision, being the coefficient of variation in the intra- and interassay studies below 17% at both low and high bile acid concentration. For other bile acids such as deoxycholate, lithocholate, and sulfolithocholate, the methods so far proposed require further improvement and need to be validated using independent techniques.

9.   COMPARISON BETWEEN METHODS

We compared values obtained with RIA to those obtained using an enzymatic method (3α Sterognost), an EIA method, and a GLC method. Table VI shows the results obtained: a good agreement was observed at high levels and poor agreement below 5 μmole/liter, where the enzymatic and GLC methods lack sensitivity. Samuelson [42] reported a comparative study for the analysis of cholic and chenodeoxycholic acid using RIA, GLC, and 3α-HSD. Their results indicate a good correlation between methods at high bile acid concentrations. However, analysis of sera from healthy subjects revealed a good correlation

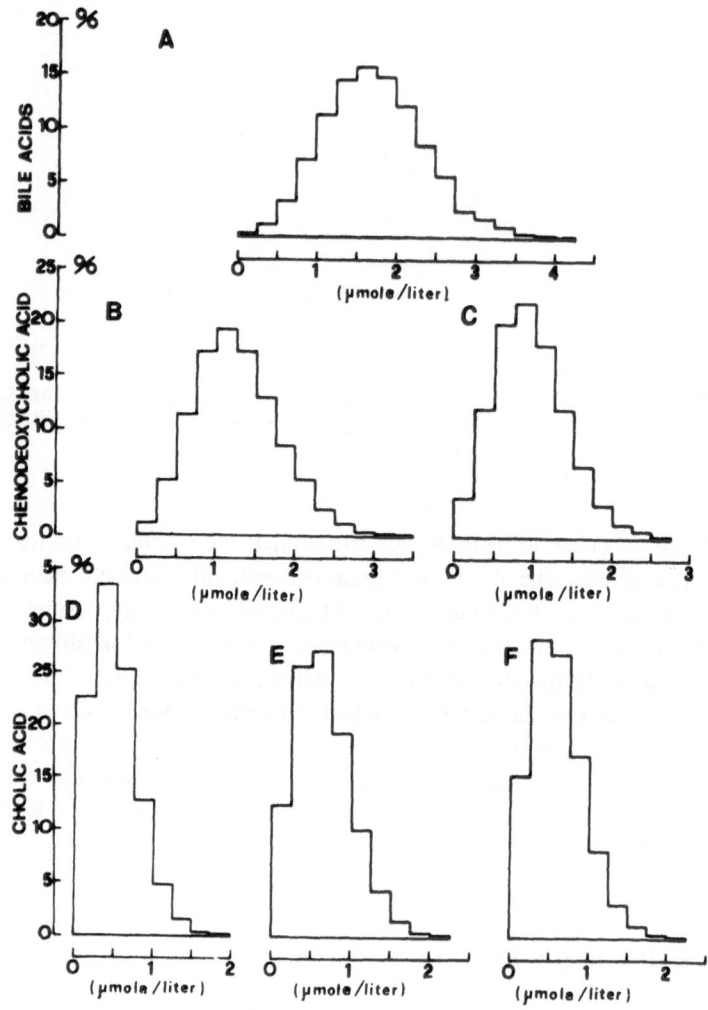

FIGURE 10. Distribution of normal values of serum bile acid levels using the different RIA procedures. (A) Becton Dickinson method. $\mu = 6.051$; $x^2 = 32.2$; $p < 0.01$. (B) ^{3}H-RIA method. $\mu = 4.35$; $x^2 = 33.7$; $p < 0.005$. (C) Nordiclab method. $\mu = 3.46$; $x^2 = 24.8$; $p < 0.01$. (D) ^{3}H-RIA method. $\mu = 1.54$; $x^2 = 23.4$; $p < 0.005$. (E) Abbott method. $\mu = 2.23$; $x^2 = 29.6$; $p < 0.001$. (F) Nordiclab method. $\mu = 1.89$; $x^2 = 22.1$; $p < 0.005$.

TABLE V. Precision of Replicate and Sample-to-Sample Assays of the Six Procedures Compared

Method	Low concentration pool				Medium concentration pool				High concentration pool			
	Interbatch		Intrabatch		Interbatch		Intrabatch		Interbatch		Interbatch	
	Mean (μmole/liter)	CV (%)	Mean (μmole/liter)	CV (%)	Mean (μmole/liter)	CV (%)	Mean (μmole/liter)	CV (%)	Mean (μmole/liter)	CV (%)	Mean (μmole/liter)	CV (%)
^{3}H-CCA	0.45	7.2	0.48	7.6	1.5	7.4	1.40	7.0	8.4	6.8	8.9	7.0
^{125}I-Abbott cholylglycine	0.40	7.8	0.44	8.1	1.4	7.8	1.55	8.1	8.0	6.5	8.2	6.8
^{125}I-Nordic Lab cholic acid	0.38	8.5	0.42	9.0	1.2	8.5	1.38	9.0	7.6	9.8	8.0	10.1
^{3}H-CCDCA	0.68	7.0	0.72	7.8	2.4	9.0	2.30	10.0	10.2	7.5	9.8	8.1
^{125}I-Nordic Lab chenodeoxychlic acid	0.55	8.1	0.60	9.2	2.1	10.0	1.85	10.7	8.1	9.2	8.8	9.8
^{125}I-Becton Dickinson	1.25	10.1	1.45	16.5	4.2	7.5	4.50	16.0	19.1	3.7	19.0	12.1

TABLE VI.   Comparison of Different Methods for Serum Bile Acid Analysis

Sample[a] (μmole/liter)	GLC	3α-HSD	RIA ³H	EIA
Glycoholic acid				
0.1	Traces	0.25	0.14	0.20
1	0.90	1.45	1.08	1.5
10	9.25	11.05	10.70	12.0
100	102	100.5	98.5	107
Glycochenodeoxycholic acid				
0.1	0.25	0.35	0.18	0.26
1	1.21	1.82	1.11	1.35
10	11.0	13.2	9.61	10.7
100	96	98	113	96

[a] Sample, charcoal-extracted human serum and increased amount of bile acids.

between RIA and GLC but no correlation between RIA and 3α-HSD. The authors conclude that this is in part due to the presence of varying amounts of secondary bile acids which can be measured with the enzymatic method but not with RIA [42].

In a previous paper we also reported on the correlation between RIA, EIA, enzymatic bioluminescence, and an HPLC method for primary bile acid (sum of conjugated cholic and chenodeoxycholic acid). The analyses were carried out on 15 normal subjects and 15 patients with liver disease (Fig. 11). There was a reasonable agreement for samples with high bile acid concentrations, but the analytical information obtained by the three methods differed. The bioluminescent procedure, using a 7α-hydroxysteroid dehydrogenase enzyme, measured both free and glycine/taurine-conjugated bile acids, whereas the immunological

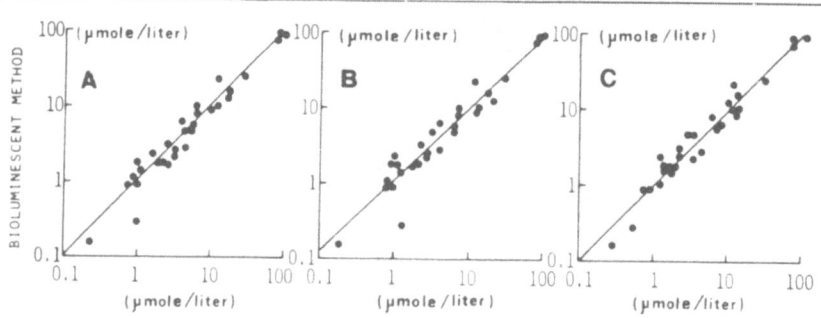

FIGURE 11.   Relationship between serum primary bile acid levels using different independent methods. (A) Radioimmunoassay ($y = 0.9997x + 0.017$; $r = 0.97$; $n = 30$). (B) Enzyme immunoassay ($y = 0.9584x + 0.040$; $r = 0.959$; $n = 30$). (C) HPLC ($y = 1.006x - 0.0077$; $r = 0.975$; $n = 30$).

methods, including RIA and EIA (see Section 10), measured only glycine- and taurine-conjugated bile acids. HPLC has the potential to measure all bile acids but the detector we used (UV at 210 mm) is only sensitive for conjugated bile acids.

10. ENZYME IMMUNOASSAY

Enzyme immunoassay has become an established tool in clinical chemistry. The assay fulfills the requirements of simplicity and sensitivity previously shown only by RIA. In contrast to RIA, EIA employs stable reagents, requires less expensive equipment, and is more suitable for automation. No radioactive compounds are required obviating the need for counting, licenses, and disposal of hazardous waste.

Recently, EIAs have been developed for serum bile acid analysis (Table VII). The underlying principle is similar to that of RIA: the "tracer" is a bile acid covalently linked with an enzyme (peroxidase, galactosidase) instead of a radioisotope. Matern et al. [43], Maeda et al. [44], and Ozaki et al. [45] developed EIA methods for cholic acid conjugates and ursodeoxycholic acid, respectively, based on a "competitive principle." The enzyme-labeled bile acid competes with the bile acids in the sample for a limited number of binding sites on the antibody. Once equilibrium is reached, the antibody-bound antigen is separated (usually by solid-phase techniques or with a second antibody), and the enzymatic activity is determined spectrophotometrically by measuring specific color-producing substrates. The absorbance is inversely proportional to the amount of bile acids present in the sample.

Baquir et al. [46] developed another EIA called "homogeneous enzyme immunoassay" for chenodeoxycholic acid conjugates. It is based on the principle that when a bile acid enzyme interacts with the specific antibody, the enzymatic activity is drastically reduced. An increased proportion of a bile acid in the sample will result in a release of enzymatic activity due to displacement of the bile acid enzyme antibody. The enzymatic activity is directly proportional to the concentration of bile acid. No separation of "bound" and "free" antigen is required, but a kinetic measurement (two point) of the enzymatric activity is needed. The homogeneous method is in principle less sensitive than the "heterogenous" enzyme immunoassay, but enough for serum bile acid analysis when applied to patients with liver disease.

At present, the only three commercially available EIA methods for bile acids are produced; one by Immunotech Corp. (Cambridge, Massachusetts), specific for cholic acid conjugates [47], and two by IFCI (Bologna, Italy), one specific for cholic acid conjugates and the other for chenodeoxycholic acid conjugates.

TABLE VII. Characteristics of the Enzymatic Immunoassay for Serum Bile Acids[a]

Reference	Specificity	Acts directly on serum	Enzyme label	Principles of the method	Sensitivity	Precision (CV %)	Normal values (µmole/liter)
43	Chl-Gly/Chl-Tau	Yes	Peroxidase	Competitive	0.5 µmole/liter	18–22	(Not reported)
45	Chn-Tau/Chn-Tau	Yes	Peroxidase	Homogeneous	50 pmole/tube	8.6	(Not reported)
44	Urs	Yes	Galactosidase	Competitive	0.8 pmole/tube	6–12.8	(Rabbit)
46	Chl/Gly/Chl-Tau	Yes	Alkaline phosphatase	Competitive	0.2 µg/ml	6–12.8	(Not reported)
47	Urs	Yes	Alkaline phosphatase	ELISA	20 pmole/tube	3.22 ± 1.28	0.27 ± 0.12
48	Chn-Gly/Chn-Tau	Yes	Peroxidase	Competitive	0.01 pmole/tube	5–8	1.1 ± 0.20
48	Chl-Gly/Chl-Tau	Yes	Peroxidase	Competitive	0.01 pmole/tube	5–8	0.54 ± 0.18
48	Lit-Gly/Lit-Tau	Yes	Peroxidase	Competitive	0.02 pmole/tube	5–9	0.12 ± 0.05

[a] Chl-Gly, cholylglycine; Chl-Tau, cholyltaurine; Chn-Gly, chenodeoxycholylglycine; Chn-Tau, chenodeoxycholyltaurine; Urs, ursodeoxycholic acid; CV, coefficient of variation.

TABLE VIII.   Serum Concentrations[a] of Different Bile Acids in Relation to Age

	Conjugated cholic acid (μmole/liter)	Conjugated chenodeoxycholic acid (μmole/liter)	Conjugated lithocholic acid (μmole/liter)	Conjugated sulfolithocholic acid (μmole/liter)
Prematures	6.46 ± 5.39	10.15 ± 4.66	0.45 ± 0.22	1.35 ± 0.31
Newborns at term	7.50 ± 4.75	8.47 ± 3.88	0.35 ± 0.15	0.42 ± 0.21
Infants				
1–2 months	4.54 ± 2.81	8.47 ± 4.22	0.35 ± 0.10	0.45 ± 0.22
3–6 months	3.39 ± 2.53	5.64 ± 4.10	0.25 ± 0.15	0.49 ± 0.21
6–12 months	2.55 ± 1.66	4.95 ± 3.78	0.22 ± 0.05	0.38 ± 0.20
Children				
1–2 years	0.93 ± 0.67	2.58 ± 1.40	0.21 ± 0.06	0.50 ± 0.48
2–3 years	0.49 ± 0.19	2.13 ± 0.69	0.20 ± 0.05	0.45 ± 0.25
3–4 years	0.51 ± 0.23	1.56 ± 0.91	0.17 ± 0.03	0.42 ± 0.20
4–6 years	0.44 ± 0.29	1.28 ± 0.73	0.18 ± 0.03	0.52 ± 0.18
6–12 years	0.43 ± 0.15	1.18 ± 0.43	0.19 ± 0.03	0.46 ± 0.17
Adults	0.48 ± 0.12	1.05 ± 0.35	0.15 ± 0.05	0.61 ± 0.18

[a] μmole/liter; mean value ± SD.

liver diseases using our ^{3}H-RIA and the EIA from Immunotech. The coefficient of correlation is high ($r = 0.96$) with a slight overestimation at low levels using the EIA procedure.

As is shown in Tables I and VIII, the sensitivity of RIA and EIA methods is at the lower limit of the normal range of bile acid levels in an adult population. This constitutes a limitation in the derivation of normal values and in monitoring slight elevations in liver disease and is not applicable to monitoring patients with bile acid malabsorption with low serum bile acid levels.

Following a previous experiment on solid-phase EIA for steroids, we recently developed an extremely sensitive EIA for cholic and chenodeoxycholic acid conjugates. The sensitivity is superior to all published methods for bile acid analysis, reaching levels of 10 fmole/tube [48]. The specific antibody was purified by salt precipitation and immobilized on polystyrene plastic beads (6.4 mm ϕ). Glycine-conjugated bile acids were coupled with horseradish peroxidase (HRP) and purified by gel exclusion chromatography.

The method is based on a competitive principle; i.e., the sample or bile acid standard and a fixed amount of the bile acid–HRP label, 20–30 pg/tube, is incubated (1 hr) with the specific antibody. At the end of incubation the solid beads (bound fraction) are washed and the bile acid–HRP activity measured by a color-producing substrate (OPD/H_2O_2) at 490 nm. Alternatively, the enzyme activity is detected by chemiluminescence using luminol/H_2O_2. Light emission occurs in a few seconds, and its variation versus time is inversely proportional to the amount of bile acid in the sample. This method fulfills the normal criteria of reproducibility, accuracy, and precision, and a comparison with RIA showed excellent agreement ($r < 0.90$).

The reagents used are very stable, and the sensitivity is 50 times greater than any existing RIA when the colorimetric detection of HRP is used: it increases up to 100 times when the chemiluminescent substrate is used. Figure 12 shows the dose–response curves for chenodeoxycholic acid conjugates obtained with a conventional RIA and with this novel solid-phase enzyme immunoassay; sensitivity is greater by a factor of 100. The use of a macrophase immobilized antibody further simplifies the assay, allowing the development of an automated system by reducing the cost of the assay. This new method will be useful in serum bile acid analysis for better defining the normal levels and slight elevation (occurring in mild liver disease), and in monitoring patients with bile acid malabsorption, whose serum bile acid levels are often reduced.

In conclusion, the EIA method seems even more sensitive than RIA, and

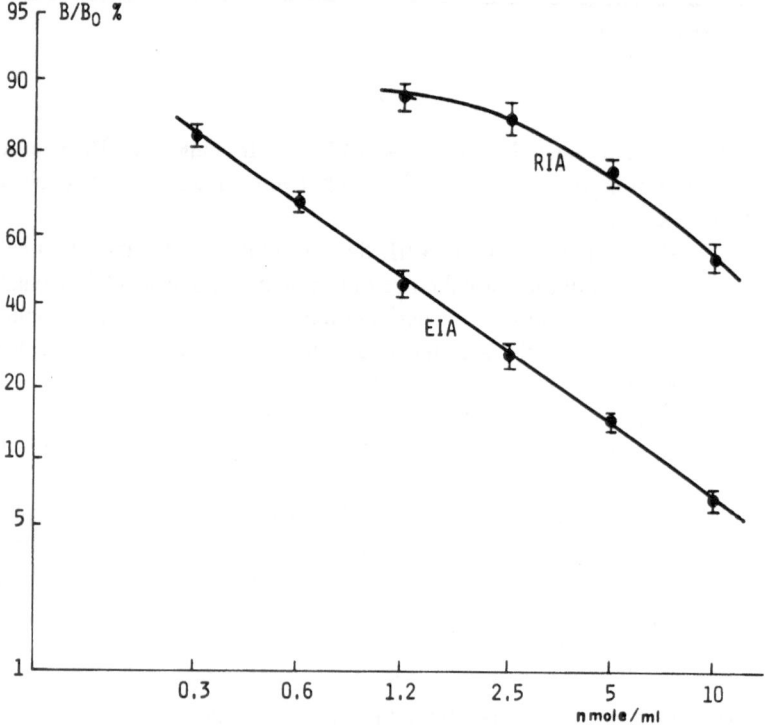

FIGURE 12. A typical dose–response curve obtained using the novel solid-phase enzyme immunoassay for serum chenodeoxycholic acid conjugated. On the right the dose–response curve using a conventional RIA.

its wider application will be extremely important in the introduction of serum bile acid measurements into clinical practice. From an analytical point of view, other immunoassays must be developed, in particular, a specific procedure for glycine, taurine, and free bile acid, in order to further investigate the most appropriate bile acid candidates for liver or intestinal function tests.

11. SERUM BILE ACIDS IN PHYSIOLOGY AND PATHOLOGY

In the last 10 years, the development of sensitive analytical methods to measure bile acids in serum has led to two main findings:

1. The physiological meaning of serum bile acid levels and the identification of their determinants.
2. The evaluation of the diagnostic usefulness of serum bile acid concentrations in liver disease and impaired enterohepatic circulation (EHC).

11.1. Serum Bile Acids as a Mirror of the Enterohepatic Circulation

The main function of the EHC (Fig. 13) is to allow appropriate cycling of the bile acid pool (which is about 2–4 g) through the liver and the intestine [49],

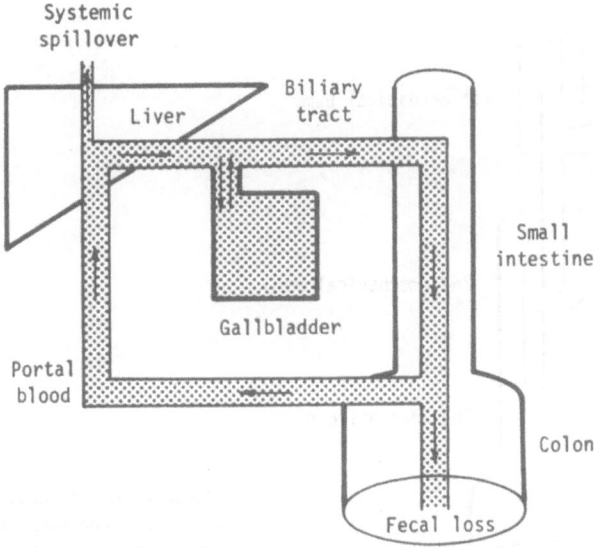

FIGURE 13. The enterohepatic circulation of bile acids.

where they regulate biliary secretion and cholesterol transport and participate in the fat absorption process [50].

This cycling is made possible by the coordinate activity of two mechanical pumps (Fig. 14) (gallbladder and intestinal motility) and two chemical pumps (liver uptake and intestinal absorption), which, therefore, represent the driving forces of the EHC in humans [49]. The amount of bile acids present in the EHC, i.e., the pool, is constant, with fecal loss (about 0.3 g/day) being balanced by hepatic *de novo* synthesis (about 0.3 g/day) [50]; however, with each entero-hepatic bile acid cycling, a fraction (from 10 to 50%, depending on the bile acid class) [51] of the pool returning to the liver escapes hepatic uptake and spills over into the systemic circulation (Fig. 15). The bile acids returning to the liver from the portal blood and from systemic circulation mix in the sinusoidal space and are cleared during their first hepatic passage. The fraction of bile acids spilling into the systemic circulation is believed to be constant [51], i.e., in-dependent of the range of loads presented to the liver [52,53]; therefore, one can assume that the amount of bile acids present in serum reflects the dynamics of the EHC [49,54,55]: higher concentrations are present when a larger amount

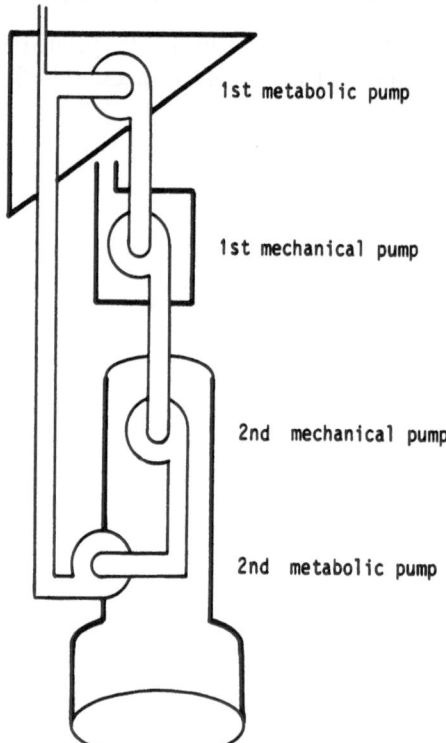

1st metabolic pump

1st mechanical pump

2nd mechanical pump

2nd metabolic pump

FIGURE 14. The driving forces of the enterohepatic circulation: metabolic pumps (hepatic transport and intestinal absorption); mechanical pumps (biliary and gastrointestinal motility).

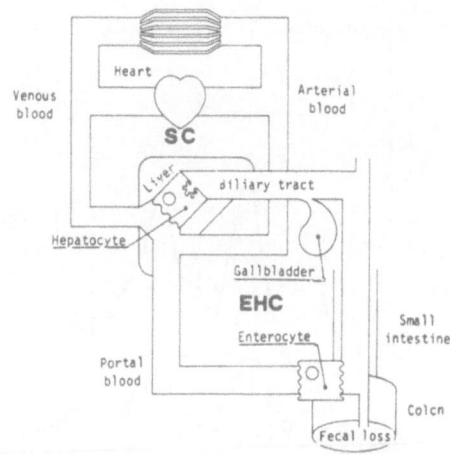

FIGURE 15. The enterohepatic (EHC) and
systemic (SC) circulations.

of bile acids gets through the intestine and, after absorption, reaches the liver, and lower concentrations when most of the bile acid pool is stored in the gallbladder. The observation of measurable serum bile acid concentrations also during fasting [56,57] indicates that the bile acid pool continuously moves within the EHC.

The evaluation of serum bile acid "circadian rhythm" in normal humans [53,54,56,141] (Fig. 16) reflects the physiological dynamics of the bile acid pool, which is differently influenced by the digestive and interdigestive motility of biliary tract and intestine.

During fasting, the interdigestive migrating myoelectric complex [58] creates a cyclical variation in the biliary bile acids which are continuously secreted into the duodenum [59]. Consequently, fasting serum bile acids fluctuate, reflecting these cyclical motor events in the biliary tract and intestine.

After food ingestion, a large amount of bile acid moves from the biliary tract to the intestine and in response an increase in serum bile acid concentration is observed (Fig. 16).

11.2. Determinants of Serum Bile Acid Levels

The results of several studies, performed both in the experimental animal and in humans, have clearly demonstrated that serum bile acid concentrations are determined by the instantaneous balance between input of bile acids from the intestine and clearance of bile acids by the liver [53]. However, since hepatic uptake is, in physiological conditions, almost constant [8,9], intestinal input is considered the main determinant of bile acid concentrations.

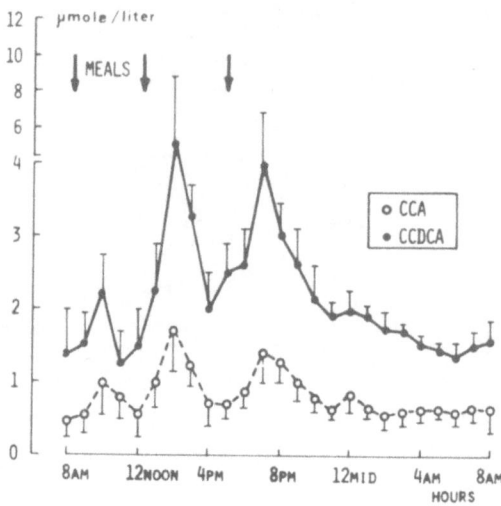

FIGURE 16. Serum levels of cholic (CCA) and chenodeoxycholic (CCDCA) acid conjugates after three meals and an overnight fast in 10 normal subjects (mean ± SD).

Intestinal input is the result of active and passive absorption of bile acids secreted into the intestine and, in turn, also reflects gallbladder contraction and intestinal motility.

Bile acids are absorbed by the intestine by two different mechanisms [60,61], depending on molecular structure and number of OH groups (Fig. 17): (1) active transport, which occurs exclusively in the distal segment of the ileum; conjugated, particularly tauroconjugated, trihydroxy bile acids undergo this modality of absorption; (2) a passive, nonionic diffusion, which is more diffuse in the proximal small intestine and colon and depends on the ability of each bile acid to remain undissociated at the pH of intestinal content. Unconjugated-, or glycine-conjugated, dihydroxy bile acids are reabsorbed by this passive modality (Fig. 17). The overall efficiency of intestinal bile acid absorption is high (~ 30%) [50].

Several lines of evidence suggest that the roles of gallbladder and intestinal motility are interrelated and influence serum bile acid levels. In the last few years we have performed several experiments [62,63], in order to elucidate the contributions of these two mechanical pumps to the EHC.

As far as the role of the intestine is concerned, if we abolish intestinal motility, during general anesthesia, or reduce it, by administration of drugs such as loperamide or atropine, serum bile acid concentrations significantly decrease. On the contrary, the reduction of intestinal transit time provoked by the administration of sorbitol or metoclopramide results in an increase in serum concen-

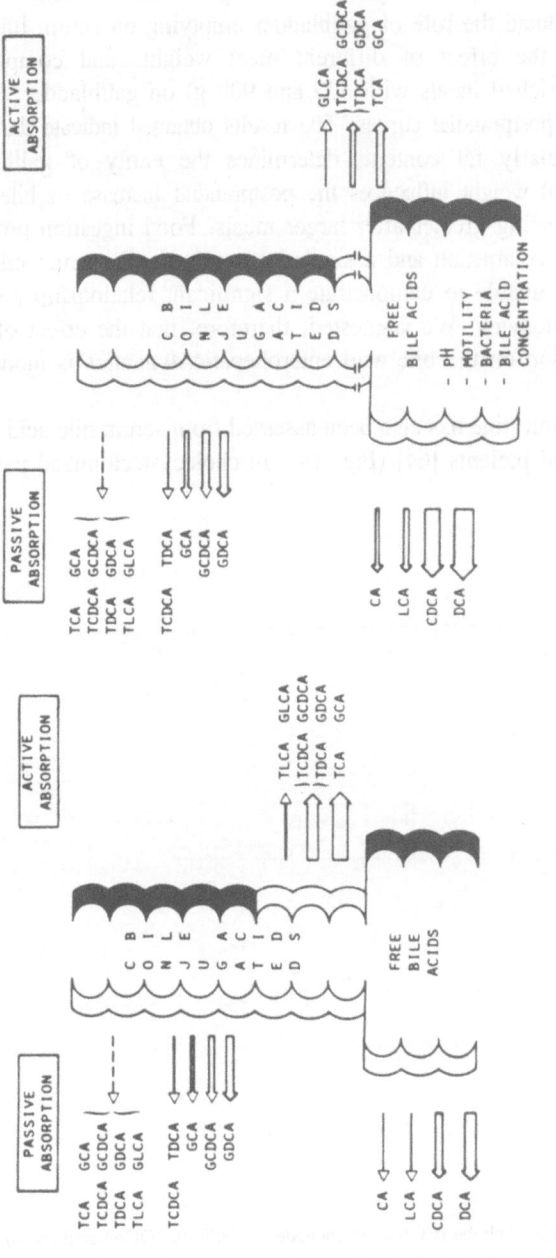

FIGURE 17. Intestinal absorption of bile acids in healthy subjects and ileal disease patients. CA, cholic acid; GCA, glycocholic acid; TCA, taurocholic acid; LCA, lithocholic acid; GLCA, glycolithocholic acid; TLCA, taurolithocholic acid; CDCA, chenodeoxycholic acid; GCDCA, glycochenodeoxycholic acid; TCDCA, taurochenodeoxycholic acid; DCA, deoxycholic acid; GDCA, glycodeoxycholic acid; TDCA, taurodeoxycholic acid).

trations of bile acids. In all the experiments, we monitored gallbladder volume by ultrasonography, and did not observe significant variations, thus excluding, in these experimental conditions, any influence of gallbladder motility.

In order to evaluate the role of gallbladder emptying on serum bile acid levels, we studied the effect of different meal weights and composition (standard and fat-enriched meals with 300 and 900 g) on gallbladder volume and serum bile acid postprandial curves. The results obtained indicate that meal composition, particularly fat content, determines the entity of gallbladder emptying, while meal weight influences the postprandial increase of bile acids in serum, the latter being greater after larger meals. Food ingestion produces a prompt gallbladder contraction and a similarly prompt rise in serum bile acid levels, but we were unable to demonstrate a significant relationship between these two early phenomena. We suggested, therefore, that the effect of gallbladder motor function on the bile acid enterohepatic dynamics is modulated by the intestine.

The role of the intestine has also been assessed from serum bile acid levels in cholecystectomized patients [64] (Fig. 18). In cholecystectomized patients,

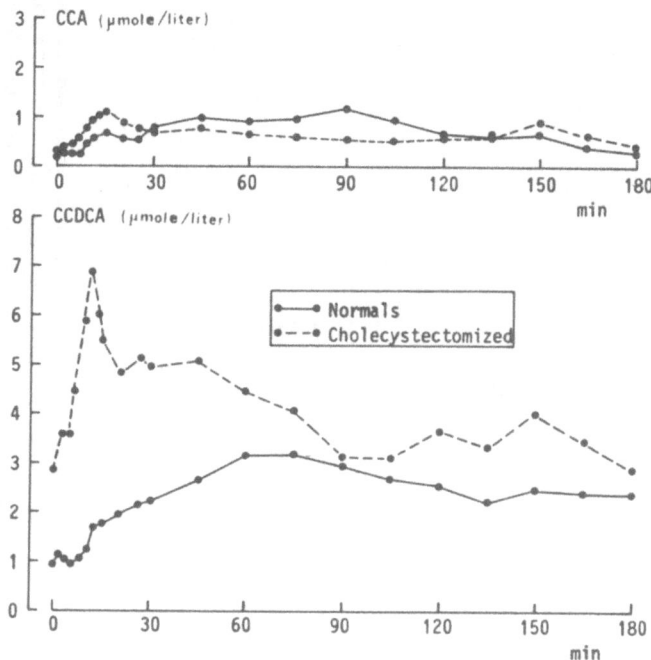

FIGURE 18. Serum levels of cholic (CCA) and chenodeoxycholic (CCDCA) acid conjugates in response to meal ingestion in cholecystectomized ($n = 10$) and control ($n = 10$) subjects (mean values).

both after a meal and after cholecystokinin infusion, serum bile acid levels display a single and early (within 30 min) peak, while normal subjects document a postprandial curve which is biphasic, with an early and small (less than ~20% of the total area under the curve of serum concentrations) peak and a second peak, which occurs later (after 60–90 min) and is greater. The areas under the curve of serum concentrations of bile acids are, however, similar in cholecystectomized and healthy subjects. This different serum profile could be explained by the larger fraction of the bile acid pool that is present during fasting in the intestine in cholecystectomized patients compared to healthy subjects.

Uptake of bile acids by the hepatocyte has been widely studied in the last 10 years. Hepatic uptake is a very efficient, carrier-mediated, active, sodium-dependent process [8,9]. Although it obeys saturation kinetics, the first-pass clearance for each bile acid is constant within a wide range of bile acid loads to the liver. Therefore, regardless of the intestinal input, this process never reaches saturation in physiological conditions. In fact, no differences were found in the rate of hepatic uptake in either fasting or postprandial conditions [52,53].

Hepatic extraction depends on bile acid structure [14,15,65]: the maximal velocity of uptake (V_{max}) is greater for trihydroxy than dihydroxy bile acids; the uptake of conjugated, particularly tauroconjugated, bile acids is more efficient than for free bile acids.

Fractional hepatic extraction is about 50–60% for the unconjugated dihydroxy bile acids, 70–80% for the conjugated dihydroxy bile acids, and 80–90% for the conjugated trihydroxy bile acids [51].

These values explain, in addition to a different modality in intestinal absorption, the higher serum concentrations of chenodeoxycholic acid than cholic acid [54] (Fig. 16).

A further determinant of hepatic uptake is represented by the extent of bile acid–albumin binding, since an inverse correlation has been documented [15,66] between the rate of bile acid uptake and the extent of bile acid–albumin binding.

A possible clinical implication of this relationship is reflected in Gilbert's syndrome, as well as in other unconjugated hyperbilirubinemic conditions. In fact, we have demonstrated [67] lower serum levels of cholic acid in these conditions compared with healthy subjects. *In vitro* studies also show that increasing bilirubin concentrations displace glycocholic acid and, to a lesser extent, glycochenodeoxycholic acid from binding to albumin. This competition between bilirubin and bile acids may explain the faster hepatic uptake of cholic acid conjugates and, hence, their lower serum levels in these conditions.

A putative carrier (receptor) for bile acids has been identified and characterized in rat liver plasma membrane fractions [68], although it is unclear whether this receptor is from canalicular, sinusoidal, or both surfaces. The relationship of this receptor to sinusoidal uptake has not been examined.

Recently, the existence of a lobular gradient for bile acid uptake has been suggested [69], revealing that the major portion of bile acids in portal blood is removed by only the first six to nine periportal hepatocytes, while the remaining cells in the lobule remove progressively less bile acids. It is not known, however, whether these findings reflect intrinsically different capacities for bile acid uptake of cells in the various regions of the lobule.

Knowledge of the determinants of serum bile acid levels makes possible a better understanding of the usefulness of this measurement in health and disease.

11.3. Serum Bile Acids in Hepatobiliary Disease

As mentioned in Section 11.2, bile acids secreted into the duodenum and efficiently absorbed by the intestine are efficiently, but partially, extracted by the liver (as for high-clearance drugs) [51]. Therefore, the fraction of bile acids spilling into the systemic circulation is believed to be constant, i.e., independent of the load presented to the liver during eating and fasting [52,53].

The major determinants of serum bile acids in physiological conditions are the hepatic clearance and, mainly, the intestinal input. Since serum bile acid concentrations, particularly their diurnal variations, may be considered a mirror of the EHC in health, it is likely that this happens also in liver and intestinal diseases. [55].

As far as liver diseases are concerned, it is well known that in patients with liver disease the concentrations of serum bile acids are frequently higher than in healthy subjects [1–3,70,71]. Factors influencing this elevation are not clearly identified: only in patients with chronic liver disease is the mechanism of increased serum bile acid levels partially elucidated [72–74].

Although it has been shown that the intestinal input is the major determinant of serum bile acid levels in these patients, the differences between normal subjects and liver disease patients are the consequence of a reduced hepatic clearance or of the presence of intrahepatic shunting. The reduced clearance results not only from reduced liver blood flow and intrahepatic shunting, but also, and mainly, from a reduced inherent capacity of the liver to remove bile acids from the blood. However, it is important to note that in patients with mild to moderate liver disease, hepatic uptake operates below saturation [75], and therefore, the first hepatic extraction, although reduced, remains constant, as in normal subjects. It is not known, however, whether the differences in serum bile acid concentration between normal subjects and liver disease patients are due to differences in the first-pass hepatic extraction or to a reduced V_{max}.

Several studies have evaluated the possibility of using serum bile acid determinations in the diagnosis of liver disease.

Evaluation of the overall diagnostic accuracy and the definition of modalities (intravenous or oral load, endogenous or exogenous load) and of timing (fasting or postprandial, or both) have been the main variables studied.

11.4. Diagnostic Accuracy

In the last few years, several studies on the diagnostic usefulness of serum bile acid determination in liver disease have been performed with conflicting conclusions.

Differences both in the assays and in statistical approach, together with differences in the populations studied, can be considered as the underlying causes of the discrepancies reported in the literature [75].

Briefly, serum bile acids are increased in liver disease and the increase possibly correlates with the severity of the disease. Several studies [76–81] have documented that the increase in serum bile acids parallels liver damage, and some authors [82,83] have proposed serum bile acid levels as a method capable of also defining the histological severity of the disease.

Serum bile acid increase, however, does not always seem to be able to discriminate between different liver diseases. In fact, serum bile acid determinations have been shown to be insensitive in fatty liver [84], in alcoholic [85,86] and posttransfusional hepatitis [87], and in the detection of injury caused by methotrexate [88], and to be useful, according to other authors, in the identification of anicteric liver disease [89], of vinyl chloride hepatotoxicity [90], and of Gilbert's syndrome [67]. There is wider agreement, on the other hand, regarding the usefulness of serum bile acids in identifying chronic liver disease [83,91,92]. Therefore, it can be assumed that serum bile acid determination provides a high specificity and a lower sensitivity [75,92].

These properties could be related to the "organ specificity" of bile acids rather than to a "functional specificity," as in most enzymes used as liver function tests. It is likely that serum bile acid evaluation may provide simultaneous information on functional hepatic mass, liver blood flow, and biliary excretory function.

We have measured serum bile acid levels, using a highly accurate and sensitive RIA method, [23] in a large number of liver disease patients and control subjects [91] (Fig. 19). In addition to serum bile acid levels, routine liver function tests were performed; results were evaluated by multivariate statistical methods, discriminant and factor analyses.

Multivariate analysis of variance documented significant differences between liver disease patients and control subjects, and also within the different classes of liver disease patients. The highest sensitivity was obtained from fasting serum bile acid determinations rather than the postprandial (2 hr after meal ingestion) determination. Fasting measurements also provided the highest percent of correct allocation within the different groups of liver disease considered.

Within the different bile acid classes, cholic acid conjugates provided the most sensitive determination, together with the combined determination of both cholic and lithocholic acid conjugates.

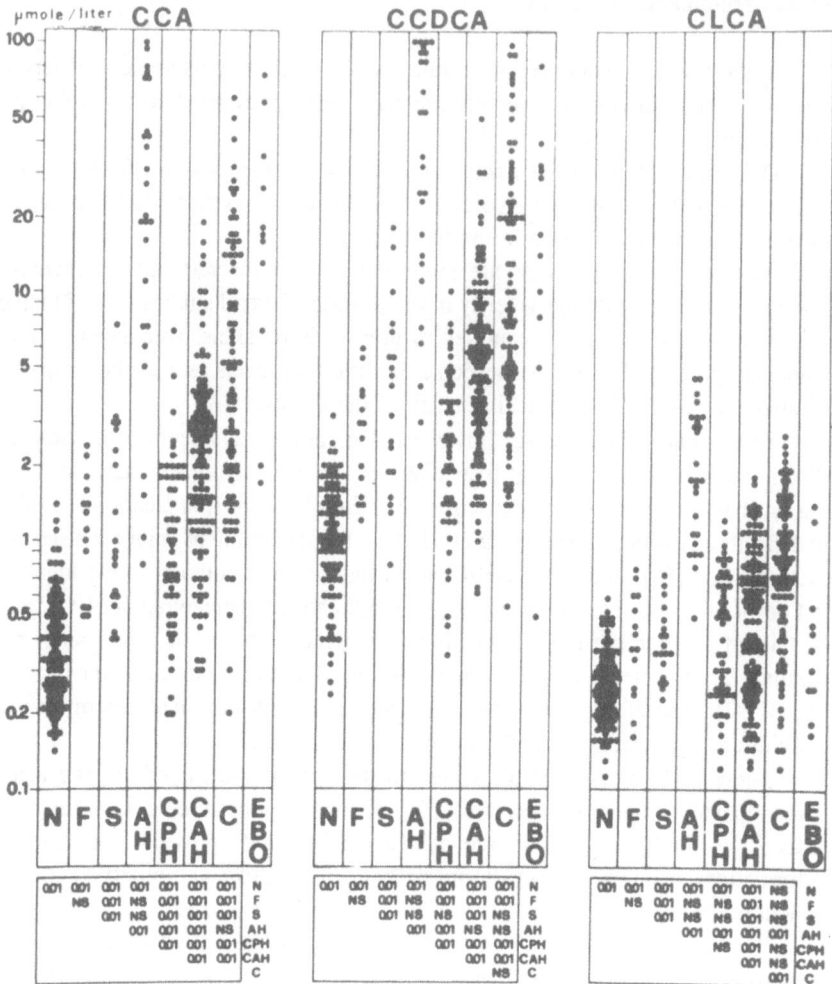

FIGURE 19. Serum fasting levels of cholic (CCA), chenodeoxycholic (CCDCA), and lithocholic (CLCA) acid conjugates in controls (N) and liver disease patients. F, hepatic fibrosis; S, hepatic steatosis; AH, acute hepatitis; CPH, chronic persistent hepatitis; CAH, chronic active hepatitis; C, cirrhosis; EBO, extrahepatic biliary obstruction.

The percentage of correct allocation was similar when considering serum bile acids and liver function tests; however, the combination of both tests provided the highest percent of correct allocation. Furthermore, this combination significantly reduced misclassification of liver disease patients, thus allowing better identification of patients needing treatment.

Factor analysis, performed to evaluate the relative variation of serum bile acid levels to liver function tests, identified two factors possibly related to cytolysis and protein synthesis; the serum bile acid concentrations highly correlated with both factors.

We therefore believe that serum bile acid determinations should play an adjunctive rather than substitutive role with respect to liver function tests, as documented by the observation that the combined use of serum bile acids and liver function tests is the most useful approach in detecting different liver diseases [91,93].

As far as the single classes of bile acids are concerned, it has been shown [91,93] that cholic acid determination produced a higher sensitivity, specificity, and predictive value than chenodeoxycholic acid and, when associated with liver function tests, allows a more accurate identification of both mild and severe disease.

11.5. Modalities of Serum Bile Acid Determination

Several bile acid tests have been proposed in order to improve the accuracy of serum bile acid determinations in detecting liver disease. An i.v. bile acid tolerance test was proposed by some authors [94,95], but results were unsatisfactory: the hepatic uptake was so rapid that the plasma disappearance was influenced by both distribution and hepatic uptake and the use of a distribution marker did not increase the sensitivity of the test.

Bile acids, as in the case of drugs with a high first-pass hepatic clearance, are not suitable to evaluate hepatic clearance which depend on blood flow and distribution volume [51].

The use of a postprandial serum determination, i.e., an endogenous load, was first proposed [96] to improve the diagnostic accuracy of fasting determination, by revealing a defective uptake of a load by a diseased liver. This method (GLC) was at first refused because of its low sensitivity; it was in fact unable to discriminate between fasting and postprandial values in controls. This problem was overcome by the use of more sensitive methods (RIA). When it was demonstrated that hepatic uptake in liver disease patients, although reduced, is far below its V_{max} and similar to uptake in healthy subjects, and the fractional hepatic clearance is constant in fasting and postprandial subjects, this invalidated the rationale of the test [75].

Furthermore, since the postprandial peaking is quite variable [97,98], depending on gallbladder and gastric emptying and on intestinal motility, an accurate evaluation would require measurement of the area under the curve of a postprandial profile, which is a much more expensive and time-consuming method.

In our opinion, the accuracy of fasting determinations is not increased by serum bile acid measurement after a meal and cerulein-induced gallbladder motility [99].

Still another approach is to present a known load of bile acids to the liver by the physiological route, i.e., portal blood, after oral ingestion. Several bile acids (chenodeoxycholic [100], cholic [101], and ursodeoxycholic acid [102]) have been used as an oral tolerance test, but its clinical utility seems to be limited, mainly because of methodological problems (e.g., influence of gastric emptying, intestinal transit time, interference with endogenous bile acids). However, owing to some peculiar characteristics regarding its passive intestinal absorption), ursodeoxycholic acid seems preferable for an oral tolerance test. Preliminary results are encouraging in both mild [103] and severe liver disease [104], but the time needed, the technical assistance required, the elevated cost, and the lack of sufficient proof of higher accuracy have limited its clinical use.

We conclude, therefore, that the fasting determination of serum bile acid concentrations seems to be preferred because of its accuracy and convenience.

11.6. Serum Bile Acids in Bile Acid Malabsorption Syndromes

An efficient intestinal absorption process preserves the bile acid pool within the EHC. In health, only about 10–15% of the total bile acid pool is lost daily in the feces, and this loss is compensated by hepatic synthesis [50]. In the presence of an interruption of the EHC due to an intestinal pathology, with or without mucosal damage, bile acids are malabsorbed (Fig. 17).

Three types of bile acid malabsorption syndromes have been identified: the first, as a consequence of ileal resection or disease [105,106]; the second, considered a primary syndrome [107,108], probably related either to increased bile acid synthesis, an overloading of the ileal transport system, or to impaired intestinal absorption of a normal bile acid load due probably to a defect in the carriers of ileal mucosa; and the third, found in uremic [109] and cholecystectomized [64] patients. The increased amount of bile acids in the colon causes alterations in the colonic water transport, and consequently diarrhea [110].

By perfusion studies [110,111], it has been demonstrated that dihydroxy bile acids induce net fluid secretion at high concentrations and, at low concentrations, block absorption of fluid and water. This effect is not present following colonic perfusion with trihydroxy bile acids.

To induce net secretion, however, bile acids must be dihydroxylated and present in elevated concentrations (>1.5 mmole) in the aqueous phase; furthermore, an appropriate pH (7.5–8.0) is necessary [112,113] (Fig. 17).

Obviously, the most frequent bile acid malabsorption syndrome is the first one; its clinical and metabolic picture depends on the length of the ileal resection or involvement. In fact, two main aspects have been described [106] (Fig. 20): a bile acid diarrhea, usually present in resections under 100 cm, and a fatty-acid diarrhea, when the ileal involvement is larger. In the first case, fecal bile acid output is high (see Chapter 12, this volume), bile acid concentrations in the aqueous phase are high, and hepatic bile acid synthesis is increased to maintain normal jejunal bile acid concentrations; therefore, steatorrhea is minimal. In

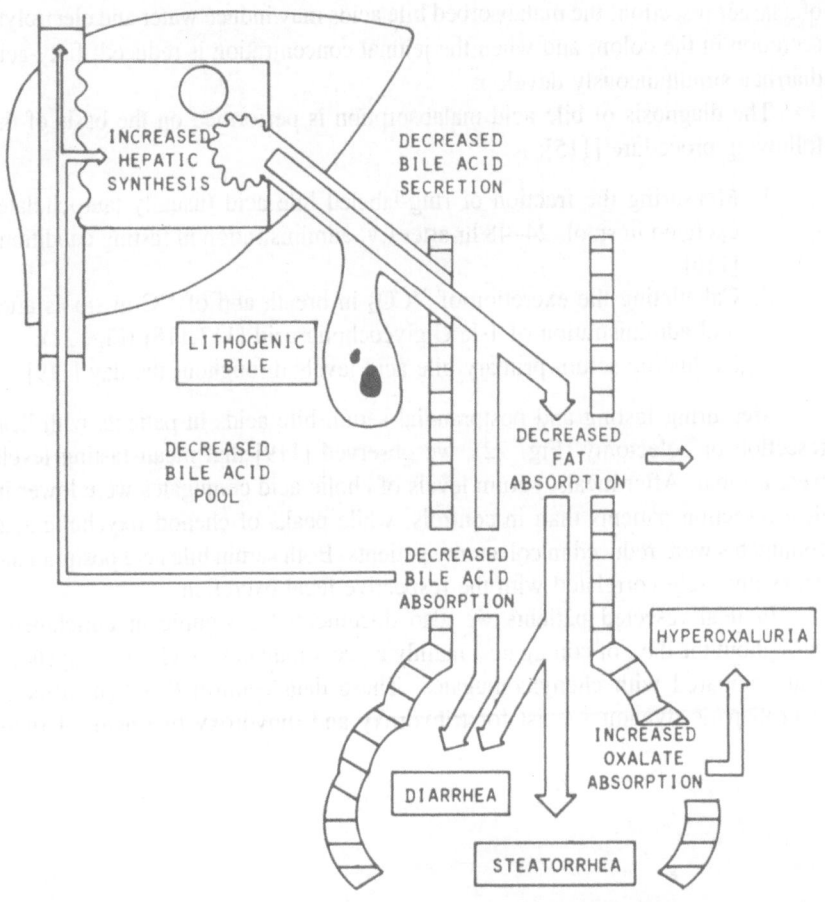

FIGURE 20. Metabolic consequence of bile acid malabsorption.

fatty-acid diarrhea, bile acid malabsorption is severe, and jejunal bile acid concentrations are usually decreased, despite the increased bile acid hepatic synthesis rate. Because of a decreased bile acid secretion, the concentration of bile acids in the fecal water of colonic content is normal. Steatorrhea is usually severe; water and electrolyte secretion is induced by hydroxy fatty acids.

Results from our laboratory [114] suggest, however, that these two syndromes are not completely separated: in fact, in cases of very short (40–50 cm) ileal involvement, little bile acid malabsorption is present; in resection of about 100 cm, bile acid malabsorption is accounted for almost only by trihydroxy bile acids, possibly because of the better intestinal conservation of dihydroxy bile acids, which undergo passive absorption in the small intestine and colon. In this case, therefore, a bile acid diarrhea may not be present. Finally, in the presence of a larger resection, the malabsorbed bile acids may induce water and electrolyte secretion in the colon, and when the jejunal concentration is reduced, fatty-acid diarrhea simultaneously develops.

The diagnosis of bile acid malabsorption is performed on the basis of the following procedure [115]:

1. Measuring the fraction of ring-labeled bile acid (usually taurocholate) excreted in stool, 24–48 hr after i.v. administration in fasting conditions [116].
2. Calculating the excretion of $^{14}CO_2$ in breath and of ^{14}C in stools after oral administration of 1-[^{14}C]glycocholic acid [117,118] (Fig. 21).
3. Evaluating serum primary bile acid levels throughout the day [119].

Measuring fasting and postprandial serum bile acids in patients with ileal resection or colectomy (Fig. 22), we observed [119] that mean fasting levels were normal. After meals, serum levels of cholic acid conjugates were lower in ileal resection patients than in controls, while peaks of chenodeoxycholic acid conjugates were reduced in colectomy patients. Both serum bile acid postprandial peaks inversely correlated with the respective fecal excretion.

In ileal resected patients we also documented a significant enrichment, throughout the day, of conjugates, mainly glycoconjugates, of chenodeoxycholic acid compared with chenylconjugates. These data confirm that two different enterohepatic dynamics exist for trihydroxy and dihydroxy bile acids, largely

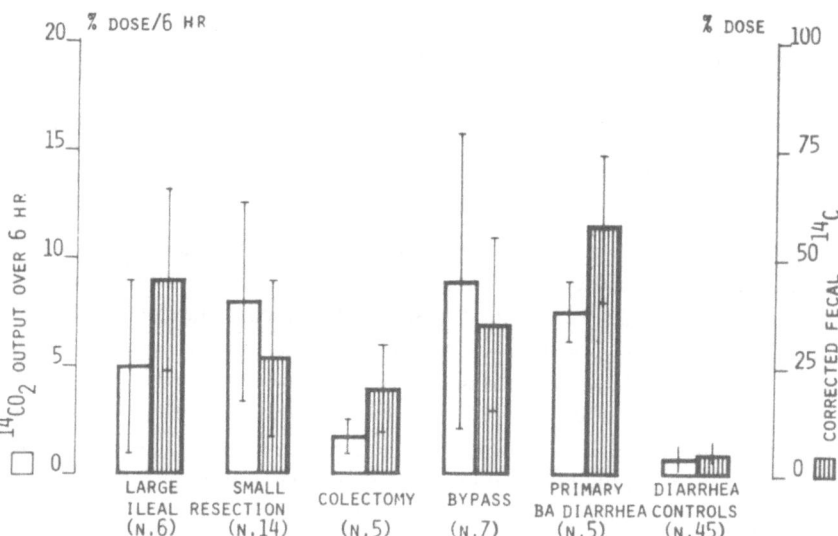

FIGURE 21. $^{14}CO_2$ output in breath and ^{14}C recovery in stools, after oral administration of 1-^{14}C-glycocholic acid, in different intestinal involvement (mean ± SD).

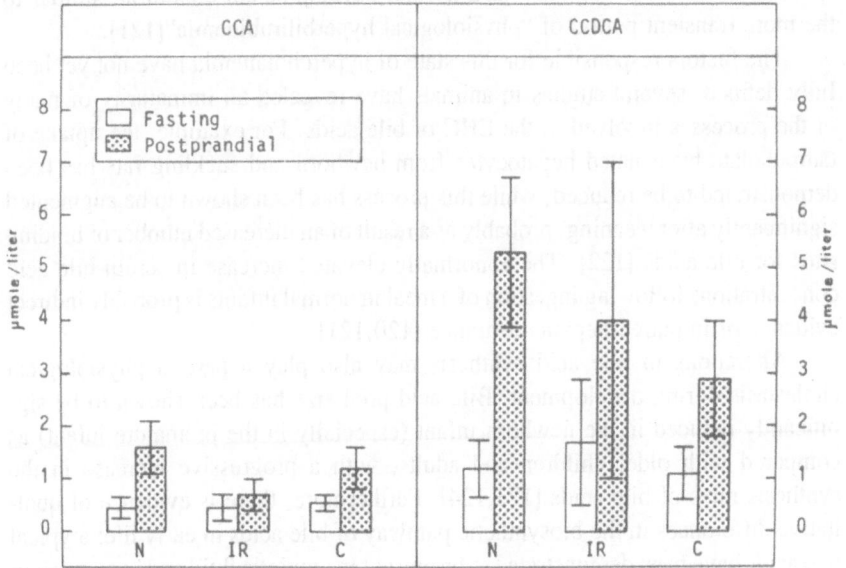

FIGURE 22. Fasting and postprandial serum levels of cholic (CCA) and chenodeoxycholic (CCDCA) acid conjugates in controls (N), ileal resection (IR), and colectomy (C) patients (mean ± SD).

determined by the intestinal absorption, and also indicate that cholic acid post-prandial peaks may be a marker of ileal disease, while those of chenodeoxycholic acid provide indirect information about colonic impairment.

Evaluation, in the morning and in the evening, of the ratio between chenyl and cholylconjugates further enhances the usefulness of this test in the diagnosis of bile acid malabsorption. It is important to note that bile acid malabsorption does not mean bile acid diarrhea; in fact, prerequisites of the occurrence of a bile acid-mediated diarrhea are the presence of a large amount of secretory bile acids in the fecal content (dihydroxy bile acid malabsorption) and a pH value keeping these bile acids in solution [112,113].

11.7. Serum Bile Acid Levels in Infancy and Childhood

The availability of a specific and sensitive RIA for different bile acid conjugates enables the characterization of serum bile acid patterns during infancy and childhood [120] (Table VIII). During the first year of life, serum concentrations of the two primary bile acids are elevated and comparable to those observed in older children and adults with liver disease. Serum levels of cholic and chenodeoxycholic acid conjugates decrease progressively, reaching adult values by the end of the first year of life. A period of "physiological cholestasis" has therefore been recognized as a normal developmental condition, similar to the more transient period of "physiological hyperbilirubinemia"[121].

The factors responsible for this state of hypercholanemia have not yet been fully defined: several studies in animals have revealed an immaturity of many of the processes involved in the EHC of bile acids. For example, the uptake of taurocholate by isolated hepatocytes from newborn and suckling rats has been demonstrated to be reduced, while this process has been shown to be augmented significantly after weaning, probably as a result of an increased number of binding sites for bile acids [122]. The abnormally elevated increase in serum bile acid concentrations following ingestion of a meal in normal infants is probably indirect evidence of impaired hepatic clearance [120,121].

Alterations in bile acid synthesis may also play a part in physiological cholestasis during development. Bile acid pool size has been shown to be significantly reduced in the newborn infant (especially in the premature infant) as compared with older children and adults, with a progressive increase in the synthesis rates of bile acids [123,124]. Furthermore, there is evidence of qualitative differences in the biosynthetic pathway of bile acids in early life: atypical bile acids have been demonstrated to be present in amniotic fluid and in meconium [125,126], some of which are well-known cholestatic agents.

These findings suggest that during fetal life, the synthesis of bile acids may occur via an alternative pathway, which is initiated through hydroxylation of the cholesterol side chain and has lithocholic and 3β-hydroxy-5-cholenoic acids as intermediates [142].

We have recently shown that serum concentrations of conjugated lithocholate are elevated in the human fetus between the fourteenth and the twenty-first weeks of gestation (1.7 ± 1.04 μmole/liter, as compared to 0.7 ± 0.3 μmole/liter in maternal serum; $p < 0.01$) [127]. Conversely, serum concentrations of deoxycholic acid are lower in the fetus, compared with the mother. We speculated that lithocholic acid could be synthesized by the fetal liver. Its cholestatic properties can be minimized by sulfation, a process that is probably reduced in fetal life, as suggested by the lower concentrations of sulfolithocholate in fetal serum compared with maternal blood, as well as by studies in animals [128]. The synthesis of this cholestatic compound (and possibly others) and the underdeveloped sulfation capacity may therefore interfere with bile acid secretion and bile flow in early life.

Owing to the peculiarities of bile acid metabolism during the first months of life and their variable developmental changes, primary serum bile acid levels have proved to be of limited value as indicators of liver dysfunction in neonates and infants.

Several studies have been carried out in an attempt to differentiate neonatal cholestasis caused by intrahepatic versus extrahepatic obstruction, but inconclusive results have been obtained [129,130]. Patients with biliary atresia tend to have higher total serum bile acid levels, as well as an increased proportion of chenodeoxycholate, compared with values obtained in patients with intrahepatic cholestasis. Conversely, in intrahepatic cholestatic syndromes, total serum bile acids tend to be lower, cholic acid is often the predominant bile acid in serum, and an increase in the cholate/chenodeoxycholate ratio in response to cholestyramine has been described. Unfortunately, for both serum bile acid concentration and composition, a considerable variability and overlap of values have been recorded, preventing clear differentiation between infants with intra- and extrahepatic cholestasis.

The determination of serum levels of conjugated sulfolithocholic acid (SLCG) seems to be more useful in the diagnosis of liver disease during the first year of life. In fact, serum SLCG concentrations in newborns and infants are similar to those of normal adults [120] and increase markedly in the presence of liver disease, as a result of sulfotransferase induction during "pathological cholestasis" [131]. Unfortunately, this test has also proved of insufficient discriminatory value as far as the cholestatic syndromes of infancy are concerned, but it has been successfully used to differentiate steatosis from cholestasis in infants during total parenteral nutrition, since abnormal levels of SLCG have been found only in children with cholestasis [132].

In older children, the use of serum bile acid determination as an indicator of liver dysfunction has confirmed the results obtained in adults: serum fasting cholic and chenodeoxycholic acid concentrations were found to be elevated in a variety of hepatobiliary disorders [133]. In adults, postprandial serum bile acid determination has been suggested to be even more sensitive than fasting values in the evaluation of liver diseases [96], but there are conflicting data in children. In a recent report, postprandial serum glycocholic acid levels were determined in 122 children with liver disease and were found more frequently abnormal than any other single liver function test. They proved even more sensitive than the combined determination of total bilirubin, aspartate transaminase, alkaline phosphatase, serum albumin, and prothrombin time; only the addition of γ-glutamyl transpeptidase increased the frequency of abnormality to that found with single serum conjugated cholic acid determination [134]. In our experience, as well as that of other authors [135], fasting serum bile acid measurement is as sensitive as postprandial determination and simpler to perform.

Unfortunately serum bile acid levels have also proved of limited help in the differential diagnosis of liver diseases in children. Some studies have reported a considerable overlap of values found in different diseases and a poor correlation between serum bile acid levels and the characteristics and severity of histological lesions [134,136]. Conversely, other authors reported that fasting serum bile acids are able to differentiate children with chronic aggressive hepatitis from those with chronic persistent hepatitis [137]. A discriminant value of serum bile acid determination has also been reported in a study that evaluated children with α-1-antitrypsin deficiency. While conventional liver function tests were often abnormal in patients with mild liver disease, primary serum bile acid levels were found to be increased only in patients with liver cirrhosis [138].

The actual advantages of serum bile acid determination over conventional liver function tests in children are probably limited to a few conditions [139]: disorders of bilirubin metabolism (in which normal bile acid levels are found), chronic hepatitis (serum bile acid levels seem to be related to disease activity), and finally, longitudinal follow-up of patients at risk for liver disease. We have found that serum chenodeoxycholic acid levels are frequently abnormal earlier than conventional liver function tests in patients with cystic fibrosis, in whom "focal" biliary cirrhosis can develop (20–25% of cases) [140]. We made the same observation in hypertransfused children with thalassemia major, and in our opinion serum chenodeoxycholic acid determination is a simple and sensitive test, which could be performed at regular intervals in patients at risk. Finally, serum cholic acid postprandial determination has been reported to be of value in the evaluation of the interruption of the EHC of bile acids caused by bowel disease [119]. Serum cholic acid levels after a meal were found to be below normal in children with ileal resection [141], Crohn's disease [142], primary bile acid diarrhea [143], and cystic fibrosis [144], and the normal expected meal-stimulated increase was absent. This test has therefore been proposed as a sensitive index of ileal involvement in children, with the advantage that radiological contrast studies and the Schilling test might be required less frequently.

REFERENCES

1. S. Sherlock and V. Walshe, *Clin. Sci.* **6**, 223 (1948).
2. D. Rudman and F. E. Kendall, *J. Clin. Invest.* **36**, 530 (1957)
3. J. B. Carey, *J. Clin. Invest* **37**, 1494 (1958).
4. S. Skrede, H. E., Solberg, J. P. Blomhoff, and E. Gjone, *Clin. Chem.* **24**, 1095 (1978).
5. M. Blum and N. Sprits, *J. Clin. Invest* **45**, 187 (1966).
6. A. F. Hofmann, *in* "Bile Acid Metabolism in Health and Disease" (G. Paumgartner and A. Stiehl, Eds.) p. 151, MTP Press, Lancaster (1977).
7. N. F. La Russo, Gastroenterology **67**, 806 (1984).
8. J. C. Glasinovic, M. Dumont, M. Duval, and S. Erlinger, *J. Clin. Invest.* **55**, 419 (1975).
9. H. Reichen and G. Paumgartner, *Am. J. Physiol.* **231**, 734 (1968).
10. J. M. Dietshy, *J. Lipid. Res.* **91**, (1968).
11. E. R. Schiff, *J. Clin. Invest.* **51**, 1351 (1972).
12. W. J. Simmonds, M. G. Korman, V. L. W. Go, and A. F. Hofmann, *Gastroenterology* **65**, 705 (1973).
13. F. Scagnolari, A. Roda, A. Fini, and B. Grigolo, *Biochim. Biophys. Acta* **791**, 274 (1984).
14. A. Roda, G. Cappelleri, R. Aldini, E. Roda, and L. Barbara, *J. Lipid Res.* **23**, 490 (1982).
15. R. Aldini, A. Roda, A. M. Morselli, G. Cappelleri, E. Roda, and L. Barbara, *J. Lipid Res.* **23**, 1167 (1982).
16. W. Kramer, H. P. Buscher, W. Gerok, and G. Kurz, *Eur. J. Biochem.* **102**, 1 (1979).

17. A. Klopstock, M. Pinto, and A. Rimon, *J. Immunol.* **92**, 515 (1664).

18. B. F. Erlanger, F. Borek, S. M. Beiser, and S. Lieberman, *J. Biol. Chem.* **228**, 713 (1957).

19. G. M. Murphy, S. M. Edkins, J. W. Williams, and D. Catty, *Clin. Chim. Acta* **54**, 81 (1974).

20. L. M. Demers and G. W. Hepner, *Clin. Chem.* **22**, 602 (1976).

21. S. Matern, R. Krieger, and W. Gerok, *Clin. Chim. Acta* **73**, 277 (1976).

22. J. W. O. Van den Berg, M. Van Blankenstein, E. P. Bosuava-Jacobs, and M. Frenkel, *Clin. Chim. Acta* **73**, 277 (1976).

23. A. Roda, E. Roda, R. Aldini, D. Festi, G. Mazzella, C. Sama, and L. Barbara, *Clin. Chm.* **23**, 2107 (1977).

24. S. W. Schalm, G. P. Van Berge Henegouwen, A. F. Hofmann, A. E. Cowen, and J. Turcotte, *Gastroenterology* **73**, 285 (1977).

25. J. G. Spenney, B. I. Hirschowitz, A. A. Mihas, and R. Gibson, *Gastroenterology* **72**, 305 (1977).

26. A. A. Mihas, J. G. Spenney, B. I. Hirschowitz, and R. Gibson, *Clin. Chim. Acta* **76**, 389 (1977).

27. S. Matern, R. Krieger, C. Hans, and W. Gerok, *Scand J. Gastroenterol.* **12**, 641 (1977).

28. A. E. Cowen, M. G. Korman, A. F. Hofmann, J. Turcotte, and J. A. Carter, *J. Lipid Res.* **18**, 692 (1977).

29. A. E. Cowen, M. G. Korman, A. F. Hofmann, J. Turcotte, and J. A. Carter, *J. Lipid Res.* **18**, 698 (1977).

30 A. Roda, E. Roda, D. Festi, R. Aldini, G. Mazzella, G. Sama, and L. Barbara, *Steroid* **32**, 13 (1978).

31. I. Makino, A. Tashiko, H. Hashimoto, S. Nakagawa, and I. Yashizawa, *J. Lipid Res.* **19**, 443 (1978).

32. O. Mentausta and O. Janne, *Clin. Chem.* **25**, 264 (1979).

33. E. Minder, G. Karlajanis, V. Schmied, P. Vitins, and G. Paumgartner, *Clin. Chim. Acta* **92**, 177 (1979).

34. P. Miller, S. Weiss, M. Cornell, and J. Dockery, *Clin. Chem.* **27**, 1968 (1981).

35. E. I. Minder, *J. Lipid Res.* **20**, 986 (1978).

36. Y. A. Baqir, J. Murison, P. E. Ross, and I. A. D. Bouchier, *J. Clin. Pathol.* **32**, 560 (1979).

37. G. J. Beckett, W. M. Hunder, and I. W. Percy-Robb, *Clin. Chim. Acta* **93**, 145 (1979).

38. A. Roda, E. Roda, R. Aldini, M. Capelli, D. Festi, C. Sama, G. Mazzella, A. Morselli, and L. Barbara, *Clin. Chem.* **26**, 1647 (1980).

39. E. G. Orbàn, J. P. Pallos, and L. T. Kocsàr, *Acta Med. Hung.* (in press).

40. S. Yamauchi, M. Kojima, and F. Nakayama, *Steroids* **41**, 165 (1983).

41. S. Yamauchi, M. Kojima, and F. Nakayama, *Chem. Pharm. Bull.* **32**, 3088 (1984).

42. K. Samuelson. *Scand. J. Clin. Invest.* **40**, 289 (1980).

43. S. Matern, K. Tietjen, H. Matern, and W. Gerok, *in* "European Labelled Immunoassays of Hormones and Drugs" (S. B. Pal-Walter de Gruyter, ed.), pp. 457–467 (1978).

44. Y. Maeda, T. Setoguchi, T. Katsuki, and E. Ishikawa, *J. Lipid Res.* **20**, 960 (1979).

45. A. Ozaki. *J. Lipid Res.* **20**, 2340 (1979).

46. Y. A. Baquir, J. Murison, P. E. Ross, and I. A. D. Bouchier, *Anal. Biochem.* **93**, 361 (1979).

47. Endab Cholylglycine EIA Kit, Immunotechnical Corporation, Cambridge, MA.

48. A. Roda, S. Girotti, S. Lodi and S. Preti, *Talanta* **31**, 895 (1984).

49. A. F. Hofmann, *Clin. Gastroenterol.* **6**, 3 (1977).

50. M. C. Carey, *in* "The Liver: Biology and Pathology" (H. Arrias, H. Popper, D. A. Schachter, and D. A. Shafritz, (eds.), p. 429, Raven Press, New York (1982).

51. G. P. Van Berge Henegouwen and A. F. Hofmann, *Eur. J. Clin. Invest.* **13**, 433 (1983).

52. B. Angelin, I. Bjorkhem, K. Einarsson, and S. Ewerth, *J. Clin. Invest.* **70**, 724 (1982).

53. N. F. LaRusso, N. E. Hoffman, M. G. Korman, A. F. Hoffmann, and A. E. Cowen, *Am. J. Dig. Dis.* **23**, 385 (1978).

54. L. Barbara, A. Roda, E. Roda, R. Aldini, G. Mazzella, D. Festi, and C. Sama, *Rendic. Gastroenterol.* **8**, 194 (1976).

55. L. Barbara, D. Festi, F. Bazzoli, R. Aldini, A. Roda, and E. Roda, *in* "Recent Advances in Bile Acid Research" (L. Barbara, R. H. Dowling, A. F. Hofmann, and E. Roda, eds., p. 29, Raven Press, New York, (1985).

56. S. W. Schalm, N. R. LaRusso, A. F. Hofmann, N. E. Hoffman, G. P. Van Berge Henegouwen, and M. G. Korman, *Gut* **19**, 1 (1978).

57. F. Baldi, F. Ferrarini, M. Salera, P. Giacomoni, S. Tovoli, M. Cassan, D. Festi, and L. Barbara, *Gastroenterol. Clin. Biol.* **7**, 721 (1983).

58. S. J. Sarna, *Gastroenterology* **89**, 894 (1985).

59. T. L. Peeters, G. Van Trappen, and J. Janssens. *Gastroenterology* **79**, 678, (1980).

60. B. Borgstom, G. Lundh, and A. F. Hofmann, *Gastroenterology* **45**, 229 (1963).

61. E. Krag and S. F. Phillips, *J. Clin. Invest.* **53**, 1686 (1974).

62. D. Festi, S. Baroncini, G. Mazzella, R. Aldini, A. Roda, E. Roda, and L. Barbara, *Ital. J. Gastroenterol.* **12**, 335 (1980).

63. L. Barbara, D. Festi, F. Bazzoli, R. Frabboni, R. Aldini, M. Malavolti, A. Roda, and E. Roda, *in* "Enterohepatic Circulation of Bile Acids and Sterol Metabolism" (G. Paumgartner, A. Sthiel, and W. Gerok, eds.), p. 139, MTP Press, Lancaster (1985).

64. E. Roda, R. Aldini, G. Mazzella, A. Roda, C. Sama, D. Festi, and L. Barbara, *Gut* **19**, 640 (1978).

65. N. E. Hoffman, J. H. Iser, and R. A. Smallwood, *Am. J. Physiol.* **229**, 298 (1975).

66. R. Aldini, A. Roda, A. M. Morselli, B. Grigolo, P. Simoni, E. Roda, and L. Barbara, *in* "Recent Advances in Bile Acid Research" (L. Barbara, R. M. Dowling, A. F. Hofmann, and E. Roda, eds.), p. 33, Raven Press, New York, (1985).

67. A. Roda, E. Roda, C. Sama, D. Festi, R. Aldini, A. M. Morselli, G. Mazzella, and L. Barbara, *Gastroenterology* **82**, 77 (1982)

68. L. Accattino and F. R. Simon, *J. Clin. Invest.* **57**, 496 (1976).

69. G. Paumgartner and A. L. Jones, *Am. J. Physiol.* **238**, G233 (1980).

70. C. R. Pennington, P. E. Ross, and I. A. D. Bouchier, *Gut* **18**, 903 (1977).

71. A. Roda, E. Roda, D. Festi, G. Mazzella, R. Aldini, and L. Barbara, *Acta Gastro-ent. Belg.* **41**, 653 (1978).

72. P. Paré, J. C. Hoefs, and M. Ashcavai, *Gastroenterology* **81**, 959 (1981).

73. R. Y. Poupon, R. E. Poupon, D. Lebrec, L. Le Quercec, and F. Darnis, *Gastroenterology* **80**, 1438 (1981).

74. H. Ohkubo, K. Okuda, S. Iida, K. Ohnishi, S. Ikawa, and I. Makino, *Gastroenterology* **86**, 514 (1984).

75. A. F. Hofmann, *Hepatology* **2**, 512 (1982).

76. L. Barbara, *Ital. J. Gastroenterol.* **10**(Suppl. 1), 8 (1978).

77. M. Van Blakenstein, M. Frenkel, J. W. O. Van den Berg. F. J. W. Ten Kate, E. P. Bosman-Jacobs, and A. C. Touw-Blommenstyn, *Dig. Dis. Sci.* **28**, 137 (1983).

78. M. Jones, S. Weinstock, R. Koretz, K. J. Lewin, J. Higgins, and G. L. Gitnick, *Gastroenterology* **78**, 1189 (1980).

79. K. Samelson, A. Aly, C. Johansson, and A. Norman, *Scand. J. Gastroenterol.* **16**, 225 (1981).

80. S. Skrede, H. E. Solberg, J. Blomhoff, and E. Gjone, *Clin. Chem.* **24**, 1095 (1978).

81. P. Tobiasson, A. Fryden, and C. Tagesson, *Scand. J. Gastroenterol.* **16**, 763 (1981).

82. M. G. Korman, A. F. Hofmann, and W. H. J. Summerskill, *N. Engl. J. Med.* **290**, 1399 (1974).

83. P. S. Monroe. A. L. Baker. J. F. Schneider. P. S. Krager, P. D. Klein, and D. Schoeller, *Hepatology* **2**, 317 (1982).

84. K. Einarsson, B. Angelin, I. Bjorkhem, and H. Glaumann, *Hepatolgoy* **5**, 108 (1985).

85. P. Tobiasson and B. Boeryd. *Scand. J. Gastroenterol.* **15**, 657 (1980).

86. H. J. Milstein, J. R. Bloomer, and G. Katskin, *Dig. Dis.* **21**, 281 (1976).

87. J. M. Mishler, L. Barbosa, L. J. Mihalko, and H. McCarter, *JAMA* **246**, 2340 (1981).

88. C. M. Lawrence, R. C. Strange, and R. Summerly, *Clin. Chim. Acta* **139**, 341 (1983).

89. J. G. Douglas, G. J. Beckett, I. A. Nimmo, N. D. C. Finlayson, and I. W. Percy-Robb, **Gut** *22*, 141 (1981).

90. G. M. Liss, R. A. Greenberg, and C. H. Tamburro, *Am. J. Med.* **78**, 68 (1985).

91. D. Festi, A. M., Morselli-Labate, A. Roda, F. Bazzoli, R. Frabboni, P. Rucci, F. Taroni, R. Aldini, E. Roda, and L. Barbara, *Hepatology* **3**, 707 (1983).

92. R. Ferraris, G. Colombatti, M. T. Fiorentini, R. Carosso, W. Arossa, and M. De La Pierre, *Dig. Dis. Sci.* **28**, 129 (1983).

93. R. Ferraris, M. T. Fiorentini, G. Galatola, P. Rolfo, and M. De La Pierre, *Ann. Intern. Med.* (1985) (in press).

94. M. G. Korman, N. F. LaRusso, N. E. Hoffman, and A. F. Hofmann, *N. Engl. J. Med.* **292**, 1205 (1975).

95. N. F. LaRusso, N. E. Hoffman, A. F. Hofmann, and M. G. Korman, *N. Engl. J. Med.* **292**, 1209 (1975).

96. N. Kaplowitz, E. Kok, and N. B. Javitt, *Jama* **225**, 292 (1973).

97. G. J. Bekett, J. C. Douglas, N. D. Finlayson, and I. W. Percy-Robb, *Digestion* **22**, 248 (1981).

98. D. Festi, *in* "Serum Bile Acids in Health and Disease" (L. Barbara, R. H. Dowling, A. F. Hofmann, and E. Roda, eds.), p. 56, MTP Press, Lancaster, (1983).

99. D. Festi, R. Aldini, R. M. Rossi, A. Roda, E. Roda, and L. Barbara, *Gastroenterology* **82**, 1227 (1982) (Abstr.).

100. P. Tobiasson, A. Fryden, and C. Tagesson. *Scand. J. Gastroenterol.* **17**, 763 (1981).

101. S. Matern, M. Haag, C. H. Hans, and W. Gerok, *in* "Bile Acids in Health and Disease" (G. Paumgartner and A. Stiehl, eds.), p. 253, MTP Press, Lancaster, (1977).

102. H. Tashiro, *Acta Hepatol. Jpn.* **20**, 369 (1979).

103. H. Ohkubo, K. Okuda, S. Iida, and I. Makino, *Gastroenterology* **81**, 126 (1981).

104. G. Miescher, G. Paumgartner, and R. Preising, *Eur. J. Clin. Invest.* **13**, 439 (1983).

105. A. F. Hofmann and J. R. Poley, *Gastroenterology* **62**, 918 (1972).

106. A. F. Hofmann, *Arch. Intern. Med.* **130**, 597 (1972).

107. E. H. Thaysen and L. Pedersen, *Gut* **17**, 963 (1976).

108. R. Aldini, A. Roda, F. Bazzoli, G. Mazzella, D. Festi, C. Sama, A. M. Morselli-Labate, E. Roda, and L. Barbara, *Ital. J. Gastroenterol.* **12**, 251 (1980).

109. S. J. Gordon, L. J. Miller, M. D. Kinsley, and O. D. Kowlessar, *Gut* **13**, 415 (1976).

110. H. S. Mekhjian, S. F. Phillips, and A. F. Hofmann, *J. Clin. Invest.* **50**, 1569 (1971).

111. H. S. Mekhjian, S. F. Phillips, and A. F. Hofmann, *Dig. Dis. Sci.* **24**, 545 (1979).

112. H. J. Binder and C. L. Rawlings, *J. Clin. Invest.* **52**, 1460 (1973).

113. B. McJunkin, H. Fromm, R. P. Sarva, and P. Amin, *Gastroenterology* **80**, 1454 (1981).

114. R. Aldini, A. Roda, D. Festi, C. Sama, G. Mazzella, F. Bazzoli, A. M. Morselli, E. Roda, and L. Barbara, *Dig. Dis. Sci.* **27**, 495 (1982).

115. L. Barbara, *in* "Bile Acids in Gastroenterology" (L. Barbara, R. H. Dowling, A. F. Hofmann, and E. Roda, eds.), p. 172, MTP Press, Lancaster, (1983).

116. M. M. Stanley and B. Nemchansky, *J. Lab. Clin. Med.* **70**, 627 (1967).

117. H. Fromm and A. F. Hofmann, *Lancet* **2**, 621 (1971).

118. A. Roda, E. Roda, R. Aldini, G. Mazzella, D. Festi, C. Sama, and L. Barbara, *Clin. Chem.* **23**, 2127 (1977).

119. R. Aldini, A. Roda, D. Festi, G. Mazzella, A. M. Morselli, C. Sama, E. Roda, N. Scopinaro, and L. Barbara, *Gut* **23**, 829 (1982).

120. L. Barbara, R. Lazzari, A. Roda, R. Aldini, D. Festi, C. Sama, A. M. Morselli, A. Collina, F. Bazzoli, G. Mazzella, and E. Roda, *Pediatr. Res.* **14**, 1222 (1980).

121. F. J. Suchy, W. F. Balistreri, J. E. Heubi, J. E. Searchy, and R. S. Levin, *Gastroenterology* **80**, 1037 (1981).

122. F. J. Suchy and W. F. Balistreri, *Pediatr. Res.* **16**, 282 (1982).

123. J. B. Watkins, P. Szczepanik, and J. B. Grould, *Gastroenterology* **69**, 706 (1975).

124. J. E. Heubi, W. F. Balistreri, and F. J. Suchy, *J. Lab. Clin. Med.* **100**, 127 (1982).

125. P. Back and K. Walter, *Gastroenterology* **78**, 671 (1980).

126. G. Délèze, G. Paumgartner, G. Karlaganis, W. Giger, M. Reinhard, and D. Sideropoulos, *Eur. J. Clin. Invest.* **8**, 41 (1978).

127. C. Colombo, A. Roda, E. Roda, M. Buscaglia, C. A. dell'Agnola, P. Filippetti, M. Ronchi, and F. Sereni, *Pediatr. Res.* **19**, 227 (1985).

128. W. F. Balistreri, L. Zimmer, F. J. Suchy, and K. E. Bove, *J. Lipid Res.* **25**, 228 (1984).

129. N. B. Javitt, J. P. Keating, J. Grand, and R. C. Harris, *J. Pediatr.* **90**, 736 (1977).

130. N. B. Javitt, K. P. Morrissey, E. Siegel, H. Goldberg, L. M. Gartner, M. Hollander, and E. Kok, *Pediatr. Res.* **7**, 119 (1973).

131. W. F. Balistreri, F. J. Suchy, M. K. Farell, and J. E. Heubi, *J. Pediatr.* **98**, 399 (1981).

132. M. K. Farell, W. F. Balistreri, and F. J. Suchy, *J. Parenteral Enteral Nutri.* **6**, 30 (1982).

133. J. D. Lloyd-Still and L. M. Demers, *Am. J. Clin. Pathol.* **71**, 444 (1979).

134. A Matsui, H. T. Bocharopoulos, A. P. Mowat, B. Portmann, and G. M. Murphy, *J. Clin. Pathol.* **35**, 1011 (1982).

135. G. P. Davidson, M. Carey, F. Morad-Hassel, J. M. Sondheimer, D. Crazier, and G. G. Forstner, *J. Clin. Pathol.* **3**, 390 419807.

136 B. Strandvik and K. Samuelson, *Scand. J. Gastroenterol.* **20**, 381 (1985).

137. P. Jara, R. Codoceo, A. Hernanz, C. Castano, C. Diaz, Y. L. Montero, and C. Varquez, *Digestion* **27**, 81 (1983).

138. A. Nemeth, K. Samuelson, and B. Strandvik, *J. Pediatr. Gastroenterol. Nutr.* **1**, 479 (1982).

139. J. E. Heubi, *J. Pediatr. Gastroenterol. Nutr.* **1**, 457 (1982).

140. C. Colombo, A. Roda, E. Roda, L. Piceni Sereni, D. Maspero, A. Giunta, and L. Barbara, *Dig. Dis. Sci.* **28**, 306 (1983).

141. K. D. R. Setchell, A. M. Lawson, E. J. Blackstock, and G. M. Murphy, *Gut* **19**, 32 (1978).

142. C. Colombo, G. Zuliani, M. Ronchi, J. Breidenstein, and K. D. R. Setchell, *Pediatr. Res.* **21**, 197 (1987).

Chapter 7

BIOLUMINESCENCE ASSAYS USING IMMOBILIZED ENZYMES IN BILE ACID ANALYSIS

J. Schölmerich, A. Roda, and M. DeLuca

1. INTRODUCTION

Although it is well established that serum bile concentrations are elevated in patients with various liver diseases [1,2], the clinical utility of their determination is still debated. Two major, incompletely resolved problems limit the usefulness of such tests: (1) the exact definition of the principal physiological determinants of serum bile acid levels and pathological events affecting such concentrations, and (2) the analytical heterogeneity of the existing serum bile acid methods currently available.

As far as the first point is concerned, we know that the momentary balance between intestinal input and hepatic uptake of bile acids has been identified as the most important physiological determinant of serum bile acid levels. An increase in serum bile acid levels postprandially or in the fasting state is considered to be specific for liver disease, reflecting decreased hepatic uptake or systemic shunting of portal venous blood. A decreased level, especially postprandially, is an indicator of bile acid malabsorption that normally reflects ileal dysfunction [3].

Abbreviations used in this chapter: NAD, oxidized form of nicotinamide-adenine dinucleotide; NADH, reduced form of nicotinamide-adenine dinucleotide; FMN, flavin mononucleotide; NADP, oxidized form of nicotinamide-adenine dinucleotide phosphate; EDTA, ethylenediaminetetraacetate.

J. Schölmerich Department of Internal Medicine, University of Freiburg, Freiburg, West Germany **A. Roda** Institute of Chemical Sciences and Departments of Internal Medicine and Gastroenterology, University of Bologna, Bologna, Italy **M. DeLuca** Department of Chemistry, University of California, San Diego, California

Because the fractional clearance of bile acids is very high (50–90%), the peripheral blood contains bile acids at very low concentrations (μM levels) compared with that of bile (mM levels). In addition, variations in serum bile acid levels related to physiological phenomena (intestinal and gallbladder motility) or to disease states are often small.

As far as the second point is concerned, the qualitative pattern of serum bile acids is very complex. Human serum contains at least 16 major bile acids: the two primary bile acids, chenodeoxycholic and cholic acid, and the two secondary bile acids, lithocholic and deoxycholic acids, which are all present in both unconjugated as well as conjugated form. In addition, lithocholyl conjugates are mostly sulfated. Several different approaches to serum bile acids analysis have been adopted in an attempt to define the clinical utility of such determinations. Gas–liquid chromatography (GLC) combined with mass spectrometry is the currently available reference method [4]. In order to measure all the bile acids present in serum, a prior class separation is required. This method is expensive, requires specialized technical ability, and is difficult to carry out on a large-scale study.

Other chromatographic procedures, such as high-performance liquid chromatography (HPLC), are still unsatisfactory, mainly because of a lack of sensitivity in detecting the separated bile acids. Using the conventional colorimetric detector at 205 nm, unconjugated bile acids are poorly detected [5,6].

The only methods widely applied since the first paper appeared in 1973 by Hofmann's group [7] are the competitive binding techniques, including radioimmunoassay (RIA) and enzyme immunoassay [8]. Although RIA exhibits adequate sensitivity, the specificity of most reported RIA is not entirely satisfactory. Unconjugated bile acids, which may constitute a considerable fraction of serum bile acids, are determined partially since antibodies mostly specific for conjugated bile acids are used in most assays reported to date [9].

Enzymatic methods have been the first biochemical methods largely applied for serum bile acid analysis [10]. These methods, based on the use of bile acid-specific hydroxysteroid dehydrogenases (HSD), such as 3α-HSD or 7α- and 12α-HSD, are potentially adequate for measuring serum bile acids because of the high specificity of the currently available enzymes. With such enzymes it is possible to measure total 3α-hydroxy bile acids, both free and glycine/taurine conjugates, which represent the majority of the physiological bile acids. The major problem in using conventional enzymatic techniques is their sensitivity. These enzymes are NAD/NADH dependent, and all the existing methods are based on the measurement of NADH production via a direct colorimetric or fluorimetric detection or by means of a coupled amplified reaction [11–13]. Despite significant improvements, the methods are not yet completely satisfactory for testing increased or decreased bile acid levels with respect to normal values. More recently, a bioluminescence detection of NADH rendered the enzymatic technique as sensitive as the immunological methods, and due to its extreme simplicity, it is suitable for large-scale clinical studies and is a potential tool in clinical chemistry [14,15].

Rather than review this field which is in its infancy and has seen limited applications, this chapter will outline the principles of the bioluminescence technique and describe our experiences in developing and applying it to the determination of bile acids in biological fluids.

2. BIOLUMINESCENCE ASSAYS

2.1. Principles

Bioluminescence assays have been reported for a variety of compounds which can be coupled to the production of reduced pyridine nucleotides [16–19]. The enzymatic reactions are the following:

$$NADH + H^+ + FMN \rightarrow NAD^+ + FMNH_2 \qquad (1)$$

$$NADPH + H^+ + FMN \rightarrow NADP^+ + FMNH_2 \qquad (2)$$

$$FMNH_2 + RCHO + O_2 \rightarrow FMN + RCOOH + H_2O + h\nu \quad (3)$$

Reactions (1) and (2) are catalyzed by specific NADH- or NADPH-dependent FMN oxidoreductases. Reaction (3) is catalyzed by bacterial luciferase. This is the oxidation of $FMNH_2$ and a long-chain aldehyde which results in the production of light. If NADH or NADPH is the limiting substrate in the coupled reaction, then light intensity is directly proportional to the concentration of this substrate.

The enzymes that catalyze these reactions have been purified from several marine bacteria. The assays described here were done with the enzymes from *Vibrio harveyi*. The luciferase was purified as described by Hastings *et al.* [20], and the oxidoreductases were purified by the method of Jablonski and DeLuca [21]. If the specificity for NADH or NADPH is not necessary, then it is possible to use a bacterial diaphorase instead of the oxidoreductase. This enzyme will use either NADH or NADPH to reduce FMN.

The assays that have been developed for bile acids utilize a third enzyme which catalyzes the oxidation of a hydroxy steroid, reaction (4).

$$OH \text{ bile acid} + NAD^+ \rightarrow O \text{ bile acid} + NADH + H^+ \qquad (4)$$

The NADH produced is measured in the coupled light-producing reactions, and in this assay the peak light obtained is proportional to the bile acid concentration. Since there are specific dehydrogenases that act only on certain hydroxyl groups, it is possible to couple each of these individual enzymes to the bioluminescence system.

2.2. Coimmobilized Enzymes

Immobilized enzymes have several advantages over the soluble forms. When one is measuring several enzymatic reactions in a coupled reaction, the coimmobilized enzymes are much more efficient than the soluble forms [22–25]. For example, the same amount of soluble NADH:FMN oxidoreductase and luciferase produces about 1% of the light for a given concentration of NADH when compared with the coimmobilized enzymes. This difference in efficiency becomes more pronounced with more enzymes in the reaction sequence [26]. In addition, when one uses either bacterial or firefly luciferase as the last enzyme in the sequence, the reactions are thermodynamically irreversible and therefore will act to pull all of the previous enzymatic reactions. Finally, it is possible to use the coimmobilized enzymes in a flow cell where samples can be pumped through repetitively. Therefore, the same enzymes can be used for multiple assays [27].

Many different procedures are available for immobilizing enzymes and these have been reviewed extensively [28]. The experiments described here have been done with enzymes immobilized onto cyanogen bromide-activated Sepharose [29]. The recovery of active enzymes on Sepharose is generally quite good, ranging from a low of 10% to as much as 90%. One disadvantage of Sepharose is that bacterial contamination is a problem after prolonged storage. This can be prevented partially by storing the Sepharose–enzyme mixture in buffers containing azide.

3. MATERIALS

3.1. Enzymes

Bacterial luciferase prepared as described [18] from *V. harveyi*, Strain B392, has been used for all assays described. However, a preparation from *V. fischeri* (Boehringer, Mannheim, Federal Republic of Germany) has been used with similar success (unpublished results). NADH:FMN-oxidoreductase from *V. harveyi* [29] was used initially and later replaced by bacterial diaphorase (Boehringer, Mannheim, Federal Republic of Germany). Several HSD were used: 3α-HSD and 7α-HSD from Sigma Chemical Company (St. Louis, MO) and 3α-HSD from Worthington Biochemical Cooperation (Freehold, NJ) and from Nyegaard and Company (Oslo, Norway). 12α-HSD was provided by the late Dr. I. A. Macdonald (Halifax, Nova Scotia, Canada) prepared from *Clostridium* group P [30–32]. The 3α-HSD from Worthington was contaminated with an aldehyde dehydrogenase and could only be used after purification with aid of a Sephadex G100 column, as described [33]. The enzymes from Sigma were contaminated to varying degrees, depending on the lot used. However, some preparations could be utilized without further purification. The 3α-HSD from Nyegaard did not contain aldehyde dehydrogenase.

Sepharose 4B was purchased from Pharmacia (Piscataway, NJ), cyanogen bromide from Eastman Chemical Company (Rochester, NY), Decanal, FMN, and glutathione from Sigma, and NAD, NADH, NADP, NADPH grade I from Boehringer (Mannheim, Federal Republic of Germany). All other chemicals used were of analytical grade.

3.2. Bile Acids

Bile acids for standard solutions were obtained from a variety of sources and had a purity of more than 98% as judged by thin-layer chromatography [34].

3.3. Samples

Serum and urine samples from patients with different hepatobiliary diseases were provided and analyzed by GLC and HPLC by Dr. G. van Berge Henegouwen (Arnheim, The Netherlands). Some sera were analyzed by RIA as well [9,14]. Bile samples extracted in methanol from the National Cooperative Gallstone Dissolution Study [35,36] analyzed by GLC were provided by Dr. A. F. Hofmann (San Diego, CA).

3.4. Instruments

An Aminco Chem Glow photometer was used for the development of the assays described. Several other instruments from other suppliers have been used with similar success. However, the choice of the instrument depends on the needs of the individual laboratory. For continuous-flow system analysis an apparatus with a two-stream flow system has been developed (Fig. 1) [37].

4. METHODS

4.1. Enzyme Immobilization

The combinations of enzymes shown in Table I were dissolved in 3.5 ml of 0.1 M sodium pyrophosphate, pH 8.0. Bovine serum albumin (BSA) was added to give a final protein concentration of 8 mg/ml. Variations of the composition of the enzyme mixture have not been studied extensively and can prob-

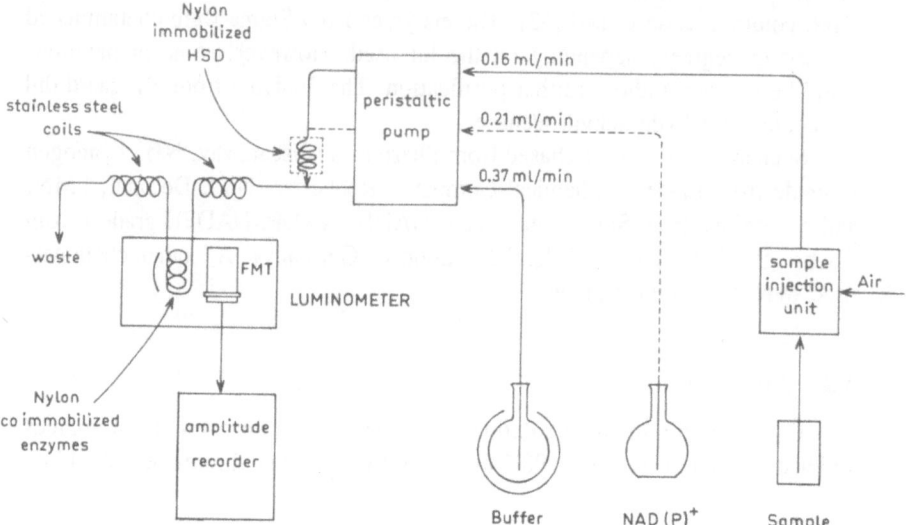

FIGURE 1. Schematic description of the continuous flow system for bioluminescence assays for bile acids using coimmobilized enzymes. The dotted lines indicate the system used with separately immobilized HSD.

TABLE I. Enzyme Composition of the Immobilized Enzyme Systems, Recovery of Enzymes after Immobilization in U/ml IES Solution, and Percent of Total Enzyme Recovered

Enzyme	Hydroxysteroid-dehydrogenase	Diaphorase	Luciferase[a]
Composition			
3α-IES	2.3 mg (10.9 U/mg)	6.0 mg (1 U/mg)	10.0 mg (6 × 10⁷ LU/mg)
7α-IES	7.0 mg (7.0 U/mg)	6.0 mg (1 U/mg)	15.0 mg (6 × 10⁷ LU/mg)
12α-IES	1.5 mg (0.1 U/mg)	10.0 mg (1 U/mg)	5.0 mg (6 × 10⁷ LU/mg)
Recovery			
3α-IES	0.04 U/ml, 34.2%	0.21 U/ml, 37.9%	0.60 × 10⁶ LU/ml, 7.1%
7α-IES	0.31 U/ml, 33.7%	0.25 U/ml, 38.0%	1.16 × 10⁶ LU/ml, 7.1%
12α-IES	0.03 U/ml, 35.7%	1.83 U/ml, 30.4%	0.54 × 10⁶ LU/ml, 7.5%

[a] LU, relative light units.

ably be used in order to improve the sensitivity of the assays further. The enzyme mixture was dialyzed against 4 liters of 0.1 M sodium pyrophosphate, pH 8.0 for 16 hr, in order to remove dithiothreitol (DTT), which is used for storage of luciferase but interferes with the immobilization procedure.

One gram of Sepharose 4B was activated with cyanogen bromide according to the method of March *et al.* [38], and 3 ml of the enzyme mixture was then added to the Sepharose and stirred for 2 hr at room temperature. The Sepharose–enzymes were washed with 200 ml 0.1 M potassium phosphate, pH 7.0, 500 ml 0.1 M potassium phosphate, pH 7.0 with 1 M NaCl, and again with 300 ml 0.1 M potassium phosphate. For preparations containing diaphorase the washing buffers contained 0.5 mM glutathione.

The washed Sepharose–enzymes were stored in 10 ml of 0.1 M potassium phosphate, pH 7.0, containing 0.02% sodium azide and 2 mM DTT. The latter was added weekly. If diaphorase was used, the storage buffer contained 0.5 mM glutathione instead of DTT, which interferes with diaphorase activity. For longer storage glycerol (10% v/v) was added. The immobilized systems were stored at 4°C for up to 3 months or rapidly frozen in liquid nitrogen and stored indefinitively at −20°C.

The recovery of enzymes after immobilization was tested using standard spectophotometric assays for HSD and diaphorase in the enzyme mixture after dialysis and in the immobilized system. The activity of luciferase was measured by injecting 100 µl FMN (0.15 mM, EDTA 0.5 mM) which had been photoreduced into a cuvette containing 400 µl 0.1 M potassium phosphate, pH 7.0, 0.0001% decanal, and 100 µl enzyme mixture (diluted 1:10–1:1000) in an Aminco Chem Glow photometer. The resulting light emission peak was used to calculate the relative light units per milliliter enzyme solution. The yield recovered after immobilization was between 30 and 40% for the different HSD, 40 and 60% for oxidoreductase, 30 and 50% for the bacterial diaphorase, and 6 and 10% for the bacterial luciferase. However, modifications of the enzyme composition may lead to better results (unpublished results). The immobilization of these enzymes onto nylon tubes has also been successfully performed [37].

4.2. Continuous-Flow Systems

In order to render these assays easier and less expensive, an automated flow system was used. Kricka *et al.* [27] developed a flow system for primary bile acids using a cell containing the three enzymes coimmobilized on Sepharose beads. Results agreed well with GLC, but the factor limiting the continuous use of the flow cell was bacterial contamination of the Sepharose. This was overcome by a new flow system that uses enzymes coimmobilized on nylon [37]. The flow system involves two streams: the primary one is a continuous flow of air into which a known volume of sample is intermittently inserted (Fig. 1).

Nylon tubes are ideal for a continuous-flow system presenting no problems, such as carryover, packing, or destruction of gel matrix. The continuous-flow system can be set up with the three enzymes coimmobilized or with two separate reactors. The first is inside the luminometer which contains the two bioluminescent enzymes, i.e., FMN oxidoreductase and luciferase; the second contains only the substrate-specific enzymes, i.e., 3α-HSD, 7α-HSD, or 12α-HSD. The last solution offers two major advantages. (1) It is possible to carry out the first catalysis at pH 9.5, which is optimal for the enzyme, and then appropriately lower the pH to 7 by mixing online with the working bioluminescence solution buffered at pH 7, thus obtaining an optimal pH for the bioluminescence enzymes. This results in a more sensitive method. (2) The use of two separate coils offers the possibility to perform multiple analyses by changing only the specific nylon coil for the analysis to be carried out. For example, for bile acids it is possible to analyze 3α-, 7α-, and 12α-hydroxy bile acids online and obtain a complete composition of the major bile acids in human serum or bile.

The nylon immobilized enzyme system can also be adapted as a detector for bile acids reverse-phase HPLC. Potentially, it will represent an extremely sensitive detector of column-separated bile acids, offering advantages of increased specificity and sensitivity, and permitting analysis of unconjugated-, glycine-conjugated, and taurine-conjugated bile acids with the same detector response.

4.3. Assay Techniques

Several assay techniques have been tested. In principle, 400 μl 0.1 M potassium phosphate, pH 7.0, containing decanal 0.0001%, 10 μl of 0.5 mM FMN, 10–50 μl of the Sepharose–enzyme suspension, and 10 μl of 5 mM NAD(P) are mixed into a tube; 10–50 μl of standard or sample is added, and the tube is vigorously shaken and placed into the luminometer.

Light emission is recorded continuously until a peak is reached after 20–40 sec (first-peak method). The time used depends largely on the enzyme composition and can be shortened by several modifications (unpublished observations). Alternatively, the shaking-and-reading procedure can be repeated until a final peak is reached (final-peak method).

The most feasible alternative is injection of 100 μl 0.5 mM NAD(P) into the tube which is already placed in the luminometer and contains assay buffer, FMN, decanal, Sepharose–enzymes, and sample or standard. The injection starts the reaction and, in addition, mixes the contents of the tube. This method has proved to be the easiest and exhibits the best reproducibility [15,39]. However, this is not possible with all instruments. Figure 2 shows the time course of light emission obtained with these different techniques. Background is obtained by measuring the light emission without sample or standard.

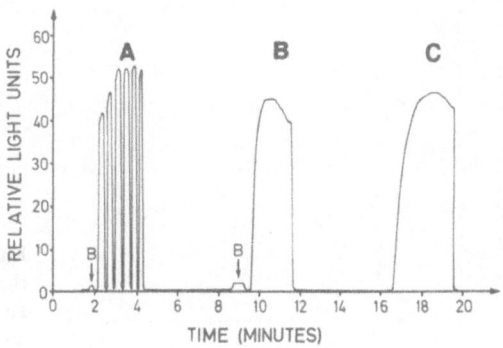

FIGURE 2. Kinetics for the different test systems described. (A) Repeated shaking until a final peak height of light emission is reached = final peak method. (B) First peak method. (C) Injection assay. 20 μmole/liter taurocholate is used in each test. B, background light emission without bile acid present. (From Ref. [39], with permission of the publisher.)

It is important to be aware that the decanal suspension is unstable, and therefore the decanal solution has to be prepared freshly before use. It can be prepared either by shaking 5 μl of decanal in 10 ml of water and then storing in ice or by dissolving 10 μl of decanal in 10 ml of ethanol. The appropriate amount is then added to the assay buffer at 1- to 2-hr intervals.

Since the descending part of the kinetic curve of the reaction is caused by sedimentation of the Sepharose beads, it is also of importance that the peak light emission is reached during the first 30 sec after injection (Fig. 3).

Results obtained by the three different test techniques on 20 serum samples revealed a good correlation ($r > 0.99$ for each comparison).

4.4. Standard Curves

Standard curves for 3α-, 7α-, and 12α-hydroxy bile acids could be constructed for concentrations from 0.2 to 1000 μmole/liter. With the enzyme

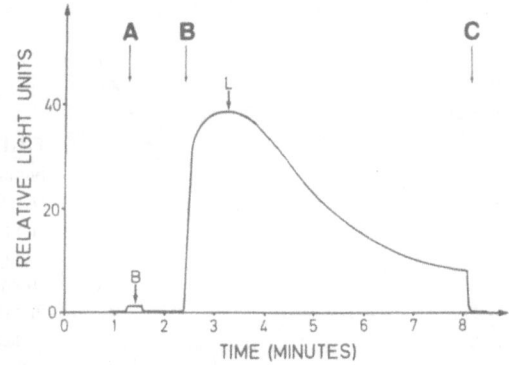

FIGURE 3. Kinetics of the injection assay. B, background light emission with injection of 100 μl NAD (0.5 mmole/liter) (A) in the absence of bile acid. L, light emission with injection of 100 μl NAD (0.5 mmole/liter) (B) in the presence of taurocholate (20 μmole/liter). The decrease of light emission reflects sedimentation of the Sepharose beads. (C) Closing of the light detector.

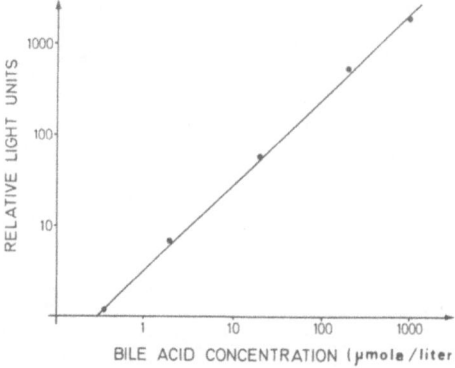

FIGURE 4. Standard curve for tauro-cholate, using the 3α-hydroxy bile acid assay ($y = 0.943x + 0.516$; $r = 0.9901$). (From Schölmerich *et al.* [15], with permission of the publisher.)

activities and the instruments used, the light output was less than twice the background at concentrations below 0.2 μmole/liter (0.5 pmole/tube). Linearity was lost at concentrations above 1000 μmole/liter (Fig. 4). While standard curves for taurine and glycine conjugates and the corresponding unconjugated bile acid were identical, significant differences were found between different bile acids when standards were prepared without albumin. These differences disappeared, however, when BSA was added to the standard solutions at a final concentration of 1.5 g/dl. Figure 5 gives an example for the 7α-HSD assay.

When 13 different bile acids or conjugates were analyzed at the same concentration, a coefficient of variation of 5.7 was found in the presence of albumin (Table II). The exact amount of BSA necessary has not yet been determined. However, kinetic studies of some of the enzymes used revealed that the affinity of 3α-HSD and 7α-HSD for different bile acids is different and can be modified by the addition of albumin (Fig. 6). Since the binding affinity of albumin to bile acids parallels the affinity of the HSD, the effect on the test

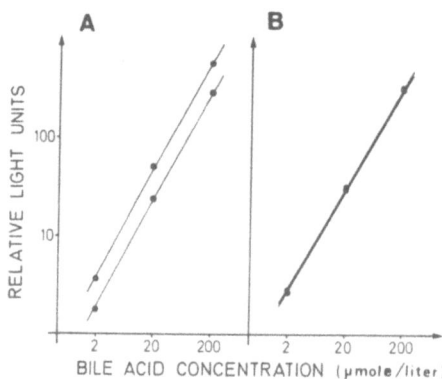

FIGURE 5. Effect of the addition of albumin on the 7α-hydroxy bile acid assay. (A) Standard curves for taurochenodeoxycholate (left; $y = 1.10x + 0.25$) and taurocholate (right; $y = 1.09x - 0.04$) in the absence of albumin; (B) identity of both curves in the presence of albumin (1.5 g/dl); taurochenodeoxycholate, $y = 1.04x + 0.14$; taurocholate, $y = 1.04x + 0.13$.

TABLE II. Light Emission with Identical Standard Concentration of Different Bile Acids[a]

Bile acid	Concentration (μmole/liter)	Light units	Variation (%)
Taurocholate	100	348	+4.5
Glycocholate	100	335	+0.6
Cholate	100	342	+2.7
Taurochenodeoxycholate	100	345	+3.6
Glycochenodeoxycholate	100	290	−12.9
Chenodeoxycholate	100	320	−3.9
Taurodeoxycholate	100	320	−3.9
Deoxycholate	100	352	+5.7
Tauroursodeoxycholate	100	359	+7.8
Ursodeoxycholate	100	323	−3.0
Glycolithocholate	100	315	−5.4
Ursocholate	100	336	+0.9
β-Muricholate	100	343	+3.0
Mean		332.9 ± 19.1	CV: 5.7%

[a] Light units represent mean of two analyses with the 3α-hydroxy bile acid assay. Variation is calculated as percent of the mean light emission for all measurements. From Ref. 15, with permission of the publisher.

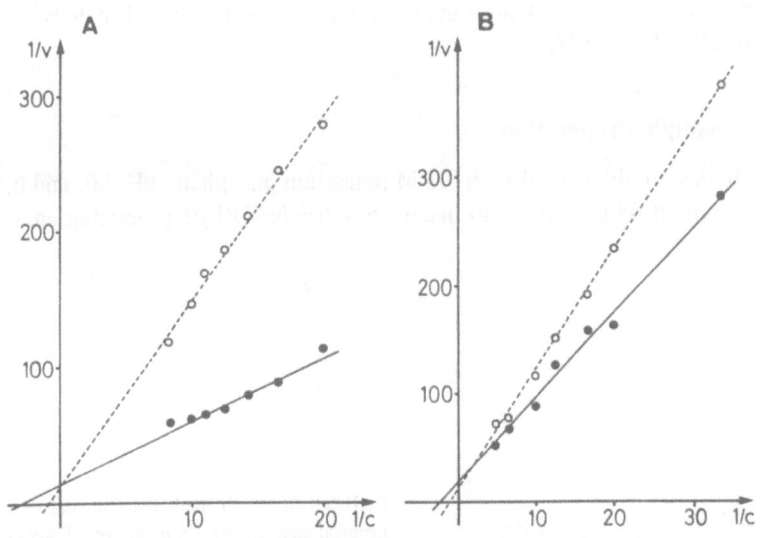

FIGURE 6. Enzyme kinetics for 3α-HSD (A) and for 7α-HSD (B). Comparison with (open circles) and without (closed circles) albumin using taurochenodeoxycholate (A) and taurocholate (B).

systems can be explained. In addition, the high affinity of the reaction product—an oxo bile acid—to albumin might be responsible for the increase of V_{max} and, thus, for the sensitivity of the assays. Therefore albumin should be used in the preparation of standard solutions and for urine and bile samples. In the latter case it is easier to add the albumin to the assay buffer.

4.5. Assay Specificity

None of the test systems studied reacted with bile acid lacking the specific hydroxyl group. Other substrates such as malate, lactate, and β-hydroxybutyrate did not give significant light emission up to concentrations of 10 mM. Androsterone did react with the 3α system as described [33]. But this compound is present in biological fluids in much smaller concentrations and therefore the interference is negligible. Cholesterol did not react with the systems, and the addition of a 3β-hydroxy bile acid resulted in a light emission of 0.28% of that obtained with the corresponding 3α-hydroxy compound, probably owing to impurity of the bile acids used.

4.6. Stability

Sepharose–enzyme systems containing oxidoreductase were stable over a period of 3 months when stored at 4°C. In contrast, systems containing diaphorase lost activity continuously. However, this could be prevented by the addition of glutathione (Fig. 7). Enzymes immobilized onto nylon coils had an activity half-life of 20–40 days [37].

4.7. Sample Preparation

Sera were diluted 1:4 with 0.1 M potassium phosphate, pH 7.0, and heated for 15 min at 68°C in order to inactivate other NAD(P)H generating enzymes.

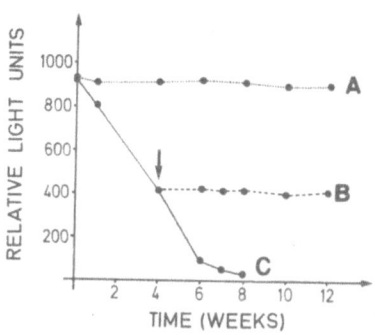

FIGURE 7. Light emission with the use of an identical standard over a period of 12 weeks for a preparation with NADH:FMN oxidoreductase (A). The same for a system containing diaphorase instead with (B) and without (C) addition of glutathione.

Urine samples were used undiluted after centrifugation (300 g, 5 min). Bile extracts were diluted 1:100–1:1000 and used without further preparation. Unmodified rat bile samples were studied after dilution (1:500). As stated in Section 4.4, it may be useful to add BSA to the samples or to the assay buffer if a differentiation of bile acids is to be obtained. However, no detailed studies on sample preparation have been performed to date.

4.8. Recovery and Precision

In order to determine the recovery of added standards, several standard solutions with different concentration were added to different samples (serum, urine, and bile) with a large range of bile acid concentrations. The recovery rates in serum ranged from 87 to 113%; in urine and bile similar values were found. Intraassay precision was analyzed by 10 repeated analyses of the same sample in 1 day. The coefficient of variation ranged between 5.2 and 8.2 for all samples studied and was independent of the sample concentration.

Interassay precision was determined by day-to-day analyses of the same sample and had the same range as intraassay precision.

Dilution of the serum samples from 1:4 up to 1:64 revealed a linear relationship between dilution and measured concentration (Fig. 8). Sample size could be changed from 10 up to 100 μl without significant influence on the results.

4.9. Calculations

Using the equations by Macdonald *et al.* [40,41]:

$$3\alpha\text{-Hydroxy bile acids} = \text{Cholic acid} \quad + \text{ chenodeoxycholic acid}$$
$$+ \text{ deoxycholic acid} \tag{5}$$

$$7\alpha\text{-Hydroxy bile acids} = \text{Cholic acid} \quad + \text{ chenodeoxycholic acid} \tag{6}$$

$$12\alpha\text{-Hydroxy bile acids} = \text{Cholic acid} \quad + \text{ deoxycholic acid} \tag{7}$$

the three major bile acids appearing in human serum and urine could be calculated. Using a similar system, bile acid groups could be estimated in rat serum and bile. However, since more than three major bile acids appear in this species [42], an additional assay, i.e., for 7β-hydroxy bile acid, would be necessary in order to calculate individual bile acids.

4.10. Assay Validations

Results obtained with these bioluminescence assays were compared with those revealed by GLC and HPLC on the same sera, urines, or bile extracts

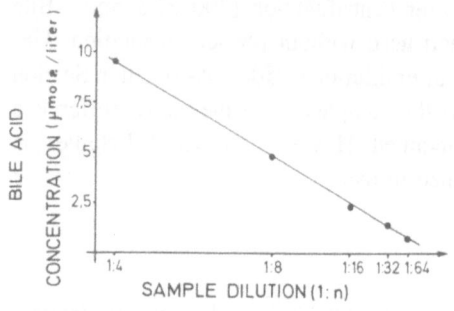

FIGURE 8. Linear relationship between sample dilution and measured bile acid concentration.

[6,43]. Furthermore, the correlation between the difference of both measurements in percent of the total mean value and that mean concentration was calculated.

5. RESULTS

5.1. Standards

Using BSA for the preparation of standards for different bile acids each of the major human bile acids, either free or as taurine or glycine conjugate, could be used for the standard curve. Figure 9 gives the curves obtained with the taurine conjugates of the three human bile acids. As shown in Table II, the use of other bile acids revealed similar results. Therefore, cholyltaurine was used for all standards used in validation experiments.

5.2. Validation

Results obtained with the three bioluminescence assays were closely correlated with those determined by GLC [14,15,39]. Figure 10 gives the comparison for 3α-hydroxy bile acids in serum, Fig. 11 that for total bile acids after enzymatic solvolysis in urine. Regardless of the sample or the assay, values obtained with either bioluminescence or GLC were always closely correlated ($r > 0.88$).

Using equations (5)–(7), concentrations of chenodeoxycholic and cholic acid were calculated for the 28 samples where all three assays had been performed. The results were again closely correlated with those measured by GLC (Table III). However, owing to multiple calculation steps, the error was increased, and thus, deoxycholate present at a concentration of less than 10% of the total could not be calculated accurately. Similar results were obtained for serum, urine, and bile extracts. However, the results with bile extracts were somewhat less satisfactory than for the other samples. This might be due to the interference of methanol with the enzyme system or to dilution problems. Pre-

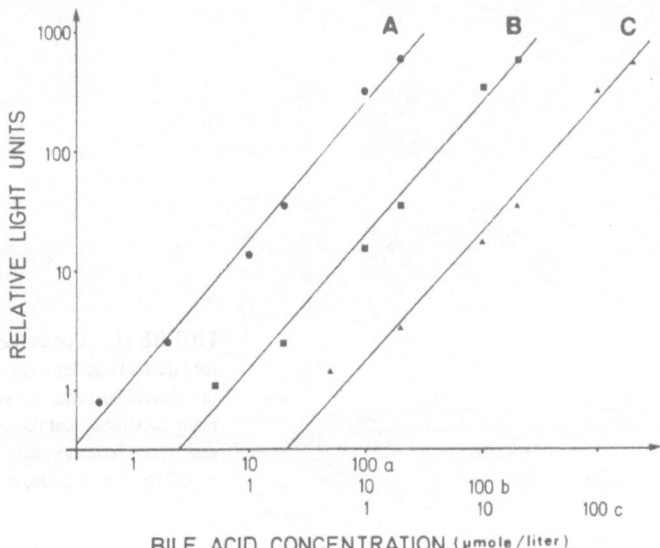

FIGURE 9. Standard curves for (A) taurocholate ($y = 1.15x + 0.12$, $r = 0.9956$), (B) tauro-chenodeoxycholate ($y = 1.10x + 0.23$, $r = 0.9920$), and (C) taurodeoxycholate ($y = 1.04x + 0.34$, $r = 0.9901$) using the 3α-hydroxy bile acid assay. Observe different scales on the horizontal axis. (From Ref. [15], with permission of the publisher.)

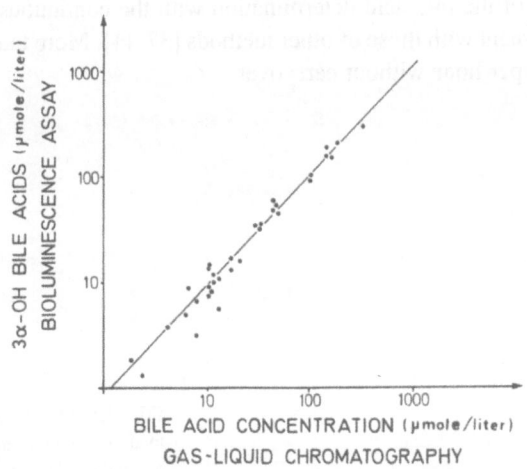

FIGURE 10. Comparison of total serum bile acid concentrations obtained by the bioluminescence assay and those obtained by gas–liquid chromatography ($y = 1.053x - 0.088$, $r = 0.9892$, $n = 33$). (From Ref. [15], with permission of the publisher.)

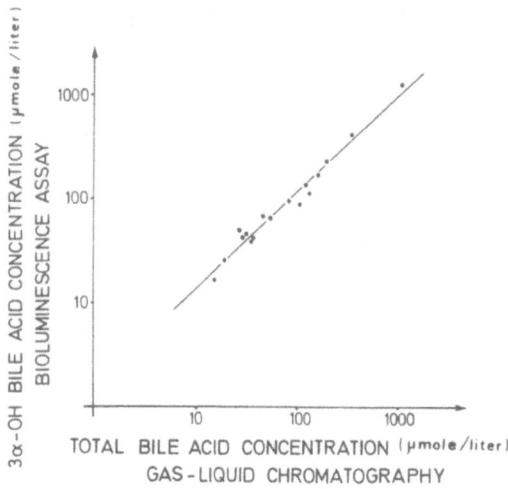

FIGURE 11. Comparison of total urine bile acid concentrations obtained by the bioluminescence assay after enzymatic solvolysis and those obtained by gas–liquid chromatography ($y = 0.923x + 0.215$, $r = 0.9806$, $n = 17$).

liminary experience with unextracted bile suggests better results and, in addition, makes the test much easier.

No correlation was found between the error as percent and the total bile acid concentration (Fig. 12). This was true for all sample types and for all three assays studied.

The use of rat serum or bile revealed a close correlation between bioluminescence and HPLC values (Table IV).

The results of the bile acid determination with the continuous-flow system are well in agreement with those of other methods [37,44]. More than 20 samples can be analyzed per hour without carryover.

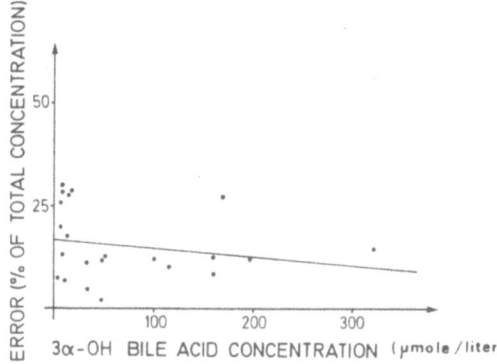

FIGURE 12. Correlation between the total bile acid concentrations (mean of bioluminescence and gas–liquid chromatography) and the error (difference as percent of the total concentration); $y = -0.02x + 16.64$, $r = -0.1770$.

TABLE III. Calculated Values for the Two Major Bile Acids in Human Serum—Bioluminescence and GLC[a]

Serum number	Cholic acid (μmole/liter)		Chenodeoxycholic acid (μmole/liter)	
	Bioluminescence	GLC	Bioluminescence	GLC
1	56.0	52.6	111.0	94.5
2	2.8	2.0	1.0	2.2
3	25.7	19.3	145.6	122.3
4	1.5	2.1	4.5	4.8
5	0	2.9	6.9	4.5
6	1.4	13.9	24.1	17.1
7	43.9	47.3	109.2	119.5
8	3.7	4.9	3.8	5.3
9	28.1	24.5	7.1	9.1
10	26.5	32.8	24.0	15.0
11	0.7	4.9	8.7	5.9
12	10.8	16.6	37.0	32.3
13	19.5	32.1	22.3	12.7
14	5.8	4.2	0	3.3
15	3.9	4.7	3.3	4.3
16	103.4	100.3	77.4	80.4
17	10.4	12.5	5.4	9.2
18	87.5	63.6	29.4	47.5
19	49.8	58.7	47.3	45.5
20	3.1	3.6	10.6	14.9
21	7.9	5.5	3.3	6.6
22	245.6	266.7	51.5	77.9
23	28.3	33.0	13.3	17.1
24	14.4	20.3	10.6	12.0
25	6.0	4.6	6.9	5.8
26	11.0	8.9	4.0	7.6
27	0.9	0.7	0.3	0.4
28	0	0	1.3	2.4
	$r = 0.8963$		$r = 0.9134$	
	$y = 1.046x - 0.173$		$y = 1.153x - 0.270$	

[a] Cholic acid and chenodeoxycholic acid in human serum as calculated from the results of the three bioluminescence assays or as measured with gas–liquid chromatography (GLC).

5.3. Some Applications

The assays described have been used on a group of patients with methotrexate treatment for psoriasis. Using an upper normal limit of 4.5 μmole/liter, all but 2 of 21 patients were correctly classified when compared with histology

TABLE IV. Bile Acids in Rat Serum—Bioluminescence and HPLC[a]

Bile acids	HPLC (μmole/liter)	Bioluminescence assay (μmole/liter)
Total bile acids	22.9	15.6
	179.0	185.8
7α-Hydroxy bile acids	5.0	4.3
	77.8	60.1
12α-Hydroxy bile acids	1.0	0.9
	50.8	63.6
Cholic acid	1.0	0.9
	50.8	50.8
α-Muricholic acid + chenodeoxycholic acid	6.0	3.4
	33.0	27.0
β-Muricholic acid + hyodeoxycholic acid	15.8	11.3
	95.2	101.2

[a] Comparison of bile acid composition of two rat sera using bioluminescence assays and high-pressure liquid chromatography (HPLC). Values for individual bile acids were calculated using three bioluminescence assays.

(J. Schölmerich *et al.*, unpublished data) (Table V). Studies on the bioavailability of chenodeoxycholic acid in gallstone dissolution have also been successful.

In addition, some experimental studies on rats were performed using these assays. Serum bile acid concentrations proved to be as effective as serum glutamate pyruvate transaminase (ALT) in detecting galactosamine-induced liver damage (Fig. 13). Furthermore, serum bile acid determinations could be used for sequential analysis during the course of the experiment, since in contrast to all other tests a very small sample size is required, which can be repeatedly taken from the animal [45].

The assays have been used for studies on bile acid transport in the perfused rat liver. They proved to be of value in the determination of bile acid uptake, secretion, and metabolism [42,46].

TABLE V. Sensitivity, Specificity, and Predictive Value of Bile Acid Determination for Methotrexate-Induced Liver Disease[a]

Sensitivity alteration	Sensitivity (%)	Specificity (%)	+ Predictive value (%)	− Predictive value (%)
Fatty degeneration	67	100	100	67
Vacuoles	100	72	28	100
Necrosis	75	75	67	82
Inflammation	70	80	78	73
Fibrosis	75	62	67	51
Total (>5 criteria)	91	89	91	89

[a] 4.5 μmole/liter was used as the upper limit of normal range.

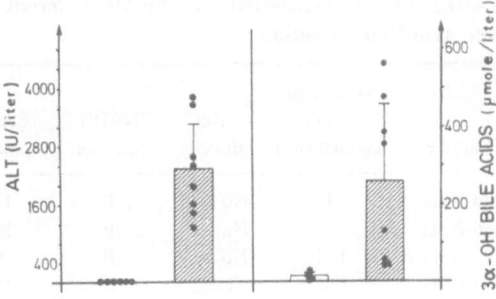

FIGURE 13. ALT activity in serum (left) and total bile acid concentration (right) for control rats and those with galactosamine-induced liver damage (shaded columns). (From Ref. [45], with permission of the publisher.)

6. DISCUSSION

6.1. The Method

6.1.1. Comparison with Other Methods

The bioluminescence methods appear to be ideal for measuring total (unsulfated) and classes of 7α- and 12α-hydroxylated bile acids in serum and biological fluids. Their sensitivity is comparable to that of competitive binding assays and better than that of all other methods reported to date other than inverse isotope dilution using mass spectrometry. The observed detection limit of the assays and the maximal sample size allow for the detection of 0.03 μmole/liter in serum.

No systematic comparison with other enzymatic methods has been done. The conventional enzymatic assays have been further improved, as has been recently described [13,47]. However, the tests described here are still two orders of magnitude more sensitive than most others reported to date (Table VI).

In contrast to other enzymatic methods, the bioluminescence assay does not give overestimated bile acid concentrations. This might be due to the fact that the typical enzymatic assay is an end-point determination which is influenced by enzyme contaminants such as malate dehydrogenase [55].

6.1.2. Technical Considerations

Interference with other serum components does not seem to have a significant role if the sera are heated prior to analysis [56]. The contamination of the HSD with aldehyde or alcohol dehydrogenases is obviously not of importance with serum and urine samples, though it may disturb the analysis of bile extracts with methanol. These contaminants can increase the background light emission and, thus, decrease sensitivity of the assays, but this can be overcome by purification of the enzymes [33].

It is crucial to have albumin in the assay system since otherwise the different affinity of the HSDs to different bile acid results in a variable light emission at

TABLE VI. Characteristics of the More Recent Enzymatic Methods for Serum Bile Acid Determinations[a]

Enzyme	Volume of serum required (ml)	Act directly	NADH detection	Sensitivity (pmole/tube)	Normal values (μmole/liter)	Reference
3α-HSD	3	No	F	1500	0–8.2	11
3α-HSD	2	No	F	1000	0.3–9.3	48
3α-HSD	1–3	No	F	630	0.9–6.3	49
3α-HSD	0.8	Yes	F	1000	3–6	50
7α-HSD	1–5	No	F	1000	1.2–6.3	51
3α-HSD	0.1	Yes	F	100	3.2–12.7	12
3α-HSD	0.2	Yes	S	2000	1–7	52
3α-HSD	0.05	Yes	S	0.3	3.03 ± 1.13	47
3α-HSD	0.2	Yes	S	1000	1–7	13
3α-HSD	0.1	Yes	F	0.5	6.3 ± 2.9	53
7α-HSD	0.01	Yes	B	0.5	nd	14
3α-HSD	0.01	Yes	B	0.2	nd	15
12α-HSD	0.01	Yes	B	1.0	nd	39
7α-HSD	0.005	Yes	C	10.0	1–2.5	37
3α-HSD	0.02	No	B	4.0	7.0	54

[a] F, fluorimetry; S, spectrophotometry; B, bioluminescence; C, continuous-flow system; nd, not determined.

identical total bile acid concentrations reflecting the bile acid composition of the sample [15]. Using albumin eliminates this problem, as is shown by the analysis of standard mixtures of three bile acids (Table VII). The effect of added BSA is explained by the change of the enzyme–substrate interaction induced by BSA in a standardized enzymatic test, which reflects the fact that bile acids have an identical relative affinity to albumin and the HSDs [57].

The use of a commercially available bacterial diaphorase and luciferase allows the preparation of a test system without enzyme purification steps and,

TABLE VII. Analysis of Mixed Standards of Bile Acids with Three Bioluminescence Assays[a]

Total bile acid concentration (μmole/liter)	3α-hydroxy assay		7α-hydroxy assay		12α-hydroxy assay	
	μmole/liter	%	μmole/liter	%	μmole/liter	%
1000	902.5	90.2	600.0	90.9	571.0	86.3
100	100.1	100.1	59.0	89.4	68.1	102.2
10	8.8	88.2	6.3	95.4	7.1	107.0

[a] Standards were mixed with cholic acid: chenodeoxycholic acid:deoxycholic acid = 1:1:1. Results are given as μmole/liter and as percent of the expected value as calculated from the mixture.

in addition, increases the sensitivity of the assay by lowering the background reading [29,39]. The immobilization onto nylon tubes or the use of flow cells [27,58] allows the automatization of the assays. In addition, a detection system for separation methods such as HPLC seems feasible and would allow the sensitive detection and differentiation of individual conjugated and unconjugated bile acids.

The limitations of the bioluminescence assays are several. First, this method is not able to differentiate between free and conjugated bile acids. However, this is a problem with the other sensitive assay, RIA, which normally detects only conjugated bile acids, and the combination of the bioluminescence assays with HPLC can certainly overcome this problem. The second limitation is that the assays cannot measure bile acids sulfated at the respective position which may be found in some patients with liver disease [59]. Third, the method can only measure hydroxy classes of bile acids. The individual bile acid can only be calculated using multiple equations, which increases the possible error. However, the overall results obtained comparing the bioluminescence assays with other methods suggest their accuracy and reliability [14,15,27,37,39,54]. The practical advantages, aside from those mentioned in the comparison with other methods, are several. The assays are extremely easy and rapid to perform—up to 30 samples can be analyzed per hour when the samples are prepared by heating and dilution. The stability of the enzyme system allows daily use until the prepared batch is used up. The application of a reusable flow cell [27,58] allows up to 700 tests with one preparation. Thus, the assays are considerably cheaper than other methods when the instrument is available. Finally, the result of the assay is available immediately.

6.2. Clinical Applications

Experiences with the clinical application of the bioluminescence assays are limited so far. Availability of this simple technique may well facilitate studies to assess the clinical utility of total bile acids or classes of hydroxy bile acid measurements. However, the published results suggest that insofar as parenchymal liver disease in adults is concerned, serum bile acid measurements offer a high degree of organ specificity but only moderate sensitivity [60–62]. Postprandial serum bile acid determinations appear to be particularly sensitive for the detection of portal systemic shunting as may occur in patients with liver cirrhosis [63,64]. In children, cholic acid conjugates correlated better with morphological evidence of liver disease than conventional liver tests [65]. The preliminary experience with the psoriasis patients treated with methotrexate suggests that the bioluminescence assays may prove to be of value for some special groups of patients with potential liver disease since the upper limit of normal values may be lower than with the conventional enzymatic assays. A study on the bioavailability of chenodeoxycholic acid in the treatment of gallstone disease revealed promising results (A. F. Hofmann, personal communication).

The extreme sensitivity of the bioluminescence assays suggests that it could be useful for the detection of bile acid malabsorption. In such patients, the postprandial increase in serum bile acid levels is lower than in healthy subjects, and the lower postprandial rise in such patients has heretofore been detectable only by RIA and could not be detected by GLC or enymatic methods [66,67].

The use of the combination of all three assays developed so far, probably in the flow system described (Fig. 1), allows determination of the pattern of the human serum bile acids. In our judgment, the pattern of serum bile acids is unlikely to be of greater value than simple determination of the total bile acid concentration. It has been shown, however, that this pattern in cholestatic liver disease differs appreciably from that present in severe hepatic necrosis [68].

The situation may be different for intestinal disease. Since conjugates of cholic acid are absorbed exclusively by active ileal transport, in contrast to the glycine conjugates of dihydroxy bile acids, which may have passive absorption in the proximal small intestine, the postprandial increase of cholates might be of particular value in detecting bile acid malabsorption caused by ileal disease or resection [69].

In summary, the bioluminescence assays described in either the standard form as injection assay or an automated form using nylon immobilization or flow cells offer some possibilities to study the diagnostic value of the determination of total or individual bile acids in health or in hepatic or ileal disease. The promise, however, has yet to be fullfilled.

6.3. Experimental Applications

Experiences with these assays in experimental work are limited. It has been shown, however, that they are useful in monitoring liver damage in small animals

TABLE VIII. Serum Bile Acids in Hamsters[a]

Bile acids	Fasting (μmole/liter)	Postprandial (μmole/liter)
3α-Hydroxy bile acids	7.7 ± 1.6	37.1 ± 8.0
7α-Hydroxy bile acids	5.9 ± 1.2	35.0 ± 4.5
12α-Hydroxy bile acids	3.6 ± 0.6	8.0 ± 2.7
Chenodeoxycholic acid	4.1 ± 1.3	25.6 ± 5.5
Cholic acid	1.6 ± 1.2	7.7 ± 2.2
Deoxycholic acid	2.0 ± 1.5	1.3 ± 0.6

[a] Values were measured using three luminescence assays. Individual bile acids were calculated using the equations of Macdonald *et al.* (41).

[45]. A number of unpublished studies on cholestatic and normal hamsters (Table VIII) and rats with selective biliary obstruction have shown the possible usefulness of the method, when only small samples are available. Furthermore, the advantage of an immediate result of bile acid analysis is of value when experimental systems such as the perfused rat liver or isolated liver cells are used [42,46]. In particular, experiments where large numbers of samples have to be analyzed for bile acid concentrations require such a simple, rapid, and reliable method.

7. SUMMARY

The bioluminescence assays, though not yet widely introduced into clinical and experimental practice, are a promising method probably facilitating the analysis of the utility of bile acid measurements. Furthermore, they add to the battery of simple and reliable tests needed for this purpose. The assays have advantages in specificity, simplicity, and economy compared to competitive binding techniques, and in sensitivity compared to GLC and conventional enzymatic techniques. In addition, the stability and potential reusability of the immobilized enzymes suggest that the method has prospects for further development, in particular, automatization.

Finally, the use of a flow system may increase the sensitivity of separation methods such as HPLC and therefore further promote the bile acid analysis of biological fluids.

Note added in proof: After preparation of the manuscript two papers have described applications of bioluminescence assays: S. Bellentani, W. G. M. Hardison, and F. Manenti, *J. Hepatol.* **2,** 525 (1985); S. S. Rossi, A. F. Hofmann, and L. M. Clayton, *J. Pharm. Sci.* **75,** 288 (1986).

REFERENCES

1. S. Sherlock and V. Walshe, *Clin. Sci.* **6,** 223 (1984).
2. D. Rudman and F. E. Kendall, *J. Clin. Invest.* **36,** 530 (1957).
3. A. F. Hofmann, N. V. LaRusso, and G. P. van Berge Henegouwen, *in* "Bile Acid Metabolism in Health and Disease" (G. Paumgartner and A. Stiehl, eds.), p. 151, MTP Press, Lancaster (1977).
4. K. D. R. Setchell and A. Matsui, *Clin. Chim. Acta* **127,** 1 (1983).
5. K. Maruyama, H. Tanimura, and Y. Hikasa, *Clin. Chim. Acta* **100,** 47 (1980).
6. A. T. Ruben and G. P. van Berge Henegouwen, *Clin. Chim. Acta* **119,** 41 (1982).
7. W. J. Simmonds, M. G. Korman, V. L. W. Go, and A. F. Hofmann, *Gastroenterology* **65,** 705 (1973).
8. S. Matern, K. Tietjen, H. Matern, and W. Gerok, *in* "Biological Effects of Bile Acids" (G. Paumgartner, A. Stiehl, and W. Gerok, eds.), p. 273, MTP Press, Lancaster (1979).

9. E. Roda, R. Aldini, M. Capelli, D. Festi, C. Sama, G. Mazella, A. M. Rorselli, and L. Barbara, *Clin. Chem.* **26,** 1647 (1980).

10. T. Iwata and K. Yamasaki, *J. Biochem. (Tokyo)* **56,** 424 (1964).

11. G. M. Murphy, B. H. Billing, and D. N. Baron, *J. Clin. Pathol.* **23,** 594 (1970).

12. F. Mashige, E. Yanagizawa, and T. Osuga, *Clin. Chim. Acta* **70,** 79 (1974).

13. F. Mashige, N. Tanaka, A. Maki, S. Kamei, and M. Yamanata, *Clin. Chem.* **27,** 1352 (1981).

14. A. Roda, L. J. Kricka, M. DeLuca, and A. F. Hofmann, *J. Lipid Res.* **23,** 1354 (1982).

15. J. Schölmerich, G. P. van Berge Henegouwen, A. F. Hofmann, and M. DeLuca, *Clin. Chim. Acta* **137,** 21 (1984).

16. M. DeLuca, "Methods in Enzymology," Volume LVIII, Academic Press, New York (1978).

17. E. Jablonski and M. DeLuca, *in* "Clincal and Biochemical Luminescence" (L. J. Kricka and T. J. N. Carter, eds.), Dekker, New York (1981).

18. M. DeLuca, *in* "Proceedings of the International Meeting on Luminescence Assays: Perspectives in Endocrinology and Clinical Chemistry" (M. Serio and M. Pazzagli, eds.), p. 115, Plenum Press, New York (1981).

19. M. DeLuca and W. D. McElroy, eds., "Bioluminescence and Chemiluminescence: Basic Chemistry and Analytical Applications," Academic Press, New York (1981).

20. J. W. Hastings, T. O. Baldwin, and M. Z. Nicoli, *in* "Methods in Enzymology 57," (M. DeLuca, ed.), p. 135, Academic Press, New York (1978).

21. E. Jablonski and M. DeLuca, *Biochemistry* **16,** 2932 (1977).

22. K. Mosbach and B. Mattiasson, *Acta Chem. Scand.* **24,** 2093 (1970).

23. B. Mattiasson and K. Mosbach, *Biochem. Biophys. Acta* **235,** 253 (1971).

24. P. Srere, B. Mattiasson, and K. Mosbach, *Proc. Natl. Acad. Sci. USA* **70,** 2534 (1973).

25. N. Siegbahr and K. Mosbach, *FEBS Lett.* **137,** 6 (1982).

26. M. DeLuca and L. J. Kricka, *Arch. Biochem. Biophys.* **226,** 285 (1983).

27. L. J. Kricka, G. K. Wienhausen, J. E. Hinkley, and M. DeLuca, *Anal. Biochem.* **129,** 392 (1983).

28. K. Mosbach, ed., "Methods in Enzymology 44," Academic Press, New York, (1976).

29. G. K. Wienhausen, L. J. Kricka, J. E. Hinkley, and M. DeLuca, *Appl. Biochem. Biotech.* **7,** 463 (1982).

30. I. A. Macdonald, E. C. Meier, D. E. Mahony, and G. A. Costain, *Biochim. Biophys. Acta* **450,** 142 (1976).

31. I. A. Macdonald, J. F. Jellett, and D. E. Mahony, *J. Lipid Res.* **20,** 234 (1979).

32. I. A. Macdonald and Y. P. Rochon, *J. Chromatogr.* **259,** 154 (1983).

33. J. Ford and M. DeLuca, *Anal. Biochem.* **110,** 43 (1981).

34. A. Roda, A. F. Hofmann, and K. J. Mysels, *J. Biol. Chem.* **258,** 6362 (1983).

35. J. J. Albers, S. M. Grundy, P. A. Cleary, D. M. Small, and J. M. Lachin, *Gastroenterology* **82,** 638 (1982).

36. A. F. Hofmann, S. M. Grundy, J. M. Lachin, S-P. Lan, R. A. Baum, R. F. Hanson, T. Hersh, N. C. Hightower, J. W. Marks, H. Mekhjian, R. A. Shaefer, R. D. Soloway, J. L. Thistle, F. B. Thomas, and M. B. Tyor, *Gastroenterology* **83,** 738 (1982).

37. A. Roda, S. Girotti, S. Ghini, B. Grigolo, G. Carrea, and R. Bovara, *Clin. Chem.* **30,** 206 (1984).

38. S. C. March, I. Parikh, and P. Cuatrecasas, *Anal. Biochem.* **60,** 149 (1974).

39. J. Schölmerich, J. E. Hinkley, I. A. Macdonald, A. F. Hofmann, and M. DeLuca, *Anal. Biochem.* **133,** 244 (1983).

40. I. A. Macdonald, C. N. Williams, and D. E. Mahony, *Anal. Biochem.* **57,** 127 (1974).

41. I. A. Macdonald, C. N. Williams, and B. C. Musial, *J. Lipid Res.* **21,** 381 (1980).

42. J. Schoelmerich, S. Kitamura, and K. Miyai, *Biochem. Biophys. Res. Commun.* **115,** 518 (1983).

43. G. P. van Berge Henegouwen, A. Reuben, and K-H. Brandt, *Clin. Chim. Acta* **54**, 249 (1974).
44. A. Roda, S. Girotti, S. Ghini, B. Grigolo, G. Carrea, and R. Bovara, *in* "Analytical Applications of Bioluminescence and Chemiluminescence" (L. J. Kricka, P. E. Stanley, G. H. G. Thorpe, and T. P. Whitehead, eds.), p. 129, Academic Press, London (1984).
45. J. Schölmerich, M. DeLuca, and M. Chojkier, *Hepatology* **4**, 639 (1984).
46. J. Schölmerich, S. Kitamura, and K. Miyai, *Hepatology* **4**, 763 (1984) (Abstr.).
47. J. C. Nicolas, J. Chaintreuil, B. Descomps, and A. Crastes de Paulet, *Anal. Biochem.* **103**, 170 (1980).
48. H. P. Schwartz, K. V. Bergman, and G. Paumgartner, *Clin. Chim. Acta* **50**, 197 (1974).
49. O. Fausa, *Scand. J. Gastroenterol.* **10**, 747 (1975).
50. P. A. Siskos, P. T. Cahill, and N. B. Javitt, *J. Lipid Res.* **18**, 666 (1977).
51. O. Fausa and B. A. Skalhegg, *Scand. J. Gastroenterol.* **12**, 44 (1977).
52. H. Steensland, *Scand. J. Clin. Lab. Invest.* **38**, 447 (1978).
53. K. Mashige, T. Osuga, N. Tanaka, K. Imai, and M. Yamanaka, *Clin. Chem.* **24**, 1150 (1978).
54. I. Styrélius, A. Thore, and I. Björkhem, *Clin. Chem.* **29**, 1123 (1983).
55. C. A. Barth and K. Wirthensohn, *J. Lipid. Res.* **22**, 1025 (1981).
56. N. B. Javitt, K. Budai, P-Y. Shan, P. A. Siskos, and P. Cahill, *in* Biological effects of bile acids (G. Paumgartner, A. Stiehl, and W. Gerok, eds.), p. 267, University Park Press, Baltimore (1978).
57. A. Roda, G. Cappelleri, R. Aldini, E. Roda, and L. Barbara, *J. Lipid. Res.* **23**, 490 (1982).
58. D. Vellom, J. Hinckley, A. Loucks, H. Egghart, and M. DeLuca, *in* "Analytical Applications of Bioluminescence and Chemiluminescence" (L. J. Kricka, P. E. Stanley, G. H. G. Thorpe, and T. P. Whitehead, eds.), p. 133, Academic Press, London (1984).
59. R. Raedsch, B. H. Lauterburg, and A. F. Hofmann, *Dig. Dis. Sci.* **26**, 394 (1981).
60. A. F. Hofmann, *Hepatology* **2**, 512 (1982).
61. D. Festi, A. M. Morselli, A. Roda, F. Bazzoli, R. Frabboni, P. Rucci, F. Taroni, R. Aldini, E. Roda, and L. Barbara, *Hepatology* **3**, 707 (1983).
62. R. Ferraris, G. Colombatti, M. T. Fiorentini, R. Carosso, W. Arossa, and M. Dela Pierre, *Dig. Dis. Sci.* **28**, 129 (1983).
63. G. A. Mannes, F. Stellaard, and G. Paumgartner, *Digestion* **25**, 217 (1982).
64. H. Porchet and J. Bircher, *Gastroenterology* **82**, 629 (1982).
65. M. Angelico, A. F. Attili, and L. Capocaccia, *Am. J. Dig. Dis.* **22**, 941 (1977).
66. W. F. Balistreri, F. J. Suchy, and J. E. Heubi, *J. Pediatr.* **96**, 829 (1982).
67. R. Aldini, A. Roda, D. Festi, G. Mazella, A. M. Morselli, C. Sama, E. Roda, N. Scopinaro, and L. Barbara, *Gut* **23**, 829 (1982).
68. K. Linnet and J. Rye-Anderson, *Clin. Chim. Acta* **127**, 217 (1983).
69. S. W. Schalm, N. F. La Russo, A. F. Hofmann, G. P. van Berge Henegouwen, and M. G. Korman, *Gut* **19**, 1006 (1978).

Chapter 8

BILE ACIDS IN EXTRAHEPATIC TISSUES

**Jacqueline Dupont, Pilar A. Garcia,
Bernhard Hennig, Nina K. Dodd, Suk Yon Oh,
Charles E. Sizer, Satindra K. Goswami, and
Walter G. Hyde**

---

1. INTRODUCTION

Unequivocal physical identification of individual nonhepatic bile acids is lacking, though there is evidence of the presence of cholanoic acids in brain [1,2], skeletal muscle, kidney, pancreas, and adipose tissues [2]. The presence of cholanoic acids in tissues raises the question of whether the compounds originated by local biosynthesis or by transfer from blood after synthesis in the liver, even though it has been stated that the liver is the sole site of bile acid formation [3] without adequate evidence.

Bile acids generally occur in nature in free and conjugated forms. Quantitative isolation of bile acids is difficult because of their physicochemical properties and the presence of other compounds in tissues, particularly the acidic pigments and fatty acids. Methods have been described for the isolation of bile acids from various sources [4–6]. Methods that utilize a vigorous hydrolysis step destroy the natural forms of the bile acid, and the separation of fatty acids from

---

Jacqueline Dupont and Pilar A. Garcia Department of Food and Nutrition, Iowa State University, Ames, Iowa **Bernhard Hennig** Departments of Nutrition and Food Science. University of Kentucky, Lexington, Kentucky **Nina K. Dodd** Department of Food and Nutrition, SVT College of Home Science, Juho, Bombay, India **Suk Yon Oh** Division of Food and Nutrition, College of Health, University of Utah, Salt Lake City, Utah **Charles E. Sizer** Department of Food and Nutrition, Iowa State University, Ames, Iowa **Satindra K. Goswami** Department of Food and Nutrition, Iowa State University, Ames, Iowa **Walter G. Hyde** Veterinary Diagnostic Laboratory, Iowa State University, Ames, Iowa *Present address for C.E.S.:* Brikpak Inc., Dallas, Texas *Present address for S.K.G.:* Institute for Basic Research, Staten Island, New York

341

bile acids, particularly lithocholic and other less polar bile acids, poses a problem. Utilization of the most recently derived methods [5,6] has not been extensive. Removal of fatty acids from samples is essential inasmuch as they can interfere with bile acid analysis by gas–liquid chromatography. This is usually done by thin-layer chromatography (TLC), and different methods have been proposed for the separation of fatty acids from bile acids [6–9].

Evidence is presented herein for identification of individual conjugated bile acids, their quantitation by high-performance liquid chromatography (HPLC) and 3-hydroxysteroid dehydrogenase assay, and the possibility of their local synthesis as suggested by the activity of cholesterol 7α-hydroxylase. The 7α-hydroxycholesterol that is formed by the action of cholesterol 7α-hydroxylase is considered to be totally committed to bile acid production [3,10]. The methods described have been applied to studies of variations in bile acid composition in human lipoproteins and leukocytes and kinetics of metabolism in rats.

Atherosclerotic plaques are characterized by the accumulation of cholesterol and its esters in the peripheral tissues. These deposits could come from three sources; increased influx from blood mediated by the increase of low-density lipoprotein (LDL), excess cholesterol synthesis in peripheral tissues, and decreased efflux from peripheral cells mediated by high-density lipoprotein (HDL). These points stress the importance of understanding the regulation of cholesterol synthesis and utilization by peripheral tissues *in vivo*. The usefulness of the leukocyte for studying cholesterol biosynthesis in peripheral tissues has been reported by Fredrickson [11], Hsia [12], and Fogelman *et al.* [13]. We have presented evidence that the leukocyte also can synthesize 7α-hydroxycholesterol [14]. An indicator of cholesterol metabolism in peripheral tissues therefore is envisioned in the leukocyte. It is our hypothesis that the leukocyte will reflect *in vivo* responses of peripheral tissues to factors affecting cholesterol metabolism. Therefore, a study was designed to use leukocytes for investigating the effects of age and diet composition on cholesterol and bile acid metabolism in peripheral tissues of adult men and the plasma lipoprotein concentrations of steroids.

Our finding that bile acid transport is via lipoproteins and not albumin in humans led to further investigation of bile acid kinetics in the rat. Cholanoic acids (bile acids) in plasma lipoproteins may originate from either hepatic or extrahepatic biosynthesis. Neither portal nor peripheral blood has been carefully studied to determine the nature of bile acid transport. In the past, it has been believed that bile acids are transported in blood attached to albumin [15,16]. Very little intestinal bile acid is transported to liver by the lymphatic system [16]. We examined the cycling of 4-[^{14}C]cholesterol and its bile acid products in lipoproteins and tissues. The results suggest that cholanoic acids may be involved in the removal of plasma sterols for excretion.

This chapter will outline the methodologies employed in these studies and discuss the results supporting the presence of bile acids in extrahepatic tissue.

2. ANALYTICAL METHODS

2.1. Chemicals

All chemicals used were of analytical grade. Free and conjugated bile acids were obtained from Supelco (Bellefonte, PA), and the sulfated bile acids were purchased from P-L Biochemicals, Inc. (Milwaukee, WI). Precoated thin-layer silica gel G plates of 0.5-mm thickness (Analtech, Inc., Newark, DE) and silica gel 60 plates of 0.25-mm thickness (E. Merck, Darmstadt, Federal Republic of Germany, distributed by VWR Scientific, Chicago, IL) were used.

2.2. Extraction Method I

One gram of lyophilized tissue was saponified with 15 ml of N NaOH in 90% ethanol at 65°C for 1 hr with occasional mixing. The sample was cooled, 5 ml of water was added, and the neutral steroids were extracted twice with 15 ml of petroleum ether. The petroleum ether extract was backwashed with 5 ml of N NaOH in 50% ethanol to remove traces of bile acids and mixed with the original extract. Ten milliliters of 50% $ZnSO_4 \cdot 7H_2O$ was added to the petroleum ether extract and mixed thoroughly in a vortex mixer, then centrifuged for 10 min at 3000 g to remove most of the pigments. The supernatant was decanted into a 500-ml round-bottom flask, and the precipitate was suspended in 15 ml of 95% ethanol containing 0.1% ammonium hydroxide and heated to 65°C for 30 min. The mixture was centrifuged again, the supernatant decanted into the same round-bottom flask. The precipitate was washed with 15 ml of 95% ethanol, and added to the other washes. The pooled extracts in the round-bottom flask was evaporated to dryness in a rotary evaporator at 50–55°C under reduced pressure. The residue was then extracted with 20 ml of chloroform–methanol (1:1), and evaporated almost to dryness under nitrogen in a water bath at 50–55°C. Any resulting precipitate was separated by centrifugation. The remaining liquid was subjected to TLC to separate fatty acids from bile acids.

2.3. Extraction Method II

One gram of lyophilized tissue was homogenized with 25–30 ml of 95% ethanol, containing 0.5% HCl, and incubated at 37°C for 1 hr [17]. The sample was centrifuged at 1000 g for 10 min, and the supernate was removed. Neutral steroids were extracted from the supernate with petroleum ether. The ethanolic residue was reduced in volume to 0.5–1 ml, by evaporation in a rotary evaporator at 50°C.

2.4. Removal of Fatty Acids

Precoated silica gel G TLC plates of 0.5-mm thickness were used. The extract was spotted as streaks on the plates, and a standard solution containing a mixture of glycocholic, lithocholic, and oleic acids was spotted at one corner of the plate. The plate was developed in a filter paper-lined chamber containing the solvent system toluene/ethanol/methanol/water/ammonium hydroxide (50:20:14:3:1, by volume) for 1 hr at room temperature (27–28°). The plate was air-dried under the hood and redeveloped in the same solvent system for 1 hr, and the plate again was dried in the hood. The entire plate except the known bile acid standard channel was covered with aluminum foil, and the exposed area sprayed with a manganous chloride spray reagent [18]. The plate was heated in an oven at 100°C for 10 min, and the area between glycocholic and lithocholic acid was marked. An alternative color reagent was 10% phosphomolybdic acid in ethanol, and heated at 105°C for less than 2 min. The bile acid area was scraped off the plate and eluted with chloroform–methanol (1:1, v/v). The extract was reduced to a volume of 0.5 ml under nitrogen at 50–55°C in a water bath.

2.5. TLC Separation of Bile Acids into Groups

Precoated plates of 0.25-mm thickness were used. In practice, a standard solution containing a mixture of glycocholic, glycolithocholic, taurocholic, taurolithocholic, cholic, and lithocholic acids was spotted at one corner of the plate, and the unknown bile acid solutions were spotted as streaks. The solvent system used was the same as reported earlier [19]. The plate was run in the solvent system ethanol/isooctane/isopropyl alcohol/ethyl acetate (25:10:10:10, by volume) for 2.5 hr and then dried, and the standard bile acid channel was sprayed with a color reagent. The glycine-conjugated bile acids lay between glycocholic and glycolithocholic, taurine-conjugated bile acids between taurocholic and taurolithocholic, and free bile acids between cholic and lithocholic acid. The zone corresponding to the different bile acid groups was scraped off the plates, and the bile acids were eluted with chloroform–methanol (1:1, v/v). The eluate was evaporated to a small volume for the identification of individual bile acids.

2.6. TLC Identification of Bile Acids

The individual bile acids were separated on TLC by different solvent systems; e.g., (1) isooctane/ethyl acetate/acetic acid (10:10:2, by volume) [20] and hexane/methyl ethyl ketone/acetic acid (63:40:6, by volume) [21] for free bile acids and (2) isoamyl alcohol/acetone/acetic acid/water (63:26:6:5, by volume) [22] and isoamyl acetate/propionic acid/n-propanol/water (4:3:2:1, by volume) [23] for glycine- and taurine-conjugated bile acids. The conjugated bile acids were hydrolyzed with 20% KOH in ethylene glycol and extracted through Amberlite XAD-2 columns [24], and free bile acids were separated by using TLC. The plate was sprayed with manganous chloride spray reagent, viewed under ultraviolet light, and compared with authentic bile acids run on the same plate. With this reagent, less than 1 µg of bile acids can be detected.

2.7. High-Pressure Liquid Chromatography

Samples recovered from TLC separation of individual bile acids were used for HPLC. Reverse-phase HPLC, coupled to an ultraviolet detector set at 193 nm, was used for identification and quantitation of conjugated bile acids [25]. Samples dissolved in the elution solvent mixture (2 propanol:8.8 mM potassium phosphate buffer, pH 2.5, 160:340) were analyzed on a 4 mm × 30 cm Bondapak/C 18 column (Waters Associates, Milford, MA) by eluting at 1 ml/min, and 1-ml fractions were collected. Quantitation was done by comparing the sample peak area with corresponding standards.

2.8. Mass Spectrometry

HPLC and TLC fractionated sample components were collected in glass screw-cap vials and concentrated to dryness. Twenty microliters of solvent was added to each vial to redissolve the sample residue, and 5 µl of each transferred to quartz capillary sample cups for solid-probe mass spectrometry. The solvent was evaporated to dryness in the capillaries by using a vacuum desiccator. Authentic compounds were used to obtain standard spectra.

Chemical ionization (CI) mass spectra were run by using a Finnigan 400 mass spectrometer and a Teknivent 29K interaction data system. Spectra were collected by using the following conditions: 70 eV accelerating voltage, 0.025 torr source pressure CI methane, source temperature 260°C, and manifold temperature 120°C.

2.9. Cholesterol 7α-Hydroxylase and Hydroxymethylglutaryl-CoA Reductase

Human leukocytes and rat tissues were analyzed for cholesterol 7α-hydroxylase and hydroxymethylglutaryl-CoA (HMG-CoA) reductase activities. The sample was homogenized, and microsomal preparations were made from the homogenates [26]. Microsomal protein was determined by the Lowry procedure

[27]. Conversion of cholesterol to 7α-hydroxycholesterol in this system has been verified [14]. HMG-CoA reductase activity was measured in microsomes by using Young and Rodwell's procedures [26].

2.10. Lipid Quantitation

Free and total cholesterol were determined by a slight modification of the enzymatic procedures of Allain *et al.* [28] and Carlson *et al.* [29]. Phospholipids were analyzed by using the method of Sandhu [30] and triglycerides by the method of Fletcher [31].

2.11. Enzymatic Assay of Total Bile Acids

Some samples were analyzed by a 3α-hydroxysteroid dehydrogenase (3α-HSD) enzyme assay [32]. 3α-HSD, obtained from Sigma Chemicals, was prepared in 0.01 M, pH 7.2, potassium phosphate buffer to give 3–4 units/ml. To 0.1 ml of sample or standard, 0.005 M BNAD and 0.6 ml of 0.1 M sodium pyrophosphate were added and the mixture was allowed to equilibrate at 25°C for 15 min. Absorbance at 340 nm was recorded; then 0.1 ml of enzyme solution was added, and the solution was allowed to equilibrate at room temperature (25°C) for 40 min. The change in absorbance was recorded at 340 nm. The concentration of unknown was determined from the standard curve. The correlation of this analysis with GLC analysis has been verified [33].

2.12. Plasma Lipoprotein Separation

Plasma lipoproteins were isolated into various density classes by repeated ultracentrifugation as previously described [34,35]. Ultracentrifugation was performed at 4°C in a Beckman L-3 model ultracentrifuge with a type 50 titanium rotor. The lipoprotein fractions were separated by ultracentrifugation sequentially at densities of $d = 1.006$, $d = 1.063$, and $d = 1.21$ g/ml achieved by adding crystalline KBr. All fractions were removed by the tube-slicing technique. Very low-density liproprotein (VLDL) was separated by centrifugation at $d < 1.006$ at 106,000 g for 18 hr, isolated, and then recentrifuged at $d = 1.006$ for 18 hr to remove excess albumin. The $d = 1.006$–1.063 g/ml density fraction LDL was obtained after centrifugation at 106,000 g for 24 hr. The $d = 1.063$–1.21 g/ml fraction HDL was separated by centrifugation at 106,000 g for 48 hr. isolated, and then recentrifuged as was VLDL. The purity of the lipoproteins was judged by paper electrophoresis [36].

Lipoproteins were lyophilized and extracted 3 times with chloroform:

methanol (2:1). All samples were dried under nitrogen and redissolved in chloroform:methanol (2:1) for further analyses.

2.13. Tissue Collection

2.13.1. Rat Tissue

In general, rats were anesthetized with ether, and approximately 12–15 ml of blood per animal was drawn from the abdominal aorta into tubes containing Na-EDTA (1 mg/ml blood). Blood was handled at 4°C and centrifuged at low speed (800 g) for 15 min for lipoprotein separations. Plasma samples from two rats at each time point were then combined for further analyses. Liver, intestine, muscle, epididymal fat pads, lung, kidney, and heart were excised, rinsed with physiological saline, and immediately frozen in liquid nitrogen and then stored at − 16°C. For analysis, samples were lyophilized. The equivalent of 2–4 g wet weight of tissue was used. For identification of bile acids organs were pooled when necessary to obtain sufficient tissue.

2.13.2. Human Plasma

In human subjects, a fasting blood sample (100 ml) was drawn by venipuncture in EDTA vacutainer tubes. A 30-ml blood sample was used for the separation of mixed leukocytes [26], 30 ml for separation of mononuclear leukocytes [37], and the rest of the sample for plasma and lipoprotein separations. The number of cells was determined by Coulter counter and differential count with Wright stain [38]. Mixed leukocytes were lyophilized and extracted with chloroform:methanol (1:1 v/v).

2.14. Experimental Designs

The details of experimental design for the human study and the rat kinetic study and their statistical analyses are included in the results of those studies.

3. APPLICATIONS

3.1. Identification and Quantitation of Cholanoic Acids

3.1.1. Extraction

Method I was used for tissues subjected to TLC identification procedures. Method II was evaluated for quantitative recovery of added glycocholic acid

TABLE I. Recovery of Added Glycocholic Acid by Extraction Method II

Tissue	[^{14}C]glycocholic acid		
	Added (cpm)	Recovered (cpm)	%
Liver	13242	11918	90
Heart	11849	10901	92
Lung	11769	10945	93
Kidney	12879	11720	91
Adipose	13018	11456	88

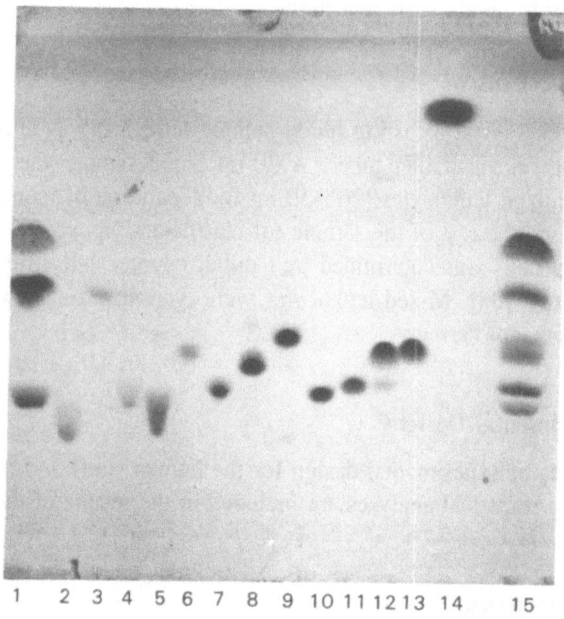

FIGURE 1. Thin-layer chromatography on silica gel G of 0.5-mm thickness developed with toluene/ethanol/methanol/water/ammonium hydroxide (50:20:14:3:1 by volume) and visualized with manganous chloride spray reagent [18]. 1, Glycocholic, lithocholic, and oleic acid (top to bottom); 2, glycolithocholic sulfate; 3, taurolithocholic acid; 4, taurolithocholic sulfate; 5, lithocholic sulfate; 6, glycolithocholic acid; 7, glycochenodeoxycholic acid; 8, taurocholic acid; 9, taurochenodeoxycholic acid; 10, cholic acid; 11, hyocholic acid; 12, hyodeoxycholic acid; 13, chenodeoxycholic acid; 14, cholesterol; 15, glycocholic, glycolithocholic, taurocholic, taurolithocholic, cholic, lithocholic, and oleic acid (top to bottom).

(Table I). From 88 to 93% of added radioactivity was recovered. Extraction method II [17] has been adopted as satisfactory for extraction.

3.1.2. TLC Isolation of Bile Acids

The TLC separation of bile acids from fatty acids is shown in Fig. 1. It is evident from this figure that bile acids are clearly separated from fatty acids and cholesterol. Cholesterol moves far ahead of both bile acids and fatty acids. All bile acids, free or conjugated, except the sulfated bile acids, lie between glycocholic and lithocholic acids. If sulfated bile acids are suspected to be in a biological specimen, then glycolithocholic sulfate and lithocholic acid should be run as standards. All bile acids form a relatively narrow band, which can be scraped from the plates after visualization with a color reagent. The solvent system should be freshly prepared, and any remaining liquid should be drained from the chamber before fresh solvent is added. This solvent system is very

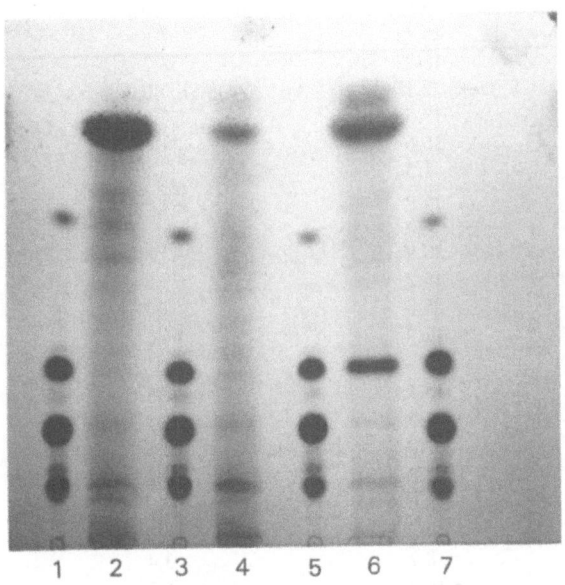

FIGURE 2. Thin-layer chromatography on silica gel 60 of 0.25-mm thickness developed with isooctane/ethyl acetate/acetic acid (10:10:2) [19] and visualized with manganous chloride spray reagent [18] and ultraviolet light. 1, Cholic, hyodeoxycholic, chenodeoxycholic, and lithocholic acid; 2, heart; 3, same as 1; 4, kidney; 5, same as 1; 6, lung; 7, same as 1. Top large bands in 2, 4, and 6 are pigment impurities.

TABLE II. Comparison of Authentic Glycocholic Acid Electron Impact Spectra with Published Data and with a Tissue Compound Isolated by TLC

m/e	Glycocholic acid		
	Reference 39	Standard	Rat kidney
465	0.4	—	—
429	8.0	—	—
414	4.3	—	—
411	8.8	—	—
396	6.2	—	—
390	2.3	—	—
372	12.3	6.8	0.8
354	13.7	13.0	1.6
336	2.9	—	—
312	6.8	6.2	1.2
299	5.0	—	0.5
294	6.3	—	1.4
281	7.5	—	1.8
271	90.6	73.8	71.6
253	100.0	57.8	70.6
244	8.8	1.6	4.9
226	21.0	9.2	14.2
211	15.5	4.0	10.4
199	18.1	7.9	11.3
147	23.6	5.0	17.4
145	28.1	7.8	28.2
130	15.2	38.2	26.9
117	32.1	100.0	42.1
76	17.5	13.9	7.6
30	36.6	NA	NA
449	1.5	—	—
431	8.6	24.4	—
416	6.5	—	—
413	30.9	42.8	—
398	45.2	75.8	—
374	6.8	9.3	—
356	7.9	64.3	—
338	3.7	3.1	—
314	15.1	52.1	—
401	2.5	3.1	—
296	7.7	2.1	—
283	3.8	0.8	—
273	18.6	7.7	—
255	44.7	20.0	—
246	8.2	3.4	—
228	14.2	6.7	—
213	31.6	12.6	—
201	18.1	7.6	—
147	24.8	3.5	—
145	21.6	12.1	—
130	46.3	37.1	—
117	100.0	100.0	—

useful to isolate the bile acids in their biological state without destroying the structure. The recovery of bile acids from the thin-layer chromatogram by enzymatic assay was more than 90%.

3.1.3. TLC Identification of Bile Acids in Tissues

The results of TLC identification of bile acids in heart, lung, and kidney are illustrated in Fig. 2. It appears that cholic acid is present in each tissue examined by TLC in the solvent system isooctane/ethyl acetate/acetic acid (10:10:2, v/v/v). It is evident from Fig. 2 that chenodeoxycholic acid is predominant in lung, while cholic acid is most abundant in heart and kidney. Traces of material were detected in bands corresponding to hyodeoxycholic and lithocholic acids.

3.1.4. Physical Identification of Bile Acids

The glycine-conjugated bile acid mass spectra obtained using low ionizing voltage were quite similar to published spectra [39] (Table II). The molecular ion and some other higher-mass ions were not evident in the spectra. This was attributed to the small sample size and also to possible interference from dissolved silica from the TLC plates.

TABLE III. Comparison of Authentic Glycine-Conjugated Cholanoic Acid Chemical Ionization Spectra with Tissue Samples Isolated by TLC

	Glycochenodeoxycholic acid			Glycocholic acid	
m/e	Standard	Adipose tissue	m/e	Standard	Adipose tissue
413	2.8	1.8	412	4.8	3.9
398	0.7	—	411	8.9	—
			373	6.3	3.8
356	11.3	5.3	372	5.6	—
338	4.1	3.5	354	24.2	14.0
			353	26.4	—
296	0.8	0.8	337	45.3	30.2
283	4.8	3.3	319	7.8	—
273	1.1	0.5	271	—	3.8
255	45.4	52.9	253	13.2	14.0
228	6.7	6.9	211	5.2	6.0
213	13.5	13.5	199	4.6	5.0
201	11.5	8.2	147	4.9	6.4
147	17.2	15.1	145	5.4	7.8
145	18.4	17.7	130	7.4	27.4
130	5.6	6.1	117	29.8	36.4
117	14.8	14.2			

Attempts to improve the spectra by using chemical ionization mass spectrometry were successful to a limited extent. The spectra generated did not resemble published electron impact (EI) spectra. However, the spectra of the reference standards and the isolated samples were quite similar, thus helping to establish the identity of the bile acids (Table III).

Neither the EI nor the CI spectra of the taurine-conjugated bile acids closely matched the spectra published in the literature (Table IV). We thought that samples were thermally degraded in the sample probe. Spectra of the commercial standards and the isolated samples were similar to each other, however, and contribute to identifying the samples.

3.1.5. Quantitation of Bile Acids

In representative samples HPLC quantitation was compared with HSD assay. Results indicate similarity between 3α-HSD and HPLC quantitations (Table

TABLE IV. Comparison of Authentic Taurine-Conjugated Cholanoic Acid Chemical Ionization Spectra with Tissue Samples Isolated by HPLC

	Taurocholic acid				Taurochenodeoxycholic acid		
m/e	Reference 39	Standard	Human leukocyte	m/e	Reference 39	Standard	Human leukocyte
379	2.8	—	—	385	—	4.6	—
353	16.7	13.1	4.6	381	2.5	—	—
338	6.3	25.6	9.2	357	—	100.0	47.4
337	—	100.0	29.6	355	100.0	25.6	13.8
335	6.9	98.0	51.3	340	55.9	5.3	11.8
320	4.9	30.9	3.9	339	—	21.7	53.2
312	0.6	—	—	337	14.9	—	—
299	5.9	—	—	322	10.0	—	—
295	2.8	—	—	314	2.7	—	—
294	5.0	—	—	301	42.0	50.0	—
271	0.9	—	—	297	4.3	—	—
253	100.0	12.5	9.9	296	5.1	—	—
244	0.9	—	—	283	12.4	—	—
243	—	2.6	2.6	273	4.0	—	—
226	6.1	2.6	4.6	257	—	9.9	19.7
211	6.4	3.9	3.9	255	80.6	9.2	13.1
125	—	61.8	63.1	246	3.5	—	—
				229	—	11.8	11.8
				228	11.5	—	—
				213	48.3	5.3	6.5
				127	—	71.0	75.7
				113	—	98.0	100.0

TABLE V. Correlation of HPLC and 3α-Hydroxysteroid Dehydrogenase Analyses of Cholanoic Acids in Rat Tissues[a]

Tissue	Total cholanoic acids (μg/g tissue)	
	HPLC	HSD
Liver	165	153.4
Heart	13.9	10.8
Lung	11.8	10.9
Kidney	32.3	31.2
Adipose	5.2	4.4

[a] $r = 0.945$.

V). Rat heart, lung, kidney, and adipose tissue as well as liver contained cholanoic acids (Table VI). They were predominantly taurine conjugated in the animals used (adult Spraque–Dawley male rats fed laboratory ration). Glycine-conjugated compounds were present and could be concentrated by TLC, as shown by the mass-spectra data. Human leukocytes were found to contain bile acids (Table VII). Glycine-conjugated cholic acid was predominant, while for chenodeoxycholic acid, the taurine conjugated form was most abundant.

3.1.6. Cholesterol 7α-Hydroxylase Activity

The presence of cholesterol 7α-hydroxylase was demonstrated in nonhepatic rat tissue, as shown in Table VII. Heart, lung, and kidney tissues exhibited

TABLE VI. HPLC Quantitation of Cholanoic Acids in Hepatic and Extrahepatic Tissues

Tissue	Total	TC[a]	TCDC[a]	TDC[a]	GC[a]	GCDC[a]	GDC[a]
Rat (μg/g tissue)							
Liver	165	115	49.5	ND[b]	ND	ND	ND
Heart	13.9	9.7	4.2	ND	ND	ND	ND
Lung	11.8	4.4	7.4	ND	ND	ND	ND
Kidney	32.3	15.4	11.3	3.4	2.2	ND	ND
Adipose	5.2	2.4	3.4	2.4	ND	ND	ND
Human (μg/ml blood)							
Mixed leukocytes	1.5	0.49	0.13	ND	0.57	0.31	0.04

[a] TC, taurocholic; TCDC, taurochenodeoxycholic; TDC, taurodeoxycholic; GC, glycocholic; GCDC, glycochenodeoxycholic; GDC, Glycodeoxycholic acids.
[b] ND, not detectable.

TABLE VII. Cholesterol 7α-Hydroxylase Activity in Microsomes of Selected Tissues

Tissue	% Cholesterol conversion[a]	nmole/40 min[a]
Liver	4.56	1.026
Heart	1.45	0.203
Lung	1.03	0.241
Kidney	1.34	0.378
Mixed leukocytes	0.247	0.113

[a]Per milligram microsomal protein.

significant levels of activity: 23–32% of the level found in liver. Human mixed leukocytes had about 10% of the activity of rat liver per milligram of microsomal protein. Later work indicated that the enzyme activity was associated with the mononuclear cells [14], so those cells were used for later work.

3.2. Human Study

3.2.1. Selection and Description of Subjects

The subjects for this study were 29 men from three different age groups (19–25, 40–50, and 60–70 years) who were students and current and former employees at Iowa State University. All were white (Caucasian background) and in apparent good health. None of the subjects were engaged in regular jogging, swimming, or other training exercises. Information describing these men is given in Table VIII.

3.2.2. Dietary Data, Blood Pressure, and Blood Sample Collection

The subjects were instructed to record their food intake in household measures for 3 consecutive days, 2 weekdays and 1 weekend day. Nutritive value of the diets was calculated by using the nutrient database of the U.S. Department of Agriculture Lipid Nutrition Laboratory [40]; the SAS (Statistical Analysis Systems) computer program was employed. Within the same week diet records were kept, blood pressure was taken using a sphygmomanometer before blood was drawn.

Paired Student *t*-tests were performed on the dietary, clinical, and biochemical data to examine effects of age. Correlation coefficients between selected variables also were calculated.

TABLE VIII.   Characteristics of Men of Three Age Groups: Mean (Range)

	Age group (years)		
Variable	19–25 (n = 10)	40–50 (n = 9)	60–70 (n = 10)
Age (years)	22.4 (19–25)	43.8 (40–50)	64.0 (61–68)
Weight (lb)	175.8 (143–199)*[a]	173.8 (136–208)*	172.6 (127–195)*
Relative weight[b]	99.5 (83–109)*	98.3 (83–108)*	102.3 (78–120)*
Blood pressure (mm Hg)			
Systolic	118.4 (94–152)*	115.5 (88–140)*	136.6 (120–156)**
Diastolic	70.2 (56–84)*	76.7 (62–98)*	78.8 (58–96)*

[a] Means with *, ** subscripts within the same row are significantly different ($p < 0.05$).

[b] Relative weight, $\% = \dfrac{\text{Actual weight}}{\text{Desirable weight for height}} \times 100$.

3.2.3.   Results of Human Study

3.2.3a. General Characteristics.   Table VIII shows the characteristics of the subjects studied. They ranged from 19 to 68 years of age. The mean weights and relative weights of the subjects in the three age groups did not differ significantly. The mean systolic pressure of men 60–70 years old was significantly greater than those for men 19–25 and 40–50 years old, but the diastolic pressure of men in the three age groups did not differ significantly.

3.2.3b. Diet Composition.   Diet composition data (Table IX) show that the total calories, fat, protein, and carbohydrate intakes of men 19–25 years old were significantly higher than those of men 40–50 years old, but did not differ significantly from those of men 60–70 years old. The mean saturated and poly-unsaturated fatty-acid intakes of the two older groups studied were significantly lower than those of the 19–25 age group, but the polyunsaturated-to-saturated ratios of the diets were similar among the three groups. The percentages of calories derived from fat, protein, and carbohydrate were not significantly different among the three age groups.

3.2.3c. Leukocytes.   There were no significant differences in the numbers or proportions of leukocytes as age increased (Table X). However, the cholesterol content of leukocytes of the oldest age group studied was significantly higher than the mean values for the two younger groups. The mean total bile acid levels were similar in all three age groups.

Our previous work [14] has shown that the mononuclear leukocytes possess

TABLE IX. Diet Composition of Healthy Men in Three Different Age Groups (mean ± SEM of 3 days)

Variable	Age group (years) 19–25 ($n = 10$)	40–50 ($n = 9$)	60–70 ($n = 10$)
Protein (g)	102 ± 9[a]	77 ± 10**	106 ± 17*
Carbohydrate (g)	310 ± 27*	208 ± 24**	258 ± 21*·**
Fat (g)			
Total	120 ± 9*	83 ± 10**	94 ± 13*·**
Saturated (S)	49*	30**	34**
Oleic	45*	30**	34**
Linoleic (P)	21*	13**	14**
P/S[b]	0.402*	0.451*	0.443*
Cholesterol (mg)	456 ± 68*	335 ± 61**	415 ± 68***
Energy			
Total (kcal)	2763 ± 179*	1927 ± 223**	2322 ± 197*·**
Protein (%)	15*	16*	18*
Carbohydrate (%)	46*	45*	46*
Fat (%)	39*	39*	36*

[a] Means with *·**·*** superscripts within the same row are significantly different ($p < 0.05$).
[b] Polyunsaturated: saturated fatty-acid ratio.

TABLE X. Total and Differential Leukocyte Count and Cholesterol and Bile Acid Concentrations in Leukocytes of Healthy Men in Three Different Age Groups (Mean ± SEM)

Variable	Age group (years) 19–25 ($n = 10$)	40–50 ($n = 9$)	60–70 ($n = 10$)
Leukocytes (total number/ml)			
Plasma	7194 ± 474*[a]	6494 ± 417*	7816 ± 694*
Mononuclear (%)	44 ± 2.7*	40 ± 3.4*	38 ± 2.9*
Granulocytes (%)	56 ± 2.7*	60 ± 3.4*	62 ± 2.9*
Cholesterol (μg/10^8 cells)			
Total	101 ± 11.9*	106 ± 14.3*	135 ± 28.1**
Free cholesterol	78 ± 10.3*	79 ± 10.8*	100 ± 18.9**
Cholesteryl ester	23 ± 3.1*	27 ± 3.8*	35 ± 9.4*
Bile acids (nmole 10^8 cells)	2.11 ± 0.164*	2.16 ± 0.252*	2.11 ± 0.178*

[a] Means with *·** superscripts within the same row are significantly different ($p < 0.05$).

TABLE XI. Hydroxymethylglutaryl-Coenzyme A Reductase[a] and Cholesterol 7α-Hydroxylase[b] Activity in Mononuclear Leukocyte Microsomes of Healthy Men in Three Different Age Groups (Mean ± SEM)

	Age group (years)					
	19–25 ($n = 10$)		40–50 ($n = 9$)		60–70 ($n = 10$)	
	HMG-CoA reductase	Cholesterol 7α-hydroxylase	HMG-CoA reductase	Cholesterol 7α-hydroxylase	HMG-CoA reductase	Cholesterol 7α-hydroxylase
Per mg microsomal protein	10.5 ± 0.54**	0.106 ± 0.0059*	13.1 ± 0.81**	0.107 ± 0.0051*	14.4 ± 1.22**	0.128 ± 0.0113*
Per 10^7 mononuclear cells	30.4 ± 3.24*	0.303 ± 0.0300*	51.4 ± 6.52***	0.419 ± 0.0549***	45.2 ± 4.10**	0.401 ± 0.0386**
Per ml blood	9.1 ± 0.49*	0.091 ± 0.0054*	12.2 ± 0.75**	0.099 ± 0.0079*	12.5 ± 0.77**	0.110 ± 0.0069*

[a] Enzyme activity is expressed as pmole of mevalonate formed per hour.
[b] Enzyme activity is expressed as percent of [14C]-cholesterol converted to [14C]-7α-hydroxycholesterol per 40 min.
: Means with *,** superscripts within the same row are significantly different ($p < 0.05$).

HMG-CoA reductase and cholesterol 7α-hydroxylase activities. HMG-CoA reductase is the rate-limiting step in the synthesis of cholesterol [41]. The mean HMG-CoA reductase activity for the 40–50 and 60–70 age groups was significantly higher than that for the youngest age group when activity was expressed as per milligram microsomal protein or per 10^7 mononuclear cells or per milliliter blood (Table XI). Cholesterol 7α-hydroxylase is the rate-limiting step of bile acid synthesis [10]. When expressed as per milligram microsomal protein or per milliliter blood, its activity was similar for the three age groups. But when values were expressed as per 10^7 mononuclear cells, the cholesterol 7α-hydroxylase activity for the 40–50 and 60–70 age groups was significantly higher than that observed in the men 19–25 years old. This particular trend was associated with a decrease in the proportion of mononuclear cells but no decrease in total number with increase in age.

3.2.3d. Plasma Lipids.   Table XII shows the concentration of various lipids in plasma. The mean plasma cholesterol values increased significantly with increases in age from 19 to 70 years, and the mean annual change was 1.57 mg cholesterol per dl plasma as estimated by regression analysis. The magnitude of increase in plasma free cholesterol and cholesteryl ester with age was similar; consequently, free–ester ratio did not change with increased age. For the plasma phospholipid, mean values were similar in all three age groups. Plasma triglycerides increased significantly for the successive age groups, with 93 mg/dl in the 19–25 group, 107 mg/dl in the 40–50 group, and 122 mg/dl in the 60–70 group. The mean values for plasma bile acids did not differ significantly among the three age groups.

TABLE XII. Cholesterol, Phospholipid, Triglyceride, and Bile Acid Levels in Plasma of Healthy Men in Three Different Age Groups (Mean ± SEM)

Variable[a] (per dl plasma)	Age group (years)		
	19–25 ($n = 10$)	40–50 ($n = 9$)	60–70 ($n = 10$)
TC (mg)	204 ± 5.4*[b]	232 ± 11.1**	282 ± 8.3***
FC (mg)	53 ± 1.8*	62 ± 4.2**	74 ± 2.1***
CE (mg)[c]	150 ± 3.9*	170 ± 7.2**	209 ± 7.1***
PL (mg)	210 ± 20.5*	208 ± 10.7*	215 ± 21.7*
TG (mg)	93 ± 3.5*	107 ± 7.2**	122 ± 5.5***
BA (μM)	0.749 ± 0.077*	0.767 ± 0.094*	0.609 ± 0.056*

[a] TC, total cholesterol; FC, free cholesterol; CE, cholesteryl ester; PL, phospholipids; TG, triglycerides; BA, bile acids.
[b] Means with different superscripts within the same row are significantly different ($p < 0.05$).
[c] Calculated CE = total cholesterol minus free cholesterol.

3.2.3e. Lipoproteins. The levels of various lipids in the different lipo-proteins are shown in Table XIII. The change in VLDL total cholesterol with advancing age was not significant, although its free cholesterol was significantly higher in the 60–70 age group than in the 19–25 age group. The LDL cholesterol increased significantly with increase in age at a mean rate of 1.29 mg/dl plasma per year. The degree of rise in LDL cholesteryl ester was proportionately greater than the increment in its free cholesterol from 19–25 to 60–70 years. There were no significant changes in HDL cholesterol and in the free–ester ratio.

All three age groups showed similar mean phospholipid concentrations in all the lipoprotein fractions analyzed. VLDL triglycerides increased signifi-cantly with advancing age, but LDL and HDL triglycerides remained unchanged. Bile acids were present in all the three lipoprotein fractions, LDL, VLDL, and HDL, and these amounts accounted for almost all the plasma bile acids. The lipo-protein bile acids tended to decrease with increase in age, but this trend was not significant.

3.2.4. Discussion of Human Study

We observed a significant increase in systolic blood pressure with increase in age, and our results are consistent with the findings of the Health and Nutrition Examination Survey (HANES) study [42]. The diastolic pressure also increased with age, but not to a significant degree. Analysis of our data has shown a high degree of correlation between systolic and diastolic pressure levels ($r = 0.64$, $p < 0.01$). An interdependence between serum cholesterol and blood pressure has often been suggested, but the relationship never has been clearly shown in humans. Our results show significant and positive relationships of plasma and LDL cholesterol with systolic blood pressure ($r = 0.52$ and 0.54, respectively, $p < 0.01$), and these trends are in agreement with the findings of Juustila [43]. Adel *et al.* [44] observed that, *in vivo,* hypertensive rats synthesize cholesterol more rapidly than do normotensive rats. Our results also show that the HMG-CoA reductase activity was high in men with high systolic pressure, and this positive association was significant ($r = 0.48$, $p < 0.01$).

No change in the total leukocyte count occurred with aging, and this is consistent with the observations of Helman and Rubenstein [45], Otani [46], and Zacharski *et al.* [47]. We observed a decline in the proportion of mononuclear cells with increase in age, and this trend also has been reported by Mackinney [48] and Reddy and Goh [49]. Friedman *et al.* [50] have reported the use of leukocyte count as a predictor of myocardial infarction, and their data suggest that the predictive power of leukocyte count is quite similar to that of serum cholesterol or blood pressure. We did not observe any association between the total leukocyte count and the plasma cholesterol or other lipids or blood pressure.

The concentration of bile acids in lipoproteins has not been reported in most

TABLE XIII. Cholesterol, Phospholipid, Triglyceride, and Bile Acid Levels (per dl Plasma) in Lipoproteins of Healthy Men in Three Different Age Groups[a]

	Age group (years)	TC (mg)	FC (mg)	CE[b] (mg)	PL (mg)	TG (mg)	BA (μM)	$\frac{FC}{CE}$	$\frac{TC}{PL}$
Very low density lipoprotein	19–25	20.4 ± 0.97*[c]	4.7 ± 0.24*	15.7 ± 0.89*	33.2 ± 2.20*	54.5 ± 2.23*	0.105 ± 0.007*	0.30*	0.61*
	40–50	21.3 ± 2.35*	5.3 ± 0.67*·**	17.0 ± 1.46*	29.8 ± 1.47*	64.4 ± 3.08**	0.104 ± 0.006*	0.31*	0.71*·**
	60–70	24.9 ± 2.09*	6.5 ± 0.57**	18.4 ± 1.53*	30.3 ± 2.36*	83.7 ± 4.18***	0.097 ± 0.055*	0.35*	0.82*·**
Low-density lipoprotein	19–25	127.1 ± 5.45*	30.4 ± 1.43*	96.7 ± 4.44*	121.9 ± 17.39*	24.4 ± 0.90*	0.423 ± 0.051*	0.31*	1.04*
	40–50	145.4 ± 6.16**	30.7 ± 2.22*	114.7 ± 5.09**	118.4 ± 8.62*	26.7 ± 1.41*	0.402 ± 0.043*	0.27*	1.38***
	60–70	193.1 ± 6.04***	41.4 ± 1.48**	151.7 ± 5.77***	139.6 ± 18.29*	24.6 ± 1.30*	0.317 ± 0.037*	0.27*	1.38***
High-density lipoprotein	19–25	47.8 ± 1.89*	10.9 ± 0.34*	36.9 ± 1.71*	54.0 ± 3.06*	13.1 ± 0.90*	0.216 ± 0.016*	0.29*	0.88*
	40–50	46.6 ± 1.27*	11.7 ± 0.77*	34.9 ± 1.24*	52.3 ± 2.03*	13.1 ± 0.94*	0.198 ± 0.013*	0.31*	0.89*
	60–70	46.3 ± 2.55*	9.8 ± 0.55*	36.5 ± 2.16*	50.0 ± 2.06*	13.2 ± 0.89*	0.189 ± 0.012*	0.27*	0.95*

[a] TC, total cholesterol; CE, cholesteryl ester; FC, free cholesterol; PL, phospholipids; TG, triglycerides; BA, bile acids.
[b] Calculated CE = total cholesterol minus free cholesterol.
[c] Mean values not sharing a common superscript are significantly different ($p < 0.05$).

literature about lipoproteins. In one study bile acids in HDL were determined [51]. An early study led to the conclusion that bile acids were bound predominantly to albumin [52], and later investigations [15,53] used the same procedure of incubating serum or pure proteins with radiotracer-labeled bile acids to reach the same conclusions.

In our study, the HMG-CoA reductase activity in the mononuclear cells increased significantly as men aged from19 to 50 years, which was followed by little change from 60 to 70 years. Our results do not agree with the observations in rats by Yamamoto and Yamamura [54], Story and Kritchevsky [55], and Dupont *et al.* [56], who reported either a decrease or no change in cholesterol synthesis in liver and some extrahepatic tissues as the animals aged. The discrepancy may be attributed partly to differences in the tissue studied or in the responses of humans compared to rats. The significant and positive association of both HMG-CoA reductase ($r = 0.49, p < 0.01$) and cholesterol 7α-hydroxylase ($r = 0.37, p < 0.05$) with age also suggests a decreased function or number of the peripheral cell receptors that bind LDL and remove cholesterol from plasma. Because it is the LDL receptor-binding mechanism that regulates cholesterol entrance and thereby synthesis within peripheral cells [57], any decrease in this activity will result in increased cholesterol synthesis and turnover by peripheral cells to meet their requirement of cholesterol for cell membrane synthesis. Goldstein and Brown [58] observed that, as the rate of growth of cultured cells (human skin fibroblasts) declines, the number of cell receptors that bind LDL also decreases. Our results also suggest this trend.

A rise in serum and LDL cholesterol with increase in age has been reported in many studies [59–64], and our results, too, confirm this trend. The increase in plasma cholesterol was almost entirely in LDL, with insignificant increases in VLDL, and Connor *et al.* [65] have made the same observation. Our results show a significant and positive association ($p < 0.01$) of plasma and LDL cholesterol levels with the HMG-CoA reductase ($r = 0.69$ and 0.54) and cholesterol 7α-hydroxylase ($r = 0.55$ and 0.46) activity in mononuclear leukocytes. This was not surprising inasmuch as it already has been demonstrated that the leukocytes of hypercholesterolemic subjects respond to incubation in a lipoprotein- or lipid-free medium with a greater-than-normal induction of HMG-CoA reductase [13]. Under steady-state conditions, *in vivo,* the leukocytes of these subjects would not synthesize more cholesterol because the higher extracellular plasma LDL cholesterol represses the HMG-CoA reductase. When plasma cholesterol levels increase, less efficient mechanisms such as scavenger cells or macrophages of the reticuloendothelial system [66] become more active in degrading the increased amount of LDL. These cells, when overloaded with cholesteryl esters, are converted to foam cells, which are classically known components of atherosclerotic plaques.

Epidemiological evidence suggests that excessive dietary intake of energy or fat may be responsible for an increase in plasma cholesterol. The results of our study show no significant correlations of energy, protein, carbohydrate, fat, or cholesterol intake with plasma cholesterol levels. Negative correlations of the energy ($r = -0.37$, $p < 0.05$) and carbohydrate intake ($r = -0.46$, $p < 0.01$) with VLDL cholesterol and of total fat intake with leukocyte cholesterol levels ($r = -0.40$, $p < 0.05$) were significant. We also observed a significant decrease in the calorie and fat intake with increase in age, and similar trends have been observed by Munro [67], McGandy *et al.* [68], and in the HANES study [69]. These trends suggest that the greater nutrient intake of the 19–25 age group of men was compensated for by their more active life-style, causing rapid turnover and low plasma lipids. It also is possible that, for healthy subjects such as the group we observed, the influence of dietary calories, fat, and cholesterol on plasma cholesterol levels is overridden. The fat intake was related positively and significantly to all lipoprotein bile acids, and this may be attributed to an increased turnover of bile acids with the increase in fat intake.

3.3. Steroid Kinetics in Rats

3.3.1. Tracer Administration and Sample Collection

A total of 27 adult male Wistar/Sprague–Dawley-derived rats weighing 503 ± 9 g were used for the study. The numbers of samples used for replication of lipoprotein and tissue analyses are given in Table XIV. The rats were caged in wire-mesh cages singly in a room with 26°C temperature, $50 \pm 10\%$ humidity, and a light–dark cycle of 0600–1800 hr light, 1800–0600 hr dark. They had free access to water and were fed a low-fat stock ration [70] *ad libitum*.

Donor rats were used to prepare radioactive LDL cholesterol. Blood was

TABLE XIV. Concentration of Cholesterol, Cholesteryl Ester, and Bile Acids in Plasma Lipoproteins and Tissues of Rats[a]

	FC	CE	BA
Lipoprotein (nmole/ml plasma)			
VLDL ($n = 11$)	57 ± 30	54 ± 33	0.30 ± 0.03
LDL ($n = 11$)	77 ± 41	96 ± 48	0.39 ± 0.03
HDL ($n = 10$)	256 ± 108	560 ± 153	0.46 ± 0.19
Tissue (nmole/g tissue)			
Liver ($n = 23$)	2900 ± 713	930 ± 240	342 ± 44
Muscle ($n = 23$)	1930 ± 220	240 ± 60	20 ± 3
Adipose ($n = 23$)	1970 ± 42	170 ± 13	12 ± 0.5

[a] FC, free cholesterol; CE, cholesteryl ester; BA, bile acids.

drawn from the abdominal aorta into tubes containing Na EDTA (1 mg/ml blood) and centrifuged at 800 *g* for 30 min. To the plasma was added 250 μCi of 4-[^{14}C]cholesterol (purified by TLC) in 200 μl of acetone–ethanol (1:1). The dissolved cholesterol was added while the plasma was being stirred over a period of 2 min. The plasma was incubated at 37°C for 2 hr and then dialyzed overnight against a 0.15 M NaCl solution containing 0.01% of EDTA, pH 7.0. The lipoprotein fractions were isolated as described, and the LDL fraction contained 81% of the added radioactivity.

Rats used for the study were injected via the tail vein with 0.25 ml of the labeled LDL containing approximately 3.5 μCi of tracer. The injections were made carefully into tail veins heated for 1 min under a lamp. Only those rats observed to have successful injections were used for the study. After a 12-hr fast, blood samples and tissues were harvested at 0.5, 1, 2, 3, 4, 8, 12, 16, 20, 24, and 28 days after injection.

3.3.2. Lipoprotein and Tissue Concentrations of Steroids

The concentrations of cholanoic acids and of free and esterified cholesterol are shown in Table XIV. Liver contained more than three times as much free as esterified cholesterol, whereas in VLDL and LDL, proportions of free and esterified cholesterol seemed to be relatively similar. On the other hand, HDL contained about twice as much cholesteryl ester as free cholesterol. Generally, cholesterol transferred to HDL from other lipoprotein fractions is rapidly esterified by lecithin-cholesterol acyl transferase (LCAT) [71]. The mass of cholanoic acids was 8.2% of liver steroids and 0.27, 0.23, and 0.056% of VLDL, LDL, and HDL steroids, respectively.

The liver, skeletal muscle, and adipose tissue values are similar to those previously reported for rats [24].

3.3.3. Cycling Patterns of LDL-4-[^{14}C]Cholesterol

The tracer LDL-[^{14}C]cholesterol was lost from LDL extremely rapidly (Fig. 3). The [^{14}C]cholesterol reappeared with highest specific activity (SA) as bile acids in gut (Fig. 4) and liver (Fig. 5). Liver free and esterified cholesterol specific activities exhibited two-component disappearance curves and were in approximate equilibrium with each other (Fig. 5). The cholanoic acid SA was higher than that of neutral sterol by approximately 10-fold for most of the time course. This agrees with the known precursor–product relationship of these compounds and can be accounted for by the relative masses of neutral and acidified sterol [72,73].

In VLDL, free and esterified cholesterol were not in equilibrium with each other until day 16 (Fig. 6). The cholesteryl ester SA of VLDL was about the same as that of liver (Fig. 5), but free cholesterol SA was much lower. This

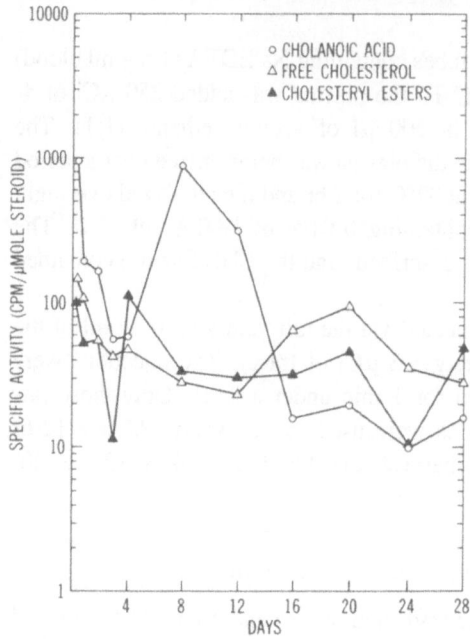

FIGURE 3. Pattern of change of LDL steroid-specific activity for 28 days after i.v. injection of LDL-4-[^{14}C]cholesterol.

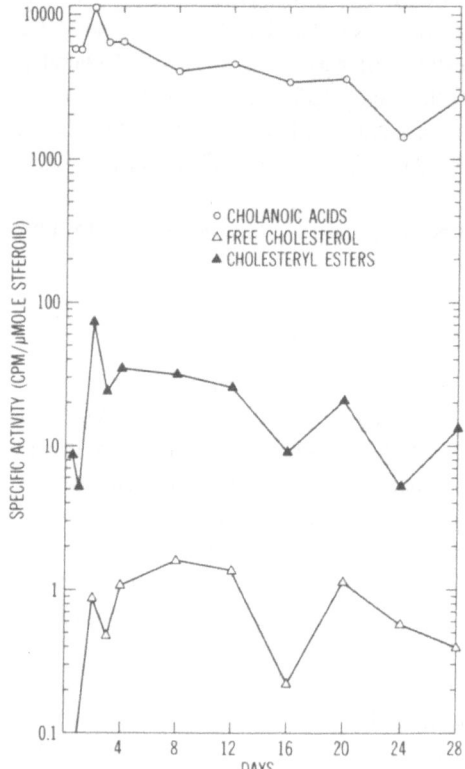

FIGURE 4. Pattern of appearance and disappearance of steroids in rat intestine after i.v. injection of LDL-4-[^{14}C]cholesterol.

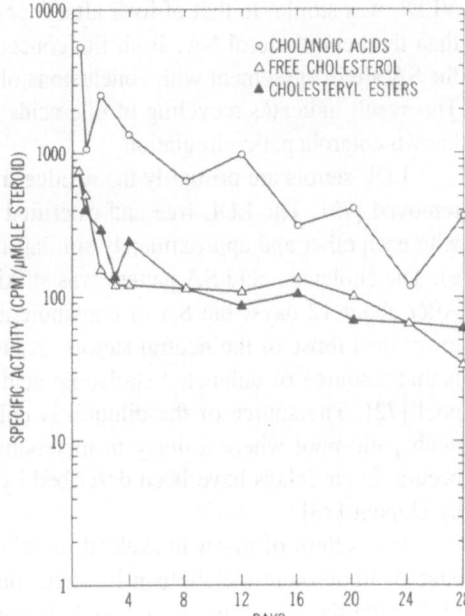

FIGURE 5. Pattern of appearance and disappearance of steroids in rat liver after i.v. injection of LDL-4-[14C]cholesterol.

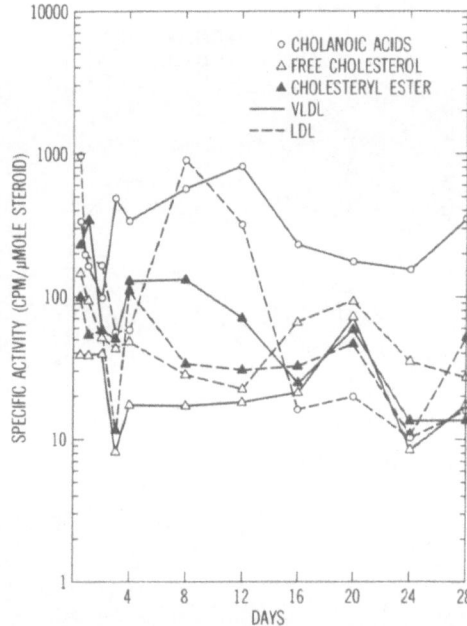

FIGURE 6. Pattern of appearance and disappearance of steroids in VLDL and LDL in rats after i.v. injection of LDL-4-[14C]cholesterol.

suggests separate source pools for the compounds. The cholanoic acid SA in VLDL was similar to that of liver after 3 or 4 days and was about 10-fold higher than the neutral sterol SA. Both the concentration of esterified cholesterol and the SA are in agreement with conclusions of Oschry and Eisenberg [74] for rats. This result indicates recycling of bile acids from liver to VLDL consonant with known enterohepatic circulation.

LDL sterols are primarily the residue from VLDL that has had triglycerides removed [75]. The LDL free and esterified cholesterol SA were in equilibrium with each other and approximately similar in SA to VLDL free cholesterol (Fig. 6). The cholanoic acid SA pattern was strikingly different in LDL from VLDL. After about 12 days, the SA of cholanoic acid dropped precipitously to values lower than those of the neutral sterols. A possible interpretation of these events is that a source of unlabeled cholanoic acid is diluting the LDL cholanoic acid pool [72]. The source of the dilution is unlikely to be the liver, but may be a nonhepatic pool where a delay in metabolism of cholesterol to cholanoic acid occurs. Such delays have been described by Schwartz *et al.* [76] and discussed by Dupont [73].

The pattern of tracer in skeletal muscle is shown in Fig. 7. Present knowledge of lipoprotein metabolism indicates that the tracer should arrive via LDL [77]. After day 1, all the steroid pools in muscle remained with constant SA for

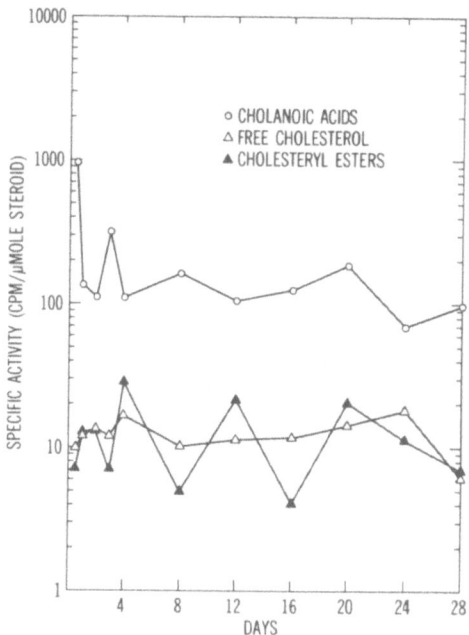

FIGURE 7. Pattern of appearance and disappearance of steroids in skeletal muscle of rats after i.v. injection of LDL-4-[^{14}C]cholesterol.

the duration of the study. This tissue could, therefore, constitute a metabolic delay pool controlling final excretion of steroids. HDL is synthesized primarily in the liver [75]. The neutral sterols of HDL were not in equilibrium with their counterparts in the liver (Fig. 8). Free cholesterol and cholanoic acids had SA patterns similar to those of VLDL (Fig. 6), but esterified cholesterol had an opposite relationship to free cholesterol in the two fractions. This probably is caused by the action of LCAT on unlabeled cholesterol being taken up by HDL for reverse cholesterol transport [78]. The cholanoic acid pattern suggests equilibration of liver and enterohepatic circulating bile acids by the eighth day.

Evidence suggesting removal for excretion of cholesterol from the plasma by HDL [79] is supported by the rapid disappearance of labeled cholesteryl esters over 20 days. The return to equilibrium with other steroids at 24 and 28 days suggests that equilibrium has been reached among all the metabolically active sterol pools in the body.

The kinetic pattern of cholanoic acids in LDL suggests a cycle of cholanoic acid sequestration by LDL, resulting in plasma clearance by the degradation of the LDL particles in the liver [77]. The lower SA of cholanoic acids and free cholesterol in LDL than in liver after 16 days is in agreement with the postulate that the LDL is finally degraded by liver, and those steroids are more likely to

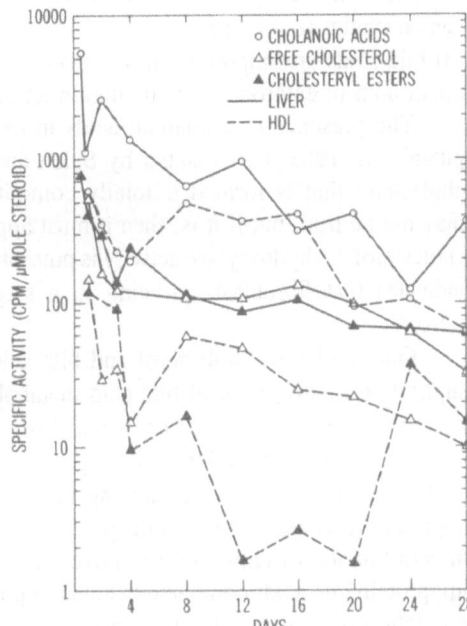

FIGURE 8. Patterns of appearance and disappearance of steroids in liver and HDL in rats after i.v. injection of LDL-4-[^{14}C]cholesterol.

be excreted via bile than retained by the liver. The very low SA of these compounds in the intestine suggests dilution by unlabeled pools.

To analyze the kinetics of the patterns of cholesterol metabolism shown, quantitative estimates of rates and pool size must be available. The data are not sufficient to construct mathematical models for analysis using programs such as simulated analysis and modeling, as done by Schwartz *et al.* [76]. It is necessary to obtain more data points over time, including tissue cholanoic acid pools and absolute excretion values.

4. CONCLUSIONS

The methodologies used in the experiments described in this chapter are becoming obsolete. Nevertheless, analysis of steroids remains a challenging task regardless of new instrumentation and products. The chemical properties of the compounds make them elusive to isolate, and their variety makes it necessary to employ more than one quantitative analysis procedure to a mixture. The mixtures are always complex. Probably it will be necessary to continue to use a variety of methods both old and new and to evaluate the results of each experiment according to its internal controls rather than absolute standards.

Several important considerations for bile acid metabolism have been developed in this chapter. They are: (1) bile acids are found in many tissues and can be synthesized in most tissues; (2) bile acids are transported via lipoproteins; (3) bile acids are important in the coordinated regulation of cholesterol metabolism both in enterohepatic circulation and in whole-body kinetics.

The presence of cholanoic acids in extrahepatic tissues is no longer disputable. In 1983 it was stated by Salen and Shefer [3] that "the 7α-hydroxycholesterol that is formed is totally committed to bile acid production." That may not be true, but if it is, then it must apply to tissues other than liver where cholesterol 7α-hydroxylase activity is present. Preliminary work in our laboratory indicates that fibroblasts growing in a serum-free medium synthesize acidic steroids.

Our studies of cholesterol and bile acid metabolism in human leukocytes suggest an involvement of bile acid metabolism in the normal turnover of cholesterol. The HMG-CoA reductase and 7α-hydroxylase activities are in synchrony and are active in directions contrary to regulation by plasma cholesterol concentration alone. The data showing kinetics of bile acid metabolism in rat lipoproteins and tissues support the premise that bile acid turnover may be involved in coordination of cholesterol turnover. The rapid turnover of the small pool of lipoprotein bile acids compared to all other pools and to cholesterol pools suggests a function possibly related to transport.

Transport of sterols through water layers adjacent to membranes remains an unknown process [80]. Bile acids have been shown to be uniquely suitable to that process in the intestine [81]. Furthermore, bile acids have been shown to be required for pancreatic cholesterol esterase [82] and carboxyl ester hydrolase from human pancreatic juice to hydrolyze esters of cholesterol and esters of vitamins A, D_3, and E [83].

The activity of lysosomal acid cholesterol esterase has been shown to require Na taurocholate for maximum activity [84,85]. While not discussing the possibility of the bile acid as a cofactor, the respective authors conclude that taurocholate is preferred in their analyses over any detergent. Taurocholate has been found, also, to stimulate lysophospholipase C, whereas detergents inhibited the activity [86].

The understanding of coordination of metabolism of cholesterol in the whole body is just beginning. Previous work has not included consideration of bile acids beyond their function in enterohepatic circulation and metabolism [87]. It is likely that future work will evolve to account for "bile acids" in extrahepatic tissues.

ACKNOWLEDGMENTS.   This work was supported in part by National Institutes of Health Grant HL 23598. Medical supervision was provided by Dr. L. Z. Furman and staff at the I.S.U. Student Health Service; technical assistance by Dr. William Runyan, James Roth, Barbara Krumhardt, Ruth Smith, and Cynthia Schriver. Mary W. Marshall supervised the computer analysis of the diets in the Lipid Nutrition Laboratory, USDA, Beltsville, MD.

REFERENCES

1. H. J. Nicholas, in "The Bile Acids" (P. P. Nair and D. Kritchevsky, eds.), Volume 3, p. 1, Plenum Press, New York (1976).
2. S. Y. Oh, and J. Dupont, *Lipids* **10**, 340 (1975).
3. G. Salen and S. Shefer, *Annu. Rev. Physiol.* **45**, 679 (1983).
4. P. Eneroth and J. Sjövall, in "Methods in Enzymology" (R. B. Clayton, ed.), Volume 15, p. 237, Academic Press, New York (1969).
5. J. M. Street, D. J. H. Trafford, and H. L. J. Makin, *J. Lipid Res.* **24**, 491 (1983).
6. K. D. R. Setchell, A. M. Lawson, N. M. Tanida, and J. Sjövall, *J. Lipid Res.* **24**, 1085 (1983).
7. S. M. Grundy, E. H. Ahrens, Jr., and T. A. Miettinen, *J. Lipid Res.* **6**, 397 (1965).
8. A. M. Weber, L. Chartrand, G. Doyon, S. Gordon, and C. C. Roy, *Clin. Chim. Acta* **39**, 524 (1972).
9. R. Spears, D. Vukusich, S. Mangat, and B. S. Reddy, *J. Chromatogr.* **116**, 184 (1976).
10. S. Shefer, S. Hauser, I. Bekersky, and E. H. Mosbach, *J. Lipid Res.* **11**, 404 (1970).
11. D. S. Fredrickson, in "The Metabolic Basis of Inherited Disease" (J. B. Standbury, J. B. Wyngaarden, and D. S. Fredrickson, eds.), p. 502, McGraw-Hill, New York (1972).
12. D. Y. Y. Hsia, *Enzyme* **13**, 161 (1972).

13. A. M. Fogelman, J. Edmond, J. Seager, and G. Popjak, *J. Biol. Chem.* **250**, 2045 (1975).
14. N. K. Dodd, C. E. Sizer, and J. Dupont, *Biochem. Biophys. Res. Commun.* **106**, 385 (1982).
15. A. Roda, G. Cappelleri, R. Aldini, E. Roda, and L. Barbara, *J. Lipid Res.* **23**, 490 (1982).
16. S. Ewerth, I. Björkhem, K. Einarsson, and L. Ost, *J. Lipid Res.* **23**, 1183 (1982).
17. J. D. Manes and D. L. Schneider, *J. Lipid Res.* **12**, 376 (1971).
18. S. K. Goswami and C. F. Frey, *J. Chromatogr.* **53**, 389 (1970).
19. S. K. Goswami and C. F. Frey, *Biochem. Med.* **17**, 20 (1977).
20. P. Eneroth, *J. Lipid Res.* **4**, 11 (1963).
21. M. N. Chavez and C. L. Krone, *J. Lipid Res.* **17**, 545 (1976).
22. E. Fujihira, N. Takahashi, A. Minato, K. Uenoyama, T. Ogiso, and S. Hirose, *Chem. Pharm. Bull. (Tokyo)* **20**, 2719 (1972).
23. A. F. Hofmann, *J. Lipid Res.* **3**, 127 (1962).
24. J. Dupont, S. Y. Oh, and P. Janson, in "The Bile Acids" (P. P. Nair and D. Kritchevsky, eds.), Volume 3, p. 17, Plenum Press, New York (1976).
25. R. Shaw, A. Smith, and W. Elliott, *Anal. Biochem.* **861**, 450 (1978).
26. N. L. Young and V. Rodwell, *J. Lipid Res.* **18**, 572 (1977).
27. O. H. Lowry, N. J. Rosebrough, A. L. Farr, and R. J. Randall, *J. Biol. Chem.* **193**, 265 (1951).
28. C. C. Allain, L. S. Poon, C. S. G. Chan, W. Richmond, and P. C. Fu, *Clin. Chem.* **20**, 470 (1974).
29. S. E. Carlson, A. D. Mitchell, and S. Goldfarb, *Biochim. Biophys. Acta* **531**, 115 (1978).
30. R. S. Sandhu, *Clin. Chem.* **22**, 1973 (1976).
31. M. J. Fletcher, *Clin. Chim. Acta* **22**, 393 (1968).
32. L. A. Turnberg and A. Anthony-Mote, *Clin. Chim. Acta* **24**, 253 (1969).
33. P. A. Siskas, P. T. Cahill, and N. B. Javitt, *J. Lipid Res.* **18**, 666 (1977).
34. R. J. Havel, H. A. Eder, and J. H. Bragdon, *J. Clin. Invest.* **34**, 1345 (1955).
35. B. Hennig and J. Dupont, *J. Nutr.* **113**, 1104 (1983).
36. W. P. Jencks and E. L. Durrum, *J. Clin. Invest.* **34**, 1437 (1955).
37. M. C. Territo, D. W. Golde, and M. J. Cline, in "Manual of Clinical Immunology" (N. R. Rose, and H. Friedman, eds.), p. 142, American Society for Microbiology, Washington, DC (1976).
38. I. Davidsohn and D. A. Nelson, in "Clinical Diagnosis," 15th ed. (I. Davidsohn and J. B. Henry, eds.), p. 149, Saunders, Philadelphia (1974).
39. R. Shaw and W. H. Elliott, *Biomed. Mass Spectrom.* **5**, 433 (1978).
40. Consumer and Food Economic Institute expansion of data published in, "Nutritive Value of American Foods in Common Units," U.S. Department of Agriculture data sets No. 456-1, 456-2, Washington, DC (1977).
41. J. M. Dietschy and M. S. Brown, *J. Lipid Res.* **15**, 508 (1974).
42. U.S. Department of Health and Human Services, Public Health Service, "Hypertension in Adults 25–74 Years of Age, United States 1971–1975," Vital and Health Statistics Series 11, No. 221, DHHS(PHS) Pub. No. 81-1671, National Center for Health Statistics, Hyattsville, MD (1981).
43. H. Juustila, *Acta Med. Scand.* **613** (Suppl.), 94 (1977).
44. H. N. Adel, Q. B. Deming, M. M. Daly, V. M. Raeff, and L. M. Brun, *J. Lab. Clin. Med.* **66**, 571 (1965).
45. N. Helman and L. S. Rubenstein, *Am. J. Clin. Pathol.* **63**, 35 (1975).
46. T. Otani, *Arch. Fr. Pediatr.* **15**, 227 (1958).
47. L. R. Zacharski, L. R. Elveback, and J. W. Linman, *Am. J. Clin. Pathol.* **54**, 148 (1971).
48. A. A. Mackinney, Jr., *J. Gerontol.* **33**, 213 (1978).
49. M. M. Reddy and K. Goh, *J. Gerontol.* **34**, 5 (1979).

50. G. D. Friedman, A. L. Klatsky, and A. B. Siegelaub, *N. Engl. J. Med.* **290,** 1275 (1974).
51. G. Middelhoff, R. Mordasini, A. Stiehl, and H. Greten, *Scand. J. Gastroenterol.* **14,** 267 (1979).
52. D. Rudman and F. E. Kendall, *J. Clin. Invest.* **36,** 538 (1957).
53. C. W. Burke, B. Lewis, D. Panvelinalla, and S. Tabaqchali, *Clin. Chim. Acta.* **32,** 207 (1971).
54. M. Yamamoto and Y. Yamamura, *Atherosclerosis* **13,** 365 (1971).
55. J. A. Story and D. Kritchevsky, *Experientia* **30,** 242 (1974).
56. J. Dupont, M. M. Mathias, A. A. Spindler, and P. Janson, *Age* **3,** 19 (1980).
57. M. S. Brown and J. L. Goldstein, *Science* **191,** 150 (1976).
58. J. L. Goldstein and M. S. Brown, *Metabolism* **26,** 1257 (1977).
59. A. Keys, O. Mickelsen, E. O. Miller, E. R. Hays, and R. L. Todd, *J. Clin. Invest.* **29,** 1347 (1950).
60. E. Y. Lawry, G. V. Mann, A. Peterson, A. P. Wysock, R. O'Connell, and F. J. Stare, *Am. J. Med.* **22,** 605 (1957).
61. I. H. Page, E. Kirk, W. H. Lewis, Jr., W. R. Thompson, and D. D. Van Slyke, *J. Biol. Chem.* **111,** 613 (1935).
62. F. J. Schilling, G. Christakis, A. Orbach, and W. H. Becker, *Am. J. Clin. Nutr.* **22,** 133 (1969).
63. L. A. Carlson, *Acta Med. Scand.* **167,** 399 (1960).
64. D. L. Adlersberg, L. E. Schaefer, A. G. Steinberg, and C. I. Wang, *JAMA* **162,** 619 (1956).
65. W. E. Conner, D. T. Witiak, D. B. Stone, and M. L. Armstrong, *J. Clin. Invest.* **48,** 1363 (1969).
66. M. S. Brown, S. K. Basu, J. R. Falck, Y. K. Ho, and J. L. Goldstein, *J. Supramol. Struct.* **13,** 67 (1980).
67. H. N. Munro, *in* "Mammalian Protein Metabolism" (H. N. Munro and J. B. Allison, eds.), p. 3, Academic Press, New York (1964).
68. R. B. McGandy, C. H. Barrows, Jr., A. Spanias, A. Meredity, J. L. Stone, and A. H. Morris, *J. Gerontol.* **21,** 581 (1966).
69. U.S. Department of Health, Education and Welfare, "Dietary Intake findings, United States 1971–74," Vital and Health Statistics Series 11, No. 202, DHEW Pub.(HRA) 77-1647, National Center for Health Statistics, Hyattsville, MD (1977).
70. R. D. Reeves and L. Arnrich, *J. Nutr.* **104,** 118 (1974).
71. G. S. Getz and R. V. Hay, *in* "The Biochemistry of Atherosclerosis," p. 151, Dekker, New York (1979).
72. J. M. Reiner, *Exp. Mol. Pathol.* **20,** 78 (1974).
73. J. Dupont, *in* "Cholesterol Systems in Insects and Animals" (J. Dupont, ed.), p. 117, CRC Press, Boca Raton, FL (1982).
74. Y. Oschry and S. Eisenberg, *J. Lipid Res.* **23,** 1099 (1982).
75. H. B. Brewer, Jr., *Klin. Wochenschr.* **59,** 1023 (1981).
76. C. C. Schwartz, M. Berman, Z. R. Vlahcevic, L. G. Halloran, D. H. Gregory, and L. Swell, *J. Clin. Invest.* **61,** 408 (1978).
77. M. S. Brown, P. T. Kovanen, and J. L. Goldstein, *Science* **212,** 628 (1981).
78. J. A. Glomset, *Prog. Biochem. Pharmacol.* **15,** 41 (1979).
79. G. Heiss, N. J. Johnson, S. Reiland, C. E. Davis, and H. A. Tyroler, *Circulation* **62,** 116 (1980).
80. M. C. Phillips, L. R. McLean, G. W. Staudt, and G. H. Rothblat, *Atherosclerosis* **36,** 409 (1980).
81. J. S. Patton and M. C. Carey, *Science* **204,** 145 (1979).
82. G. B. Vahouny, S. Weersing, and C. R. Treadwell, *Biochim. Biophys. Acta.* **98,** 607 (1965).

83. D. Lombardo and O. Guy, *Biochim. Biophys. Acta* **611,** 147 (1980).
84. N. J. Haley, *J. Lipid Res.* **21,** 961 (1980).
85. R. D. Kenagy and E. L. Bierman, *Biochim. Biophys. Acta* **754,** 174 (1983).
86. S. Huterer and J. R. Wherrett, *Biochim. Biophys. Acta* **794,** 1 (1984).
87. S. K. Goswami, *in* "Cholesterol Systems in Insects and Animals" (J. Dupont, ed.), p. 97, CRC Press, Boca Raton, FL (1984).

Chapter 9

TISSUE-BOUND BILE ACIDS

Nabila Turjman and Padmanabhan P. Nair

1. INTRODUCTION

In the last decade, bile acids have attracted the attention of nutritional epidemiologists because of the suspected association between dietary characteristics, fecal steroids, and carcinoma of the colon. The bile acids, cholic and chenodeoxycholic acids, formed in the liver are the primary end products of cholesterol metabolism in higher animals. In the liver, they are conjugated to glycine and taurine and/or sulfated to form water-soluble products that are secreted into the bile. A significant amount of bile salts and neutral sterols that reach the small bowel are reabsorbed and returned to the liver via the enterohepatic circulation, where they exercise negative feedback control on the synthesis of cholesterol as well as on its conversion to bile acids [1]. The neutral sterols (both exogenous and endogenous) and bile salts reaching the distal ileum and large bowel are further metabolized to a number of secondary products under the influence of intestinal microflora. Fecal bile acids and neutral sterols are therefore a complex mixture of closely related compounds, representing the combined effects of hepatic and gastrointestinal metabolism.

Bile acids are present in the tissues either in the soluble form, extractable with ethanol/ammonia, or firmly bound to tissues and released only by enzymatic

---

The systematic nomenclature of the bile acids referred to in this article is as follows: cholic acid, 3α,7α,12α trihydroxy-5β-cholanoic acid; chenodeoxycholic acid, 3α,7α dihydroxy-5β-cholanoic acid; deoxycholic acid, 3α,12α dihydroxy-5β-cholanoic acid; lithocholic acid, 3α hydroxy-5β-cholanoic acid; NELL, N-ε-lithocholyl lysine.

Nabila Turjman and Padmanabhan P. Nair   Lipid Nutrition Laboratory, Beltsville Human Nutrition Research Center, Agricultural Research Service, U.S. Department of Agriculture, Beltsville, Maryland and the Department of Biochemistry, The Johns Hopkins University School of Hygiene and Public Health, Baltimore, Maryland   *Present address for N.T.:* Department of Microbiology and Immunology, Duke University, Durham, North Carolina

hydrolysis with cholanoylamino acid hydrolase (also termed cholylglycine hydrolase, E.C. 3.5.1.24). The first bile acid isolated in the tissue-bound form was lithocholic acid in pathological specimens of human livers from obese patients undergoing bypass surgery [2,3]. Tissue-bound lithocholic acid (TBL) was shown to be elevated in livers of rats treated with the carcinogen methylazoxymethanol, as compared to those treated with saline along [4]. TBL, detected for the first time in tissues outside the digestive tract, was found to be somewhat uniformly distributed among normal and neoplastic human mammary tissues [5,6] and in neoplasms of the uterus, kidney, lung, and the colon [7].

2. METHODOLOGY

2.1. Isolation Procedures

The original method [2,8] for the isolation and assay of TBL has served as a basis for further refinement in more recent studies [4]. The procedure involved a preliminary solvent extraction of soluble bile acids followed by enzymatic release of bound bile acids from the residue for quantitation by gas–liquid chromatography (GLC).

Generally, about 100 mg of liver tissue is homogenized with 95% ethanol/0.1% ammonia [9], a solvent system that effectively extracts bile salts from tissues. The residue, obtained after repeated extractions with this solvent system, is considered to be free of extractable bile salts. Treatment of this residue under optimum conditions with crude cholanoylamino acid hydrolase (from *Clostridium perfringens* ATCC 19574) [10] releases tissue-bound bile acids, which are then isolated and quantified by standard procedures involving thin-layer chromatography (TLC) and GLC.

Normal livers from experimental animals (rats) yield significant amounts of free bile acids, including lithocholic acid [4,9]. However, in cholestatic liver disease there is a profound alteration in sterol and bile acid metabolism, and higher concentrations of lithocholic acid would be expected [11]. In our experience, in liver injury associated with bowel bypass surgery for obesity, hepatic tissue was essentially devoid of any soluble lithocholate, but contained significant amounts of TBL [2,3]. Since, in the liver, sulfation of lithocholate is an active process involving the detoxification of this bile acid, efforts to quantify lithocholic acid sulfate in the soluble and bound forms revealed that the sulfate form is almost exclusively localized in the soluble fraction. Maruyama and co-workers [12], using computerized GLC–mass spectrometry–selected-ion monitoring, have also found most of the lithocholic acid sulfate in the soluble fraction, whereas TBL consisted of nonsulfated lithocholic acid. The evidence from several laboratories indicates that hepatic TBL is perhaps the consequence of the activation of a latent alternate enzymatic pathway capturing lithocholic acid that escapes sulfation [13]. Gelb and co-workers [14], following similar procedures, have validated the methods for the determination of soluble and tissue-bound bile acids in colon cancer and in colonic polyps, although they found no correlation between histological data and bile acid composition or concentration.

A modification of the original procedure has increased the sensitivity of the procedure severalfold [4,15]. This method entailed the use of a proteolytic enzyme prior to hydrolysis with cholanoylamino acid hydrolase (Fig. 1). The tissue is first converted to acetone-dry powders to provide stable and homogeneous aliquots for analysis. Aliquots of the acetone-dry powder are repeatedly extracted with 95% ethanol/0.1% ammonia, and the insoluble residue from these extractions is predigested with trypsin for 2 hr at 37°C [4], after which the enzyme is heat-inactivated. The mixture is then hydrolyzed with cholanoylamino acid hydrolase in the presence of disodium EDTA and β-mercaptoethanol. The lithocholic acid released is extracted from an acidified medium into ethyl acetate,

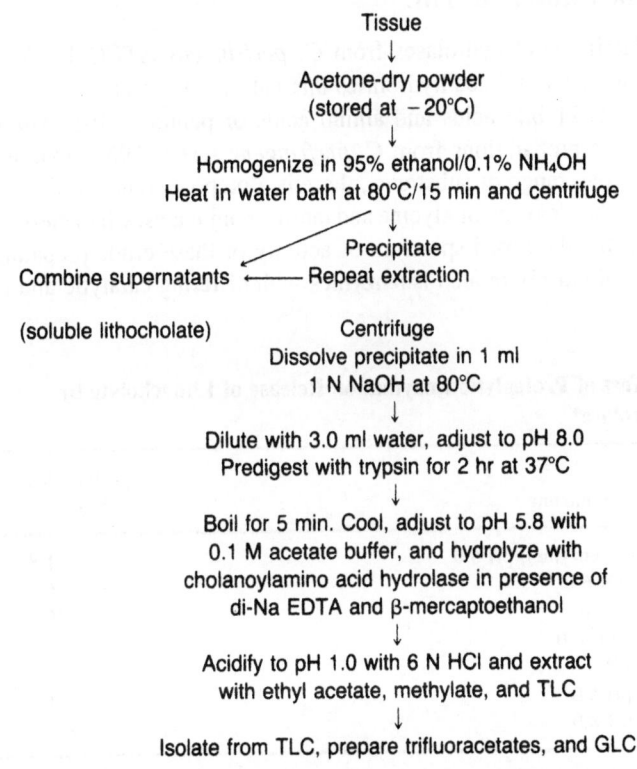

Figure 1. Analysis of tissue-bound lithocholate.

methylated, purified by TLC, and then quantified by GLC as the corresponding trifluoroacetate.

Predigestion of tissue residue with proteases appeared to improve the yield of TBL, presumably by enhancing steric accessibility of the sites of bound lithocholic acid to the action of clostridial cholanoyl amino acid hydrolase [4]. Table I summarizes the results obtained using several of these proteolytic enzymes at pH 6.0. Trypsin was the most effective enzyme, and its efficiency was further enhanced at its optimum pH of 8.0.

The enzymatic cleavage of TBL from fibrous and connective tissue such as breast, muscle, and uterus requires further modification. In this instance, tissue residues were subjected to a series of two sequential digestions with collagenase and trypsin. Treatment of tissues with collagenase prior to tryptic digestion increased the yield of TBL by about 2.5-fold (Table II). Omission of the collagenase step may therefore result in serious underestimates of TBL in fibrous tissues.

2.2. Enzymatic Cleavage of TBL

Cholanoylamino acid hydrolases from *C. perfringens* ATCC 19574 constitute a group of closely related hydrolases that catalyze the cleavage of the C-N bond between C-21 bile acids and amino acids or peptides [10]. Although most crude enzyme preparations from *C. perfringens* ATCC 19574 exhibit activity against a broad range of substrates, there is a preponderance of the isoenzymes catalyzing the cleavage of glycine and taurine conjugates. Characterization studies revealed that the broad spectrum of activity of these crude preparations represented several closely related isoenzymes with differing catalytic and ther-

TABLE I.   Effect of Proteolytic Enzymes on Release of Lithocholate by Clostridial Hydrolase[a]

Incubation mixture	µg lithocholic acid released/g dry wt
Control (no proteolytic enzyme), pH 6.0	1.8
Protease (2 mg/ml), pH 6.0	36
Pepsin (2 mg/ml), pH 6.0	46
Peptidase (2 mg/ml), pH 6.0	24
Pronase (2 mg/ml), pH 6.0	27
Trypsin (2 mg/ml), pH 6.0	71
Trypsin (2 mg/ml), pH 8.0	149

[a] The procedure outlined in Fig. 1 was employed in these assays.

TABLE II.   Enzymatic Release of TBL from Breast Tissues[a]

Nature of enzymes used	μg TBL/g of dry tissue
Control, hydrolase only[b]	6.69
Trypsin followed by hydrolase[c]	18.1
Collagenase followed by hydrolase[d]	3.7
Collagenase, trypsin and hydrolase[e]	47.4

[a] Aliquots, 100 mg each of acetone-dry powder, were extracted twice with 95% ethanol/0.1% ammonium hydroxide. The residues were dissolved in 1 ml of 1 N NaOH and pH was adjusted for optimal activity of the enzyme system in use.

[b] The pH was adjusted to 5.8 with 0.1 M acetate buffer, and hydrolysis was carried out with 0.3 ml of cholylglycine hydrolase (activity, not less than 10K units/ml) in the presence of di-Na EDTA and β- mercaptoethanol for 2 hr at 37°C.

[c] The pH was initially adjusted to 8.0 and predigested with 3 mg trypsin for 2 hr at 37°C. Trypsin was inactivated by boiling for 5 min.

[d] The conditions were the same as in [c] except that trypsin was replaced with 3.0 mg of collagenase.

[e] Predigestion with 3.0 mg collagenase for 30 min. boiling, and a second digestion with trypsin as in [c] for 90 min.

mal stability characteristics [16]. In view of the pleiotropic character of this group of hydrolases, it is essential to observe certain precautions to preserve the activity of the relatively unstable form of the enzyme that catalyzes the release of bile acids from tissue-bound form [17]. In recent studies involving the use of N-ε-lithocholyl lysine (NELL) as a substrate, we have pointed out that this enzyme (in distinction to those that hydrolyze glycine and taurine conjugates) is unstable at temperatures about 0°C and as a general rule should be stored at below −20°C in small aliquots to preserve its activity [17].

2.3.   Isolation of N-ε-Lithocholyl Lysine from Tissues by Nonenzymatic Hydrolysis

An alternative to the enzymatic cleavage of lithocholic acid from tissue-bound form is the isolation of the intact amino acid (lysine) residue to which the bile acid is linked. Fresh tissue (100 mg) is homogenized with 2.0 ml of a mixture of 6 N HCl:propionic acid (1:1) and hydrolyzed under partial vacuum at 110°C for 3 hr in a vacuum hydrolysis tube [18]. The hydrolyzate is transferred to a 50-ml round-bottomed flask, and the solution is evaporated to dryness in a rotary evaporator. The residue is extracted with ethanol, and the pooled ethanolic extract is subjected to TLC [2]. Standard NELL [19] is run alongside to locate the zone corresponding to NELL, visualization being carried out using a spray of 8-hydroxy-1,3,6-pyrenetrisulfonic acid followed by ultraviolet visualization. Since, among the isomeric lithocholyl lysines only NELL is ninhydrin positive, microquantitation as its ninhydrin chromogen is feasible.

REFERENCES

1. H. Danielsson, *in* "The Bile Acids" (P. P. Nair and D. Kritchevsky, eds.), Volume 2, p 1, Plenum Press, New York (1973).
2. P. P. Nair, A. I. Mendeloff, M. Vocci, J. Bankoski, M. Gorelik, G. Herman, and R. Plapinger, *Lipids* **12,** 922 (1977).
3. P. P. Nair, R. Solomon, J. Bankoski, and R. Plapinger, *Lipids* **13,** 966 (1978).
4. N. Turjman, A. I. Mendeloff, C. Jacob, C. Guidry, and P. P. Nair, *J. Steroid Biochem.* **14,** 1237 (1981).
5. N. Turjman and C. Jacob, *Fed. Proc.* **40,** 1683 (1981) (Abstr.).
6. P. P. Nair, W. Daughtrey, R. N. Patnaik, A. I. Mendeloff, and N. Turjman, VI International Bile Acid Meeting, Falk Symposium No. 29, Freiburg, W. Germany, October 1980.
7. N. Turjman and C. Jacob, VI International Bile Acid Meeting, Falk Symposium No. 29, Freiburg, W. Germany, October 1980.
8. P. P. Nair, *in* "The Bile Acids" (P. P. Nair and D. Kritchevsky, eds.), Volume 3, p. 29, Plenum Press, New York (1976).
9. T. Okishio, P. P. Nair, and M. Gordon, *Biochem. J.* **102,** 654 (1967).
10. P. P. Nair, M. Gordon, and J. Reback, *J. Biol. Chem.* **242,** 7 (1967).
11. N. Carulli, M. Ponz de Leon, P. Loria, F. Zironi, and R. Iori, *in* "Frontiers in Gastrointestinal Research" (P. Gentilini and M. U. Dianzani, eds.), Volume 8, p. 217, Karger, Basel (1984).
12. I. Maruyama, T. Maeyama, R. Kumashiro, K. Tanikawa, T. Shinka, and I. Matsumoto, VI International Bile Acid Meeting, Falk Symposium No. 29, Freiburg, W. Germany October 1980 (Abstr. No. 72).
13. N. Turjman, G. Kessie, and P. P. Nair, *Fed. Proc.* **42,** 1925 (1983) (Abstr. No. 983).
14. A. M. Gelb, C. K. McSherry, J. R. Sadowsky, and E. H. Mosback, *Am. J. Gastroenterol.* **77,** 314 (1982).
15. N. Turjman and P. P. Nair, *Cancer Res.* **41,** 3761 (1981).
16. R. Patnaik and M. Sokolow, *Gastroenterology* **68,** 966 (1975) (Abstr.).
17. P. P. Nair, G. Kessie, and V. Flanagan, *J. Lipid Res.* **27,** 905 (1986).
18. F. Westzll and H. Hesser, *Anal. Biochem.* **61,** 610 (1964).
19. P. P. Nair, G. Kessie, and N. Turjman, *J. Steroid Biochem.* **23,** 573 (1985).

Chapter 10

SERUM BILE ACIDS

G. M. Murphy

1. INTRODUCTION

The advances in methodology described elsewhere in this volume (Chapters 1,2,5,6,7) have led to a veritable explosion of data on peripheral bile acid concentrations. This chapter considers those data in the light of the known influences on serum bile acids in health and disease. In particular, the role played by the analytical method used will be stressed. In line with most of the previous literature, the diagnostic and prognostic value of serum bile acid measurements in liver disease will be examined. Finally, the usefulness of serum bile acid determinations in nonhepatic diseases will be reviewed.

But first, two apologies—one for treating serum, plasma, and systemic as synonymous and the other to the authors of the numerous recent scholarly reviews of this subject [1–10]. Inevitably, therefore, I am guilty of much plagiarism and omission.

2. SERUM BILE ACIDS AND THE DIFFERENT ENTEROHEPATIC BILE ACID CYCLES IN NORMAL HUMANS

Modern serum bile acid measurements have highlighted the fact that the different bile acids in humans have different enterohepatic circulations. They do not travel together as a single entity or bolus through one anatomically defined enterohepatic circulation under the impetus of gallbladder contraction and ileal absorption. The various bile acids are distributed diferently during fasting, are

G. M. Murphy   Gastroenterology Unit, Department of Medicine, Guy's Campus, United Medical and Dental Schools, London, England

absorbed by different mechanisms from different intestinal sites postprandially, are extracted from the portal vein by the liver with different efficiencies, and may have different excretory routes. It is probably more realistic to consider that there are as many "enterohepatic circulations" as there are bile acids. And each of these aspects of bile acid metabolism has an effect on serum bile acid concentrations.

2.1. Bile Acid Distribution

During fasting the bile acid pool in normal humans is largely confined to the gallbladder, biliary tree, and intestine. Results of modern studies show some agreement on the distribution of the bile acid pool in the fasting and postprandial states, but there are problems of interpretation. Although hepatic bile is stored and concentrated in the gallbladder [11], not all of the hepatic bile secreted during fasting enters the gallbladder. Estimates of the distribution of bile between the gallbladder and the intestine vary widely and are clearly methodologically dependent. Using cholescintigraphy and ^{99m}Tc-HIDA it has been shown that during an overnight fast about 75% of the hepatic bile has been stored in the gallbladder [12]. Using intubation procedures and also non-bile acid biliary markers, other workers have shown that during the night only 40–50% of hepatic bile may enter the gallbladder [13–15], in accord with earlier studies [16–19]. Apart from the obvious difficulties of interpreting the results of studies using intubation procedures, it must also be borne in mind that the distribution of the bile acid pool cannot necessarily be inferred from the distribution of some other biliary marker. Using a gamma-labeled bile acid ([^{75}Se]taurohomocholate), it has been found that about 50% of the fasting bile acid pool is in the gallbladder [20]. Independent of gallbladder contraction which may occur spontaneously [21], the issue is further complicated in that there appears to be a cyclic variation in the concentration of the bile acids in the bile which is continuously secreted into the duodenum; this variation follows the phases of the interdigestive migrating motor complexes [22].

2.2. Absorption of Bile Acids from the Intestine into the Portal Blood

Bile acids may be absorbed to some extent from all regions of the gastrointestinal tract, including the stomach and the large intestine [23]. The mechanisms of absorption and the particular bile acids preferentially absorbed vary from region to region, and because of this it may be misleading to regard even the small intestine as a single compartment [24]. Active transport of bile acids is still believed to be confined to the distal small bowel and to be greatest in the terminal ileum [25]. Not all bile acids are actively absorbed equally well—

trihydroxy bile acids, for example, appear to be selectively absorbed when compared to dihydroxy bile acids [26–28]. In recent years with the advent of bile acid treatment to dissolve gallstones, the passive absorption of bile acids, particularly from the jejunum, has received much attention. Passive absorption of bile acids also appears to be selective and favors glycine-rather than taurine-conjugated bile acids, and nonpolar species rather than polar ones. In the fasting state the input of bile acids from the gut is less than that postprandially, and differences as to site of absorption may not modify the peripheral bile acid composition so dramatically [29–31].

2.3. Bile Acids in the Portal Blood

There have been few studies of the bile acid content of human portal blood, but portal blood bile acid concentrations of up to 1000 μmole/liter have been reported (compared with <10 μmole/liter in the peripheral circulation). There are also differences in bile acid composition. These qualitative and quantitative differences between the bile acids in the portal vein and those in the peripheral circulation are due to the fact that the liver handles the different bile acids with different efficiency. Also, bile acids that are not extracted from the portal vein on first pass are subsequently presented to the liver by the systemic circulation. Therefore, to asess the effects of the liver in relation to peripheral bile acid concentrations we must consider at least two processes: hepatic extraction of bile acids from the portal vein and hepatic extraction of bile acids from the systemic circulation.

2.4. Hepatic Extraction from the Portal Vein

The hepatic extraction of bile acids from portal blood has received much attention [32–34] and has been shown to be a sodium-dependent carrier-mediated process [35,36]. The factors that influence portal blood bile acid clearance include the bile acid concentration, state of conjugation, degree of hydroxylation, and extent of protein binding.

Increases in bile acid concentration have been shown to stimulate Na^+, K^+-ATPase activity and to produce an increase in the number of bile acid carriers (or binding sites) in the liver surface membrane of hepatoma cells [37,38]. However, whether such effects could be produced by the changes in portal blood concentrations that occur in humans after meals is not known.

With regard to the state of conjugation of the bile acids, many studies have shown that conjugated bile acids are cleared from the portal vein more efficiently than unconjugated bile acids. About 90% of one conjugated cholate is removed from the portal blood during one passage through the human liver, compared with about 70% unconjugated cholate [39] and 50–60% unconjugated cheno-deoxycholate [40,41]. This suggests that the more polar bile acids are cleared more efficiently.

The role of protein in the hepatic uptake of bile acids is more controversial. Many studies have shown marked differences in the affinities of the different bile acids for albumin [42–44]. Unconjugated bile acids bind more tightly than conjugated and glycine-more tightly than taurine-conjugated. These findings, together with the observation than albumin competitively inhibited hepatic uptake of taurocholate but had little effect on that of cholate, suggest that an albumin–bile acid complex is itself involved in the uptake process [45]. Other workers, however, have found the albumin binding of the different bile acid fractions to be essentially the same and have shown that the hepatic uptake of taurine-conjugated bile acids is more efficient than that of glycine-conjugated, which is, in turn, more efficient than that of unconjugated bile acids [46]. Whatever the explanation for these discrepancies in the findings of different studies, it has been noted that bile acids are not the only ligands that may bind to albumin and that concentrations of other substances such as fatty acid may modify albumin–bile acid interactions [47].

2.5. Hepatic Uptake of Bile Acids from the Systemic Circulation

At any time the bile acids in the systemic circulation are a mixture of "new" (those bile acids which have not been extracted from the portal blood on their first pass and have yet to be re-presented to the liver) and "old" (which refers to those bile acids already present in the systemic circulation). Since in normal humans the hepatic transport mechanisms are considered to be operating well below their capacity, the absolute amount extracted will depend on the liver blood flow; the relative uptakes of the different bile acids will depend on their intrinsic clearance. As with the bile acids in portal blood, the intrinsic clearance of the bile acids in the systemic circulation is largely influenced by their structure. Following their intravenous injection, conjugated bile acids are cleared more rapidly than unconjugated and trihydroxy more rapidly than dihydroxy bile acids [42,48,49]. So the modifying effect of one passage of systemic bile acids through the liver is to enrich the circulation in dihydroxy and unconjugated bile acids.

2.6. Compartmental Models and Peripheral Bile Acid Concentrations

In attempting to define the factors responsible for the distribution of bile acids in normal humans, many workers have used compartmental models and analyses. For example, to explain and define the plasma disappearance of intra-venously administered cholate, a five-compartment model was suggested [50].

This model did not, however, include a hepatic sinusoid compartment. It has been argued that it is important to include a sinusoidal compartment when considering the plasma disappearance of a bile acid with a high first-pass clearance [51]. This is because complete mixing of the injected bile acid within the plasma compartment may not occur if the first pass is high, and the assumption of instantaneous mixing following intravenous injection may not be justifiable and may lead to errors [52]. It seems essential to postulate a well-mixed sinusoidal compartment one fraction of which passes into the liver and the other directly into the peripheral plasma compartment [51].

3. SERUM BILE ACID MEASUREMENTS AND THE DYNAMICS OF THE NORMAL ENTEROHEPATIC CIRCULATIONS

3.1. Radioimmunoassay

The application of radioimmunoassay to serum bile acid concentrations has made a large contribution to our knowledge of peripheral bile acid dynamics (see Chapter 6). Radioimmunoassay offers the advantage of ease of application together with high sensitivity. It thus provides a convenient method for monitoring the changes induced in serum bile acid concentrations by bile acid enterohepatic cycling. Consequently, radioimmunoassay has often been hailed as the method of choice for routine bile acid measurements [53–55]. The specificity of such measurements has received relatively little attention. Many authors have validated their assays by applying techniques, such as gas–liquid chromatography, to the same samples. These other techniques may be of entirely different specificity; for example, gas–liquid chromatography assays often include unconjugated species as well as conjugated. Interference from other bile acids, particularly at low or normal concentrations [56,57], has seldom been examined. In spite of these difficulties, radioimmunoassay has nevertheless made a considerable contribution to our practical knowledge of the enterohepatic circulation of bile acids in normal subjects.

In 1974 LaRusso and colleagues followed the dynamics of serum-conjugated cholate [58]. It was subsequently shown that following a liquid meal, serum-conjugated chenodeoxycholate concentrations increased before, and to higher values than, conjugated cholate concentrations [59]. Similar results were obtained by other groups, and although the responses to lunch and dinner were not always as consistent as those after breakfast, it was clearly demonstrated that the enterohepatic cycling of cholate conjugates differed from that of chenodeoxycholate conjugates [30]. The possible mechanisms responsible for this difference were discussed in Section 2.2 and are associated with the different sites of intestinal absorption [60,61] and the different hepatic extractions of these bile acids. It has also been shown that the rhythm of the enterohepatic cycling is determined by eating and that this rhythm will vary considerably according to the nature of the meals [62,63]. Thus, the slower gastric emptying or intestinal transit of solid meals has little effect on the absorption of glycochenodeoxycholate but does delay the absorption of conjugates of cholic acid from the terminal ileum [64]. However, not only were different types of meals used in some of these studies, but the radioimmunoassays employed differed considerably in their specificity, some including taurine conjugates, others not. Using assays designed to measure primarily glycine conjugates, early and late increases in concentrations following meals have been observed in some subjects for both chenodeoxycholate and cholate, but not for deoxycholate. One possible explanation proposed for the lack of a later increase in glycodeoxycholate has been simply slower absorption of the bile acid. Another explanation has been that the glycodeoxycholate pool, which is smaller than that of the primary bile acids, may be absorbed completely from the jejunum. It is clear, however, that many factors other than polarity must be taken into account when considering bile acid intestinal absorption. It might be predicted from physicochemical chemical considerations alone that ursodeoxycholic acid and its glycine conjugate would be predominantly absorbed in the ileum [65]. However, with the use of radioimmunoassay methods, it has been shown tht ursodeoxycholate may be absorbed from the proximal small bowel [66].

3.2. Gas–Liquid Chromatography/Mass Spectrometry

The results obtained using radioimmunoassay have been largely confirmed using gas–liquid chromatography/mass spectrometry (GC–MS). Thus, cheno-deoxycholic and deoxycholic acid concentrations (measured after deconjugation) in normal subjects increased following meals before those of cholic acid [67]. In this study the increase in deoxycholic acid was somewhat greater than that of cholate and demonstrated both early and late peaks, thus disagreeing somewhat with the results of radioimmunoassay. Indeed, it was argued that both dihydroxy bile acids short-circuited the hepatic–ileal circuit. The slow return of concentrations to baseline values was attributed to continuing enhanced small-intestinal input caused by recirculation of the bile acid pool. Later results support this view of a short-circuited enterohepatic cycle of deoxycholic as well as cheno-deoxycholic acid [41].

3.3. High-Pressure Liquid Chromatography

At the time this manuscript was completed there had been few studies of fasting and postprandial concentrations of glycine-conjugated and taurine-conjugated bile acids in normal subjects using HPLC, although it is evident that with improvements in the sensitivity of this technique, this situation will change (see Chapter 2). Using a high-pressure liquid chromatographic/enzymatic assay, Linnet [68] observed that it was glycine-conjugated rather than taurine-conjugated bile acids that were preferentially absorbed in the proximal small bowel. In contrast to the results of these studies, glycocholic acid concentrations increased more rapidly than those of glycochenodeoxycholic acid. Linnet observed that the rapid decrease of the elevated glycocholate concentrations was in accord with its higher hepatic clearance [68].

3.4. Unconjugated Serum Bile Acid Concentrations

Until recently little attention was given to unconjugated serum bile acid concentrations. One obvious explanation for this is that the available methodology was incapable of quantitatively distinguishing serum unconjugated bile acids. In normal sera the measurements of even the conjugated bile acid concentrations were usually made at the limits of sensitivity; hence unconjugated bile acids were not detected and consequently were assumed not to be present. In sera from patients with cholestatic liver disease the unconjugated bile acid concentrations were not detectable against the background of elevated conjugated bile acid concentrations—and again were assumed not to be present [69,70]. The majority of bile acid radioimmunoassays measured conjugated bile acids, and few, if any, antisera specific for unconjugated bile acids have been described. Lately the situation has changed; methods for the separation of conjugated and unconjugated bile acids have improved (see Chapter 1), and much information is accumulating on unconjugated serum bile acid concentrations. Many studies have confirmed the early gas–liquid chromatographic observations of Makino *et al.* [71] that significant quantities of unconjugated bile acids may be found in fasting sera from normal subjects [31, 72, 73–76]. In common with conjugated bile acids the unconjugated fraction increases within 2 hr after meals [31,77]. It is generally accepted that the normal liver conjugates the majority of unconjugated bile acids presented to it and that deconjugation is the first step in secondary bile acid formation in the intestinal lumen. Unconjugated serum bile acid concentrations may thus prove to be an index of the absorption from the intestine of newly formed secondary bile acids [78]. If the site of formation of secondary bile acids is normally the distal ileum, then presumably the early peak in unconjugated serum bile acid concentrations is due either to the absorption of unconjugated bile acids already present in the intestine or to the absorption of unconjugated bile acids newly formed from conjugated bile acids already present in the intestine [77,78]. This early peak in unconjugated serum bile acids in normal subjects after meals is thus compatible with the notion of some of the bile acid pool being sequestered in the intestine during fasting. The further metabolism and/or absorption of this sequestered pool may be brought about by the increase in colonic motility occasioned by the ingestion of food [79,80].

3.5. Protein-Bound Serum Bile Acid Concentrations

The importance of bile acid binding to plasma proteins in the maintenance of serum bile acid concentrations is not clear. Early studies suggested that binding to plasma proteins was a mechanism that inhibited the surface-active effects of bile acids in the peripheral circulation [81–88]. Recently attention has turned to the effects of protein binding on the hepatic extraction of bile acids from the peripheral circulation. Most authors suggest that binding to albumin impairs hepatic extraction [87,88], but some studies have indicated that protein binding actually enhances plasma bile acid uptake into the hepatocyte [89,90]. The various bile acids are bound with differing affinities to the plasma proteins. Although different techniques for assessing protein binding have been used, there is a consensus of opinion that unconjugated bile acids are more tightly protein bound than conjugated bile acids [83,87,91]. The most important plasma protein is albumin, although globulins and apolipoproteins have also been incriminated [83,87,92].

It has been suggested [93] that different affinities of the various bile acids for plasma albumin account for the increased concentrations of unconjugated bile acids found in the plasma of patients with bacterial overgrowth [93,246] and the reduced concentrations of glycocholic acid found in plasma from subjects with Gilbert's syndrome [88]. Although usually those bile acids more tightly bound to albumin disappear from plasma more slowly than those with less affinity, clearly there are exceptions. Taurodehydrocholic acid, for example, is one of the least protein bound and yet it has the slowest plasma disappearance rate [42,94]. There is little information on the distribution of endogenous bile acids between bound and nonbound fractions. It seems, however, at least in normal sera, that results obtained using radioactive bile acids and either pooled [94] or individual sera [43] are similar to those obtained for endogenous serum bile acids [44].

4. SERUM BILE ACIDS AND LIVER DISEASE

Because of the obvious, although complex, interaction between plasma bile acid concentrations and "bile acid liver function," the measurement of serum bile acids has long tempted clinicians as a possible simple, but sensitive and reliable index of liver disease. The constraints put on the model, outlined in Section 2.5, of normal bile acid metabolism by liver disease will be different in various liver diseases and at different stages of each disease. Peripheral bile acid concentrations are altered not only by abnormalities in the hepatic uptake, transport, synthesis, and biliary excretion of bile acids, but also by changes in hepatic blood flow and portal systemic shunting.

4.1. Total 3α-Hydroxy Bile Acid Concentrations

Of all the methods used for measuring serum bile acids, enzymatic determination of total 3α-hydroxy bile acids is clearly the one that lends itself most readily to application to large numbers of samples. Until recently this method has involved tedious extraction procedures, but many improvements have now been described. Total 3α-hydroxy bile acids may be measured in human serum without extraction [95–98], with automation and continuous-flow techniques [99–102], using bioluminescence [103,104] and even electrochemical detection [105]. In spite of these advances, clinical chemists are still reluctant to add this assay to their package of routine liver function tests. This hesitation is not simply because of economic reasons, but is due rather to the controversy that remains with regard to the sensitivity and specificity of this measurement both analytically and clinically.

Analytical problems with the enzymatic assay have received much attention, but it is worth emphasizing that there is a wide range of normal values in the literature, upper limits of normal being 5–21 μmole/liter. With few exceptions, the normal values obtained for fasting serum bile acids with the enzymatic approach are far in excess of those obtained by gas–liquid chromatography and GC–MS. This has raised doubts about the analytical sensitivity and specificity of the enzyme procedure [106–110] and subsequently its accuracy at low concentrations. There is also considerable controversy about the clinical sensitivity and specificity of the enzymatic assay.

Until recently the majority of authors suggested that the measurement of serum bile acids provided a sensitive index of liver function [111–124]. Modern applications of the enzymatic assay have thrown some doubt on that observation [125–132]. One way of defining clinical sensitivity is to consider it simply as the percentage of results in samples from patients with biopsy-proven liver damage that are above the upper limit of the normal range. Defined this way, sensitivity in the different studies has varied from 40 to 80% for fasting total serum bile acids. The two extremes are not incompatible. In primary biliary cirrhosis the diagnostic sensitivity of total serum bile acid concentrations may be 90–100% [128,133], whereas in patients with alcoholic liver disease it may be only 40% [126,128,129]. The observations made in the early literature that normal serum bile acids may be found in acute hepatitis and obstructive jaundice [134,135] have also been repeatedly confirmed [128,136].

Attempts to improve the sensitivity of serum bile acid assays have included using measurements following meals, exogenously induced gallbladder contraction, and oral administration of ursodeoxycholic acid (UDCA). With regard to serum bile acids following gallbladder contraction, many authors have found an increased sensitivity [111,112,126,128,137,138], but others have not [119,121,124,131,139–141]. One obvious explanation for these different results is the large variation between individuals in their serum bile acid response to gallbladder contraction. This problem may be overcome by assessing serum bile acid profiles rather than a single serum bile acid determination [142]. Some workers, for example, have used an oral dose of UDCA to standardize the bile acid input into the portal vein, estimated the area of serum bile acid curves, and found that use of the maximum value improved the detection rate [118,143]. This approach has been criticized because of the need for prolonged blood sampling [144].

Most of the recent studies have also compared the value of serum bile acid assays in the detection of liver disease to that of the more "routine" plasma analysis, the so-called liver function tests. Any assessment of these comparisons is made a little difficult by the fact that different groups use different routine tests; not everyone, for example, includes γ-glutamyl transferase. Nevertheless, it is probably fair to say that serum bile acid measurements are usually regarded (but not by everybody; see Refs. 126,128,145) as superior to the routine liver function tests. One thing on which all authors appear to agree is that serum bile acid measurements provide a more sensitive test of hepatobiliary damage than does serum bilirubin, an observation of importance in Gilbert's syndrome, as discussed in Section 3.5.

4.2. The Specificity of Serum Bile Acid Measurements for Liver Disease

Although the "sensitivity" of serum bile acid measurements as an index of liver disease has received much attention, little has been given to the specificity of these assays. It is clear, however, that increased total 3α-hydroxy serum bile acid concentrations are not confined to patients with liver disease. In 1969, Lewis *et al.* described patients with small-bowel bacterial overgrowth and ileal resection who had elevated unconjugated serum bile acid concentrations [93], observations which were recently corroborated [246]. These concentrations were ascribed to the normally less efficient hepatic clearance of unconjugated bile acids from the peripheral circulation and the increased input of unconjugated bile acids into the portal vein in such patients. Although it may not be correct to exclude liver disease without liver biopsy evidence, this early observation did indicate that raised total 3α-hydroxy serum bile acid could not be expected to be a specific indicator for liver disease. Raised concentrations of total serum bile acid concentrations in patients without liver disease have since been observed by many groups of workers [93,125,146]. A preliminary communication has suggested that even raised conjugated serum bile acid concentrations are not confined to liver disease [128]. Elevated concentrations of conjugated cholate and conjugated chenodeoxycholate were found in patients with coeliac disease (again, liver disease cannot be excluded as biopsies were not performed). The issue is further complicated by lack of absolute analytical specificity in the conjugated bile acid assays, although at the concentrations encountered it is unlikely that this is the explanation [56]. A more probable cause of increased serum bile acid concentrations in these patients would be an increased bile acid input into the portal vein. The latter might be the consequence of a larger-than usual bile acid pool [147] associated with an abnormally low bile acid excretion [148,149].

4.3. Elevated Peripheral Bile Acid Concentrations in Liver Cirrhosis

Several groups have investigated the origins of the raised peripheral bile acid concentrations found in liver cirrhosis. It is known that both bile acid synthesis and biliary bile acid output are reduced in this situation [150–152]. Consequently, it would be anticipated that intestinal input of bile acids into the portal vein might be reduced in liver cirrhosis. Hence, if serum bile acids are elevated in patients with cirrhosis, then that elevation might reflect impaired hepatic uptake of bile acids from the portal vein. Pare *et al.* [129] measured total 3α-hydroxy bile acids and also, using radioimmunoassay, glycocholic and glycolithocholate sulfate concentrations in the portal vein, hepatic vein, and peripheral vein of eight patients with alcoholic cirrhosis. All the patients appeared to have reduced liver capacity to transport bile acids, as indicated by the ratio of hepatic vein bile acid to portal vein bile acid concentrations. This ratio did not alter when the portal vein bile acid content was increased by cholecystokinin-induced gallbladder emptying. That is, even these cirrhotic livers were functioning below their maximum capacity to transport bile acids (assuming that the only factor to change with cholecystokinin was the portal vein bile acid load). There was a significant correlation between the portal vein load and the peripheral bile acid concentrations, which indicates that intestinal input is still an important factor in the determination of serum bile acid concentrations of patients with liver cirrhosis. The reduced ratio of portal vein to peripheral vein bile acid concentrations indicates extrahepatic shunting is also a factor in the elevated serum bile acids in cirrhosis. The shunting implies that there is no direct relationship between portal vein extraction and serum bile acids [153,154] and to some extent explains why serum bile acid concentrations often fail to correlate with liver status assessed by other liver function tests [114,118] or histology [115].

Direct measurement of the hepatic extraction of ^{14}C bile acids given as an intravenous bolus, and thus avoiding the influences of extrahepatic portal systemic shunting, has indicated that extraction ratios decrease as the severity of the liver disease (as indicated by the serum albumin) increases [115]. Ohkubo *et al.* [155] used an oral dose of UDCA to study the hepatic transport of bile acids in patients with cirrhosis. They found significant correlations of the bile acid concentrations in the portal vein and superior mesenteric vein with those in the peripheral vein and again stressed the importance of the intestinal input. The UDCA area under the curve was abnormal in all 10 patients studied and reflected intestinal absorption, intra- and extrahepatic shunting, and hepatic extraction. The authors argued that the fractional extraction ratio of UDCA was largely determined by the effective hepatic blood flow rather than reduced liver cell function. This is similar to the view expressed by Marigold *et al.* [94], but contrary to that of others [156,157]. The mechanisms of increased serum bile acid concentrations in liver disease are clearly complex and involve factors not directly related to liver cell function.

4.4. Serum Bile Acid Hydroxylation Patterns in Liver Disease

The synthesis of the primary bile acids, cholic acid and chenodeoxycholic acid, takes place only in the liver, and it has often been suggested that measurement of their serum concentrations would give useful information in hepatobiliary disease. Rudman and Kendall in 1957 [158] observed that the increase in the dihydroxy fraction of the serum bile acids was greater than that in the trihydroxy in patients with cirrhosis. This observation of low ratios of serum trihydroxy/dihydroxy bile acids in patients with cirrhosis has frequently been repeated [134,159]. The application of gas–liquid chromatography to serum bile acid determination showed that low values of the ratio were usually associated with low concentrations of deoxycholic acid [71,160,161].

The mechanism of the reduction of serum deoxycholic acid concentrations in cirrhosis has received much attention. Some workers favor reduced intestinal bacterial activity [162], whereas others argue in favor of a selective impairment in cholic acid synthesis [134,163–165]. Whatever the mechanism, the diagnostic usefulness of the ratio of the serum concentrations of the two primary bile acids is disputed. Carey [159] suggested that the trihydroxy/dihydroxy ratio distinguished those patients with predominantly hepatocellular liver disease (ratio <1) from those whose liver disease was mainly obstructive in nature (ratio >1). Most groups have found low values for this ratio in patients with cirrhosis [119,134,139,166,167], but question the diagnostic reliability. Here again, however, methodological considerations are important. Whereas it is sound to equate the trihydroxy fraction with cholic acid, it may not be valid to equate the dihydroxy fraction with chenodeoxycholic acid unless a gas–liquid chromatography technique demonstrates a low or absent deoxycholic acid. Then again, the necessary hydrolysis of the samples might hide the fact that the Trihydroxy/Dihydroxy ratio in the glycine fraction (presumably because glycine conjugates are more susceptible to bacterial degradation) may be different from that in the taurine fraction [69,70].

5. GAS CHROMATOGRAPHY–MASS SPECTROMETRY AND MONOHYDROXY BILE ACIDS IN LIVER DISEASE

The technique of GC–MS has two distinct advantages over the more classical methods for serum bile acid measurement—definitive identification and sensitivity. For these reasons, GC–MS, more than any other technique, has been used in the search for serum evidence of unusual bile acids in liver disease (see Chapter 5).

However, many authors emphasize the need for caution in the interpretation of results obtained by GC–MS. Its application to serum bile acid determination requires extraction and isolation of the bile acids and preparation of their volatile derivatives. This extensive preparation work may lead to artifact formation. For example, the epimerization of 3α-hydroxy groups to 3β-hydroxy groups may be produced by the action of intestinal flora [168–170] or by strong alkaline hydrolysis of 3α-sulfates [171,172]. Use of strong acids in extraction or methylation may result in the introduction of double bonds into the bile acid ring structure [173,174]. The GC–MS analysis of monohydroxy bile acids presents particular difficulties because, compared with di- and trihydroxy bile acids, monohydroxy bile acids have low solubilities in aqueous solutions [175,176]. Although it may be easy to circumvent this problem by using alkaline alcohol solutions, these in turn may create difficulties with resin extraction procedures [173,177]. In spite of these many difficulties, the use of GC–MS in the field of bile acids has resulted in a huge expansion of our knowledge in both normal and disease states.

In 1967, a new pathway for bile acid synthesis in rat liver was described [178]. This pathway began, rather than ended, with oxidation of the cholesterol side chain and included among its intermediates two unsaturated steroid carboxylic acids, 3β-hydroxy-5-cholenoic acid and 3β-hydroxy-5-cholestenoic acid. Subsequently, the taurine conjugate of 3β-hydroxy-5-cholenoic was shown to be capable of producing cholestasis in animal models [179]. In 1969, Schaffner and Popper [180] hypothesized that cholestasis was associated with the production of monohydroxy bile acids within the liver cell. They argued that one of the consequences of a hypoactive, hypertrophic smooth endoplasmic reticulum would be reduced hydroxylation activity. Thus, the alternate pathway of bile acid synthesis would be favored. This hypothesis was responsible for a massive expansion in the search for monohydroxy bile acids in the peripheral circulation of patients with cholestatic liver disease. The method used was almost invariably GC–MS.

In 1971, Makino *et al.* studied the monohydroxy bile acids found in urine from four patients with extrahepatic biliary atresia [181]. The unusual monohydroxy bile acids found were *allo*-lithocholic and 3β-hydroxy-5-cholenoic acid. The presence of the latter bile acid was taken to be indicative of the alternate pathway and its importance in liver disease. Less experienced hands confirmed these findings in biliary atresia, but also reported the presence of another unsaturated monohydroxy bile acid in pooled sera from patients with primary biliary cirrhosis [182]. The detection of a diunsaturated bile acid, 3β-hydroxychol-5,22 dienoic acid, perhaps emphasizes the danger of accurately defining artifacts by GC–MS. However, the monounsaturated bile acid cholenoic acid has been found in the plasma of patients with liver disease in many studies [183–187,247]. Using capillary GC–MS, 3β-hydroxy-5-cholenoic acid was shown to be the major monohydroxy component in the sera of adult patients with severe intrahepatic cholestasis. Unsaturated C-20, C-22, and C-27 steroid carboxylic acids were also found [188]. The immediate precursor of 3β-hydroxy-5-cholenoic acid in the alternate pathway is cholestenoic acid, a C-27 unsaturated steroid acid.

Not all the unusual bile acids that have been found in plasma from patients with liver disease can be readily incorporated into alternate pathways of hepatic bile acid synthesis. In 1976, 6- and 7α-hydroxy bile acids which had not previously been found in human plasma were identified in cholestatic plasma [186]. This finding emphasized the need for caution in assessing the origin of serum bile acid concentrations.

One obvious question is whether or not the unusual bile acids mentioned above are found in normal sera. There is, however, relatively little information on the bile acid composition of normal sera. This is understandable, in that in normal serum, bile acid concentrations are much lower than those usually found in patients with liver disease. With modern GC–MS techniques claiming high sensitivity, there appears to be little or no agreement on bile acid composition. This, again, is perhaps not too surprising. Few authors have attempted to provide a complete definitive analysis of normal concentrations. As emphasized by Danielson and Sjövall [189], most workers select which bile acid ions to monitor, and not all apply the same selection criteria to both normal samples and those from patients with liver disease. Another explanation that has been proposed for a wide variation in normal values has been differences in the selection of normal subjects [190].

Nevertheless, in spite of these reservations, it is clear that those bile acids which at one time were thought to be specifically associated with cholestatic liver disease may be present in normal sera and in the unconjugated fraction [31]. This raises the issue of whether these monohydroxy bile acids are in any way hepatotoxic in man?

Lithocholic acid has been shown to be an effective hepatotoxic agent in many animal species. Man is exceptional in that the human liver appears to be less susceptible to lithocholate. In contrast to the results of the numerous animal studies, there have been no reports of lithocholate-induced cholestasis in humans, although lithocholate *per se* has been fed to humans [191,192] and its precursor chenodeoxycholic acid has been widely used for the dissolution of human gall-stones. The mechanism by which the human liver detoxifies lithocholate is believed to be sulfation [193]. It has been shown that about 20% of the lith-ocholate formed in the colon is absorbed [194,195], and it might be expected that the majority of this would be sulfated, excreted in bile, and not reabsorbed [28,196–199]. The intestinal input of lithocholate sulfate into the portal vein is therefore low, and one might not expect a postprandial response. A lack of postprandial response has been shown in normal subjects [64,200] and in patients with liver disease [129,201], but postprandial rises were recorded in two patients with intermittent intrahepatic cholestasis [202]. Less than 10% of the total serum bile acids were sulfated in 40 patients with severe cholestasis, whereas 50% of the corresponding urinary bile acids were sulfated [203]. Subsequent to that study it was shown that bile acid sulfation may also occur in the kidney [204]. Not only is there competition between sulfated and nonsulfated bile acids at the intestinal mucosa [28,199] and at the liver plasma membrane [205], but some workers consider that this competition extends to biliary bile acid excretion [206–208]. However, as recently suggested, another explanation for the relatively slight decrease in serum bile acid sulfate concentrations following biliary drainage might be preferential accumulation of the sulfates in the tissues [209]. Thus although bile acid sulfation is increased in liver disease [203,210–212], the degree of increase, as indicated by serum concentrations, is variable [213,214]. The relatively low biliary excretion of sulfates may simply reflect increased synthesis of primary bile acids [209].

Glucuronidation has also been proposed as a defense mechanism against potentially hepatotoxic bile acids, but it appears to be quantitatively less important than sulfation [215–217]. There is correspondingly less data on serum bile acid glucuronides [218–220]. Bile acid glucuronidation may take place, like sulfation, in the liver, kidney, and small-intestinal mucosa [221,222]. Unconjugated bile acids may be the preferred substrates [222]. Some groups have found reduced serum bile acid glucuronides in cirrhosis [220], and others have shown reduced bile acid glucuronyl transferase activity in the cirrhotic liver [222]. Still other workers have observed that where biliary total bile acid excretion is maintained, there are increased urinary bile acid sulfates and glucuronides [223].

Within the last few years, it has been shown that sulfation of lithocholic acid does not necessarily inhibit its cholestatic properties [224,225]. The same has been observed with lithocholate glucuronide [226]. A reduced total bile acid pool also apparently enhances the cholestatic properties of these esters [227].

One hypothesis recently put forward is that the cholestatic potential of a bile acid may be related not simply to its insolubillity in water, but to its ability to bind calcium ions [226]. Other workers have found that sulfation may impair hepatic uptake and thus increase urinary excretion of the more hydrophilic species

[228,229]. These preliminary observations have led to renewed interest in this field and in particular in the etiological role of altered bile acid metabolism in cholestatic liver disease [230].

6. SERUM BILE ACID COMPOSITION IN OTHER CIRCUMSTANCES

The origins of serum bile acid concentrations may be simple, but the results of the many interactions affecting serum bile acid composition are difficult to predict. One would not necessarily therefore expect to find a direct relationship between biliary bile acid composition and peripheral bile acid concentrations. Nevertheless, there are certain circumstances in which serum bile acid composition does appear to reflect that in bile.

6.1. Serum Bile Acids during Bile Acid Feeding

Several groups have proposed that measurement of serum bile acid concentrations would enable one to predict the composition of the biliary bile acid pool [231–234]. The regression equations between the amount of any bile acid (expressed as a percentage of the total bile acids) in the fasting serum and the amount in the corresponding fasting bile have been derived for the common bile acids and used to predict the biliary bile acid composition associated with a particular serum bile acid composition. Thus, it has been possible to obtain information on the biliary bile acid composition during bile acid feeding solely from observations of the serum bile acids [232,234]. However, although there is good agreement between groups of subjects in these studies, agreement between individual pairs of samples is not so good. Again, the impact of different methodologies on the data cannot be ignored and may help explain the different findings in this field. Serum UDCA concentrations (conjugated and unconjugated, radioimmunoassay) have been claimed to be better predictors of compliance with UDCA feeding than serum conjugates of chenodeoxycholic acid (conjugated alone, radioimmunoassasy) were for biliary chenodeoxycholic acid levels [232]. Other workers have found that although there appears to be no significant relationship between fasting serum bile acid (as determined by radioimmunoassay) and primary bile acid pool sizes, the use of "areas under the curve" or of peak serum levels following cholecystokinin infusion did provide a reasonable estimate of primary bile acid pool size [233]. This suggests that serum bile acid composition can be used as an index of biliary bile acid composition in situations where bile acid liver function and intestinal absorption are normal, i.e., when the predominant factor determining serum bile acid composition is input into the portal vein from the intestine.

6.2. Serum Bile Acids during Pregnancy

There is general agreement that serum bile acid composition changes during pregnancy. However, the changes are small, and even the application of similar methods has yielded different results. Early studies found that serum bile acid concentrations remained within normal limits throughout pregnancy [235,236]. However, using capillary gas chromatography, serum bile acids were found to rise, and the predominant bile acid in the last trimester of pregnancy was cholic acid [237]. But using radioimmunoassay, chenodeoxycholic acid was found to show an increase, although small [238]. A detailed study of fasting and 2-hr postprandial concentrations of conjugated cholic and conjugated chenodeoxycholic acids from 14 weeks until 6 weeks postdelivery, using radioimmunoassay, found no significant changes in chenodeoxycholic acid concentrations, but an increase in the fasting concentrations of conjugated cholate, which began at about 34 weeks' gestation and continued through pregnancy [239]. This then raises the question of whether normal pregnancy is associated with mild sub-clinical cholestasis [240]. Early in human pregnancy, the bile acid pool is increased [241]; there is a shift toward taurine rather than glycine conjugation, an increase in the proportion of cholate in bile, and a corresponding decrease in chenodeoxycholate [242]. These changes may be accompanied by reduced Brom-sulphalein clearance [240]. Although the factors responsible for these changes have not been elucidated, alterations of gallbladder function may be involved [243]. In pregnancy complicated by cholestasis, the changes in both biliary bile acids [242] and serum bile acids [235,238,244,245] are much more dramatic. Cholic acid may comprise up to 90% of the total biliary bile acids [242], and there are large increases in both serum cholate and chenodeoxycholate concentrations. Thus, in pregnancy, elevated serum bile acids may precede the appearance of any cholestatic symptoms and be a genuine early sign when compared with the more "routine" liver function tests [244,245].

7. CONCLUSION

The recent methodological advances in this field have resulted in reinterpretation and reassessment of the stages involved in normal bile acid enterohepatic cycling and their impact on serum bile acid concentrations. It has become clear that serum bile acids can help elucidate the dynamics of bile acid metabolism in health and disease. However, any consideration of serum bile acid data must take into account the limitations of the methodology used to produce it.

ACKNOWLEDGMENTS. Thanks are due to Professor Hermon Dowling and all the members of the Gastroenterology Unit, United Medical Schools, for helpful discussion during the preparation of this manuscript and to Lorraine Byrne and Cathy Weeks, who so patiently typed and retyped it.

REFERENCES

1. G. P. Van Berge-Henegouwen and A. F. Hofmann, *Eur. J. Clin. Invest.* **13**, 433 (1983).
2. G. P. Van Berge-Henegouwen, A. T. Ruben, and W. Van der Brock, *Neth. J. Med.* **26**, 309 (1983).
3. W. Berry and J. Reicken, *Semin. Liver Dis.* **3**, 330 (1983).
4. "Bile Acids in Hepatobiliary and Gastrointestinal Disease," International Symposium, IRC Press, Oxford, Washington DC (1984).
5. T. Osuga, Y. Matsuzaki, M. Imawari, K. Mitamura, and J. Shoda, *in* "Advances in Bile Acid and Bile Alcohol Research in Japan" (S. Ikawa, H. Kawasaki, and N. Kasbara, eds.), pp. 305–314, Tokyotanabesyoji, Tokyo (1984).
6. "Serum Bile Acids in Health and Disease," *in* "Workshops in Bile Acid Research" (L. Barbara, R. H. Dowling, A. Hofmann, and E. Roda, eds.), pp. 9–80, MTP Press, Lancaster (1983).
7. R. Y. Poupon, *Gastroenterol.-Clin. Biol.* **5**, 769 (1981).
8. "Editorial," *Lancet* **2**, 1136 (1982).
9. A. Pagani, A. Panzo, and T. Galante, *Minerva Med.* **75**, 99 (1984).
10. G. Paumgartner, G. A. Mannes, and F. Stellaard, *in* "Bile Acids in Gastroenterology," (L. Barbara, R. H. Dowling, A. F. Hofman, and E. Roda, eds.), pp. 217–222, MTP Press, Lancaster (1983).
11. H. O. Wheeler, *Am. J. Med.* **51**, 588 (1971).
12. E. A. Schaffer, P. McOrmond, and H. Duggan, *Gastroenterology* **79**, 899 (1980).
13. O. G. Bjornsson, T. E. Adrian, J. Dawson, R. F. McCloy, G. R. Greenberg, S. R. Bloom, and V. S. Chadwick, *Eur. J. Clin. Invest.* **9**, 293 (1979).
14. H. Y. Mok, K. von Bergmann, and S. M. Grundy, *Gastroenterology* **78**, 1023 (1980).
15. G. P. Van Berge and A. F. Hofmann, *Gastroenterology* **75**, 879 (1978).
16. V. L. W. Go, A. F. Hofmann, and W. H. J. Summerskill, *Gastroenterology* **58**, 321 (1970).
17. T. C. Northfield and A. F. Hofmann, *Gut* **16**, 1 (1975).
18. W. C. Duane and K. C. Hanson, *J. Lab. Clin. Med.* **92**, 859 (1978).
19. K. Von Bergmann, H. Y. Mok, and S. M. Grundy, *Gastroenterology* **71**, 932 (1976).
20. R. Jazrawi, A. Lanzini, S. Mellor, and T. Northfield, Stockholm World Congress of Gastroenterology (1982) (Abstr.).
21. E. Paumgartner and T. Sauerbruch, *in* "Clinics in Gastroenterology," Volume 12 (M. Classen and H. W. Schreiber, eds.), pp. 3–24, Saunders, London (1983).
22. T. L. Peeters, G. Van Trappen, and J. Janssens, *Gastroenterology* **79**, 678 (1980).
23. J. M. Dietschy, *J. Lipid. Res.* **9**, 297 (1968).
24. A. F. Hofmann, N. F. LaRusso, and G. P. Van Berge-Henegouwen, *in* "Bile Acid Metabolism in Health and Disease" (G. Paumgartner and A. Stiehl, eds.), MTP Press, Lancaster (1977).
25. F. A. Wilson, *Am. J. Physiol.*, **241**, G83 (1981).
26. L. Lack and I. M. Weiner, *Am. J. Physiol.* **210**, 1142 (1966).
27. K. W. Heaton and L. Lack, *Am. J. Physiol.* **214**, 485 (1968).
28. T. S. Low-Beer, M. P. Tyor, and L. Lack, *Gastroenterology* **56**, 721 (1969).
29. N. F. LaRusso, N. E. Hoffman, M. C. Korman, and A. F. Hofmann, *Am. J. Dig. Dis.* **23**, 385 (1978).

30. M. Ponz de Leon, G. M. Murphy, and R. H. Dowling, *Gut* **19**, 32 (1978).
31. K. D. R. Setchell, A. M. Lawson, E. J. Blackstock, and G. M. Murphy, *Gut* **23**, 637 (1982).
32. L. R. Schwartz, R. Burr, M. Schwenck, E. Pfaff, and H. Grein, *Eur. J. Biochem.* **55**, 617 (1975).
33. J. Reichen and G. Paumgartner, *Gastroenterology* **68**, 132 (1975).
34. S. Matern and W. Gerok, *Rev. Physiol. Biochem. Pharmacol* **85**, 126 (1979).
35. L. R. Schwartz, M. Schwenck, E. Pfaff, and H. Grein, *Eur. J. Biochem.* **71**, 369 (1976).
36. J. Reichen and G. Paumgartner *Am. J. Physiol.* **231**, 734 (1976).
37. D. C. Ballard and F. R. Simon, *Gastroenterology* **79**, 1002 (1980) (Abstr.).
38. F. R. Simon, E. M. Sutherland, and M. Gonzalez *J. Clin. Invest.* **70**, 401 (1982).
39. I. T. Gilmore and R. P. H. Thompson, *Clin. Sci.* **60**, 65 (1981).
40. G. P. Van Berge Henegouwen and A. F. Hofmann, *Gastroenterology* **73**, 300 (1977).
41. B. Angelin, I. Björkhem, and K. Einarsson, *J. Clin. Invest.* **70**, 724 (1982).
42. A. E. Cowen, M. G. Korman, A. F. Hofmann, and P. J. Thomas, *Gastroenterology* **68**, 1567 (1975).
43. A. Chitranukroh and B. H. Billing, *Clin. Sci.*, **65**, 77 (1983).
44. P. Loria, S. M. Smith, G. M. Murphy, and R. H. Dowling *Clin. Sci.* **62**, 50 pp. (1981) (Abstr.).
45. M. S. Anwer and D. Hegner, *Hoppe-Seyler's Z. Physiol. Chem.* **360**, 515 (1979).
46. A. Roda, G. Capelleri, R. Aldini, E. Roda, and L. Barbara, *J. Lipid Res.* **23**, 490 (1982).
47. G. J. Beckett, P. Armstrong, and I. W. Percy-Robb, *Biochim. Biophys. Acta* **664**, 602 (1981).
48. L. R. Engelking, S. Barnes, C. A. Dasher, D. C. Naffel, and B. I. Hirschowitz, *Clin. Sci.* **57**, 499 (1979).
49. N. E. Hoffman, D. E. Donald, and A. F. Hofmann, *Am. J. Physiol.* **229**, 214 (1975).
50. W. Horak, R. Waldram, I. M. Murray-Lyon, E. Schuster, and R. Williams, *Gastroenterology* **71**, 809 (1976).
51. A. F. Hofmann, G. Molino, M. Milanese, and G. Belforte, *J. Clin. Invest.* **71**, 1003 (1983).
52. M. G. Korman, N. F. LaRusso, N. E. Hoffman, and A. F. Hofmann, *N. Engl. J. Med.* **292**, 1205 (1975).
53. A. Roda, E. Roda, R. Albini, D. Festi, B. Mazella, C. Sarna, and L. Barbara *Clin. Chem.* **23**, 1107 (1977).
54. E. Minder, G. Karlaganis, U. Schmick, P. Vitrins, and G. Paumgartner, *Clin. Chim. Acta* **92**, 177 (1977).
55. S. Matern, R. Krieger, C. Hans, and W. Gerok, *Scand. J. Gastroenterol.* **12**, 641 (1977).
56. D. G. Sampson, G. M. Murphy, L. M. Cross, and D. Cathy, *Anal. Lett.* **12**, 927 (1979).
57. J. W. O. Van der Berg, M. Van Blackenstein, and E. P. Bosman-Jacobs, *Clin Chim Acta* **73**, 277 (1976).
58. N. F. LaRusso, M. G. Korman, N. E. Hoffman, and A. F. Hofmann, *N. Engl. J. Med.* **291**, 689 (1974).
59. S. W. Schalm, W. F. LaRusso, M. G. Korman, A. E. Cowe, N. E. Hoffman, A. Carter, J. Turcok, and A. F. Hofmann, *Clin. Res.* **23**, 396 (1975) (Abstr.).
60. G. Hislop, A. F. Hofmann, and L. T. Schoenfield, *J. Clin. Invest.* **46**, 1070 (1967) (Abstr.).
61. S. W. Schalm, N. F. LaRusso, A. F. Hofmann, N. E. Hoffman, G. P. Van Berge-Henegouwen, and M. G. Korman, *Gut* **19**, 1006 (1978).
62. S. D. Ladas, P. E. T. Isaacs, G. M. Murphy, and G. E. Sladen, *Gut* **25**, 405 (1984).
63. J. N. Baxter, J. S. Grime, M. Critchley, and R. Shields, *Gut*, **26**, 342 (1984).
64. G. W. Hepner and L. M. Demers, *Gastroenterology* **72**, 499 (1977).

65. H. Igimi, and M. C. Carey, *J. Lipid. Res.* **22**, 254 (1981).
66. D. C. Ruppin, M. F. Myszor, G. M. Murphy, and R. H. Dowling, *Clin. Sci.* **63**, 55 pp. (1981) (Abstr.).
67. B. Angelin and I. Björkhem, *Gut* **18**, 606 (1977).
68. K. Linnet, *Gut* **24**, 249 (1983).
69. G. M. Murphy, A. Ross, B. H. Billing, *Gut* **13**, 201 (1972).
70. G. Neale, B. Lewis, K. Weaver, and D. Panveliwalla, *Gut* **12**, 145 (1971).
71. I. Makino, S. Nakagawa, and K. Mashino, *Gastroenterology* **56**, 1033 (1969).
72. A. Bremmelgaard and B. Almé, *Scand. J. Gastroenterology* **15**, 593 (1980).
73. H. Kimura, N. Suzuki, T. Sato, J. Goto, and T. Nambara, *Jpn. J. Clin. Chem.* **8**, 126 (1979).
74. S. Matern, K. E. Titjen, O. Fackler, R. Hinger, R. Herz, and W. Gerok, *in* "Biological Effects of Bile Acids" (G. Paumgartner, A. Stiehl, and W. Gerok, eds.), pp. 109–118, MTT Press, Lancaster (1979).
75. S. Okuyama, N. Kokubun, S. Higashidate, D. Uemura, and Y. Hirata, *Chem. Lett. (Jpn.)* **12**, 1443 (1979).
76. S. M. Smith, M. Myszor, K. D. R. Setchell, and G. M. Murphy *J. Clin. Pathol.*, **36**, 235 (1984).
77. M. F. Myszor, D. C. Ruppin, G. M. Murphy, and R. H. Dowling, *Clin. Sci.* **63**, 56 pp. (1982) (Abstr.).
78. K. D. R. Setchell, A. M. Lawson, E. J. Blackstock, and G. M. Murphy, *Clin. Sci.* **60**, 76 (1981) (Abstr.).
79. F. W. Smith and M. H. Sleisenger, *in* "Gastrointestinal Disease," Volume 2 (M. H. Sleisenger and J. Fordtran, eds.), pp. 1523–1548 Saunders, Philadelphia (1978).
80. A. Hirst and A. Newton, *J. Physiol.* **47**, 54 (1913).
81. E. J. Cohn, *Blood* **3**, 471 (1948).
82. R. M. Burton, and U. Westphal, *Metabolism* **21**, 253 (1972).
83. D. Rudman and F. E. Kendall, *J. Clin. Invest.* **36**, 538 (1957).
84. R. A. Cooper and J. H. Jandl, *J. Clin. Invest.* **47**, 809 (1968).
85. H. O. Green, J. Moritz, and L. Lack, *Biochim. Biophys. Acta* **231**, 550 (1981).
86. B. Catanese, R. Lisuani, and D. Piccmelli, *Biochem. Pharmacol.* **18**, 1707 (1969).
87. C. W. Burke, B. Lewis, D. Panveliwalla, and S. Tabaqchali, *Clin. Chim. Acta* **32**, 207 (1979).
88. A. Roda, E. Roda, C. Sarna, D. Festi, R. Aldini, A. M. Rorselli, G. Mazella, and L. Barbara, *Gastroenterology* **82**, 77 (1982).
89. R. A. Weisiger, J. Gollan, and R. Ockner, *Science* **211**, 1048 (1981).
90. E. L. Forker and B. A. Luxon, *J. Clin. Invest.* **67**, 1517 (1981).
91. A. Roda, G. Cappelleri, R. Aldini, E. Roda, and L. Barbara, *J. Lipid Res.* **23**, 490 (1982).
92. H. Abberger, H. Bickel, H. P. Buscher, K. Fuchte, W. Gerok, W. Kramer, and G. Kurz, *in* "Bile Acids and Bile Lipids." (G. Paumgartner, A. Stiehl, and W. Gerok, eds.), pp. 233–246, MTP Press, Lancaster, (1980).
93. B. Lewis, D. Panveliwalla, S. Tabaqchali, and D. P. Wootton, *Lancet* **1**, 219 (1969).
94. J. H. Marigold, A. J. Bull, I. T. Gilmore, D. T. Coltart, and R. P. H. Thompson, *Clin. Sci.* **63**, 197 (1982).
95. P. A. Siskos, P. J. Calvill, and N. B. Javitt, *J. Lipid Res.* **18**, 666 (1977).
96. F. Mashigo, N. Tanaka, A. Maki, S. Kamei, and M. Yamanaka, *Clin. Chem.* **27**, 1352 (1981).
97. M. Y. Qureshi, S. M. Smith, and G. M. Murphy, *J. Clin. Pathol.* **37**, 317 (1984).
98. E. Dworsky, L. Corneliussen, K. Wold, and M. Krutnes, *in* "Bile Acids in Hepatobiliary and Gastrointestinal Diseases," pp. 25–36, IRL Press, Oxford, Washington, DC (1984).

99. D. H. Steensland, *Scand. J. Clin. Lab. Invest.* **38**, 447 (1978).

100. A. Roda, S. Girotti, S. Ghini, B. Grigolo, G. Correa, and R. Bovara, *Clin. Chem.* **30**, 206 (1984).

101. F. Mashige, T. Osuga, W. Tanake, K. Imai, and M. Yamanaka, *Clin. Chem.* **24**, 1150 (1978).

102. D. Clements, S. Smith, and E. Elias, *in* "Bile Acids in Hepatobiliary and Gastronintestinal Diseases," pp. 141–144, IRL Press, Oxford, (1984).

103. J. Schoelmerick, G. P. Van-Berge Henegouwen, A. F. Hofmann, and M. De Luca, *Clin. Chim. Acta* **137**, 21 (1984).

104. A. Roda, L. J. Kricka, M. De Luca, and A. F. Hofmann, *J. Lipid Res.* **23**, 1354 (1982).

105. S. Kamada, M. Maeda, and A. Tsuji, *J. Chromatog.* **239**, 773 (1982).

106. R. Engert and M. D. Turner, *Anal. Biochem.* **51**, 399 (1973).

107. M. G. Bolt and J. L. Boyer, *in* "Bile Acid Metabolism in Health and Disease" (G. Paumgartner and A. Stiehl, eds.), pp. 285–292, MTP Press, Lancaster, (1977).

108. A. Bruusgaard, H. Sorensen, O. T. Gilhuus-Moe, and B. A. Skalhegg, *Clin. Chim. Acta* **77**, 387 (1977).

109. P. Hansson, S-A. Hedstorm, B. Hultborg, and M. Olin, *Clin. Chim. Acta* **125**, 241 (1982).

110. G. M. Murphy, *in* "Bile Acids in Hepatobiliary and Gastrointestinal Diseases," pp. 45–49, IRL Press, Oxford, Washington, (1984).

111. N. Kaplowitz, E. Kok, and N. B. Javitt, *JAMA* **225**, 292 (1973).

112. S. Barnes, G. A. Gallo, D. B. Trash, and J. S. Morris, *J. Clin. Pathol.* **28**, 506 (1975).

113. N. F. LaRusso, N. E. Hoffman, A. F. Hofmann, and M. G. Korman, *N. Engl. J. Med.* **292**, 1202 (1975).

114. C. Hirayamu, T. Irisa, K. Arimura, and M. Nakamura, *Acta Hepato-Gastroenterol.* **23**, 385 (1976).

115. H. J. Milstein, J. R. Bloomer, and G. Klatskin, *Am. J. Dig. Dis.* **21**, 281 (1976).

116. M. Angelico, A. F. Attili, and L. Capocaccia, *Dig. Dis.* **22**, 941 (1977).

117. N. B. Javitt, *in* "Clinics in Gastroenterology 6." (G. Paumgartner, ed.), pp. 219–226, Saunders, London, Philadelphia, Toronto, (1977).

118. T. Osuga, K. Mitamura, F. Mashige, K. Imai, *Clin. Chem. Acta* **75**, 81 (1977).

119. C. R. Pennington, P. E. Ross, and I. A. D. Bouchier, *Gut* **18**, 903 (1977).

120. S. Matern and W. Gerok, *Acta Hepato-Gastroenterol* **26**, 185 (1979).

121. S. Skrede, H. E. Solberg, J. P. Blomhoff, and E. Gjone *Clin. Chem.* **24**, 1095 (1978).

122. A. Stiehl, E. Ast, and P. Czygan, *Inn. Med.* **5**, 14 (1978).

123. K. Kobayashi, R. M. Allen, J. R. Bloomer, and G. Klatskin, *JAMA* **241**, 2043 (1979).

124. P. S. Monroe, A. Z. Baker, J. F. Schneider, P. S. Krager, P. D. Klein, and D. Schoeller, *Hepatology* **2**, 317 (1982).

125. K. Samuelson, A. A. Johansson, and A. Norman, *Scand. J. Gastroenterol.* **16**, 225 (1981).

126. R. Alm, J. Carlson, and S. Eriksson, *Scand. J. Gastroenterol.* **17**, 213 (1982).

127. H. Rickers, E. Christensen, T. Arnfred, U. Dige, and E. Hess Thaysen, *Scand. J. Gastroenterol.* **17**, 565 (1982).

128. P. E. T. Isaacs, G. M. Murphy, L. M. Matthews, D. Sampson, and R. H. Dowling, *Clin. Sci.* **56**, 28 pp. (1979) (Abstr.).

129. P. Pare, J. C. Hoeffs, and M. Ashcavai, *Gastroenterology* **81**, 959 (1981).

130. R. Ferraris, G. Colombatti, M. T. Fiorentini, R. Carosso, R. Grossa, and M. De La Pierre, *Dig. Dis. Sci.* **28**, 127 (1983).

131. P. Tobiasson and B. Boeryd, *Scand. J. Gastroenterol.* **15**, 657 (1980).

132. J. G. Douglas, G. T. Beckett, I. A. Nimmo, N. D. Finlayson, and I. W. Percy-Robb, *Gut* **22**, 141 (1981).

133. R. Raedsch, B. A. Lanterburg, and A. F. Hofmann, *Dig. Dis. Sci.* **26**, 394 (1981).
134. S. Sherlock and V. Walshe, *Clin. Sci.* **6**, 223 (1948).
135. M. Rautureau, B. Chevrel, and J. Cavoli *Rev. Med. Chir. Mal Foil.* **42**, 167 (1967).
136. B. T. Starkey and V. Marks, *Clin. Chim. Acta* **119**, 165 (1982).
137. R. Testa, E. Betti, F. Dillepiare, M. Randazzo, A. Piccioto, G. Mausi, E. Scopinaro, V. Savarino, M. Sanguineti, and G. Celle, *Pathologica* **74**, 201 (1982).
138. G. A. Mannes, F. Stellaard, and G. Paumgartner, *Digestion* **25**, 217 (1982).
139. K. Linnet and J. Rye Anderson, *Clin. Chim. Acta* **127**, 217 (1983).
140. D. Festi, M. A. Morselli Labate, A. Roda, F. Bazzoli, R. Frabboni, P. Rucci, F. Taroni, R. Aldini, E. Roda, and L. Barbara, *Hepatology* **3**, 707 (1983).
141. M. Van Blankenstein, M. Frenkel, J. W. van der Berg, F. J. Ten-Kate, E. A. Bosman-Jacobs, and A. L. Toun-Blommestein, *Dig. Dis. Sci.* **28**, 137 (1983).
142. H. Pohlablen, H. Losgen, G. Brunner, E. Schmidt, and F. W. Schmidt, *in* "Bile Acids in Hepatobiliary and Gastrointestinal Disease," pp. 121–125, IRL Press, Oxford, (1984).
143. T. Osuga, Y. Matsuzaki, M. Imawari, K. Mitamura, and J. Shoda, *in* "Advances in Bile Acids and Bile Alcohol Research in Japan" (S. Ikawa, H. Kawasaki, and N. Kaibara, eds.), pp. 305–314, Tokyotanabestjoji, Tokyo, (1984).
144. D. Festi, *in* "Workshops in Bile Acid Research" (L. Barbara, R. H. Dowling, A. Hofmann, and E. Roda, eds.), pp. 56–57, MTP Press, Lancaster, (1983).
145. A. Chitranukroh, J. R. Heady, S. Rosalla, J. A. Summerfield, and B. H. Billing, *in* "Bile Acids in Hepatobiliary and Gastrointestinal Diseases," pp. 121–125, IRL Press, Oxford (1984).
146. K. Samuelsson, A. Ally, A. Johnsson, and A. Norman, *Scand. J. Lab. Invest* **39**, 511 (1979).
147. K. W. Heaton *in* "Clinics in Gastroenterology," (G. Paumgartner, ed.), pp. 69–89, Saunders, London, Philadelphia, Toronto, (1977).
148. T. A. Miettinen, *Lancet* **2**, 358 (1968).
149. R. Gilberg and H. Andersson, *Scand. J. Gastroenterol.* **12**(Suppl **45**), 24 (1977) (Abstr.).
150. Z. R. Vlachevic, M. F. Prugh, D. H. Gregory, and L. Swell, *Clin. Gastroenterol.* **6**, 25 (1977).
151. A. Stiehl, E. Ast, P. Cygan, W. Frohling, R. Paedsch, and D. Kommerill, *Gastroenterology* **74**, 572 (1978).
152. L. A. Turnberg and G. Grahame, *Gut* **11**, 126 (1979).
153. P. M. Huot, and D. Maclean, *Gastroenterology* **76**, 1285 (1979).
154. L. Linblad, K. Lundholm, and T. Schersten, *Scand. J. Gastroenterol.* **12**, 395 (1977).
155. H. Ohkubo, K. Okuda, S. Iida, K. Ohnishi, S. Ikawa, and I. Makino, *Gastroenterology* **86**, 514 (1984).
156. R. Y. Poupon, R. E. Poupon, K. Lebrec, L. L. Qnernec, and F. Darnis, *Gastroenterology* **80**, 1438 (1981).
157. Y. Poupon, *Gastroenterology* **88**, 221 (1985) (Letter).
158. D. Rudman and F. E. Kendall, *J. Clin. Invest.* **37**, 1494 (1958).
160. M. A. Eastwood, W. L. Mowbray, R. P. H. Thompson, and R. Williams, *Br. J. Nutr.* **24**, 1029 (1970).
161. D. H. Sandberg, J. Sjövall, K. Sjövall, and D. A. Turner, *J. Lipid Res.* **6**, 182 (1965).
162. R. C. Knodell, M. D. Kinsey, E. C. Boedeker, and D. P. Collin, *Gastroenterology* **71**, 196 (1976).
163. W. C. McCormick, III, C. C. Bell, Jr., L. Swell, and Z. R. Vlachevic, *Gut* **14**, 895 (1973).
164. B. Angelin, K. Einarsson, and K. Hellström, *Dig. Dis. Sci.* **23**, 1115 (1978).
165. Z. R. Vlachevic, M. Goldman, O. C. Shwark, J. Gustafsson, and L. Swell, *Hepatology* **1**, 146 (1981).
166. J. R. Bloomer, R. M. Allen, and G. Klatskin, *Arch. Intern. Med.* **126**, 57 (1976).

167. A. Bremmelgaard and B. Almé, *Scand. J. Gastroenterol.* **15**, 573 (1980).

168. P. Eneroth, R. Gordon, R. Ryhage, and J. Sjövall, *J. Lipid Res.* **7**, 511 (1966).

169. B. Almé, A. Bremmelgaard, J. Sjövall, and P. Thomassen, *J. Lipid Res.* **18**, 339 (1977).

170. S. Hirano, N. Masuda, and H. Oda, *J. Lipis Res.* **22**, 735 (1981).

171. R. H. Palmer and M. G. Bolt, *J. Lipid Res.* **12**, 671 (1971).

172. M. Maeda, H. Ohama, H. Takeda, M. Tabe, M. Mantu, and T. Namishisa, *J. Lipid Res.* **25**, 14 (1984).

173. T. Beppu, Y. Seyama, T. Kasama, and T. Yamaka, *J. Biochem.* **89**, 1963 (1981).

174. T. Harano, C. Fujita, K. Harano, and K. Yamasaki, *Steroids* **30**, 393 (1977).

175. P. M. Small and W. H. Admirand, *Nature* **221**, 265 (1969).

176. M. C. Carey, S. J. Wu, and J. B. Watkins, *Biochim. Biophys. Acta* **575**, 16 (1979).

177. J. Yanagisawa, H. Ichimiya, M. Nagai, and F. Nakayama, *J. Lipid Res.* **25**, 750 (1984).

178. K. A. Mitropoulos and N. B. Myant, *Biochem. J.* **103**, 472 (1967).

179. N. B. Javitt and S. Enerman, *J. Clin. Invest.* **47**, 1002 (1968).

180. F. Schaffner and H. Popper, *Lancet* **2**, 355 (1969).

181. I. Makino, J. Sjövall, A. Norman, and B. Strandvik, *FEBS Lett.* **15**, 161 (1971).

182. G. M. Murphy, F. H. Jansen, and B. H. Billing, *Biochem. J.* **129**, 491 (1972).

183. G. Délèze and G. Paumgartner, *Helv. Paediatr. Acta* **32**, 29 (1977).

184. P. Back, J. Sjövall, and K. Sjövall, *Med. Biol.* **52**, 31 (1974).

185. L. G. Linarelli, C. N. Williams, and M. J. Phillips, *J. Paediatrics* **81**, 484 (1972).

186. J. A. Summerfield, B. H. Billing, and C. H. L. Shackleton, *Biochem. J.* **154**, 507 (1976).

187. E. I. Minder, G. Karlaganis, and G. Paumgartner, *J. Lipid Res.* **20**, 986 (1979).

188. R. Lester, J. ST, Pyrek, and J. M. Little, *Gastroenterology* **82**, 1113 (1982) (Abstr.).

189. H. Danielsson and J. Sjövall, *Annu. Rev. Biochem.* **44**, 233 (1975).

190. I. Björkhem and O. Falk, *Scand. J. Lab. Invest.* **43**, 163 (1983).

191. G. A. Bray, T. F. Gallagher, *J. Clin. Invest.* **47**, 11a (1968).

192. H. Gerds, in "Bile Acids in Human Disease" (P. Back and W. Gerok, eds.), pp. 203–207, F. K. Schattauer Verlag, Stuttgart, New York (1982).

193. R. H. Palmer, *Proc. Natl. Acad. Sci.* **58**, 1047 (1967).

194. R. N. Allan, J. L. Thistle, A. F. Hofmann, J. A. Carter, and P. Yu, *Gut* **17**, 405 (1976).

195. R. N. Allan, J. L. Thistle, and A. F. Hofmann, *Gut* **17**, 413 (1976).

196. A. E. Cowen, M. G. Korman, A. F. Hofmann, O. W. Cass, *Gastroenterology* **69**, 59 (1975).

197. A. E. Cowen, M. G. Korman, A. F. Hofmann, O. W. Cass, and S. B. Coffin, *Gastroenterology* **69**, 67 (1975).

198. A. E. Cowan, M. G. Korman, A. F. Hofmann, and P. J. Thomas, *Gastroenterology* **69**, 77 (1975).

199. E. H. Dewitt and L. Lack, *Am. J. Physiol.* **238**, 34 (1980).

200. H. J. Wildgrube, H. Stang, D. Schiller, M. Winckler, J. Weber, H. Campana, and G. Mauritz, *J. Clin. Chem. Clin. Pathol.* **20**, 319 (1982).

201. C. B. Campbell, C. McGuffie, L. W. Powell, R. K. Roberts, and A. W. Stewart, *Am. J. Dig. Dis.* **23**, 509 (1978).

202. R. J. Vonk, C. M. A. Bijleveld, F. Kuipers, C. M. F. Kneepkens, J. Fernandez, and A. van Zauten, *Hepatology* **2**, 729 (1982) (Abstr.).

203. A. Stiehl, *Eur. J. Clin. Invest.* **4**, 59 (1974).

204. J. A. Summerfield, J. L. Gollan, and B. H. Billing, *Biochem. J.* **156**, 339 (1976).

205. T. C. Bartholomew and B. H. Billing, *Biochem. Biophys. Acta* **754**, 101 (1983).

206. A. Eklund, A. Norlander, and A. Norman, *Eur. J. Clin. Invest.* **10**, 349 (1980).

207. J. S. Dooley, C. Bartholomew, J. A. Summerfield, and B. H. Billing, *Clin. Sci.* **67**, 61 (1984).

208. D. P. Cleland, T. C. Bartholomew, and B. H. Billing, *Hepatology* **4**, 477 (1984).

209. H. Takikawa, T. Beppu, and Y. Seyama, *Gut* **26**, 38 (1984).
210. A. Stiehl, D. L. Earnest, and W. H. Admirand, *Gastroenterology* **68**, 534 (1975).
211. G. P. van Berge Henegouwen, K. H. Brandt, H. Eyssen, and G. Parmentier, *Gut* **17**, 861 (1976).
213. I. Makino, H. Hashimoto, K. Shinozaki, K. Yoshino, and S. Nagawa, *Gastroenterology* **68**, 545 (1975).
214. J. A. Summerfield, J. Cullen, S. Barnes, and B. H. Billing, *Clin. Sci. Mol. Med.* **52**, 51 (1977).
215. P. Back, K. Spaczynski, and W. Gerok, *Hoppe-Seyler's Z. Physiol. Chem.* **355**, 749 (1974).
216. W. Frohling and A. Stiehl, *in* "Advances in Bile Acid Research" (S. Maton, J. Hackenschmidt, P. Back, and W. Gerok, eds.), pp. 153–156, F. K. Schattauer Verlag, Stuttgart, New York (1974).
217. L. Capocaccia, M. Angelico, A. Attili, D. Alvaro, A. DeSantis, and M. Marin, *in* "Bile Acids in Gastroenterology" (L. Barbara, R. H. Dowling, A. F. Hofmann, and E. Roda, eds), pp. 209–216, MTP Press, Lancaster (1983).
218. P. Back, *Hoppe-Seyler's Z. Physiol. Chem.* **357**, 213 (1976).
219. A. Stiehl, M. Becker, P. Czygan, W. Frohling, B. Kommerell, *Eur. J. Clin. Invest.* **10**, 307 (1980).
220. H. Takikawa, H. Otsuko, T. Beppu, Y. Seymama, and T. Yamakawa, *Digestion* **27**, 189 (1983).
221. H. Matern, S. Matern, C. Schelzig, and W. Gerok, *FEBS Lett.* **118**, 251 (1980).
222. S. Matern, H. Matern, E. W. Parthmann, and W. Gerok, *J. Clin. Invest.* **74**, 402 (1984).
223. R. Raedsch, A. Stiehl, S. Walker, P. Czygan, and B. Kammerell, *in* "Bile Acids and Cholesterol in Health and Disease" (G. Paumgartner, A. Stiehl, and W. Gerok, eds.), pp. 107–110, MTP Press, Lancaster (1982).
224. N. P. Dorvil, I. M. Yousef, B. Tuchweber, and C. C. Roy, *Am. J. Clin. Nutr.* **37**, 221 (1983).
225. I. M. Yousef, B. Tuchweber, R. J. Vonk, D. Masse, M. Ander, and C. C. Roy, *Gastroenterology* **80**, 233 (1981).
226. D. G. Oelberg, V. Chari, M. Little, E. W. Adcock, and R. Lester, *J. Clin. Invest.* **73**, 1507 (1984).
227. F. Kuipers, R. Havinga, and R. J. Vonk, *Clin. Sci.* **68**, 127 (1985).
228. M. A. O'Donovan, C. B. Hartman, M. C. Carey, and J. L. Gollan, *Gastroenterology* **86**, 1198 (1984) (Abstr.).
229. M. A. O'Donovan, J. L. Gollan, M. C. Carey, A. M. Lawson, and K. D. R. Setchell, *Gastroenterology* **88**, 1683 (1985) (Abstr.).
230. S. Erlinger, *Gastroenterology* **87**, 1396 (1984) (Comment).
231. M. J. Whiting, and J. McK. Watts, *Gastroenterology* **78**, 220 (1980).
232. F. Bazzoli, H. Fromm, A. Roda, A. N. Kishore, E. Boda, L. Barbara, and P. Amin, *Dig. Dis. Sci.* **30**, 650 (1985).
233. S. P. J. Van der Werf, G. P. van Berge Henegouwen, and W. van der Brock, *J. Lipid Res.* **26**, 168 (1985).
234. M. J. Whiting, R. H. L. Down, J. McK. Watts, *Ann. Clin, Biochem,* **20**, 336 (1984).
235. K. Sjövall and J. Sjövall, *Clin. Chim. Acta* **13**, 207 (1966).
236. K. Samuelsson and P. A. Thomasson, *Acta Obstet-Gynaecol. Scand.* **59**, 417 (1980).
237. T. Laatkainen and A. Hesso, *Clin. Chim. Acta* **64**, 63 (1975).
238. T. Heikkinen, O. Maentausta, P. Ylostalo, and O. Janne, *Br. J. Obstet. Gynaecol.* **83**, 240 (1981).
239. I. C. Julton, J. G. Douglas, D. J. R. Hutchon, and G. J. Beckett, *Clin. Chim. Acta* **130**, 171 (1983).
240. M. J. Simcock and F. M. W. Forster, *Med. J. Anesth.* **25**, 917 (1967).

241. F. Kern, G. T. Everson, B. De Mark, C. McKinley, R. Showalter, W. Erfling, D. Z. Braverman, P. Szczepanik-van Leeuwen, and P. D. Klein, *J. Clin. Invest.* **68,** 1229 (1981).

242. T. Laatikainen, P. Lehtonen, and A. Hesse, *Clin. Chim. Acta* **85,** 145 (1978).

243. G. Radberg and J. Svanik, *Gut* **27,** 10 (1986).

244. T. Laatikainen and E. Ikone, *Obstet. Gynaecol.* **50,** 313 (1977).

245. J. Heikkinen, O. Maentausta, P. Ylostalo, and O. Janne, *Eur. J. Obstet. Gynaecol. Reprod. Biol.* **14,** 153 (1982).

246. K. D. R. Setchell, D. L. Harrison, J. M. Gilbert, and G. M. Murphy, *Clin. Chim. Acta* **152,** 297 (1985).

247. T. C. Bartholomew, J. A. Summerfield, B. H. Billing, A. M. Lawson, and K. D. R. Setchell, *Clin. Sci.* **63,** 65 (1982).

Chapter 11

URINARY BILE ACIDS

Peter Back

1. INTRODUCTION

Bile acids do not occur to a significant extent in the urine in health. Methods to prove their existence in the urine of healthy subjects were unavilable until the late 1960s. In principle, this picture of a quantitatively negligible renal excretion is still true. However, methods have been developed to determine the tiny amounts present even in health, and an enormous complexity of the bile acid spectrum has been disclosed.

Since collection of urine is a harmless noninvasive procedure, any safe information to be obtained from the detailed analysis of urine may be of practical clinical importance. So far, the spectrum of the urinary bile acids has been recognized to be more detailed and complicated than the spectrum of the biliary and the serum bile acids. Though it is not yet fully known to which extent the metabolism in different organs contributes to this complexity, it is assumed that compensatory enzymatic activities mainly of the liver are responsible for the metabolic diversity found. These compensatory enzymatic activities are seemingly latent, widely overshadowed in quantity by the normal metabolic activities of the liver that are engaged in providing an amphipathic bile for digestive purposes.

However, these latent activities are most important under two circumstances: during the prenatal development period and in hepatobiliary disease. Teleologically, the synthesis of hydrophilic catabolites is required under both these conditions. This review therefore deals mainly with the progress made in the knowledge of bile acid derivatives that are able to pass over into the watery phase.

Peter Back   *Hospital of the University of Freiburg, Freiburg im Breisgau, West Germany*

405

2. METHODOLOGY

2.1. Liquid–Liquid Extraction

The extraction of urine acidified to pH 1 with n-butanol yields bile acid conjugates quantitatively [1]. However, the evaporation of n-butanol is time consuming, emulsions may form, and the residues obtained contain many contaminants. The separation of unconjugated bile acids and glycine conjugates from taurine conjugates may be achieved by first extracting the acidified urine with ethyl acetate and thereafter with n-butanol [2]. Extraction by means of quaternary ammonium added to organic solvent was studied using tetraheptyl-ammonium chloride [3], and the extraction of taurine-conjugated bile acids was achieved by ion-pair formation with tetrabutyl- and tetrapentylammonium [4].

Countercurrent extraction was developed very early [5], but was not suitable for routine analyses. Generally, liquid–liquid extractions lead to extracts that contain numerous other lipids and compounds of similar polarity. Further purification steps are required, as described earlier [6] and reviewed by Sjövall and Setchell in this volume, Chapter 1.

2.2 Liquid–Solid Extraction

The introduction of the neutral cross-linked polystyrene polymer Amberlite XAD-2 into extraction of steroid conjugates from urine [7] was soon appreciated as suitable for the extraction of bile acids from biological fluids [8]. Recoveries from plasma or serum were shown to be quantitative [8], and the detailed precautions to be observed with respect to pH and flow rate in the extraction of bile acids from urine of healthy persons were thoroughly evaluated by Almé et $al.$ [9]. Taking into account pitfalls in processing the urine, which may contain precipitates at neutral or acid pH leading to losses of monohydroxy bile acids [10], the Amberlite XAD-2 extraction method is the most carefully studied extraction method up to now. The batch procedure using Amberlite XAD-7 [11] has not been studied to the same extent. In some cases, a higher capacity may be required than is available with these adsorbents. However, 1 μmole of steroid per 1 g of resin can be reproducibly extracted [7,12].

More recently, methods using reverse-phase octadecylsilane-bonded silica have further simplified extractions of bile acids from aqueous medium. Octadecylsilane-bounded silica cartridges behave as polar adsorbents and therefore exclude nonpolar lipids. Sep-Pak cartridges extract 10 μmole of steroid per cartridge from 100 ml of urine [13]. Several reports deal with elution conditions using either Sep-Pak or Bond Elut cartridges [14–16]. The capacity to extract glycocholic and glycochenodeoxy-cholic acids from urine with Sep-Pak was found to be high and quantitative up to 30 μmole of bile acid per sample (20 ml of urine) [17].

Ion-pair liquid–gel extraction of bile acids using hydroxyalkylated Sephadex-LH 20 containing 10% (w/w) of C_{12}–C_{14} hydroxyalkyl chains (Lipidex 1000) has been demonstrated as alternative to liquid–solid extraction procedures [18,19]. Addition of solid decyltrimethylammonium bromide to urine up to a final concentration of 0.03 M resulted in a 96–99% extraction. This method has not yet attained wider application. However, it has the advantage of having still higher capacity than the aforementioned liquid–solid extraction methods. The columns may be reused for long periods. The amounts of solids extracted from urine by Lipidex 1000 and Amberlite XAD-2 seem to be comparable, being in the range of 1–2 mg/ml of urine [19].

In summary, different techniques have been developed guaranteeing prepurification of bile acids from urine with negligible losses up to high concentrations. In some of these techniques the remaining problem is quantitative extraction of minute amounts. Therefore, it seems advisable to adjust adsorbent bed sizes and amounts according to the individual analytical problem. Further enrichment of bile acids out of the desorbed solids may be achieved by extraction with 85% ethanol [9].

2.3. Separation into Subclasses

Until now, several subclasses of bile acids have been discriminated: neutral compounds; unconjugated, "free" bile acids; glycine conjugates; taurine conjugates; monosulfated glycine conjugates; monosulfated taurine conjugates; glucuronides of unconjugated bile acids; glucuronides of conjugated bile acids.

The reported occurrence of bile acids as neutral derivatives [9] may be artifactual, since cation-exchange chromatography on Amberlyst A-15 in H^+ form, in order to convert the salts to the protonated forms [9], may lead to esterification [17]. Conjugation with other amino acids such as ornithine [20], arginine [21], and ciliatine [22] has been reported; amino acids other than glycine and taurine have, however, not been found by investigation of the amino acid composition of the urinary bile acid fractions [17].

Di- and trisulfates of bile acids, proposed to occur in cholestasis [23], have not been found in subsequent studies of urine samples of healthy persons [9] and patients suffering from extrahepatic cholestasis [17]. The investigation of the nature of urinary bile acid conjugates performed by Hedenborg and Norman [17] also excluded the possibility of the simultaneous occurrence of sulfate esters and glucuronides of cholic and chenodeoxycholic acid conjugates.

2.3.1. Unconjugated Bile Acids

Unconjugated bile acids are eluted as sodium salts together with conjugated bile acids on Sephadex LH-20 6 g [12] and 4g columns [24], respectively, using chloroform/methanol 1 : 1 (v/v), containing 0.01 M NaCl. Free bile acids can be separated from conjugates subsequently by thin-layer chromatography (e.g., Ref. 25), which, however, is not suitable for studies involving quantitative determinations. More preferable is anion-exchange chromatography on lipophilic diethylaminohydroxypropyl Sephadex LH-20 (DEAP-Sephadex LH-20) [9] after previous conversion of the salts to the acid forms. Almé et al. have demonstrated the presence of unconjugated bile acids in the urine even in healthy subjects; cholic acid predominated in this fraction. Irregularly encountered were norcholic acid, 1,3,12-trihydroxy-5β-cholanoic acid, 3α,6β,12α-trihydroxy-5β-cholanoic acid, deoxycholic acid, and hyocholic acid. Rarely, and in low amounts, were encountered 3α,6α,12α-trihydroxy-5-β-cholanoic acid [26], formerly tentatively designated as 3β,7β,12α-trihydroxy-5α-cholanoic acid [9], and 3β,7β,12α-trihydroxy-5β-cholanoic acid. It can be concluded that with the exception of deoxycholic acid, mainly trihydroxylated bile acids occur as unconjugated bile acids in normal urine.

In cholestasis unconjugated bile acids may make up a large proportion of the total urinary bile acids excreted. Under these conditions tetrahydroxylated bile acids and chenodeoxycholic acid may contribute to this fraction (9). This is true also in pregnancy [27].

In healthy persons ($n = 6$) the excretion of unconjugated bile acids was found to be between 0 and 10% of total bile acids excreted (glucuronides of bile acids not included); in instances of elevated total bile acid excretion ($n = 5$) 0–4.9% [9]. When glucuronides were considered, unconjugated bile acids were found in three healthy subjects in the range of 0.2–6.5% [26].

2.3.2. Glycine-Conjugated Bile Acids

Glycine-conjugated bile acids have been determined in most studies together with unconjugated bile acids and taurine-conjugated bile acids [28–32]. Most of these studies dealt with the problem of the occurrence of sulfated bile acids, and the techniques applied did not discriminate between the different nonsulfated conjugate classes. From a clinical point of view it may not seem urgent to subdivide the class of nonsulfated bile acids further. However, it is in principle important to know the contribution of individual components, when one tries to calculate renal clearances (Section 5).

The separation of sodium salts of glycine conjugates and taurine conjugates is feasible by chromatography on Sephadex LH-20 4g columns [9], if small fractions of the eluate with the solvent mixture chloroform/methanol containing 0.02 M NaCl 60 : 40 (v/v) are collected. This procedure, however, may be time consuming. The separation of glycine from taurine conjugates and sulfates is well effected by anion-exchange chromatography on diethylaminohydroxypropyl Sephadex LH-20 [9]. Glycine conjugates will elute with 16.5 ml/meq DEAP

LH-20 of 0.3 M ammonium acetate buffer of apparent pH 5.0 in 72% ethanol from a column containing approximately 1 meq of the exchanger. This procedure has been reproduced severalfold [17,33]. Small amounts of bile acid glucuronides were stated to occur in this fraction [17]. Since glucuronides are stable in alkaline hydrolysis, they will not interfere with the ultimate analysis of the amounts of bile acids in the glycine-conjugated fraction.

In the urine of healthy subjects, cholic acid, 1,3,12-trihydroxy-5β-cholanoic acid, and 3α,12β-dihydroxy-5β-cholanoic acid were found as regular components of this fraction [9]. Inconsistently encountered were allo-cholic acid, 3β,7β,12α-trihydroxy-5β-cholanoic acid, 3α,7β,12α-trihydroxy-5β-cholanoic acid, and deoxycholic acid. In two subjects traces of hyocholic acid were detected, and in only one person chenodeoxycholic acid.

The percentages found in healthy subjects were up to 50% of the total (glucuronides not included) [9]. When glucuronides were considered [26], up to 27.9% were found as glycine conjugates ($n = 3$). Particularly noteworthy is the fact that glycine conjugates showed lowest relative percentages in persons who otherwise excreted relatively high amounts of sulfates or glucuronides [9,26].

In instances of elevated bile acid excretion into the urine, glycine conjugates were found to contain in some cases tetrahydroxylated bile acids, hyocholic acid, deoxycholic acid, and chenodeoxycholic acid [9]. The relative percentage of glycine conjugates ranged between 5 and 12% of total bile acids excreted under pathological conditions of liver disease. Within the group of less polar bile acids excreted with the urine, the glycine conjugates constitute the main subclass.

2.3.3. Taurine-Conjugated Bile Acids

Whereas glycine conjugates are separable from taurine conjugates by Sephadex LH-20 chromatography [9], interference occurs between taurine conjugates and monosulfates of glycine-conjugated bile acids in this system [9]. Separation by anion-exchange chromatography was found to be the method of choice, since with a volume of 11.0 ml/meq DEAP Sephadex LH-20 of ammonium acetate buffer (0.15 M, apparent pH 6.6) in 72% ethanol, taurine conjugates could be eluted with no overlap to monosulfated bile acids [9].

The taurine-conjugated fraction of urinary bile acids is small in healthy subjects. A mean value of 2.8% can be calculated from the data of Almé et al. [9], glucuronides not included. Some individuals excrete only one or two bile acids within this fraction. Most often noted were hyocholic and deoxycholic acid. Rarely present were cholic acid, 3α,6α,12α-trihydroxy-5β-cholanoic acid and 3α,12β-dihydroxy-5β-cholanoic acid. Only in one instance was chenodeoxycholic acid found.

Under pathological conditions of cholestasis the relative proportion of the taurine-conjugated fraction is increased [9]; a mean value of 12.3% of total bile acids excreted can be calculated. Under these circumstances tetrahydroxy bile acids and 1,3,12-trihydroxy-5β-cholanoic acid may also be found in this fraction [9]. The occurrence of tetrahydroxylated bile acids and of trihydroxylated compounds carrying hydroxyl groups at C_1 or C_6 was noted as a particular feature of the taurine-conjugated fraction in liver disease [34].

2.3.4. Sulfate Esters of Bile Acids

Although solvolysis of steroid sulfates had been shown to be a reliable cleavage procedure leaving intact the stereochemistry of the functional group of steroids [35], this sulfate ester cleavage procedure was not applied in studies on urinary bile acids until 1971, when it was first routinely used in a study on urinary bile acids excreted in extrahepatic biliary atresia [36]. Under this condition, when infants lack an efficient enterohepatic circulation of bile acids, large amounts of conjugated tri-, di-, and monohydroxy bile acids are excreted in the urine [37].

The demonstration that the sulfate ester of glycolithocholate can be excreted by renal pathways in the rat [38] and the high recovery of radioactivity from human urine after administration of labeled cholic and chenodeoxycholic acids, if solvolysis was introduced into the workup [36,39], were the first strong indicators that "sulfates" of bile acids are excreted into the urine. In the course of these studies, 3β-hydroxy-5-cholenoic acid was found to be the main monohydroxy bile acid excreted in extrahepatic biliary atresia, and it was thought to be excreted as a sulfate ester [40].

Using Sephadex LH-20 column chromatography, originally applied in a study of metabolites of monohydroxy bile acids in the bile of the rat [24], it was possible to further enrich the fraction of urinary bile acid sulfates from patients with liver diseases [28,41].

By means of infrared (IR) spectrometry, the sulfate ester bond, indicated by bands in the fingerprint region characteristic for $>\!S = O$ and $>\!SO_2$ [42], was demonstrated for isolated glycochenodeoxycholic acid sulfate using synthetic material as reference [29]. Similar chemical proof was obtained for sulfochenodeoxycholate isolated from the urine of cirrhotic patients [32].

Although it was shown that chromatography on Sephadex LH-20 separates synthetic mono, di-, and trisulfates of bile acids [9,43], polysulfates have not been found in the urine of patients [9,17]. The requirement of sodium as counterion is a prerequisite for chromatography on these columns [44], and it was soon realized that ion-exchange effects, leading to "memory" fractions, may occur on these columns, if they are eluted with solvents of increasing polarity. Furthermore, Sephadex LH-20 columns are sensitive to overloading.

These disadvantages, particularly the partial overlap in elution of taurine conjugates with monosulfates of glycine conjugates [9,25], are overcome by column chromatography on DEAP Sephadex LH-20 [9], which has been studied and applied most extensively of several lipophilic anion exchangers synthesized [45,46]. Since the anion-exchange capacity of this material is high (the commercially available form has an amine content of about 1.2 mmole/g dry gel), small columns containing about 1 g are sufficient to take up about 300 μmole of bile acid [9]. This is adequate for aliquots of 10–50 ml of urine even if the content of bile acid sulfates is high. By increasing the pH to 7.6 of the 0.34 M ammonium acetate buffer, the sulfates will be eluted [9]. However, most of the glucuronides of bile acids are also eluted within this fraction [17]. Further subfractionation with acetate buffers by stepwise increase in the pH from 7.0 to 8.0 and by increasing the concentration from 0.17 to 0.35 M allows separation of these two classes of polar compounds [17].

DEAP Sephadex LH-20 was claimed to give good subseparation of bile acid conjugates, glucuronides, sulfates, and disulfates [47]. However, others have not found satisfying subseparation of glucuronides and conjugates with this system [17]. The sodium acetate eluted within some fractions may pose problems in further quantitative processing.

Using the more basic groups of the piperidinium ion, piperidinohydroxypropyl Sephadex LH-20 (PHP-LH-20) was employed in the subseparation of free, glucuronidated, and sulfated bile acids after previous cleavage of the glycine and taurine bonds [48]. Sulfates were eluted with 1% $(NH_4)_2CO_3$/70% ethanol from this material, whose basic group has an apparent pK_b of about 11.

The sulfate ester fraction of bile acids was shown in most instances to be the most important bile acid fraction in the urine.

Early workers found more than 50% of the urinary bile acids as sulfates in disease states [28–32,49], and it was soon realized that sulfation of individual bile acids in the urine is inversely related to the degree of hydroxylation [31,39]. Monohydroxy bile acids are exclusively excreted as sulfate esters; the degree of sulfation of dihydroxy bile acids varies according to the stereochemistry of the hydroxyl groups; trihydroxy bile acids are generally less sulfated; and tetrahydroxy bile acids are virtually absent from this fraction [9,34,50,51].

In healthy persons sulfated bile acids are also the main constituents of the urinary bile acid fraction [9]; in another study, however, only 5–15% were found to be sulfated [23].

The methods available do not yet allow fractionation of bile acid sulfates into glycine- and taurine-conjugated bile acid sulfate esters. The sulfate ester fraction typically contains 3β-hydroxy-5-cholenoic acid, allo-lithocholic acid, and lithocholic acid [40,42]. Some dihydroxy bile acids have also been found exclusively in this fraction: 3β,12α-dihydroxy-5-cholenoic acid, ursodeoxycholic acid, 3β,12α-dihydroxy-5β-cholanoic acid, allo-chenodeoxycholic acid, and 3β,7α-dihydroxy-5β-cholanoic acid. There is also another unsaturated dihydroxy bile acid not yet further identified [9]. Chenodeoxycholic acid is usually the most important bile acid of the sulfate fraction. However, it also occurs to a small extent in the nonsulfated fractions.

Many other dihydroxy bile acids, such as hyodeoxycholic acid, can be found in the nonsulfated fractions as well as the sulfate fraction; for more detailed information the work of Almé et al. [9,26] may be consulted.

It is interesting that only a few trihydroxy bile acids occur in the sulfate fraction: cholic acid is important under cholestatic conditions; $3\alpha,6\alpha,12\alpha$-trihydroxy-5β-cholanoic acid [26], allo-cholic acid, and 1,3,12-trihydroxy-5β-cholanoic acid can be minor components [9]. All of these are excreted as nonsulfated compounds.

Though sulfation at positions other than C-3 have been described [12,138,168], recent work confirms the widely held opinion that for cholic and chenodeoxycholic acid the esterification occurs only at position C-3 [9,17].

2.3.5. Glucuronic Acid-Coupled Bile Acids

An investigation of the polar bile acid fraction of urine obtained by means of Sephadex LH-20 chromatography revealed that part of the assumed "sulfates" was not cleaved by solvolysis in acetone/ethanol/HCl [25,52]. Therefore, the polar compounds remaining were subjected to treatment with bovine liver β-glucuronidase in order to test whether glucuronic acid-coupled conjugates might have been present. This enzymatic cleavage indeed liberated an additional fraction for analysis [25]. Evidence for the presence of bile acid glucuronides was further provided by selective application of enzymatic cleavage or solvolysis to the polar bile acid fraction and determination of the liberated bile acid fraction after rechromatography on Sephadex LH-20 [53–55]. The inhibition of β-glucuronidase cleavage of postulated bile acid glucuronides was reported to be possible with glucaro-1,4-lactone [56,138].

After chemical synthesis of glucuronides of different mono-, di-, and trihydroxy bile acids [57], mass spectrometric data obtained from peracetylated methyl esters of bile acid glucuronides led to positive identification of an intact glucuronide of a dihydroxy bile acid, suggested to represent chenodeoxycholic acid [58]. The methylated and acetylated bile acid glucuronide fraction was further separated by thin-layer chromatography in a system of 1,2-dichloroethane/diethylether 8 : 2 (v/v) [58].

In a patient suffering from malabsorption [59] and in healthy adults [26] the glucuronide fraction of bile acids was isolated together with the monosulfate fraction from DEAP LH-20. Selective solvolysis and subsequent rechromatography as well as methylation of the glucuronides were used to separate glucuronides by chromatography on Lipidex 1000 [26]. 6α-Hydroxylated bile acids were found to account for the majority of the aglycone moieties in the patient with malabsorption, but glucuronic acid-coupled cholic and chenodeoxycholic acids were also found. That glucuronidation of bile acids containing a 6α-hydroxyl group is a metabolic pathway in humans is supported by findings in cholecystectomized patients, who received ^{14}C-labeled hyodeoxycholic acid orally. These patients, fitted with T-tube drainages, excreted 30–84% of the label as urinary hyodeoxycholic acid glucuronide [167]. In the urine of the healthy subjects almost all of the bile acid glucuronides were found to be otherwise unconjugated [26]. The subseparation of dimethyl esters of bile acid glucuronides on Lipidex 1000 gave evidence for about 10 different compounds occurring in the urine. The glucuronyl linkage in hyodeoxycholic acid was found at position C-6, whereas in cholic and chenodeoxycholic acids it was at the C-3 position.

The mass spectrometric data obtained for the dimethylester of the major pertrimethylsilylated bile acid glucuronides [26] resembled those of the acetylated methyl esters of synthetic dihydroxy bile acid glucuronides [57], taking into account the differences resulting from losses of trimethylsilanol groups versus acetic acid and the differences resulting from the specific fragmentation pattern described for trimethylsilylether derivatives of glucuronides of steroids [60]. The position of the glycosidic linkage of bile acid glucuronides cannot be determined from mass spectra [26,57,58]; the identity of the aglycone can be disclosed only after cleavage of the glycosidic bond.

In the procedure using piperidinohydroxypropyl Sephadex LH-20 (PHP-LH-20), originally proposed for separations of subclasses of conjugates of bile acids in serum [48], the question of the conjugation pattern of bile acid glucuronides could not be answered, since the peptide bonds were cleaved enzymatically [61] before application of the bile acid mixture onto the column [62]. The use of small columns containing 110 mg of PHP Sephadex LH-20 [62,63] with an ion exchange capacity of 0.5 meq/g [64] may require special attention to be paid to the total amount of bile acids applied to these columns in analyses of urine with high concentrations of bile acids. After the elution of unconjugated bile acids with 0.2 M acetic acid in 90% ethanol, the glucuronides were reported to be eluted with 0.4 M formic acid in 90% ethanol. With this elution program it was not possible to separate glycine conjugates from free bile acids and glucuronides and, on the other hand, taurine conjugates from glucuronides and sulfates [48].

As yet there appears to be no fully satisfactory method for the complete separation of glucuronides of bile acids in a single chromatographic step that would allow the isolation of double conjugates. Owing to the uncertainties involved in the recovery from biochemical cleavage steps, the quantitative determination of glucuronides of bile acids in the urine still poses problems. On the other hand, the quantitation of other groups of bile acids may be disturbed by their presence, especially if they are labile under the conditions of alkaline hydrolysis and solvolysis [48]. The variability of the reported percentages of glucuronides of bile acids in the urine of healthy and diseased subjects may be a reflection of methodological problems still encountered. The contribution of bile acid glucuronides to total urinary bile acids in healthy subjects has been reported to be as high as 35.8% in one study [26] and 12.9% in another [62]. They may be excreted in higher amounts in liver diseases. However, their relative contribution to total urinary bile acids was found to be 21.2% [25], 7.0% [54], 7.5% [62], and 9.1% [65]. It seems clear that glucuronides of bile acids represent a minor fraction of the total urinary bile acids.

All of major bile acids have been reported to be present in the glucuronide fraction of urinary bile acids; particularly interesting is the preponderance of 6α-hydroxylated compounds in healthy subjects [26,59]. Exact proof for the agly-cone moieties, as well as for the sugar moiety, is, however, hampered by the quite uniform fragmentation pattern in mass spectrometry [17,26,57]. The question of the occurrence of monohydroxy bile acid glucuronides [25,26,54,58,62,65] is not yet fully settled.

3. BILE ACIDS DETECTED IN THE URINE

3.1. Bile Acid Groups

Urine of healthy subjects and of patients with liver disease has usually been investigated using small aliquots owing to the limited capacities of the equipment employed for the analyses. The question of detection of minor compounds may therefore not be fully answered at present. In the future techniques may be developed for separating greater amounts of urinary bile acids, and for the exact determination of minor components, chemical isolation of the individual components will be required. Even with the use of mass spectrometric detection systems, not all components encountered can be identified at present, and structural elucidation will probably require supplementary techniques, such as nuclear magnetic resonance spectrometry. This is mainly due to the lack of available reference compounds for comparing physicochemical characteristics with unknowns.

3.1.1. Monohydroxy Bile Acids

Monohydroxy bile acids were not positively identified in human urine in early studies [66,67], and bile acids excreted in the urine of surgically jaundiced rats were shown to consist of polyhydroxylated compounds [1,80,81]. However, the feeding of radioactively labeled lithocholate to rats led to renal excretion of 6% of the label, whereas rats fed the sulfate conjugate of glycolithocholate excreted 24% of the label in the urine [38].

Since monohydroxy bile acids of unknown structure, and differing from lithocholic acid and iso-lithocholic acid, were observed after solvolysis of urinary bile acids in children with extrahepatic biliary atresia [36], solvolysis was also applied in the workup procedure by Makino *et al.* [40]. This mass spectrometric study established the identity of monohydroxy bile acids in human urine. Surprisingly, it was 3β-hydroxy-5-cholenoic acid and not lithocholic acid, that was found to be the main monohydroxy bile acid excreted. The second most important monohydroxy bile acid found was allo-lithocholic acid. Lithocholic acid was present in small amounts in only one of the four children investigated. This infant excreted chenodeoxycholic acid as the main urinary bile acid. This finding was confirmed by using a different gas chromatographic systems [68–70]. 3β-Hydroxy-5-cholenoic acid was also demonstrated to be a main monohydroxy bile acid excreted in the urine of adult patients with various forms of cholestatic liver disease [47,71]. In the latter patients, lithocholic acid was consistently found, and the presence of 3β-hydroxy-5-cholenoic acid together with lithocholic acid was suggested to reflect an altered cholesterol and bile acid metabolism in the cholestatic condition [40,68,70,71]. This suggestion was based on the finding that the major secondary bile acid, deoxycholic acid, was virtually absent in the urine of completely obstructed patients [41], as well as on previous observations of a mitochondrial origin of 3β-hydroxy-5-cholenoic acid [103].

The uncertainties involved in identification of monohydroxy bile acids by means of gas chromatography alone are reflected by the demonstration of non-sulfated lithocholic acid in patients' urine, whereas combined gas chromatographic–mass spectrometric studies revealed that monohydroxy bile acids are excreted exclusively as sulfate esters [9]. To what extent monohydroxy bile acids occur as glucuronides, as has been proposed [25,26,54,62,72], will probably be decided only after chemical isolation.

3.1.2. Dihydroxy Bile Acids

In early investigations, chenodeoxycholic acid was the only dihydroxy bile acid shown to be excreted into the urine [66]. In almost all patients with liver disease, chenodeoxycholic acid sulfate is the major bile acid to be found [29,32]. On the other hand, in healthy subjects [9] and in normal pregnancy [73], deoxycholic acid is the predominant dihydroxy bile acid. In cases of recurrent cholestasis of pregnancy deoxycholic acid has also been observed to be the major dihydroxy bile acid excreted with the urine [9,74]. Deoxycholic acid is found mainly in the sulfate fraction.

Other dihydroxy bile acids appear of minor importance. In healthy subjects there are very small amounts of allo-deoxycholic acid, exclusively found as sulfate [9]; inconsistently allo-chenodeoxycholic acid and hyodeoxycholic acid can be identified as sulfates [9]. 3β-Hydroxylated dihydroxy compounds, such as 3β,12α-dihydroxy-5β-cholanoic acid and 3β,12α-dihydroxy-5α-cholanoic acid, are also found as sulfates. Ursodeoxycholic acid is excreted as sulfate. Considerable amounts of $3\alpha,12\beta$-dihydroxy-5β-cholanoic acid are excreted as the glycine conjugate [9].

In liver disease all these dihydroxy bile acids may be found in the urine [9]. However, in obstructive jaundice the diminution of deoxycholic acid may be striking [71]. To identify all minor components is impossible for laboratories not equipped with computerized gas chromatography–mass spectrometry. The obvious gas chromatographic pattern consists of chenodeoxycholic acid, deoxycholic acid, ursodeoxycholic acid, and $3\beta,12\alpha$-dihydroxy-5-cholenoic acid [17,71].

Hyodeoxycholic acid was identified in small amounts in the taurine conjugate fraction [9] and in the sulfate fraction of patients with intrahepatic cholestasis of pregnancy [73,74]. It was proposed to be excreted in the urine of children with intrahepatic cholestasis [65]. The chemical ionization mass spectrometry pattern provided in this study, however, allows no definitive discrimination from the pattern of other dihydroxy bile acid derivatives. Hyodeoxycholic acid and its isomer $3\alpha,6\beta$-dihydroxy-5β-cholanoic acid were, however, safely identified, though in very small amounts, in the urinary dihydroxy bile acid profile of a patient suffering from primary biliary cirrhosis [75].

$3\beta,7\alpha$-Dihydroxy-5β-cholanoic acid was detected in the urine of cholestatic patients [91] and was demonstrated to occur as a sulfate in one case of recurrent cholestasis of pregnancy [73], as well as in a case of primary biliary cirrhosis and in congenital intrahepatic cholestasis [9].

Besides hyodeoxycholic acid and chenodeoxycholic acid, several still unidentified cholanoic acid "diols" seem to be present as glucuronides in the urine of healthy subjects and of patients [26].

3.1.3. Trihydroxy Bile Acids

Cholic acid, the main trihydroxy bile acid of mammals, was shown to be excreted in the urine in early investigations [66]. In the surgically jaundiced rat, the occurrence of 6β-hydroxylated bile acids was observed. The rat was shown to mainly hydroxylate in order to metabolize and renally excrete radioactively labeled, administered chenodeoxycholic acid [1,80,81].

With the introduction of solvolysis procedures into the workup for urinary bile acids [36], it was shown that almost all the radioactivity from administered 24-[^{14}C]cholic acid could be recovered from the urine of infants with extrahepatic biliary atresia. In these patients 55–77% of the isotope was excreted in the urine over a 4-day period. Most of these patients excreted cholic acid as the main bile acid in the urine [40].

Subfractionation of the bile acid conjugates into subclasses of nonsulfated and sulfated bile acids revealed that the degree of sulfation of cholic acid varied considerably in the urine of patients with different liver diseases [29]. In patients with cirrhosis of the liver, 24% of cholic acid was reported to be sulfated [32].

Glucuronides of conjugates of cholic acid have been tentatively identified in the urine of patients with extrahepatic obstruction [17].

In health, cholic acid is found mainly in the free and glycine- and taurine-conjugated bile acid fraction [9,62]. Sulfates are present in relatively low percentages [9,62], and glucuronides of cholic acid were reported to represent a mean of 13.6% of total cholic acid excreted [62].

Other trihydroxy bile acids occasionally detected in health are of great quantitative importance in disease. Some of the trihydroxy bile acids carry a 6-hydroxy group. Hyocholic acid can be found in almost all samples [9,49]. 6α- and 6β-hydroxylated metabolites with 3α,12α-dihydroxy structures are present [9,26], and evidence was also obtained for trihydroxy bile acids with the 3,6β,7-trihydroxy structure [9,84]. Interesting is the occurrence of a 1,3,12-trihydroxy bile acid [9], which in later studies was positively identified as 1β,3α,12α-trihydroxy-5β-cholanoic acid by nuclear magnetic resonance spectrometry [76,85]. Allo-cholic acid, 3α,7β,12α-trihydroxy-5β-cholanoic acid and 3β,7β,12α-trihydroxy-5β-cholanoic acid may be found occasionally in very small amounts [9], and the fraction of unconjugated bile acids often contains norcholic acid.

Hyocholic acid was one of the first unusual hydroxylated bile acids to be recognized in patients suffering from hepatobiliary diseases [77]. It is mainly found in the nonsulfate fraction of urinary bile acids in liver diseases [9,84,91], however, its monosulfate [9] and glucuronide [26] conjugates have also been described. It becomes an important bile acid in the urine of patients with intrahepatic cholestasis during treatment with phenobarbital [87].

In a patient with cerebrotendinous xanthomatosis receiving treatment with ursodeoxycholic acid, ω-muricholic acid was identified in the urine by capillary gas chromatography [89].

Unequivocal identification of the many trihydroxy bile acids encountered in disease requires optimal gas chromatographic separation systems [78,79], and further knowledge can be expected if capillary gas chromatography coupled to mass spectrometric systems is used in analyses. Even with this methodology, uncertainties may continue with regard to the chemical identity of cholanoic acid "triols," and the only method of structure elucidation may remain chemical isolation of the substances in question.

3.1.4. Tetrahydroxy Bile Acids

Since the studies on bile acids of reptiles, fishes, and amphibians [82,83], the existence of tetrahydroxylated bile acids of the coprostanic acid series is known (e.g., varanic acid, arapaimic acid).

The finding of tetrahydroxylated bile acids of the 5β-cholanoic acid series in the urine of patients with liver disease [84] was a completely new observation. These bile acids were identified as C-1- and C-6-hydroxylated cholic acids. Since their elution properties were almost identical in gas chromatography, their separate determination was possible only by computerized gas chromatography–mass spectrometry, taking into account the different fractional contribution of specific fragment ions [9]. The final proof for their structures came from the comparison with synthetic $3\alpha,6\alpha,7\alpha,12\alpha$-tetrahydroxy-$5\beta$-cholanoic acid obtained from authentic $3\alpha,6\alpha$-dihydroxy,7-oxo-12α-acetoxy-5β-cholanic acid [85] and from a nuclear magnetic resonance (NMR) study after isolation of the individual tetrahydroxy bile acids that confirmed and established their structures as $3\alpha,6\alpha,7\alpha,12\alpha$-tetrahydroxy-$5\beta$-cholanoic acid and $1\beta,3\alpha,7\alpha,12\alpha$-tetrahydroxy-$5\beta$-cholanoic acid [86]. The occurrence of $3\alpha,6\beta,7\alpha,12\alpha$-tetrahydroxy-$5\beta$-cholanoate, suggested to be a metabolite of cholic acid in the urine of patients with intrahepatic cholestasis [85], was also confirmed by NMR spectrometry after its isolation as a methyl ester [86].

These bile acids were not found in the urine of healthy subjects [9]. However, they occurred in the urine of normal pregnant women [73] and in pregnant women with intrahepatic cholestasis of pregnancy [9,73,74]. They have been noted to be particularly prominent in the urinary bile acid profile of cholestatic patients treated with phenobarbital [34], and a study following the time course of their excretion during the institution and maintenance of phenobarbital therapy revealed that their excretion can be stimulated by this drug under various conditions of intrahepatic cholestasis [87] (see Section 3.4.5).

Other tetrahydroxy bile acids seem to be present in low quantities in the urine of patients with liver diseases [9,73,74]; they may be traceable if isotopically labeled cholic acid, the presumed precursor, is administered to patients [34]. A tetrahydroxylated bile acid, suggested to represent a 2,3,6,7-tetrahydroxy compound, was found in the urine of healthy newborns in considerable quantity [92]. Other sites of hydroxylation in tetrahydroxylated bile acids differing from the positions at C-1, C-2, C-3, C-6, C-7, and C-12 are not yet established. For further information the work of Bremmelgaard and Sjövall [34] should be consulted.

There are also small amounts of a side chain (C_{23}) hydroxylated cholic acid, detected in the urine in primary biliary cirrhosis and in congenital intrahepatic cholestasis [9]. Patients with cerebrotendinous xanthomatosis may also excrete this bile acid [89].

Tetrahydroxylated bile acids have not been found in the sulfate or glucuronide fraction of urinary bile acids. They are mainly present in the taurine conjugate fraction [9]. A partial subseparation of different tetrahydroxy bile acids can be achieved by Sephadex LH-20 chromatography [85], chromatography on neutral aluminium oxide (activity grade V) [88], thin-layer chromatography [86,87], and capillary gas chromatography (OV–101) [170].

3.1.5. Other Bile Acids

Oxo bile acids have been identified in the glycine conjugate and sulfate ester fractions of urine in health and in disease [9,119] (see Section 3.4).

C_{27} and C_{29} bile acids have been identified by combined gas chromatography–mass spectrometry in infants with coprostanic acidemia [90] (see Section 3.5).

3.2. Bile Acid Pattern in the Urine of Healthy Adults

Prior to 1968 virtually no bile acids were found in the urine of normal subjects with the methods then available [66]. In normal persons the cumulative urinary excretion of label resulting from administration of [^{14}C]cholate was found to be less than 0.12% [93].

By the inclusion of a solvolysis into the workup procedure to allow for the detection of more polar conjugates, the total excretion of bile acids was found to be 0.46–0.97 μmole/day [30]. In this study, cholic acid, chenodeoxycholic acid, and deoxycholic acid were detected; the percentage of sulfated bile acids ranged from 50 to 81%. Only trace concentrations, less than 0.2 mg/liter, even after solvolysis, were found in 15 healthy subjects [49]. With the use of isolation methods carefully designed and evaluated for the highest possible recoveries and employing computerized gas chromatography–mass spectrometry, the urinary excretion of bile acids was found to range between 6.38 and 10.98 μmole/24 hr in five healthy subjects [9]. The lower limit for quantification was 0.01 μmole/24 hr, about 4 μg/24 hr. The percentage of sulfated bile acids ranged between 43 and 87%, but glucuronides of bile acids were not included. In five normal primiparous women the total bile acid excretion was found to range between 7 and 30 μmole/24 hr during the third trimester [73]. The mean value, 16 μmole/24 hr, was found to be higher than the mean value of 7 μmole/24 hr found for normal nonpregnant women in the study of Almé et al. [9]. If glucuronides were considered [26], two healthy women excreted 7.03 and 11.29 μmole/24 hr glucuronides, amounting to 12 and 17%, respectively, of the total. A large proportion of the glucuronides was shown to be 6-hydroxylated compounds. Interestingly, the 1-hydroxylated derivative of deoxycholic acid was found to be a main bile acid in the urine during pregnancy. Certainly, studies of more subjects are required in order to establish the significance of these findings. The occurrence of 10–14 different compounds resembling bile acid glucuronides was reported [26].

With the use of group separations on Sephadex LH-20 and enzymatic cleavage procedures to liberate free bile acids, a total of 3.3 mg/24 hr of mono-, di-, and trihydroxy bile acids was encountered in healthy children [65]. Sulfated bile acids were reported to contribute 45% and glucuronides up to 3% of the urinary bile acids in these children. 6-Hydroxylated bile acids were not encountered. These were also not detected in a recent study, demonstrating specific separation of conjugates, sulfates, and glucuronides, where a mean total urinary bile acid excretion of 1.9 mg/24 hr, 60% as sulfates, 13% of glucuronides, was reported [62]. Since thin-layer chromatography was used to establish the purity of the fractions [94], further investigations using current separation techniques seems justified.

3.3.   Bile Acids in the Urine of the Newborn

Bile acids have been found to be excreted in the urine of 2-week-old normal infants [95,97], but bile acids were below the detection level in the urine of infants aged 2 months or more [96]. A recent study, however, using gas chromatography–mass spectrometry, demonstrates the occurrence of bile acids within the range of 0.4–1.2 μmole/24 hr in children 5 months to 7 years of age [51].

Cholic acid was found to predominate in the urine; occasionally chenodeoxycholic acid was detected. The excretion of bile acids was shown to be temporary [97], and there was no accompanying rise in the concentration of unconjugated bilirubin in the serum. Cholic acid was found to be the main urinary bile acid of the newborn at a mean value of 0.34 μmole/24 hr [92]. On the basis of body weight and body surface, its excretion markedly exceeds the urinary excretion found in children aged 9–14 years [98] and in adults [9].

Between 46 and 70% of cholic acid was present in the taurine conjugate fraction and 4–35% in the sulfate fraction. Some unconjugated cholic acid was present. Chenodeoxycholic acid constituted about 10–20%. 3β-Hydroxy-5-cholenoic acid was found to be excreted to the extent of 0.04 μmole/24 hr (mean value). Its excretion relative to body weight and body surface was seemingly not increased when compared with the values found in older children or adults [92].

Particularly noteworthy is the fact that tetrahydroxylated bile acids are excreted in the urine of the newborn. Their sum may exceed the amounts of cholic acid encountered [92]. These studies demonstrate that the newborn infant excretes in the urine bile acids that are not normally found in the urine of the healthy adult. Hyocholic acid, demonstrated to occur in the urine of the newborn, is also virtually absent from the urinary bile acid spectrum of healthy adults [9]. These findings support the hypothesis that during the developmental phase [33] and in the perinatal period, differences of bile acid metabolism are found compared with the bile acid metabolism at later ages. Summarized, the main differences are that taurine conjugation and sulfate esterification appear to be important in the early phase of life and that atypical hydroxylations occur to a significant extent.

The organ site of atypical hydroxylations in the newborn has not yet been determined although probably the liver, since it is known to be able to hydroxylate steroids in the 1β-position [99] and bile acids in 6α-position [100]. Furthermore, fetal liver microsomes have been reported to conduct steroid hydroxylations in the 2α- and 2β-position [101,102]. The detection of otherwise hydroxylated bile acids seems still possible. Recent analysis of human fetal gallbladder bile [172] which revealed a range of 6α- and 1β-hydroxylated bile acids indicates the fetal liver as the site of formation of these bile acids during development.

3.4. Urinary Bile Acids in Hepatobiliary Disease

3.4.1. Cholestasis of Infancy

3.4.1a. Extrahepatic Biliary Atresia. This syndrome presents a unique possibility to study bile acid metabolism in the cholestatic syndrome without interference of the intestinal flora. In this condition monohydroxy bile acids have been found to be excreted with the urine [40], mainly 3β-hydroxy-5-cholenoic acid and allo-lithocholic acid, whereas lithocholic was mostly absent. The origin of 3β-hydroxy-5-cholenoic acid was suggested to be not directly from cholesterol, since less than 5% of the amount of radioactivity present in the di- and trihydroxycholanoate fractions could be recovered in the monohydroxylated bile acids after administration of 4-[14C]cholesterol to a patient. However, mitochondrial cholesterol may be the sterol source responsible for the biogenesis of 3β-hydroxy-5-cholenoic acid, since disrupted rat liver mitochondria have been shown to convert cholesterol to this bile acid [103]. Liver cell subfractionation studies using biopsy material obtained at explorative laparotomy from a patient with extrahepatic biliary atresia indicated that the microsomal fraction supplemented with cytosolic supernatant converted more 3β-hydroxy-5-cholenoic acid to allo-lithocholic acid than to lithocholic acid [108].

In extrahepatic biliary atresia cholic acid and chenodeoxycholic acid were found to be the main bile acids excreted [40]. However, the gas chromatographic pattern of the bile acid fraction obtained after solvolysis and alkaline hydrolysis indicated that even in this disease, a more complex composition of urinary bile acids prevails [69,108]. Other workers had already reached this conclusion by demonstrating several metabolites with chromatographic features of bile acids after administration of 4-[14C]cholesterol to a patient with extrahepatic biliary atresia [36]. 24-[14C]Lithocholic acid was found to be transformed to a small extent into more polar compounds [109], which, however, were not further characterized. Most of the lithocholic acid administered was excreted as the sulfate esters of tauro- and glycolithocholic acid [110], as was the label of 24-[14C]chenodeoxycholic acid. Cholic acid, on the other hand, was found to be excreted in the form of the glycine conjugate [110]. Sulfation seems to be unimpaired [114].

3.4.1b. Intrahepatic Cholestasis of Infancy (Neonatal Hepatitis Syndrome).

In comparison to extrahepatic biliary atresia, the mean daily values for the excretion of cholic and chenodeoxycholic acids showed no differences in the infants with neonatal hepatitis, and the ratio of cholic to chenodeoxycholic acid was not discriminatory [111]. After cessation of overt jaundice, the total urinary excretion of bile acids reverted to 0.6–3.7 μmole/24 hr in infants with intrahepatic cholestasis [96].

With regard to the conjugation pattern, in addition to glyco- and taurocholic acid at least three more conjugates of cholic acid were isolated during early studies [112]. In retrospect, these conjugates may now be assumed to have represented sulfates and/or glucuronides of conjugates of cholic acid.

A diminution of the urinary excretion of bile acid sulfates was noted in cases of intrahepatic cholestasis due to paucity of bile ducts, when compared with cases of hepatitis [113] and when compared with infants with extrahepatic biliary atresia [104], though in the latter study the difference was not statistically significant.

Monohydroxy bile acids, particularly 3β-hydroxy-5-cholenoic acid, were regularly encountered in infantile forms of intrahepatic cholestasis [51], but were not discriminatory between biliary atresia and neonatal hepatitis. As in adults, the excretion rate of 3β-hydroxy-5-cholenoic acid was related to the excretion rate of chenodeoxycholic acid [105].

The occurrence of hydroxylated derivatives of cholic acid was realized only in recent years [51,113]. In patients with intrahepatic cholestasis due to paucity of interlobular bile ducts the excretion of tetrahydroxy bile acids was found to amount to a mean percentage of 19.7% of the total in the presence of a low mean excretion of bile acid sulfates (16.2%). On the other hand, in patients with hepatitis, tetrahydroxy bile acids were found to be excreted to the extent of 4.8%, whereas sulfated bile acids amounted to 46.2% [113]. The relatively high proportion of tetrahydroxylated bile acids was suggested to reflect a biochemical difference within the disease entity of dysplasia of intrahepatic bile ducts in contrast to the conditions encountered in the hepatitis syndrome. Sulfation and hydroxylation may be interpreted as alternatives of bile acid metabolism occurring at different rates in these two patient groups, which cannot easily be distinguished on clinical grounds.

A recent study, performed prospectively over 2–8 years, gave some indication for the prognostic value of determining tetrahydroxy bile acids in cholestatic syndromes of infancy caused by α_1–antitrypsine deficiency. Infants initially exhibiting a good capacity of polyhydroxylation showed a more favorable course in their liver disease [51].

In the most severe form of intrahepatic cholestasis, familial progressive intrahepatic cholestasis (Byler's disease), hyocholic acid was an important constituent of the urinary bile acid spectrum [106,107], whereas, interestingly, tetrahydroxy bile acids were not detected in the urine [107]. Whether this finding has implications in terms of etiology of this disease remains an intriguing question.

3.4.1c. Erythroblastosis Fetalis. The prognosis of jaundice occurring in the newborn due to maternal Rh isoimmunization is good, provided it is not complicated by kernicterus. In this condition urinary bile acids have been studied, and the ratio of cholic to chenodeoxycholic acid was found to be greater than 5 in cases with elevation of conjugated bilirubin in the serum [115].

3.4.2. Acute Hepatitis

In acute viral hepatitis the urinary bile acid excretion seems to correlate with serum bilirubin values [71]. The complexity of the bile acid excretion occurring in hepatitis was soon recognized, as well as the requirement of gas chromatographic–mass spectrometric means for the identification of single components [71]. Most workers using gas chromatography alone restricted quantitative measurements to the main primary and secondary bile acids [23,28,29]. The regular occurrence of deoxycholic acid in the urine of patients with acute hepatitis was noted, in contrast to the observations in patients with extrahepatic obstruction [41,71]. The monohydroxy bile acid, 3β-hydroxy-5-cholenoic acid, was identified in most cases of viral hepatitis. However, it occurred in lower amounts than in extrahepatic obstruction [41,71]. These findings were extended using computerized gas chromatography–mass spectrometry [34,116], and many unusual bile acids could be identified in addition to the main primary and secondary bile acids. An evaluation of the significance of the occurrence of all the minor components is still impossible.

With respect to the conjugation pattern of cholic acid, this bile acid occurred in the urine as glycine conjugate (11–56%), as taurine conjugate (14–42%), and as sulfate ester (2–67%) [116]. In most patients small amounts of unconjugated cholic acid were found [34]. In contrast to these findings, chenodeoxycholic acid and deoxycholic acid were predominantly excreted as sulfate esters. In 11 of 12 patients studied the relative percentage of sulfates of chenodeoxycholic acids varied between 84 and 100% [116]. Determinations of bile acid glucuronides have yet to be carried out in acute viral hepatitis.

When compared with the bile acid pattern of patients with acute viral hepatitis, patients with toxic hepatitis (halothane, α-methyl DOPA) show slight differences in their urinary bile acid profile. Allo-cholic acid was less frequently found in the sulfate fraction, and 1-hydroxylated derivatives of deoxycholic acid and chenodeoxycholic acid were more often encountered [116]. One patient, reported to have toxic hepatitis due to intake of rifampicin and isoniazid, excreted relatively large amounts of 3β-hydroxy-5-cholenoic acid (23.9% of the total) and the urinary bile acid excretion of this patient exceeded the upper limit of normal (approximately 10 μmole/24 hr).

3.4.3. Chronic Hepatitis and Cirrhosis of the Liver

There are few data available regarding the urinary excretion of bile acids in chronic forms of hepatitis. Detailed studies discriminating between diverse forms of chronic hepatitis in terms of etiology or stage are lacking. Furthermore, no attempt has yet been made to elucidate the bile acid profile to the same extent as for acute forms of hepatitis. Cholic acid was detected merely as a nonsulfated compound; chenodeoxycholic acid and deoxycholic acid were mainly found in the sulfate fraction [30]. Great variations in the degree of sulfation, ranging from 28.7 to 100%, were reported, probably reflecting the heterogeneity of the groups of patients studied.

The first investigations of urinary bile acids excreted by cirrhotic patients were directed at establishing the presence of bile acid sulfates [30,32]. Only the main bile acids, cholic acid, chenodeoxycholic acid, deoxycholic acid, and lithocholic acid were determined. Although the clinical stages of the patients showed great variations, from stable inactive cirrhosis to cirrhosis with active alcoholic hepatitis, the principle was confirmed that the degree of sulfation is inversely related to the degree of hydroxylation. Detailed inspection of the bile acid pattern [34] revealed a similar complexity as is found in acute hepatitis [91,116]. The data of the Swedish workers included 10 patients with portal cirrhosis and five patients with primary biliary cirrhosis. In the patients with portal cirrhosis the total urinary bile acid excretion was found to range from normal, to values as high as 308 μmole/24 hr, indicating great variations of the clinical stages within the group. This diversity was equally well reflected by the differences in serum bilirubin values, although far less by the activities of alanine aminotransferase in the serum. Total urinary bile acid excretion seems to be loosely correlated to serum bilirubin values in portal cirrhosis, but has a fair correlation with serum bilirubin values in primary biliary cirrhosis.

With respect to the profile of metabolites, the relatively high percentage of norcholic acid observed in some of the patients, who otherwise excreted low amounts of bile acids, is noteworthy. This bile acid was found exclusively in unconjugated form [34]. In portal cirrhosis, a disease characterized by derangement of hepatic lobular structure, the proportion of unconjugated bile acids in the urine is generally increased: 13% in portal cirrhosis and 8.3% in primary biliary cirrhosis. With respect to sulfation, no impairment was observed with sulfates accounting for between 48.8 and 72.2%, while no bile acids were found as disulfates [34]. Bile acids with hydroxy groups in the C-1 and C-6 positions were relatively more abundant in cases with a low overall bile acid excretion. The absolute amounts of tetrahydroxy bile acids excreted may serve as an index of the hydroxylating function of the liver, since these are found almost exclusively in the urine. It seems worthwhile to investigate why most patients with cirrhosis of the liver excrete small amounts of tetrahydroxy bile acids in the urine (range 0.07–3.27 μmole/24 hr) [116]. In this study [116] 1 of 15 patients had a very high excretion (27.85 μmole/24 hr). Drug intake may be important to establish since this may account for these high levels of tetrahydroxy bile acids [87]. The presence of unusual trihydroxylated bile acids, such as 1,3, 12-, 3α,6α,7α-, and 3α,7β,12α-trihydroxy cholanoic acids, has also been reported [117,118], with hyocholic acid consistently present.

In patients with decompensated stages of cirrhosis of the liver, an increased excretion of several oxo bile acids, such as 3α-hydroxy-12-oxo-, 3α,12α-dihydroxy-7-oxo-, 3α,7α-dihydroxy-12-oxo-, and 3α-hydroxy-7,12-dioxocholanoic acids has been reported [119]. In this study a great number of bile acids (up to 74.4%) were unidentified. Since strong alkaline hydrolysis over 6 hr had been performed in this study, a reinvestigation directed towards the occurrence of oxo bile acids, applying milder conditions of hydrolysis, seems justifiable.

Early reports dealing with the determination of bile acid glucuronides in the urine of cirrhotic patients stress the importance of this fraction [56]. Up to 55% of the total bile acids excreted in the urine of these patients, who simultaneously suffered from cholestatic symptoms, were estimated as glucuronides. With the use of different techniques for separating sulfates and glucuronides, only 1.8–18% were found [62], and forthcoming technology [17] may be of help in further studies.

Though considerable progress has been made with respect to the characterization of urinary bile acid excretion in cirrhosis of the liver, no comprehensive data on all subclasses of compounds are available as yet. It seems advisable to define more strictly different etiologies and stages of the disease in future analyses.

3.4.4. Extrahepatic Biliary Obstruction

Extrahepatic biliary obstruction may lead to very high urinary excretion of bile acids in adults. Some of the highest reported values are 85 mg/24 hr (approximately 215 μmole/24 hr) [71], 166 μmole/24 hr [116], and 78 mg/24 hr (approximately 195 μmole/24 hr) [62]. The spectrum obtained seems to be simpler than that found in acute hepatitis, at least in terms of the relative quantity of unusual bile acids present [34]. An interruption of the enterohepatic circulation of bile acids is mainly reflected by decreased urinary deoxycholic acid excretion [71]. However, lithocholic acid may be present [71,116]. The main bile acids in the urine are cholic and chenodeoxycholic acids, which together comprise 61.4–93.4% [116]. The third major component is 3β-hydroxy-5-cholenoic acid;

mean percentages of 12% [71], 8.4% [116], and 6% [62] have been found. There is no consistent conjugation pattern for cholic acid, whereas chenodeoxycholic acid is excreted mainly as a sulfate. The pattern of nonsulfated bile acids comprises small amounts of unconjugated norcholic acid. Within the glycine and taurine conjugates, allo-cholic acid and relatively small amounts of tetrahydroxylated bile acids, not exceeding 10% of the total unsulfated bile acids, are found [34,116]. This fraction regularly contains hyocholic acid. Interestingly, 1-hydroxylated derivatives are lacking in most patients [116]. The pattern of the bile acids found in the sulfate fraction reveals the occurrence of 3β-hydroxylated bile acids, dihydroxy bile acids of the 5α- and 5β-cholanoate series, and the monohydroxy bile acids [116]. Generally speaking, the paucity of atypical hydroxy bile acids seems to be a characteristic of extrahepatic biliary obstruction.

3.4.5. Intrahepatic Cholestasis

3.4.5a. Recurrent Intrahepatic Cholestasis of Pregnancy. Only a few patients have been studied with the condition of intrahepatic cholestasis of pregnancy. However, these investigations are exemplary in attaining the highest possible degree of precision. This is reflected by the narrow standard deviations of four parallel determinations performed in a urine sample of one patient reported by Almé *et al.* [9]. Cholic acid and chenodeoxycholic acid are major components of the urine. As in normal pregnancy, deoxycholic acid is a predominant compound present, and this was confirmed by Thomassen [74] in other patients with this condition. The 1-hydroxylated derivative of deoxycholic acid, 1β,3α,12α-trihydroxy-5β-cholanoic acid, is quantitatively important. Early changes during the course of intrahepatic cholestasis of pregnancy involve the appearance of tetrahydroxylated bile acids. Both groups of bile acids, tetrahydroxylated bile acids, and 1-hydroxylated deoxycholic acid occur mainly, if not exclusively, in the nonsulfate fraction. Another finding is the occurrence of the 12-oxo derivative of deoxycholic acid, which makes up approximately 4% of the bile acids near term [74]. This bile acid has also been found in patients with intermittent intrahepatic cholestasis [34], and not in more serious forms of liver disease. In contrast to those forms of cholestasis, it is noteworthy that the main monohydroxy bile acid found was lithocholic acid, and not 3β-hydroxy-5-cholenoic acid [9,34].

3.4.5b. Intermittent Intrahepatic Cholestasis of Unknown Etiology. Familial benign recurrent intrahepatic cholestasis may be related to intrahepatic cholestasis of pregnancy and to cholestasis due to the intake of oral contraceptives [120]. In the most detailed study on urinary bile acids [34,116] in this syndrome, exclusion of other forms of cholestatic liver disease that pose difficulties for diagnosis, such as unrecognized drug intake should be considered. With these premises in mind, the profile of urinary bile acids found in several intrahepatic forms of cholestasis may show similarity. It was found that male adult patients with this syndrome excreted a great variety of bile acids in the urine. Some trihydroxy bile acids were identified that were not present in acute hepatitis, cirrhosis of the liver, or extrahepatic obstruction, these included $3\alpha,7\beta,12\alpha$-, $3\beta,7\beta,12\alpha$-, and $3\alpha,6\beta,12\alpha$-trihydroxy-5β-cholanoic acid [34]. Quantitatively as well as qualitatively the pattern encountered resembles that found in recurrent cholestasis of pregnancy. Since glucuronides of bile acids have not yet been determined in this condition, it is possible that some additional trihydroxy bile acids may be identified in this condition.

3.4.5c. Other Forms of Intrahepatic Cholestasis and the Effect of Pheno-barbital. In adult patients with acute forms of intrahepatic cholestasis as a possible sequelae of intake of drugs or hormones, cholic and chenodeoxycholic acids predominated; however, tetrahydroxylated bile acids as well as hyocholic acid were found as major constituents of the nonsulfated bile acid fraction of the urine [87].

The daily excretion of these C-1 and C-6 hydroxylated bile acids was found to be stimulated by phenobarbital (Fig. 1). This effect is temporary, lasting for several weeks, until the plasma concentrations of the respective precursors, the primary bile acids [85], have dropped to lower levels or even into the normal range. Patients may, however, respond differently with respect to the time course and extent of alteration of the bile acid metabolism. Under phenobarbital treatment the relative contribution of tetrahydroxylated bile acids was found to reach 60–70% of total nonsulfated bile acids in the urine, even with normalization of the total amounts of bile acids excreted. A corresponding observation had been made previously in patients with intermittent intrahepatic cholestasis of unknown etiology, where bile acids hydroxylated at C-1 and C-6 were found to be relatively more abundant in two patients during treatment with phenobarbital [34]. Bile and sulfate excretion was not stimulated by phenobarbital.

Since bile acids, if converted to tetrahydroxylated compounds, can be directly removed from the body by renal excretion, the induction of atypical hydroxylations by phenobarbital has been tried as a therapeutic regimen in patients with drug-induced jaundice, with hypoplasia of the bile ducts, and with sclerosing cholangitis [87,122]. Similar patients were studied earlier [121].

The pattern of urinary bile acids may be governed by metabolic derangements that are not yet fully understood, and this can be influenced by drug treatment under certain circumstances.

Other workers have not found a stimulation of atypical hydroxylations of bile acids during phenobarbital treatment in children with persistent intrahepatic cholestasis (aged 1–3 years) and children with intrahepatic biliary hypoplasia (aged 4–5 years) [65]. A slight rise of the urinary excretion of bile acid glucuronides was observed, possibly indicating an effect of phenobarbital on glucuronyl transferases. Such an effect of phenobarbital on the urinary excretion of bile acid glucuronides has been described in an adult patient suffering from

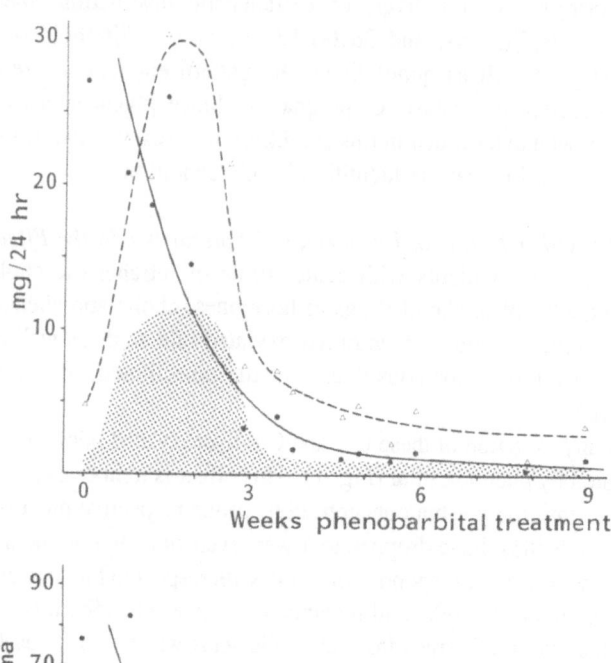

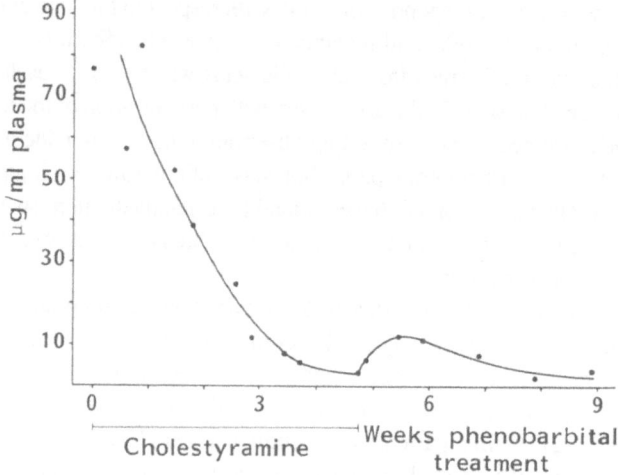

FIGURE 1. Effect of phenobarbital on the excretion of urinary bile acids and the concentration of plasma bile acids in a patient with intrahepatic cholestasis due to sclerosing cholangitis. Phenobarbital dosage: 5.3 mg/kg body weight daily. (A) Bile acid excretion with the urine. △------△ total 1- and 6-hydroxylated bile acid conjugates; ●——● cholic acid conjugates; shaded area, hyocholic acid conjugates. (B) Bile acid concentration in the plasma.

chronic intrahepatic cholestasis [123], but this has not been consistently observed in other patients [124]. This was explained by the transient nature of the elevation of urinary bile acid glucuronide excretion, which falls according to the decrease in the concentration of the respective conjugates in the plasma [124]. An effect of phenobarbital on the excretion of bile acid sulfates was not observed, nor was an effect on the excretion of cholic and chenodeoxycholic acids observed [125].

Early studies dealing with the effect of phenobarbital had shown that the urinary excretion of 24-[^{14}C]cholic and 2,4-[^{3}H]chenodeoxycholic acids did not change during administration of phenobarbital to patients with intrahepatic bile duct hypoplasia [126]. It was stated that the specific radioactivity of the urinary bile acids was higher during this period, though, on the other hand, it was shown that the slope of the specific radioactivity–decay curve of plasma cholate and plasma chenodeoxycholate was steeper than in the controls [126]. Assuming that an increased oxidative metabolism of bile acids had taken place in these patients, these results might be explained by distribution of the radioactive label between the precursors and their hydroxylated metabolites in the urine [85].

3.4.6. Hepatoma

3β-Hydroxy-5-cholenoic acid was found to be a major bile acid in the urine of patients with hepatoma [127]. In 10 cases investigated the mean percentage excreted was found to be 15.7 ± 11.5%. Though this was higher than in metastatic cancer of the liver without jaundice, no statistically significant difference could be found between these cancers. A high excretion of hyocholic acid was also observed in a patient with hepatoma [118].

3.4.7. Portacaval Shunt

In six patients undergoing elective portacaval anastomosis the urinary bile acid excretion was measured before and after the operation [128]. The authors determined cholic acid, chenodeoxycholic acid, deoxycholic acid, and lithocholic acid. It was found that the total urinary bile acid excretion increased significantly over preoperative values, mainly owing to an increase in nonsulfated cholic acid. In rat experiments, portacaval anastomosis led to an approximately 10-fold increase in bile acid excretion in the urine as measured by radioimmunoassay [129].

3.5.   Other Diseases Associated with Alterations in Urinary Bile Acid Excretion

3.5.1.   Malabsorption

In a patient with a malabsorption syndrome, who was reported to have a heavily increased smooth surface endoplasmic reticulum by electron microscopic examination of the liver tissue, 12.4 μmole/24 hr of bile acids was found to be excreted in the urine. A relatively high percentage of taurine conjugates, the presence of tetrahydroxy cholanoates, and an unusually high proportion of 6α- and 6β-hydroxylated bile acids were found, suggesting an abnormality in bile acid metabolism [59]. Further investigation revealed that glucuronides, mainly of 6-hydroxylated bile acids, were excreted. The glucuronides were found to be otherwise unconjugated and were characterized as their dimethyl ester pertrimethylsilyl ethers by computerized combined gas chromatography–mass spectrometry. The simultaneous occurrence of fragments typical for TMS ethers of steroid glucuronides [60] and for dihydroxy- and trihydroxycholaonoate structures was demonstrated in the gas chromatographically separated compounds. Within the intact molecules the sugar moiety and the aglycons could not, however, be positively identified by mass spectrometry alone. This was possible for the aglycons after enzymatic or chemical (periodate) cleavage of the glucuronides followed by hydrolysis of the resulting formates [26,59].

3.5.2. Zellweger Syndrome

Coprostanic acidemia, present in a few children with cholestatic liver disease [130,131], is probably related to the cerebrohepatorenal syndrome (Zellweger syndrome). In the full syndrome trihydroxycoprostanic, dihydroxycoprostanic, and varanic acids were found in the urine [132], and a quantitative account disclosed a complex composition in this condition [133]. The overall excretion of bile acids in urine was found to be elevated only in a patient with overt jaundice. Most of the bile acids were present in the nonsulfate fraction; trihydroxycoprostanic acid was the main constituent. Dihydroxycoprostanic acid was found only in the fully expressed syndrome, whereas varanic acid was detected only in the serum in this series. A considerable portion of the gas chromatographic peaks, suggested to represent bile acids, remained unidentified [133]. In a recent report [134] dealing with the metabolism of [^{3}H]5β-cholestane-3α,7α,12α-triol in a patient with Zellweger syndrome, more than half the radioactivity excreted overnight, 15 hr after the infusion was associated with a tetrahydroxylated 5β-cholestanoic acid with all hydroxy groups in the steroid nucleus. In a recent GC–MS study of the urine from Zellweger patients these tetrahydroxycholestanoic acids were identified as 6α- and 1β-hydroxy derivatives of 3α,7α,12α-trihydroxy-5β-cholestan-26-oic acid [173]. In two cases of coprostanic acidemia a 3α,7α,12α-trihydroxy-5β-C_{29}-dicarboxylic acid was identified [133]. The urinary excretion of C_{24} bile acids was low; cholic acid, chenodeoxycholic acid, and hyocholic acid were consistently detected; deoxycholic acid, ursodeoxycholic acid, and 3β-hydroxy-5-cholenoic acid inconsistently found [133].

4.   ORIGIN OF THE ATYPICAL BILE ACIDS IN THE URINE

With the advent of very advanced technology the last decade of bile acid research has seen considerable progress in detecting unusual or "atypical" bile acids in the urine. These developments have not yet come to a standstill, since if larger quantities of urine are analyzed new compounds will continue to be discovered. This is exemplified by studies on urinary tetrahydroxycholanoates [85], by the investigation of bile acid glucuronides [26], and by recent findings in Zellweger syndrome [133,134,173]. Any discussion of the origin of bile acids in the urine will therefore be preliminary, and only main areas of study can be discussed.

4.1.   Conjugation Reactions

Coupling of the microsomally formed coenzyme A thioesters to the amino acids glycine and taurine takes place in the cytoplasmic compartment of the liver cell [135].

Esterification with sulfuric acid has been demonstrated to occur in several organs. Bile acid sulfuryl transferases have been found in human liver and small intestine and in kidney tissue of the rat [136,137]. The synthesis of bile acid monosulfates was also shown to be conducted by isolated perfused rat kidney [138] and by perfused rat liver [139].

Also, the synthesis of bile acid glucuronides seems not to be restricted to liver tissue [140], since closely similar apparent affinities for different bile acid substrates were found for microsomal UDP glucuronosyltransferase activities obtained from kidney and duodenum [141]. Adrenal tissue showed no activity with bile acids. Others have not found glucuronides of chenodeoxycholic acid in the system of the isolated perfused rat kidney after perfusion with 24-[^{14}C]chenodeoxycholate, though some breakdown of the polar compounds obtained, after incubation with β-glucuronidase was noted [138].

Any discussion of the significance of sulfation and glucuronidation of bile acids must consider the quantitative aspects of these mechanisms, particularly in terms of the different substrates. Since the excretory routes of some of these conjugates are mainly the renal pathways, the amounts per time ultimately found in the urine may approximate maximal coupling rates, which must be studied in relation to substrate supply.

4.2.   Atypical Hydroxylations

Whereas the urinary excretion of free 6β-hydroxycortisol was noticed as early as 1960 in the urine of neonates [142] and polyhydroxylations of steroids became well-known metabolic events occurring characteristically in the prenatal developmental phase [143], unusual hydroxylated bile acids were first positively identified in adult patients with hepatobiliary diseases [77,84].

6α-Hydroxylation of taurolithocholate had been shown to be catalyzed by human liver microsomes [100], and cytochrome P-450 dependency was demonstrated for this reaction [144].

Although the organ sites of atypical nuclear hydroxylations other than 6α-hydroxylation have not yet been established, it is conceivable that 1β- and 6β-hydroxylations also take place in the liver. This assumption is corroborated by the demonstration of the occurrence of 1β- and 6α-hydroxylated bile acids in the plasma of patients without kidneys [145]. The finding of C-1 and C-6 hydroxylated bile acids in human fetal gallbladder bile during early gestation [172] and the demonstration that the microsomal fraction from human fetal liver can 1β-hydroxylate taurodeoxycholic [174] add further support that these reactions mainly occur in the liver. Since urinary excretion of 1β- and 6β-hydroxylated bile acids has been found to be stimulated by phenobarbital [86,87], the involvement of microsomal monooxygenase systems is likely. It can be expected that still other positions at the steroid nucleus, e.g., C-15, may be found to be hydroxylated [85]. However, no *in vitro* study has as yet appeared to show the location, specificity, and kinetic properties of those enzyme systems. The problem of multiplicity of hydroxylations seems to be related to the current fundamental question of whether microsomal P-450s are few in number and of broad substrate specificity or enormously diversified with very specific substrate requirements [146].

Since evidence for multiple hydroxylations at the steroid nucleus was obtained from studies on the urinary bile acid composition of healthy neonates [51] and meconium of healthy newborns [33], it may be speculated that the mechanisms of P-450 multiplicity may unfold in the diseased state by repression-derepression regulation [147], or DNA intragenic rearrangements combined with RNA processing [146], similar to that seen for other multigene families.

Another explanation may be that zonation effects within the liver parenchyma come into play in the diseased state. The biotransformation systems are assumed to be situated preferentially in the perivenous zone [148], which may become increasingly involved in bile acid metabolism under certain conditions of hepatobiliary disease. The perivenous zone seems to constitute a reserve capacity for bile acid excretion [149] and is also the preferential cell location for the increase of smooth endoplasmic reticulum [150] and of inducible cytochrome P-450 [171] during administration of phenobarbital.

With atypical hydroxylation reactions, quantitative experiments have to be carried out before their significance for the "detoxification" of different substrates can be evaluated. The effects of phenobarbital application leading to resolution of epiphenomena of cholestasis, such as accumulation of bile acids and bilirubin in the plasma, seems to work best in patients who already have a relatively high spontaneous rate of formation of atypically hydroxylated bile acids. A capacity momentum of the monooxygenase systems may be involved in the question of whether bile acids may take part in pathogenic events in hepatobiliary diseases [151].

4.3. Alterations of the Side Chain

The persistence of the full side chain of cholesterol in the C_{27} bile acids detected in the urine of infants with Zellweger syndrome or coprostanic acidemia may be explained by a defective β-oxidative cleavage in these patients [134]. Liver peroxisomes had been shown to convert 3α,7α,12α-trihydroxy-5β-cholestanoic acid into cholic acid in the rat [152]. In Zellweger syndrome the absence of peroxisomes and biochemical defects of mitochondria have been described [153]. The occurrence of a C_{29}-dicarboxylic bile acid in this syndrome was suggested to reflect a defective degradation of some other sterol than one derived from cholesterol, possibly β-sitosterol [133]. Norcholic acid, observed as an unconjugated compound in the urine of cirrhotic patients [34], may be formed from 23-hydroxycholic acid [9].

4.4. Epimerizations

Epimers of the main bile acids with α-oriented hydroxyl groups may arise by bacterial oxidoreductions during passage through the intestinal tract [154–157] and subsequent hepatic metabolism. However, even in extrahepatic biliary atresia with no enterohepatic circulation of bile acids, ursodeoxycholic acid was found to be excreted in the urine [69]. During follow-up from the fifth to the eleventh months of life its relative contribution to the total bile acids excreted in urine was studied in one patient several times and it was suggested that hepatic 7β-hydroxylation may be a primary reaction [68,69].

4.5. Unsaturated Bile Acids

3β-Hydroxy-5-cholenoic acid, which is usually the main urinary monohydroxy bile acid found in cholestatic conditions, may arise by mitochondrial synthesis from cholesterol, starting with degradation of the side chain [103]. It is a bile acid normally found in the urine of healthy adults [9]. It may originate from a mitochondrial pool of cholesterol, since it was not labeled to a considerable extent by exogenously administered radioactively labeled cholesterol [40]. It was suggested that any 3β-hydroxy-5-cholenoic acid formed in the liver may be sulfated and excreted under normal conditions [9]. Another unsaturated bile acid identified in the urine is 3β,12α-dihydroxy-5-cholenoic acid, which may be a metabolite of 3β-hydroxy-5-cholenoic acid, since a positive correlation was found between the urinary excretion of these two compounds as sulfates [34]. 3β-Hydroxy-5-cholenoic acid may be excreted as a conjugate coupled with glucuronic acid [54,62]. Phenobarbital may influence the mode of conjugation, favoring glucuronide excretion, in contrast to sulfate excretion [54].

5. THE RENAL CLEARANCE OF BILE ACIDS

Early studies suggested a very effective proximal tubular reabsorption of bile acid conjugates [158]. This was confirmed by stopflow analyses in dogs, and the transport maximum of the proximal tubular reabsorption was found to amount to 200 μg/ml of the glomerular filtration rate [159]. In view of this enormously high transport capacity of filtered bile acids [160,161], it remains uncertain exactly how bile acids ultimately reach the urine, whether by tubular secretion [162,163,169] or by diffusion processes [164]. The latter hypothesis may be corroborated by the fact that there is virtually no concentration gradient between plasma and urine for total bile acids.

Considerations of overall clearance disregard the question of the mechanisms involved in taking into account only the measurable variables of supply (concentrations in plasma) and output (amount excreted in the urine per time period). Using the conventional formula for the renal clearance of a given solute

$$\frac{\text{Conc. of solute in the urine} \times \text{vol. of urine per time}}{\text{Conc. of solute in plasma}} = \text{Clearance}_{\text{solute}}$$

a term for the permeability of the solute can be derived, if it is assumed that clearance is proportional to permeability (cm/min) and to the filtration surface (cm^2). In a given subject, filtration surface can be set unity. This method of consideration seems to be a possibility for evaluating the present data in terms of permeabilities of different bile acids and bile acid groups under pathophysiological conditions in a single subject. The great scattering of "clearance" values observed [23,49] may in part be due to interindividual variations in the filtration surface. The unidirectional permeability of an individual bile acid should give a constant value in a patient, assumed to have constant renal function over a certain observation period. In order to confirm this constant value, it is desirable that the bile acid load, i.e., the concentration of the bile acid in the plasma, differs during the observation period.

Some workers have found relatively constant values for apparent clearances of nonsulfated bile acids [165] and variable apparent clearances for sulfated bile acids that were directly related to the total urinary bile acid output. Their findings support the hypothesis that an increased filtered load of nonsulfated bile acids inhibits reabsorption of bile acid sulfates [165]. Others have observed that with increasing concentrations of nonsulfated bile acids in the plasma, saturation of their renal excretion ensued (i.e., variable apparent clearance), whereas the urinary excretion of sulfates correlated with their concentrations in the plasma [164]. These findings were interpreted as pointing to uni- and bidirectional diffusion processes, possibly at more distal sites of the tubular system where a diffusion gradient between plasma and urine may be present.

So far, it is established only that the "clearances" of sulfated bile acids are higher than the "clearances" of nonsulfated bile acids [23,49,169], but this has been disputed [164].

Very high apparent permeability coefficients could be estimated for conjugates of tetrahydroxylated bile acids [87], exceeding by far those found for bile acid sulfates. Since the protein binding of bile acids had been shown to be inversely related to the number of hydroxyl groups present at the steroid nucleus [166], tetrahydroxylated bile acids may be filtered to a proportionally larger extent than less polar bile acids. It is not known whether they are reabsorbed as well as are the conjugates of cholic acid [159]. Furthermore, their hydrophilicity seems to be high owing to the special steric arrangement of the additional hydroxy groups [86,87,122], thereby favoring a distribution into the aqueous phase.

The beneficial effect seen on the plasma disappearance of bile acids in intrahepatic cholestasis, if atypical hydroxylations are induced, may serve as a strong argument for the assumption that owing to the stimulation of additional hydroxylations, and to a lesser extent of glucuronidation, therapeutically effective excretion of bile acids by renal pathways may be achieved.

6. CONCLUSION

The urinary profile of bile acids is in part still enigmatic, and the diversity found may lead to the conclusion that it is of no real value to the clinician as a routine test of liver function. However, there may indeed be clues in the detailed analyses of bile acids in the urine, and valuable information may accrue relating to structural, subcellular, and biochemical derangements of the hepatobiliary system.

So far, the number of patients thoroughly investigated is small, and larger groups must be studied. The disease entities and stages require more strict definition, so that analyses of urinary bile acids may be accepted as clinically useful. Rapidity of measurements and use of routine methods are practical demands, and further progress in understanding the biochemical and physiological reasons for the occurrence of the diversity of bile acids in the urine seems to be equally important.

ACKNOWLEDGMENT AND DEDICATION.   This work was supported by Deutsche Forschungsgemeinschaft grants Ba 253/3-6, Ge 110/11, and SFB 154. This chapter is dedicated to Prof. Dr. W. Gerok, to whom I owe thanks for many years of friendly tutelage, on the occasion of his 60th birthday.

REFERENCES

1. T. A. Mahowald, J. T. Matschiner, S. L. Hsia, E. A. Doisy, Jr., W. H. Elliott, and E. A. Doisy, *J. Biol. Chem.* **225**, 795 (1957).
2. T. Alström and A. Norman, *Acta Med. Scand,* **191**, 521 (1972).
3. A. F. Hofmann, *J. Lipid Res.* **8**, 55 (1967).
4. B. Fransson and G. Schill, *Acta Pharm. Suec.* **12**, 417 (1975).
5. E. H. Ahrens, Jr., and L. C. Craig, *J. Biol. Chem.* **195**, 763 (1952).
6. P. Eneroth and J. Sjövall, in "The Bile Acids" (P. P. Nair and D. Kritchevsky, eds.), Volume 1, p. 121, Plenum Press, New York (1971).
7. H. L. Bradlow, *Steroids* **11**, 265 (1968).
8. I. Makino and J. Sjövall, *Anal. Lett.* **5**, 341 (1972).
9. B. Almé, A. Bremmelgaard, J. Sjövall, and P. Thomassen, *J. Lipid Res.* **18**, 339 (1977).
10. J. Yanagisawa, H. Ichimiya, M. Nagai, and F. Nakayama, *J. Lipid Res.* **25**, 750 (1984).
11. S. Barnes and A. Chitranukroh, *Ann. Clin. Biochem.* **14**, 235 (1977).
12. J. F. Pageaux, B. Duperray, D. Anker, and M. Dubois, *Steroids* **34**, 73 (1979).
13. C. H. L. Shackleton and J. O. Whitney, *Clin. Chim. Acta* **107**, 231 (1980).
14. J. O. Whitney and M. M. Thaler, *J. Liq. Chromatogr.* **3**, 545 (1980).
15. M. Axelson and B. L. Sahlberg, *Anal. Lett.* **14**, 771 (1981).
16. K. D. R. Setchell and J. Worthington, *Clin. Chim. Acta* **125**, 135 (1982).
17. G. Hedenborg and A. Norman, *Scand. J. Clin. Lab. Invest.* **44**, 725 (1984).
18. A. Dyfverman and J. Sjövall, in "Biological Effects of Bile Acids" (G. Paumgartner, A. Stiehl, and W. Gerok, eds.), p. 281, MTP Press, Lancaster (1979).
19. A. Dyfverman and J. Sjövall, *Anal. Biochem.* **134**, 303 (1983).
20. L. Peric-Golia and R. S. Jones, *Proc. Soc. Exp. Med.* **110**, 327 (1962).
21. J. M. Yousef and M. M. Fisher, *Can. J. Physiol. Pharmacol.* **53**, 880 (1975).
22. M. Tamari, M. Ogowa, and M. Kametaka, *J. Biochem.* **80**, 371 (1976).
23. A. Stiehl, *Eur. J. Clin. Invest.* **4**, 59 (1974).
24. T. Cronholm, I. Makino, and J. Sjövall, *Eur. J. Biochem.* **26**, 251 (1972).
25. P. Back, K. Spaczynski, and W. Gerok, *Hoppe-Seyler's Z. Physiol. Chem.* **355**, 749 (1974).
26. B. Almé and J. Sjövall, *J. Steroid Biochem.* **13**, 907 (1980).
27. P. A. Thomassen, *Eur. J. Clin. Invest.* **9**, 425 (1979).
28. I. Makino, K. Shinozaki, and S. Nakagawa, *Lipids* **8**, 47 (1973).
29. I. Makino, K. Shinozaki, S. Nakagawa, and K. Mashimo, *J. Lipid Res.* **15**, 132 (1974).
30. I. Makino, H. Hashimoto, K. Shinozaki, K. Yoshino, and S. Nakagawa, *Gastroenterology* **68**, 545 (1975).
31. P. Back, *Verh. Dtsch. Ges. Inn. Med.* **79**, 950 (1973).
32. A. Stiehl, D. L. Earnest, and W. H. Admirand, *Gastroenterology* **68**, 534 (1975).
33. P. Back and K. Walter, *Gastroenterology* **78**, 671 (1980).
34. A. Bremmelgaard and J. Sjövall, *Eur. J. Clin. Invest.* **9**, 341 (1979).
35. S. Burstein and S. Lieberman, *J. Am. Chem. Soc.* **80**, 5235 (1958).
36. A. Norman and B. Strandvik, *J. Lab. Clin. Med.* **78**, 181 (1971).
37. A. Norman and B. Strandvik, *Acta Paediatr. Scand.* **62**, 253 (1973).

38. R. H. Palmer, *J. Lipid Res.* **12**, 680 (1971).

39. B. Strandvik, *Opuscula Med.* (Suppl. 29), 1 (1973).

40. I. Makino, J. Sjövall, A. Norman, and B. Strandvik, *FEBS Lett.* **15**, 161 (1971).

41. P. Back, *Z. Gastroenterol.* **11**, 477 (1973).

42. W. Otting, *in* "Biochemisches Taschenbuch" (H. M. Rauen, ed.), Part 1, p. 880, Springer Verlag, Berlin (1964).

43. A. Stiehl, *in* "Bile Acids in Human Diseases" (P. Back and W. Gerok, eds.), p. 73, F. K. Schattauer Verlag, Stuttgart, New York (1972).

44. J. Sjövall and R. Vihko, *Acta Chem. Scand.* **20**, 1419 (1966).

45. J. Ellingboe, B. Almé, and J. Sjövall, *Acta Chem. Scand.* **24**, 463 (1970).

46. B. Almé and E. Nyström, *J. Chromatogr.* **59**, 45 (1971).

47. A. Stiehl, R. Raedsch, G. Rudolph, P. Czygan, and S. Walter, *Clin. Chim. Acta* **123**, 275 (1982).

48. H. Takikawa, H. Otsuka, T. Beppu, Y. Seyama, and T. Yamakawa, *J. Biochem.* **92**, 985 (1982).

49. G. P. van Berge Henegouwen, K. H. Brandt, H. Eyssen, and G. Parmentier, *Gut* **17**, 861 (1976).

50. A. Stiehl, *in* "Trends in Hepatology" (L. Bianchi, W. Gerok, and H. Popper, eds.), MTP Press, Lancaster (1985).

51. A. Németh and B. Strandvik, *Scand J. Clin. Lab. Invest.* **44**, 387 (1984).

52. R. H. Palmer, *Proc. Natl. Acad. Sci. USA* **58**, 1047 (1967).

53. P. Back, *Verh. Dtsch. Ges. Inn. Med.* **80**, 415 (1974).

54. P. Back, *in* "Advances in Bile Acid Research" (S. Matern, J. Hackenschmidt, P. Back, and W. Gerok, eds.), p. 149, F. K. Schattauer Verlag Stuttgart, New York (1975).

55. W. Fröhling and A. Stiehl, *in* "Advances in Bile Acid Research" (S. Matern, J. Hackenschmidt, P. Back, and W. Gerok, eds.), p. 153, F. K. Schattauer Verlag Stuttgart, New York (1975).

56. W. Fröhling and A. Stiehl, *Eur. J. Clin. Invest.* **6**, 67 (1976).

57. P. Back and D. V. Bowen, *Hoppe-Seyler's Z. Physiol. Chem.* **357**, 219 (1976).

58. P. Back, *Hoppe-Seyler's Z. Physiol. Chem.* **357**, 213 (1976).

59. B. Almé, Å. Nordén, and J. Sjövall, *Clin. Chim. Acta* **86**, 251 (1978).

60. S. Billets, P. S. Lietman, and C. Fenselau, *J. Med. Chem.* **16**, 30 (1973).

61. P. P. Nair, M. Gordon, and J. Reback, *J. Biol. Chem.* **242**, 7 (1967).

62. H. Takikawa, T. Beppu, and Y. Seyama, *Gastroenterol. Jpn.* **19**, 104 (1984).

63. H. Takikawa, H. Otsuka, T. Beppu, Y. Seyama, and T. Yamakawa, *Digestion* **27**, 189 (1983).

64. J. Goto, F. Hasegawa, H. Kato, and T. Nambara, *Clin. Chim. Acta* **87**, 141 (1978).

65. A. Stiehl, M. Becker, P. Czygan, W. Fröhling, B. Kommerell, H. W. Rotthauwe, and M. Senn, *Eur. J. Clin. Invest.* **10**, 307 (1980).

66. J. A. Gregg, *Am. J. Clin. Pathol.* **49**, 404 (1968).

67. T. Usui, S. Nakasone, and M. Kawamoto, *Yonago Acta Med.* **10**, 250 (1966).

68. P. Back, H. Schumacher, and W. Gerok, *Verh. Dtsch. Ges. Inn. Med.* **78**, 1300 (1972).

69. P. Back, *Helv. Med. Acta* **37**, 193 (1973).

70. P. Back, *Klin. Wochenschr.* **51**, 926 (1973).

71. P. Back, *Clin. Chim. Acta* **44**, 199 (1973).

72. P. Back, *in* "Liver and Bile" (L. Bianchi, W. Gerok, and K. Sickinger, eds.), p. 121, MTP Press Lancaster (1977).

73. P. A. Thomassen, *Eur. J. Clin. Invest.* **9**, 425 (1979).

74. P. A. Thomassen, *Eur. J. Clin. Invest.* **9**, 417 (1979).

75. B. Almé, A. Bremmelgaard, J. Sjövall, and P. Thomassen, *in* "Advances in Bile Acid Research" (S. Matern, J. Hackenschmidt, P. Back, and W. Gerok, eds.), p. 145, F. K. Schattauer Verlag, Stuttgart, New York (1975).

76. K. Carlström, D. N. Kirk, and J. Sjövall, *J. Lipid Res.* **22**, 1225 (1981).

77. G. P. van Berge Henegouwen, Galzuren en Cholestase. Thesis, Drukkerij Elerie, Bennekon, The Netherlands (1974).

78. M. Makita and W. W. Wells, *Anal. Biochem.* **5**, 523 (1963).

79. P. Back, J. Sjövall, and K. Sjövall, *Med. Biol.* **52**, 31 (1974).

80. J. T. Matschiner, T. A. Mahowald, S. L. Hsia, E. A. Doisy, Jr., W. H. Elliott, and E. A. Doisy, *J. Biol. Chem.* **225**, 803 (1957).

81. S. L. Hsia, J. T. Matschiner, T. A. Mahowald, W. H. Elliott, E. A. Doisy Jr., S. A. Thayer, and E. A. Doisy, *J. Biol. Chem.* **225**, 811 (1957).

82. G. A. D. Haslewood, "Bile Salts," Methuen, London (1967).

83. G. A. D. Haslewood and L. Tökés, *Biochem. J.* **126**, 1161 (1972).

84. B. Almé, A. Bremmelgaard, J. Sjövall, and P. Thomassen, *in* "The Liver. Quantitative Aspects of Structure and Function" (R. Preisig, J. Bircher, and G. Paumgartner, eds.), p. 227, Editio Cantor, Aulendorf (1976).

85. A. Bremmelgaard and J. Sjövall, *J. Lipid Res.* **21**, 1072 (1980).

86. P. Back, H. Fritz, and C. Populoh, *Hoppe-Seyler's Z. Physiol. Chem.* **365**, 479 (1984).

87. P. Back, *Klin. Wochenschr.* **60**, 541 (1982).

88. P. T. Clayton, D. P. R. Muller, and A. M. Lawson, *Biochem. J.* **206**, 489 (1982).

89. B. J. Koopman, B. G. Wolthers, J. C. van der Molen, G. T. Nagel, R. J. Waterreus, and H. J. G. H. Oosterhuis, *Clin. Chim. Acta* **142**, 103 (1984).

90. G. G. Parmentier, G. A. Janssen, E. A. Eggermont, and H. J. Eyssen, *Eur. J. Biochem.* **102**, 173 (1979).

91. J. A. Summerfield, B. H. Billing, and C. H. L. Shackleton, *Biochem. J.* **154**, 507 (1976).

92. B. Strandvik and S. Å. Wikström, *Eur. J. Clin. Invest.* **12**, 301 (1982).

93. M. Blum and N. Spritz, *J. Clin. Invest.* **45**, 187 (1966).

94. H. Takikawa, H. Otsuka, T. Beppu, Y. Seyama, and T. Yamakawa, *J. Biochem.* **92**, 985 (1982).

95. A. Norman, B. Strandvik, and Ö. Ojamäe, *Acta Paediatr. Scand.* **61**, 571 (1972).

96. A. Norman and B. Strandvik, *Acta Paediatr. Scand.* **62**, 264 (1973).

97. A. Norman and B. Strandvik, *Acta Paediatr. Scand.* **62**, 161 (1973).

98. A. Eklund, A. Norman, and B. Strandvik, *Scand. J. Clin. Invest.* **40**, 595 (1980).

99. J. J. Schneider and N. S. Bhacca, *J. Biol. Chem.* **241**, 5313 (1966).

100. D. Trülzsch, J. Roboz, H. Greim, P. Czygan, J. Rudick, F. Hutterer, F. Schaffner, and H. Popper, *Biochem. Med.* **9**, 158 (1974).

101. B. P. Lisboa and J-C. Plasse, *Eur. J. Biochem.* **31**, 378 (1972).

102. B. P. Lisboa and J-C. Plasse, *Steroids Lipids Res.* **3**, 142 (1972).

103. K. A. Mitropoulos and N. B. Myant, *Biochem. J.* **103**, 472 (1967).

104. Y. Tazawa, M. Yamada, M. Nakagawa, T. Konno, and K. Tada, *Acta Paediatr. Scand.* **73**, 392 (1984).

105. Y. Tazawa and T. Konno, *Acta Paediatr. Scand.* **71**, 91 (1982).

106. F. Stellaard, J. B. Watkins, P. Szczepanik-van Leeuwen, and D. Alagille, *Gastroenterology* **77**, A 42 (1977).

107. Y. Tazawa, M. Yamada, M. Nakagawa, T. Konno, and K. Tada, *J. Pediatr. Gastroenterol. Nutr.* **4**, 32 (1985).

108. P. Back, *Klin. Wochenschr.* **51**, 926 (1973).

109. A. Norman and B. Strandvik, *Acta Paediatr. Scand.* **63**, 92 (1974).

110. A. Norman and B. Strandvik, *Acta Paediatr. Scand.* **63**, 97 (1974).

111. A. Norman and B. Strandvik, *Acta Paediatr. Scand.* **62**, 253 (1973).

112. A. Norman, B. Strandvik, and R. Zetterström, *Acta Paediatr. Scand.* **58**, 59 (1969).

113. P. Back and H. J. Zeisel, 16th Meeting of the Eur. Assoc. for the Study of the Liver, Lisbon, Sept. 3–5, No. 102 (1981).

114. H. Nittono, N. Natatsu, K. Obinata, T. Watanabe, S. Niijima, H. Sasaki, O. Arisaka, K. Yamada, and H. Kato, *in* "Advances in Bile Acid and Bile Alcohol Research in Japan" (S. Ikawa, H. Kawasaki, and N. Kaibara, eds.), p. 179, Tokyotanabesyoji, Tokyo (1984).

115. A. Norman and B. Strandvik, *Acta Paediatr. Scand.* **62,** 161 (1973).

116. A. Bremmelgaard and J. Sjövall, Supplementary publication No. SUP 90040, *Eur. J. Clin. Invest.* **9,** 341 (1979).

117. Y. Amuro, E. Hayashi, T. Endo, K. Higashino, and S. Kishimoto, *Clin. Chim. Acta* **127,** 61 (1983).

118. Y. Amuro, K. Higashino, T. Endo, T. Endo, and E. Hayashi, *Proc. Jpn. Soc. Med. Mass Spectrom.* **8,** 61 (1983).

119. Y. Amuro, T. Endo, K. Higashino, K. Uchida, and Y. Yamamura, *Clin. Chim. Acta* **114,** 137 (1981).

120. A. G. F. de Pagter, G. P. van Berge Henegouwen, J. A. ten Bokkel Huininck, and K. H. Brandt, *Gastroenterology* **71,** 202 (1976).

121. J. R. Bloomer and J. L. Boyer, *Ann. Inter. Med.* **82,** 310 (1975).

122. P. Back, *in* "Trends in Hepatology" (L. Bianchi, W. Gerok, and H. Popper, eds.), p. 53, MTP Press, Lancaster (1985).

123. P. Back, *Verh. Dtsch. Ges. Inn. Med.* **80,** 415 (1974).

124. P. Back, *in* "Liver and Bile" (L. Bianchi, W. Gerok, and K. Sickinger, eds.), p. 121, MTP Press, Lancaster (1977).

125. W. Fröhling, A. Stiehl, P. Czygan, M. Liersch, B. Kommerell, H. W. Rotthauwe, and M. Becker, *in* "Bile Acid Metabolism in Health and Disease" (G. Paumgartner and A. Stiehl, eds.), p. 101, MTP Press, Lancaster (1977).

126. A. Stiehl, M. M. Thaler, and W. H. Admirand, *Pediatrics* **51,** 992 (1973).

127. I. Makino, and S. Nakagawa, *in* "Advances in Bile Acid Research" (S. Matern, J. Hackenschmidt, P. Back, and W. Gerok, eds.), p. 135, F. K. Schattauer Verlag, Stuttgart, New York (1975).

128. L. Capocaccia, A. F. Attili, A. Cantafora, F. Bracci, L. Paciscopi, C. Puoti, U. Pièche, and M. Angelico, *Dig. Dis. Sci.* **26,** 513 (1981).

129. M. Grün, E. Richter, and W. D. Heine, *Hepato-gastroenterol.* **29,** 232 (1982).

130. H. Eyssen, G. Parmentier, F. Compernolle, J. Boon, and E. Eggermont, *Biochim. Biophys. Acta* **273,** 212 (1972).

131. R. F. Hanson, J. N. Isenberg, G. C. Williams, D. Hachey, P. Szezepanik, P. D. Klein, and H. L. Sharp, *J. Clin. Invest.* **56,** 577 (1975).

132. R. F. Hanson, P. Szczepanik-van Leeuwen, G. C. Williams, G. Grabowski, and H. L. Sharp, *Science* **203,** 1107 (1979).

133. G. G. Parmentier, G. A. Jansson, E. A. Eggermont, and H. J. Eyssen, *Eur. J. Biochem.* **102,** 173 (1979).

134. B. F. Kase, I. Björkhem, P. Hågå, and J. I. Pedersen, *J. Clin. Invest.* **75,** 427 (1985).

135. P. G. Killenberg, *J. Lipid Res.* **19,** 24 (1978).

136. L-J. Chen, T. J. Imperato, and R. J. Bolt, *Biochim. Biophys. Acta* **522,** 443 (1978).

137. L. Lööf and B. Wengle, *Scand. J. Gastroenterol.* **14,** 513 (1979).

138. J. A. Summerfield, J. L. Gollan, and B. H. Billing, *Biochem. J.* **156,** 339 (1976).

139. M. Liersch and A. Stiehl, *Z. Gastroenterol.* **12,** 131 (1974).

140. H. Matern, S. Matern, C. Schelzig, and W. Gerok, *FEBS Lett.* **118,** 251 (1980).

141. S. Matern, H. Matern, E. H. Farthmann, and W. Gerok, *J. Clin. Invest.* **74,** 402 (1984).

142. R. A. Ulstrom, E. Colle, J. Burley, and R. Gunville, *J. Clin. Endocrinol.* **20,** 1080 (1960).

143. F. L. Mitchell, *Vitam. Horm.* **25,** 191 (1967).

144. P. Czygan, H. Greim, D. Trülzsch, J. Rudick, F. Hutterer, F. Schaffner, O. Rosenthal, H. Popper, and D. Y. Cooper, *Biochim. Biophys. Acta* **354,** 168 (1974).

145. A. Bremmelgaard, Scand. *J. Clin. Lab. Invest.* **43**, 603 (1983).

146. D. W. Nebert and M. Negishi, *Biochem. Pharm.* **31**, 2311 (1982).

147. P. Back, *Gastroenterology* **78**, 1651 (1980).

148. K. Jungermann and N. Katz, *in* "Metabolic Compartmentation" (H. Sies, ed.), p. 411, Academic Press, London (1982).

149. A. L. Jones, D. L. Schmucker, R. D. Adler, R. K. Ockner, and J. S. Mooney, *in* "The Liver. Quantitative Aspects of Structure and Function" (R. Preisig, J. Bircher, and G. Paumgartner, eds.), p. 36, Editio Cantor Aulendorf (1976).

150. J. Wanson, P. Drochmans, C. May, W. Penasse, and A. Popowski, *J. Cell Biol.* **66**, 23 (1975).

151. G. Rothe, M. Renner, and P. Back, *Hepatology* **4**, 780 (1984).

152. J. I. Pedersen and J. Gustafsson, *FEBS Lett.* **121**, 345 (1980).

153. S. Goldfischer, C. L. Moore, A. B. Johnson, A. J. Spiro, M. P. Valsamis, H. K. Wisniewski, R. H. Ritch, W. T. Norton, I. Rapin, and L. M. Gartner, *Science* **182**, 62 (1973).

154. J. Sjövall, *Acta Chem. Scand.* **13**, 711 (1959).

155. P. Eneroth, B. Gordon, R. Ryhage, and J. Sjövall, *J. Lipid Res.* **7**, 511 (1966).

156. V. C. Aries and M. J. Hill, *Biochim. Biophys. Acta* **202**, 535 (1970).

157. T. Setoguchi and T. Katsuki, *in* "Advances in Bile Acid and Bile Alcohol Research in Japan" (S. Ikawa, H. Kawasaki, and N. Kaibara, eds.), p. 233, Tokyotanabesyoji, Tokyo (1984).

158. D. Rudman and F. E. Kendall, *J. Clin. Invest.* **36**, 530 (1957).

159. I. M. Weiner, J. E. Glasser, and L. Lack, *Am. J. Physiol.* **207**, 964 (1964).

160. L. Lack and I. M. Weiner, *in* "The Bile Acids," Volume 2 (P. P. Nair and D. Kritchevsky, eds.), p. 50, Plenum Press, New York (1973).

161. S. Barnes, J. L. Gollan, and B. H. Billing, *Biochem. J.* **166**, 65 (1977).

162. G. R. Zins and I. M. Weiner, *Am. J. Physiol.* **215**, 840 (1968).

163. A. Chitranukroh, G. Taggart, and B. H. Billing, *Clin. Sci.* **68**, 63 (1985).

164. P. Back and W. Gerok, *in* "Bile Acid Metabolism in Health and Disease" (G. Paumgartner and A. Stiehl, eds.), p. 93, MTP Press, Lancaster (1977).

165. J. A. Summerfield, J. Cullen, S. Barnes, and B. H. Billing, *Clin. Sci. Mol. Med.* **52**, 51 (1977).

166. D. Rudman and F. E. Kendall, *J. Clin. Invest.* **36**, 538 (1957).

167. E. Sacquet, M. Parquet, M. Riottot, A. Raizman, P. Jarrige, C. Huguet, and R. Infante, *J. Lipid Res.* **24**, 604 (1983).

168. R. Raedsch, B. H. Lauterburg, and A. F. Hofmann, *Dig. Dis. Sci.* **26**, 394 (1981).

169. C. L. Corbett, T. C. Bartholomew, B. H. Billing, and J. A. Summerfield, *Clin. Sci.* **61**, 773 (1981).

170. A. Stiehl, R. Raedsch, G. Rudolph, U. Gurdert-Remy, and M. Senn, *Hepatology* **5**, 492 (1985).

171. K. Ohnishi, A. Mishima, and K. Okuda, *Hepatology* **2**, 849 (1982).

172. C. Colombo, G. Zuliani, M. Ronchi, J. Breidenstein, and K. D. R. Setchell, *Pediatr. Res.* **21**, 197 (1987).

173. A. M. Lawson, M. J. Madigan, D. Shortland, and P. T. Clayton, *Clin. Chim. Acta* **161**, 221 (1986).

174. J. Gustafsson, S. Andersson, and J. Sjövall, *Biol. Neonate* **47**, 26 (1985).

Chapter 12

FECAL BILE ACIDS

K. D. R. Setchell, J. M. Street, and J. Sjövall

1. INTRODUCTION*

Fecal bile acid excretion represents an important pathway for the elimination of cholesterol from the body. Extensive transformation of primary bile acids occurs due to the action of the intestinal microflora [1] resulting in a complex array of metabolic products being formed within the lumen which are subsequently excreted. Since the enterohepatic circulation of bile acids is extremely efficient, most of the bile acid pool is conserved by intestinal reabsorption; however, a small percentage of the bile acids escapes the normal reabsorption mechanism and is therefore excreted. In the steady state this fecal loss will be approximately equal to the daily synthesis, and therefore, accurate quantitative determination of fecal bile acid excretion provides important information about bile acid kinetics and cholesterol homeostasis. Qualitative analysis, on the other hand, provides insight into intraluminal events involving bacteria and bile acid interaction, since microbial activity will largely influence the pattern of fecal bile acid excretion. Diet, drugs, gut transit time, intraluminal environment, gallbladder emptying, and hepatic function are other factors that will have a bearing on fecal bile acid excretion, and these are discussed in some detail.

The quantity of bile acids excreted in feces is small (<500 mg/day) relative to the mass of stools and the amounts of the many interfering compounds that are present, and this presents considerable problems in their reliable estimation.

---

* Abbreviations used are as follows: NAD, oxidized form of nicotinamide-adenine dinucleotide; NADH, reduced form of nicotinamide-adenine dinucleotide.

K. D. R. Setchell and J. M. Street   Department of Pediatric Gastroenterology and Nutrition, Clinical Mass Spectrometry Laboratories, Children's Hospital Medical Center, Cincinnati, Ohio
J. Sjövall   Department of Physiological Chemistry, Karolinska Institute, Stockholm, Sweden

This chapter will describe the methods that have been reported for fecal bile acid analysis, attempting to give a critical assessment of available techniques. As far as possible, we have attempted to provide a comprehensive review of fecal bile acid excretion as it relates to health and disease.

2. METHODS FOR DETERMINATION OF FECAL BILE ACIDS

2.1. Early Techniques

One of the earliest procedures for determination of fecal bile acids involved a precipitation step using zinc hydroxide followed by titration [2]. This approach which was used by others [3–6] was reported to give an incomplete recovery of bile acids evidenced from experiments in which the endogenous bile acid pool was labeled by administration of [14C] cholesterol and [14C] bile acids. In addition, interference from titratable contaminants that were not removed during the workup was observed [7]. Attempts to improve on these methods included the introduction of a charcoal adsorption step to remove acidic impurities [8] or the use of silicic acid chromatography to isolate bile acids from sterols and fatty acids [6,9].

Spectrophotometry was used for bile acid analysis [10,11]. However, the methods were only sensitive to dihydroxy and trihydroxy bile acids and did not detect monohydroxy bile acids, which comprise a high proportion of the total bile acids in human feces. The high normal ranges that were reported illustrate the deficiencies of this detection technique.

Although many of these early procedures "appeared" to give reliable data, this was probably accounted for by the fact that the poor recovery of bile acids through the method was compensated for by an overestimation in the "end-point" method of detection.

Using the argument that the specific activity of fecal bile acids is equal to that of plasma cholesterol after the administration of radioactively labeled cholesterol, this procedure, because of its simplicity, was used by many laboratories for determination of fecal bile acids [12]. Studies by Lindstedt showed that the specific activity curve for [14C] cholesterol in plasma crossed the curves for cholic and deoxycholic acids in bile at 3–5 and 8–12 days, respectively, in healthy subjects [13]. Similar equilibration times were found by others [14], but the equilibration of plasma cholesterol and cholic acid was much longer, 12–16 days, in hypercholesterolemic subjects [15]. The crossover of the specific activity curves of plasma cholesterol and fecal bile acids showed wide variations, 7–21 days, in five patients on a liquid formula diet [7]. Despite these drawbacks, this approach gave fecal bile acid measurements in reasonable agreement with the gas–liquid chromatography (GLC) techniques published at the time [7,16,17].

2.2. Techniques Based on Gas–Liquid Chromatography

Gas-liquid chromatography has been the most widespread technique of choice for the determination of bile acids in feces, and over the last 20 years the general procedures described fall into five basic methods, with numerous modifications reported. These are reviewed below and summarized in Table I.

2.2.1. Method of Grundy, Ahrens, and Miettinen (1965)

Since its description, the method of Grundy et al. [7] has formed the basis for most of the fecal bile acid measurements described in the literature. In the original method, lipids, bile acids, and steroids were extracted by homogenization of the stools in an equal volume of water. From this homogenization, an aliquot (1 g) was taken, and 24-[^{14}C] deoxycholic acid was added as a recovery marker. Sterol esters were subjected to mild saponification for 1 hr by refluxing in 1 N sodium hydroxide in 90% ethanol. Neutral sterols were then removed by partitioning into petroleum ether, and after backwashing of the ether with 1 N NaOH in 50% ethanol to remove residual bile acids, this extract could be used for fecal neutral steroid determination [22]. In order to hydrolyze conjugated bile acids, rigorous saponification of the aqueous phase was carried out after addition of 10 N NaOH at 15 psi/3 hr. Unconjugated bile acids were then extracted into chloroform/methanol (2/1 v/v) after acidification of the aqueous phase to pH 2 with HCl. The removal of fatty acids and acidic pigments was effected by chromatography on columns of florisil. After preparation of the methyl esters by 5% HCl in dry methanol, preparative thin-layer chromatography (TLC) was carried out in two solvent systems to further remove fatty acids from bile acids. The bile acid methyl esters were recovered from the TLC plates and trimethylsilyl ethers prepared for GLC analysis. Columns packed with DC-560 or SE-30 liquid phase were routinely used for total bile acid determination, but where specific separation and quantification of the individual bile acids was required, Hi-eff 8BP or QF-1 columns proved more satisfactory. Recoveries of radioactive deoxycholic acid ranged from 86 to 96%, but not surprisingly, it was shown that 3-oxo- and 3,12-dioxo-cholanoic acids were partially destroyed during the saponification steps. Numerous variations of this technique have followed, usually in attempts to decrease the lengthy analysis time of the method, but few major modifications have been described, and in most instances the sequence of steps has been rearranged and minor changes in solvents and GLC conditions reported [23–29]. For example, Soxhlet extraction in ethanol of the original stool sample has been used [30], or lyophilized stools have been extracted directly with chloroform/methanol mixtures [23] or subjected to immediate saponification [24]. Column chromatography on florisil has been used in place of the TLC step [31]. Conversely, additional TLC steps have been included to improve the specificity of the assay [32]. Omission of the alkaline hydrolysis step has been used as a

TABLE I. Methods and Main Steps in the Analysis of Fecal Bile Acids by GLC[a]

Method	Reference	Pretreatment	Extraction	Purification steps	Derivatization	Internal standard	GLC liquid phase
Grundy et al., 1965	7	Homogenization, mild and rigorous saponification	Chloroform/methanol (2/1 v/v)	1. Neutral steroids removed by pet. ether 2. Florisil and/or TLC	Methyl ester –TMS ether	5α-Cholestane	DC-560 SE-30 Hi-eff 8BP
Ali et al., 1966	18	Homogenization, lyophilized	Ethylene chloride/methanol (3/1 v/v)	Extraction of lipids with pet. ether	Methyl ester–TFA; silicic acid chromatography	Methyl-CDCA	QF-1
Evrard and Janssen, 1968	19	Homogenization, lyophilized	Acetic acid toluene added after extraction	1. Saponification 2. Sterols extracted by pet. ether 3. Bile acids extracted by ether 4. Chromate oxidation	Methyl ester	23-Nor-DCA	JXR
Eneroth et al., 1968	20	Homogenization	Hot chloroform/methanol (1/1 v/v)	1. Saponification in KOH/dioxan 2. Ether extraction 3. Silicic acid chromatography	Methyl ester –TFA	3-Oxo-12α-hydroxy -5β-cholanoic and 3α-acetoxy- 5β-cholanoic acids	QF-1
Setchell et al., 1983	21	Homogenization	Reflux with 1. 90% ethanol 2. 80% ethanol 3. chloroform/methanol (1/1 v/v)	1. Lipidex 1000 and Bond Elut 2. Amberlyst A-15 3. DEAP-LH-20 4. Enzymatic or alkaline hydrolysis of conjugate fractions	Methyl ester –TMS ether Lipidex 5000	Coprostanol	OV-1 fused silica capillary column

[a] TMS, trimethylsilyl; TFA, trifluoroacetate; CDCA, chenodeoxycholic acid; DCA, deoxycholic acid; pet. ether, petroleum ether.

means for determining only unconjugated bile acids [33], which constitute the largest proportion of the total fecal bile acids excreted by normal humans. Subdivision of the fecal extract and differential hydrolysis and solvolysis of the samples has also been used as a method of determining the relative proportions of unconjugated bile acids, sulfates, and total nonsulfated species [34,35].

2.2.2. Method of Ali, Kuksis, and Beveridge (1966)

The method of Ali *et al.* [18] essentially applied a similar sequence of steps to that of Grundy *et al.* [7]. Methanol/ethylene dichloride mixture (1/3 v/v) was employed for the initial prolonged (24 hr) Soxhlet extraction. Neutral lipids were removed by extraction into petroleum ether, and following acidification to pH 2, bile acids were extracted. The remaining residues were combined and hydrolyzed in 2 N NaOH for 4 hr at 120°C and 15 psi, and after reextraction of bile acids, the total bile acid extract was further purified by silicic acid chromatography of the methyl esters to remove the bulk of fatty acids. Further purification and separation of bile acids into groups based on the number and type of functional groups was achieved on TLC. Gas–liquid chromatography analysis of the trifluoroacetate derivative was then performed on a 1% QF-1 column. No standards were employed to take account of procedural losses, and the methyl ester of chenodeoxycholic acid was used as the internal standard for quantification by GLC. Since this primary bile acid can be excreted in significant quantities in pathological conditions, these authors analyzed the sample both before and after addition of methyl chenodeoxycholate, thus doubling the time of analysis. Huang *et al.* [36,37] subsequently utilized [^{14}C] cholic acid, which was added to the feces for recovery determination, and quantification by GLC was performed using 5α-cholestane as the internal standard.

2.2.3. Method of Evrard and Janssen (1968)

Of all the methods developed for fecal bile acids, that of Evrard and Janssen [19] represented perhaps the most significant deviation from the original technique of Grundy *et al.* [7]. Extraction of bile acids and neutral steroids was effected by toluene after prior heating of the sample with glacial acetic acid and with the addition of 23-nordeoxycholic acid as an internal standard for later quantification. Saponification was carried out by heating with 20% KOH in ethylene glycol, and lipids removed by partitioning into petroleum ether. After acidification, unconjugated bile acids were extracted with diethyl ether and methyl esters prepared using ethereal diazomethane [38]. This extract was oxidized with chromic acid in order to convert all hydroxy cholanoates to their respective oxo analogs, which were analyzed by GLC. The method was originally intended to be a rapid alternative for determination of total fecal bile acids, and for this reason the authors did not consider the inability to separate individual oxo bile acids or to distinguish bile acid isomers to be a disadvantage. The use of 23-nordeoxycholic acid as an internal standard afforded a means of correcting for procedural losses in the extraction and derivatization steps and simplified quantification.

Slight modifications of this method have included the use of lyophilized feces, alternative internal standards, radiolabeled bile acids for recovery determinations, and different GLC conditions [39–41]. Introduction of a borohydride reduction step and determination of acidic steroids by a 3α-hydroxysteroid dehydrogenase enzyme has also been described [42]. Applying the extraction method of Evrard and Janssen [19], a simplified method was described [43] in which enzymatic hydrolysis was carried out and bile acids quantified as methyl ester-TMS ethers. During the extraction procedure acetates will be formed and cholylglycine hydrolase will not cleave the amide bond of a bile acid conjugate when acetyl groups are in the ABCD rings of the molecule [44]. Clearly, this simplification results in problems because it was claimed that lithocholic, chenodeoxycholic, and cholic acids were the major bile acids in feces, and numerous degradation products were found.

2.2.4. *Method of Eneroth, Hellström, and Sjövall (1968)*

In a critical series of publications Eneroth *et al.* [20,45,46] described and extensively evaluated a method for fecal bile acid analysis and applied this in a relatively detailed investigation of the bile acid composition of human feces. These studies also represented some of the earliest applications of mass spectrometry for the identification of bile acids in feces. In the development of the method, subjects were given labeled bile acids so that fecal samples contained a wide range of endogenously labeled bile acid metabolites. This feature, not employed in the validation of previously described methods, was important because extraction efficiencies cannot be measured accurately by simple addition of radiolabeled bile acids to fecal samples [21]. Bile acids were extracted by a continuous and prolonged Soxhlet extraction with hot chloroform/methanol (1/1 v/v). The extract was saponified with 4 N KOH in dioxan under conditions that do not hydrolyze bile acid conjugates, and after acidification bile acids were extracted for 16 hr into diethyl ether. Bile acids were purified by silicic acid chromatography prior to GLC analysis of the methyl ester and methyl ester-trifluoroacetate derivatives on QF-1 columns. 3-Oxo-12α-hydroxy-5β-cholanoic and 3α-acetoxycholanoic acids were used as internal standards for quantitative GLC analysis. Quantitative recoveries of endogenously labeled and added bile acids were achieved during the extraction stages, and the recovery of [^{14}C] cholic acid metabolites was 95.6%, while that for chenodeoxycholic acid metabolites was found to be lower (77.3%).

2.2.5. Method of Setchell, Lawson, Tanida, and Sjövall (1983)

Very few improvements in fecal bile acid analysis appeared in the literature until recent years when methods utilizing liquid–solid and liquid–gel extraction/chromatography steps and capillary column GLC were introduced. Setchell *et al.* [21] described an extensive method for detailed analysis of the profile of bile acids in feces, taking particular attention to incorporate procedures that would create minimal artifactual formation or destruction of the metabolites and permit a high degree of flexibility for use in either research or routine applications. Extraction of bile acids and related compounds from fecal homogenates (equivalent to 1/500–1/1000 of daily stool collection) was achieved by sequential refluxing for short periods in 90% ethanol, 80% ethanol, and chloroform/methanol (1/1 v/v). The combined extracts were dried and desalted after reconstitution in 0.01 M acetic acid by passage through a column of the lipophilic gel Lipidex 1000 and a cartridge of octadecylsilane bonded silica (Bond Elut or Sep-Pak C_{18}).

Since the behavior of these two adsorptive chromatography systems is complementary [47], this stage affords quantitative recovery of a wide range of nonpolar and polar lipophilic compounds from the extract while eliminating salts, water-soluble polar compounds, and other contaminants such as pigments. Cation exchange chromatography on SP-Sephadex or Amberlyst A-15 in the $[H^+]$ form was used to remove interfering cations prior to group separations of neutral compounds, including sterols, unconjugated bile acids, glycine-, taurine-, and sulfate-conjugated bile acids, by anion exchange chromatography on the lipophilic gel diethylaminohydroxypropyl Sephadex LH-20 (DEAP-LH-20 or Lipidex-DEAP). With minor modifications, solvent systems were adapted from the method originally described for separation of urinary bile acid conjugates [48]. Since in most instances most bile acids are unconjugated, this scheme allows isolation of this fraction without exposure to adverse pH conditions or prior saponification steps, as was used in previous techniques. For routine purposes a single solvent reflux in 80% ethanol would provide a satisfactory extraction of the bulk of fecal bile acids [21], and collection of the unconjugated bile acid fraction only from DEAP-LH-20 would enable a relatively rapid analytical method suitable for most clinical purposes. Furthermore, information on neutral sterol excretion from the same analysis is possible by direct derivatization of the neutral fraction, or where there is interest in sterol or steroid conjugates, this can be attained by using the alternative DEAP-LH-20 buffer systems described previously for neutral steroids [49]. Enzymatic hydrolysis of the glycine and taurine conjugates was carried out. In the absence of suitable commercial sources of bacterial sulfatase enzymes, bile acid sulfates were subjected to solvolysis and alkaline hydrolysis. Bile acid esters which have been reported to occur in fecal extracts [50,51] would be isolated in the fraction containing neutral sterols or unconjugated bile acids, depending on the site of esterification. Their determination, if required, is possible after mild saponification of these fractions and chromatographic separation of the resulting unconjugated bile acids from sterols on Lipidex 1000 or DEAP-LH-20.

In an earlier version of this method developed by the same group [52], bile acids and sterols were adsorbed to the neutral cross-linked polystyrene resin Amberlite XAD-2 and sterols eluted by washing the column with hexane. Procedures based on this general principle were adopted by others [53–55]. The extraction of bile acids by Amberlite XAD-2 resin is extremely flow dependent, and in our earlier experiences, washing with hexane did not give a quantitative recovery of lipids and frequently resulted in variable recoveries of monohydroxy bile acids. Furthermore, the limited capacity of the resin makes this a less robust purification step than the alternative combination of Lipidex 1000 and a Bond Elut cartridge. Amberlite XAD-2, erroneously described as an anion exchange resin, was reported to be advantageous in the final purification of fecal extracts for bile acid analysis. However, in this method it was introduced after methylation of the hydrolyzed extract [56]. Amberlite XAD-2 has also been employed for extraction of bile acids after their separation into conjugate groups and following their hydrolysis [53]. However, in this method sulfated bile acids were hydrolyzed enzymatically using a sulfatase enzyme that has limited specificity toward bile acid sulfates.

In a recent minor modification of the technique of Setchell *et al.* [21], a Soxhlet extraction step was used and neutral sterols removed by prolonged extraction into petroleum ether before bile acids were extracted for 6 hr using chloroform/methanol mixture (1/1 v/v) [57]. It is difficult to understand the claim by these authors that their alternative extraction procedure, representing a minor modification of the method of Eneroth *et al.* [20], was less time consuming than the extraction procedure of Setchell *et al.* [21], since it involved a 12-hr extraction compared with 5 hr by the original method.

2.3. Gas–Liquid Chromatography–Mass Spectrometry

Very few specific direct applications of mass spectrometry (MS) to fecal bile acid analysis have been described, but the technique commands particular importance in bile acid methodology primarily as the means of characterizing the peaks in GLC profiles. With the advent of high-resolving capillary columns, which are rapidly replacing conventional packed columns, the use of GLC–MS will assume increasing importance, but its high cost will probably continue to preclude its application to routine fecal bile acid determinations.

Early methods for compound identification usually relied on peak shift techniques using either different derivatives or liquid phases. In the earliest examples of combined GLC–MS, Eneroth et al. [45,46] described the characterization of mono-, di-, and trihydroxy bile acids in human feces. Electron impact ionization (22.5 eV) of the bile acids as their methyl ester and methyl ester–trifluoroacetates yielded spectra suitable for identification of the principal fecal bile acids and their isomers, including several oxo bile acids [45,46].

In view of the complex composition of bile acids in feces, repetitive scanning MS is necessary to provide a satisfactory analysis of minor metabolites. This approach, used extensively by Sjövall and co-workers for both qualitative and quantitative analysis of urinary bile acids [48], is well suited to analysis of fecal bile acid extracts. With the use of capillary columns, relatively fast scanning is required to acquire sufficient scans over the narrow peaks. Informative reviews on MS of bile acids have been written which provide detailed information on mass spectral interpretation of bile acid structures [58–63], and from these, series of mass/charge (m/z) values can be selected and ion current chromatograms plotted for general or specific bile acid structures [48,54,64]. A typical example of this is seen in Fig. 1, which shows reconstructed ion current chromatograms for ions specific for a number of oxo bile acids identified in human feces [65]. By selection of specific ions and suitable internal or external standards, it is possible to quantify the amount of bile acid in the extract, but few examples have been reported of quantification in this way. In general, quantification has been performed from peak height or peak area response in the gas chromatograms. Quantitative determination of methyl ether methyl esters of fecal bile acids has been carried out using the deuterated external standards [^{2}H$_6$] lithocholic, [^{2}H$_9$] deoxycholic, [^{2}H$_9$] chenodeoxycholic, and [^{2}H$_{12}$] cholic acids prepared separately by derivatization with reagents labeled with deuterium. This approach was used in early selected ion monitoring (SIM) methods for steroids and prostaglandins (see references in Ref. 66). It is not to be recommended because of the possibility of incomplete derivatization or exchange of labeled and unlabeled derivative groups. Selected ion monitoring of the principal fecal bile acids was compared with repetitive magnetic scanning, for quantification, with the conclusion that the former approach provides increased sensitivity but no improvement in precision [53].

A SIM technique for 3-methylcholanthrene in stools, with a sensitivity equivalent to 17 ng/g stool, was recently described. This is probably the only description of the specific application of SIM–MS to a problem related to fecal bile acid excretion. This carcinogen has been postulated to be important in the etiology of colonic carcinogenesis because of its possible formation by bacterial degradation of bile acids. The technique used a high-pressure liquid chromatography (HPLC) method for the isolation of 3-methylcholanthrene and GLC–MS

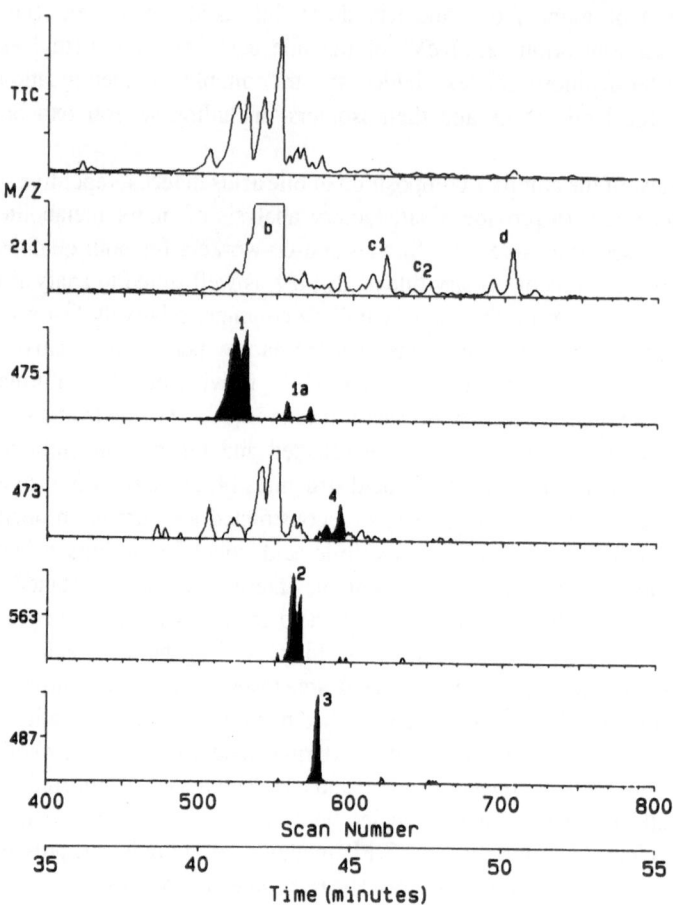

FIGURE 1. Total ion current (TIC) and selected ion current chromatograms, obtained in the GC–MS analysis of oximes of unconjugated bile acids in feces. The ions were selected to represent the 3-oxime TMS-4-ene structure (*m/z* 211), molecular ions of oxime TMS derivatives of methyl oxocholanoates (*m/z* 475), oxocholenoates (*m/z* 473), and hydroxyoxocholanoates (*m/z* 563) and M-89 for the derivative of dioxocholanoates (m/z 487). The following bile acids were identified: (1) 3-oxo-5β-cholanoic; (1a) 3-oxo-5α-cholanoic; (2) 3-oxo-12α-hydroxy-5β-cholanoic; (3) 3,12-dioxo-5β-cholanoic; (4) 3-oxo-4-cholenoic. Peaks indicated by the letters b, c1, c2, and d represent oxime derivatives of neutral sterols. [Reproduced from Ref. [65] with permission of authors.]

with selected ion monitoring of the ions m/z 268 and m/z 270 for the [^{14}C]–labeled analog [67].

2.4. Enzymatic Methods

Methods involving enzymatic determination of bile acids, using either spectrophotometry or fluorometry as the end point for quantification, were developed primarily out of a need to improve the ease and speed of analysis (Table II). Many techniques have been described [68–77], and although satisfactory correlations with GLC methods have been claimed, these methods generally only give a crude measurement of total fecal bile acid excretion which may be suitable for the detection of clinical conditions involving excessive bile acid loss [78].

The enzyme 3α-hydroxysteroid dehydrogenase (E.C. 1.1.1.50) isolated from *Pseudomonas testosteroni* was first applied to the measurement of 3α-hydroxy bile salts in blood [79]. This enzyme catalyzes the oxidation of 3α-hydroxy to 3-oxo steroids with a concomitant reduction in NAD. In the early assays the generation of NADH was monitored from the absorbance at 340 nm [68–72,74]; however, by the later inclusion of diaphorase (E.C. 1.6.4.3), the nonfluorescent compound resazurin is converted to the fluorescent resorufin and the fluorescence monitored at 580 nm after excitation at 565 nm [73]. More recently, a more convenient spectrophotometric assay was developed where nitrotetrazolium blue reagent was reduced to give the blue-colored diformazan and the cascade reaction monitored spectrophotometrically at 540 nm [75]. Although this assay was mainly developed for serum bile acid analysis [80], it has been applied to fecal bile acid determination [75]; however, it was shown to be 10–30 times less sensitive than the enzymatic fluorometric procedure. The use of a higher wavelength minimizes interference due to pigments and reduces the need for sample purification. The assay was reported to be suitable for human and animal feces. Recoveries of bile acids added to fecal extracts were quantitative, and the inprecision of the assay was <10%. Many studies show a good correlation between the fluorometric/photometric methods and GLC techniques [70,73,74,78]. However, Setchell *et al.* [81], in comparing an enzymatic method (Enzabile) with GLC-MS, found higher values with the former technique, a linear relationship, and significant scatter of the data.

Despite the use of an enzymatic method for the end-point determination, some form of extraction continues to be necessary, and solvent extractions with and without saponification or partial purification have been adopted from the early GLC methodologies (Table I). Drawbacks of the enzymatic methods include the inability to distinguish bile acids other than those with a free 3α-hydroxyl group (e.g., 3β-epimers, 3-oxo, and conjugated species which comprise a significant proportion of the fecal bile acids) and the limited sensitivity and spec-

TABLE II. Techniques for Fecal Bile Acid Determination by Enzymatic Methods

Method	Reference	Sample preparation	Enzyme method	Absorbance reading[a]	Normal range
Woodbury and Kern, 1971	68	NaOH hydrolysis (1.5 hr, 120°C, 16 psi) / Diethyl ether extraction / TLC separation of lipids	3α-Hydroxysteroid dehydrogenase	340 nm	—
Sheltawy and Losowsky, 1975	69	Aqueous NaOH hydrolysis / Ethyl acetate extraction / TLC separation of lipids	3α-Hydroxysteroid dehydrogenase	340 nm	245–880 μmole/day (n = 11)
De Wael *et al.*, 1977	70	KOH/ethylene glycol hydrolysis / Diethyl ether extraction	3α-Hydroxysteroid dehydrogenase	340 nm	630 and 1110 μmole/day (n = 2)
Ferraris *et al.*, 1980	71	Methanol extraction	Sterognost Kit / 3α-Hydroxysteroid dehydrogenase / Diaphorase / Resazurin	565 nm (E) / 580 nm (F)	40–390 mg/day (n = 12)
Crowell and MacDonald, 1980	72	KOH/glycerol autoclaved / Ether extraction of lipids / Ether/HCl extraction bile acids	3α-Hydroxysteroid dehydrogenase / 7α-Hydroxysteroid dehydrogenase / 12α-Hydroxysteroid dehydrogenase	340 nm	32–64 μmole/g / 19–52 μmole/g / 32–71 μmole/g dried wt (n = 18)
Beher *et al.*, 1981	73	Hot ethanol extraction (30 min) / Aqueous NaOH hydrolysis (150° C/2 hr)	3α-Hydroxysteroid dehydrogenase / Diaphorase / Resazurin	565 nm (E) / 580 nm (F)	194–263 mg/day (n = 3)
Malchow-Møller *et al.*, 1982	74	Ethanol/acetic acid extraction / Saponification in NaOH / Pet. ether removal of lipids / Ethyl acetate extraction	Sterognost Kit / 3α-Hydroxysteroid dehydrogenase / Diaphorase / Resazurin	565 nm (E) / 580 nm (F)	412 ± 63 μmole/day
Beher *et al.*, 1983	75	Continuous soxhlet extraction with chloroform:methanol (2:1 v/v) / NaOH hydrolysis	3α-Hydroxysteroid dehydrogenase / Diaphorase type II-L / Nitro blue tetrazolium	540 nm	

[a] E, excitation; F, fluorescence.

ificity of the assay. However, for clinical diagnosis these drawbacks may not necessarily limit the utility of the assay [78].

Crowell and MacDonald [72] used 7α-hydroxysteroid dehydrogenase (E.C. 1.1.1.159) and 12α-hydroxysteroid dehydrogenase (E.C. 1.1.1.176) prepared from *Escherichia coli* [82,83] and the 3α-hydroxysteroid dehydrogenase for the determination of 3α-, 7α-, and 12α-hydroxy bile acids in extracts of feces. The absorbance of reduced NAD was read at 340 nm and Triton X-100 added to inhibit nonspecific turbidity. Values for 12α-hydroxy bile acid excretion exceeded those for other bile acids in many samples, suggesting the presence of interfering compounds. Again, these enzymatic methods will not detect oxo groups or protected hydroxyl groups. Attempts to quantify the bile acids after sodium borohydride reduction surprisingly led to lower values for 3α-, 7α-, and 12α-hydroxy bile acids, providing no information about the amounts of oxo bile acids present [73]. This highlights the problem of accurately interpreting data for fecal bile acids based on enzymatic determination.

2.5. High-Pressure Liquid Chromatography Analysis

High-pressure liquid chromatography is an ideal tool for the determination of polar and nonvolatile compounds. Its application to bile acid analysis has been reviewed in detail in Chapter 4 [84]. In general, it is more suited to the determination of conjugated rather than unconjugated bile acids, since the latter group have negligible UV absorbance. Since the majority of fecal bile acids are unconjugated, this presents a particular problem when HPLC is applied. However, this can be overcome by converting the carboxylic acid group to a suitable chromophore, such as a phenacyl derivative, thereby allowing UV detection at 254 nm [85–87] or using a fluorescing label [88–91].

In an early application of HPLC to fecal bile acid determination, bile acids were separated as their paranitrobenzyl esters on a micro-Partisil column using isocratic elution with 2% isopropanol in isooctane and detection by the UV absorbance at 254 nm. A satisfactory separation of standards of 5β-cholanoic,3-oxo-5β-cholanoic, iso-lithocholic, and lithocholic acids was demonstrated, and the method was used for isolation of radiolabeled metabolites of iso-lithocholic acid formed from [14C] taurolithocholic acid by human fecal microorganisms [85]. In a later report from this group [92] a gradient elution with tetrahydrofuran in hexane allowed the separation of a wide range of microbial metabolites of sulfolithocholic acid and provided evidence for its conversion to several nonpolar metabolites, including 5β-chol-3-enoic, and fatty acid esters of both lithocholic and iso-lithocholic acid. The structure of the unsaturated metabolite was confirmed by electron impact ionization MS [92].

High-pressure liquid chromatography was recently used to measure bile acids in feces after their extraction and separation into conjugate groups by the lipophilic gel piperidinohydroxypropyl Sephadex LH-20, although sulfates were excluded by this method. Using Hyamine 1622 as a counter ion, HPLC separation was carried out on a Bondapak C_{18} reversed-phase column in a solvent system of methanol/phosphate buffer (0.5 M, pH7)/acetonitrile with detection at 254 nm [93]. There appears to be no real advantages in using this methodology since the workup of samples described is as extensive as that used for the GLC methods, yet the latter technique provides more detailed and reliable information about the composition of bile acids in feces. This is exemplified by some of the chromatograms shown where the amounts of the primary bile acids (cholic and chenodeoxycholic) appear to be in excess of the secondary ones (deoxycholic and lithocholic) for a normal subject. Although quantitative data were not given, these authors stated that the limit of sensitivity was 1–2 μg, the precision for fecal samples 11% (CV), and that conjugated bile acids accounted for less than 5% of the bile acids in each sample.

2.6. Isotopic Methods

Radioactive bile acid tracers were used early in studies of the enterohepatic circulation and for the determination of bile acid kinetics. After administration of a labeled bile acid, its specific activity in bile declines exponentially as newly synthesized nonradioactive bile acids are formed [94]. On the assumption that synthesis equals fecal loss, if synthesis is constant and assuming regularity in fecal excretion, then the specific activity of the radioactive bile acids in the stool should also decline exponentially. Lindstedt and Norman derived a mathematical description of fecal bile acid excretion using values for isotope excretion following injection of a labeled bile acid [95].

To account for the irregularities in defecation, a nonabsorbable marker is required [96], such as chromium sesquioxide [97,98]. Unless bile acids are quantified or specific radioactivity is determined [94], data on pool size and synthesis rates cannot be calculated since the isotope excretion alone provides only a measure of turnover.

To determine fecal bile acid loss in patients with ileal resection, Woodbury and Kern [68] compared the rates of isotope excretion for $^{57}CrCl_3$ and [^{14}C]cholic acid. The difference in rates of excretion of the two isotopes gave an indication of the proportion of cholic acid absorbed and was referred to as the "coefficient of cholic acid absorption." In patients with massive ileal resection or disease, the excretion rates for the two isotopes were almost identical, indicating no absorption of the bile acid, while in less extensive forms of ileal disease the $^{57}CrCl_3$ was excreted more rapidly than [^{14}C]cholic acid. The coefficient of absorption was inversely related to fecal bile acid excretion ($r = 0.86$) measured by an enzymatic method.

Earlier studies of isotope excretion used only the bile acid [99,100] and not the nonabsorbable marker. To ensure good mixing of the two isotopes in the proximal small bowel, the [^{14}C]cholic acid was given intravenously several hours earlier than the oral administration of ^{57}CrCl$_3$. In normal subjects less than 20% of the administered bile acid isotope will be excreted within 24 hr, whereas in ileal disease greater proportions are lost in the same period [68,99–103]. Glycocholate labeled with [^{14}C]glycine has also been used more specifically for a clinical assessment of the state of bacterial deconjugation in the small bowel by analysis of expired ^{14}CO$_2$. In the "breath test" Fromm and Hofmann [104] also sampled feces and, using combustion [105], determined the amount of isotope excreted. The importance of measuring fecal excretion of [^{14}C]bile acid has been highlighted and indicates that results for breath ^{14}CO$_2$ alone can be misleading [104,106]. In a study of feces collected over 5 days following the administration of [^{14}C]taurocholate [107], the values for isotope excretion by combustion of whole feces agreed with those obtained following extraction of bile acids by the method of Grundy et al. [7] and showed increased isotopic excretion in patients with ileal disease.

In recent years a ^{75}Se-tagged taurohomocholic acid was introduced as an alternative to [^{14}C]-labeled bile acids for the determination of bile acid loss [108]. Turnover and fecal excretion of this γ-emitting compound were compared with values for [^{14}C]taurocholate, and the two compounds were found to behave in an identical manner [109,110]. The use of a γ-ray counter was simpler than determining the β-emitting isotope [111].

Under normal conditions bile acids are eliminated from the body almost exclusively by fecal excretion. Determination of bile acid production by Lindstedt's isotope dilution method [94] has therefore been widely used as an alternative to the chemical analysis of fecal bile acids. The relative reliability of the two fundamentally different methods has been the subject of considerable controversy. Early observations indicated that bile acid production determined by Lindstedt's method was higher than chemically determined fecal bile acid excretion and that this result might be species dependent [1]. In a comparison of bile acid excretion in nine subjects simultaneously using the method of Lindstedt [94] and a GLC method [112], values determined by the latter technique were 18.1–44.2% lower, irrespective of whether [^{3}H]- or [^{14}C]-labeled bile acids were used or whether β-sitosterol or chromium sesquioxide was used as a fecal flow marker [28]. Tangedahl et al. [113] also found higher values with the isotope dilution method [94] than with the GLC method [7]. These studies were discussed in detail by Davidson et al. [114] with special reference to possible overestimation of bile acid production due to the use of [^{3}H]chenodeoxycholic acid whose label was not stable in vivo [113]. They found good agreement between the isotope dilution method [94] using [^{14}C]-labeled cholic and chenodeoxycholic acids and the GLC method [7] when applied to studies of normoglyceridemic subjects, but no correlation between the two methods when hypertriglyceridemic patients were studied [114]. Duane *et al.*, also using [^{14}C]-labeled bile acids, studied five healthy subjects on 13 occasions and found that the isotopic method gave synthesis values that averaged 16.3% lower than the chemical method [115]. Using a one-sample determination of pool size [116], where the specific activity at 12 hr post-isotope administration is divided into the amount of isotope given, measurements of bile acid synthesis by the isotopic method were on an average 5.6% higher than the fecal bile acid excretion determined chemically [115].

The method based on labeling of plasma cholesterol and fecal bile acids by injection of [^{14}C]cholesterol (see Section 2.6) has also been compared with the chemical method for determination of fecal bile acid [7,16,17]. A good agreement was obtained when calculations were based on the specific activity of plasma cholesterol 56 hr before the midpoint in time of collection of fecal pools [17]. In this study an equilibration time of 4–6 weeks was allowed after injection of [^{14}C]cholesterol. This is probably necessary to allow for individual variations in equilibration times.

Comparisons between estimates of cholesterol synthesis and excretion by isotope kinetic and chemical methods have also been carried out. It is outside the scope of this chapter to discuss these studies, but the results of the methods are in reasonable agreement provided that an appropriate model for cholesterol metabolism is used, and markers for fecal flow [97] and sterol losses [117] are included [118]. Determinations of bile acid synthesis from the rate of production of ^3H_2O from [24,25-^{3}H]cholesterol have given values in agreement with chemical determination of bile acids in feces in normolipidemic subjects [119]. In analogy with the isotope dilution method using [^{14}C]-labeled bile acids, the values for bile acid synthesis were higher than those for chemically determined bile acid output in hypertriglyceridemic subjects [119].

Although several studies have indicated a good agreement between values obtained with Lindstedt's method [94] and GLC analyses [7], this may be limited to healthy subjects [114,119] and possibly depend on dietary factors. Discrepancies may have several reasons (see Ref. 114). The validity of the isotope dilution method [120] depends on a steady-state situation, equilibration of the entire administered dose of tracer with the endogenous pool(s), single exponential decay of specific activity, and absence of reentry of label into the pool it has left. The timing of administration of isotope may be important since bile acid synthesis rate varies with fasting and rate of enterohepatic circulation, which can affect the rate of dilution of isotope. Steady-state conditions may be difficult to attain, especially with respect to pool composition, since part of the bile acids are in the large bowel undergoing partly reversible bacterial metabolism.

On the other hand, the difficulty of accurately determining fecal bile acid excretion by chemical methods may offer a more simplistic answer to the observed discrepancies between the two techniques. A prerequisite for accurate chemical analysis of fecal bile acid output is demonstration of steady-state conditions and high recoveries of chromium sesquioxide as a marker for fecal flow [117]. Overestimation by inclusion of non-bile acid components is possible, as is the loss or neglect of hitherto unknown, labile or derivatized bile acid structures. A possible loss of bile acids through bacterial degradation of the steroid skeleton, as implied for neutral sterols in some patients [117] but not found in other species [121], is not supported by studies addressing this question using [^{14}C]-labeled bile acids [7,117]. However, only a small number of patients on special diets have been studied, and unexplained loss of [^{14}C]deoxycholic acid has been observed in rabbits [122].

2.7. General Problems in Fecal Bile Acid Analysis

Basic problems encountered in the analysis of fecal bile acids are discussed in several articles and are reviewed here [7,18–21,81,112]. Individual steps in the preparation of biological samples for analysis of bile acids are discussed in detail in Chapter 1 [123].

Whether fecal bile acids determined in excreted stools reflect the state of bile acid metabolism in the gastrointestinal tract is difficult to determine, since metabolic degradation of bile acids by bacteria may continue after excretion. To minimize the possibility of such changes occurring, freshly voided feces require immediate freezing and storage at $-20°C$, and only then can the fecal bile acid profiles be considered to reflect metabolic events within the intestinal lumen. However, the bacterial metabolism of bile acids consists of both reversible (e.g., oxidoreduction) and irreversible (e.g., deconjugation, dehydroxylation) reactions, and one may expect the relative concentrations of irreversibly formed metabolites to increase from cecum to rectum. The composition of bile acids in different segments of the large intestine and the distribution of metabolites between aqueous and solid phases in the contents cannot be determined by fecal bile acid analysis.

Storage of fecal samples, which is frequently unavoidable, also leads to questions concerning the stability of endogenous compounds. Fecal fats and lipids, for example, have been shown to be stable for up to 2 days when stored at $4°C$, but hydrolysis of fatty acid esters will occur at this temperature [124], and transesterification is a possibility that could involve the bile acids. Therefore, if these classes of compounds are to be analyzed, stool samples should be immediately frozen following their excretion. No studies have addressed in detail the question of long-term storage stability of the bile acids. Bile acid composition of feces stored at $-20°C$ was shown to be similar to that for feces homogenized in alcohol and stored at $0°C$ [36]. When fecal samples requiring bile acid analysis are to be stored for long periods before their analysis, it may be preferable to perform an initial extraction step and store the dried extract [21].

The first step required in the analysis of feces invariably involves homogenization of the stool collections. Homogenization has been performed in many solvents, including water, methanol, chloroform/methanol mixtures, and other organic solvents. However, homogenization using cold water is, in our opinion, the most convenient and least hazardous technique. It has been our experience that the most efficient homogenization is achieved by a stepwise dilution of the stools with small quantities of cold distilled water [21]. Although the direct extraction of whole stools or freeze-dried samples with organic solvents has been used extensively, individual stools are not homogeneous with respect to their qualitative and quantitative composition [81]. This finding indicates the need for homogenization in order to sample representative aliquots for analysis [81]. These results of detailed analyses are at variance with earlier reports using other methods, which have suggested that the "large bowel is a good mixing organ" and that random stools are adequate for fecal bile acid determination [24,78]. Visual inspection of stools indicates that the large bowel does not mix the contents that efficiently, and it is therefore not surprising to find chemical heterogeneity in stool composition [81].

A great variety of solvents have been used for the initial extraction of bile acids from the wet, lyophilized, or homogenized aliquots of stools. While vigorous alkaline hydrolysis as described by Grundy *et al.* [7] has been widely used for its convenience, this method should now be replaced by milder extraction techniques. Alkaline ethanol extraction, which leads to a mild saponification of lipids and hydrolysis of bile acid esters, gives an improved recovery of sulfated bile acids [125] compared with chloroform/methanol (1/1 v/v) or acetic acid extraction [19,43]. With the latter procedure, however, acetoxy derivative formation and hydrolysis of sulfates is an undesirable drawback. Prolonged refluxing in chloroform/methanol mixtures, which is time consuming [20], and extraction using methanol/hydrochloric acid [126] may lead to esterification of bile acids. Several unidentified bile acid esters have been found in fecal samples; however, in these early studies, the bulk of bile acids were first extracted with acetone to minimize the possibility of esters being formed during prolonged refluxing with chloroform/methanol [50,51]. Ethanol and aqueous ethanol mixtures have been used in several studies [122,127], but incomplete recoveries have been observed, presumably owing to the strong binding of bile acids to bacteria or dietary residue [127–129].

Complete recovery of bile acids is best achieved by multiple extraction steps [21,127]. Extraction of the bulk of bile acids into ethanol or aqueous ethanol, followed by a second extraction of nonpolar metabolites into chloroform/methanol (1/1 v/v), although more time consuming, ensures a quantitative recovery [21]. On a practical point, during the evaluation of extraction efficiency it should be realized that simple addition of radiolabeled bile acids to fecal samples does not give a true reflection of the efficiency of the extraction of endogenous bile acids. Only when the endogenous pool is labeled by prior administration of radiolabeled bile acids can an accurate assessment of the recovery of fecal bile acids be made [7,20,21,122,127]. This is exemplified from previous studies showing that the additional chloroform/methanol (1/1 v/v) extraction stage recovered 1.6–11.8% of endogenous radiolabeled bile acids from human feces [21] and 2.6–6.7% from rat feces [127] not extractable into other more polar solvents. Furthermore, while the recovery of a selection of radiolabeled bile acids added to fecal samples was >90% by a single aqueous ethanol extraction, the recovery of radiolabeled endogenous bile acids by the same procedure was shown to be only 76.0–89.9% [21], highlighting the deficiencies of this approach to method evaluation.

Modifications of the extraction procedure are required when feces are obtained from patients receiving sequestrants such as cholestyramine [130,131]. This anion exchange resin avidly binds bile acids and those with the lowest pKa; i.e., taurine conjugates and sulfates are bound the strongest. Organic extraction is not sufficient to displace the bile acids from the sequestrant, and it is necessary to introduce a competing ion to the solvent. For example, when 2.5 mg of taurocholate was added to rat feces containing cholestyramine, extraction using toluene–glacial acetic acid [19] recovered only 32%; however, when 0.5 M HCl was used, the recoveries of taurolithocholate, taurodeoxycholate, and taurocholate were quantitative [131]. Alternatively, the residue after a conventional extraction may be extracted with 0.2 M ammonium carbonate in 80% ethanol, known to elute bile acids from strong anion exchangers having a polystyrene matrix [130,132].

Following the extraction step, a crude extract is obtained which generally requires considerable purification before bile acids can be analyzed. Direct measurements on these extracts have been carried out, however, by less specific methods of measurement, such as enzymatic techniques. Neutral sterols present the major problem in GLC analyses because their GLC characteristics are so similar to those of bile acid derivatives and coelution can create difficulties in the identification and quantification of bile acids. Liquid–liquid partitioning with petroleum ether or n-hexane has been used for the removal of nonpolar lipids from aqueous alkaline and acidic solutions; however, partial losses of nonpolar bile acids such as lithocholate have been reported [8]. When 5% potassium carbonate–petroleum ether was used to remove lipids from alcoholic extracts, as much as 30% of the total bile acids was extracted into the petroleum ether phase [18,133], although this was probably the result of artifact formation of ethyl esters during prior prolonged refluxing in alcohol [20]. A small proportion of bile acids in feces has been reported to be esterified with an alcohol at the carboxyl group [50,51,134] or with a fatty acid at C-3 [135,136], and liquid–liquid partitioning into nonpolar solvents may result in a substantial loss of these derivatives. When solvent extraction is used, a small amount of alcohol should be added to aqueous acidic solutions to prevent micellar solubilization of lipids, which inhibits their partitioning into the extraction solvent.

Lipophilic gels, such as Lipidex 1000 or 5000, offer an alternative approach to the removal of unwanted lipids. Their high capacity (50 mg/ml gel bed) for absorbing nonpolar lipids [137] from aqueous solutions and the ability to perform reverse-phase chromatography in aqueous alcohols once lipids have been absorbed (see Chapter 1) have made this gel useful for the retention of monohydroxy sterols from biological samples [21,138]. In the method for fecal bile acids described by Setchell *et al.* [21], this gel was used in combination with octadecylsilane-bonded silica (Bond Elut or Sep-Pak) cartridges primarily as a means of removing inorganic salts and some pigments from the organic extract, neutral sterols being separated at a later stage by a liophilic anion exchange gel.

A variety of methods have been employed for the removal of fatty acids from fecal extracts, although this step is less essential than the removal of sterols because the majority of fatty-acid derivatives elute considerably faster than bile acid derivatives during GLC. Given some of the limitations described earlier, fatty acids may be extracted into petroleum ether from an acidified aqueous solution [139], preferably in the presence of ethanol, but TLC has been commonly employed. Preparative TLC, of both the free bile acids and the methyl ester derivatives in either single- or dual-solvent systems, has been reported [7,69]. The recovery of bile acid methyl esters is better than that of free acids after this step. Overloading of the TLC plates, particularly with feces rich in lipid, and contamination of the GLC profile by impurities from the silica gel are frequent drawbacks of this technique. Column chromatography on florisil [7] or silicic acid [18,20], of the methyl esters [18] or the free acids [7,20], has also been employed to remove fatty acids and pigments. Liquid–solid extraction using resins such as Amberlite XAD-2 affords a method of adsorbing fatty acids and bile acids, and after washing with acidic water, elution of the resin with hexane permits the removal of fatty acids [140]. Similarly, neutral sterols have been separated from bile acids by this approach [52–54].

Final isolation of fecal bile acids from an aqueous matrix can be achieved by solvent, solid, or gel extraction. Unconjugated bile acids are extractable from acidic aqueous solution by organic solvents such as diethyl ether or ethyl acetate. A low pH (<3) is necessary to suppress ionization, and under these conditions unconjugated bile acids and nonpolar glycine conjugates will be extracted into diethyl ether [50]. Solvolysis of some nonamidated sulfates (see Chapter 1) will also occur in diethyl ether under acid conditions [141], which may or may not be an advantage of this approach. For more polar bile acids including taurine conjugates, *n*-butanol has been used [50,127], but evaporation of large volumes of this solvent is tedious and artifact formation at low pH is a possibility.

Liquid–solid extraction by Amberlite XAD-2 resin [142] or using commercially available octadecylsilane-bonded silica cartridges [143] offers a more attractive method of isolating bile acids from aqueous solution [47]. Furthermore, unconjugated bile acids may be extracted from mildly acidic (pH 3–4) solution by Lipidex 1000 [21,47,144,145], while the inclusion of an ion-pair reagent permits the extraction of bile acid conjugates [146].

In general, human feces contains predominantly unconjugated bile acids; however, during antibiotic or drug therapy and in certain diseases, the proportion of conjugated bile acids may be significantly increased. Until recently little was done to determine the contribution of bile acid conjugates in feces because most of the methods incorporated a hydrolytic step at an early stage of the procedure.

Alkaline hydrolysis has been carried out at high temperature and pressure in aqueous or aqueous/alcohol mixtures, and ethylene glycol has also been used. Inevitably there may be artifact formation by these rigorous conditions [51,147–149]. Although the recovery of the principal primary and secondary bile acids having only hydroxyl functions is satisfactory using rigorous alkaline hydrolysis, a poor recovery of oxo bile acids occurs [51,147]. Lepage *et al.* [149] demonstrated the vulnerability of oxo bile acids to alkaline hydrolysis when alcohol was present in the mixture. In a study of 11 different oxo bile acids, none of the bile acids tested with a 3-oxo group could be recovered following saponification in 95% ethanol or 95% methanol; however, only minor by-products were formed when aqueous alkali was employed. 12-Oxo and 7-oxo metabolites of hydroxylated bile acids were less susceptible, but 3,7-dioxo-, 3,12-dioxo-, and 3,7,12-trioxo-5β-cholanoic acids were extensively converted by reduction to hydroxy-oxo products in the presence of alcohols, although chemical changes were less pronounced with aqueous alkaline hydrolysis [149].

Enzymatic hydrolysis is not without problems since the presence of hydroxyl groups in the bile acid nucleus is a prerequisite for enzymatic hydrolysis, and products of esterification or acetylation reactions, which can occur during solvent extractions, will not be hydrolyzed by the enzyme [44].

Since oxo bile acids are present in significant amounts in human feces [20,21], rigorous hydrolysis procedures should be avoided if possible. These can be avoided if the bulk of unconjugated bile acids are isolated separately from the conjugated species, as described in recent methods [21,54,57,64], and the necessary hydrolytic steps can be performed separately on the individual conjugate groups.

Although bile acid sulfates can now be measured directly by fast-atom-bombardment ionization MS [62] in bile and serum samples, the sensitivity is presently inadequate to detect sulfated bile acids in small aliquots of feces [81], and therefore the analysis of this fraction of fecal bile acids can only be performed following a solvolytic and hydrolytic step. Several options are available for solvolysis [48,125,141,150–156] (see Chapter 1). Methanol/acetone/HCl is frequently used but is relatively time consuming [125,150], and partial formation of acetonides of *cis*-glycol structures, as exist in hyocholic and β-muricholic acids, will result. Parmentier and Eyssen [150] showed that the rate of solvolysis at 37°C was faster than at ambient temperature; sulfates at positions C-3 and C-12 of cholic acid were completely solvolyzed after 3 hr and 6 hr, respectively, while the C-7 sulfate required 12–18 hr to achieve complete solvolysis. Similarly, in a comparison of four different solvolytic methods for the sulfate conjugates of chenodeoxycholic acid [151], yields in excess of 90% were obtained when methanol/acetone/HCl was used; however, the C-7 sulfate was inefficiently solvolysed (17% recovery) when ethanol/ethyl acetate/H_2SO_4 was employed [48,152]. A drawback of the latter technique is the partial formation of acetates and ethyl esters, which necessitate an alkaline hydrolysis. Dimethoxypropane/HCl has been employed for the simultaneous solvolysis and methylation [153–155] of bile acids, but few studies have attempted to investigate the conditions necessary for quantitative solvolysis [141,151]. In addition, the formation of acetonide derivatives of *cis*-glycols occurs with this reagent. Studies of Burstein and Lieberman [156] led to the description of a hydrolysis–solvolysis procedure for conjugated bile acid 3-sulfates [141] in which solvolysis occurred during ether extraction of acidified (pH 1) aqueous extracts, but the technique was not evaluated for amidated and other polar bile acid sulfates, and alkaline hydrolysis must precede the solvolytic procedure.

When GLC is used for the analysis of the final extract, it is necessary to convert the involatile bile acids to appropriate volatile derivatives and to select a suitable liquid phase and internal standard for quantification. The most appropriate combination of derivative and liquid phase is one that permits adequate separation of the components in the mixture at temperatures at which the liquid phase is thermally stable and under conditions where minimum adsorption occurs. At present, with the introduction of chemically bonded fused silica capillary columns, adsorption effects and thermal stability are less of a consideration than with conventional packed columns. Because of the complexity of bile acids in fecal extracts, the most appropriate choice of column for their analysis is a high-resolution capillary column with a nonselective liquid phase, such as OV-1 or SE-30 and other thermally stable methyl silicones. Numerous derivatives and GLC liquid phases have been employed for bile acid analysis, but it is outside the scope of this chapter to discuss their relative merits in detail. For more information the reader is directed to several reviews on this subject and references therein [47,59,61,62,157].

3. FECAL BILE ACID EXCRETION IN HEALTH

3.1. Normal Patterns of Fecal Bile Acid Excretion in Adults

Relative to numbers of quantitative studies, there are few reports that characterize in detail the qualitative patterns of fecal bile acid excretion. The principal secondary bile acid, deoxycholic acid, was identified in human feces by Lindstedt [158] and confirmed by Carey and Watson [159]. Lithocholic, its 3β-isomer, and 3α-hydroxy-12-oxo-5β-cholanoic acids were found by Heftmann et al. [160] and confirmed along with chenodeoxycholic, 3β,12α-dihydroxy-5β-cholanoic, and 3β-hydroxy-12-oxo-5β-cholanoic acids by Danielsson et al. [161]. Cholic [162], 3-oxo-5β-cholanoic [1,51], 3α,12α-dihydroxy-7-oxo-5β-cholanoic, and 3α,7β,12α-trihydroxy-5β-cholanoic acids were also identified over two decades ago [163]. In a series of papers, Eneroth et al. [20,45,46] described the characterization of a series of mono-, di-, and tri-substituted cholanoic acids, and despite some limitations of the methods outlined earlier, these investigations stand even today as some of the most detailed qualitative analyses of fecal bile acids. The identifications were based on GLC–MS analysis of methyl esters, methyl ester trifluoroacetates, and methyl ester trimethylsilyl ether derivatives. The GLC retention times and peak shifts of the compounds formed as products of oxidation and reaction with dimethylhydrazine were valuable in assigning the stereochemistry of some of the isomeric bile acids. These studies and those more recently employing the higher chromatographic resolving power of capillary columns [21] reveal the tremendous complexity of fecal bile acid excretion, which in agreement with previous studies of the metabolism of labeled bile acids [50,51,161] shows feces to contain a mixture of isomeric hydroxy- and oxo-cholanoic acids (Table III). In the normal healthy subject, bile acids lacking a 7α-hydroxyl group and with a 3α- or 3β-hydroxyl, either with or without a substituent at C-12, are the predominant bile acids excreted. Such metabolites are formed from biliary bile acids by the action of bacterial microflora in deconjugating and 7α-dehydroxylating primary bile acids during their passage through the gastrointestinal tract. Although Gustafsson et al. [166] isolated a single microorganism capable of carrying out dehydroxylation, oxidation, and reduction of the hydroxyl groups on the steroid nucleus, a variety of bacteria that colonize the colon and distal ileum possess enzyme activity toward steroids and bile acids [167,168].

Table III gives a comprehensive list of bile acids that have been identified in human adult feces. Although the early studies failed to identify the presence of 5α-isomers of bile acids, this may be explained by the difficulty of resolving the relatively low concentrations of these allo bile acids from other quantitatively more important bile acid metabolites. With the advent of capillary columns, with superior resolving power, a large number of isomeric bile acids can be determined

TABLE III. Bile Acids Identified in the Feces of Adult Human Subjects

Bile acid structure (trivial name)	Reference[a]
No hydroxyls	
5α-Cholanoic	18,54
5β-Chol-3-enoic	54,64,136,164
Monohydroxy bile acids	
3α-Hydroxy-5β-cholanoic (lithocholic)	18,20,21,45,47,54,64,65,112,136,160,161,164
3α-Hydroxy-5α-cholanoic (allo-lithocholic)	164
3β-Hydroxy-5β-cholanoic (iso-lithocholic)	18,20,21,45,47,54,64,136,160,161,164
3β-Hydroxy-5α-cholanoic	164
3β-Hydroxy-5-cholenoic	21
7β-Hydroxy-5β-cholanoic	18
12β-Hydroxy-5β-cholanoic	54
12α-Hydroxy-3-cholenoic	64
12α-Hydroxy-2-cholenoic	64
Dihydroxy bile acids	
3α,6α-Dihydroxy-5β-cholanoic (hyodeoxycholic)	164
3β,6α-Dihydroxy-5β-cholanoic	164
3α,7α-Dihydroxy-5β-cholanoic (chenodeoxycholic)	21,45,47,54,64,112,161,164
3α,7β-Dihydroxy-5β-cholanoic (ursodeoxycholic)	18,21,45,54,64,112
3β,7α-Dihydroxy-5β-cholanoic	18,21,45,54,64
3α,12α-Dihydroxy-5β-cholanoic (deoxycholic)	18,20,21,45,47,54,64,112,159,161,164
3α,12α-Dihydroxy-5α-cholanoic (allo-deoxycholic)	64,164
3β,12α-Dihydroxy-5β-cholanoic	18,20,21,45,112,161,164
3β,12α-Dihydroxy-5α-cholanoic	47,164
3α,12β-Dihydroxy-5β-cholanoic	18,45,54,112
3α,12β-Dihydroxy-5α-cholanoic	164
3β,12β-Dihydroxy-5β-cholanoic	18,45
3β,12β-Dihydroxy-5α-cholanoic	64,164
Trihydroxy bile acids	
3α,6α,7α-Trihydroxy-5β-cholanoic (hyocholic)	54,112
3α,6β,7α-Trihydroxy-5β-cholanoic (muricholic)	164
3α,7α,12α-Trihydroxy-5β-cholanoic (cholic)	18,20,21,46,54,64,162,164

3α,7α,12α-Trihydroxy-5α-cholanoic (allo-cholic)	46
3β,7α,12α-Trihydroxy-5β-cholanoic	21,46,47,64
3α,7β,12α-Trihydroxy-5β-cholanoic	21,46,54,64,164
3β,7β,12α-Trihydroxy-5β-cholanoic	46,64
Oxo bile acids	
3-Oxo-4-cholenoic	65
3-Oxo-5β-cholanoic	45,54,64,65,161
3-Oxo-5α-cholanoic	65
12α-Hydroxy-3-oxo-5β-cholanoic	18,20,45,65,164
7α-Hydroxy-3-oxo-5β-cholanoic	45,54,64,112
7α,12α-Dihydroxy-3-oxo-5β-cholanoic	164
3,12-Dioxo-5β-cholanoic	45,54,64
7-Oxo-5β-cholanoic	54
3α-Hydroxy-7-oxo-5β-cholanoic	18,21,45,54,112
3α,12α-Dihydroxy-7-oxo-5β-cholanoic	20,21,46,54,112,164
12-Oxo-5β-cholanoic	54
3α-Hydroxy-12-oxo-5β-cholanoic	18,20,21,45,65,160,161,164
3β-Hydroxy-12-oxo-5β-cholanoic	20,21,45,161
3α-Hydroxy-12-oxo-5α-cholanoic	164
3α,7α-Dihydroxy-12-oxo-5β-cholanoic	46,47
Esters	
3β-Hydroxy-5β-cholanoic, 3-palmitoleate	136
3β-Hydroxy-5β-cholanoic, 3-palmitate	136
3β-Hydroxy-5β-cholanoic, 3-stearate	136
3β-Hydroxy-5β-cholanoic, 3-oleate	136
Glycine and taurine conjugates	
3β-Hydroxy-5α-cholanoic	81
3β-Hydroxy-5β-cholanoic (iso-lithocholic)	54
3α-Hydroxy-5β-cholanoic (lithocholic)	54,55,57,81,165
12α-Hydroxy-5α-cholanoic	55
12β-Hydroxy-5β-cholanoic	54
3α,6α-Dihydroxy-5β-cholanoic (hyodeoxycholic)	81
3α,7α-Dihydroxy-5β-cholanoic (chenodeoxycholic)	54,55,57,165
3α,12β-Dihydroxy-5β-cholanoic	54
3α,12β-Dihydroxy-5α-cholanoic	55,81

(continued)

TABLE III. (*Continued*)

Bile acid structure (trivial name)	Reference[a]
3β,12α-Dihydroxy-5α-cholanoic	81
3β,12β-Dihydroxy-5α-cholanoic	81
3α,12α-Dihydroxy-5β-cholanoic (deoxycholic)	54,55,57,81,165
3α,6α,7α-Trihydroxy-5β-cholanoic (hyocholic)	54
3α,7α,12α-Trihydroxy-5β-cholanoic (cholic)	54,55,165
3α,7β,12α-Trihydroxy-5β-cholanoic	54
3-Oxo-5β-cholanoic	54
12α-Hydroxy-3-oxo-5β-cholanoic	81
3α,12α-Dihdroxy-7-oxo-5β-cholanoic	54
12-Oxo-5β-cholanoic	54
3,12-Dioxo-5β-cholanoic	54
Sulfates	
3β-Hydroxy-5-cholenoic	21
3β-Hydroxy-5β-cholanoic (iso-lithocholic)	21,54,55,81
3β-Hydroxy-5α-cholanoic (allo-lithocholic)	81
3α-Hydroxy-5β-cholanoic (lithocholic)	21,54,55,64,81,165
3α,7α-Dihydroxy-5β-cholanoic (chenodeoxycholic)	21,54,55,81,165
3α,7β-Dihydroxy-5β-cholanoic (ursodeoxycholic)	54,55
3α,12β-Dihydroxy-5α-cholanoic	55,81
3α,12β-Dihydroxy-5β-cholanoic	54
3α,12α-Dihydroxy-5β-cholanoic (deoxycholic)	21,54,55,64,165
3β,12α-Dihydroxy-5β-cholanoic	21,64
3α,7α,12α-Trihydroxy-5β-cholanoic (cholic)	21,54,55,165
3α,7β,12α-Trihydroxy-5β-cholanoic	21,54
7α-Hydroxy-3-oxo-5β-cholanoic	54
12α-Hydroxy-3-oxo-5β-cholanoic	81

[a] Only references related to the characterization of bile acids in the feces from normal adult subjects using GLC or MS procedures are given. We acknowledge that many other publications report the presence of the bile acids, deoxycholic, lithocholic, chenodeoxycholic, and cholic acids, but these are not included and can be found in the various citations given in Table IV.

in a single analysis. For example, all four isomers of the 3-hydroxycholanoic acid and six of the eight theoretically possible isomers of 3,12-dihydroxycholanoic acid could be identified and quantified from the capillary column GLC analysis (Fig. 2).

Bile acids having no hydroxyl groups or with unsaturation have been rarely detected and only in trace amounts in human feces [21,54,136,165,169]. These facts are significant, given the supposed association between bile acids and colonic carcinogenesis (see Section 4.1). Norman and Palmer [51], in studies on the fate of ingested [^{14}C]lithocholic acid, first recognized the presence of an unidentified, unsubstituted, and unsaturated bile acid metabolite in human feces and bile. No evidence for 5β-cholanoic acid was obtained in these studies.

The presence of esterified bile acids in feces has long been recognized [50,51] from observations that a significant proportion of the radiolabeled metabolites of administered [^{14}C]cholic and [^{14}C]lithocholic acids was recovered from feces as saponifiable derivatives. Although the existence of ethyl esters of bile acids has been reported [134], some caution should be taken in interpreting observations of C-24 esterified bile acids since such products can result from exposure to alcohols during extraction steps. The natural occurrence of fatty acyl derivatives of lithocholic acid metabolites in intestinal contents has been clearly demonstrated [136], and this may represent a biological detoxification mechanism because these metabolites will be much less soluble and therefore expected to be less well absorbed from the gastrointestinal tract. The major ester formed from sulfolithocholic acid by human microflora was identified by GLC–MS as the palmitate of iso-lithocholic acid, and lesser amounts of the palmitoleyl, stearyl, and oleyl esters were also found [136]. Since most methods employed for fecal bile acid analysis incorporated a saponification step, information on this type of bile acid metabolite in human feces is lacking, except for the original observations of Norman and Palmer [51]. In this context, application of the relatively nondestructive procedures of Setchell *et al.* [21] for fecal bile acid analysis which utilize lipophilic gel chromatography would permit this group of bile acids to be isolated and, following saponification, to be identified. This is supported by the finding of Korpela *et al.*, using DEAE Sephadex for the group separation of bile acid derivatives [170]. They found nonpolar hydrolyzable derivatives of dihydroxy bile acids in the neutral fraction but did not attempt to characterize these derivatives. Clearly, further work is required to evaluate the biological importance of this type of metabolite.

Quantitatively, the majority of bile acids excreted by healthy adults are in the unconjugated form owing to deconjugation during passage through the small intestine and colon [168]. An extensive literature exists on quantitative fecal bile acid excretion in health and disease, which is highlighted in later sections. Figures given in Table IV are an attempt to bring together many of the studies that have been reported, measuring daily fecal bile acid excretion, irrespective of the

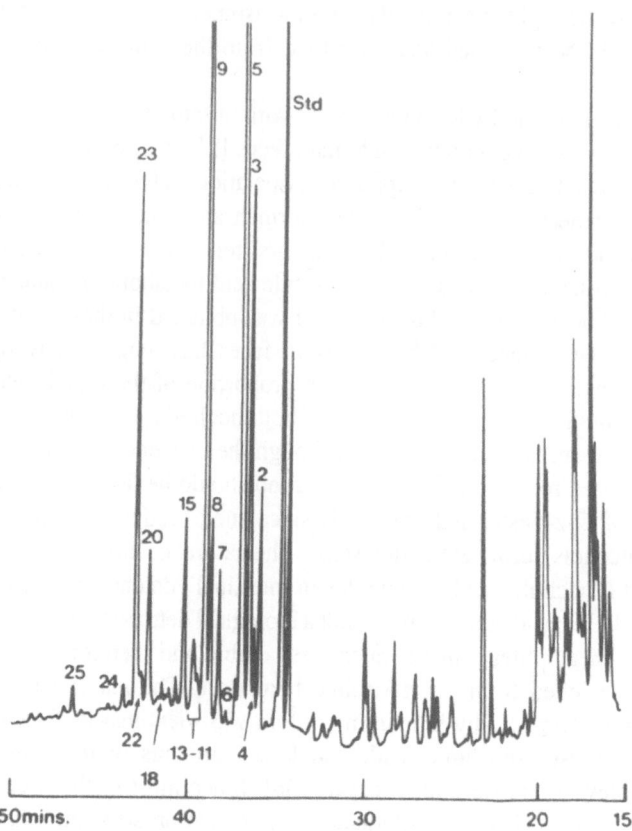

FIGURE 2. Capillary column GLC analysis of the methyl ester–trimethylsilyl ether derivatives of the unconjugated bile acids isolated from the feces of a healthy adult. GLC analysis was performed on a 25-m DB-1 fused silica column using temperature-programmed operation from 225°C to 295°C with increments of 2°C/min after initial and final isothermal periods of 5 min and 20 min, respectively. Helium was used as carrier gas (flow approx. 2 ml/min) and identification of the compounds listed was achieved by mass spectrometry. The following bile acids were identified. (1) 5β-chol-3-enoic; (2) 3α-hydroxy-5α-cholanoic; (3) 3β-hydroxy-5β-cholanoic; (4) monohydroxy bile acid; (5) 3α-hydroxy-5β-cholanoic; (6) 3α,12α-dihydroxy-5α-cholanoic; (7) 3α,12β-dihydroxy-5β-cholanoic; (8) 3β,12α-dihydroxy-5β-cholanoic; (9) 3α,12α-dihydroxy-5β-cholanoic; (10) 3β-hydroxy-5α-cholanoic; (11) 3α,7α-dihydroxy-5β-cholanoic; (12) 12α-hydroxy-3-oxo-5β-cholanoic; (13) 3α,7α,12α-trihydroxy-5β-cholanoic; (14) 3α,6β,7α-trihydroxy-5β-cholanoic; (15) 3α,6α-dihydroxy-5β-cholanoic; (16) 3β,6α-dihydroxy-5β-cholanoic; (17) dihydroxy-monoxo bile acid; (18) 3β,12α-dihydroxy-5α-cholanoic; (19) trihydroxy bile acid; (20) 3β-hydroxy-12-oxo-5β-cholanoic; (21) trihydroxy bile acid; (22) 3β,12β-dihydroxy-5α-cholanoic; (23) 3α-hydroxy-12-oxo-5β-cholanoic; (24) 7α,12α-dihydroxy-3-oxo-5β-cholanoic; (25) dihydroxy-monooxo bile acid.

TABLE IV. Quantitative Fecal Bile Acid Excretion in Adult Subjects in Health and Disease

Reference		n	Age (years)	Sex (M/F)	Sample size (days)	Diet and drugs	Analytical methods	Total daily bile acid excretion[b]
Normals								
Ali et al., 1966 [18]	Middle-aged men	3	—	3/0	4	Fat-free diet	TLC/GLC	130–650 mg
Eneroth et al., 1968 [20]	Normal	10	—	—	NS	Regular solid or formula diet	GLC	22.4–404 mg
Evrard and Janssen, 1968 [19]	Normal	5	—	—	3–4	Ad lib	GLC	127–290 (218) mg
Forman et al., 1968 [172]	Normal	20	—	—	1	Not stated	TLC/ Fluorometry	196–460 (292) mg
Reddy and Wynder, 1973 [173]	Normal Americans	17	28–60	—	2	Western diet	GLC	256 ± 34 mg
	American vegetarians	12	30–62	—	2	Lactovegetarian	GLC	133 ± 15 mg
Sheltawy and Losowsky, 1975 [69]	No GI/liver disease	11	—	—	2	Not stated	Enzymatic	100–360 (202) mg 3.3 ± 0.3 mg/kg
Huang et al., 1976 [174]	Normal	5	20–35	5/0	1	Hospital diet	GLC	293 ± 69 mg
De Wael et al., 1977 [70]	No GI/liver disease	9	—	—	3	Controlled	Enzymatic	410–1210 μmole
Wynder, 1979 [175]	American Seventh-Day Adventists	11	30–75	—	2	Lactovegetarian	GLC	54 ± 16 mg
	Japanese migrants	21	35–45	—	2	Typical Japanese	GLC	83 ± 8 mg
	Chinese migrants	11	28–45	—	2	Typical Chinese	GLC	54 ± 13 mg
Bilheimer et al., 1979 [176]	Normal	14	50	14/0	4	Controlled	Enzymatic	4.9 ± 2.3 mg/kg

(continued)

TABLE IV. *(Continued)*

Reference	Subjects	n	Age (years)	Sex (M/F)	Sample size (days)	Diet and drugs	Analytical methods	Total daily bile acid excretion[b]
Podesta et al., 1980 [35]	Normal	15	17–84	8/7	3	Not stated	GLC	603 ± 71 mg
Reddy et al., 1980 [177]	Normal	30	24–41	15/15	3	Mixed Western	GLC	2.77 ± 0.15 mg/kg
Tanida et al., 1981 [54]	Japanese	8	24–49	8/0	3	Ad lib	GLC-MS	128–366 μmole
Malchow-Møller et al., 1982 [74]	Normal	16	22–42	—	2	Ad lib	Enzymatic	412 ± 63 μmole
	No GI/liver disease	25	21–89	—	3	100 g fat/15 g fiber	Enzymatic	343 ± 63 μmole
Setchell et al., 1987 [81]	Normal	8	21–35	4/4	5	Ad lib	GLC-MS	81 ± 24 mg
Korpela et al., 1986 [170]	Normal	5	23–30	5/0	3	Mixed Western	GLC	472 ± 21.8 μmole
Intestinal disease								
Miettinen, 1968 [388]	Controls	19	—	—	—	Not stated	—	220 ± 70 mg
	Celiac disease	5	—	—	—	Not stated	—	172 ± 457 mg
Miettinen, 1971 [349]	Normal	6	29 ± 2	—	2	Low cholesterol	GLC	232 ± 45 mg
	Ulcerative colitis	5	28 ± 6	3/2	2		GLC	300 ± 52 mg
Miettinen, 1971 [263]	Normal	13	—	—	3	Not stated	GLC	238 ± 25 mg
								3.8 ± 0.2 mg/kg
	Incomplete ileal resection	3	51–65	1/2	3	Low cholesterol	GLC	215 ± 36 mg
								3.1 ± 0.8 mg/kg
	Complete ileal resection	10	10–69	4/6	3	Low cholesterol	GLC	1765 ± 282 mg
								32.9 ± 5.7 mg/kg
Miettinen and Peltokallio, 1971 [423]	Normal	—	—	—	3	Not stated	GLC	<436 mg; <5.6 mg/kg
	Colectomy/short ileal resection	16	20–68	8/8	3	Low cholesterol	GLC	341 ± 68 mg; 5.8 ± 1.2 mg/kg
		16	20–68	8/8	3	+ Na taurocholate (3 g)	GLC	1161 ± 81 mg/kg

Reference	Condition	n	Age	M/F	SS	Diet	Method	Value
Woodbury and Kern, 1971 [68]	Diarrhea and ileal resection <100 cm	4	28-44	0/4		Not stated	Enzymatic	363-1140 mg
	Pancreatic insuff. and diarrhea	1	61	1/0	1	Not stated	Enzymatic	574 mg
	Diarrhea and ileal resection >100 cm	5	20-71	2/3	1	Not stated	Enzymatic	1377-4034 mg
Hofmann and Poley, 1972 [380]	Small-ileal resection	6	13-60	3/3	1	LCT	GLC	1400-6500 mg
	Small-ileal resection	6	13-60	3/3	1	LCT + cholestyramine	GLC	2600-5900 mg
	Small-ileal resection	6	13-60	1/2	1	MCT	GLC	500-3000 mg
	Small-ileal resection	6	13-60	1/2	1	MCT + cholestyramine	GLC	1200-4100 mg
	Large-ileal resection	3	44-67	1/2	1	LCT	GLC	2000-2800 mg
	Large-ileal resection	3	44-67	1/2	1	LCT + cholestyramine	GLC	2300-7800 mg
	Large-ileal resection	3	44-67	1/2	1	MCT	GLC	900-1700 mg
	Large-ileal resection	3	44-67	1/2	1	MCT + cholestyramine	GLC	1500-4000 mg
Allan et al., 1974 [378]	Normal	7	—	—	5	Not stated	GLC	799 ± 117 mg
	Postvagotomy (no diarrhea)	7	—	—	5	Not stated	GLC	2538 ± 632 mg
	Postvagotomy (diarrhea)	7	—	—	5	Not stated	GLC	1030 ± 482 mg
Findlay et al., 1974 [369]	Normal	6	28-36	—	7	Ad lib	GLC	295.7 ± 32.5 mg
	Normal	6	28-36	—	7	+ 10 g bran/day	GLC	352.2 ± 45.5 mg
	Diverticular disease	7	30-84	—	7	Ad lib	GLC	273.1 ± 69.2 mg
	Diverticular disease	7	30-84	—	7	+ 10 g bran/day	GLC	240.9 ± 84.6 mg
Mitchell et al., 1974 [259]	Normal (ITT 24-113 hr)	10	—	—	1	Not stated	GLC	222 ± 99; 104-358 mg
	Ileal resection >40 cm (ITT 24-73 hr)	9	32-78	5/4	1	Not stated	GLC	3155 ± 1197; 1611-4600 mg
Campbell and McIvor, 1975 [31]	Normal + non-GI disease	26	—	—	3	Ad lib	GLC	371 ± 128 mg (115-685) mg
	Ileal resection	16	—	—	3	Ad lib	GLC	2360 ± 1110 mg (250-6320) mg

(continued)

TABLE IV. (*Continued*)

Reference	Subjects	n	Age (years)	Sex (M/F)	Sample size (days)	Diet and drugs	Analytical methods	Total daily bile acid excretion[b]
Huang et al., 1976 [36]	Normal	5	—	5/0	—	Prison diet	GLC	4.2 ± 1.0 mg/kg
	Normal challenged with *Shigella*	5	—	5/0	—	Prison diet	GLC	5.14 ± 1.0 mg/kg
Krag and Krag 1976 [362]	Crohn's disease with diarrhea	6	19–63	3/3	—	Not stated	Enzymatic	600–4500 mg
	Crohn's disease no diarrhea	5	23–36	3/2	—	Not stated	Enzymatic	100–200 mg
Duncombe et al., 1977 [379]	Normal	4	—	—	—	Not stated	GLC	204–620 (473) mg
	Postvagotomy diarrhea	6	—	—	—	Not stated	GLC	745–1168 (1024) mg
Faloon et al., 1977 [421]	Preoperative	16	—	—	—	Not stated	GLC	206 ± 34 mg
	Jejunoileostomy	16	—	—	—	Constant regulated diet 1100 cal	GLC	621 ± 96 mg
Huigbregtse et al., 1977 [425]	Ileostomy	10	25–73	6/4	1	Hospital diet	Enzymatic	196–1070 mg
		10	25–73	6/4	1	Hospital diet	Enzymatic	164–1230 mg
Nelson et al., 1977 [363]	Normal	2	—	—	5	Normal	GLC	160–430 μmole
		2	—	—	5	elemental diet	GLC	680–130 μmole
	Noncholerreic diarrhea	4	—	—	5	Normal	GLC	270–1010 μmole
		4	—	—	5	elemental diet	GLC	40–680 μmole
	Cholerrheic diarrhea	6	—	—	5	Normal	GLC	6370 ± 1640 μmole
		6	—	—	5	elemental diet	GLC	2700 ± 1120 μmole
Roller and Kern, 1977 [214]	Acquired pancr. insuff.	9	32–73	7/2	3	No treatment	Enzymatic	14.0 ± 4.0 mg/kg
	Acquired pancr. insuff.	9	32–73	7/2	3	+ cotazym	Enzymatic	6.4 ± 2.0 mg/kg

Reference	Condition	N	Age	Sex	No.	Diet	Method	Value
Van Blankenstein et al., 1977 [397]	Crohn's ileal resection/disease	18	—	—	—	Not stated	Enzymatic	180–3600 (1220) mg
	Ileal resection (non-Crohn)	5	—	—	—	Not stated	Enzymatic	880–3100 (1900) mg
	Pelvic radiotherapy (diarrhea)	8	—	—	—	Not stated	Enzymatic	200–2320 (1000) mg
	Thyroid carcinoma (diarrhea)	2	—	—	—	Not stated	Enzymatic	1100–2000 mg
	Other GI disorders	11	—	—	—	Not stated	Enzymatic	190–1640 mg
	Normal	10	—	—	—	Not stated	Enzymatic	140–640 mg
Podesta et al., 1978 [34]	Normal	14	—	—	—	Not stated	GLC	680 ± 106 mg
	Diarrhea/no malabsorb.	7	—	—	—	Not stated	GLC	890 ± 145 mg
	Diarrhea/malabsorb.	12	—	—	—	Not stated	GLC	1751 ± 331 mg
	Cholelithiasis	5	—	—	—	Not stated	GLC	845 ± 64 mg
Tarpila et al., 1978 [370]	Diverticular disease	10	35–64	5/5	3	Normal diet	GLC	475 ± 82 mg
	Diverticular disease	10	35–64	5/5	3	+ Fiber suppl. (6 m)	GLC	384 ± 61 mg
	Diverticular disease	10	35–64	5/5	3	+ Fiber suppl. (12 m)	GLC	240 ± 39 mg
	Diverticular disease	12	35–64	6/6	3	Normal diet	GLC	330 ± 48 mg
	Diverticular disease	12	35–64	6/6	3	Normal diet	GLC	320 ± 49 mg
	Diverticular disease	12	35–64	6/6	3	Normal diet	GLC	346 ± 89 mg
Hutcheon et al., 1979 [384]	Normal	—	—	—	3	Ad lib	GLC	238 ± 25 mg
	Postcholecystectomy diarrhea	2	48–51	1/1	3	Ad lib	GLC	1299, 754 mg
Flynn et al., 1980 [262]	Normal	17	—	10/7	5	Ad lib	GLC	226.2 ± 37 mg
	Diverticular disease	21	—	—	5	Ad lib	GLC	132 ± 17.3 mg
	Diverticular disease	21	—	—	5	Fiber suppl. (1 mo)	GLC	217.1 ± 35.6 mg
	Diverticular disease	21	—	—	5	Fiber suppl. (1 yr)	GLC	125.2 ± 21.0 mg
Ferraris et al., 1980 [71]	Normal	12	20–68	9/3	3	Not stated	Enzymatic	163 ± 100 mg
	Cholegenic diarrhea	3	—	—	3	Not stated	Enzymatic	1500 ± 533 mg
Huang et al., 1978 [37]	Normal	6	—	6/0	<1	Not stated	GLC	4.1 ± 1.2 mg/kg
	Traveler's diarrhea	6	—	6/0	<1	Not stated	GLC	7.4 ± 1.4 mg/kg

(continued)

TABLE IV. (*Continued*)

Reference	Subjects	n	Age (years)	Sex (M/F)	Sample size (days)	Diet and drugs	Analytical methods	Total daily bile acid excretion[b]
Kay et al., 1980 [424]	Conventional ileostomy	5	23–50	2/3	3	Controlled	GLC	5.6 ± 1.4 mg/kg
	Continent ileostomy + resection	5	18–30	2/3	3	Controlled	GLC	7.6 ± 1.0 mg/kg
		5	16–51	2/3	3	Controlled	GLC	19.8 ± 5.9 mg/kg
Pasanen et al., 1980 [432]	Normal	7	—	—	3	Not stated	GLC	4.2 ± 0.3 mg/kg
	Pancreatic insuff.	9	—	—	3	Low cholesterol	GLC	9.5 ± 1.5 mg/kg
Beher et al., 1981 [73]	Normal	3	—	—	3	Not stated	Enzymatic	194–263 mg
	Low serum HDL	2	—	—	3	Not stated	Enzymatic	86, 126 mg
	Diarrhea	2	—	—	3	Not stated	Enzymatic	302, 630 mg
McJunkin et al., 1981 [245]	Normal	3	31–44	2/1	1–3	Ad lib	GLC	701 ± 130 (677–723) µmole
	Diarrhea	7	22–69	4/3	1–3	Ad lib	GLC	774 ± 255 (55–2182) µmole
	Small ileal resection	9	30–58	1/8	1–3	Ad lib	GLC	3723 ± 783 (805–7387) µmole
	Large ileal resection	7	29–63	4/3	1–3	Ad lib	GLC	5834 ± 1098 (1256–10651) µmole
	Jejunoileal bypass	8	29–44	3/5	1–3	Ad lib	GLC	4300 ± 1004 (1432–9326) µmole
Rutgeerts et al., 1981 [358]	Normal	23	18–68	17/6	3	Ad lib	Enzymatic	275 ± 11.3 mg
	Irritable bowel (diarrhea)	14	—	—	3	Ad lib	Enzymatic	254 ± 43.8 mg
Aldini et al., 1982 [264]	Controls	8	19–38	6/4	3	No drugs	GLC	213 ± 71 mg
	Crohn's ileal resection >100 cm	6	28–44	4/2	3	No drugs	GLC	2819 ± 1971 mg
	Crohn's ileal resection or disease <100 cm	8	28–74	3/4	3	No drugs	GLC	667 ± 411 mg
	Colectomy	5	32–48	3/2	3	No drugs	GLC	484 ± 180 mg
Vuoristo and Miettinen, 1982 [241]	Normal	8	25 ± 3	4/4	—	Low cholesterol	GLC/TLC	11.3 ± 2.4 µmole/kg / 710 ± 220 µmole
	Celiac disease	30	35 ± 10	17/13	—	Low cholesterol	GLC/TLC	11.5 ± 7.0 µmole/kg / 640 ± 440 µmole
	Celiac disease	13	—	—	—	Low cholesterol	GLC/TLC	13.1 ± 9.4 µmole/kg / 710 ± 480 µmole

					+ Gluten free	GLC/TLC	
	Celiac disease	13	—				11.8 ± 5.1 μmole/kg
							730 ± 280 μmole
Fiasse et al., 1983 [260]	Normal	10	23–48	2/8	3	Ad lib	GLC	710 ± 120 μmole
	Crohn's	45	14–65	26/19	3	Hospital diet	GLC	1690 ± 270 μmole
	+ Ileal resection <50 cm	30	15–64	21/9	3	Hospital diet	GLC	4540 ± 530 μmole
	+ Ileal resection 50–100 cm	19	19–61	12/7	3	Hospital diet	GLC	5840 ± 540 μmole
	+ Ileal resection >100 cm	14	22–58	6/8	3	Hospital diet	GLC	5580 ± 740 μmole
Blake et al., 1983 [270]	Normal	12	—	—	1	Ad lib	Enzymatic	800 μmole
	Vagotomy	16	—	—	1	Ad lib	Enzymatic	803 ± 57 μmole
	Cholecystectomy and vagotomy	12	—	—	1	Ad lib	Enzymatic	1197 ± 135 μmole
Koga et al., 1984 [360]	Normal	6	19–24	3/3	2	Elemental low fat	GLC	192.1 ± 33.0 mg
	Crohn's disease	9	16–28	5/4	2	Elemental low fat	GLC	491.5 ± 50.3 mg
	Crohn's disease	9	16–28	5/4	2	+ Butter fat	GLC	738.8 ± 91.8 mg
Tanida et al., 1984 [165]	Normal Japanese	13	32–74	11/2	3	Ad lib	GLC–MS	93.8–712.3 (386.7) μmole
	Adenomatous polyps	13	34–78	10/3	3	Ad lib	GLC–MS	55.0–837.6 (346.9) μmole
Hikasa et al., 1984 [169]	Normal	14	63 ± 11	5/9	3	Ad lib	GLC–MS	470 ± 231 μmole
	Colon cancer	14	63 ± 8	5/9	3	Ad lib	GLC–MS	297 ± 402 μmole
Vuoristo and Miettinen, 1985 [387]	Normal	7	22–32	4/3	3	Low cholesterol	GLC	263 ± 33 mg
	Untreated celiac disease	11	20–49	4/7	3	Solid food (10g fat/day)	GLC	260 ± 44 mg
Breuer et al., 1985 [182]	Celiac disease	7	16–54	2/5	3	Gluten-free diet	GLC	233 ± 23 mg
	Celiac disease	7	16–54	2/5	3	Pre-gluten-free diet	GLC	330 ± 84 mg
	Normal	21	—	15/6	3	Hospital diet	GLC	87.6 ± 14.5 mg
	Colon cancer	23	—	8/15	3	Hospital diet	GLC	43.1 ± 5.1 mg
Tanida et al., 1986 [350]	Normal	12	<50	11/1	3	Ad lib	GLC	215 ± 148 μmole
	Ulcerative colitis (active)	14	14–47	6/8	3	Low residue	GLC	445 ± 392 μmole
Inborn errors								
Salen and Grundy, 1973 [448]	Normolipemic	2	71–73	2/0	4	Liquid formula	GLC	210 and 306 mg
	Type II hypercholesterolemia	3	37–58	2/1	4	Liquid formula	GLC	116–234 mg
	CTX	2	32–46	1/1	4	Liquid formula	GLC	93 ± 28 and 136 ± 46 mg

(continued)

TABLE IV. *(Continued)*

Reference	Subjects	n	Age (years)	Sex (M/F)	Sample size (days)	Diet and drugs	Analytical methods	Total daily bile acid excretion[b]
Miettinen, 1980 [452]	Normal	13	—	—	3	Not stated	GLC	638 ± 67 µmole
	Phytosterolemia	1	27	1/0	3	Solid food	GLC	606–858 µmole
	Phytosterolemia	1	32	1/0	3	Liquid formula	GLC	180 µmole
Hyperlipidemia								
Lewis and Myant, 1967 [205]	Normal	4	43–72	1/3		Normal	Isotopic	123–274 mg
	Normal	4	43–72	1/3		Normal	Isotopic	165–354 mg
	Type II	3	49–69	1/2		Ad lib or low-cholesterol diet	Isotopic	95–222 mg
	Type II	3	49–69	1/2		Ad lib or low-cholesterol diet	Isotopic	109–265 mg
Miettinen et al., 1967 [476]	Normocholesterolemia	19	—	10/9	2	Controlled	GLC	220 ± 16 mg
	Hypercholesterolemia	8	31–50	1/7	2	Controlled	GLC	154 ± 14 mg
Miettinen, 1968 [481]	Hypercholesterolemia	5	31–56	—	3	Low cholesterol	GLC	65–231 mg
Moore et al., 1969 [422]	Hypercholesterolemia + Ileal bypass	3	—	—	3–4	Low cholesterol	Isotopic	285–620 (399)
		5	—	—	3–4	Not stated	Isotopic	1435–3101 (1934)
Grundy et al., 1971 [415]	Type II (severe)	4	14–28	1/3	3	Liquid formula	GLC	84–224 mg
	After ileal exclusion	4	14–28	1/3	3	Liquid formula	GLC	768–1778 mg
	Type II	7	14–83	4/3	3	Low-cholesterol diet	GLC	110–646 mg
	Type II	7	14–83	4/3	3	+ cholestyramine	GLC	620–2428 mg
Grundy et al., 1972 [208]	Type I	2	11	2/0	4	Liquid formula	GLC	49–116 mg
	Type II	5	57–72	3/2	4	Liquid formula	GLC	130–343 mg
	Type III	3	20–68	3/0	4	Liquid formula	GLC	144–324 mg
	Type IV	3	48–60	2/1	4	Liquid formula	GLC	297–1132 mg
	Type V	5	48–60	3/2	4	Liquid formula	GLC	249–421 mg

Nazir et al., 1972 [485]	Type II	4	16–56	0/4	3	Controlled	GLC	73–277 mg
	Type II	4	16–56	0/4	3	+ Cholestyramine (12 g/day)	GLC	546–654 mg
Miettinen and Aro, 1972 [469]	Normal	13	—	7/6	—	Controlled	GLC	238 ± 25 mg; 3.8 ± 0.2 mg/kg
	Hypercholesterolemia	39	—	22/17	—	Controlled	GLC	181 ± 11 mg; 2.8 ± 0.2 mg/kg
	Hypertriglyceridemia	12	—	—	—	Controlled	GLC	381 ± 91 mg; 4.7 ± 0.9 mg/kg
Moutafis and Myant, 1973 [486]	Type II	1	41	1/0	3–10	Low cholesterol	Isotopic	315 ± 29 mg
	Type II	1	41	1/0	3–10	+ 2.7 g cholesterol	Isotopic	416 ± 33 mg
Sodhi and Kudchodhar, 1973 [470]	Hypercholesterolemia	8	33–56	2/6	3	Controlled	GLC	141 ± 75 mg
	Hypercholesterolemia and hypertriglyceridemia	8	30–47	7/1	3	Controlled	GLC	443 ± 133 mg
Nestel and Hunter, 1974 [275]	Normal	8	29–50	6/2	2	Controlled	GLC	267 ± 70 mg 3.88 ± 0.8 mg/kg
	Hypercholesterolemia	8	32–55	3/5	2	Controlled	GLC	199 ± 69 mg 2.99 ± 0.8 mg/kg
	Overweight (hypertriglyceridemia)	6	39–53	5/1	2	Ad lib	GLC	507 ± 97 mg 6.07 ± 0.8 mg/kg
	Overweight (normolipemic)	6	22–38	3/3	2	Ad lib	GLC	429 ± 76 mg 3.87 ± 0.51 mg/kg
Sedaghat et al., 1975 [487]	Hyperlipidemic	4	52–72	3/1	4	Liquid formula	GLC	160–328 (237) mg
Simons and Myant, 1975 [483]	Type IV/diabetes	1	—	1/0	4–6	Ad lib	GLC	931 ± 38 mg
	Type V/diabetes	1	—	1/0	4–6	Ad lib	GLC	1034 ± 11 mg
	Type IIb	1	—	1/0	4–6	Ad lib	GLC	947 ± 84 mg
Miller et al., 1976 [473]	Normals	5	17–28	5/0	2	Ad lib or low-cholesterol diet	Isotopic	171 ± 435 mg
	Type IIa	6	35–61	6/0	2	Ad lib or low-cholesterol diet	Isotopic	338 ± 478 mg
	Type IIb	2	47–48	0/2	2	Ad lib or low-cholesterol diet	Isotopic	305 ± 359 mg
	Type IV	2	52–56	1/1	2	Ad lib or low-choleterol diet	Isotopic	459 ± 563 mg
Subbiah et al., 1976 [28]	Various hyperlipoproteinemias	9	30–62	6/3	1	Solid food	GLC	108.9–1274.2 mg

(continued)

TABLE IV. (*Continued*)

Reference	Subjects	n	Age (years)	Sex (M/F)	Sample size (days)	Diet and drugs	Analytical methods	Total daily bile acid excretion[b]
Angelin et al., 1976 [474]	Normal	14	54 ± 2	7/7	7–10	Low cholesterol	Isotopic	1000 ± 120 µmole
	Type IIa	19	56 ± 1	12/7	7–10	Low cholesterol	Isotopic	1000 ± 80 µmole
	Type IIb	12	54 ± 3	5/7	7–10	Low cholesterol	Isotopic	1130 ± 110 µmole
	Type IV	23	51 ± 2	20/3	7–10	Low cholesterol	Isotopic	2310 ± 220 µmole
Miettinen and Lempinen, 1977 [482]	Type II	12	26–50	8/4	3	Low-cholesterol diet	GLC	193 ± 31 mg (120–472) mg
	Type II	12	26–50	8/4	3	Cholestyramine (32 g/day)	GLC	1434 ± 176 mg (593–2311) mg
	After ileal resection	12	26–50	8/4	3	Not stated	GLC	2414 ± 264 mg (741–4020) mg
Lin and Connor, 1980 [475]	Normal	1	31	1/0	7	Liquid formula (low cholesterol)	GLC	184 ± 14.2 mg; 2.8 ± 0.2 mg/kg
	Normal	1	31	1/10	7	Liquid formula (high cholesterol)	GLC	214 ± 29.3 mg; 3.2 ± 0.4 mg/kg
	Type IIa	1	67	0/1	7	Liquid formula (low cholesterol)	GLC	162 ± 25.4 mg; 3.7 ± 0.6 mg/kg
	Type IIa	1	67	0/1	7	Liquid formula (high cholesterol)	GLC	225 ± 52.8 mg; 4.9 ± 0.3 mg/kg
Sodhi et al., 1980 [279]	Normals	53	43 ± 12	—	—	Not stated	GLC	295 ± 134 mg
	Hypercholesterolemic	50	42 ± 17	—	—	Not stated	GLC	182 ± 72 mg
	Hypertriglyceridemia	21	48 ± 11	—	—	Not stated	GLC	421 ± 225 mg
	Hypercholesterolemia/ hypertriglyceridemia	40	47 ± 12	—	—	Not stated	GLC	390 ± 211 mg
Thompson et al., 1980 [478]	Type IIa homozygote	2	15–25	2/0	—	Low cholesterol	GLC	551–577 mg
	Type IIa heterozygote	2	31–47	2/0	—	High cholesterol	GLC	281–583 mg

Reference	Group	n	Age	M/F	No.	Diet	Method	Value
Anderson et al., 1984 [488]	Type IIa or IIb	10	34–66	10/0	3	Controlled	GLC	109 ± 37 mg
	Type IIa or IIb	10	34–66	10/0	3	+ Oat bran	GLC	180 ± 43 mg
	Type IIa or IIb	10	34–66	10/0	3	Controlled	GLC	154 ± 37 mg
	Type IIa or IIb	10	34–66	10/0	3	+ Bean suppl.	GLC	108 ± 20 mg
Liver disease								
Erb et al., 1972 [492]	Normal	6	—	—	1	Not stated	GLC	197.6 ± 53.8 mg
	Cholestasis	4	—	—	1	Not stated	GLC	(4.1) mg
	Acute hepatitis	6	—	—	1	Not stated	GLC	(18.3) mg
Miettinen, 1972 [382]	Normal	13	—	—	—	Not stated	GLC	238 ± 25 mg; 3.8 ± 2 mg/kg
	Cirrhosis (no obstruction)	6	26–60	1/5	—	Low cholesterol	GLC	128 ± 19 mg; 2.2 ± 0.4 mg/kg
	Cirrhosis (obstruction)	8	19–62	4/4	—	Low cholesterol	GLC	95 ± 17 mg; 1.6 ± 0.3 mg/kg
Saudek, 1977 [493] Hepatitis B	Incubation period	1	61	0/1	4	Controlled	GLC	347 ± 127.7 mg
	Preicteric period	1	61	0/1	4	Controlled	GLC	243 ± 10.1 mg
	Icteric period	1	61	0/1	4	Controlled	GLC	121.4 ± 27 mg
Endo et al., 1979 [269]	Normal	16	—	—	2	Not stated	GLC/GLC–MS	145 ± 48 mg
	Benign recurrent intrahepatic cholestasis (anicteric)	1	32	1/0	2	Not stated	GLC/GLC–MS	430 ± 107 mg
Von Bergmann et al., 1979 [489]	Normal	14	—	14/0	4	Mixed solids/liquid formula	GLC	4.8 ± 0.5 mg/kg
	Normal	3	—	3/0	4	Liquid formula	GLC	5.7 ± 0.8 mg/kg
	Normal	3	—	3/0	4	+ cholestyramine	GLC	24.9 ± 4.5 mg/kg
	Alcoholic cirrhosis	8	44–61	8/0	4	Liquid formula	GLC	4.6 ± 0.6 mg/kg
	Alcoholic cirrhosis	8	44–61	8/0	4	+ cholestyramine	GLC	19.8 ± 3.9 mg/kg
Amuro et al., 1981 [268]	Normal	5	—	—	1	Normal (Japanese)	GLC/GLC–MS	135.3 ± 23.3 mg
	Liver cirrhosis	13	42–63	13/0	1	Normal (Japanese)	GLC/GLC–MS	5.3–178.4 (50.6) mg
Amuro et al., 1981 [267]	Normal	12	35–60	12/0	3	Normal (Japanese)	GLC	142.6 ± 38.3 mg
	Mild cirrhosis	8	42–53	8/0	3	Normal (Japanese)	GLC	77.9 ± 55.6 mg
	Advanced cirrhosis	8	42–76	8/0	3	Normal (Japanese)	GLC	29.8 ± 31.5 mg

(continued)

TABLE IV. *(Continued)*

Reference	Subjects	n	Age (years)	Sex (M/F)	Sample size (days)	Diet and drugs	Analytical methods	Total daily bile acid excretion[b]
Kesaniemi et al., 1981 [490]	Normal relatives	17	39 ± 3	9/9	2	Low cholesterol	GLC	4.7 ± 0.5 mg/kg
	Primary biliary cirrhosis	9	51 ± 4	1/8	2	Low cholesterol	GLC	3.3 ± 0.4 mg/kg
	Chronic active hepatitis	24	43 ± 4	8/16	2	Low cholesterol	GLC	3.3 ± 0.4 mg/kg
Kesaniemi et al., 1982 [491]	Normal	7	25 ± 1	4/3	2	Controlled	GLC	4.2 ± 0.38 mg/kg
	Chronic cholestasis	10	46 ± 5	2/8	2	Controlled	GLC	1.7 ± 0.4 mg/kg
Diabetes								
Bennion and Grundy, 1977 [521]	Pima Indians with maturity-onset diabetes (poor control)	6	19–63	1/5	4	Controlled diet	GLC	226–761 (415) mg
	Good control	6	19–63	1/5	4	Controlled diet	GLC	122–422 (261) mg
Saudek and Brach, 1978 [522]	Diabetes (poor control)	4	60–84	1/3	2–4	Controlled	GLC	138–720 (379) mg
	Diabetes (good control)	4	60–84	1/3	2–4	Controlled	GLC	106–477 (287) mg
	Diabetes + type V HLP (poor control)	1	43	1/0	2–4	Controlled	GLC	1083 mg
	Diabetes ± type V HLP (good control)	1	43	1/0	2–4	Controlled	GLC	1051 mg
Abrahams et al., 1982 [523]	Diabetes (poor control)	13	42–70	13/0	40	Controlled diet	GLC	518 ± 89 mg
	Diabetes (good control)	13	42–70	13/0	39	Controlled diet	GLC	496 ± 51 mg

[a] Values for daily fecal bile acid excretion in adults, published prior to 1966 are not included but these are reviewed in Ref. 7. GI, gastrointestinal; LCT, long-chain triglycerides; MCT, medium-chain triglycerides; ITT, intestinal transit time; pancr. insuff., pancreatic insufficiency; SS, single stool.

[b] Values are expressed as range with mean in parentheses, mean ± SD, or mean ± SEM, and the reader is directed to the reference for more detail.

method of analysis. In most instances bile acid excretion is quoted as mg(μmole)/ day or mg(μmole)/kg body wt/day, and few examples report a detailed breakdown of the individual bile acids. Since a great variety of assay methods have been employed and the number of patients, dietary status, and collection procedures differ markedly between studies, comparison of the individual studies is often difficult; however, as a general rule the less specific the method of assay, the higher the quoted normal range. This is evident when the values for fecal bile excretion determined by the early titration and spectroscopic techniques (reviewed in Grundy et al. [7]) are compared with the GLC methods developed later. In general, though, total fecal bile acid excretion in healthy adults has been quoted to range between 22 and 650 mg/day, with the average range for excretion from all the reported studies being 200–300 mg/day.

This wide range in values probably reflects individual variations in bile acid excretion due to variability in diet between the subjects studied, since it is clear from Table IV that alteration in diet greatly influences fecal bile acid excretion. Few attempts have been made to determine the variability in fecal bile acid excretion within subjects from day to day. Mower et al. [171] measured total bile acid excretion (mg/g dry weight feces) in 16 japanese men on two separate occasions 10 days apart and found greater than 100% difference in the values for two-thirds of the subjects studied, including six subjects whose excretion differed by fourfold and one by 20-fold.

In a detailed study [81] all feces excreted were collected from two healthy subjects over a period of 12 days, and 20 individual unconjugated bile acids and the principal conjugated bile acids were quantified in each collection. Although bile acid excretion (mg/day) was relatively constant in one subject, in the other considerable variation was observed. These variations, in both subjects, were more pronounced when data were expressed in terms of bile acid excretion in mg/g dry weight and mg/g wet weight (Fig. 3). This presumably reflects the differences in water content and fecal weights between the collections and suggests that bile acid excretion based on concentrations is less meaningful than the absolute amount excreted from day to day. This in part may explain the wide variability found by Mower et al. [171] because values were expressed as mg/g dry weight. Analysis of the individual bile acids revealed marked variation in their relative proportions from day to day, and this was particularly apparent for deoxycholic acid and the other 3,12-dihydroxy-5β-cholanoic acid isomers.

Data on quantitative excretion of the individual fecal bile acids in healthy subjects reported by different groups are difficult to assimilate, but relevant references are indicated in Table IV. The accuracy of some of the reported values may, however, be questionable, unless the homogeneity of the peaks in gas chromatograms was confirmed by MS or extensive peak shift techniques, and very few studies have incorporated these criteria for peak identification.

Lithocholic and deoxycholic acids are quantitatively the major bile acids

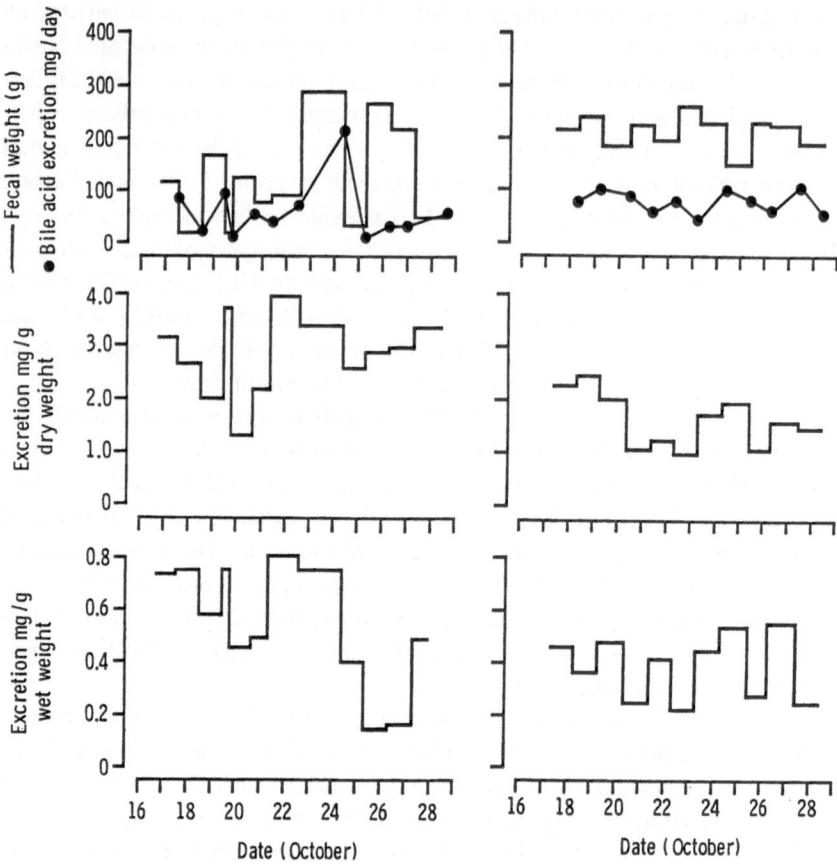

FIGURE 3. Variation in fecal weight (g) expressed as total weight and total bile acid excretion (mg/day) over a prolonged period in two healthy subjects (upper panel). Bile acid excretion is also expressed in terms of the concentrations mg/g dry weight of stool (middle panel) and mg/g wet weight of stool (lower panel) [Reproduced from Ref. 81].

excreted and proportionally account for about 20–35% and 30–55%, respectively, of the total fecal bile acids excreted [81]. It is significant that many of the early studies reported excretion values for bile acids of unknown structure, although using more recent techniques, a great many more isomeric species have now been characterized and quantified [21,53,54,164]. The proportions of chenodeoxycholic and cholic acids excreted are in general low in the normal subject; however, in conditions that may decrease transit time or alter the microflora, these "primary" bile acids may constitute a significant proportion of the total, at the expense of 7α-dehydroxylated metabolites [81]. The major oxo bile acids found in feces are 3α-hydroxy-12-oxo-5β-cholanoic and 3β-hydroxy-12-oxo-5β-cholanoic acids, which can account for up to 17% of the total bile acids excreted. In a semiquantitative analysis of oxo bile acids as a group, following their specific isolation as oxime derivatives on a cation exchange gel, the 3-oxo bile acids were found to constitute 1–20% of the amounts of lithocholic acid in the feces of eight healthy adults (c.f. Fig. 1) [65]. Concentrations of 3-oxo-5β-cholanoic, 12α-hydroxy-3-oxo-5β-cholanoic, and 3,12-dioxo-5β-cholanoic acids were 27 ± 16, 37 ± 30, and 2–10 mg/g wet weight, respectively, in the feces of normals, while 3-oxo-4-cholenoic and 3-oxo-5α-cholanoic acids were present but at concentrations too low to be quantified [65].

All the classical procedures for fecal bile acid determination incorporate a hydrolytic step; therefore, information about the state of conjugation of bile acids is lost. Although Norman and Palmer [51] demonstrated the presence of small proportions of glycine- and taurine-conjugated bile acids and sulfation has been shown to be an important pathway in the metabolism of lithocholic acid [178,179], it is only in recent years that conjugated fecal bile acids have received any attention; this followed reports that up to 30–40% of the total bile acids in the feces of healthy subjects were sulfate conjugates [34,35]. The methods employed in these studies relied on differential hydrolysis of two portions of the same sample, an inaccurate technique that frequently may give negative results, and since GLC chromatograms were not published and MS confirmation was not used, it is impossible to know whether these findings may be due to nonspecific interference in the chromatographic peaks quantified. Specific group isolation of the individual bile acid sulfates, which is possible using lipophilic ion exchangers [21,48], is at present the only satisfactory method of determining this fraction of bile acids. With these procedures Setchell et al. [21,81] showed GLC chromatograms of each individual conjugate group (Fig. 4) and with the use of MS identified and quantified a large number of sulfated bile acids in the feces of healthy subjects (Table III). In seven subjects studied conjugated bile acids accounted for <6% of the total fecal bile acids, with the sulfates representing <3%. Quantitative excretion was less than 10 mg/day. Although earlier Islam et al. [180] could not identify bile acid sulfates in feces, Salvioli and Salati [181] found they accounted for 10.3% of the total fecal bile acids and that following chenodeoxycholic acid therapy this fraction increased to account for 22.3% of the total. The former study used Sephadex LH-20 for the isolation of lithocholic acid sulfate. In a comprehensive quantitative study of fecal bile acid excretion in Japanese subjects using the techniques of Setchell et al. [21], sulfated bile acids accounted for between 1 and 10% of the total [165,169]. Similarly, in separate studies of 15 and 21 healthy subjects, Breuer et al. [55,182] found that bile acid sulfates comprised 4.4 ± 1.2% and 3.1 ± 0.9% (mean ± SD), respectively, of the total fecal bile acids.

A variety of bile acid sulfates have been identified (Table III), but in general,

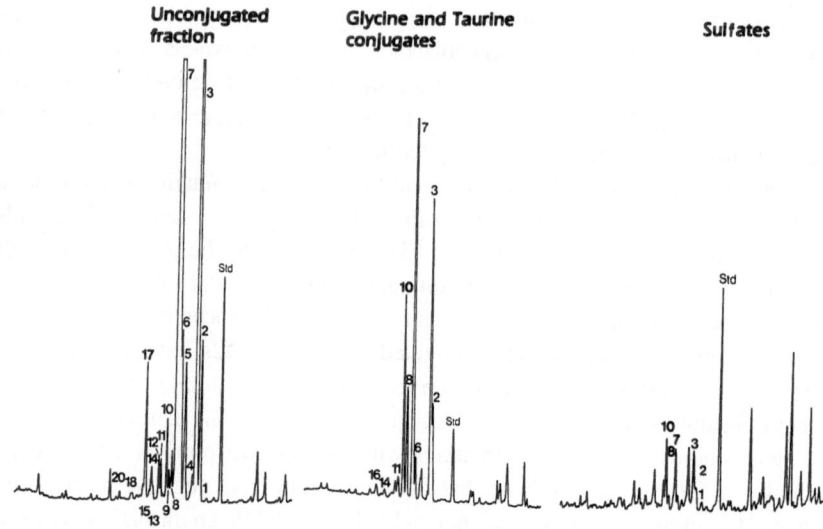

FIGURE 4. Gas chromatographic analysis of the bile acids excreted in the feces of a healthy adult. Bile acids were separated into (1) unconjugated bile acids, (2) glycine and taurine conjugates combined, and (3) sulfates, using lipophilic gel chromatography on Lipidex-DEAP. After deconjugation and/or preparation of methyl ester–trimethylsilyl ether derivatives, analysis was performed on a 30-m DB-1 chemically bonded fused silica column using temperature programmed operation 225°–295°C with increments of 2°C/min. The internal standard (Std), coprostanol, was added, and a list of those bile acids which were identified by mass spectrometry are as follows: (1) 3α-hydroxy-5α-cholanoic; (2) 3β-hydroxy-5β-cholanoic; (3) 3α-hydroxy-5β-cholanoic; (4) unknown monohydroxy bile acid; (5) 3α,12β-dihydroxy-5β-cholanoic; (6) 3β,12α-dihydroxy-5β-cholanoic; (7) 3α,12α-dihydroxy-5β-cholanoic; (8) 3α,7α-dihydroxy-5β-cholanoic; (9) 12α-hydroxy-3-oxo-5β-cholanoic; (10) 3α,7α,12α-trihydroxy-5β-cholanoic; (11) 3α,6α-dihydroxy-5β-cholanoic; (12) 3β,6α-dihydroxy-5β-cholanoic; (13) dihydroxy-monooxo bile acid; (14) 3β,12α-dihydroxy-5α-cholanoic; (15) 3α-hydroxy-12-oxo-5α-cholanoic; (16) 3β,12β-dihydroxy-5α-cholanoic; (17) 3α-hydroxy-12-oxo-5β-cholanoic; (18) unknown bile acid (19) 7α,12α-dihydroxy-3-oxo-5β-cholanoic; (20) unknown bile acid; (21) dihydroxy-monooxo bile acid. *Note:* In the original publication [81] of these chromatograms, several peaks in the conjugate fractions were incorrectly numbered, e.g. in the glycine/taurine and sulfate fractions peak 9 should have been no. 10 (cholic acid) while peak 6 in the sulfate fraction should have been no. 7 (deoxycholic acid). These assignments of structures correct the earlier errors.

in healthy subjects lithocholic and deoxycholic are the major ones found in this fraction. As is the case with neutral steroid sulfates, there is often a preponderance of 3β- and 5α-isomers.

Glycine- and taurine-conjugated bile acids have been identified in the feces of healthy adults [21,54,55,57,81], but generally comprise a relatively small proportion of the total bile acids excreted. Not unexpectedly, chenodeoxycholic and cholic acids can be found in this fraction, and it should be emphasized that in conditions such as diarrhea, where the gut transit times are markedly decreased, or in antibiotic therapy, the proportions of these conjugates may increase [21,47]. A list of bile acids that have been recognized as taurine and/or glycine conjugates is given in Table III and a typical GLC profile of the conjugate fractions is shown in Fig. 4.

3.2. Bile Acids in Children and Infants

There have been no detailed studies published on the development of the qualitative pattern of bile acids in the feces of neonates and infants. Since the fetal gut is sterile [183], the spectrum of secondary bile acids identified in meconium (see Section 3.3) is presumably due to transplacental passage of maternal bile acids. Sharp et al. [184] could not detect secondary bile acids in pooled stool samples collected from approximately 100 normal newborns after the first meconium passage. Gut microflora appears to be rapidly established postnatally [183], and the products of bacterial action, the secondary bile acids, appeared only in the pooled stool samples collected at day 5 of life. Since this study relied on pooling of a large number of samples, any individual variability could not be assessed. Several groups have reported the virtual absence of deoxycholic acid from gallbladder bile or duodenal contents of infants and children up to the age of 12 months [185–188], and it was early suggested [185] that this may be due to decreased microbial transformation of the primary bile acids during the first year of life. In support of this, Tazawa et al. [189] found cholic and chenodeoxycholic acids to be the major bile acids present in the feces of 20 children, including 12 normals aged 4–46 weeks, and demonstrated only trace amounts of other secondary bile acids in some specimens. A similar GLC analysis of fecal samples from five infants during the period from birth to 56 days reported that secondary bile acids accounted for less than 10% of the total fecal bile acids present [190]. Finley and Davidson [191] did not present information on the fecal qualitative patterns in the normal 2-week-old infants they studied, but a single published representative GLC profile shows chenodeoxycholic and cholic acids to be the predominant bile acids [191]. These results differ from the findings of Huang et al. [174] and our own observations [192]. Huang et al. [174] found a significantly higher proportion of cholic acid and lower amounts of deoxycholic and lithocholic acids ($p < 0.001$) in fecal samples

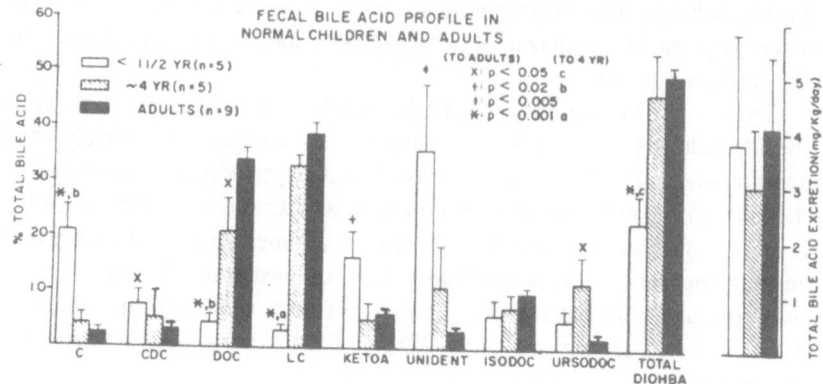

FIGURE 5. Comparison of the fecal bile acid profiles of normal adults with infants and children. All values are (mean ± SEM) percentage of total bile acids. Total bile acid excretion is expressed as mg/kg of body weight per day. Note that only five subjects (inmate volunteers) were studied for total daily bile acid excretion. C, cholic; CDC, chenodeoxycholic; DOC, deoxycholic; LC, lithocholic; KETOA, oxo-hydroxy bile acids including 7-oxo-deoxycholic, 12-oxo-lithocholic, 7-oxo-lithocholic, and 3,12-dioxo-cholanoic acid; UNIDENT, unidentified bile acids: ISODOC, 3β,12α-dihydroxy-5β-cholanoic acid; URSO, ursodeoxycholic; total DIOHBA, bile acids having two hydroxy groups in the sterol nucleus. [Reproduced from Ref. [174], with permission.]

of infants under 1.5 years of age compared with adult fecal bile acid profiles (Fig. 5). However, a large proportion of the fecal bile acids in the children (36.1 ± 11.9%) was unidentified. Furthermore, oxo bile acids were also present at proportionally higher levels than those seen in the adult.

In a study of fecal bile acid excretion [192] of three newborns over the first 7 days of life (Fig. 6), cholic and chenodeoxycholic acids were the predominant species identified (cholic acid levels being greatly in excess of chenodeoxycholic acid levels), in accordance with the observations of Huang *et al.* [174] and consistent with the observation that the duodenal bile of newborns contains mainly cholic acid [185–187,193,194]. When fecal bile acids were subdivided into conjugate groups, it was evident that in the first 2 days of life, bile acid conjugates predominate, but on day 3 of life, the microflora is sufficiently well developed to enable deconjugation of the primary bile acids to occur, so that cholic acid chenodeoxycholic acids are excreted mainly in the unconjugated form.

While qualitative profiles of fecal bile acids in newborns differ from those found in adults, absolute quantitative levels also differ. Table V summarizes the reported values of quantitative bile acid excretion in health and disease in newborn infants and children.

In a comparison of the fecal bile acid excretion of normal adults and children [174], daily excretion by adults was eight- to ninefold greater than by children

TABLE V. Quantitative Total Fecal Bile Acid Excretion in Infants and Children[a]

Reference	Subjects	n	Age	Sex (M/F)	Number of days	Dietary status	Analytical method	Daily fecal excretion[b]
Lewis and Myant, 1967 [205]	Type II hyperlipidemia	1	10 yr	0/1	5–10	*Ad lib* and corn oil (low cholesterol)	Isotopic	127 ± 13 mg
Leyland, 1970 [206]	Type II hyperlipidemia	1	10 yr	0/1	5–10		Isotopic	157 ± 22 mg
	No GI disease	16	2 mo–14 yr	—	3	Not stated	GLC	10–85 mg
	Biliary atresia	5	—	0/1	3	Not stated	GLC	<3 mg
Shurpalekar et al., 1971 [207]	Normal	10	10–12 yr	0/1	3	Controlled	—	71–82.8 (78.5) mg
	Normal	10	10–12 yr	0/1	3	+ Cholesterol (4 g/d)	—	133.8–162.1 (146.9) mg
	Normal	10	10–12 yr	0/1	3	+ Cellulose (100 g/d) and cholesterol (4 g/d)	—	192.1–231.6 (212.7) mg
Grundy et al., 1972 [208]	Type I hyperlipidemia	2	11 yr	2/0	3	Liquid formula, no fat, low cholesterol	GLC	49 ± 8 mg and 116 ± 31 mg
	Type I hyperlipidemia	2	11 yr	2/0	3	As above + clofibrate	GLC	39 ± 10 mg and 64 ± 27 mg
	Type I hyperlipidemia	2	11 yr	2/0	3	High fat, high cholesterol	GLC	67 ± 14 mg and 114 ± 48 mg
	Type I hyperlipidemia	2	11 yr	2/0	3	As above + clofibrate	GLC	57 ± 18 mg and 110 ± 26
Weber et al., 1972 [76]	Premature	16	—	—	3	Not stated	Enzymatic	7.3 ± 4.4 mg/kg
	Normal	18	4 mo–5 yr	—	3	Not stated	Enzymatic	6.1 ± 2.7 mg/kg
Weber et al., 1973 [197]	Normal	18	4 mo–5 yr	—	3	Not stated	Enzymatic	110.0 ± 11.0 mg/m²
	Celiac	12	(2.4y)	—	3	Not stated	Enzymatic	140.1 ± 29.3 mg/m²
	CF/pancreatic insuff.	24	4 mo–12 yr	—	3	Antibiotics (5 pt)	Enzymatic	743.2 ± 55.3 mg/m²
	Ileal resection	6	2 mo–9 yr (2.5y) <8m	—	3	Not stated	Enzymatic	957.7 ± 195.1 mg/m²
	Exocrine pancreatic hypoplasia	1	5 yr	—	3	Enzyme suppl.	Enzymatic	345.3 mg/m²
	No treatment	1	5 yr	—	3	No treatment	Enzymatic	494.6 mg/m²
	Chronic relapsing pancreatitis	2	—	—		Not stated	Enzymatic	88.3 ± 172 mg/m²

(continued)

TABLE V. (*Continued*)

Reference	Subjects	n	Age	Sex (M/F)	Number of days	Dietary status	Analytical method	Daily fecal excretion[b]
Nestel, 1974 [209]	Normal	12	15–18 yr	—	2	Constant diet	GLC	200 ± 84 mg
	Normal plasma cholesterol/anorexic	1	14 yr	0/1	2	Constant diet	GLC	197 mg
	Hypercholesterolemia/anorexic	4	15–16 yr	0/4	2	Constant diet	GLC	30–100 mg
Roy et al., 1975 [210]	Premature (1.3–1.8 kg birth wt)	25	7–36 days	—	3	MCT diet	Enzymatic	3.0 ± 0.6 mg/kg
		25	7–36 days	—	3	LCT diet	Enzymatic	7.8 ± 1.2 mg/kg
Carter et al., 1975 [211]	Normal	5	—	—	—	Not stated	Enzymatic	3.33 ± 0.94 mg/kg
	Hyperlipidemia type II (homozygous)	2	—	—	—	Not stated	Enzymatic	1.65 ± 0.37 mg/kg
	Hyperlipidemia type II (heterozygous)	11	—	—	—	Not stated	Enzymatic	4.2 ± 1.92 mg/kg
Gaze et al., 1975 [212]	Normal	10	1–7 yr	—	1	Not stated	Enzymatic/	14.3–23.1 (18.0) μmole/kg
	Diarrhea/Hirshsprung's/Soave pullthrough	4	1–6.5 yr	—	1	Not stated	Fluorimetric	25.8–38.0 μmole/kg
	Colostomy/Hirshsprung's	1	13 mo	—	1	Not stated	Fluorimetric	8.0 μmole/kg
Goodchild et al., 1975 [213]	Normal	11	6 mo–14 yr (6yr)	5/6	1	Not stated	Enzymatic	14.3–23.1 (18.0) μmole/kg
	Cystic fibrosis	29	3 mo–21 yr (9yr)	17/12	1	Not stated	Enzymatic	0.8–134.0 (47.6) μmole/kg
	Pancreatic insuff.	4	2–16 yr (10yr)	3/1	1	Not stated	Enzymatic	2.8–58.0 (30.9) μmole/kg
Huang et al., 1976 [174]	Normal	5	4–18 mo	3/2	1			
	Normal	5	44–65 mo	5/0	1			

Reference	Group	n	Age			Treatment	Method	Value
Potter and Nestel, 1976 [190]	Normal	4	5–10 days	—		Cow's milk	GLC	1.1–2.2 (1.7) mg/kg
	Normal	5	3 wk	—		Cow's milk	GLC	2.5–4.6 (3.6) mg/kg
	Normal	5	5–7 wk	—		Soy milk	GLC	3.4–10.8 (6.8) mg/kg
	Normal	3	8 wk	—		Cow's milk	GLC	1.0–5.6 (2.8) mg/kg
Weber et al., 1976 [198]	Normal	25	4 mo–15 yr (5.2yr)	—	3	Not stated	Enzymatic	109.8 ± 9.8 mg/m²
	Cystic fibrosis	43	2 mo–20 yr (3.9yr)	—	3	No enzyme therapy	Enzymatic	751.1 ± 48.3 mg/m²
	Cystic fibrosis no steattorrhea	6	—	—	3	No enzyme therapy	Enzymatic	133.4 ± 15.9 mg/m²
	Cystic fibrosis	6	—	—	3	Not stated	Enzymatic	999.9 ± 202 mg/m²
	Cystic fibrosis	6	—	—	3	Enzyme suppl.	Enzymatic	543.5 ± 72.1 mg/m²
	Cystic fibrosis	7	—	—	3	Enzym. + Na bicarb.	Enzymatic	500.4 ± 58.6 mg/m²
	Cystic fibrosis	7	—	—	3	Enzym. ± MCT diet	Enzymatic	457.7 ± 77.8 mg/m²
Roller and Kern, 1977 [214]	Cystic fibrosis	8	11–24 yr (18yr)	8/0	3	No treatment	Enzymatic	11.67 ± 1.96 mg/kg
	Cystic fibrosis	8	11–24 yr (18yr)	8/0	3	Enzyme supple.	Enzymatic	12.50 ± 2.02 mg/kg
Watkins et al., 1977 [199]	Cystic fibrosis	6	4 mo–3.5 yr	—	3	Controlled	GLC	545 ± 166 mg; 1170 ± 236 mg/m²
	Cystic fibrosis	6	4 mo–3.5 yr	—	3	Enzyme suppl.	GLC	513 ± 146 mg; 1165 ± 335 mg/m²
Smalley et al., 1978 [215]	Normal	11	6 mo–14 yr (6yr)	5/6	3	Ad lib	Enzymatic	18 ± 3 μmole/kg
	Cystic fibrosis/ steatorrhea	16	3–12 yr (7yr)	6/10	3	Ad lib	Enzymatic	74 ± 37 μmole/kg
		16	3–12 yr (7yr)	6/10		Enzyme suppl.	Enzymatic	62 ± 29 μmole/kg
		16	3–12 yr (7yr)	6/10		Reduced LCT(+ MCT)	Enzymatic	56 ± 26 μmole/kg
		16	3–12 yr (7yr)	6/10		Reduced LCT(+ MCT) + enzyme suppl.	Enzymatic	35 ± 18 μmole/kg
			3–12 yr (7yr)					

(continued)

TABLE V. *(Continued)*

Reference	Subjects	n	Age	Sex (M/F)	Number of days	Dietary status	Analytical method	Daily fecal excretion[b]
	Cystic fibrosis/ no steatorrhea	2	6 mo–6 yr	—	3	Ad lib	Enzymatic	15–19 μmole/kg
Bilheimer et al., 1979 [176]	Hyperlipidemia Type II (homozygous)	7	5–21 yr (12yr)	3/4	4	Controlled	Enzymatic	30–406 mg; 4.6 ± 3.4 mg/kg
	Hyperlipidemia Type II (heterozygous)	4	8–31 yr (16yr)	1/6	4	Controlled	Enzymatic	94–377 mg; 5.3 ± 2.1 mg/kg
Jonas et al., 1979 [216]	Normal	4	18–13 wk	—	3	Ad lib	Enzymatic	13.5 ± 3.1 μmole/kg
	Acute diarrhea, recovery stage	6	15–10 wk	—	3	Varied diets	Enzymatic	33.9 ± 11.6 μmole/kg
Martin and Nestel, 1979 [217]	Normal	5	5–12 yr	—	8	Low cholesterol	GLC	60–198 mg; 3.8 ± 1.4 mg/kg
	Hyperlipidemia Type IIa	9	5–18 yr	—	8	Low cholesterol	GLC	31–147 mg; 4.0 ± 2.1 mg/kg
Nestel et al., 1979 [195]	Normal	10	3–16 mo	—	8	Low cholesterol; High PUFA soy milk	GLC; GLC	11–139 mg; 1.8–11.1 (5.3) mg/kg
	Normal	10	3–16 mo	—	8	High cholesterol; Low PUFA (cow's milk)	GLC	13–36 mg; 1.1–5.3 (2.7) mg/kg
Schwarz et al., 1979 [218]	Normal	1	3 yr	1/0	3	Controlled	GLC	54 ± 6 mg; 3.3 ± 0.5 mg/kg
	Hyperlipidemia Type II (homozygous)	1	13 mo	1/0	3	Controlled	GLC	106 mg; 12.0 mg/kg
Boyle et al., 1980 [200]	Cystic fibrosis and steatorrhea	1	3 yr	1/0	3	Controlled	GLC	233 ± 42 mg; 15.5 ± 3.2 mg/kg
	Cystic fibrosis and steatorrhea	8	12–15 yr (16yr)	—	3	Enzyme suppl.	Enzymatic	4700 ± 900 μmole/m²
	Cystic fibrosis and stetorrhea	8	12–25 yr (16yr)	—	3	+ Cimetidine	Enzymatic	4200 ± 100 μmole/m²
Finley and Davidson, 1980 [191]	Premature infants (1–2.5 kg birth wt)	18	10–16 days	—	3	Cow's milk formula	GLC	2.3 ± 1.1 mg/kg

Reference	Condition	n	Age			Treatment	Method	Values[b]
Heubi et al., 1980 [201]	No GI disease	5	5 mo–9 yr	—		Not stated	Enzymatic	91 ± 24 mg/m²
	Ileal resection at <3 days	7	5 mo–9 yr	3/4		Not stated	Enzymatic	205–1412 mg/m²
Fondacaro et al., 1982 [202]	Cystic fibrosis	5	—	—		Antibiotics	Enzymatic	214.0 ± 48.4 mg/m²
	Ileostomy	5	—	—		No treatment	Enzymatic	90.8 ± 21.9 mg/m²
Heubi et al., 1982 [203]	Congenital bile acid malabsorption	2	53 and 64 mo	—	3	Not stated	Enzymatic	960 and 991 mg/m²
Zavoral et al., 1982 [204]	Normal siblings	8	7–16 yr (12.7yr)	—	3	Controlled	GLC	177 ± 20 mg; 4.05 ± 0.84 mg/kg; 120 ± 15 mg/m²
	Type IIa hypercholesterolemia	12	3–13 yr (9.5yr)	—	3	Controlled	GLC	143 ± 18 mg; 4.41 ± 0.56 mg/kg; 124 ± 15 mg/m²
Colombo et al., 1983 [196]	Normal	6	4–11 yr	—	3	Controlled	GLC	60.59 ± 24.63 mg; 2.8 ± 1.1 mg/kg
	Cystic fibrosis with no liver disease	16	4–17 yr	—	3	Enzyme suppl.	GLC	196.2 ± 158.5 mg; 7.8 ± 5.6 mg/kg
	Cystic fibrosis and liver disease	2	5–11 yr	—	3	Enzyme suppl.	GLC	11.6–139.5 mg; 0.4–9.3 mg/kg
Tazawa et al., 1984 [189]	Intractable diarrhea	2	20–26 mo	—	3	i.v. hyperalimentation (HA)	GLC	1261 and 1877 μmole/m²
	Chronic diarrhea	6	2–11 mo	—	3	i.v. HA or milk form.	GLC	173 ± 97 μmole/m²
	No GI disease	12	1–11 mo	—	3	Not stated	GLC	167 ± 86 μmole/m²
Colombo et al., 1984 [219]	Normal	10	3–13 yr	5/5	3	Not stated	GLC	2.79 ± 1.1 mg/kg
	CF/pancreat. insuff.	4	—	—	3	Not stated	GLC	11.2 ± 3.8 mg/kg
	CF/no pancreat. insuf.	3	—	—	3	Not stated	GLC	2.2 ± 0.45 mg/kg
Darling et al., 1985 [220]	Cystic fibrosis	10	2–16 yr	—	3	Enzyme suppl. and placebo	GLC	401.8 ± 3.6 mg
	Cystic fibrosis	10	2–16 yr	—	3	Enzyme suppl. and taurine	GLC	372.6 ± 58.7 mg
Setchell et al., 1985 [221]	Cystic fibrosis (no liver disease)	7	3–17 yr	3/7	3	Enzyme suppl.	GLC	7.53 ± 4.14 mg/kg

[a] PUFA, polyunsaturated fatty acids; LCT, long-chain triglycerides; MCT, medium-chain triglycerides; Enzy, pancreatic enzyme supplement; GI, gastrointestinal; HA, hyperalimentation; i.v., intravenous.

[b] Values are expressed : s range with mean in parentheses, mean ± SD. or mean ± SEM. and the reader is directed to the relevant reference for more detail.

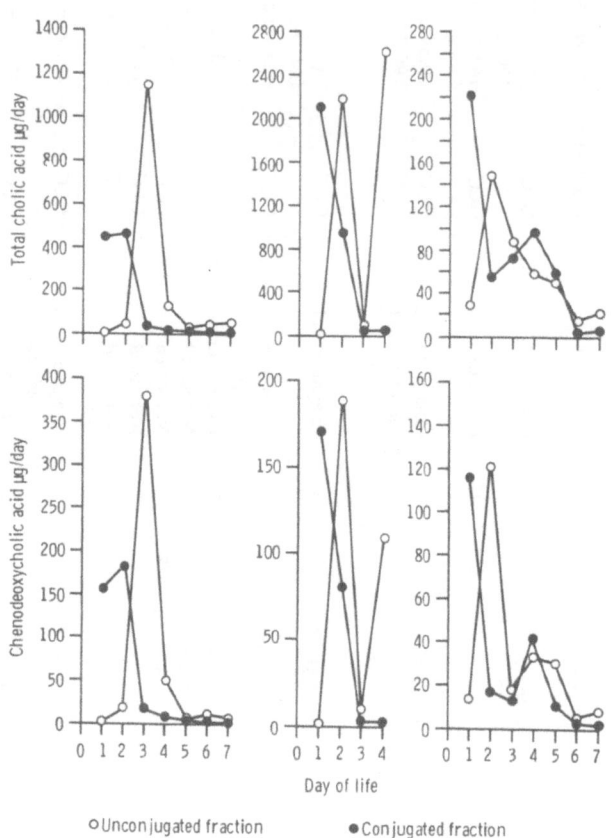

FIGURE 6. Changes in the fecal excretion of unconjugated and conjugated forms of cholic and chenodeoxycholic acids (mg/g feces) in the meconium and stools of three normal newborn infants during the first week of life. Note that the concentrations of cholic acid exceed those of chenodeoxycholic acid and that by day 3 the proportion of the unconjugated bile acid exceeds the conjugated species, providing evidence for rapid development of bacterial deconjugation after birth. (Setchell *et al.*, unpublished data.)

less than 18 months old. The differences disappeared when data were expressed in terms of mg/kg body wt per day [174]. These observations have been confirmed by others [190,195,196], and additionally it has been indicated that body surface area should be taken into account when expressing data [189,197–204].

In general, total bile acid excretion reported for children 5 years of age is less than 85 mg/day (Table V), which is significantly lower than for older children and adults (see Table IV). For all newborns and children, bile acid excretion is relatively constant when values are based on body weight (mean values 3–6 mg/kg body wt per day) and is similar to that of adults.

Obvious differences arise in newborn infants on different dietary regimes. For example, soy-based formula feeding leads to significantly increased total fecal bile acid excretion compared to cow's milk, and a concomitant decrease in the plasma cholesterol level occurs [190,195]. These differences may be explained by the higher intake of polyunsaturated fatty acids associated with the soy milk formula. To our knowledge only two studies have measured bile acid excretion in premature infants; one used an enzymatic technique [76] and the other used GLC [191]. The former study reports values (7.3 ± 4.4 mg/kg body wt per day) markedly higher than the latter (2.3 ± 1.1 mg/kg body wt per day), but the differences probably reflect the different methodologies. On the basis of these limited studies, it is not clear whether there are differences in fecal bile acid excretion between normal-term newborns and premature infants. Roy *et al.* [210] studied 25 premature neonates and found an improvement in fat malabsorption and an increased fecal bile acid loss with a diet high in medium-chain triglycerides compared to a diet with long-chain triglycerides.

Quantitative data on the excretion of individual bile acids in the feces of normal newborns and infants is extremely limited and scattered; and it is therefore difficult to make comparisons between studies because of the variability in the age of subjects studied. The most useful information to date can be found in the publication of Huang *et al.* [174] in which the percent composition of the individual bile acids is reported for subjects in the age range 4–18 months and 44–65 months and these are compared with data for normal adults (Fig. 5).

3.3. Meconium

Several studies have been carried out to characterize bile acids present in meconium [184,222,223], and a list of bile acids that have been identified is given in Table VI. As meconium is sterile *in utero,* the presence of the secondary bile acids, such as deoxycholic acid, which are products of bacterial metabolism, is presumed to be due to transplacental passage of bile acids rather than fetal hepatic synthesis. The presence of other bile acids with hydroxylation at C-1, C-6, and C-23, the allo bile acids, 3β-hydroxy-5-cholenoic, and norcholic acid [227] probably reflects fetal bile acid metabolism; for example, 1β-hydroxylation

TABLE VI.   Bile Acids Found in Human Meconium

Bile acid structure	State of conjugation[a]	Reference
Monohydroxy		
3α-hydroxy-5β-cholanoic acid	S	184,222,224–227
3β-hydroxy-5-cholenoic acid	T,S	222,224–227
Dihydroxy		
3α, 7α-dihydroxy-5β-cholanoic acid	U,G,T,S	184,222,226,227
3α, 12α-dihydroxy-5β-cholanoic acid	NS,S	184,222,225,227
3α, 6α-dihydroxy-5β-cholanoic acid	NS	227
3β, 12α-dihydroxy-5-cholenoic acid	S	227
3α, 7β-dihydroxy-5β-cholanoic acid	—	222
unidentified unsaturated dihydroxy	NS,S	227
Trihydroxy		
3α, 7α, 12α-trihydroxy-5β-cholanoic acid	U,G,T,S	184,222,225–227
3α, 6α, 7α-trihydroxy-5β-cholanoic acid	U,G,T,S	226,227
3α, 7α, 12α-trihydroxy-5α-cholanoic acid	NS,S	227
3β, 7β,12α-trihydroxy-5β-cholanoic	S	227
Tetrahydroxy		
1, 3, 7, 12-tetrahydroxy-5β-cholanoic acid	U,T,S	192,227,661
3, 7, 12, 23-tetrahydroxy-5β-cholanoic acid	NS	192,227,661
Oxo bile acids		
3α-hydroxy-7-oxo-5β-cholanoic acid	—	226
Nor bile acids		
3α, 7α, 12α-trihydroxy-24-nor-5β-cholan-23-oic acid	U,S	227
Bisnor bile acids		
(20S)3α-hydroxy-23,24-bisnor-5β-cholan-22-oic	—	228
(20S)3β-hydroxy-23,24-bisnor-5-chol-en-22-oic	—	228
(20R)3β-hydroxy-23,24-bisnor-5-chol-en-22-oic	—	228
Cortolic acids		
3β-hydroxy-pregn-5-en-21-oic	—	228
Etianic acids		
3β-hydroxy-5-androst-ene-17β-carboxylic	S/Gluc	223,662
3α-hydroxy-5α-androstane-17β-carboxylic	S/Gluc	223,662
3α-hydroxy-5β-androstane-17β-carboxylic	S/Gluc	223,662
3β-hydroxy-5α-androstane-17β-carboxylic	S/Gluc	223,662

[a] Gluc, glucuronide; S, sulfate; T, taurine; U, unconjugated; G, glycine; NS, nonsulfated.

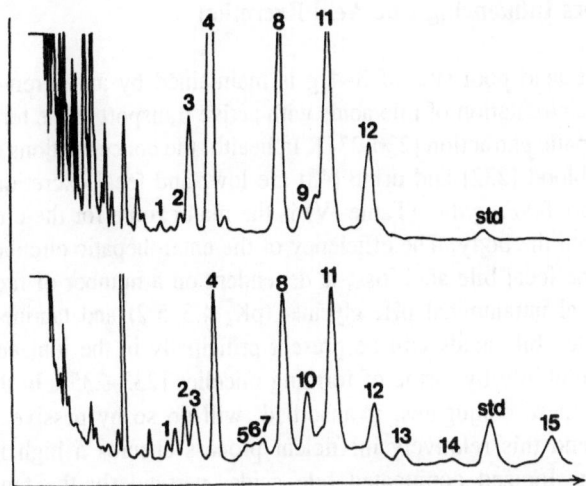

FIGURE 7. GLC analysis of the nonsulfated and sulfated bile acids of meconium. Bile acids identified are as follows: (1) norcholic; (2) allo-cholic; (3) 1,3,7,12-tetrahydroxy-5β-cholanoic; (4) cholic; (5) allo-deoxycholic; (6) compound b; (7) 3β,7α-dihydroxy-5β-cholanoic; (8) hyocholic; (9) 3α,7β,12α-trihydroxy-5β-cholanoic; (10) deoxycholic; (11) chenodeoxycholic; (12) compound a; (13) 3β,12α-dihydroxy-5β-cholanoic; (14) lithocholic; (15) 3β-hydroxy-5-cholanoic; (std.) internal standard (ursodeoxycholic). (Chromatograms are reconstructed and peaks numbered for clarity from the published figure of Ref. [227], with permission.)

by human fetal hepatic microsomes was recently confirmed [229]. Gas–liquid chromatograms of the nonsulfated and sulfated bile acids in meconium, published in one of the most detailed studies [227], are shown in Fig. 7. Bile acids occur in meconium predominately as taurine and sulfate conjugates [192,222,225,227], but the contribution, if any, of bile acid glucuronides has yet to be determined. In newborns from healthy pregnancies, 53–84% of the total bile acids in meconium were sulfate conjugates [225]. Meconium from three infants with cholestatic mothers had elevated levels of bile acids, with a markedly increased level in one case where the mother had a severe degree of cholestasis [225].

A series of saturated and unsaturated monohydroxy steroid acids containing 20, 21, and 22 carbon atoms have been isolated from meconium and identified by MS [223,228,662,663]. The C_{20} acids (etianic acids) have been reported to be present as sulfate or glucuronide conjugates in similar or greater concentrations than those of lithocholic and 3β-hydroxy-5-cholenoic acids in meconium. The accumulation of these short-chain acids in meconium may arise from the high concentrations of steroidal hormones in the fetus acting as a substrate for side-chain oxidation; however, since their biosynthesis, metabolism, and function are unclear, further work is necessary to determine their origin.

3.4. Factors Influencing Bile Acid Excretion

The bile acid pool size of 3–4 g is maintained by an extremely efficient enterohepatic circulation of bile acids with active transport at the terminal ileum and rapid hepatic extraction [230,231]. In health, the concentrations of bile acids in systemic blood [232] and urine [48] are low, and fecal excretion, variously reported as 50–800 mg/day (Table IV) is the major route for the elimination of bile acids from the body. The efficiency of the enterohepatic circulation, which will determine fecal bile acid loss, is dependent on a number of factors.

At normal intraluminal pH, glycine- (pK_a 4.3–5.2) and taurine- (pK_a 1.8–1.9) conjugated bile acids will be present principally in the ionized form with high water solubility by virtue of forming micelles [233–235]. In the proximal small bowel these conjugates, if absorbed, will do so by passive ionic diffusion [236], and this relatively inefficient process ensures a high intraluminal concentration. Ionized conjugated bile acids, particularly the taurine conjugates, will be favored by the active transport process found in the terminal ileum [237], which is the major site of bile acid reabsorption in humans. The unconjugated bile acids (pK_a 5.0–6.0) will be protonated to a larger extent and will therefore be absorbed by a nonionic passive diffusion process [236]. This process is at least 5–6 times more efficient than diffusion of charged particles [236]. Any decrease in small-bowel luminal pH, therefore, or the presence of large quantities of unconjugated bile acids will cause precipitation and adsorption of the bile acids to dietary residues or bacteria. Maintenance of the enterohepatic circulation is therefore dependent not only on the intraluminal pH, but also on the distribution of the microflora within the gastrointestinal tract. When excessive amounts of unconjugated bile acids are present, for example, as in conditions of bacterial overgrowth and in certain diseases where the luminal pH is decreased (e.g., cystic fibrosis), the patency of the enterohepatic circulation may be threatened and excessive spillage of bile acids into the colon may result.

Differences in the physicochemical properties of conjugated bile acids suggest that any alteration in the glycine/taurine conjugate ratio may give rise to changes in fecal bile acid excretion [238]. A switch to a higher proportion of taurine conjugates in the bile acid pool might lead to an increased fecal bile acid excretion because of the greater dependence on the active ileal reabsorption mechanism and because of exposure to greater numbers of bacteria present in the distal ileum [239]. On the other hand, glycine conjugates are more susceptible to bacterial deconjugation [240], and therefore any increase in these bile acids in the enterohepatic circulation may augment fecal losses. When the glycine/taurine conjugate ratio was significantly decreased in six healthy male subjects by feeding taurine [240], bile acid pool size and fecal bile acid excretion decreased only slightly. Individual variations were large, and therefore firm conclusions could not be drawn from these data, but it appears that a change in the conjugation pattern probably has little effect on fecal bile acid excretion.

The most crucial factor in the conservation of the bile acid pool is the active transport of bile acids at the terminal ileum. Resection or dysfunction in this region will seriously compromise the integrity of the enterohepatic circulation. Compensatory hyperfunction of the terminal ileum may occur when other sections of the intestine are diseased, as occurs in some cases of celiac sprue [241]. Bile acids passing into the colon undergo extensive deconjugation, dehydroxylation, and oxidoreduction, which leads to the complex profile seen in feces [20,21] and the predominance of unconjugated species [21]. Colonic absorption of bile acids has been demonstrated [236,242,243], and despite bile acid binding to dietary residue and intestinal microorganisms, it may contribute significantly to the conservation of the bile acid pool in health, since numerous unconjugated and secondary bile acids have been observed in normal peripheral serum [244]. The binding of bile acids to dietary residue or bacteria [127–129] and the distribution of different bile acids in the aqueous phase [245] will both be influenced by pH. These are important factors which will affect fecal bile acid excretion and will be discussed later.

Changes in small-bowel transit time would be expected to affect the extent of bacterial degradation and the residence time in the region of active absorption. Gut transit times are affected by bowel resection, conditions inducing diarrhea, and dietary changes, particularly in fiber content. Despite significant increases in intestinal transit time induced by the ingestion of the nonabsorbable sugar mannitol, relatively little effect on the fecal excretion rate of orally administered radiolabeled cholic acid was demonstrated, indicating that gut transit is a minor factor in causing excessive fecal bile acid loss [99,246]. This finding is further illustrated in a series of papers on infectious enteritis [36,37,247,248] in which decreased intestinal transit time was not always associated with an alteration in qualitative or quantitative fecal bile acid profiles. In later studies, total bile acid synthesis rate was measured during acute sorbitol treatment and showed a mean reduction of approximately 27% accompanied by a decrease in primary bile acid synthesis and an increase in the fractional turnover of cholic acid [249]. Hardison et al. [250] also used sorbitol and atropine (to increase gut transit time) and concluded that acute changes in intestinal transit time are associated with acute changes in biliary bile acid excretion rate, although neither of these later studies directly examined fecal bile acid excretion.

Intestinal microflora metabolize bile acids by a number of reactions, the principal one being hydrolysis of the amide bond of the conjugates, and this reaction generally precedes 7α-dehydroxylation and oxidoreduction at other positions [167,168,251,252]. Various strains of bacteria have different abilities to perform these transformations, so that changes in the microflora in the gut may alter both the quantitative and qualitative pattern of fecal bile acid excretion. In the case of the newborn infant, colonization of the gastrointestinal tract normally starts within a few days postnatally [183]. The method of delivery, diet, and gestational age influence the type and time scale of microflora development [253]. Differences are found between vaginal- and cesarean-delivered infants. Cesarean section, prematurity, and breast feeding have been shown to influence numbers and frequency of detection of anaerobic species in the gastrointestinal tract [253], and the effect of these factors on qualitative and quantitative fecal bile acid excretion in the newborn is unknown.

A number of mechanisms control the degree of proliferation and types of microflora present in the gut. Decreased small-intestinal motility, as observed in blind-loop syndrome, and increased intraluminal pH, as seen in achlorhydric or partial-gastrectomy patients, will increase bacterial growth in the normally sparsely populated regions of the stomach and proximal small intestine [254]. Despite observations of altered bile acid metabolism in bacterial overgrowth [255,256], analysis of fecal bile acids has not to our knowledge been reported. The ability to deconjugate has been observed with many intestinal microorganisms *in vitro* [168,251] and appears to be relatively insensitive to inhibition. *In vivo* examination of specimens of intestinal fluid demonstrates a close relationship between the presence of anaerobic bacteroides and unconjugated bile acids [254]. No correlation was found between bile acid deconjugation and other microorganisms [254]. Although unconjugated bile acids represent only a small proportion of biliary output, they comprise >90% of the bile acids found in in the stools of healthy subjects [21] and in most disease states. In contrast, the ability to 7α-dehydroxylate is much less common in intestinal organisms [167,168,251], and *in vitro* this reaction has been shown to be inhibited at pH less than 6.5 [257] and by increased concentrations of bile acids [258,259]. Increased levels of primary bile acids with concomitant reduction in secondary metabolites have been reported in a wide variety of disorders listed in Table VII and are particularly associated with the presentation of diarrhea. Dihydroxy bile acids, deoxycholic and chenodeoxycholic acids, have been shown to induce secretion or inhibit absorption of electrolytes and water in the colon [272]. Although deoxycholic acid exhibits a greater potency than chenodeoxycholic acid at normal colonic pH [272], chenodeoxycholic acid is more soluble [273]. The relationship between fecal pH and the proportion of bile acids in the aqueous phase is discussed in Section 4.9. Microbial transformations also affected by the intraluminal pH include interconversion of ursodeoxycholic acid, 7-oxo-lithocholic acid, and chenodeoxycholic acid [274].

Fecal bile acid output is in many cases dependent on the body weight, sex, and age of the subject, so that adequate numbers of matched controls are necessary for proper evaluation of experimental results, yet frequently these considerations are neglected. The effect of dietary calorific intake on fecal bile acid excretion is discussed in Section 5.4. The effects of obesity are similar to those

TABLE VII.   Conditions in Which an Increase in the Fecal Primary Bile Acid: Secondary Bile Acid Ratio Has Been Described

Condition	Reference
1. Crohn's disease	260, 261
2. Diverticular disease	262
3. Ileal resection	245, 260, 261, 263, 264
4. Liver cirrhosis	265–268
5. Benign recurrent cholestasis	269
6. Diarrhea associated with	
1. Vagotomy	270
2. Uremia	33
3. Hypogammaglobulinemia	271
4. Soave pullthrough operation	212

seen with high calorie intake. Total bile acid excretion and synthesis rate is high in overweight normolipidemic subjects, but is comparable to that of healthy subjects when corrected for body weight [275,276]. Hellström and Einarsson [277] criticize the expression of excretion in terms of mg/kg body weight per day on the premise that normal adult body weight correlates poorly with fecal bile acid excretion, an observation that is at variance with earlier reports [278]. A compilation of data from several studies has shown a correlation between body weight and fecal bile acid output in normolipidemic and hyperlipidemic subjects [279]. When analyzed in terms of populations, characterized by plasma lipid levels, the correlation was statistically significant in all groups except hypercholesterolemic subjects. It is of interest that the correlation was better in normolipidemia ($r = 0.54$) than in hyperlipidemias ($r = 0.21$–0.48). Low body weight or weight loss in obese subjects is usually associated with decreased bile acid excretion [275,280–282], although the correlation is poor and correction for body weight does not always bring excretion figures to within the expected normal range. Slightly higher values for total fecal bile acids when expressed as mg/day, mg/kg body weight per day, or mg/m^2 per day have been reported in males compared with females in health, obesity, and hyperlipidemia, although the differences were usually not statistically significant [278,279,281]. A study of 165 Japanese men in Hawaii aged 43–74 found no correlation between total or individual fecal bile acid excretion and the weight, height, or age of the subject [171]. With regard to age, Van der Werf et al. [283] showed a higher fractional turnover of cholic acid in elderly persons (55–75 years) compared to younger subjects (20–30 years), although pool sizes, synthesis rates of cholic acid, and gut transit times were similar. In contrast, Einarsson et al. [284] did not find the fractional turnover to be age dependent and found a significant negative correlation between age and total bile acid synthesis. Higher deoxycholate intestinal input rates have been reported in the older subject [283], but how this affected fecal bile acid excretion was not investigated. Finally, drugs and dietary components have been found to affect bile acid metabolism and increase fecal excretion. Several excellent reviews on these effects have been published [285–287], and specific effects are discussed in Section 5.

4. FECAL BILE ACID EXCRETION IN DISEASE

4.1. Colon Cancer

Probably more has been written on the subject of fecal bile acids and colonic carcinogenesis than on any other area in which bile acids have been implicated in gastrointestinal disease. For this reason, no attempt has been made to comprehensively review all aspects of this subject, and the reader is directed to numerous reviews published elsewhere and references therein covering the epidemiological and environmental factors [288–294], dietary effects [175,185,295–305], experimental animal models for colon cancer [304,306–312], and bacteriology [308,313–321].

In one of the earliest studies of fecal bile acid excretion, Hill *et al.* [316] studied 44 patients with cancer of the large bowel and found 82% of them to have higher fecal bile acid concentrations than controls; 70% of these were shown to have a concomitant presence of nuclear dehydrogenating Clostridia in the stool which compared with only 9% of the 90 patients studied with other diseases. These findings and those from metabolic epidemiological studies gave support to the hypothesis that high concentrations of bile acid metabolites produced by anerobic bacteria may act as carcinogens or cocarcinogens in initiating and/or promoting tumor growth [313].

In a further study, Reddy and Wynder [322], using essentially similar methodology in 31 patients with colon cancer, found elevated total fecal bile acid concentrations (19.87 ± 0.94 SEM mg/g dry wt) compared with controls (10.85 ± 0.76 SEM mg/g dry wt) and a group of patients with other digestive diseases (9.51 ± 1.13 SEM mg/g dry wt). These two key studies, taken together with the many studies of fecal bile acid excretion in subjects from populations at different risks for colon cancer (Table VIII), gave impetus to the idea of a connection between bile acids and colon carcinogenesis. Subsequent studies, however, have failed to corroborate these early findings; moreover, fecal bile acid concentrations in the important studies performed since have been shown to be either lower than or not significantly different from those of control groups [169,182,328–331] (Fig. 8). The reason for the disparity between these data is difficult to explain, but it may relate to differences in methods of fecal collection, nutritional status of the patients, or analytical methodology. Interestingly, in a

TABLE VIII. Total Fecal Bile Acid Concentrations in Normal Subjects from Populations Differing in Risk for Colon Rectal Cancer and in Patients with and at Risk for Carcinogenesis

Reference	Subjects	n	Age (mean)	Sex (M/F)	Sample size (days)	Diet and drugs	Analytical methods	Concentrations
Normals								
Hill and Aries, 1971 [24]	Normal English	22	—	—	Single random stool	Western mixed	GLC	6.13 ± 0.88 mg/g dry wt
	Normal Scottish	14	—	—		Western mixed	GLC	6.18 ± 0.90 mg/g dry wt
	Normal Indians	10	—	—		Rice diet	GLC	0.51 ± 0.16 mg/g dry wt
	Normal Ugandans	11	—	—		Matoke	GLC	0.54 ± 0.09 mg/g dry wt
Hill, 1974 [315]	Normal	60	—	—	Single random stool	Not stated	GLC	6.0 mg/g dry wt
	Vegans	20	—	—		Not stated	GLC	3.5 mg/g dry wt
Crowther et al., 1976 [323]	Normals, Hong Kong (high income)	19	—	—	Single stool	Not stated	GLC	4.74 mg/g dry wt
	Normals, Hong Kong (average income)	24	—	—	Single stool	Not stated	GLC	3.3 mg/g dry wt
	Normals, Hong Kong (low income)	21	—	—	Single stool	Not stated	GLC	2.17 mg/g dry wt
IARCIM Study, 1977 [324]	Normals, Finland. Spring–Autumn	30	55–64	30/0	Single stool	Ad lib but recorded	GLC	8.3 ± 2.6 mg/g dry wt
		26	55–64	26/0	Single stool	Ad lib but recorded	GLC	7.3 ± 1.4 mg/g dry wt
	Normals, Denmark. Spring–Autumn	30	55–64	30/0	Single stool	Ad lib but recorded	GLC	8.5 ± 3.7 mg/g dry wt
		26	55–64	32/0	Single stool	Ad lib but recorded	GLC	8.2 ± 1.7 mg/g dry wt
Mower et al., 1978 [171]	Normal Japanese in Hawaii	165	43–74	165/0	Single stool	Ad lib but recorded	GLC	10.96 ± 0.67 mg/g dry wt
Reddy et al., 1978 [325]	Normals, Kuopio, Finland	15	Middle age		1	Finnish diet	GLC	4.9 ± 0.42 mg/g dry wt
	Normals, New York	15	Middle age		2	Western diet	GLC	11.7 ± 0.54 mg/g dry wt
Mower et al., 1979 [326]	Hawaii Japanese immigrants	247	30–74	165/82	Single random stool	Japanese/Western	GLC	0.48 ± 61.8 (8.61) mg/g dry wt
	Normal Japanese	122	30–74	70/52		Japanese	GLC	0.1–95.3 (7.45) mg/g dry wt

(continued)

TABLE VIII. (*Continued*)

Reference	Subjects	n	Age (mean)	Sex (M/F)	Sample size (days)	Diet and drugs	Analytical methods	Concentrations
Tanida et al., 1981 [54]	Normal Japanese	8	24–49	8/0	3	*Ad lib*	GLC-MS	5.2–>50.6 µmole/g dry wt
Turjman et al., 1981 [327]	Lactoovarian SDA	21	39–65	—	3	*Ad lib* but recorded	GLC	3.98 ± 0.69 mg/g dry wt
	Nonvegetarian SDA	25	39–65	—	3	*Ad lib* but recorded	GLC	4.61 ± 0.57 mg/g dry wt
	Vegans, SDA	18	39–65	—	3	*Ad lib* but recorded	GLC	2.17 ± 0.31 mg/g dry wt
	Normals	22	39–65	—	3	*Ad lib* but recorded	GLC	6.46 ± 0.82 mg/g dry wt

Colon cancer and related conditions

Reference	Subjects	n	Age (mean)	Sex (M/F)	Sample size (days)	Diet and drugs	Analytical methods	Concentrations
Hill et al., 1975 [316]	Colon cancer	44	(62.7)	24/20	Single	Not stated	GLC	2–>10 (6.7) mg/g dry wt
	Upper GI disease	50	(53.7)	23/27	stool	Not stated	GLC	1–9.7 (4.2) mg/g dry wt
	Non-GI disease	28	(53)	18/10	Single	Not stated	GLC	1.2–>10 (5.1) mg/g dry wt
	Nonmalignant large-bowel disease	12	(53.5)	11/1	stool	Not stated	GLC	1.8–7.3 (5.2) mg/g dry wt
Reddy and Wynder, 1977 [322]	Normals	34	30–64	16/18	1–2	Mixed Western diet	GLC	10.85 ± 0.06 mg/g dry wt
	Colon cancer	31	45–65	15/16	1–2	Mixed Western diet	GLC	19.87 ± 0.94 mg/g dry wt
	Adenomatous polyps	13	30–65	8/5	1–2	Mixed Western diet	GLC	16.79 ± 0.77 mg/g dry wt
	Other digestive dis.	9	38–54	5/4	1–2	Mixed Western diet	GLC	9.51 ± 1.13 mg/g dry wt
	Normal Japanese	12	28–38		1–2	Japanese/Western mixed	GLC	6.02 ± 0.21 mg/g dry wt
Mudd et al., 1978 [328]	Normals	57	—	—	Single	*Ad lib*	Enzymatic	25.6–32.2 µmole/g dry wt
	Colorectal adenoma	17	49–75	8/9	stool	*Ad lib*	Enzymatic	25.1 ± 3.3 µmole/g dry wt
	Ulcerative colitis	11	23–75	7/4	Single stool	*Ad lib*	Enzymatic	26.9 ± 3.3 µmole/g dry wt
Moskovitz et al., 1979 [329]	Normal	23	(63)	15/9	Single	*Ad lib* but recorded	GLC	8.9 ± 1.3 mg/g dry wt
	Colon cancer	15	(71)	(71)	stool	*Ad lib* but recorded	GLC	5.8 ± 1.0 mg/g dry wt
Mudd et al., 1980 [330]	Colon cancer	19	—		Single	Not stated	Enzymatic	26.3 ± 2.5 µmole/g dry wt
	Normal	19	—		stool	Not stated	Enzymatic	28.1 ± 2.7 µmole/g dry wt

Reference	Subjects	n	Age		Stools	Diet	Method	Value
Murray et al., 1980 [331]	Non-GI disease	36	(65)	20/16	Single stool	Ad lib	Enzymatic	20.5 ± 2.2 μmole/g dry wt
	Colon cancer	37	65–68	19/16	Single stool	Ad lib	Enzymatic	11.8 ± 0.7 μmole/g dry wt
Hikasa et al., 1984 [169]	Normal Japanese	14	63 ± 11	5/9	3	Ad lib	GLC–MS	7.0–31.7 mg/g dry wt
	Colon cancer (Japanese)	14	63 ± 8	5/9	3	Ad lib	GLC–MS	5.8–22.9 mg/g dry wt
Tanida et al., 1984 [165]	Normals, Japanese	13	34–78	10/3	3	Ad lib	GLC–MS	1.56–13.0 (6.7) mg/g dry wt
	Adenomatous polyps	13	32–74	11/2	3	Ad lib	GLC–MS	0.76–20.3 (6.6) mg/g dry wt
Breuer et al., 1985 [182]	Normal	21	(49)	15/6	3 Single stools	Ad lib	GLC	3.80 ± 0.70 mg/g dry wt
	Colon cancer	23	(62)	8/15	3 Single stools	Normal diet	GLC	3.03 ± 0.34 mg/g dry wt
Owen et al., 1986 [332]	Normal	36	—	—	—	Not stated	GLC	6.71 ± 0.86 mg/g dry wt
	Colon cancer	34	—	—	—	Not stated	GLC	7.40 ± 1.12 mg/g dry wt
Breast cancer								
Murray et al., 1980 [354]	Normal	36	—	—	Single stool	Ad lib	GLC	20.5 ± 1.9 μmole/g dry wt
	Breast cancer	30	—	0/30	stool	Ad lib	GLC	15.6 ± 1.8 μmole/g dry wt
Papatestas et al 1982 [355]	Normal	71	—	0/71	3 Single stools	Not stated	GLC	13 ± 14 mg/g dry wt
	Breast cancer	78	—	0/78	stools	Not stated	GLC	17 ± 15 mg/g dry wt
Owen et al 1986 [332]	Normal	36	—	—	—	Not stated	GLC	6.71 ± 0.86 mg/g dry wt
	Breast cancer	16	—	0/16	—	Not stated	GLC	6.45 ± 0.84 mg/g dry wt
Ulcerative colitis								
Reddy et al., 1977 [29]	Normals	40	(44)	21/19	1–2	Mixed Western	GLC	10.8 ± 0.68 mg/g dry wt
	Ulcerative colitis	15	(34)	6/9	1–2	Mixed Western	GLC	12.2 ± 0.71 mg/g dry wt
	Relatives of patients	15	(38)	6/9	1–2	Mixed Western	GLC	11.54 ± 0.68 mg/g dry wt
	Other GI diseases	15	(45)	23/22	1–2	Mixed Western	GLC	10.59 ± 1.14 mg/g dry wt
Tanida et al., 1986 [350]	Normals	12	<50	11/1	3	Ad lib	GLC	12.4 ± 13.3 μmole/g dry wt
	Ulcerative colitis	14	14–47	6/8	3	Low residue	GLC	17.2 ± 9.2 μmole/g dry wt
Familial polyposis/Gardner's syndrome								
Watne and Core, 1975 [346]	Normal	7	—	—	—	Not stated	GLC	(7.26) mg/g dry wt
	F. polyposis/Gardner's	7	—	—	—	Not stated	GLC	(6.4) mg/g dry wt
Reddy et al., 1976 [27]	Familial polyposis	8	17–50	6/2	2	Mixed Western	GLC	8.31 ± 2.22 mg/g dry wt
	Relatives, controls	10	9–54	—	2	Mixed Western	GLC	9.67 ± 3.06 mg/g dry wt
	Normals	17	20–50	—	2	Mixed Western	GLC	10.7 ± 0.61 mg/g dry wt

" Figures in parentheses denote mean values. Non-GI disease refers to control subjects having no evidence of gastrointestinal disease. Ad lib refer to no dietary restrictions and subjects consuming usual diet. SDA, Seventh-Day Adventist.

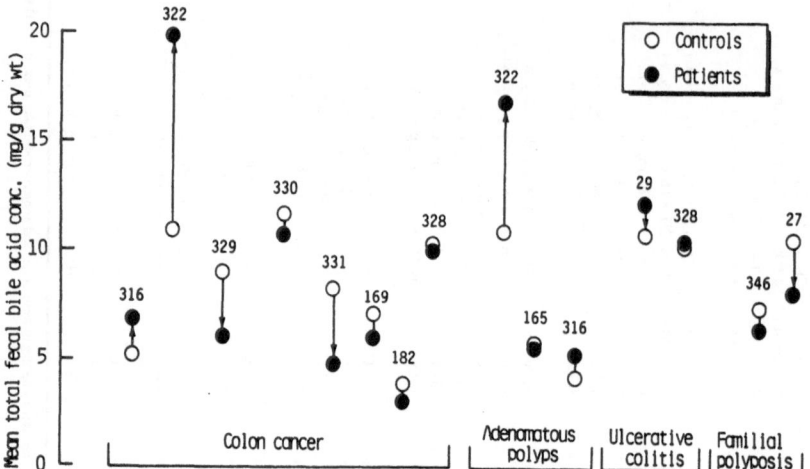

FIGURE 8. Graphic representation of the mean values for fecal bile acid concentration (mg/g dry wt feces) reported by various groups for healthy controls and patients with colon cancer or patients with diseases classified as high risk for development of colon cancer. Numbers refer to reference number of publication in bibliography.

study of 14 patients with colon cancer living in Japan, a population with a low incidence of the disease, total fecal bile acid concentration was shown to be lower in the patient group compared with a group of healthy Japanese controls [169]. In the study of Hikasa *et al.* [169] and also in a separate study of healthy Japanese subjects and patients with adenomatous polyps [54,165], concentrations of total fecal bile acids (approx. 6 mg/g dry wt) were comparable to figures reported by many groups for healthy Westerners (Fig. 8). It is therefore difficult to understand why these values, using the most sophisticated techniques to date, do not support the earlier figures showing average concentrations for Japanese subjects to be 0.5 mg/g dry wt feces [333]. Indeed, while Mower *et al.* [326] demonstrated that fecal bile acid concentrations in Japanese immigrants to Hawaii, consuming a more Westernized type of diet, were higher than for healthy Japanese living in Japan, they were in close agreement to those found in the most recent studies, [54,165,169]. In a similar study to that of Hikasa *et al.* [169] and using a slight modification [55] of the method described by Setchell *et al.* [21], bile acid concentrations in 23 patients with colon cancer were compared with those in 21 healthy controls [182]. Concentrations of total fecal bile acids were not significantly different between the patients and controls, and interestingly, these values for Westerners were noticeably lower than those found by Hikasa *et al.* [169] and Tanida *et al.* [54,165] for Japanese subjects.

These discrepancies appear to cast some doubt on the reliability of some of the earlier studies of fecal bile acid concentrations from populations at differing risk for colon cancer. Figure 9 summarizes the reported average fecal bile acid

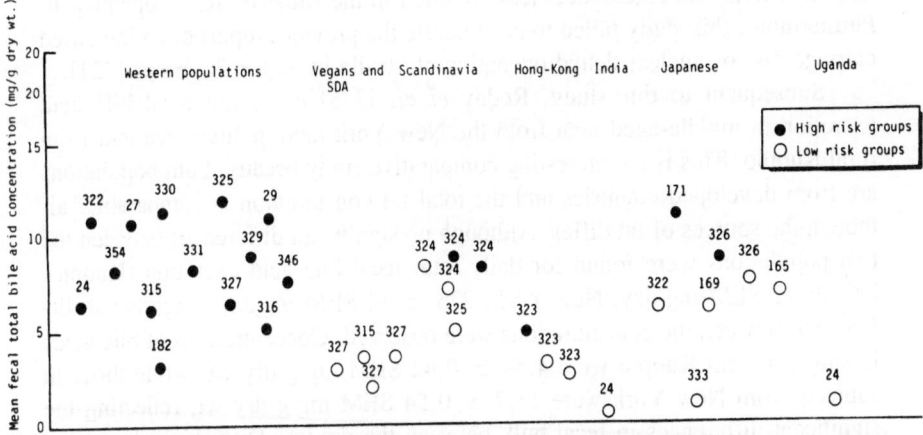

FIGURE 9. Mean fecal bile acid concentrations (mg/g dry wt feces) reported for subjects in populations or groups differing in the incidence rates of, and therefore risk for, colon cancer. Numbers refer to relevant reference number of publication in bibliography.

concentrations for healthy subjects in different populations or in groups having different incidence rates for the disease. When the combined studies are examined, it is clear that there is little evidence to substantiate the general concept that bile acid excretion by subjects living in Japan, who are at lower risk for the disease, is lower than that of healthy subjects in the Western world.

Further anomalies also exist in the reported literature involving population studies. Crowther et al. [323] examined fecal bile acid excretion in three groups of Hong Kong Chinese, classified according to their socioeconomic groups. Epidemiological study of the area showed the high-income group to have the highest incidence of colon and breast cancer, and the fecal bile acid concentration determined from random stools in this group (4.74 mg/g dry wt) was greater than for the low-income group (2.17 mg/g dry wt). While differences in fecal flora were noted, the earlier reported higher prevalence of nuclear dehydrogenating clostridia was not apparent in the higher-risk group.

Prompted by the reported differences in fecal bile acid concentrations in subjects from populations with known differences in incidence rates for colon cancer, the International Agency for Research in Cancer Intestinal Microecology (IARCIM) group carried out a study of subjects in rural Kuopio (Finland) and Copenhagen (Denmark). These two developed communities have similar lifestyles but show a fourfold (males) and 2.8-fold (females) difference in the incidence of colorectal cancer [324]. Fecal bile acids were measured on two occasions in the spring and autumn. Mean concentrations, which ranged from 7.3 to 8.5 mg/g dry wt, were found to be not significantly different between the two populations, except for a small difference in the autumn values. Again, differences in the bacterial flora were observed, with Bacteroides more common and eubacteria and enterococci less common in the subjects from Copenhagen. Furthermore, this study failed to corroborate the previous report of an increased carriage rate of nuclear dehydrogenating clostridia in high-risk groups [321].

Subsequent to this study, Reddy *et al.* [325] compared fecal bile acid excretion in middle-aged men from the New York metropolitan area and from rural Kuopio. This is an interesting comparative study because both populations are from developed countries and the total fat consumption is comparable, although the sources of fat differ. Although no significant differences between the two populations were found for daily total fecal bile acid excretion (Kuopio, 277 ± 22 SEM mg/day; New York, 275 ± 14 SEM mg/day), significant differences between the concentrations were observed. Concentrations of bile acids in subjects from Kuopio were 4.59 ± 0.42 SEM mg/g dry wt, while those in subjects from New York were 11.7 ± 0.54 SEM mg/g dry wt, reflecting the significant differences in fecal bulk between the groups. Disturbing, however, is the gross difference between the fecal bile acid concentrations reported for subjects living in Kuopi in this study [325] and the values previously reported by the IARCIM group [324] (8.3 ± 2.6 and 7.3 ± 1.4 mg/g dry wt for the same population in spring and autumn, respectively).

Seventh-Day Adventists comprise an interesting group because of their unique dietary practices and the prevailing low incidence of colorectal cancer in this population [173,327]. Fecal bile acid concentrations have been measured by GLC in this population and were shown to be significantly lower than in healthy Americans, either as concentrations [327] or as daily excretion [173]. In the study of the strict vegans, although the concentration of deoxycholic acid in the feces was lower than in stools from subjects in the general population, this bile acid constituted 52% of the total bile acids, which was higher than the proportion observed for the general population (41%). In the study of Reddy and Wynder [173], the same trend was apparent, although the proportion of deoxycholic in the stool of the American Seventh-Day Adventists was 44% compared to 41% for the control group. Deoxycholic acid excretion (mg/day) by American vegetarians, on the other hand, comprised only 24% of the total bile acids while lithocholic accounted for 17%. These further examples of differences between studies from different groups of workers highlight the problem of interpreting the role of bile acids in the etiology of colon carcinogenesis.

Within the limitations of the early methodology, attempts have been made to measure individual bile acid concentrations in the stools of patients with colon cancer, but in most instances early published data are restricted to the principal primary and secondary bile acids. While considerable attention was given to the increased concentrations of deoxycholic acid excreted in the feces of colon cancer patients compared with controls, findings that were not corroborated in later studies, close analysis of the data indicates that proportional to the total bile acids excreted, there is no significant difference from the control groups. For

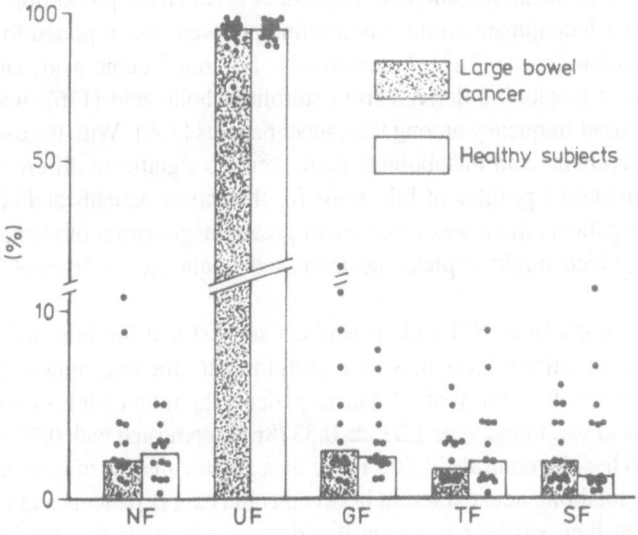

FIGURE 10. Distribution of total fecal bile acids within various conjugate groups for healthy Japanese subjects and Japanese patients with colon cancer. Each dot represents one subject and the columns indicate median values in each group. No statistical differences were found between the patients and healthy controls. NF, neutral fraction; UF, unconjugated fraction; GF, glycine-conjugated fraction; TF, taurine-conjugated fraction; SF, sulfate fraction. (Reproduced from Ref. [169]).

example, Reddy and Wynder [322] reported elevated concentrations of deoxycholic acid (6.96 ± 0.41 SEM mg/g dry wt) in colon cancer patients compared with controls (3.76 ± 0.43 SEM mg/g dry wt), but in both groups deoxycholic represented 35% of the total bile acids excreted. By analogy, lithocholic acid represented 32% of the total bile acids in the cancer group compared with 29% for the controls. Similarly, Moskovitz et al. [329] showed that the deoxycholic acid concentration in the feces of colon cancer patients (2.5 ± 0.5 mg/g dry wt) was lower than in controls (3.8 ± 0.8 mg/g dry wt), but proportionally in both groups it represented 43% of the total bile acids excreted.

In the most detailed published analytical study of fecal bile acid excretion in colon cancer, Hikasa et al. [169], using methods described by Setchell et al. [21], studied 14 cancer patients and 14 healthy Japanese controls. Fecal bile acid excretion calculated from 3-day stool collections was expressed as total bile acid/day, and as concentrations relative to fecal wet weight and to fecal dry weight. In addition, bile acids were separated into conjugate classes to determine the importance, if any, of the mode of conjugation in colonic carcinogenesis (Fig. 10). This study indicated fecal bile acid excretion to be lower in colon cancer patients than in controls, irrespective of how the data were calculated. Greater than 90% of the bile acids were excreted in the unconjugated form, and there were no significant quantitative differences between the proportions of bile acids within each conjugate group. Similar findings were also reported for Westerners with colon cancer [182]. Qualitatively, 5β-chol-3-enoic acid, an unsaturated bile acid metabolite derived from sulfolithocholic acid [136], was identified in increased frequency among the cancer patients [169]. With the exception of chenodeoxycholic acid metabolites, there were no significant differences between the qualitative profiles of bile acids for the cancer patients and controls. In the cancer patients there was evidence of greater degradation of chenodeoxycholic acid, which might explain the increased frequency of 5β-chol-3-enoic acid found.

Recent reports from Hill and co-workers suggest that the lithocholic acid/deoxycholic acid ratio in feces may be a useful marker for screening of patients with colon cancer. In a study of 34 cancer patients the ratio of lithocholic acid/deoxycholic acid was found to be 1.91 ± 0.33 (SEM) compared with 0.90 ± 0.09 (SEM) for 16 healthy controls [332]. These data and the finding of no significant difference in total bile acid excretion between colon cancer patients and controls do not support their earlier contention that deoxycholic acid and total bile acid excretion is elevated in this group of patients. Indeed, the finding of a significant shift in the ratio of lithocholic acid/deoxycholic acid would imply that there is altered hepatic synthesis of the primary bile acids chenodeoxycholic and cholic acid in colorectal cancer patients, but this has not been substantiated from bile acid kinetic studies. Furthermore, the increased ratio of lithocholic acid/deoxycholic acid is not corroborated by others [165,182,329,334]. It is of interest that in a study examining serum bile acids in colorectal cancer patients, lithocholic acid concentrations were significantly elevated compared with controls [334].

4.2. Adenomatous Polyps

Epidemiological studies have suggested a strong link between patients with adenomatous polyps and cancer of the large bowel [335–338], and it is generally accepted that patients with this condition are at greater risk for colorectal cancer than the normal population [339]. For this reason and because colon cancer is considered to proceed through an adenoma–carcinoma sequence [340,341], fecal bile acid measurements have been made in patients with adenomatous polyps. In the study by Hill and co-workers [316] of patients with large-bowel cancer, a group of 12 patients with nonmalignant large-bowel disease, including diverticulosis and polyps, were found to have total fecal bile acid concentrations lower than that found in the cancer patients and not significantly different from a group of patients with nongastrointestinal diseases used as controls. Reddy and Wynder [322] analyzed the stools of 13 patients (age range 30–65 years) with adenomatous polyps and observed higher concentrations for total fecal bile acids (16.79 ± 0.77 SEM mg/g dry wt) compared with a similarly aged group of healthy controls (10.85 ± 0.76 SEM mg/g dry wt). These values were, however, lower than for colon cancer patients. While concentrations of deoxycholic and lithocholic acids were significantly higher than in controls, these bile acids represented 36% and 32% of the total bile acids present, which was almost identical to the composition in the stools of cancer patients. Analysis of the values showed that "other unidentified" bile acids constituted 27% of the total bile acids present.

The most detailed clinical and analytical investigations of fecal bile acid excretion in patients with adenomatous polyps have been carried out by Tanida *et al.* [165,342], although these were of patients living in Japan, where the incidence of colorectal cancer is low. Wide variations in fecal bile acid excretion were observed in 13 patients and 13 closely matched controls irrespective of whether data were expressed as μmole/day, μmole/g dry wt, or mg/g wet weight. The mean total bile acid excretion of the healthy controls (387 μmole/day) was not significantly different from that of the patient group (349 μmole/day), although the median values were greater. These values are consistent with previous reports for normal Japanese subjects determined by different methodologies, but do not support the previous findings of Reddy and Wynder [322] or Hill *et al.* [316] for Western subjects with adenomatous polyps, in whom increased fecal bile acid concentrations were reported. Consistent with the study of Japanese subjects with colon cancer, the conjugation profile indicated that quantitatively the unconjugated bile acids comprised >90% of the total bile acids, and there were no significant differences between the patients with adenomatous polyps and healthy controls with respect to state of bile acid conjugation. Comparisons of individual bile acids within each conjugate fraction for both groups were unremarkable and showed no major quantitative differences. This was also the case for microbiological analysis of stools reported elsewhere [343]. Comparison of the ratios of individual bile acids for each subject gives an indication of the extent of dehydroxylation and/or oxidoreduction, and an increased ratio (but not statistically significant) of lithocholic/chenodeoxycholic acid and of deoxycholic/cholic acids in the patient group suggested a greater degree of 7α-dehydroxylation of bile acids in the patients with adenomatous polyps.

These studies were extended to address the question of whether differences in fecal bile acid excretion were dependent on polyp size, distribution, multiplicity, and degree of dysplasia, because it had previously been reported that there was a correlation between adenomatous polyp size and both fecal bile acid excretion and presence of nuclear dehydrogenating clostridia [344]. In a detailed and extensive study of fecal bile acid excretion in 33 patients with adenomatous polyps, which addressed the above factors, Tanida *et al.* [342] concluded that the highest fecal bile acid excretion was found in patients with the most polyps and the greatest severity of dysplasia, while in those patients with the largest polyps fecal bile acid excretion tended to be lower. Statistical analysis of the data, however, revealed no significant differences, and a large variation was found for the extent of dehydroxylation and oxidoreduction. It was concluded that there was little evidence to support a direct role for bile acids in the development of adenomatous polyps in the Japanese population.

4.3. Familial Polyposis

Familial polyposis is an inherited disorder with a high malignancy potential [345] occurring with an incidence of 1 in 8300–10,000 births in the Western world and Japan. In most instances it is believed that patients with familial polyposis will develop colorectal cancer unless the colon is removed prophylactically, and for this reason this condition has commanded interest from the point of view of fecal bile acid excretion [27,346–348].

Reddy *et al.* [27] studied eight patients (age range 17–50 years) and 10 relatives of these patients (age range 9–54 years) as well as 17 healthy nonrelated controls. Although there were no significant differences between the three groups of subjects for total neutral sterol excretion, of interest was the finding of decreased microbial metabolism of cholesterol in the patients with familial polyposis. Conversion of cholesterol to its usual metabolites, coprostanol and coprostanone, was minimal (approx. 3%) in five of the eight patients, while in the family-related subjects and nonrelated controls cholesterol conversion to the extent of 78–99% was observed. No major differences were noted between the groups for the concentrations of total fecal bile acids; however, greater concentrations of cholic acid and lesser concentrations of lithocholic acid and the unidentified "other" bile acids were reported. The ratio of total primary bile acids to secondary bile acids was slightly increased in the patients with familial polyposis, suggesting less dehydroxylation.

These latter findings were also reported by Watne and Core [346]. A group of seven subjects with familial polyposis and/or Gardner's syndrome were studied. The reason for the decreased microbial activity in familial polyposis is difficult to explain, since in the restricted studies that have been carried out on fecal flora in these patients and controls, no major differences have been reported [27,347]. It has been shown that *in vitro*, mixed cultures from fecal contents of patients and controls were able to convert cholesterol to coprostanol to the same extent. The differences in the fecal bile acid and steroid profiles between polyposis patients and controls are difficult to explain since fecal pH and transit time were reportedly similar between groups.

These results are surprising and not at all consistent with the expected trend for fecal bile acid excretion, given the hypothesis that bile acid metabolism and fecal flora may be important factors in the development of carcinogenesis.

4.4. Ulcerative Colitis

Patients with ulcerative colitis are at increased risk for the development of colon cancer, and akin to patients with adenomatous polyps and familial polyposis, fecal bile excretion has been determined in this population. Diarrhea is a frequent symptom of ulcerative colitis, and while bile acids may induce such conditions (see Section 4.8), Miettinen [349] found fecal bile acid excretion in five patients with ulcerative colitis and diarrhea to be within the normal range (159–463 mg/day). Since cholestyramine had little effect on the diarrhea, this study indicated that bile acids probably play little role in the diarrhea associated with ulcerative colitis of the large bowel.

Reddy and Wynder [322], in similar studies to those previously carried out for familial polyposis patients [27], examined fecal bile acid concentrations in 15 patients with ulcerative colitis and compared the values with those of 15 relatives of these patients and 40 healthy controls. All of the patients had colitis for more than 10 years and an intact colon. No significant differences were observed for the concentrations of total and individual bile acids or for the ratio of primary/secondary bile acids between the groups; however, differences in neutral sterol excretion were noted. Total neutral sterol concentrations were significantly higher (41.4 ± 7.0 SEM mg/g dry wt) in the patients compared with the relatives (18.9 ± 1.8 SEM mg/dry wt) and controls (18.1 ± 0.9 SEM mg/g dry wt).

These results were unexpected for this "high risk for colon cancer" group of patients, and it was suggested by Reddy and Wynder [322] that patients in this category fall into three distinct groups. Compared with controls, patients with adenomatous polyps excrete high levels of cholesterol metabolites and bile acids (not, however, supported by other workers [169,182,328–331]), patients with familial polyposis excrete similar levels of neutral sterols and bile acids but more unchanged cholesterol, and patients with ulcerative colitis excrete higher levels of neutral sterols and similar levels of bile acids.

In a recent study of 14 Japanese patients with ulcerative colitis at an active phase of the disease, fecal bile acids were measured by a slight modification of the method of Setchell et al. [21] and compared with healthy controls [350]. Fecal bile acid profiles of the patient group compared to the controls were characterized by statistically significant increases in concentrations and higher proportions of primary bile acids and their amidated conjugates. These differences in fecal bile acid excretion are at variance with earlier reports [322,349] but may be explained by the younger age group of the patients and the fact that they were studied during an active phase of the disease. This study also found that the more severe the disease activity, the greater was the bile acid excretion [350], in accord with data for fecal [^{14}C]bile acid excretion during the breath test [351].

4.5. Breast Cancer

Despite the obvious parallels in the epidemiology of colon cancer and breast cancer [352,353], very few studies have been performed to assess the importance of fecal bile acid metabolism in patients with breast cancer [332,354,355], yet great emphasis has been given to the role of steroid hormones. Murray *et al.* [354] studied 30 patients with histologically confirmed breast cancer, and total fecal bile acid was measured in single stool samples by an enzymatic method after their extraction using the procedures described by Evrard and Janssen [19]. Total fecal bile acids were significantly lower in the patients with breast cancer (15.0 ± 1.8 SEM μmole/g dry wt feces) compared to the values obtained for a group of 36 patients (both sexes) undergoing minor surgery but with no gastrointestinal disease (20.5 ± 1.9 SEM μmole/g dry wt feces). Microbiological studies of these samples revealed significant differences between the groups in the carriage rates of *Clostridium paraputificum* and nuclear dehydrogenating clostridia. These two microorganisms were identified in 62% and 58%, respectively, of the patients with breast cancer, compared with 31% and 15%, respectively, of the controls. The overall results of this study are in close agreement with the findings by the same group for patients with colon cancer [331].

The results of two later studies [332,355] failed to support the previous finding of a lower bile acid excretion in breast cancer patients [354]. In a group of 78 patients with breast cancer, total fecal bile acid excretion (17 ± 13 SD mg/g dry wt stool) is reported to be significantly higher than for closely matched controls (11 ± 8 SD mg/g dry wt of stool), but little information on the methodology was given. These differences, however, were shown not to be due to differences between the dietary intake of total calories, fat, or cholesterol between the groups.

In the most recent report, Owen *et al.* [332] found total fecal bile acid concentrations for 16 breast cancer patients to be not significantly different from those in healthy controls or colon cancer patients, but the lithocholic acid:deoxycholic acid ratio was significantly higher in the breast cancer group compared with the controls.

4.6. Crohn's Disease

Crohn's disease is a chronic inflammatory disorder which can involve any part of the alimentary tract, although the distal ileum and colon are the most commonly affected sites. It may also appear at any time of life, but is most common in adolescents and young adults. The inflammation that characterizes this disorder extends through all layers of the gut wall and involves the adjacent mesentery and lymph nodes. Since the terminal ileum may be involved, malabsorption of fat and bile acids is a common, but variable finding in Crohn's disease. Fecal excretion of intravenously administered or ingested [^{14}C]cholic or [^{14}C]glycocholic acids [99,356–358] and endogenous fecal bile acids [260,359,360] has been shown to be abnormally increased in 30–100% of Crohn's disease patients studied. Farivar *et al.* [361] estimated bile acid absorption by administering [^{3}H]ring-labeled taurocholic acid and [^{14}C]glycocholic acid. Twelve patients of the 31 studied had an abnormal [^{14}C] breath test, but only six revealed evidence of increased fecal bile acid loss. These inconsistent findings probably reflect variation in the extent of the inflammatory involvement of the ileum. Fecal bile acid excretion in Crohn's disease fails to correlate significantly with the length of ileal involvement, possibly because of difficulties in identifying affected areas or the variability within different parts of the affected ileal segment [357]. A similar lack of correlation has been observed in ileal-resected patients. In a comparative study of 11 patients with mild Crohn's disease, five without and six with diarrhea, markedly increased daily fecal bile acid excretion was observed in patients with diarrhea [362]. This report suggests that when Crohn's disease presents with diarrhea, malabsorption of bile acids may be involved in its pathogenesis. Evaluation of fecal bile acid excretion may also be compromised by variabilities in diet, since increased dietary fat enhanced fecal bile acid excretion significantly in patients with Crohn's disease, but not in control subjects [360]. This observation is similar to that seen in ileal resection, where a low-fat diet led to decreased fecal bile acid excretion [363]. In contrast to patients with ileal involvement, subjects with Crohn's disease confined to the colon generally had normal or moderately increased fecal excretion of ingested [^{14}C]glycocholic [364]. Studies in which total fecal bile acid excretion has been measured in Crohn's disease are listed in Table IV.

As found with ileal resection, many of the patients with Crohn's disease had an increased proportion of the fecal primary bile acids, cholic and chenodeoxycholic acids, and a decrease in deoxycholic and lithocholic acids [260,261]. Although these changes were statistically significant, they were not as dramatic as those seen in ileal-resected patients.

4.7. Diverticulosis

Diverticulosis, an acquired deformity of the colon, is thought to be due to increased intraluminal pressure acting at points of weakness in the bowel wall. It has been suggested that a lack of dietary fiber may cause or exacerbate the disorder [365–367]. Institution of a high-fiber diet lowers colonic pressure and reduces transit time, thereby relieving the symptoms of constipation, diarrhea, and pain [368,369].

Total daily fecal bile acid excretion (Table IV) has been variously described as being the same as [40,369], greater than [370], and less than [262] values found for control subjects, although significant differences observed were small.

The possibility that variability may be due to differences in wet or dry stool weight is not supported by figures adjusted for these variables [262,369]. The concentration of bile acids in the stool of patients with diverticulosis is usually not statistically significantly different from fecal concentrations in normal controls [262,369,370].

The inclusion of increased fiber in the diet of diverticular patients for periods of up to 12 months decreases daily fecal bile acid excretion and concentration, although the changes are not statistically significant [262,369] or are of marginal significance only ($p < 0.02$) [370]. Decreased daily fecal bile acid excretion occurs despite an increase in fecal wet weight, a result that is in accord with changes observed in normal subjects with similar dietary bran supplementation (see Section 5.1).

It is of interest that a positive correlation ($r = 0.66$) between the fecal lithocholic acid concentration and the incidence of abnormally rapid colonic myoelectrical rhythm was reported in patients with diverticular disease, while a negative correlation ($r = -0.62$) between deoxycholic acid excretion and the incidence of 6–9 cycles/min slow-wave electrical activity was also observed. Treatment with bran for long periods significantly decreased the colonic myoelectrical activity in these patients but did not alter concentrations or total daily excretion of lithocholic and deoxycholic acids in the stool. No correlation between myoelectrical activity and fecal bile acid output, however, was found in normal subjects. The importance of these findings is uncertain, but they may provide support for the concept of diverticular disease as a motility disorder [371]. The pattern of individual fecal bile acids is similar to that seen in healthy controls and is not significantly altered by increased dietary fiber [40,262,370].

4.8. Diarrhea and Malabsorption

Many studies have suggested that endogenous bile acids have a natural laxative effect and may in certain circumstances induce diarrhea. Possible mechanisms for these effects, reviewed elsewhere [372,373], include increases in colonic motility, alterations in electrolyte and water transport, and membrane damage due to the detergent action of bile acids. As a consequence, bile acid excretion has been studied in a wide variety of disorders characterized by diarrhea and malabsorption (Table IX). Reported values for daily fecal bile acid excretion are summarized in Table IV; it should be noted that diarrhea *per se* is not accompanied by an increased fecal bile acid excretion, as shown in patients with a number of diarrheal disorders.

In a series of reports on acute diarrhea due to infectious enteritis, it was suggested that the fecal bile acid profile reflected the mechanism of diarrhea [36,37,247]. In acute shigellosis diarrhea, in which overt mucosal invasion occurs, the percentage contribution of the secondary bile acids, deoxycholic and

TABLE IX.   Conditions Involving Diarrhea and Malabsorption in Which Fecal Bile Acid Excretion Has Been Evaluated

	Reference
1. Intestinal dysfunction	
a. Intractable diarrhea in infancy	189, 203, 373–377
b. Irritable bowel	375, 433
c. Crohn's disease	260, 360, 362, 397
d. Acute infectious enteritis	36, 37, 247, 248
e. Post-infectious enteritis	216
f. Celiac sprue	197, 241, 386–388
2. Pancreatic disease	
a. Cystic fibrosis	196–199, 202, 210, 213, 215, 219, 221
b. Acquired pancreatic insufficiency	214, 278, 432
3. Iatrogenic	
a. Ileal resection	68, 259, 263, 380, 398, 399
b. Ileal bypass	245, 414–416
c. Postcholecystectomy	270, 384
d. Postvagotomy	270, 378
e. Radiation enteritis	393
f. Exogenous bile acid administration	181, 513, 654–656, 660
g. Antibiotics	578–580
h. Phototherapy	392
4. Other	
a. Renal failure	33
b. Diabetes	381
c. Hypogammaglobulinemia	271
d. Cirrhosis	382, 383

lithocholic acids, to the total bile acids in feces was decreased significantly, and a concomitant increase in the percentage of cholic and chenodeoxycholic acids occurred [36]. In contrast, diarrhea induced by the noninvasive enterotoxigenic bacteria, *Escherichia coli* and *Vibrio cholerae*, is characterized by an unchanged pattern of bile acid excretion and 7-dehydroxylation [37,247,248]. Significantly decreased intestinal transit time was found in both types of diarrheal disorders, but changes in total bile acid excretion (expressed as mg/kg body weight per day) were not statistically different from controls. The increase in fecal primary bile acids and concomitant decrease in secondary bile acids in shigellosis suggest that a reduced interaction between the secreted bile acids and colonic bacterial flora takes place, which may be explained by the reduced intestinal transit time. However, the absence of changes in bile acid profiles in diarrhea due to *E. coli* and *V. cholerae* suggests that transit time is not an important factor. It is possible that inhibition of bacterial metabolism by toxic products of *Shigella flexneri* may account for the differences in bile acid excretion in this type of diarrhea.

Bile acid malabsorption has been implicated in the pathogenesis of disturbed fat absorption following infectious enteritis. Significantly increased fecal bile acid excretion was demonstrated in six children with steatorrhea and persistent failure to thrive during recovery from acute gastroenteritis [216]. Transient ileal dysfunction associated with excess bile acid loss and consequent impaired fat digestion was postulated. The profile of individual bile acids was not studied.

Changes in total bile acid excretion and/or the pattern of fecal bile acids have been implicated in the pathogenesis of chronic diarrhea in a number of conditions. Continuous or episodic diarrhea is a side effect observed in a small percentage of patients treated for duodenal ulcer with truncal vagotomy and drainage. Total daily fecal bile acid excretion has been reported to be significantly increased in patients with postvagotomy diarrhea compared with healthy controls [270,378,379]; however, in all these studies the number of patients studied was small and the scatter in bile acid excretion large. In the study of Allan *et al.* [378] total fecal bile acid excretion in seven patients postvagotomy, but without diarrhea, showed no significant differences from the normal control group, which was also the case in a separate study [270]. Critical analysis of the data from the former study [378], which concluded that there was no statistical difference between fecal bile acid excretions of postvagotomy patients with or without symptoms of diarrhea, reveals that three of the patients with postvagotomy diarrhea had fecal bile acid excretion values in excess of 3000 mg/day while in the other four patients values ranged from 1000 to 1500 mg/day. These small numbers probably account for the conflicting conclusions. In a study that subdivided patients into groups based on the severity of the diarrhea [270], it was again concluded that overall there was no significant difference in the fecal bile acid excretion between postvagotomy patients and controls, but on days in which the diarrhea was severe, a statistically significantly higher excretion was found. Cholecystectomy, in association with vagotomy, led to significantly elevated fecal bile acid excretion and concentration over and above that seen in postvagotomy diarrhea alone [270]. A single study [378] that examined the excretion of individual bile acids found a highly significant increase in the excretion of chenodeoxycholic acid in postvagotomy diarrhea patients compared with normal controls ($p < 0.005$), and since the fecal aqueous dihydroxy bile acid concentration has been shown to be of etiological important in diarrhea [245,270,380], this may be the determining factor in the development of postvagotomy diarrhea. The therapeutic value of the bile acid sequestrant cholestyramine [378,381–383] and the increased risk of developing diarrhea when, in addition, cholecystectomy was carried out [270] also imply that bile acids are involved in the pathogenesis of this disorder.

Intractable diabetic diarrhea has also been shown to respond to treatment with the bile acid sequestrant cholestyramine [381], although fecal bile acid excretion has not been studied. It has been suggested that in this condition, neuropathy of the vagal fibers occurs with consequent effects on gallbladder contraction and volume, similar to that seen with vagal resection. While cholecystectomy has been shown to exacerbate the bile acid excretion and increase fecal weight in postvagotomy patients, diarrhea is not a common symptom of cholecystectomy alone. However, in two patients with diarrhea following cholecystectomy, fecal bile acid excretion measured by [^{14}C]glycocholate was markedly increased and responded favorably to treatment, suggesting the diarrhea was bile acid-mediated [384].

Diarrhea is a common cause of infant morbidity in the Third World. When it becomes protracted, a vicious cycle, difficult to treat, of malabsorption, malnutrition, and failure to thrive may result [373]. Despite increasing evidence for the involvement of a disturbance in bile acid metabolism in many diarrheal conditions, few conditions have provided information about fecal bile acid profiles in children with protracted infantile diarrhea. Elevated concentrations of bile acids in feces, increased fecal loss of administered labeled bile acids, and a beneficial response to cholestyramine administration have been reported in patients with intractable diarrhea of previously unknown etiology [203,374]. These patients, who were without apparent ileal disease, were shown to have reduced *in vitro* uptake of taurocholic acid by ileal tissue obtained by biopsy [203], and it was suggested that this was due to a rare inborn error in which the ileal transport system was absent or displayed a reduced capacity.

Other cases of apparent primary bile acid malabsorption in children have been reported with uniform presentation of increased fecal bile acid excretion and intractable diarrhea, often present since birth [189,375–377]. Few studies have examined bile acid patterns other than total fecal bile acid estimation in stool homogenates and in the fecal aqueous phase. In a study of two patients with greatly increased total fecal bile acid excretion and intractable diarrhea, the proportion of unconjugated bile acids in the stool homogenates was 2 and 6%, respectively [189]. This compares with the usual 95% deconjugation seen in normal children and in five of the other six children with chronic diarrhea studied by this group. A study of 47 infants with acute, subacute, and intractable diarrhea failed to find conjugated bile acids in the feces, although total bile acid excretion was significantly higher in infants with diarrhea and especially in those with intractable diarrhea [385]. No qualitative difference in the excreted bile acids was observed in the groups studied; however, the TLC/fluorimetric methods used in this study are relatively crude and hence the results should be interpreted with some caution.

The mechanism of diarrhea seen in celiac sprue is not fully understood, although bile acid malabsorption may contribute in some cases [386]. Celiac sprue is characterized by a gluten-induced villous atrophy of the small intestinal mucosa with potential or actual malabsorption of virtually all nutrients. Since the disease is frequently confined to the proximal small intestine, bile acid reabsorption by the terminal ileum may be normal or even enhanced [387]. In cases where the disease extends to, and involves, the ileum, impaired absorption of conjugated bile salts may further aggravate the diarrhea caused by malabsorption of other dietary components. Normal or low fecal excretion of intravenously injected [^{14}C]cholic acid was found in 11 patients with untreated celiac disease without diarrhea, whereas five patients with diarrhea exhibited increased fecal excretion [386]. Several studies have found normal or only moderately increased amounts of bile acids in the feces from untreated celiac patients [197,241,387,388]. Low-Beer *et al.* [389] demonstrated a threefold increase in the total bile acid pool size (mean of 9.2 g versus 3.1 g for controls) and a significant increase in the taurocholate pool in celiac patients compared with healthy controls, yet the daily turnover of taurocholate was not significantly different and correlated closely with its half-life. These studies showed that the taurocholate pool was metabolized more slowly [388,389] in celiac disease and led to the suggestion that sluggish gallbladder function accounted for the greatly increased bile salt pool size. Furthermore, compensatory hyperfunction of the ileum probably results in the improved ileal conservation of bile acids observed [387]. Tropical sprue, an intestinal malabsorption syndrome, presents with steatorrhea in most patients. An increased rate of fecal excretion of intravenously injected [^{14}C]cholic acid was observed in four of seven patients with tropical sprue [390].

Diarrhea is a common gastrointestinal complaint associated with the uremic state, and Gordon *et al.* [33] analyzed fecal bile acids in random stools from a single patient with uremic watery diarrhea. In comparison to fecal bile acid profiles from a control subject and an azotemic patient without diarrhea, markedly decreased proportions of deoxycholic, lithocholic, and 3β,12α-hydroxy-5β-cholanoic acid and a concomitant increase in primary bile acids was found. Although quantitative fecal bile acid excretion was not assessed, the qualitative profile and high proportions of conjugated bile acids (75%) found suggested that bile acids may be involved in the pathogenesis of uremic diarrhea.

Gastrointestinal symptoms such as diarrhea and steatorrhea are common in patients with hypogammaglobulinemia. Giardiasis, which also presents with diarrheal symptoms, is frequently associated with this condition. The qualitative pattern of fecal bile acids was studied in a single patient with hypogammaglobulinemia and symptoms of watery diarrhea without steatorrhea [271]. Identification of some of the bile acids was tentative, but increased primary bile acids were found. However, their presence may have been due to changes in the microbial flora consequent to antibiotic use. Administration of the bile acid sequestrant cholestyramine controlled the diarrhea in this condition, but nonspecific binding of other intraluminal components could also account for the cessation of symptoms.

Phototherapy is used widely as a means of reducing serum bilirubin concentration in neonates with unconjugated hyperbilirubinemia. Watery diarrhea has been observed as a side effect in approximately 10% of treated infants [391]. Significantly higher bile salt concentrations in the aqueous component of the stools were found in infants during phototherapy compared to before and after treatment and to normal nontreated neonates used as controls [392]. This suggests that bile acids may be a factor in the pathogenesis of diarrhea associated with phototherapy of the jaundiced newborn.

Diarrhea is also associated with pelvic irradiation therapy for malignant tumors. Seventeen of twenty irradiated patients were found to have an increased fecal excretion of intravenously injected radiolabeled cholic acid [393]. All patients with diarrhea had significantly increased excretion compared to a control group and also compared to irradiated patients with no symptoms of diarrhea. Introduction of a low-fat diet proved beneficial to these patients, with an eradication of diarrhea and a significant reduction in bile acid malabsorption determined from fecal radiolabeled cholic excretion [394].

Hirschsprung's disease, or congenital colonic dilatation, is frequently treated with a Soave pull-through operation [395], a potential side effect of which is chronic diarrhea [396]. Fecal bile acids measured in four such patients [212] were moderately increased, and a reduced proportion of deoxycholic acid was found in three of the children, suggesting decreased 7α-dehydroxylation similar to that seen in other diarrheal conditions (Table VII).

4.9. Ileostomy and Ileal Resection

Since bile acids are chiefly absorbed in the terminal ileum by an active transport process, it is not surprising that resection of this region results in bile acid malabsorption. Increased excretion of an oral dose of [^{14}C]cholic acid [99,246] or [^{14}C]taurocholic acid [397] and markedly increased fecal bile acid excretions have been reported in ileal resection in many studies [68,259,263,380,398]. The physiological consequences of bile acid malabsorption may be divided into two types, depending on the response to cholestyramine treatment. The distinction between the consequences of small and large ileal resection was first made by Hofmann and Poley [380] and is summarized in Table X. Owing to failure of the normal feedback inhibition of bile acids returning to the liver from the gut, the hepatic bile acid synthesis rate will respond to compensate for the increased fecal bile acid output; therefore, with increased losses, synthesis may increase up to 10 to 15-fold [259,380]. When the fecal pH is normal (6.14–6.57) or high, the bile acids will be more soluble in the aqueous phase, and assuming that fecal pH reflects intracolonic environment, increased loss of bile acids into the colon will result in watery secretory diarrhea [245]. A close correlation ($r = 0.90$) has been found between fecal bile acid excretion and stool water content [68]. An alkaline fecal pH (6.81–8.50) is a

TABLE X. Syndromes Associated with Ileal Resection (Related to Residual Length)[a]

	Bile acid-mediated diarrhea	Fatty-acid-mediated diarrhea
Amount of residual functional small intestine	Large (<100 cm nonfunction)	Small (>100 cm nonfunction)
Fecal bile acid output	High	High
Fecal aqueous phase (bile acid)	High (pH dependent)	Low
Hepatic bile acid synthesis	Compensates for losses	Decompensated
Bile acid pool size	Normal	Decreased
Intraluminal bile acid concentration	Normal	Decreased
Steatorrhea	Minimal	Marked
Treatment with cholestyramine	Responsive	"Nonresponsive"
Treatment with ↓ LCT, ↑ MCT	No change in diarrhea	Improved
Postulated mechanism (secretagogue)	Bile acids	Fatty acids

[a] LCT, dietary long-term triglycerides; MCT, medium-chain triglycerides. (Reproduced from Ref. [399] and modified from Ref. [380]).

common occurrence in small ileal resections, whereas in large resections, fecal pH tends to be more acidic (5.44–6.51), possibly because of decreased Cl^-/HCO_3^- exchange, which normally takes place in the ileum and proximal colon [245]. Gastric acid hypersection [400,401] and formation of short-chain organic acids by the action of intestinal bacteria on malabsorbed carbohydrates [402] may also acidify the colonic environment. At abnormally acidic pH, bile acids, in particular those which are unconjugated or glycine-conjugated, will tend to precipitate, and bile acid-mediated diarrhea is unlikely. Although the hypothesis that intraluminal pH is a major determinant in the mechanism of diarrhea in ileal resection is attractive, a significant difference between fecal pH in large ileal resection and healthy controls was not established [245], and a more recent study found no significant difference in fecal pH between large and small ileal resections [264]. Very few studies have analyzed the concentration of bile acids in the fecal aqueous supernatant in intestinal resection. In their original study, Hofmann and Poley [380] examined four patients with small ileal resections (<100 cm) and three patients with large resections (>100 cm of terminal ileum) and found a high concentration of bile acids in the aqueous phase of the stool from all of the patients with small resections, but also in one with a large resection. Cholestyramine is effective in controlling diarrhea when the bile acid concentrations in the aqueous phase are high [380,403,404]. When bile acid loss is great, the capacity of the liver to synthesize new bile acids can no longer compensate, and the bile acid pool size will be reduced, leading to a marked decrease in the intraluminal bile salt concentration below the critical micellar concentration

[263,380]. Resultant malabsorption of fat will induce diarrhea by the action of fatty acids, and then cholestyramine therapy alone is not effective in controlling the diarrhea [380]. Decreasing dietary fat and substituting medium-chain tri-glycerides, which are absorbed reasonably well in the absence of bile acids, will improve the diarrhea [380,405] and may decrease bile acid loss in the stool [406].

Bile acid malabsorption may be minimal in cases of very short ileal resection or where sections of the terminal ileum as short as 35–50 cm have been left intact even when the rest of the proximal ileum has been removed [263,264]. This may be explained by adaptive hyperfunction in the retained segment [407]. Possibly because of this and/or other unknown variables, no significant corre-lation between the excretion of ingested [^{14}C]glycocholic acid [408] or the total bile acid excretion [68,260] and length of ileal resection in individual patients has been established.

Compared to healthy controls, most patients with ileal resection have an increased proportion of fecal primary bile acids (Table VII) with reduction or absence of the usual secondary bile acids, deoxycholic and lithocholic acids [245,260,261,263,264]. After ileal resection the intestinal microflora does not differ quantitatively or qualitatively from that of normal subjects [259,409], and in these patients fecal bile acid composition is essentially unchanged by treatment with antibiotics and sulfasalazine [260]. It is possible that the increased proportion of excreted primary bile acids is due to inhibition of 7α-dehydroxylation by high intraluminal bile acid levels, since this has been demonstrated *in vitro* by fecal homogenates exposed to increased concentrations of bile acids [258,259]. De-conjugation may also be affected by the decreased intestinal transit time observed in patients with cholerrheic enteropathy [99,259], although total fecal bile acid levels are only marginally increased with large changes in intestinal transit time induced by mannitol-mediated diarrhea [99,246]. *In vitro* experiments, however, have shown that feces from a patient with cholerrheic enteropathy rapidly de-conjugated primary bile acids, but 7α-dehydroxylation was possible only after a 48-hr incubation [259]. Therefore, the qualitative pattern of individual bile acids may be affected even though changes in total fecal bile acid levels are minimal. The degree of deconjugation of bile acids has not, to our knowledge, been reported in ileal resection.

In an attempt to improve the steatorrhea caused by intraluminal bile acid deficiency found in large ileal resection, several programs of bile acid feeding in these patients have been used. Sodium taurocholate reduced fat excretion but was accompanied by watery diarrhea due to the cathartic effect of its metabolite deoxycholic acid [410]. The use of ursodeoxycholic acid, a relatively nonca-thartic bile acid, has had mixed results. For example, two studies have reported no change in stool frequency, weight, and fat after ingestion of ursodeoxycholic acid [411] or tauroursodeoxycholic acid [412], whereas in another report ursodeoxycholic acid strikingly improved diarrhea and/or fat absorption in a number of patients [413].

Some of the effects of ileal resection in children are similar to those seen in adults, with greatly increased fecal bile acid excretion accompanied by high bile acid concentrations in the fecal aqueous supernatant. However, fecal fat and fecal weight in these children were only moderately increased, suggesting compensatory intestinal hyperfunction [201].

Jejunoileal bypass surgery is occasionally carried out in patients with morbid obesity who fail to respond to dietary treatment [414]. Diarrhea is an invariable side affect, and inevitably both bile salt and fat malabsorption have been implicated in its pathogenesis. Significantly increased fecal bile acid excretion similar to that seen in ileal resection has been reported in most patients with jejunoileal bypass [245,414–416] or jejunoileostomy [274,417]. Fromm *et al.* [414] found all patients with jejunoileal bypass to have diarrhea and steatorrhea, yet found bile acid excretion in excess of the upper limit of control values in only 50% of patients, and the concentration of dihydroxy bile acids in the fecal aqueous phase was found to be considerably lower than that known to induce diarrhea (1.5 mM/liter [272]). The latter finding may reflect similar circumstances to those found in large ileal resection, where precipitation of bile acids due to low fecal pH (4.41–6.4) or dilution of colonic contents by fluids delivered from the small intestine results [245]. Although the severity of the diarrhea in patients with jejunoileostomy decreased markedly after the first 6 months, both excessive fecal fat and bile acid excretion persist for at least several years [414,417]. This suggests that jejunal adaptation does not result in significant changes in fat or bile acid absorption.

A comparison of the effects of a 3:1 and 1:3 jejunoileal ratio in the bypass operation found a lower fecal bile acid excretion with the 1:3 than the 3:1 ratio bypass [418]. This surprising finding supports kinetic studies that describe a lower bile acid synthesis rate with the 1:3 ratio than with the 3:1 [418,419]. Although this anomaly might be explained by increased deconjugation of bile acids by bacterial overgrowth commonly associated with the bypass operation, no significant difference in the occurrence of deconjugated bile acids in the jejunal contents has been reported [419]. Therefore, it was postulated that a "functioning upper jejunum is a prerequisite for stimulation of bile acid synthesis in response to bile acid depletion" [418,420].

A close correlation between fecal fat and fecal bile acid pattern has been reported in jejunoileostomy patients [421]. Moderate fecal fat (<24 g on a 65-g daily intake) was shown to be associated with fecal excretion of predominantly secondary bile acids, whereas increased primary bile acids were observed when fecal fat excretion was elevated (>24 g/day). As for ileal-resected patients, an increased proportion of primary bile acids (20–92%) was observed in all jejunoileostomy subjects studied [245,414,421].

Partial ileal bypass has also been used in the treatment of hypercholesterolemia. Serum cholesterol is markedly reduced in most patients [421,422], although no demonstrable correlation with increased fecal bile acids or total steroids is observed [421]. The hypocholesterolemic effect of the partial ileal bypass was shown to be maintained in most patients when examined after 1 year.

Total colectomy and ileostomy without extensive ileal resection may be performed in cases of ulcerative colitis or multiple polyposis [423,424], and fecal bile acid excretion is normal or moderately increased in these patients [264,423–425]. This suggests either that the terminal ileum adapts to prevent bile acid losses or that colonic absorption of bile acids is normally minimal and the colon plays little role in maintaining the total bile acid pool. Significant bile acid malabsorption may occur, however, when ileal resection or the construction of a continent ileostomy reservoir is also carried out [424,426,427]. In the construction of the reservoir, about 45 cm of terminal ileum is used, and resulting changes in bacterial flora [428] and intestinal villi [429] may interfere with bile acid absorption in this region. Since deconjugation is increased and the pH of the ileal effluent may be reduced in continent, compared with conventional, ileostomy, precipitation of bile acids may contribute to their increased loss [424]. Apart from deconjugation, further bacterial modification of bile acids has been shown to be minimal in conventional ileostomy and in most cases of continent ileostomy, where secondary bile acids such as deoxycholic and lithocholic acids are reported to be negligible or undetectable in ileostomy effluent [264,423,424,430,431].

4.10. Inborn Errors of Metabolism Involving Bile Acid Metabolism

4.10.1. Cerebrotendinous Xanthomatosis and Phytosterolemia

Cerebrotendinous xanthomatosis (CTX), first described by Van Bogaert et al. in 1937, is a rare autosomal recessive lipid storage disease characterized by extensive tendon xanthomas, cataracts, and a number of symptoms indicative of neurological dysfunction [434]. Plasma cholesterol levels are low [435], but in contrast cholestanol is elevated in plasma and numerous tissues, including the brain [436]. The primary biochemical abnormality is subnormal bile acid synthesis, which leads to considerable quantities of polyhydroxylated bile alcohols being excreted in urine, bile and feces [437–443]. Evidence has been found from in vitro and in vivo studies that the primary enzymatic defect in CTX is a deficiency of mitochondrial 26-hydroxylase [444–447].

Owing to the cholesterol side-chain hydroxylation defect in CTX disease, bile acid synthesis is impaired and fecal excretion is markedly reduced [441,442,448]. Neutral sterol profiles of feces, on the other hand, are characterized primarily by the presence of large amounts of polyhydroxylated bile alcohols, particularly those hydroxylated at the C-25 position either alone or in combination with hydroxylations at C-22, C-23, and C-24. Interestingly, no evidence for intestinal 7α-dehydroxylation of bile alcohols is found, since all bile alcohols that have been identified in feces (Fig. 11) have nuclear hydroxyl groups at positions C-3, C-7, and C-12. Whether this reflects substrate specificity of the bacterial 7α-dehydroxylation enzyme or inhibition by large intraluminal amounts of bile alcohols is presently unclear. In a report of fecal bile acid and neutral sterol excretion in two patients (the GLC chromatograms of one are shown in Fig. 11), no 7α-dehydroxylated bile alcohols were identified in the feces [442].

An alternative hypothesis that the metabolic defect in CTX is one of a deficiency in 24-hydroxylation of 5β-cholestane-3α,7α,12α-25-terol has been suggested [449]. This was based on the lower rate of conversion of 5β-cholestane-3α,7α,12α-25-tetrol to cholic acid in CTX patients compared to controls and the low activity of hepatic 24-hydroxylase enzyme [449]. This, however, would not be supported by the large amounts of 24-hydroxylated bile alcohols excreted in urine and feces of CTX patients [440–442] or by the greater reduction in chenodeoxycholic acid compared to cholic acid biosynthesis.

The reason for the greater proportion of cholic acid compared to chenodeoxycholic acid in bile [436,445], supported by the finding of greater amounts of cholic acid than chenodeoxycholic acid in feces [442], may be explained by the C-25 hydroxylation pathway, since it has been shown that 5β-cholestane-3α,7α-diol is less efficiently 25-hydroxylated than 5β-cholestane-3α,7α,12α-triol [450]. Cholic acid production would therefore be less affected than chenodeoxycholic acid production, as is the case in CTX disease.

Phytosterolemia, a condition described by Bhattacharyya and Connor [451], also presents with extensive tendon xanthomas but normal plasma cholesterol levels. Elevated amounts of plant sterols have been detected in blood and tissues [451,452], and fecal bile acid excretion has been reported to be normal or slightly less than normal [452]. Unlike CTX disease, the qualitative fecal bile acid profile is essentially normal, and bile alcohols do not constitute a significant proportion of the neutral fecal sterols (Setchell *et al.*, unpublished observations).

4.10.2.   Cystic Fibrosis

Cystic fibrosis (CF) is an autosomal recessively transmitted disease characterized by obstructive lesions throughout multiple organ systems and disturbances of mucous and electrolyte secretions. The specific genetic defect remains obscure but produces a diversity of symptoms, the principal one being chronic obstructive pulmonary disease and exocrine pancreatic insufficiency. Fat and protein maldigestion and clinically significant intestinal malabsorption of a variety of substrates including bile acids is common in patients with pancreatic

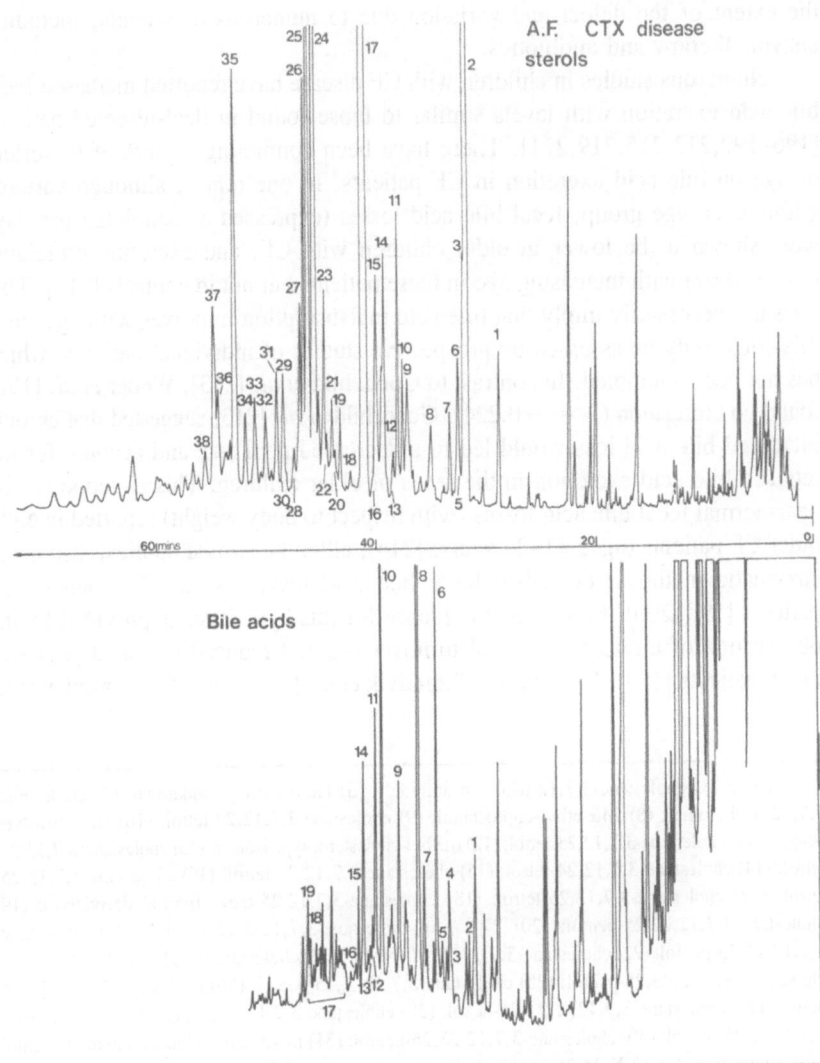

FIGURE 11. Bile alcohol and bile acid excretion in the feces of a patient with cerebrotendinous xanthomatosis. Typical capillary column GLC profiles are shown for the methyl ester–trimethylsilyl ether derivatives of bile acids (lower panel) and trimethylsilyl ether derivatives of neutral sterols (upper panel) in the feces. Separation of compounds was acheived using a 25-m DB-1 capillary column and temperature-programmed analysis 225°–295°C with increments of 2°C/min after an initial isothermal period of 5 min and final isothermal period of 20 min. Neutral sterols and bile acids were identified by GLC–MS, and because of the lack of availability of authentic standards, the stereochemistry of many of the bile alcohols is uncertain, although the position of the hydroxyls can be assigned accurately from the fragmentation patterns (X denotes unknown position of substituent). The following compounds are indicated: *Neutral sterols:* (1) coprostanol; (2) cholesterol; (3)

involvement. The study of CF patients is complicated by the wide diversity in the extent of the defect and variation due to numerous treatments, including enzyme therapy and antibiotics.

Numerous studies in children with CF disease have reported increased fecal bile acid excretion with levels similar to those found in ileal-resected patients [196–199,213,215,219,221]. There have been conflicting reports of the effect of age on bile acid excretion in CF patients. In one report, although variable within each age group, fecal bile acid losses (expressed as μmole/kg per day) were shown to be lower in older children with CF, and excretion correlated ($r = -0.67$) with increasing age in these patients but not in controls [213]. This does not necessarily imply that bile acid malabsorption improves with age since this could only be assessed by prospective studies of individual patients, which has not been attempted. In contrast to Goodchild *et al.* [213], Weber *et al.* [198] found no correlation ($r = -0.224$). Goodchild *et al.* [213] suggested that chronic increased bile acid loss would led to a decreased pool size and account for the reduced bile acid excretion in the feces of older children. This is supported by near-normal fecal bile acid losses (with respect to body weight) reported in eight older CF patients (aged 11–24 years) [214], although similar studies, also using enzymatic methods, described large bile acid losses as seen in younger CF patients [198–200]. More direct evidence for this hypothesis is provided by the observation of a raised fractional turnover rate but reduced bile acid pool size in CF patients [199]. By contrast, Strandvik *et al.* [453] recently reported normal

---

cholestanol; (5) cholestene-3,7,12-triol (unsaturated); (6) lathosterol + unknown; (7) cholestene-3,7,12-triol isomer; (8) 24α-ethyl-coprostanol; (9) cholestane-3,7,12,23-tetrol; (10) sitosterol-type isomer; (11) cholestane-3,7,12,23-tetrol; (12) methyl-cholesterol-type isomer; (13) cholestane-3,7,12,24-tetrol; (14) cholestane-3,7,12,24-tetrol; (15) cholestane-3,7,12,24-tetrol; (16) cholestane-3,7,12,25-tetrol; (17) cholestane-3,7,12,25-tetrol; (18) cholestane-3,7,12,25-tetrol (partial derivative); (19) cholestane-3,7,12,22,25-pentol; (20) 27-nor-5β-cholestane-3,7,12,24,25-pentol; (21) cholestane-3,7,12,22,25-pentol; (22) cholestane-3,7,12,22,25-pentol; (23) cholestane-3,7,12,23,25-pentol; (24) cholestane-3,7,12,23,25-pentol; (25) cholestane-3,7,12,X,25-pentol; (26) cholestane-3,7,12,24,25-pentol; (27) cholestane-3,7,12,X,22,25-hexol; (28) cholestane-3,7,12,X,25-pentol; (29) cholestane-3,7,12,X,25-pentol; (30) cholestane-3,7,12,25,26-pentol; (31) mixture of unknown hexol + pentol; (32) cholestane-3,7,12,X,24,25-hexol; (33) cholestane-3,7,12,X,24,25-hexol; (34) cholestane-3,7,12,22,24,25-hexol; (35) cholestane-hexol (unknown structure); (36) cholestane-hexol (unknown structure); (37) cholestane-3,7,12,X,23,25-hexol; (38) cholestane-hexol (unknown structure). *Bile acids:* (1) norcholic; (2) cholesterol; (3) 3β,12α-dihydroxy-5β-cholanoic; (4) 3β,7α,12α-trihydroxy-5β-cholanoic; (5) 3α,7α-dihydroxy-5β-cholanoic (chenodeoxycholic); (6) 3α,7α,12α-trihydroxy-5β-cholanoic (cholic); (7) monooxo-monohydroxy bile acid; (8) 3α,6α,12α-trihydroxy-5β-cholanoic; (9) trihydroxycholestanoic; (10) 7-oxo-3α,12α-dihydroxy-5β-cholanoic; (11) 7-oxo-3α,12α-dihydroxy-5α-cholanoic; (12) trihydroxy bile acid; (13) 12-oxo-3α,7α-dihydroxy-5β-cholanoic; (14) 5β-cholestane-3,7,12,25-tetrol isomer; (15) tetrahydroxy-cholanoic (side-chain hydroxy); (16) tetrahydroxy-cholanoic; (17) trace of bile alcohols; (18) 3α,7α,12α,25-tetrahydroxy-5β-cholestanoic; (19) 3α,7α,12α-trihydroxy-5β-cholest-24-enoic.

concentrations of bile acids in the bile of five CF patients aged 17–24 years despite marked fat malabsorption. Bile acid kinetics were determined by the technique of Lindstedt [94], and the cholic and chenodeoxycholic acid pool sizes ranged from 320 to 2743 mg and from 496 to 1008 mg, respectively. Synthesis rates for these bile acids ranged from 193 to 1442 mg/day and from 88 to 658 mg/day, respectively, which were normal or slightly raised compared with rates for healthy controls [453]. The two contrasting results [199,453] may reflect differences in the age range of the patients studied.

Liver disease, which is likely to be of greater significance among older patients, may compromise bile acid synthesis and reduce pool size and fecal excretion, since CF patients with overt liver disease tend to have lower fecal bile acid losses [213,215]. Conversely, children with CF with pulmonary involvement alone have been reported to have normal fecal bile acid and fat excretion [198,219].

The exact mechanism leading to increased fecal bile acid loss in CF is not clear. It has been suggested that unhydrolyzed triglyceride or other maldigested dietary components may interfere with bile acid reabsorption by intraluminal nonspecific binding. Several studies report good correlations between fecal losses of bile acids and fat in untreated children with CF [197,198] or concomitant reductions in both when pancreatic enzyme therapy is introduced. Similar responses are seen in association with reduction in dietary long-chain triglycerides or with low-fat diets [197,198,215]. Goodchild *et al.*, on the other hand, found no correlation between fecal fat and bile acid excretion in CF, but the patients studied were of a wide age range (3 months–21 years) and patients with liver disease were not excluded [213]. Furthermore, a recent study by Colombo *et al.* [196] also found no correlation, despite the exclusion of CF patients with liver disease. In studies in which fecal fat excretion was reduced following institution of enzyme therapy [199,214] or enzyme therapy supplemented with cimetidine [20], bicarbonate [198], or taurine [220], no significant reduction in fecal bile acid excretion was reported. Patients with acquired pancreatic insufficiency responded more favorably to enzyme therapy and showed a concomitant reduction in bile acid and fat excretion [214]. Significantly decreased fecal bile acid excretion has been observed in children with CF treated with antibiotics (cloxacillin, gentamycin, and carbenicillin), despite continued high fecal fat losses during treatment [454]. This finding, however, was not supported in a later study [202].

A primary ileal mucosal cell defect in bile acid uptake has also been suggested as a possible cause of bile acid malabsorption in CF [202]. In a single report using intestinal biopsy specimens from CF patients and patients with ileostomies, ileal uptake of taurocholate was significantly reduced in the tissue from CF patients [202]. This study does not exclude a local environmental effect at the brush border, although electron micrographs of biopsy tissue were structurally similar in CF and control patients. Furthermore, an ileal defect in bile acid transport is unlikely, since CF patients with isolated pulmonary involvement or with some remaining pancreatic function have normal fecal bile acid excretion [198], while adult patients with acquired pancreatic insufficiency due to other causes also show increased fecal fat and bile acid losses [214,278,425].

Insufficient neutralization of hydrochloric acid or reduced pancreatic bicarbonate could contribute to bile acid loss via precipitation of bile acids under conditions of low intraluminal pH [455]. Precipitation would be exacerbated by the high glycine/taurine ratio found in CF patients [456,457]. This is supported by the improvement in steattorhea seen when the glycine/taurine bile acid ratio was reduced by dietary supplementation with taurine [220], although no significant change in bile acid excretion was found. Similar effects with sodium bicarbonate or cimetidine have also been reported, suggesting that factors other than bile acid precipitation are involved in bile acid malabsorption [199,200].

With regard to the individual patterns of fecal bile acids in CF, studies are limited. Interpretations of qualitative profiles are compromised by chronic and intermittent usage of antibiotics. Increased fecal primary bile acids have been found in CF patients with and without pancreatic involvement following discontinuation of antibiotic treatment [198,219]. A reduced capacity of fresh stool samples from CF patients, with and without pancreatic involvement, to deconjugate and dehydroxylate bile acids [458] may reflect modification of the intestinal microflora by prior antibiotic treatment. Weber *et al.* [198] reported the degree of 7α-dehydroxylation to increase following various treatments (enzyme therapy, sodium bicarbonate, low-fat diet), but not to normalize. A later study from the same group found normal levels of primary bile acids in the feces of CF children treated with pancreatic enzyme, and only in those children treated by antibiotics were high proportions of primary bile acids found [454]. In contrast to healthy controls, significant levels of ursodeoxycholic acid ($10.25 \pm 6.2\%$) were identified in CF patients with pancreatic involvement, in addition to the increased proportion of primary bile acids [221].

In summary, the inconsistencies in the numerous studies of fecal bile acids in cystic fibrosis are probably due to the heterogeneity in the manifestation of the disease, variability in diet, or the effects of chronic and intermittent antibiotic treatment. Many of the studies are of small numbers of patients, and clearly, further, more detailed studies may assist in the elucidation of the underlying mechanism involved in fat and bile acid malabsorption in this inborn error.

4.10.3. *Peroxisomal Defects*

Cerebrohepatorenal syndrome (Zellweger syndrome), initially described in 1964, is a lethal condition of infancy characterized by various neurological, dysmorphic, and cholestatic symptoms [459]. More recently, it has been suggested that the disease may present these symptoms to a variable degree with the possibility that Zellweger disease may be the more severe expression of a larger disease entity characterized by virtual absence of peroxisomes and coprostanic acidemia [447,460–463]. A defect in the peroxisomal cleavage of the steroid side chain, not related to mitochondrial 26-hydroxylation, in the biosynthetic pathway for bile acids is postulated [447,463]. This is supported by the identification of considerable amounts of dihydroxy and trihydroxy coprostanic acids in bile, serum, urine, and feces [447,461,464]. These bile acids accounted for 7.5–15.7% and 15.0–75.3%, respectively, in the feces of three children with neurological, dysmorphic, and cholestatic symptoms of variable degree [461].

4.11. Hyperlipidemia

4.11.1. Primary Hyperlipidemia

Hyperlipidemia, or primary elevation of plasma lipids, is a blanket term encompassing a broad heterogeneous mixture of disorders. These have been provisionally classified according to the elevation of cholesterol and/or triglycerides, the abnormalities in the plasma lipoproteins, and the characteristics of their inheritance. The allocation of type is loose since apparently different genetic disorders may express the same pattern of lipids and lipoproteins. Conversely, abnormalities of the same apparent genotype may present with different patterns. However, the nomenclature of hyperlipidemias does provide a simple code for differentiating the lipid and lipoprotein patterns. Biochemical characteristics may then be measured and possible metabolic interrelationships thereby elucidated. Hyperlipoproteinemias have been classified [465,466] according to increased levels of chylomicrons (type I), LDL (type IIa), LDL + VLDL (type IIb), floating β-lipoproteins (type III), VLDL (type IV) or chylomicrons and VLDL (type V) of which types II, IV, and V are the most common. More comprehensive reviews of the classification of primary hyperlipidemia may be found elsewhere [277,279,467]. In a survey of cholesterol balance studies carried out between 1965 and 1976 [279], data are analyzed by groups according to whether the subjects were hypercholesterolemic, hypertriglyceridemic, both hypertriglyceridemic and hypercholesterolemic, or normolipemic. The survey showed that fecal excretion of bile acids in normotriglyceridemic patients with hypercholesterolemia was consistently and significantly lower than that of normolipidemic controls ($p < 0.0001$). In comparison, hypertriglyceridemic patients had consistently greater fecal bile acid excretion, although when corrected for body weight, values were not significantly different from those of normolipidemic controls. Patients with elevated plasma cholesterol and triglycerides had similar fecal bile acid excretions to those seen in hypertriglyceridemia alone. Some previous studies report similar findings [275,468–470]. However, other studies, most of which used kinetic methods, have not observed these relationships, particularly with respect to a decrease in bile acid excretion in hypercholesterolemia [471–475]. An early suggestion that hypercholesterolemia may be caused by a block in bile acid synthesis from cholesterol [476] is now known to be unfounded. It is more probable that these differences between hypercholesterolemia and hypertriglyceridemia are due to a decrease and increase, respectively, in cholesterol availability for catabolism [279].

A more detailed breakdown of fecal bile acid excretion in various hyperlipoproteinemias (Table IV) reveals more subtle differences in the manifestation of these disorders. However, many of the studies measuring fecal bile acid excretion in hyperlipoproteinemia are difficult to assess because of the absence of, or inadequate numbers of, control subjects, or confusion in the classification of lipoproteinemia. The effect of obesity and dietary conditions are also important since the caloric value and content of fiber, fat, cholesterol, and carbohydrates may all interfere with bile acid excretion [477]. To compensate for the effect of obesity, several authors present their data with respect to body weight (mg/kg body wt per day).

4.11.1a. Type I—Chylomicronemia. Fecal bile acid excretion was determined in two boys with type I hyperlipoproteinemia [208]. In one child values for six determinations were within the normal range for children reported elsewhere (see Table V), and in the second patient excretion was slightly raised but similar to that for children given high-cholesterol diets [207]. In both patients, a high-fat, high-cholesterol diet had no significant effect on fecal acidic steroid excretion [208].

4.11.1b. Type II—Hyperlipoproteinemia. It is probable that more than one genetic defect can be expressed as the type II phenotype. All are characterized by an increased concentration of plasma LDL and hence elevated cholesterol. Two main subgroups have been designated as type II: type IIa, familial hypercholesterolemia and polygenic (essential) hypercholesterolemia, is characterized by normal levels of plasma VLDL, whereas type IIb, combined hyperlipoproteinemia, has elevated VLDL concomitant with an increased LDL.

Data concerning cholesterol and bile acid excretion in homozygous familial hypercholesterolemia (type IIa) are limited. Several studies have found apparently normal fecal bile acid excretion in these patients, although none [478,479] or only a small number [176,480] of controls were included in these studies. Relatively low daily fecal bile acid values were found in studies of two homozygous children by Grundy *et al.* [415], but a possible contributory factor was the low body weight of the patients. Despite large increases in fecal bile acid excretion with cholestyramine therapy and ileal exclusion, these patients did not respond with a decrease in plasma cholesterol level.

In 36 heterozygous familial hypercholesterolemic subjects (type IIa), fecal bile acid excretion (180 ± 12 mg/day) was reported as being subnormal, although statistical analysis was not reported. This value is significantly less than the excretion levels found in type IIa essential hypercholesterolemia and type IIb mixed or type IV hyperlipoproteinemias [477]. Steroid balance studies by Miettinen *et al.* in eight members of a family with familial hypercholesterolemia showed fecal bile acid excretion to be significantly lower than in the control subjects [476]. Other studies from the same group [481,482] and from other groups [204,480] have reported a decreased fecal bile acid output to be a frequent, but not consistent, characteristic in these patients.

Type IIa essential hypercholesterolemia patients generally have normal fecal bile acid output [205,477], whereas excretion in type IIb mixed hyperlipoproteinemia may be normal or moderately raised [473,477,483]. Several studies, which do not differentiate between the various primary disorders resulting in the type IIa phenotype, report no apparent differences for fecal bile acid excretion between controls and patients in this heterogenous group [208,473,475].

4.11.1c. Type III—Familial Hyperlipoproteinemia. Very few studies of fecal bile acid excretion have been reported for this uncommon hyperlipidemia [208,477]. In the small number of patients studied, low or normal total bile acid excretion has been recorded, and detailed studies of individual fecal bile acids were not described [208,477].

4.11.1d. Type IV—Hypertriglyceridemia. Increased fecal bile acid excretion is a frequent, but not consistent, finding in patients with type IV hyperlipidemia [208,473,477,480,483]. This inconsistency may reflect the heterogenous nature of the type IV pattern, which has been observed in nonfamilial forms, familial hypercholesterolemia, and familial hypertriglyceridemia [277]. Kinetic studies have shown that the increased fecal bile acid excretion seen in these patients is principally due to accelerated synthesis of cholic acid [471,474]. These elevated levels may be due in part to increased body weight, which is commonly associated with hypertriglyceridemia. However, some lean hypertriglyceridemic patients may have a high fecal bile acid excretion and some obese patients a normal output [208,477,483].

4.11.1e. Type V—Hypertriglyceridemia. Grundy *et al.* [208] reported normal or moderately increased fecal bile acid excretion in five patients presenting with type V hyperlipidemia. In a later report, a single individual had large fecal bile acid losses, greater than 1000 mg/day [483]. This excessive excretion could not be attributed to obesity in the patient or method of analysis, and the small number of patients studied would not exclude individual variation, as seen in type IV subjects, as a possible reason for the discrepancy. Kinetic studies in five patients with the type V pattern have also demonstrated a wide variability in bile acid formation between individuals [471].

4.11.2. Bile Acid Profiles in Hyperlipoproteinemias

Despite numerous studies using GLC methods of analysis, few reports have included details of fecal bile acid profiles in hyperlipoproteinemia. Kinetic studies have shown that cholic acid synthesis is frequently subnormal in hypercholesterolemia type IIa and IIb, although total bile acid synthesis is within the normal range [471]. In contrast, bile acid synthesis was above normal in most patients with type IV hyperlipoproteinemias, chiefly because of increased cholic acid synthesis [471].

Lithocholic acid was identified as the major bile acid in the feces of eight hypercholesterolemia patients with normal triglyceride levels (type IIa). In contrast, subjects with high plasma cholesterol and triglycerides, which include type IIb, type IV, and type V hyperlipoproteinemias, had increased excretion of both lithocholic and deoxycholic acids, but the latter was the major fecal bile acid excreted [470,483,484]. These findings are supported by kinetic studies indicating that patients with high plasma cholesterol and normal triglyceride levels have a depressed synthesis of cholic acid compared with that of chenodeoxycholic acid, while in hypertriglyceridemic subjects, cholic acid synthesis exceeds that of chenodeoxycholic acid synthesis [471].

4.11.3. Secondary Hyperlipoproteinemia

Hyperlipoproteinemia may also be associated with a number of diseases, including alcoholism, diabetes mellitus, nephrotic syndrome, hypothyroidism, and obesity. Fecal bile acid excretion in these conditions is discussed later.

4.12. Liver Disease

Fecal bile acid excretion has been studied in a number of liver diseases, such as cirrhosis [265,266,268,382,489], including primary biliary cirrhosis [382,490,491], chronic active hepatitis [382,490,491], sclerosing cholangitis [491], benign recurrent intrahepatic cholestasis [269], acute viral hepatitis [492,493], and obstructive icteric liver disease [492]. Most patients with severe forms of liver disease have significantly decreased fecal bile acid excretion [267,268,382,490–492]. These patients also fail to increase bile acid synthesis in response to fecal bile acid loss induced by cholestyramine treatment [382,383].

In contrast to severe liver disease, a study of eight patients with moderately advanced but stable alcoholic cirrhosis found fecal bile acid excretion to be not significantly different from a control group, and cholestyramine treatment in these patients induced a fourfold increase in excretion, in common with that seen in subjects with normal liver function [489]. When bile duct obstruction or hypoplasia is present, fecal bile acids are not detectable [494] or are very low [492]. Phenobarbitol has been shown to markedly increase fecal bile acid excretion in a patient with congenital biliary atresia while reducing serum bile acid concentrations [495]. This effect was also found in three of four normal subjects treated with phenobarbital [496].

Despite the reduced biliary bile acid secretion and impaired micelle formation [497,498] in severe liver disease, steatorrhea is mild [382,491]. This is not unexpected since satisfactory fat absorption has been observed even in biliary fistula patients where complete exclusion of bile flow occurs [499].

Other abnormalities of bile acid metabolism that are frequently associated with liver disease include a raised plasma and biliary chenodeoxycholate/cholate ratio with a decrease or absence of deoxycholate [500–502]. In feces, an increased proportion of primary bile acids and a decrease in deoxycholic acid have been reported in a significant proportion of patients with cirrhosis [266–268]. Lithocholic acid excretion may also be reduced, although more frequently it is unchanged, and an increased proportion of other unidentified secondary bile acids is commonly found [265–268]. The decreased proportion of deoxycholic acid in feces may be due to lack of substrate (cholic acid) for gut bacterial 7α-dehydroxylation; however, this is not supported by the findings of significant amounts of cholic acid in the feces of these patients, and it is more probable that there is a reduced transformation of cholic acid to deoxycholic acid in the colon [266]. This is supported by results from a number of experiments. Administration of large amounts of cholic acid to such patients leads to a marked increase in biliary cholic acid, but little change in biliary deoxycholic acid [265,502]. In addition, *in vitro* fecal cultures from three cirrhotic patients were shown to have a marked inability to convert cholic acid to deoxycholic acid, compared with stools from controls [266]. Furthermore, rectal administration of [^{14}C]cholic acid to two patients with severe cirrhosis and low biliary deoxycholic acid levels indicated that 7α-dehydroxylation was less efficient than for two other patients with less advanced cirrhosis and normal biliary deoxycholic acid levels [265].

Markedly increased fecal bile acid excretion was reported in a single cirrhotic patient with cholestyramine suppressible diarrhea [382,383], and it was suggested that the diarrhea that occasionally presents in cirrhosis may be bile acid induced. Increased fecal bile acid excretion was observed in the anicteric phase in another patient with benign recurrent intrahepatic cholestasis [269]. Total fecal bile acids were threefold greater than normal, and an increased proportion of primary bile acids with virtual disappearance of lithocholic and deoxycholic acids was seen. A number of unusual bile acids were also present but not identified. In a similar study of the anicteric phase of acute hepatitis B, moderately diminished fecal bile acids were observed, and as jaundice developed in this patient, fecal bile acid excretion markedly decreased [493].

4.13. Cholelithiasis

Epidemiological studies have indicated that the prevalence of gallstone disease in developed societies is greater than 10% [503], and increased risk for the disease is associated with many factors that have influence on bile acid synthesis, biliary secretion, and bile composition [504,505]. No attempt has been made in this chapter to comprehensively cite the vast literature pertaining to bile acids and cholelithiasis, and the reader is therefore directed to reviews, books, or articles on the subject and references therein [506–509], including a chapter from Volume 3 in this series [510].

The bile of cholesterol gallstone patients compared to healthy subjects is more frequently saturated with cholesterol, and similarly the bile acid pool size is abnormally small [511]. Any increase in fecal bile acid excretion, due to increases in bile acid recycling or impairment in reabsorption, might lead to a decrease in the bile acid pool size and biliary secretion and thus supersaturation of bile with cholesterol. It is of interest therefore, in this connection, that ileal resection or dysfunction, long-term treatment with cholestyramine, and cystic fibrosis are associated with an increased incidence of cholelithiasis [456,512].

Examination of the literature, however, reveals little evidence to support an increase in fecal bile acid excretion in patients with cholelithiasis compared with controls [510,513,514], although there is considerable variation in reported composition. Danzinger *et al.* [513] measured fecal bile acid excretion in seven women with gallstones and six healthy women and suggested that the fecal bile acid composition was probably not different in cholelithiasis from that of healthy persons, although the proportion of $3\alpha,12\beta$-dihydroxy-5β-cholanoic (tentatively identified) was claimed to be higher in the patient group. Since the reported proportions of the individual bile acids in the feces of patients with cholesterol gallstone are extremely variable and the number of studies that have been performed is small, it is not possible to determine whether there are any qualitative changes of significance in cholelithiasis [181,513–515]. Bile acid sulfates have been measured in patients with cholelithiasis and are reported to comprise less than 13% in most cases [181,515], similar to the percentage reported for normal subjects [21,54,55,169,182,342].

Oral bile acid therapy has been effectively employed for the treatment of cholesterol gallstones and has been extensively studied (see Section 5.6). Agents other than bile acids, which are capable of enhancing bile acid synthesis, secretion, and pool size, but do not affect cholesterol secretion, have been evaluated for gallstone dissolution. Phenobarbital has been reported to have these effects and thus to decrease biliary cholesterol saturation [496,515–518].

Fecal bile acid composition was reported by Redinger [515] in patients with cholesterol gallstones, before and after treatment with phenobarbital, and in control subjects. Phenobarbital increased the excretion of total bile acids in feces by twofold, which paralleled the increased production rate of bile acids as determined by the Lindstedt technique [94]. Statistically significant increases in amounts of nonsulfated cholic, deoxycholic, and lithocholic acids and cholic acid sulfate were also found, while the proportion of chenodeoxycholic acid decreased.

The increased fecal excretion of cholic and deoxycholic acid probably results from the increased production and secretion of cholic acid, while the decreased proportion of chenodeoxycholic acid in the stools may reflect enhanced reabsorption and cycling frequency. Long-term phenobarbital treatment in gallstone patients reduced the cholesterol saturation of bile but did not effect dissolution after a 2-year period. Since cholic acid is less effective as a dissolution agent than the dihydroxy bile acids, this may be explained by the greater effect of phenobarbital on cholic acid rather than chenodeoxycholic synthesis. In normal subjects this drug was shown to have no influence on biliary lipid composition or bile acid pool size, suggesting an underlying metabolic difference between healthy subjects and gallstone patients, or that it is only effective in cases where there is a reduced bile acid pool size.

4.14. Diabetes

Diabetics are at increased risk for atherosclerosis [519] and cholelithiasis [520]. Cholesterol balance studies, including measurements of bile acid excretion, have been carried out in an attempt to elucidate the pathogenic mechanism involved in this disorder. Two independent assessments found fecal bile acid excretion to be significantly decreased during insulin treatment compared with patients with uncontrolled hyperglycemia [521,522]. All these studies found an association of increased bile acid synthesis with diabetes, although no change was found in a third study [523]. If fecal bile acid excretion is decreased, with no change or an increase in neutral steroid excretion, as may occur during insulin control of diabetes, bile may become more lithogenic and the risk of gallstone formation will increase. This suggestion is supported by the finding of increased cholesterol saturation of gallbladder bile during insulin treatment of diabetes [521,523]. Intractable diarrhea is a complication in a minority of diabetic patients, and the possibility that this symptom may be bile acid induced was discussed in Section 4.8.

5. EFFECT OF DIET AND DRUGS ON FECAL
  BILE ACID EXCRETION

5.1. Dietary Fiber

The role of fiber in the quantitative and qualitative composition of bile acids in stools has provoked much interest over the past 10–15 years. Initially the observations of Burkitt *et al.* [524] and Trowell [525] led to the proposal that some diseases prevalent in Western society, but not in Third World countries, may be due to a decreased intake of dietary fiber. Interest in the interrelationship of fiber and bile acid metabolism has sprung from the possible involvement of bile acids in the pathogenesis of colon cancer and also their interdependence with cholesterol and lipid metabolism [526]. Furthermore, the hypocholesterolemic action of some dietary fibers is well documented [287].

Experiments to define the interrelationship between fiber and bile acid excretion are hampered by problems in accurately assessing fiber intake, since isolation techniques are crude and classification of fiber type is often arbitrary [527]. Furthermore, fiber content of foodstuffs may change with methods of cultivation, types of grain, and vegetable or fruit variety, and the administration of extracted fiber as part of a diet may bear little relation to its effects when taken in its natural form. Dietary fiber is composed of a diverse group of substances generally derived from plant cell walls and possessing a wide variety of physicochemical characteristics [528,529]. It would be expected, therefore, that dietary fiber will vary excretion by a number of mechanisms, including physical binding of the bile acids to the fiber, affecting bacterial flora and intestinal transit time, increasing fecal bulk, and shielding bile acids from reabsorption.

Bile acid binding by a number of types of fiber has been demonstrated *in vitro* [128–130,530–532] and ostensibly *in vivo* [130] by identification of bile acids present in the stool residue after centrifugation. The possibility that bile acids are merely trapped within the fiber matrix is not precluded. The extent to which bile acids are adsorbed to fiber has been shown to be a function of pH, osmolarity, type of bile acid and fiber, and degree of micelle formation. Guar gum and lignin are the most effective adsorbents of bile acids *in vitro*, and alfalfa, bran, and cellulose are progressively less effective [531,532]. Binding is greater at acid pH and increased bile acid concentration, but it is inhibited by micelle formation, particularly at higher pH [530]. Extent of adsorption of individual bile acids is variable, and no consistent pattern based on number of hydroxyl groups or state of conjugation has been identified [130,530–532]. Adsorption of bacteria to fiber may also occur in the gut and·thereby alter microbial transformation. Hill *et al.* [314] compared bacteria in stool samples from six different populations and found larger numbers of anerobic organisms in specimens from Western countries than from those in Africa and Asia. In controlled experiments conflicting results have been reported with either no change [533] or an increased number [534] of anerobes in the feces after augmentation of fiber with dietary wheat bran. A number of other studies reviewed by Hill [535] report no differences in fecal bacterial flora in various populations or under controlled conditions using fiber supplementation; however, only the principal species of bacteria were characterized. Different fibers have different effects on stool bulk and intestinal transit time; both factors may have a bearing on fecal bile acid excretion.

In an early report by Williams and Holmsted [536], the effect of a wide variety of foods on fecal bulk was examined. More recent studies have shown that dietary augmentation with wheat bran [537–540] generally increases fecal wet weight considerably, whereas pectin [541,542] causes no significant change. When measured, intestinal transit time was usually unchanged with all fiber types, although some studies with wheat bran [540,543], oat fiber [544], and mixed fibers [545] found it to be significantly decreased. Inconsistencies may be due to differences in the quantitative intake of fiber since the increment was particularly large in these experiments. The type of colonic microflora and intestinal transit time may affect the degree of fiber digestion and hence influence its intraluminal effects [546]. More detailed reviews of the effect of fiber have been published elsewhere [528,547,548].

A vast literature exists on the effects of various types of fiber on bile acid excretion in the feces. These include wheat bran [537–540,543,549,550], oat bran [488,544,551,552], pectin [541,542,553], soybean flour [550,554–556], corn bran [553], cellulose [537,538], mixed high-fiber foodstuffs [538], alfalfa seeds [557], legumes [488,558], guar [559], bagasse [533,560], gum arabic [561], Metamucil [562,563], and plantago fiber [564]. The effect of these fibers on total daily bile acid excretion is summarized in Table XI.

In most studies where the diet of normals was augmented with wheat bran, total fecal bile acid excretion was unchanged (Table XI). Two exceptions, however, reported significantly increased excretion, but these used large increments of dietary fiber and reported significantly decreased intestinal transit times [540,543]. Two of the studies with low-fiber wheat bran augmentation observed a rebound effect, so that fecal bile excretion increased significantly in the period immediately after the fiber diet had been discontinued [538,539]. Significantly decreased bile acid concentrations in the stool were observed in all cases, presumably because of the increase in fecal bulk [537–540,543]. Serum cholesterol levels and the profile of individual fecal bile acids observed were essentially unchanged with increased wheat fiber intake [538,539,549].

Augmentation of the diet with oat bran [552,563] or rolled oats [544] resulted in increased fecal bile acid excretion which was statistically significant in most

TABLE XI.   Effects of Various Dietary Fibers on the Excretion of Total Fecal Bile Acids

Supplementary fiber source	Amount (g/day)	Added dietary fiber (g/day)	No. subjects	Exptl. period (weeks)	Change in daily fecal bile acid excretion	p value[a]	Reference
Wheat bran	16	—	8	3	No change	—	537
	16	—	8	3	+4% to +11%	NS	538
	20	—	6	5	+57%	NS	550
	25 (hard)	—	3	4	−9% to +42%	NS	550
	25 (soft)	—	3	4	−38 to +29%	NS	550
	23–35	—	6	3	+7%		539
	35	—	7	4–8	−15% to +98%	NS	549
	39	13	5	1	No change	—	533
	—	28	6	3	+40%	<0.01	543
	—	31	4	3	+100%	<0.0005	540
	100	36	6	6			565
Oatmeal	125	9	10	7	+38%	<0.01	544
	100	23	8[b]	1.5	+54%	<0.001	552
Oatbran	100	38	20[b]	3	+65%	NS	488
Cornbran	25	—	4	4	+6% to +108%	NS	550
Pectin	15	—	8	2.5	+6% to +11%		542
	15	—	9	3	+33%	<0.02	541
	36	—	7	2	+36%	<0.05	566
	40–50	—	2	2	+57%	<0.005	553
Soybean pulp	54	21	6	3	+21%	<0.01	556
Soybean	25	—	4	4	−29% to +35%	NS	550
Soy polysaccharide	25	11.3	14	2.5	+8%	NS	555
Mixed beans	115	38	20[b]	3	−30%	<0.01	488
Bagasse	10.5	9	19	12	+50%	<0.001	533
Chick pea	—	—	30	55	+29%	<0.001	558
Guar	36	—	7	2	+84%	<0.05	566
Gum arabic	25	—	5	3	−10%	NS	561
Metamucil	19.2	9.6	2	6	+202%	<0.005	562
Various	60	16	8[b]	4	No change	—	545

[a] p value of statistical significance; NS = p > 0.05.
[b] Hypercholesterolemic patients included.

subjects studied, including normals and hypercholesterolemic patients. No significant changes in neutral steroids were observed, although serum cholesterol levels decreased.

The addition of moderate amounts of pectin to the diet, in general, increases total fecal bile acid excretion (Table XI), although large individual variations are observed [542]. Larger increases and significantly decreased serum cholesterol concentrations were reported with a high pectin intake [553].

Soybean fiber significantly increased fecal bile acid excretion [556], whereas no change was observed with soy polysaccharide, a fiber-rich product from soybean [555]. Excretion actually decreased ($p < 0.05$) when a mixture of pinto and navy beans was added to the diet of 10 hypercholesterolemic men. A striking change in mean fecal bile acid excretion was observed with the mucilage derived from blond psyllium seed (Metamucil), although large variations occurred [562] and serum cholesterol levels were significantly reduced ($p < 0.001$).

The pattern of individual fecal bile acids is generally reported to be unchanged with increased dietary fiber in normal subjects [538,539,549,553] and diverticular patients [262,370], although the analytical methods used in most studies are limited to the determination of lithocholic, deoxycholic, chenodeoxycholic, and cholic acids only. An increased proportion of primary bile acids was reported in a single study in which the diet was enriched with natural high-fiber foods [567].

5.2. Bile Acid Sequestrants

Anion exchange resins such as cholestyramine, colestipol, and polidexide act by absorbing bile acids in the gut lumen, thereby preventing their absorption and increasing by severalfold their excretion in the feces (Tables IV and XII) with an associated decrease in serum cholesterol levels of between 20 and 30% [287]. Various sequestrants have been evaluated in the treatment of hypercholesterolemia, alone or in combination with other hypolipemic drugs such as clofibrate [568], neomycin [569], or nicotinic acid [570]. Table XII summarizes the effect of these anion exchange resins on fecal bile acid excretion in hyperlipidemic patients. The accuracy of the values for fecal bile acids following cholestyramine treatment may be questionable since problems of inefficient extraction of bile acids from the cholestyramine resin excreted in the stools can occur [132] (see Section 2.7). Increments observed with cholestyramine administration to normal subjects are clearly higher than those observed in patients with heterozygous hypercholesterolemia type IIa and IIb, normolipidemic coronary heart disease, or hypertriglyceridemia [287] (Table XII). Miettinen [287] compared heterozygous hypercholesterolemic patients with healthy controls matched for age and body weight and found cholestyramine to increase bile acid excretion to a significantly greater extent in the normal controls. The cholestyramine

TABLE XII. Effects of Sequestrants on Total Fecal Bile Acid Excretion in Hyperlipidemia

Reference	Type of hyperlipidemia	n	Sequestrant	Period (days)	Amount (g)	Daily bile acid excretion[a]		
						Before	During	Ratio[b]
Moutafis and Myant, 1969 [479]	II	1	Cholestyramine	216	15–20	190 mg	1560 mg	8.2
	II	1	Cholestyramine	216	7–20	315 mg	1407	4.5
Grundy et al., 1971 [415]	II Homozygotes	2	Cholestyramine	24–56	10–24	139–141 mg	641–914 mg	4.6–6.5
	II Heterozygotes	5	Cholestyramine	24–56	10–24	115–645 (282) mg	620–2428 (1413) mg	3.0–10.0 (6.0)
Myant, 1972 [571]	IIa	1	Cholestyramine	Long-term	16–30	195 ± 28 mg	1222 ± 10 mg	6.3
Miettinen, 1971 [572]	Normal	13	Cholestyramine	10	32	235 ± 25 mg	—	8.6
	Coronary patients	5	Cholestyramine	10	32	332 ± 78 mg	—	3.7
	Normolipemic							
	Hypercholesterolemia	2	Cholestyramine	10	32	241 ± 37 mg	—	5.1
	Hypertriglyceridemia	6	Cholestyramine	10	32	326 ± 60 mg	—	3.3
Nazir et al., 1972 [485]	IIa	4	Cholestyramine	12–15	12	73–277 (143) mg	546–645 (597) mg	2.4–7.5 (5.2)
Simons and Myant, 1977 [573]	Normal/IIa/IIb	4	Cholestyramine	56–123	16–28	203–310 (273) mg	1289–3136 (2262) mg	6.4–10.1 (8.2)

Reference	Type	Drug	n					
Miettinen and Lempinen, 1977 [482]	IIa/IIb	Cholestyramine	13	10–12	32	193 ± 31 mg	1434 ± 176 mg	7.4
Moutafis et al., 1977 [480]	Normal	Cholestyramine	2	21–28	20–30	3.9–4.7 mg/kg	28.3–28.6 mg/kg	6.0–7.3
	IIa Heterozygotes	Cholestyramine	4	21–28	20–30	3.8–5.3 mg/kg	23.9–38.2 mg/kg	4.7–6.6 (6.2)
	IIa Homozygotes	Cholestyramine	3	21–28	20–30	5.1–7.8 mg/kg	36.9–76.8 mg/kg	7.2–9.8 (8.3)
Miettinen, 1979 [569]	Hypercholesterolemia	Cholestyramine	9	25	32	203 ± 20 mg	1497 ± 170 mg	7.4
Miettinen, 1981 [287]	Normal	Cholestyramine	7	—	32	3.9 ± 0.4 mg/kg	36.2 ± 3.8 mg/kg	9.3
	IIa Heterozygotes	Cholestyramine	46	—	32	1.8 ± 4.4 mg/kg[c]	13.6–21.8 mg/kg[c]	5.0–7.6[c]
	IIb	Cholestyramine	6	—	32	2.9 ± 0.5 mg/kg	16.5 ± 4.0 mg/kg	5.7
	IIa Homozygotes	Cholestyramine	2	—	16–24	1.7–7.1 mg/kg	35.5–69.2 mg/kg	20.8; 9.7
Miller et al., 1973 [574]	IIa/IIb/IV	Colestipol	8	41–59	15	305–478 (399) mg	1074–2129 (1640) mg	2.9–5.8 (4.1)
Grundy and Mok, 1977 [568]	Various	Colestipol	14	28–42	20	428 ± 47 mg	1816 ± 291 mg	4.2
Miettinen, 1978 [575]	Hypercholesterolemia	Colestipol	6	8–24	15	256 ± 33 mg	1199 ± 220 mg	4.7
Simons and Myant, 1977 [573]	Normal/IIa/IIb	Polidexide	5	56–123	3–30	277–310 (278) mg	851–1954 (1467) mg	4.2–6.3 (5.2)

[a] Values are expressed as range with mean in parentheses: mean ± SD or mean ± SEM and the reader is directed to the relevant reference for more details.

[b] The ratio of bile acid excretion during cholestyramine treatment/control period excretion.

[c] Range of quoted means.

response in homozygous patients is frequently, but not consistently, very high [287,480]. In contrast to other hyperlipidemic disorders, treatment with cholestyramine did not significantly reduce serum cholesterol concentrations [287,480,570]. However, these patients were younger than other patients evaluated [287,570], and comparable age-matched controls were not available.

Treatment with cholestyramine in conjunction with nicotinic acid reduced plasma cholesterol levels in four homozygous hypercholesterolemic patients aged 11–18. In two of these patients in whom bile acid and total steroids were measured, excretion decreased with nicotinic acid therapy, and experimental evidence suggested this was due to an inhibition of cholesterol synthesis [570]. In adults with type IIa heterozygous hyperlipidemia, fecal bile acid response to cholestyramine appears to be independent of age [287].

Polidexide and colestipol cause similar increases in bile acid excretion to that found with cholestyramine (Table XII); in all subjects, and with all of the ion exchangers, bile acid excretion increases to a maximum with increasing doses of sequestrant. However, significantly higher fecal bile acid excretion is observed with ileal resection or bypass [482], showing that sequestrants do not completely inhibit bile acid reabsorption. This is borne out by the finding that biliary deoxycholate secretion is reduced but not eliminated during sequestrant treatment [287]. Further support of this is found from a study of 13 patients with severe hypercholesterolemia in whom fecal bile acid excretion increased sevenfold (1434 $\pm$ 176 mg/day) during cholestyramine treatment (Table XII), but when an ileal bypass procedure was carried out on these patients, fecal bile acid excretion increased still further to 2414 $\pm$ 264 mg/day ($p < 0.001$) [482]. The corresponding hypocholesterolemic action was also significantly greater with the bypass than with cholestyramine alone. The hypocholesterolemic action of sequestrants may be enhanced by addition of certain drugs. For example, fecal bile acid excretion is significantly increased in hypercholesterolemic patients when cholestyramine is used in combination with neomycin, compared to the sequestrant alone [569]. Clofibrate, which inhibits cholesterol synthesis, when used with colestipol, leads to a lower fecal bile acid excretion than the sequestrant alone, but enhances the hypocholesterolemic action of the sequestrant [568]. The combination with mevinolin, which inhibits HMG–CoA reductase, is more effective than any other combinations in this respect [576].

Cholestyramine is commonly used in the control of cholerrheic diarrhea, secondary to small ileal resection [380,403,404] and ileal dysfunction [203,374], and in a number of disorders in which diarrhea is believed to be bile acid mediated. The latter includes postvagotomy [379,381], postcholecystectomy [270,384], diabetes [381], hypogammaglobulinemia [271], and cirrhosis [382,383]. Few of these studies, however, measured fecal bile acid excretion after cholestyramine therapy. Hofmann and Poley [380] found that the elevated fecal bile acid excretion following ileal resection could be increased still further by cholestyramine treatment in some, but not all, patients. Bile acid concentrations in the aqueous phase of the stool were greatly decreased owing to binding to the sequestrant, and the improvement in diarrhea observed in patients with small-ileal resection was attributed to this change. *In vivo* the resin has been shown to bind taurine conjugates more effectively than glycine conjugates, as would be expected from their pK_a, values and dihydroxy bile acids more effectively than trihydroxy bile acids [532,577]. Detailed analyses of individual bile acids in the feces following sequestrant treatment have not, to our knowledge, been reported.

5.3. Antibiotics

A number of studies have shown that administration of antibiotics such as clindamycin, lincomycin, neomycin, and kanamycin cause steatorrhea and, in some patients, diarrhea and pseudomembranous colitis [578–580]. In addition, small doses of neomycin and paramycin have been shown to be effective in decreasing serum cholesterol levels [581–583], thus providing a possible treatment for hyperlipidemia [287]. The mechanism of action is unclear, although their ability to act by altering the intestinal microflora [11], behaving as bile acid "sequestrants," or precipitating bile acids in the small intestine [583,584] has been suggested. Alteration of the intestinal microflora seems unlikely since similar antibiotics do not show the same hypocholesterolemic effects and the hypocholesterolemic effect has also been shown to occur in germ-free animals [585]. An increased fecal excretion of bile acids was observed in several early studies, although relatively nonspecific methods of analysis were used [6,11,586]. Similar effects are found with the analog N-methyl neomycin, which has no antibiotic effect [587], suggesting that a nonspecific sequestrant action may be the likely explanation of this phenomenon, consistent with experiments *in vivo* and *in vitro* which show that bile acids are precipitated by neomycin from samples of bile and intestinal contents [583,584,588,589]. Small doses of neomycin inconsistently result in increases in fecal bile acid output, suggesting that bile acid sequestration or precipitation may not be the primary reason for the reduction in serum cholesterol concentrations [287,487,569]. Large doses of neomycin lead to moderate, but significantly increased, fecal bile acid output in patients with obesity, cirrhosis of the liver, intestinal bypass, and familial hypercholesterolemia [23,569,590,591], but the increment is lower than that seen with sequestrants such as cholestyramine. Transient large increases in fecal excretion of bile acids have been observed in the initial period after neomycin administration in some patients [487].

Neomycin has been shown to decrease the proportion of deoxycholic acid in the stool [590]. Analysis of individual fecal bile acids in patients with cystic fibrosis has revealed a lower percentage of secondary bile acids during antibiotic treatment [219,454]. Similarly, in five patients with normal gastrointestinal function, deoxycholic acid was reduced or eliminated from the bile following administration of this antibiotic [592]. Marked inhibition of 7α-dehydroxylation of [^{14}C]cholic acid was demonstrated after oral administration of neomycin, kanamycin, and chloramphenicol in a number of patients with various types of hyperlipoproteinemias [582], and significantly reduced serum cholesterol concentrations were observed in those patients with 7α-dehydroxylation inhibition. When no hypocholesterolemic effect was found, no gross changes in qualitative fecal bile acid pattern were detected [582]. In a study of a single patient given tetracycline [593], cholic and chenodeoxycholic acids became the major bile acids excreted, and only small amounts of conjugated bile acids were found in feces. Korpela *et al.* found decreased concentrations of saponifiable derivatives and increased primary bile acids and conjugates in feces during treatment with oxytetracycline [170].

5.4. Dietary Factors Influencing Bile Acid Metabolism

5.4.1. Quantity of Diet

As mentioned earlier, obesity is associated with increased cholesterol synthesis and elevated bile acid production. Excessive calorie intake *per se* has been associated with augmented biliary bile acid excretion in two nonobese Caucasian men [282]. However, we can find no reports of fecal bile acid excretion under these conditions. A reduced calorie intake causes immediate and marked reduction in fecal bile acid excretion in obese hyperlipemic subjects [594–596], before any significant changes in body weight occur. The reduction is greater for deoxycholic acid than for lithocholic acid [595,596], suggesting a selective inhibition of cholic acid synthesis. This effect may be secondary to a reduced cholesterol synthesis [282,594,595]. Hypolipemic drugs are less effective at increasing fecal bile acid excretion in patients undergoing caloric reduction [278,594] than in patients given a normal diet.

5.4.2. Dietary Fat and Carbohydrates

The interrelationship of dietary fat and bile acids has been studied extensively because of the implication of high fat/meat diets in the pathogenesis of colon cancer (discussed in Section 4.1) and the well-established hypocholesterolemic action of polyunsaturated fat [286]. Many epidemiological surveys, for example, have shown that population groups consuming a high-fat Western diet have higher fecal excretion of neutral and acidic steroids than groups consuming diets low in fat, e.g., Refs. 24 and 175. Early studies using relatively imprecise methodology for bile acid estimation generally failed to show an increase in fecal bile acid output when dietary saturated fat was increased [2,140,597].

Some recent experiments have reported significantly increased total fecal bile acid excretion, principally of the secondary bile acids deoxycholic and lithocholic acids, in normal subjects ingesting high-saturated fat/high-meat diets [42,177,314]. However, a detailed study of a large number of volunteers has compared the effect of varying the amount of total fat (22–40% of total caloric intake) and/or the polyunsaturated fatty acid ratio (0.2–1.7) in the diet, and no significant differences in fecal bile acid output were observed between these dietary regimes [598].

Consistently increased fecal bile acids, in association with elevated plasma triglyceride levels, have been observed with diets low in fat but rich in sucrose, when compared with diets enriched with polyunsaturated or saturated fatty acids [599]. The association of augmented triglyceride and fecal bile acid output has also been observed in type IV hyperlipoproteinemia (see references cited in Section 4.11.1d), obesity [279], and cholestyramine treatment [287].

Substitution of polyunsaturated for saturated fat in the diet of normal subjects has been reported to cause an increase [16,597,600–604], decrease [605], or no change [598,606–608] in fecal bile acid excretion despite consistent reduction of plasma cholesterol and, where reported, triglyceride levels. Some early studies report large subject and sample variations, and therefore no conclusions could be drawn from the results [18,609]. The variability may be due to differences in a number of factors, for example, the types of saturated fat, since medium-chain triglycerides found in coconut oil [602,609,610] are metabolized differently from long-chain fats; the proportion of cholesterol [597,611] and plant steroids [597]; non-steady-state conditions; and differences in fiber content, particularly with liquid formula diets.

In hypercholesterolemic patients the effect of polyunsaturated fat, compared to saturated fat, has been shown to be similar to that seen in some studies of normolipemic subjects, i.e., reduced serum cholesterol levels and unchanged fecal bile acid or neutral steroid excretion [607]. By contrast, when type IV and V patients are fed a diet high in polyunsaturated fats, a reduction of plasma cholesterol and triglyceride levels occurs in most cases, with an associated increase in neutral steroid and bile acid excretion [612].

5.4.3. Dietary Cholesterol

An increased intake of cholesterol as high as 30-fold does not significantly augment bile acid excretion in the feces of normal subjects [613,614]. This suggests that changes in bile acid synthesis and excretion are not compensatory mechanisms protecting against dietary cholesterol overloading and potential hypercholesterolemia in humans. A study by Lin and Connor, of 21–25 weeks' duration, in a normal and a hypercholesterolemic subject reported moderate, but significant, increases in fecal bile acids in both, concomitant to increased cholesterol absorption [475]. The Masai tribe of East Africa present an unusual biochemical pattern since despite a high-fat, high-cholesterol, and low-fiber diet, they have surprisingly low levels of serum cholesterol and β-lipoprotein. Decreasing the cholesterol intake in this population led to a lower, but not statistically different, fecal bile acid excretion [615].

5.4.4. Ethanol

Moderate consumption of ethanol may lead to small changes in plasma triglycerides and cholesterol levels in normal subjects, accentuate preexisting hypertriglyceridemia, or infrequently cause marked hyperlipemia (see Ref. 616 and references therein). Total fecal bile acid excretion is unchanged in normal subjects when moderate amounts of alcohol are consumed [616,617], whereas although increases occur in most hyperlipidemic or obese patients, the changes are not statistically significant [616,617].

5.4.5. Calcium

In some studies high calcium intake has been reported to cause a significant reduction in serum cholesterol levels [618,619] and to increase fecal bile acid excretion in most patients [619–621]. This effect appears to be dependent on the type of dietary fat ingested [621].

5.4.6. Sucrose Polyester

Sucrose polyester, a nonabsorbable mixture of various fatty acid esters of sucrose, has been shown to decrease plasma cholesterol, but not triglyceride levels, in humans and to have variable and inconsistent effects on total fecal bile acid excretion, which overall were not significant [622,623].

5.5. Absorbable Hypolipidemic Drugs

A number of absorbable agents, including clofibrate, nicotinic acid, tibric acid, thyroid and growth hormones, and probucol, have been found to have hypolipidemic effects [277,287], although the mechanisms of action have not been well elucidated.

5.5.1. Clofibrate

Miettinen [287], in a comprehensive review of studies carried out on the effects of this drug up to 1981, described data indicating that plasma triglyceride levels and, to a lesser extent, plasma cholesterol are decreased by clofibrate treatment in most hyperlipoproteinemia patients. Assignment of the type of lipoproteinemia in this review [287] was in some cases arbitrary, but in general, clofibrate was reported to reduce fecal bile acid excretion, although the decrease was frequently not statistically significant [208,498,568,624,625]. Results are variable for type II hyperlipoproteinemia patients since significantly increased bile acid excretion has been reported in several studies for this group of patients [484,624,626]. Grundy *et al.* [208] observed a significant mean reduction of 70 ± 18 mg/day ($p < 0.01$) in fecal bile acids by clofibrate treatment in various types of hyperlipidemia. The largest reductions were seen in type IV hypertriglyceridemia, whereas a small, or no, decrease was seen in type III and I. Changes in fecal bile acid excretion in type II and V hyperlipoproteinemias were not significant in 9 of 12 patients studied [208].

As discussed previously, increased fecal bile acid excretion observed in hypertriglyceridemia patients has been shown by isotope methods to be due principally to increases in cholic acid synthesis [471]. Cholic acid synthesis was lowered from 1076 ± 243 mg/day to 333 ± 85 mg/day in type IV patients after treatment with clofibrate [627]. In a separate study of various hyperlipidemic patients, however, clofibrate frequently increased, or had no affect on, deoxycholic acid excretion, while the lithocholic:deoxycholic ratio decreased, suggesting that cholic acid synthesis was not affected by the treatment [484]. This discrepancy may be due to a variable effect of clofibrate treatment in different hyperlipoproteinemia types.

5.5.2. Nicotinic Acid

The effects of nicotinic acid in hyperlipoproteinemia appear to be similar to, but more potent than, those seen with clofibrate. In general, nicotinic acid reduces plasma levels of both LDL and VLDL, although LDL values may actually increase in some type IV and V patients [628,629]. Although a number of sterol balance studies have indicated changes in cholesterol and bile acid synthesis (for reviews see Refs. 277,287) relatively few fecal bile acid analyses have been carried out. An early study reported nicotinic acid treatment to cause a marked increase in fecal bile acid and neutral sterol excretion [630]. In type II hyperlipoproteinemia, nicotinic acid treatment increased fecal bile acid output in four of five patients studied, but the change was statistically significant in only two of these subjects [481]. In a later sterol balance study of seven hypercholesterolemia patients, mean fecal bile acid excretion was increased after nicotinic acid, but the change was not significant. Other investigators also observed inconsistencies in fecal bile acid excretion, with nicotinic acid [629,631].

5.5.3. Hormones

A number of hormones are known to alter cholesterol and bile acid metabolism (see Refs. 278,287,477,632 and references therein.)

5.5.3a. Thyroid Hormones.   Thyroxine and its derivatives have a marked effect on serum cholesterol and triglyceride levels; both are frequently elevated in hypothyroidism and decreased in hyperthyroidism [278]. Fecal bile acid excretion, however, is usually within normal limits, although values may be slightly depressed in hypothyroidism and marginally elevated in hyperthyroidism [278,633]. The diarrhea sometimes seen in hyperthyroidism is characterized by markedly increased bile acid excretion and is responsive to cholestyramine therapy [278]. Normalization of hyperlipidemia in hypothyroid patients by treatment with thyroid hormone inconsistently increases fecal bile acid excretion, but total fecal steroid excretion is usually enhanced [278,633].

D-Thyroxine has been used in the treatment of hypercholesterolemia [287,634]. This treatment in euthyroid subjects with familial hypercholesterolemia (type II) induced similar results to those seen in hypothyroid patients. Plasma cholesterol concentrations decreased and total fecal steroid excretion increased in all patients studied [634].

5.5.3b. Growth Hormone.   Growth hormone deficiency usually presents with marked hypercholesterolemia [635,636] and reduced postprandial intraluminal bile salt concentrations [637]. Treatment of these patients with growth hormone has been reported to have no effect on plasma lipoprotein or cholesterol levels [635,638], although conflicting results were presented in one study [636]. Intraluminal bile acid concentrations are normalized [637], and mean fecal bile acid excretion with respect to body weight was significantly increased in a group of 16 children studied [638]. Growth hormone administration to hypercholesterolemic adults have been shown to lower serum cholesterol [639], but bile acid excretion has not been studied under these circumstances.

5.5.4. Other Agents

Probucol is a hypocholesterolemic agent that reduces serum cholesterol while having little or no effect on triglyceride levels [640], and it is therefore useful in patients with type II hyperlipoproteinemia. In type IIa patients, probucol caused a transitory significant increase in fecal bile acids in a transitional period during which serum cholesterol levels decreased [641]. Once a steady state had been reached, fecal bile acid output decreased to the pretreatment values [641]. Nestel and Billington [642] also found a significant increase ($p < 0.02$) in fecal bile acids in type IIa and IIb patients treated with probucol for 8 weeks. In contrast, in a long-term study lasting 2–6 months of drug treatment, and where fecal bile acid excretion was measured over a 6-week period, no change in fecal excretion of neutral, acidic, or total steroids was reported [643].

Tibric acid is used in the treatment of hypertriglyceridemia since its principal action is to reduce serum triglycerides; however, the excretion of acidic steroids in the feces was unchanged in a single study reporting its effects [640].

5.6. Exogenous Bile Acid Therapy

For more than a decade the oral administration of bile salts has been applied to the treatment of cholesterol gallstones, and a large literature exists on this topic. Reviews of the subject can be found in Refs. 509,510,664–647 and in various articles cited in Ref. 648.

Chenodeoxycholic acid was the first bile acid to be used in the clinical treatment of cholesterol gallstones [649,650], and later ursodeoxycholic acid was introduced and shown to be a more effective agent [651–653]. Relative to the studies that have been carried out on the pharmacokinetics and clinical utility of these bile acids, data on fecal bile acid excretion and composition are limited [181,513,654–656].

In a study of five patients with cholelithiasis [513] the composition, but not the amount, of bile acids was determined before and after treatment with chenodeoxycholic acid (1.5–4.0 g/day). Lithocholic acid in feces varied between 13.8 and 35.2%, and deoxycholic acid accounted for between 39.8 and 55% of the total bile acids before treatment. Following chenodeoxycholic acid therapy, chenodeoxycholic acid was the predominant fecal bile acid excreted in three of five patients studied, and in all patients the combined proportion of deoxycholic and cholic acids was less than 10.1%, compared to 60.3–73.9% in the control period, indicating a marked suppression of cholic acid synthesis by the exogenous bile acid. The two patients showing the lowest extent of 7α-dehydroxylation were also those on the highest doses of bile acid and exhibited the worst diarrhea [513]. Since chenodeoxycholic acid reduces water and electrolyte secretion in the colon [272], the diarrhea, in this case, is bile acid induced and similar to that found in patients with ileal resection.

With the introduction of ursodeoxycholic acid, inevitably comparisons with chenodeoxycholic acid were made, particularly with respect to their metabolism [181,654–656]. Salvioli and Salati [181] examined the effects of short-term feeding of ursodeoxycholic and chenodeoxycholic acids on the bile acid pool size and on fecal bile acid loss. Twelve nonobese patients were fed a standard diet with a daily intake of 500 mg cholesterol. After a 10-day period, six patients were given 15 mg/kg body wt of ursodeoxycholic followed by [14C]ursodeoxycholic, and the other six patients were given comparable doses of chenodeoxycholic and [14C]chenodeoxycholic. Fecal bile acids were isolated into nonsulfated and sulfated fractions using Sephadex LH-20, and an enzymatic method was used to determine total bile acids, while GLC was used to determine the individual bile acids. Fecal bile acid excretion in the control period for two groups of subjects was 8.12 ± 2.02 μmole/kg per day and 7.85 ± 2.22 μmole/kg per day, respectively, for nonsulfated bile acids and 1.27 ± 0.21 μmole/kg per day and 0.91 ± 0.17 μmole/kg per day, respectively, for the sulfates, and it was reported that the composition of the individual bile acids was similar in both fractions. The proportion of sulfated bile acids in the control period was 13.5% and 10.3%, respectively, for the two groups of patients, which was slightly higher than the values reported for healthy subjects [21,54,55,169,182,342]. Following bile acid administration, fecal bile acid excretion significantly increased, and total bile acid loss attained levels of 23.94 μmole/kg per day following ursodeoxycholic acid and 36.12 μmole/kg per day following chenodeoxycholic acid. A significant increase in the relative proportion of sulfates was also seen (17.4% for ursodeoxycholic acid and 22.3% for chenodeoxycholic acid therapy). Lithocholic acid was found to be the predominant bile acid ($70.4 \pm 13.8\%$ for nonsulfated and $76.8 \pm 12.6\%$ for sulfated) at the expense of deoxycholic, which decreased in the stool following chenodeoxycholic acid therapy, and this was markedly greater than that found following ursodeoxycholic acid therapy ($44.6 \pm 12.2\%$ for sulfated and $37.7 \pm 8.7\%$ for nonsulfated fractions).

In an earlier study by Fedorowski *et al.* [655] similar findings were observed. Fecal bile acid composition was examined in four patients given 1 g/day of chenodeoxycholic and ursodeoxycholic acids. Following chenodeoxycholic acid, the proportion of lithocholic acid ($69 \pm 7\%$) in the stools was much higher than that found following ursodeoxycholic acid administration ($44.3 \pm 19.1\%$). The findings of these two groups differ slightly from those of Makino and Nakagawa [654], where lithocholic acid comprised 60% of the fecal bile acids following ursodeoxycholic acid administration. All the studies indicate that ursodeoxycholic acid can be 7β-dehydroxylated in the gastrointestinal tract, but the differences in fecal bile acid excretion following chenodeoxycholic acid and ursodeoxycholic acid therapy can probably be explained by a slower rate of bacterial 7β-dehydroxylation compared with 7α-dehydroxylation [657]. Alternatively, it has been suggested that the lower percentage of lithocholic acid in the feces of patients receiving ursodeoxycholic is the result of continuous secretion of bile acids derived from cholic acid synthesis, which is less impaired by ursodeoxycholic than by chenodeoxycholic acid therapy [656]. Small amounts of 7-oxo-lithocholic acid have been found in the feces following the administration of both bile acids, but a significantly greater proportion of this bile acid arises from the use of ursodeoxycholic [181,655]. The role of bacterial flora in the interconversion of chenodeoxycholic and ursodeoxycholic acids and in the metabolism of 7-oxo-lithocholic acid has been studied by several groups [274,657–659].

Phenobarbital administration affects biliary lipid secretion in patients with cholesterol gallstones [515,516]. Cholic acid production rate and pool size are increased in these patients, and the proportion of deoxycholic acid excreted in feces is elevated, yet phenobarbital has no effect on stone dissolution. Coyne *et al.* [518] examined the effect of chenodeoxycholic acid and phenobarbital in combination for the treatment of cholesterol gallstones and showed effective dissolution, but no data on fecal bile acid excretion were reported.

In an experiment designed to determine whether bile acid feeding leads to an increased absorption of cholesterol, Mok *et al.* [660] studied the effects of chenodeoxycholic acid and Bilron (a commercial preparation of cholic, deoxycholic, and chenodeoxycholic acids) in eight obese subjects undergoing weight reduction. Measurements of biliary secretion and fecal neutral steroid and bile acid excretion were used to estimate the extent of cholesterol and bile acid absorption, and despite an increased flux of bile acids through the gastrointestinal tract, no significant increase in cholesterol absorption occurred during exogenous bile acid therapy.

6. CONCLUSIONS

The technical and analytical developments in recent years have significantly increased our ability to study the excretion of bile acids in feces. The complexity of the fecal bile acid mixture has been emphasized by recent analytical studies but has not been adequately considered in most clinical studies. Little is known about biological effects of the variety of positional isomers, stereoisomers, ketones, esters, and unsaturated derivatives of bile acids that can be found in feces. Most of these have been formed through the action of intestinal microflora with or without participation of intestinal or hepatic enzymes during enterohepatic circulation. The degree of absorption and potential interaction with intestinal cells varies for the different bile acid metabolites. Biological effects depend on the chemical structure, and the steroid hormones illustrate how profound effects can result from small variations in the position and orientation of hydroxyl groups and other substituents in a steroidal molecule.

These considerations indicate the importance of analyzing individual compounds rather than (or in addition to) groups of peaks in a chromatogram, or nonspecific responses in enzyme or immunoassays. Better and simpler methods are required to permit such analyses to be carried out in clinical studies. These methods should be based on separations of molecules, e.g., in chromatographic or MS systems. Developments in liquid chromatography and tandem MS are promising, but gas chromatography has also been much improved. The combination of chromatographic and MS methods is very powerful, but simplifications and computerized interpretation are needed for future applications in clinical research. The analytical methods should take into account the existence of labile structures, oxo compounds, fatty-acid esters, and other derivatives of bile acids that are frequently lost in the common methods used in clinical studies. This may be one reason for discrepancies between isotope kinetic and chemical analyses. The possibility of degradation of the steroid skeleton should not be completely discounted, and the products of degradation of cholesterol and plant sterols, if occurring, should be identified.

It will also be important to study the physical form in which the different chemical structures exist. It is of obvious interest to know whether or not a compound is in solution in an aqueous phase, adsorbed to fibers or bacterial surfaces, or present inside the bacteria. These conditions will influence the potential for absorption and interaction with cells in the intestinal tract.

Many controversies may be resolved with an increased control of experimental conditions. The problems to be studied should be carefully defined and individual factors isolated as far as possible. This is obviously very difficult in studies of patients, but it is also clear that many apparent discrepancies are due to lack of control of one or several factors influencing the fecal excretion and analysis of bile acids.

ACKNOWLEDGMENT. The authors wish to acknowledge the invaluable secretarial assistance of Mary Jo McCarthy in the preparation of the manuscript and particularly in the compilation of the tables and references to the text. Without her patience and careful help one of us (K.D.R.S.) would have undoubtedly gone insane before completion of the manuscript.

REFERENCES

1. H. Danielsson, *Adv. Lipid Res.* **1**, 335 (1963).
2. H. Gordon, B. Lewis, L. Eales, and J. F. Brock, *Lancet* **2**, 1299 (1957).
3. B. Lewis, *J. Lab. Clin. Med.* **3**, 316 (1957).
4. G. L. Curran, D. L. Azarnoff, and R. E. Bollinger, *J. Clin. Invest.* **38**, 1251 (1959).
5. H. Engelberg, *Proc. Soc. Exp. Biol. Med.* **102**, 365 (1959).
6. G. A. Goldsmith, J. G. Hamilton, and O. N. Miller, *Arch. Intern. Med.* **105**, 512 (1960).
7. S. M. Grundy, E. H. Ahrens, Jr., and T. A. Miettinen, *J. Lipid Res.* **6**, 397 (1965).
8. H. G. Roscoe and M. J. Fahrenbach, *Anal. Biochem.* **6**, 520 (1963).
9. O. N. Miller, J. G. Hamilton, and G. A. Goldsmith, *Am. J. Clin. Nutr.* **10**, 285 (1962).
10. E. H. Mosbach, H. J. Kalinsky, E. Halpern, and F. E. Kendall, *Arch. Biochem. Biophys.* **51**, 402 (1954).
11. R. C. Powell, W. T. Nones, R. S. Harding, and J. B. Vacca, *Am. J. Clin. Nutr.* **11**, 156 (1962).
12. L. Hellman, R. S. Rosenfeld, W. Insull, Jr., and E. H. Ahrens Jr., *J. Clin. Invest.* **36**, 898 (1957).
13. S. Lindstedt, *Clin. Chim. Acta* **7**, 1 (1962).
14. R. S. Rosenfeld and L. Hellman, *Arch. Biochem. Biophys.* **97**, 406 (1962).
15. S. Lindstedt and E. H. Ahrens, Jr., *Proc. Soc. Exp. Biol. Med.* **108**, 286 (1961).
16. S. M. Grundy and E. H. Ahrens, *J. Clin. Invest.* **45**, 1503 (1966).
17. B. J. Kudchockar, H. S. Sodhi, and L. Horlick, *Clin. Chim. Acta* **41**, 47 (1972).
18. S. S. Ali, A. Kuksis, and J. M. R. Beveridge, *Can. J. Biochem.* **44**, 957 (1966).

19. E. Evrard and G. Janssen, *J. Lipid Res.* **9**, 226 (1968).
20. P. Eneroth, K. Hellström, and J. Sjövall, *Acta Chem. Scand.* **22**, 1729 (1968).
21. K. D. R. Setchell, A. M. Lawson, N. Tanida, and J. Sjövall, *J. Lipid Res.* **24**, 1085 (1983).
22. T. A. Miettinen, E. H. Ahrens, and S. M. Grundy, *J. Lipid Res.* **6**, 411 (1965).
23. A. Rubulis, M. Rubert, and W. W. Faloon, *Am. J. Clin. Nutr.* **23**, 1251 (1970).
24. M. J. Hill and V. C. Aries, *J. Pathol.* **104**, 129 (1971).
25. M. T. R. Subbiah, *J. Lipid Res.* **14**, 692 (1973).
26. B. S. Reddy, J. H. Weisburger, and E. L. Wynder, *J. Nutr.* **105**, 878 (1975).
27. B. S. Reddy, A. Mastromarino, C. Gustafson, M. Lipkin, and E. L. Wynder, *Cancer* **38**, 1694 (1976).
28. M. T. R. Subbiah, N. E. Tyler, M. D. Buscaglia, and L. Marai, *J. Lipid Res.* **17**, 78 (1976).
29. B. S. Reddy, C. W. Martin, and E. L. Wynder, *Cancer Res.* **37**, 1697 (1977).
30. B. I. Cohen, R. F. Raicht, G. Salen, and E. H. Mosbach, *Anal. Biochem.* **64**, 567 (1975).
31. C. B. Campbell and W. E. McIvor, *Pathology* **7**, 157 (1975).
32. M. Takahashi, R. F. Raicht, A. N. Sarwal, E. H. Mosbach, and B. I. Cohen, *Anal. Biochem.* **87**, 594 (1978).
33. S. J. Gordon, L. J. Miller, L. J. Haeffner, M. D. Kinsey, and O. D. Kowlessar, *Gut* **17**, 58 (1976).
34. M. T. Podesta, G. M. Murphy, G. E. Sladen, N. F. Breuer, R. H. Dowling, in "Biological Effects of Bile Acids" (G. Paumgartner, A. Stiehl, and W. Gerok, eds.), p. 245, MTP Press, Lancaster, (1978).
35. M. T. Podesta, G. M. Murphy, and R. H. Dowling, *J. Chromatogr.* **182**, 293 (1980).
36. C. T. Huang, W. E. Woodward, R. B. Hornick, J. T. Rodriguez, and B. L. Nichols, *Am. J. Clin. Nutr.* **29**, 949 (1976).
37. C. T. Huang, J. N. Udall, M. Merson, and B. L. Nichols, *Am. J. Clin. Nutr.* **31**, 626 (1978).
38. H. Schlenk and J. L. Gellerman, *Anal. Chem.* **32**, 1412 (1960).
39. W. D. Mitchell, J. M. Findlay, R. J. Prescott, M. A. Eastwood, and D. B. Horn, *Gut* **14**, 348 (1973).
40. J. A. Goy, M. A. Eastwood, W. D. Mitchell, J. L. Pritchard, and A. N. Smith, *Am. J. Clin. Nutr.* **29**, 1480 (1976).
41. M. S. Sian and A. J. H. Rains, *Postgrad. Med. J.* **55**, 180 (1979).
42. J. H. Cummings, H. S. Wiggins, D. J. Jenkins, H. Houston, T. Jivraj, B. S. Drasar, and M. J. Hill, *J. Clin. Invest.* **61**, 953 (1978).
43. O. J. Roseleur and C. M. Van Gent, *Clin. Chim. Acta* **82**, 13 (1978).
44. P. P. Nair, M. Gordon, and J. Reback, *J. Biol. Chem.* **242**, 7 (1967).
45. P. Eneroth, B. Gordon, R. Ryhage, and J. Sjövall, *J. Lipid Res.* **7**, 511 (1966).
46. P. Eneroth, B. Gordon, and J. Sjövall, *J. Lipid Res.* **7**, 524 (1966).
47. K. D. R. Setchell, in "Bile Acids in Gastroenterology" (L. Barbara, R. H. Dowling, A. F. Hofmann, and E. Roda, eds.), p. 1, MTP Press, Lancaster, (1982).
48. B. Almé, A. Bremmelgaard, J. Sjövall, and P. Thomassen, *J. Lipid Res.* **18**, 339 (1977).
49. K. D. R. Setchell, B. Almé, M. Axelson, and J. Sjövall, *J. Steroid Biochem.* **7**, 615 (1976).
50. A. Norman, *Br. J. Nutr.* **18**, 173 (1964).
51. A. Norman and R. H. Palmer, *J. Lab. Clin. Med.* **63**, 986 (1964).
52. K. D. R. Setchell, A. M. Lawson, J. Sjövall, and N. Tanida, Presented at Falk Symposium No. 29. Bile Acids and Lipids, Falk Foundation, Freiburg, West Germany, p. 71 (1982) (Abstr.).
53. G. A. De Weerdt, R. Beke, and F. Barbier, *Biomed. Mass Spectrom.* **7**, 515 (1980).
54. N. Tanida, Y. Hikasa, M. Hosomi, M. Satomi, I. Oohama, and T. Shimoyama, *Gastroenterol. Jpn.* **16**, 363 (1981).

55. N. Breuer, P. Dommes. R. Tandon, and H. Goebell, *J. Clin. Chem. Clin. Biochem.* **22,** 623 (1984).
56. N. Turjman and C. Jacob, *Am. J. Clin. Nutr.* **40,** 957 (1984).
57. R. W. Owen, M. H. Thompson, and M. J. Hill, *J. Steroid Biochem.* **21,** 593 (1984).
58. J. Sjövall, P. Eneroth, and R. Ryhage, *in* "The Bile Acids Chemistry, Physiology and Metabolism" (P. P. Nair and D. Kritchevsky, eds.), Volume 1, p. 209, Plenum Press, New York (1971).
59. W. H. Elliott, *in* "Biomedical Applications of Mass Spectrometry" (G. R. Waller, ed.), p. 291, Wiley-Interscience, New York, London (1972).
60. P. A. Szczepanik, D. L. Hachey, and P. D. Klein, *J. Lipid Res.* **17,** 314 (1976).
61. W. H. Elliott, *in* "Biomedical Applications of Mass Spectrometry, First Supplementary Volume" (G. R. Waller and O. C. Dermer, eds.), p. 229, Wiley-Interscience, New York, London (1980).
62. J. Sjövall, A. M. Lawson, and K. D. R. Setchell, *in* "Methods in Enzymology" (J. H. Law and H. C. Rilling, eds.), Volume III, Academic Press, New York, p. 63 (1985).
63. K. D. R. Setchell, and A. M. Lawson, *in* "Mass Spectrometry—Applications in Clinical Biochemistry" (A. M. Lawson, ed.), Walter-de-Gruyter, (1987) (in press).
64. G. A. De Weerdt, H. Verdievel, R. Beke, and F. Barbier, *in* "Proceedings of First International Symposium on Bile Acids in Hepatobiliary and Gastrointestinal Diseases," p. 185, IRL Press, Oxford (1984).
65. R. Tandon, M. Axelson, and J. Sjövall, *J. Chromatogr.* **302,** 1 (1984).
66. K. D. R. Setchell, *in* "Hormones in Blood" (C. H. Gray, V. H. T. James, eds.), Volume 3, p. 63, Academic Press, New York, London (1979).
67. W. C. Duane, J. C. Behrens, S. G. Kelly, and A. S. Levine, *J. Lipid Res.* **25,** 523 (1984).
68. J. F. Woodbury and F. Kern, Jr., *J. Clin. Invest.* **50,** 2531 (1971).
69. M. J. Sheltawy and M. S. Losowsky, *Clin. Chim. Acta* **64,** 127 (1975).
70. J. De Wael, C. E. Raaymakers, and H. J. Endeman, *Clin. Chim. Acta* **79,** 465 (1977).
71. R. Ferraris, R. Carosso, M. T. Fiorentini, M. De La Pierre, M. Marcellino, M. Eandi and S. R. Gamalero, *Boll. Soc. Int. Biol. Sper.* **56,** 15 (1980).
72. M. J. Crowell and I. MacDonald, *J. Lipid Res.* **26,** 1298 (1980).
73. W. T. Beher, S. Stradnieks, G. T. Lin, and J. Sanfield, *Steroids* **38,** 281 (1981).
74. A. Malchow-Møller, S. Arffmann, N. La Russo, and E. Krag, *Scand. J. Gastroenterol.* **17,** 331 (1982).
75. W. T. Beher, S. Stradnieks, and G. J. Lin, *Steroids* **41,** 729 (1983).
76. A. M. Weber, L. Chartrand, G. Doyan, S. Gordon, and C. C. Roy, *Clin. Chim. Acta* **39,** 524 (1972).
77. W. V. Reimold and R. Katterman, *Gastroenterologie* **12,** 341 (1974).
78. H-D. Kalek, F. Stellaard, W. Kruis, and G. Paumgartner, *Clin. Chim. Acta* **140,** 85 (1984).
79. T. Iwata and K. Yamasaki, *J. Biochem.* **56,** 424 (1964).
80. F. Mashige, N. Tanaka, A. Maki, S. Kamei, and M. Yamanaka, *Clin. Chem.* **27,** 1352 (1981).
81. K. D. R. Setchell, J. A. Ives, G. C. Cashmore, and A. M. Lawson, *Clin. Chim. Acta* **162,** 257 (1987).
82. I. A. MacDonald, C. N. Williams, and D. E. Mahony, *J. Lipid Res.* **16,** 244 (1975).
83. I. A. MacDonald, J. F. Jellett, and D. E. Mahony, *J. Lipid Res.* **20,** 234 (1979).
84. T. Nambara and J. Goto, *in* "The Bile Acids: Methods and Applications" (K. D. R. Setchell, D. Kritchevsky, and P. P. Nair, eds.), Volume 4, Plenum Press, New York, pp. 43–64 (1988).
85. B. Shaikh, N. J. Pontzer, J. E. Molina, and M. I. Kelsey, *Anal. Biochem.* **85,** 47 (1978).
86. F. Stellaard, J. L. Hachey, and P. D. Klein, *Anal. Biochem.* **87,** 359 (1978).
87. G. Mingrone, A. V. Greco, and S. Passi, *J. Chromatogr.* **183,** 277 (1980).
88. S. Okuyama, D. Uemura, and Y. Hirata, *Chem. Lett.* **1979,** 461.

89. J. Goto, S. Komatsu, N. Goto, and T. Nambara, *Chem. Pharm. Bull.* **29,** 899 (1981).
90. J. Goto, N. Goto, F. Shamsa, M. Saito, S. Komatsu, K. Suzaki, and T. Nambara, *Anal. Clin. Acta* **147,** 397 (1983).
91. S. Kamada, M. Maeda, and A. Tsuji, *J. Chromatogr.* **272,** 29 (1983).
92. M. I. Kelsey, K-K. Hwang, S-K. S. Huang, and B. Shaikh, *J. Steroid Biochem.* **14,** 205 (1981).
93. L. R. Ferguson, G. W. Rewcastle, J. M. Lello, P. G. Alley, and R. N. Seelye, *Anal. Biochem.* **143,** 325 (1984).
94. S. Lindstedt, *Acta Physiol. Scand.* **40,** 1 (1957).
95. S. Lindstedt and A. Norman, *Acta Physiol. Scand.* **38,** 121 (1956).
96. R. M. Donaldson, Jr., and R. F. Barreras, *J. Lab. Clin. Med.* **68,** 484 (1966).
97. J. Davignon, W. J. Simmonds, and E. H. Ahrens, Jr., *J. Clin. Invest.* **47,** 127 (1968).
98. L. G. Whitby and D. Lang, *J. Clin. Invest.* **39,** 854 (1960).
99. W. E. Meihoff and F. Kern, Jr., *J. Clin. Invest.* **47,** 261 (1968).
100. M. M. Stanley and B. Nemchausky, *J. Lab. Clin. Med.* **70,** 627 (1967).
101. W. I. Austad, M. P. Tyor, and L. Lack, *Gastroenterology* **52,** 638 (1967).
102. A. F. Hofmann, *Gastroenterology* **52,** 752 (1967).
103. K. Lenz, *Digestion* **15,** 147 (1977).
104. H. Fromm and A. F. Hofmann, *Lancet* **2,** 621 (1971).
105. M. H. Griffiths and A. Mallinson, *Anal. Biochem.* **22,** 465 (1968).
106. J. H. B. Scarpello and G. E. Sladen, *Gut* **18,** 742 (1977).
107. P. Horchner, M. van Blakenstein, J. Boot, M. Frenkel, R. de Groot, T. Hoyset, E. P. Jacobs, and A. J. Looy, *Folia Med. Neerl.* **15,** 186 (1972).
108. G. S. Boyd, M. V. Merrick, R. Monks, and I. L. Thomas, *J. Nucl. Med.* **22,** 720 (1981).
109. H. Delhez, J. W. O. Van den Berg, M. Van Blankenstein, and J. H. Meerwaldt, *Eur. J. Nucl. Med.* **7,** 269 (1982).
110. H. Nyhlin, M. V. Merrick, M. A. Eastwood, and W. G. Brydon, *Gastroenterology* **84,** 63 (1983).
111. H. Nyhlin, G. Brydon, A. Danielsson, and S. Westman, *Hepatogastroenterology* **31,** 187 (1984).
112. M. T. R. Subbiah, *Ann. Clin. Lab. Sci.* **3,** 362 (1973).
113. T. N. Tangedahl, A. F. Hofmann, and B. A. Kottke, *J. Lipid Res.* **20,** 125 (1979).
114. N. O. Davidson, P. Samuel, S. Lieberman, S. P. Shane, J. R. Crouse, and E. H. Ahrens, Jr., *J. Lipid Res.* **22,** 620 (1981).
115. W. C. Duane, D. E. Holloway, S. W. Hutton, D. J. Corcoran, and N. A. Haas, *Lipids* **17,** 345 (1982).
116. W. C. Duane, R. D. Adler, L. J. Bennion, and R. L. Ginsberg, *J. Lipid Res.* **16,** 155 (1975).
117. S. M. Grundy, E. H. Ahrens, Jr., and G. Salen, *J. Lipid Res.* **9,** 374 (1968).
118. S. M. Grundy and E. H. Ahrens, Jr., *J. Lipid Res.* **10,** 91 (1969).
119. N. O. Davidson, H. L. Bradlow, E. H. Ahrens, Jr., R. S. Rosenfeld, and C. C. Schwartz, *J. Lipid Res.* **27,** 183 (1986).
120. A. F. Hofmann and S. A. Cummings, *in* "Bile Acids in Gastroenterology" (L. Barbara, R. H. Dowling, A. F. Hofmann, and E. Roda, eds.), p. 75, MTP Press, Lancaster, (1982).
121. R. W. St. Clair, N. D. M. Lehner, and T. E. Hamm, *Lipids* **10,** 25 (1975).
122. K. Hellström and J. Sjövall, *J. Lipid Res.* **3,** 397 (1962).
123. J. Sjövall, and K. D. R. Setchell, *in* "The Bile Acids: Methods and Applications" (K. D. R. Setchell, D. Kritchevsky, and P. P. Nair, eds.), Volume 4, Plenum Press, New York (1988).
124. D. R. Wybenga and J. A. Inkpen, *in* "Clinical Chemistry" (R. J. Henry, D. C. Cannon, and J. W. Winkelman, eds.), p. 1421, Harper and Row, New York, London (1974).

125. H. J. Eyssen, G. G. Parmentier, and J. A. Mertens, *Eur. J. Biochem.* **66**, 507 (1976).
126. A. van den Ende, C. E. Radecker, W. M. Mairuhu, and A. P. van Zanten, *Clin. Chim. Acta* **121**, 95 (1982).
127. B. E. Gustafsson, and A. Norman, *Br. J. Nutr.* **23**, 429 (1969).
128. T. Midtvedt and A. Norman, *Acta Pathol. Microbiol. Scand* **80**, 202 (1972).
129. F. Kern Jr., J. J. Birkner, and V. S. Ostrower, *Am. J. Clin. Nutr.* **31**, S175 (1978).
130. G. W. Kuron and D. M. Tennent, *Fed. Proc.* **20**, 268 (1961).
131. J. D. Manes and D. L. Schneider, *J. Lipid Res.* **12**, 376 (1971).
132. D. H. Sandberg, J. Sjövall, and D. A. Turner, *J. Lipid Res.* **6**, 182 (1965).
133. W. S. Harris, M. J. J. Myher, and M. T. R. Subbiah, *J. Chromatogr.* **131**, 437 (1977).
134. M. I. Kelsey and S. A. Sexton, *J. Steroid Biochem.* **7**, 641 (1976).
135. M. I. Kelsey and S. A. Sexton, *J. Chromatogr.* **133**, 327 (1977).
136. M. I. Kelsey, J. E. Molina, S.-K. S. Huang, and K.-K. Huang, *J. Lipid Res.* **21**, 751 (1980).
137. B. Egestad, T. Curstedt, and J. Sjövall, *Anal. Lett.* **15**, 243 (1982).
138. K. D. R. Setchell and A. Matsui, *Clin. Chim. Acta* **127**, 1 (1983).
139. A. Antonis and I. Bersohn, *Am. J. Clin. Nutr.* **11**, 142 (1962).
140. J. Dupont, S. Y. Oh, and P. Janson, *in* "The Bile Acids" (P. P. Nair and D. Kritchevsky, eds.), Volume 3, p. 17, Plenum Press, New York (1976).
141. G. P. Van Berge Henegouwen, R. N. Allan, A. F. Hofmann, and P. Y. S. Yu, *J. Lipid Res.* **18**, 118 (1977).
142. I. Makino and J. Sjövall, *Anal. Lett.* **5**, 341 (1972).
143. K. D. R. Setchell and J. Worthington, *Clin. Chim. Acta* **125**, 135 (1982).
144. A. Dyfverman and J. Sjövall, *in* "Biological Effects of Bile Acids" (G. Paumgartner, A. Stiehl, and W. Gerok, eds.), p. 281, MTP Press, Lancaster, (1978).
145. A. Dyfverman and J. Sjövall, *Anal. Lett.* **6**, 485 (1978).
146. A. Dyfverman and J. Sjövall, *Anal. Biochem.* **134**, 303 (1983).
147. P. Eneroth and J. Sjövall, *in* "The Bile Acids, Chemistry, Physiology and Metabolism" (P. P. Nair and D. Kritchevsky, eds.), Volume 1, p. 121, Plenum, Press, New York (1971).
148. O. J. Roseleur and C. M. Van Gent, *Clin. Chim. Acta* **66**, 269 (1976).
149. G. Lepage, A. Fontaine, and C. C. Roy, *J. Lipid Res.* **19**, 505 (1978).
150. G. G. Parmentier and H. Eyssen, *Steroids* **26**, 721 (1975).
151. B. I. Cohen, K. Budai, and N. B. Javitt, *Steroids* **37**, 6321 (1981).
152. L. Kornel, *Biochemistry* **4**, 444 (1965).
153. R. Galeazzi and N. B. Javitt, *J. Clin. Invest.* **60**, 693 (1977).
154. R. Shaw and W. H. Elliott, *J. Lipid Res.* **19**, 783 (1978).
155. A. van Faassen, F. M. Nagengast, M. Hectors, W. J. M. van den Broek, A. W. M. Huijbregts, S. D. J. van der Werf, G. P. van Berge Henegouwen, and J. H. M. van Tongeren, *Clin. Chim. Acta.* **152**, 231 (1985).
156. S. Burstein and S. Lieberman, *J. Biol. Chem.* **233**, 331 (1958).
157. J. M. Street, D. J. H. Trafford, and H. L. J. Makin, *J. Lipid Res.* **24**, 491 (1983).
158. S. Lindstedt, *Arkiv Kemi* **11**, 145 (1957).
159. J. B. Carey and C. J. Watson, *J. Biol. Chem.* **216**, 847 (1963).
160. E. Heftmann, E. Weiss, H. K. Miller, and E. Mosettig, *Arch. Biochem. Biophys.* **84**, 324 (1959).
161. H. Danielsson, P. Eneroth, K. Hellström, S. Lindstedt, and J. Sjövall, *J. Biol. Chem.* **238**, 2299 (1963).
162. M. Jenke and F. Bandow, *Z. Physiol. Chem.* **249**, 16 (1937).
163. J. G. Hamilton, *Arch. Biochem. Biophys.* **101**, 7 (1963).
164. K. D. R. Setchell, J. M. Gilbert, and A. M. Lawson, *Br. Med. J.* **286**, 1750 (1983).
165. N. Tanida, Y. Hikasa, T. Shimoyama, and K. D. R. Setchell, *Gut* **25**, 824 (1984).

166. B. E. Gustafsson, T. Midtvedt, and A. Norman, *J. Exp. Med.* **123**, 413 (1966).
167. I. A. MacDonald, V. D. Bokkenheuser, J. Winter, A. M. McLernon, and E. H. Mosbach, *J. Lipid Res.* **24**, 675 (1983) (Review).
168. P. B. Hylemon, *in* "Sterols and Bile Acids" (H. Danielsson, and J. Sjövall, eds.), p. 331, Elsevier, Amsterdam, (1985).
169. Y. Hikasa, N. Tanida, T. Ohno, and T. Shimoyama, *Gut* **25**, 833 (1984).
170. J. T. Korpela, T. Fotsis, and H. Adlercreutz, *J. Steroid Biochem.* **25**, 277 (1986).
171. H. F. Mower, R. M. Ray, G. N. Stemmermann, A. Nomura, and G. A. Glober, *J. Nutr.* **108**, 1289 (1978).
172. D. T. Forman, C. Phillips, W. Eiseman, and C. B. Taylor, *Clin. Chem.* **14**, 348 (1968).
173. B. S. Reddy and E. L. Wynder, *J. Natl. Cancer Inst.* **50**, 1437 (1973).
174. C. T. Huang, J. T. Rodriguez, W. E. Woodward, and B. L. Nichols, *Am. J. Clin. Nutr.* **29**, 1196 (1976).
175. E. L. Wynder, *Cancer* **43**, 1955 (1979).
176. P. W. Bilheimer, N. J. Stone, and S. M. Grundy, *J. Clin. Invest.* **64**, 524, (1979).
177. B. S. Reddy, D. Hanson, S. Marget, L. Matthews, M. Sibaschnig, C. Sharma, and B. Sim, *J. Nutr.* **110**, 1880 (1980).
178. R. H. Palmer, *Proc. Natl. Acad. Sci. USA* **58**, 1047 (1967).
179. R. H. Palmer and M. G. Bolt, *J. Lipid Res.* **12**, 671 (1971).
180. M. A. Islam, R. F. Raicht, and B. I. Cohen, *Anal. Biochem.* **112**, 371 (1981).
181. G. Salvioli and R. Salati, *Gut* **20**, 698 (1979).
182. N. F. Breuer, P. Dommes, S. Jaekel, and H. Goebell, *Dig. Dis. Sci.* **30**, 852 (1985).
183. L. S. Mata, M. L. Mejicanos, and F. Jimenez, *Am. J. Clin. Nutr.* **25**, 1380 (1972).
184. H. L. Sharp, J. Peller, J. B. Carey, Jr., and W. Krivit. *Pediatr. Res.* **5**, 274 (1971).
185. J. C. Encrantz and J. Sjövall, *Clin. Chim. Acta* **4**, 793 (1959).
186. J. R. Poley, J. C. Dower, C. A. Owen, Jr., and G. B. Stickler, *J. Lab. Clin. Med.* **64**, 838 (1964).
187. A. M. Bongiovanni, *J. Clin. Endocrinol.* **25**, 678, (1965).
188. A. Norman, B. Strandvik and O. Ojamäe, *Acta Pediatr. Scand.* **61**, 571 (1972).
189. Y. Tazawa, Y. Masaaki, M. Nakagawa, H. Suzuki, Y. Igarashi, T. Konno, and K. Tada, *J. Pediatr. Gastroenterol. Nutr.* **3**, 378 (1984).
190. J. M. Potter and P. J. Nestel, *Am. J. Clin. Nutr.* **29**, 546 (1976).
191. A. J. Finley and M. Davidson, *Pediatrics* **65**, 132 (1980).
192. L. Zimmer-Nechemias, W. F. Balistreri, and K. D. R. Setchell, *Gastroenterology* **92**, 1795 (1987) (Abstr.).
193. P. T. Clayton, D. P. R. Muller, and A. M. Lawson, *Biochem. J.* **206**, 489 (1982).
194. A. M. Lawson, K. D. R. Setchell, P. T. Clayton, and D. P. R. Muller, *Proceeding of 31st Annual Conference on Mass Spectrometry and Allied Topics*, American Society of Mass Spectrometry, Boston, p. 510 (1983).
195. P. J. Nestel, A. Poyser, and T. J. Boulton, *Am. J. Clin. Nutr.* **32**, 2177 (1979).
196. C. Colombo, A. Roda, E. Roda, L. Piceni-Sereni, D. Maspero, A. M. Giunta, and L. Barbara, *Dig. Dis. Sci.* **28**, 306 (1983).
197. A. M. Weber, C. C. Roy, C. L. Morin, and R. Lasalle, *N. Engl. J. Med.* **289**, 1001 (1973).
198. A. M. Weber, C. C. Roy, L. Chartrand, G. Lepage, O. L. Dufour, C. L. Morin, and R. Lasalle, *Gut* **17**, 295 (1976).
199. J. B. Watkins, A. M. Tercyak, P. Szczepanik, and P. D. Klein, *Gastroenterology* **73**, 1023 (1977).
200. B. J. Boyle, W. B. Long, W. F. Balistreri, S. J. Widzer, and N. Huang, *Gastroenterology* **78**, 950 (1980).

201. J. E. Heubi, W. F. Balistreri, J. C. Partin, W. K. Schubert, and F. J. Suchy, *J. Lab Clin. Med* **95**, 231 (1980).
202. J. D. Fondacaro, J. E. Heubi, and F. W. Kellogg, *Pediatr. Res.* **16**, 494 (1982).
203. J. E. Heubi, W. F. Balistreri, J. D. Fondacaro, J. C. Partin, and W. K. Schubert, *Gastroenterology* **83**, 804 (1982).
204. J. H. Zavoral, D. C. Laine, L. K. Bale, D. L. Weelik, R. D. Ellefson, K. Kuba, W. Krivit, and B. A. Kottke, *Am. J. Clin. Nutr.* **35**, 1360 (1982).
205. B. Lewis and N. B. Myant, *Clin. Sci.* **32**, 201 (1967).
206. C. Leyland, *Arch. Dis. Child.* **45**, 714 (1970).
207. K. S. Shurpalekar, T. R. Doraiswamy, O. E. Sundaravalli, and M. N. Rao, *Nature* **232**, 554 (1971).
208. S. M. Grundy, E. H. Ahrens, Jr., G. Salen, P. H. Schreibman, and P. J. Nestel, *J. Lipid Res.* **13**, 531 (1972).
209. P. J. Nestel, *J. Clin. Endocrinol. Metab.* **38**, 325 (1974).
210. C. C. Roy, M. Ste. Marie, L. Chartrand, A. Weber, H. Bard, and B. Doray, *J. Pediatr.* **86**, 446 (1975).
211. G. A. Carter, W. E. Connor, and A. K. Bhattachayra, *Circulation* **51**, 11 (1975).
212. H. Gaze, G. M. Murphy, R. Nelson, J. J. Corkery, and C. M. Anderson, *Arch. Dis. Child.* **50**, 243 (1975).
213. M. C. Goodchild, G. M. Murphy, A. M. Howell, S. A. Nutter, and C. M. Anderson, *Arch. Dis. Child.* **50**, 769 (1975).
214. R. J. Roller and F. Kern, *Gastroenterology* **72**, 661 (1977).
215. C. A. Smalley, G. A. Brown, M. E. Parkes, H. Tease, V. Brookes, and C. M. Anderson, *Arch. Dis. Child.* **53**, 477 (1978).
216. A. Jonas, S. Avigad, A. Diver-Haber, and D. Katznelson, *J. Pediatr.* **95**, 366 (1979).
217. G. M. Martin and P. J. Nestel, *Clin. Sci.* **56**, 377 (1979).
218. K. B. Schwarz, J. Witzum, G. Schonfeld, S. M. Grundy, and W. E. Connor, *J. Clin. Invest.* **64**, 745 (1979).
219. C. Colombo, A. Roda, E. Roda, L. Piceni-Sereni, A. Brega, R. Fugazza, and A. Giunta, *J. Pediatr. Gastroenterol. Nutr.* **3**, 556 (1984).
220. P. B. Darling, G. Lepage, C. Leroy, P. Masson, and C. C. Roy, *Pediatr. Res.* **19**, 578 (1985).
221. K. D. R. Setchell, P. Smethurst, A. M. Giunta, and C. Colombo, *Clin. Chim Acta* **151**, 297 (1985).
222. P. Back and K. Ross, *Hoppe-Seyler's Z. Physiol. Chem.* **354**, 83 (1973).
223. J. St. Pyrek, R. Lester, E. W. Adcock, and A. T. Sanghvi, *J. Steroid Biochem.* **18**, 341 (1983).
224. J. B. Watkins, and E. R. Brown, *in* "Bile Acid Metabolism in Health and Disease" (G. Paumgartner, and A. Stiehl, eds.), p. 65, MTP Press, Lancaster (1977).
225. T. J. Laatikainen, P. J. Lehtonen, and A. E. Hesso, *J. Lab Clin. Med.* **91**, 185 (1978).
226. R. Lester, J. St. Pyrek, E. W. Adcock, A. Grinberg, J. Boros, and A. Sanghvi, *Gastroenterology* **79**, 1034 (1980) (Abstr.).
227. P. Back and K. Walter, *Gastroenterology* **78**, 671 (1980).
228. J. St. Pyrek, R. Sterzycki, R. Lester, and E. W. Adcock, *Lipids* **17**, 241 (1982).
229. J. Gustafsson, S. Andersson, and J. Sjövall, *Biol. Neonate* **47**, 26 (1985).
230. L. Lack and I. M. Weiner, *in* "The Bile Acids, Chemistry, Physiology and Metabolism" (P. P. Nair, and D. Kritchevsky, eds.), Volume 2, p. 33, Plenum Press, New York, London (1973).
231. A. F. Hofmann, *Clin. Gastroenterol.* **6**, 3 (1977).
232. G. P. Van Berge Henegouwen and A. F. Hofmann, *Eur. J. Clin. Invest.* **13**, 433 (1983).

233. D. M. Small, *in* "The Bile Acids, Chemistry, Physiology and Metabolism" (P. P. Nair and D. Kritchevsky, eds.), Volume 1, p. 249, Plenum Press, New York, London (1971).
234. M. C. Carey and H. Igimi *in* "Bile Acids and Lipids" (G. Paumgartner, A. Stiehl, and W. Gerok eds.), p. 123, MTP Press, Lancaster (1981).
235. M. C. Carey, *in* "Sterols and Bile Acids" (H. Danielsson and J. Sjövall, eds.), p. 345, Elsevier, Amsterdam (1985).
236. J. M. Dietschy, H. S. Salomon, and M. D. Siperstein, *J. Clin. Invest.* **45**, 832 (1966).
237. L. Lack and I. M. Wiener, *Am. J. Physiol.* **200**, 313 (1961).
238. W. G. Hardison and S. M. Grundy, *Gastroenterology* **84**, 617 (1983).
239. T. C. Northfield, and I. McColl, *Gut* **14**, 513 (1973).
240. G. W. Hepner, J. A. Sturman, A. F. Hofmann, and P. J. Thomas, *J. Clin. Invest.* **52**, 433 (1973).
241. M. Vuoristo and T. A. Miettinen, *Eur. J. Clin. Invest.* **12**, 285 (1982).
242. P. Samuel, G. M. Saypol, E. Meilman, E. H. Mosbach, and M. Chafizadeh, *J. Clin. Invest.* **47**, 2070 (1968).
243. H. S. Mekhjian, S. F. Phillips, and A. F. Hofmann, *Dig. Dis. Sci.* **24**, 545 (1979).
244. K. D. R. Setchell, A. M. Lawson, E. J. Blackstock, and G. M. Murphy, *Gut* **23**, 637 (1982).
245. B. McJunkin, H. Fromm, R. P. Sarva, and P. Amin, *Gastroenterology* **80**, 1454 (1981).
246. F. Kern Jr., J. J. Woodbury, and W. E. Meihoff, *Tr. Am. Clin. Climatol. Assoc.* **82**, 61 (1970) (Review).
247. C. T. Huang, M. M. Levine, G. S. Daoud, D. R. Nalin, and B. L. Nichols, *Am. J. Clin. Nutr.* **33**, 40 (1980).
248. C. T. Huang, M. M. Levine, G. S. Daoud, D. R. Nalin, and B. L. Nichols, *Lipids* **17**, 612 (1982).
249. W. C. Duane, *J. Lab. Clin. Med.* **91**, 969 (1978).
250. W. G. M. Hardison, N. Tomaszewski, and S. M. Grundy, *Gastroenterology* **76**, 568 (1979).
251. S. Hayakawa, *Adv. Lipid Res.* **11**, 143 (1973) (Review).
252. G. L. Simon and S. L. Gorbach, *Gastroenterology* **86**, 174 (1984) (Review).
253. S. S. Long and R. M. Swenson, *J. Pediatr.* **91**, 298 (1977).
254. S. L. Gorbach and S. Tabaqchali, *Gut* **10**, 963 (1969).
255. S. Tabaqchali and C. C. Booth, *Lancet* **2**, 12 (1966).
256. I. H. Rosenberg, W. G. Hardison, and D. M. Bull, *N. Engl. J. Med.* **276**, 1391 (1967).
257. V. C. Aries and M. J. Hill, *Biochim. Biophys Acta* **202**, 535 (1970).
258. M. H. Floch, W. Gershengoren, S. Elliott, and H. M. Spiro, *Gastroenterology* **61**, 228 (1971).
259. W. D. Mitchell, J. M. Findlay, R. Macrae, M. A. Eastwood, and R. Anderson, *Digestion* **11**, 135 (1974).
260. R. Fiasse, H. J. Eyssen, J. P. Leonard, and C. H. Dive, *Eur. J. Clin. Invest.* **13**, 185 (1983).
261. W. D. Mitchell and M. A. Eastwood, *Scand J. Gastroenterol.* **7**, 29 (1972).
262. M. Flynn, J. Hyland, P. Hammond, C. Darby, and I. Taylor, *Br. J. Surg.* **67**, 629 (1980).
263. T. A. Miettinen, *Eur. J. Clin. Invest.* **1**, 452 (1971).
264. R. Aldini, A. Roda, D. Festi, C. Sama, G. Mazzella, F. Bazzoli, A. M. Morselli, E. Roda, and L. Barbara, *Dig. Dis. Sci.* **27**, 495 (1982).
265. T. Yoshida, W. C. McCormick III, L. Swell, and Z. R. Vlahcevic, *Gastroenterology* **68**, 335 (1975).
266. R. G. Knodell, M. Kinsey, E. C. Boedeker, and D. P. Collin, *Gastroenterology* **71**, 196 (1976).
267. Y. Amuro, T. Endo, K. Higashino, K. Uchida, and Y. Yamamura, *Gastroenterol. Jpn.* **16**, 506 (1981).
268. Y. Amuro, T. Endo, K. Higashino, K. Uchida, and Y. Yamamura, *Clin. Chim. Acta.* **114**, 137 (1981).

269. T. Endo, K. Uchida, Y. Amuro, K. Higashino, and Y. Yamamura, *Gastroenterology* **76**, 1002 (1979).
270. G. B. Blake, T. L. Kennedy, and S. T. K. McKelvey, *Br. J. Surg* **70**, 177 (1983).
271. G. J. Gleich and A. F. Hofmann, *Am. J. Med.* **51**, 281 (1971).
272. H. S. Mekhjian, S. F. Phillips, and A. F. Hofmann, *J. Clin. Invest.* **50**, 1569 (1971).
273. S. P. Borriello, "Bacteria and Gastrointestinal Secretion and Motility," *Scand. J. Gastroenterol.* **Suppl. 93**, 115–121 (1984).
274. H. Fromm, R. P. Sarva, and F. Bazzoli, *J. Lipid Res.* **24**, 841 (1983).
275. P. J. Nestel and J. D. Hunter, *Aust. NZ J. Med.* **4**, 491 (1974).
276. T. M. Mabee, P. Meyer, L. Den Besten, and E. E. Mason, *Surgery* **79**, 460 (1976).
277. K. Hellström and K. Einarsson, *Clin. Gastroenterol.* **6**, 103 (1977) (Review).
278. T. A. Miettinen, *in* "The Bile Acids, Chemistry, Physiology and Metabolism" (P. P. Nair and D. Kritchevsky, eds.), Volume 2, p. 191, Plenum Press, New York, London (1973).
279. H. S. Sodhi, B. J. Kudchodkar, and D. T. Mason, *Adv. Lipid Res.* **17**, 107 (1980) (Review).
280. M. M. Stanley, *Metabolism* **19**, 865 (1970).
281. T. A. Miettinen, *Circulation* **44**, 842 (1971).
282. L. J. Bennion and S. M. Grundy, *J. Clin. Invest.* **56**, 996 (1975).
283. S. D. J. Van der Werf, W. W. M. Huibregts, H. L. M. Lamers, G. P. Van Berge Henegouwen, and J. H. M. Van Tongeren, *Eur. J. Clin. Invest.* **11**, 425 (1981).
284. K. Einarsson, K. Nilsell, B. Leijd, and B. Angelin, *N. Engl. J. Med.* **313**, 277 (1985).
285. T. S. Low-Beer, *Clin. Gastroenterol.* **6**, 165 (1977) (Review).
286. R. Paul, C. S. Ramesha, and J. Ganguly, *Adv. Lipid Res.* **17**, 155 (1980) (Review).
287. T. A. Miettinen, *Adv. Lipid Res.* **18**, 65 (1981) (Review).
288. E. L. Wynder and T. Shigamatsu, *Cancer* **20**, 1520 (1967).
289. R. Doll, *Br. J. Cancer* **23**, 1 (1969).
290. E. L. Wynder, T. Kajitani, S. Ishikawa, H. Dodo, and A. Takano, *Cancer* **23**, 1210 (1969).
291. D. P. Burkitt, *Cancer* **28**, 3 (1971).
292. W. Haenszel, J. W. Berg, M. Segi, M. Kurihara, and F. B. Locke, *J. Natl. Cancer Inst.* **51**, 1765 (1973).
293. D. P. Burkitt, *J. Natl. Cancer Instit.* **54**, 3 (1975).
294. M. J. Hill, *in* "Colonic Carcinogenesis" (R. A. Malt and R. C. N. Williamson, eds.), p. 73, MTP Press, Lancaster (1981).
295. O. Gregor, R. Toman, and F. Prugova, *Gut* **10**, 1031 (1969).
296. M. J. Hill, *Digestion* **11**, 289 (1974).
297. M. J. Hill, *Cancer Res.* **35**, 3398 (1975) (Review).
298. A. R. P. Walker, *Am. J. Clin. Nutr.* **29**, 1417 (1976).
299. A. R. Walker and D. P. Burkitt, *Am. J. Dig. Dis.* **21**, 910 (1976).
300. D. Kritchevsky and J. A. Story, *Curr. Concepts Nutr.* **6**, 41 (1977).
301. E. L. Wynder and B. S. Reddy, *Curr. Concepts Nutr.* **6**, 55 (1977).
302. C. T. Huang, G. S. Gopalakrishna, and B. L. Nichols, *Am. J. Clin. Nutr.* **31**, 516 (1978) (Review).
303. G. A. Spiller, *in* "Dietary Fiber in Health and Disease" (G. Vahouny and D. Kritchevsky, eds.), p. 237, Plenum Press, New York (1982).
304. B. S. Reddy, *in* "Dietary Fiber in Health and Disease" (G. Vahouny and D. Kritchevsky, eds.), p. 265, Plenum Press, New York (1982).
305. D. Kritchevsky, *Cancer Res.* **43**, 2491s (1983).
306. T. Narisawa, T. Sato, M. Hayakawa, A. Sakuma, and H. Nakano, *Jpn. J. Cancer Res. (Gann)*, **62**, 231 (1971).
307. E. L. Wynder and B. S. Reddy, *J. Natl. Cancer Inst.* **50**, 1099 (1973).
308. B. S. Reddy. J. H. Weisburger, and E. L. Wynder, *J. Natl. Cancer Inst.* **52**, 507 (1974).

309. B. S. Reddy, S. Mangat, A. Sheinfil, J. H. Weisburger, and E. L. Wynder, *Cancer Res.* **37**, 2132 (1977).

310. T. Narisawa, B. S. Reddy, and J. H. Weisburger, *Gastroenterol. Jpn.* **13**, 206 (1978).

311. R. Abraham, T. A. Barbolt, and J. B. Rodgers, *Exp. Mol. Pathol.* **33**, 133, (1980).

312. B. I. Cohen and R. F. Raicht, *Cancer Res.* **41**, 3759 (1981).

313. V. Aries, J. S. Crowther, B. S. Drasar, M. J. Hill, and R. E. O. Williams, *Gut* **10**, 334 (1969).

314. M. J. Hill, J. S. Crowther, B. S. Drasar, G. Hawksworth, V. Aries, and R. E. O. Williams, *Lancet* **1**, 95 (1971).

315. M. J. Hill, *Am. J. Clin. Nutr.* **27**, 1475 (1974).

316. M. J. Hill, B. S. Drasar, R. E. O. Williams, T. W. Meade, A. G. Cox, J. E. P. Simpson, and B. C. Morson, *Lancet* **2**, 535 (1975).

317. M. J. Hill, *Cancer* **36**, 2387 (1975).

318. A. Mastromarino, B. S. Reddy, and E. L. Wynder, *Am. J. Clin. Nutr.* **29**, 1455 (1976).

319. M. J. Goldberg, J. W. Smith, and R. L. Nichols, *Ann. Surg.* **186**, 97 (1977).

320. D. Vargo, M. Moskovitz, and M. H. Floch, *Gut* **21**, 701 (1980).

321. S. P. Borriello, B. S. Drasar, A. Tomkins, and M. J. Hill, *J. Clin. Pathol.* **36**, 93 (1983).

322. B. S. Reddy and E. L. Wynder, *Cancer* **39**, 2533 (1977).

323. J. S. Crowther, B. S. Drasar, M. J. Hill, R. Maclennan, D. Magnin, S. Peach, and C. H. Teoh-Chan, *Br. J. Cancer* **34**, 191 (1976).

324. International Agency for Research on Cancer Intestinal Microecology Group, *Lancet* **2**, 207 (1977).

325. B. S. Reddy, A. Hedges, K. Laakso and E. L. Wynder, *Cancer Lett.* **4**, 217 (1978).

326. H. F. Mower, R. M. Ray, R. Shoff, G. N. Stemmermann, A. Nomura, G. A. Glober, S. Kamiyama, A. Shimada, and H. Yamakawa, *Cancer Res.* **39**, 328 (1979).

327. N. Turjman, C. Guidry, B. Jaeger, A. I. Mendeloff, B. Calkins, R. L. Phillips, and P. P. Nair, *in* "Colon and Nutrition" (H. Kasper and H. I. Goebell, eds.), p. 291, MTP Press, Lancaster (1981).

328. D. G. Mudd, S. T. McKelvey, J. M. Sloan, and D. T. Elmore, *Acta Gastroenterol Belg.* **41**, 241 (1978).

329. M. Moskovitz, C. White, R. N. Barnett, S. Stevens, E. Russell, D. Vargo, and M. H. Floch, *Dig. Dis. Sci.* **24**, 746 (1979).

330. D. G. Mudd, S. T. D. McKelvey, W. Norwood, D. T. Elmore, and A. D. Roy, *Gut* **21**, 587 (1980).

331. W. R. Muray, A. Blackwood, J. M. Trotter, K. C. Calman, and C. MacKay, *Br. J. Cancer* **41**, 923 (1980).

332. R. W. Owen, P. J. Henly, M. H. Thompson, and M. J. Hill, *J. Steroid Biochem.* **24**, 391 (1986).

333. M. H. Thompson, *in* "Colonic Carcinogenesis" (R. A. Malt and R. C. N. Williamson, eds.), p. 49, MTP Press, Lancaster (1981).

334. W. J. O. Kurtz, U. Leuschner, *Tokai J. Exp. Clin. Med.* **8**, 59 (1983).

335. C. B. Bremner and L. V. Ackerman, *Cancer* **26**, 991, (1970).

336. P. Correa, E. Duque, C. Cuello, and W. Haerszel, *Cancer* **9**, 86 (1972).

337. G. N. Stemmermann and R. Yatani, *Cancer* **31**, 1260 (1973).

338. P. Correa, J. P. Strong, A. Reif, and W. D. Johnson, *Cancer* **39**, 2258 (1977).

339. W. L. Buntain, W. H. ReMine, and G. M. Farrow, *Surg. Gynaecol Obstet.* **134**, 499 (1972).

340. B. C. Morson, *Proc. Roy. Soc. Med.* **67**, 451 (1974).

341. M. J. Hill, B. C. Morson, and H. J. R. Bussey, *Lancet* **1**, 245 (1978).

342. N. Tanida, Y. Hikasa, T. Shimoyama, and K. D. R. Setchell, *Jpn. J. Cancer Res.* (*Gann*) **76**, 104 (1985).

343. Y. Hikasa, N. Tanida, T. Ohno, and T. Shimoyama, *Scand. J. Gastroenterol.* **17**,(Suppl. 78), A299 (1982) (Abstr.).

344. M. J. Hill, *in* "Banbury Report 7, Gastrointestinal Cancer: Endogenous Factors" (W. R. Bruce, P. Correa, M. Lipkin, S. R. Tannenbaum, and T. D. Wilkins, eds.), p. 365, Cold Spring Harbor Laboratory, Cold Spring Harbor, NY (1981).

345. E. W. Naylor and E. Lebenthal, *Dig. Dis. Sci.* **25**, 945 (1980).

346. A. L. Watne and S. K. Core, *J. Surg. Res.* **19**, 157 (1975).

347. E. Bone, B. S. Drasar, and M. J. Hill, *Lancet* **1**, 1117 (1975).

348. P. Nair and N. Turjman, *Dis. Colon Rectum* **26**, 629 (1983).

349. T. Miettinen, *Gut* **12**, 632 (1971).

350. N. Tanida, Y. Hikasa, D. Dodo, K. Sawada, A. Kawaura, and T. Shimoyama, *Gastroenterologica Japonica* **21**, 245 (1986).

351. D. Lenz, *Scand. J. Gastroenterol.* **11**, 769 (1976).

352. E. L. Wynder, *in* "Prognostic Factors in Breast Cancer" (A. P. M. Forrest and P. B. Kunkler, eds.), p. 32, Williams and Wilkins, Baltimore, MD (1968).

353. B. S. Drasar and D. Irving, *Br. J. Cancer* **327**, 167 (1973).

354. W. R. Murray, A. Blackwood, K. C. Calman, and C. MacKay, *Br. J. Cancer* **42**, 856 (1980).

355. A. E. Papatestas, D. Panvelliwalla, P. I. Tartter, S. Miller, D. Pertsemlidis, and A. H. Aufses, Jr., *Cancer* **49**, 1201 (1982).

356. H. Andersson, S. Filipsson, and L. Hulten, *Scand. J. Gastroenterol.* **13**, 249 (1978).

357. K. Samuelson, C. Johansson, and A. Norman, *Scand. J. Clin. Lab. Invest.* **39**, 511 (1979).

358. P. Rutgeerts, Y. Ghoos, G. Vantrappen, and H. Eyssen, *Eur. J. Clin. Invest.* **11**, 199 (1981).

359. W. F. Balistreri, F. J. Suchy, and J. E. Heubi, *J. Pediatr.* **96**, 582 (1980).

360. T. Koga, T. Nishida, H. Miwa, M. Yamamoto, K. Kaku, T. Yao, and M. Okumura, *Dig. Dis. Sci.* **29**, 994 (1984).

361. S. Farivar, H. Fromm, D. Schindler, B. McJunkin, and F. W. Schmidt, *Am. J. Clin. Pathol.* **73**, 69 (1980).

362. E. Krag and B. Krag, *Scand. J. Gastroenterol.* **11**, 481 (1976).

363. L. M. Nelson, H. A. Carmichael, R. I. Russel, and S. T. Artherton, *Gut* **18**, 792 (1977).

364. P. Rutgeerts, Y. Ghoos, and G. Vantrappen, *Gut* **20**, 1072 (1979).

365. D. P. Burkitt and H. C. Trowell, "Refined Carbohydrate Foods and Disease," Academic Press, London (1975).

366. N. S. Painter, "Diverticular Disease of the Colon," Heinemann, London (1975).

367. J. S. S. Gear, A. Ware, P. Fursdon, J. I. Mann, D. J. Nolan, A. J. M. Brodribb, and M. P. Vessey, *Lancet* **1**, 511 (1979).

368. N. S. Painter, A. Zalmeida, and K. W. Colebourne, *Br. Med. J.* **2**, 137 (1972).

369. J. M. Findlay, A. N. Smith, W. D. Mitchell, A. J. Anderson, and M. A. Eastwood, *Lancet* **1**, 146 (1974).

370. S. Tarpila, T. A. Miettinen, and L. Metsaranta, *Gut* **19**, 137 (1978).

371. M. A. Eastwood, D. A. Watters, and A. N. Smith, *Clin. Gastroenterol.* **11**, 545 (1982) (Review).

372. E. Krag, *in* "Biological Effects of Bile Acids" (Falk Symposium 26) (G. Paumgartner, A. Stiehl, and W. Gerok, eds.), MTP Press, Lancaster (1978).

373. W. F. Balistreri, *in* "Chronic Diarrhea in Children" (E. Leberthal, ed.), p. 347, Vevey/Raven Press, New York (1984).

374. W. F. Balistreri, J. C. Partin, and W. K. Schubert, *J. Pediatr* **90**, 21 (1977).

375. E. Hess-Thaysen and L. Pederson, *Gut* **17**, 965 (1978).

376. J. M. Sondheimer, P. Szczepanik, and J. F. Perreault, *Pediatr. Res.* **12**, 442 (1978) (Abstr.).

377. A. Weber, C. C. Roy, L. Chartrand, C. L. Morin, and M. Van Caillie, *Union Med. Can.* **103**, 2089 (1974).

378. J. G. Allan, V. P. Gerskowitch, and R. I. Russell, *Br. J. Surg.* **61**, 516 (1974).

379. V. M. Duncombe, T. D. Bolin, and A. E. Davis, *Gut* **18**, 531 (1977).

380. A. F. Hofmann and J. R. Poley, *Gastroenterology* **62**, 918 (1972).

381. J. R. Condon, V. Robinson, M. I. Suleman, V. S. Fan, and M. D. McKeown, *Br. J. Surg.* **62**, 309 (1975).

382. T. A. Miettinen, *Gut* **13**, 682 (1972).

383. T. A. Miettinen, *Helv. Med. Acta* **37**, 113 (1973).

384. D. F. Hutcheon, T. M. Bayless, and T. R. Gadacz, *JAMA* **241**, 823 (1979).

385. M. Emilfork-Soto, G. Duffau-Toro, D. L. Bascur, and A. M. Urbina, *Bol. Med. Hosp. Infant Mex.* **39**, 177 (1982).

386. R. Gillberg and H. Andersson, *Scand. J. Gastroenterol.* **15**, 969 (1980).

387. M. Vuoristo and T. A. Miettinen, *Gastroenterology* **88**, 134 (1985).

388. T. A. Miettinen, *Lancet* **2**, 358 (1968).

389. T. S. Low-Beer, K. W. Heaton, E. N. Pomare, and A. E. Read, *Gut* **14**, 204 (1973).

390. C. R. Kapadia and S. J. Baker, *Aust. NZ J. Med.* **3**, 260 (1973).

391. E. John, *Aust. Pediatr. J.* **11**, 53 (1975).

392. M. Berant, E. Diamond, R. Brik, and S. Yurman, *Acta. Paediatr. Scand.* **72**, 853 (1983).

393. H. Andersson, I. Bosaeus, and C. Nyström, *Acta. Radiol. Oncol. Radiat. Phys. Biol.* **17**, 312 (1978).

394. I. Bosaeus, H. Andersson, and C. Nyström, *Acta. Radiol. Oncol. Radiat. Phys. Biol.* **18**, 460 (1979).

395. F. Soave, *Arch. Dis. Child.* **39**, 116 (1964).

396. T. Ehrenpreis, *Am. J. Dig. Dis.* **16**, 1032 (1971).

397. M. Van Blankenstein, T. Hoyset, P. Horchner, M. Frendel, and J. H. P. Wilson, *Neth. J. Med.* **20**, 248 (1977).

398. A. F. Hofmann and S. M. Grundy, *Clin Res.* **13**, 254 (1965).

399. W. F. Balistreri, J. E. Heubi, and F. J. Suchy, *J. Pediatr. Gastroenterol. Nutr.* **2**, 105 (1983).

400. M. S. Klein and S. J. Winawer, *Am. J. Gastroenterol.* **61**, 470 (1974).

401. A. Cortot, C. R. Fleming, and J. R. Malageleda, *N. Engl. J. Med.* **300**, 79 (1977).

402. A. M. Dawson, C. A. Holdsworth, and J. Webb, *Exp. Biol. Med.* **117**, 97 (1964).

403. A. F. Hofmann and J. R. Poley, *N. Engl. J. Med.* **281**, 397 (1969).

404. W. G. Thompson, *Dis. Colon. Rectum* **18**, 304 (1975).

405. W. Bochenek, J. B Rodgers, Jr., and J. A. Balint, *Ann. Intern. Med.* **72**, 205 (1970).

406. H. Andersson, *Nutr. Metab.* **20**, 254 (1976).

407. R. L. Porus, *Gastroenterology* **48**, 753 (1965).

408. A. Papazian, Y. Minaire, L. Descos, C. Andre, M. Melange, and J. Vignal, *Hepato-gastroenterol.* **28**, 106 (1981).

409. A. Mallory, D. Savage, and F. Kern Jr., *Gastroenterology* **64**, 34 (1973).

410. W. G. M. Hardison and I. H. Rosenberg, *N. Engl. J. Med.* **277**, 337 (1967).

411. N. F. La Russo, J. Turcotte, and J. L. Thistle, *Gastroenterology* **76**, 1181 (1979) (Abstr.).

412. Y. Peled, S. Bar-Meir, H. H. Rotmensch, A. Tiomny, and T. Gilat, *Isr. J. Med. Sci.* **18**, 812 (1982).

413. A. W. Huijbregts, T. M. Cox, J. Hermsen, G. P. Van Berge Henegouwen, A. Van Schaik, and V. S. Chadwick, *Neth. J. Med.* **24**, 108 (1981).

414. H. Fromm, R. P. Sarva, M. M. Ravitch, B. McJunkin, S. Farivar, and P. Amin, *Metabolism* **32**, 1133 (1983).

415. S. M. Grundy, E. H. Ahrens, Jr., and G. Salen, *J. Lab Clin. Med.* **78**, 94 (1971).

416. L. Wise and T. Stein, *Surg. Gynecol. Obstet.* **142**, 686 (1976).

417. W. W. Faloon, M. S. Flood, S. Aroesty, and C. D. Sherman, *Am. J. Clin. Nutr.* **33**, 431 (1980).

418. T. I. Sorensen, *Scand. J. Gastroenterol.* **17,** 577 (1982) (Review).
419. T. I. Sorensen, *Scand. J. Gastroenterol.* **17,** 473 (1982).
420. T. I. A. Sorensen and E. Krag, *Scand. J. Gastroenterol.* **11,** 491 (1976).
421. W. W. Faloon, A. Rubulis, J. Knipp, C. D. Sherman, and M. S. Flood, *Am. J. Clin. Nutr.* **30,** 21 (1977).
422. R. B. Moore, I. D. Frantz, Jr., and H. Buchwald, *Surgery* **65,** 98 (1969).
423. T. A. Miettinen and P. Peltokallio, *Scand. J. Gastroenterol.* **6,** 543 (1971).
424. R. M. Kay, Z. Cohen, K. P. Siu, C. N. Petrunka, and S. M. Strasberg, *Gut* **21,** 128 (1980).
425. K. Huibregtse, F. Hoek, G. T. Sanders, and G. N. Tytgat, *Eur. J. Clin. Invest.* **7,** 137 (1977).
426. H. Andersson, S. Fasth, S. Filipsson, R. Hellberg, L. Hulten, L. O. Nilsson, S. Nordgren, and N. G. Kock, *Scand. J. Gastroenterol.* **14,** 551 (1979).
427. L. O. Nilsson, H. Andersson, L. Hulten, R. Jagenburg, N. G. Kock, H. E. Myrvold, and B. Philipson, *Gut* **20,** 499 (1979).
428. A. Brandberg, N. G. Koch, and B. Philipson, *Gastroenterology* **63,** 413 (1972).
429. B. Philipson, A. Brandberg, R. Jagenburg, N. Kock, I. Lager, and C. Ahren, *Scand. J. Gastroenterol.* **10,** 145 (1975).
430. I. W. Percy-Robb, W. A. Telfer-Brunton, J. C. Gould, K. N. Jalan, J. P. McManus, and W. Sircus, *Scand. J. Gastroenterol.* **6,** 625 (1971).
431. I. W. Percy-Robb, K. N. Jalan, J. P. A. McManus, and W. Sircus, *Clin. Sci.* **41,** 371 (1971).
432. A. V. Pasanen, S. Tarpila, and T. A. Miettinen, *Scand. J. Gastroenterol.* **15,** 503 (1980).
433. M. Flynn, P. Hammond, C. Darby, J. Hyland, and I. Taylor, *Digestion* **22,** 144 (1981).
434. L. Van Bogaert, H. J. Scherer, and E. Epstein, *in* "Une forme célébrale de cholestérinoae géneralisée," Masson, Paris (1937).
435. J. R. Schimschock, E. C. Alvord, Jr., and P. D. Swanson, *Neurology* **18,** 688 (1968).
436. G. Salen, *Ann. Intern. Med.* **75,** 843 (1971).
437. S. Shefer, B. Dayal, G. S. Tint, G. Salen, and E. H. Mosbach, *J. Lipid Res.* **16,** 280 (1975).
438. G. Salen and E. H. Mosbach *in* "The Bile Acids, Chemistry, Physiology and Metabolism" (P. P. Nair and D. Kritchevsky, eds.), Volume 3, p. 115, Plenum Press, New York (1976).
439. T. Hoshita, M. Yasuhara, M. Une, A. Kibe, E. Itoga, S. Kito, and T. Kuromoto, *J. Lipid Res.* **21,** 1015 (1980).
440. B. G. Wolthers, M. Volmer, J. van der Molen, B. J. Koopman, A. E. J. de Jaeger, and R. J. Waterreus, *Clin. Chim Acta* **131,** 53 (1983).
441. K. Shimazu, M. Kuwabara, M. Yoshii, K. Kihira, H. Takeuchi, I. Nakano, S. Ozawa, M. Onuki, Y. Hatta, and T. Hoshita, *J. Biochem.* **99,** 477 (1986).
442. J. Sjövall, K. D. R. Setchell, A. M. Lawson, G. Karlaganis, and S. Skrede, *Biomed. Chromatogr.* (1988) (submitted for publication).
443. B. Egestad, P. Petersson, S. Skrede, and J. Sjövall, *Scand. J. Clin. Lab. Invest.* **45,** 443 (1985).
444. T. Setoguchi, G. Salen, G. S. Tint, and E. H. Mosbach, *J. Clin. Invest.* **53,** 1393 (1974).
445. H. Oftebro, I. Björkhem, S. Skrede, A. Schreiner, and J. I. Pedersen, *J. Clin. Invest.* **65,** 1418 (1980).
446. I. Björkhem, O. Fausa, G. Hopen, H. Oftebro, J. I. Pederson, and S. Skrede, *J. Clin. Invest.* **71,** 142 (1983).
447. R. F. Hanson, P. Szczepanik-van Leeuwen, G. C. Williams, G. Grabowski, and H. L. Sharp, *Science* **203,** 1107 (1979).
448. G. Salen and S. M. Grundy, *J. Clin. Invest.* **52,** 2822 (1973).
449. G. Salen, S. Shefer, F. W. Cheng, B. Dayal, A. K. Batta, and G. S. Tint, *J. Clin. Invest.* **63,** 38 (1979).
450. I. Björkhem, J. Gustafsson, G. Johansson, and B. Persson, *J. Clin. Invest.* **55,** 478 (1975).

451. A. K. Bhattacharyya and W. E. Connor, *J. Clin. Invest.* **53**, 1033 (1974).

452. T. A. Miettinen, *Eur. J. Clin. Invest.* **10**, 27 (1980).

453. B. Strandvik, B. Angelin, and K. Einarsson, 2nd Joint Meeting NASPG/ESPGAN, New York (1985) (Presentation).

454. C. C. Roy, G. Delage, A. Fontaine, L. Robitaille, L. Chartrand, A. Weber, and C. L. Morin, *Am. J. Clin. Nutr.* **32**, 2404 (1979).

455. P. L. Zentler-Munro, W. J. F. Fitzpatrick, J. C. Batten, and T. C. Northfield, *Gut* **25**, 500 (1984).

456. C. C. Roy, A. M. Weber, C. L. Morin, J. C. Combes, D. Nussle, A. Megevand, and R. Lasalle, *N. Engl. J. Med.* **297**, 1301 (1977).

457. J. T. Harries, D. R. R. Muller, J. P. K. McCollum, A. Lipson, E. Roma, and A. P. Norman, *Arch. Dis. Child.* **54**, 19 (1979).

458. D. Lefebvre, S. Ratelle, L. Chartrand, and C. C. Roy, *Experientia* **33**, 616 (1977).

459. P. Bowen, C. S. N. Lee, H. Zellweger, and R. Lindenberg, *Bull. Johns Hopkins Hosp.* **114**, 402 (1964).

460. S. Goldfischer, C. L. Moore, A. B. Johnson, A. J. Spiro, M. P. Valsamis, H. K. Wisniewsi, R. H. Ritch, W. T. Norton, I. Rapin, and L. M. Gartner, *Science* **182**, 62 (1973).

461. G. G. Parmentier, G. A. Janssen, E. A. Eggermont, and H. J. Eyssen, *Eur. J. Biochem.* **102**, 173 (1979).

462. I. Björkhem, *in* "Sterols and Bile Acids" (H. Danielsson and J. Sjövall, eds.), p. 231, Elsevier, Amsterdam (1985).

463. I. Björkhem, B. F. Kase, and J. I. Pedersen, *Scand. J. Clin. Lab. Invest.* **45**, Suppl 177, 23 (1985).

464. H. Eyssen, E. Eggermont, J. Van Eldere, J. Jaeken, G. Parmentier, and G. Janssen, *Acta Pediatr. Scand.* **74**, 539 (1985).

465. D. S. Frederikson, R. I. Levy, and R. S. Lees, *N. Engl. J. Med.* **276**, 32, 94, 148, 215, 273 (1967).

466. J. L. Beaumont, L. A. Carlson, G. R. Cooper, Z. Fejfar, D. S. Frederikson, and T. Shrasser, *Bull. WHO* **43**, 891 (1970).

467. J. M. Felts, and L. L. Rudel, *in* "Hypolipidemic Agents" (D. Kritchevsky, ed.), p. 151, Springer/Verlag, Berlin, Heidelberg, New York (1975).

468. B. A. Kottke, *Circulation* **40**, 13 (1969).

469. T. A. Miettinen and A. Aro, *Scand J. Clin. Lab. Invest.* **30**, 85 (1972).

470. H. S. Sodhi and B. J. Kudchodkar, *Clin. Chim. Acta* **46**, 161 (1973).

471. K. Einarsson, K. Hellström, and M. Kallner, *J. Clin. Invest.* **54**, 1301 (1974).

472. K. Einarsson and K. Hellström, *Eur. J. Clin. Invest.* **2**, 225 (1972).

473. N. E. Miller, P. J. Nestel, and P. Clifton-Bligh, *Atherosclerosis* **23**, 535 (1976).

474. B. Angelin, K. Einarsson, K. Hellström, and M. Kallner, *Clin. Sci. Mol. Med.* **51**, 393 (1976).

475. D. S. Lin and W. E. Connor, *J. Lipid Res.* **21**, 1042 (1980).

476. T. A. Miettinen, R. Pelkonen, E. A. Nikkilä, and O. Heinonen, *Acta Med. Scand.* **182**, 645 (1967).

477. T. A. Miettinen, *in* "Hypolipidemic Agents" (D. Kritchevsky, ed.), p. 109, Springer-Verlag, Berlin, Heidelberg, New York (1975).

478. G. R. Thompson, N. B. Myant, D. Kilpatrick, C. M. Oakley, M. J. Raphael, and R. E. Steiner, *Br. Heart J.* **43**, 680 (1980).

479. C. D. Moutafis and N. B. Myant, *Clin. Sci.* **37**, 443 (1969).

480. C. D. Moutafis, L. A. Simons, N. B. Myant, P. W. Adams, and V. Wynn. *Atherosclerosis* **26**, 329 (1977).

481. T. A. Miettinen, *Clin. Chim. Acta* **20**, 43 (1968).

482. T. A. Miettinen and M. Lempinen, *Eur. J. Clin. Invest.* **7**, 509 (1977).

483. L. A. Simons and N. B. Myant, *Clin. Chim. Acta* **65**, 117 (1975).
484. B. J. Kudchodkar, H. S. Sodhi, D. L. Horlick, and D. T. Mason, *Clin. Pharmacol. Ther.* **22**, 154 (1977).
485. D. J. Nazir, L. Horlick, B. J. Kudchodkar, and H. S. Sodhi, *Circulation* **46**, 95 (1972).
486. C. D. Moutafis and N. B. Myant, *Alterosclerosis* **17**, 305 (1973).
487. A. Sedaghat, P. Samuel, J. R. Crouse, and E. H. Ahrens, Jr., *J. Clin. Invest.* **55**, 12 (1975).
488. J. W. Anderson, L. Story, B. Sieling, W. J. Chen, M. S. Petro, and J. Story, *Am. J. Clin. Nutr.* **40**, 1146 (1984).
489. K. Von Bergmann, H. Y. Mok, W. G. Hardison, and S. M. Grundy, *Gastroenterology* **77**, 1183 (1979).
490. Y. A. Kesaniemi, T. A. Miettinen, and M. P. Salaspuro, *Gut* **22**, 579 (1981).
491. Y. A. Kesaniemi, M. P. Salaspuro, M. Vuoristo, and T. A. Miettinen, *Gut* **23**, 931 (1982).
492. W. Erb, R. Krohl, J. Schreiber, and J. Wildgrube, *Z. Gastroenterol.* **10**, 85 (1972).
493. C. D. Saudek, *Am. J. Med.* **63**, 453 (1977).
494. P. Back, *Klin. Wochenschr.* **51**, 926 (1973).
495. A. Stiehl, M. M. Thaler, and W. H. Admirand, *Pediatrics* **51**, 992 (1973).
496. M. E. Miller and P. J. Nestel, *Clin. Sci. Mol. Med.* **42**, 257 (1973).
497. B. W. D. Badley, G. M. Murphy, I. A. D. Bouchner, and S. Sherlock, *Gastroenterology* **58**, 781 (1970).
498. T. A. Miettinen and M. Siurala, *Scand. J. Gastroenterol.* **6**, 527 (1971).
499. H. P. Porter, D. R. Saunders, G. Tytget, O. Brunser, and C. E. Rubin, *Gastroenterology* **60**, 1008 (1971).
500. J. Sjövall, *Clin. Chim. Acta* **5**, 33 (1960).
501. Z. R. Vlachevic, I. Buhac, C. C. Bell Jr., and L. Swell, *Gut* **11**, 420 (1970).
502. S. Mehta, J. E. Struthers Jr., M. D. Kaye, and J. L. Naylor, *Gastroenterology* **67**, 674 (1974).
503. F. J. Ingelfinger, *Gastroenterology* **55**, 102 (1968).
504. L. J. Bennion and S. M. Grundy, *N. Eng. J. Med.* **299**, 1221 (1978).
505. L. Capocaccia, G. Ricci, F. Angelico, M. Angelico, and A. F. Attila, "Epidemiology and Prevention of Gallstone Disease," MTP Press, Lancaster, Boston, The Hague, Dorddrecht (1984).
506. W. H. Admirand and D. M. Small, *J. Clin. Invest.* **47**, 1043 (1968).
507. D. M. Small, *Adv. Intern Med.* **16**, 243 (1970).
508. R. H. Palmer *in* "The Bile Acids, Chemistry, Physiology and Metabolism" (P. P. Nair and D. Kritchevsky, eds.), Volume 2, p. 153, Plenum Press, New York (1973).
509. R. H. Dowling, A. F. Hofmann, and L. Barbara, "Workshop on Ursodeoxycholic Acid," MTP Press, Lancaster (1978).
510. J. W. Marks, G. C. Bonorris, and L. J. Schoenfield, *in* "The Bile Acids, Chemistry, Physiology and Metabolism" (P. P. Nair and D. Kritchevsky, eds.), Volume 3, p. 81, Plenum Press, New York (1976).
511. Z. R. Vlachevic, C. C. Bell, Jr., I. Buhac, J. T. Farrar, and L. Swell, *Gastroenterology* **59**, 165 (1970).
512. K. W. Heaton and A. E. Read, *Br. Med. J.* **3**, 494 (1969).
513. R. G. Danzinger, A. F. Hofmann, J. L. Thistle, and L. J. Schoenfield, *J. Clin. Invest.* **52**, 2809 (1973).
514. I. Makino and S. Nakagawa, *J. Lipid Res.* **19**, 723 (1978).
515. R. N. Redinger, *Lipids* **14**, 277 (1979).
516. R. N. Redinger and D. M. Small, *J. Clin. Invest.* **52**, 161 (1973).
517. R. N. Redinger, *Gastroenterology* **66**, 763 (1974).
518. M. J. Coyne, G. G. Bonnoris, A. Chung, L. J. Goldstein, D. Lahana, and L. J. Schoenfield, *N. Engl. J. Med.* **292**, 604 (1975).

519. R. J. Santen, P. W. Willis, and S. S. Fajans, *Arch. Intern. Med.* **130,** 833 (1972).

520. M. M. Lieber, *Ann. Surg.* **135,** 394 (1952).

521. L. J. Bennion and S. M. Grundy, *N. Engl. J. Med.* **296,** 1365 (1977).

522. C. D. Saudek and E. L. Brach, *Diabetes* **27,** 1059 (1978).

523. J. J. Abrahams, H. Ginsberg, and S. M. Grundy, *Diabetes* **31,** 903 (1982).

524. D. P. Burkitt, A. R. P. Walker, and N. S. Painter, *JAMA* **229,** 1068 (1974).

525. H. Trowell, *Rev. Eur. Etudes Clin. Biol.* **17,** 345 (1972).

526. G. V. Vahouny and D. Kritchevsky, "Dietary Fiber in Health and Disease" Plenum Press, New York, London (1982).

527. M. A. Eastwood and J. A. Robertson, *J. Hum. Nutr.* **32,** 53 (1978).

528. R. M. Kay and S. M. Strasberg, *Clin. Invest. Med.* **1,** 9 (1978).

529. D. A. T. Southgate, *in* "Dietary Fiber in Health and Disease (G. V. Vahouny and D. Kritchevsky, ed.), p. 1, Plenum Press, New York, London (1982).

530. M. Eastwood and L. Mowbray, *Am. J. Clin. Nutr.* **29,** 1461 (1976).

531. D. Kritchevsky, *Am. J. Clin. Nutr.* **30,** 979 (1977).

532. G. V. Vahouny, R. Tombes, M. M. Cassidy, D. Kritchevsky, and L. L. Gallo, *Lipids* **15,** 1012 (1980).

533. R. L. Walters, I. M. Baird, P. S. Davies, M. J. Hill, B. S. Draser, D. A. T. Southgate, J. Green, and B. Morgan, *Br. Med. J.* **2,** 536 (1975).

534. H. M. Fuchs, S. Dorfman, and M. H. Floch, *Am. J. Clin. Nutr.* **29,** 1443 (1976).

535. M. J. Hill, *in* "Colon and Nutrition" (H. Kasper and H. Goebell, eds.), p. 37, MTP Press, Lancaster, Boston, The Hague (1982).

536. R. D. Williams and W. Holmsted, *J. Nutr.* **11,** 433 (1936).

537. M. A. Eastwood, T. Hamilton, J. R. Kirkpatrick, and W. D. Mitchell, *Proc. Nutr. Soc.* **32,** 22A (1973).

538. M. A. Eastwood, J. R. Kirkpatrick, W. D. Mitchell, A. Bone, and T. Hamilton, *Br. Med. J.* **4,** 392 (1973).

539. R. M. Kay and A. S. Truswell, *Br. J. Nutr.* **37,** 227 (1977).

540. J. H. Cummings, M. J. Hill, T. Jivraj, H. Houston, W. J. Branch, and D. J. Jenkins, *Am. J. Clin. Nutr.* **32,** 2086 (1979).

541. R. M. Kay and A. S. Truswell, *Am. J. Clin. Nutr.* **30,** 171 (1977).

542. J. K. Ross and J. E. Leklem, *Am. J. Clin. Nutr.* **34,** 2068 (1981).

543. J. H. Cummings, M. J. Hill, D. J. A. Jenkins, J. R. Pearson, and H. S. Wiggins, *Am. J. Clin. Nutr.* **29,** 1468 (1976).

544. P. A. Judd and A. S. Truswell, *Am. J. Clin. Nutr.* **34,** 2061 (1981).

545. T. L. Raymond, W. E. Connor, D. S. Lin, S. Warner, M. M. Fry, and S. L. Connor, *J. Clin. Invest.* **60,** 1429 (1977).

546. J. H. Cummings, *in* "Colon and Nutrition" (H. Kaspar and H. Goebell, eds.), p. 91, MTP Press, Lancaster, Boston, The Hague (1982).

547. T. P. Almy, *Curr. Concepts Nutr.* **4,** 155 (1976) (Review).

548. J. L. Kelsay, *Am. J. Clin. Nutr.* **31,** 142 (1978) (Review).

549. A. W. Huijbregts, G. P. Van Berge Henegouwen, M. P. Hectors, A. Van Schaik, and S. D. Van der Werf, *Eur. J. Clin. Invest.* **10,** 451 (1980).

550. E. W. Bell, E. A. Emken, L. M. Klevay, and H. H. Sandstead, *Am. J. Clin. Nutr.* **34,** 1071 (1981).

551. M. J. Kretsch, L. Crawford, and D. H. Calloway, *Am. J. Clin. Nutr.* **32,** 1492 (1979).

552. R. W. Kirby, J. W. Anderson, B. Sieling, E. D. Rees, W. J. Chen, R. E. Miller, and R. M. Kay, *Am. J. Clin. Nutr.* **34,** 824 (1981).

553. T. A. Miettinen and S. Tarpila, *Clin. Chim. Acta* **79,** 471 (1977).

554. G. D. Calvert, L. Blight, R. J. Illman, D. L. Topping, and J. D. Potter, *Br. J. Nutr.* **45**, 277 (1981).

555. A. C. Tsai, E. L. Mott, G. M. Owen, M. R. Bennick, G. S. Lo, and F. H. Steinke, *Am. J .Clin. Nutr.* **38**, 504 (1983).

556. T. F. Schweizer, A. R. Bekhechi, B. Koellreutter, S. Reimann, D. Pometta, and B. A. Bron, *Am. J. Clin. Nutr.* **38**, 1 (1983).

557. M. R. Malinow, P. McLaughlin, and C. Stafford, *Experientia* **36**, 562 (1980).

558. K. S. Mathur, M. A. Khan, and R. D. Sharma, *Br. Med. J.* **1**(583), 30 (1968).

559. D. J. A. Jenkins, *in* "Drugs Affecting Lipid Metabolism" (R. Fumagalli, D. Kritchevsky, and R. Paoletti, eds.), p. 339, Elsevier/North Holland, Amsterdam (1980).

560. I. M. Baird, R. L. Walters, D. S. Davies, M. J. Hill, B. S. Drasar, and D. A. Southgate, *Metabolism* **26**, 117 (1977).

561. A. H. McLean Ross, M. A. Eastwood, J. R. Anderson, and D. M. W. Anderson, *Am. J. Clin. Nutr.* **37**, 368 (1983).

562. D. T. Forman, J. E. Garvin, J. E. Forestner, and C. B. Taylor, *Proc. Soc. Exp. Biol. Med.* **127**, 1060 (1968).

563. M. M. Stanley, D. Paul, D. Gacke, and J. Murphy, *Gastroenterology* **65**, 889 (1973).

564. S. Tarpila and T. A. Miettinen, *Scand. J. Gastroenterol.* **12**, 105 (1977).

565. D. J. A. Jenkins, M. J. Hill, and J. H. Cummings, *Am. J. Clin. Nutr.* **28**, 1408 (1975).

566. D. J. A. Jenkins, *in* "International Conference on Altherosclerosis" (I. A. Carlson, R. Paoletti, C. R. Sirtoli, and G. Weber, eds.), p. 173, Raven Press, New York (1978).

567. I. H. Ullrich, H. Y. Lai, L. Vona, R. L. Reid, and M. J. Albrink, *Am. J. Clin. Nutr.* **34**, 2054 (1981).

568. S. M. Grundy and H. Y. Mok, *J. Lab Clin. Med.* **89**, 354 (1977).

569. T. A. Miettinen, *J. Clin. Invest.* **64**, 1485 (1979).

570. C. D. Moutafis and N. B. Myant, *in* "Metabolic Effects of Nicotinic Acid and Its Derivatives," p. 659, Huber, Bern, Stuttgart, Vienna (1971).

571. N. B. Myant, "Advances in Experimental Medicine and Biology," Volume 26, p. 137, Plenum Press, New York, London (1972).

572. T. A. Miettinen, *Ann. Clin. Res.* **3**, 313 (1971).

573. L. A. Simons and N. B. Myant, *Adv. Exp. Med. Biol.* **82**, 188 (1977).

574. N. E. Miller, P. Clifton-Bligh, and P. J. Nestel, *J. Lab Clin. Med.* **82**, 876 (1973).

575. T. A. Miettinen, *in* "International Conference on Atherosclerosis" (L. A. Carlson, R. Paoletti, C. R. Sirtoli, and G. Weber, eds.), p. 193, Raven Press, New York (1978).

576. M. S. Brown, and J. L. Goldstein, *Science* **232**, 34 (1986).

577. D. A. Cook, L. M. Hagerman, and D. L. Schneider, *Proc. Soc. Exp. Biol. Med.* **139**, 70 (1972).

578. J. G. Bartlett, *John Hopkins Med. J.* **149**, 6 (1981).

579. E. D. Jacobson, R. B. Chodos, and W. W. Faloon, *Am. J. Med.* **28**, 524 (1960).

580. J. G. Bartlett, *Rev. Infect. Dis.* **1**, 530 (1979).

581. P. Samuel and A. Steiner, *Proc. Soc. Exp. Biol. Med.* **100**, 193 (1959).

582. P. Samuel, C. M. Holtzman, E. Meilman, and I. Sekowski, *Circ. Res.* **33**, 393 (1973).

583. W. W. Faloon, I. C. Paes, D. Woolfolk, H. Nankin, K. Wallace, and E. N. Haro, *Ann. NY Acad. Sci.* **132**, 879 (1966).

584. G. R. Thompson, J. Barrowman, L. Gutierrez, and R. H. Dowling, *J. Clin. Invest.* **50**, 319 (1971).

585. H. Eyssen, H. Vanderhaeghe, and P. De-Somer, *Fed. Proc.* **30**, 1803 (1971).

586. W. W. Faloon, A. Rubulis, and M. Rubert, *Clin. Res.* **17**, 158 (1969). (Abstr.).

587. J. F. Van Den Bosch and P. J. Claes, *Progr. Biochem. Pharmacol.* **2**, 97 (1976).

588. G. R. Thompson, M. MacMahon, and P. Claes, *Eur. J. Clin. Invest.* **1**, 40 (1979).

589. T. A. Miettinen, *in* "Proceedings of The Sixth International Congress of Pharmacology,"

Volume 4, p. 149, Finnish Pharmacological Society, Pergamon Press, Oxford, New York (1976).

590. J. G. Hamilton, O. N. Miller, and G. A. Goldsmith, *Fed. Proc.* **19,** 187 (1960).
591. D. Schwob, A. Rubulis, E. C. Lim, C. D. Sherman, and W. W. Faloon, *Am. J. Clin. Nutr.* **25,** 987 (1972).
592. W. G. Hardison and I. H. Rosenberg, *J. Lab Clin. Med.* **74,** 564 (1969).
593. J. R. Waldbaum, W. T. Beher, R. J. Priest, and S. Stradnieks, *Henry Ford Hosp. Med. J.* **30,** 160 (1982).
594. T. A. Miettinen, *Clin. Chim. Acta* **19,** 341 (1968).
595. B. J. Kudchodkar, H. S. Sodhi, D. T. Mason, and N. O. Borhani, *Am. J. Clin. Nutr.* **30,** 1135 (1977).
596. H. S. Sodhi and B. J. Kudchodkar, *Adv. Exp. Med. Biol.* **82,** 290 (1977).
597. R. B. Moore, J. T. Anderson, H. L. Taylor, A. Keys, and I. D. Frantz, *J. Clin. Invest.* **47,** 1517 (1968).
598. J. H. Bruusgaard, M. B. Katan, and J. G. A. J. Hautvast, *Eur. J. Clin. Invest.* **13,** 115 (1983).
599. H. M. Whyte, P. J. Nestel, and E. S. Pryke, *J. Lab Clin. Med.* **81,** 818 (1973).
600. P. D. Wood, R. Shioda, and L. W. Kinsell, *Lancet* **2,** 604 (1966).
601. H. S. Sodhi, P. D. S. Wood, G. Schlerf, and L. W. Kinsell, *Metabolism* **16,** 334 (1967).
602. W. E. Connor, D. T. Witiak, D. B. Stone, and M. L. Armstrong, *J. Clin. Invest.* **48,** 1363 (1969).
603. R. Nicolaysen, *Nord. Med.* **86,** 865 (1971) (Review).
604. P. J. Nestel, N. Havenstein, H. M. Whyte, T. J. Scott, and L. J. Cook, *N. Engl. J. Med.* **288,** 379 (1973).
605. P. Hill, B. S. Reddy, and E. L. Wynder, *J. Am. Diet. Assoc.* **75,** 414 (1979).
606. N. Spritz, E. H. Ahrens, and S. M. Grundy, *J. Clin. Invest.* **44,** 1482 (1965).
607. S. M. Grundy and E. H. Ahrens, *J. Clin. Invest.* **49,** 1135 (1970).
608. P. J. Nestel, N. Havenstein. T. W. Scott, and L. J. Cook, *Aust. NZ J. Med.* **4,** 497 (1974).
609. J. Avigan and D. Steinberg, *J. Clin. Invest.* **44,** 1845 (1965).
610. R. L. Jackson, O. D. Taunton, J. D. Morrisett, and A. M. Gotto, Jr., *Circ. Res.* **42,** 447 (1978) (Review).
611. P. J. Nestel, N. Havenstein, Y. Homme, T. W. Scott, and L. J. Cook, *Metabolism* **24,** 189 (1975).
612. S. M. Grundy, *J. Clin. Invest.* **55,** 269 (1975).
613. J. D. Wilson and C. A. Lindsey, Jr., *J. Clin. Invest.* **44,** 1805 (1965).
614. E. Quintao, S. M. Grundy, and E. H. Ahrens, Jr., *J. Lipid Res.* **12,** 233 (1971).
615. K. J. Ho, K. Biss, B. Mikkelson, L. A. Lewis, and C. B. Taylor, *Arch. Pathol.* **91,** 387 (1971).
616. P. J. Nestel, L. A. Simons, and Y. Homma, *Am. J. Clin. Nutr.* **29,** 1007 (1976).
617. J. R. Crouse and S. M. Grundy, *J. Lipid Res.* **25,** 486 (1984).
618. H. Yacowitz, A. I. Fleischman, and M. L. Bierenbaum, *Br. Med. J.* **1,** 1352 (1965).
619. D. J. Nazir and M. A. Mishkel, *Clin. Chim. Acta* **62,** 117 (1975).
620. W. D. Mitchell, T. Fyfe, and D. A. Smith, *J. Atheroscler. Res.* **8,** 913 (1968).
621. A. K. Bhattacharyya, C. Thera. J. R. Anderson, F. Grande, and A. Keys, *Am. J. Clin. Nutr.* **22,** 1161 (1969).
622. J. R. Crouse and S. M. Grundy, *Metabolism* **28,** 994 (1979).
623. C. J. Glueck, R. J. Jandacek, M. T. Ravi Subbiah, L. Gallon, R. Yunker, C. Allen, E. Hogg, and P. M. Laskarzewski, *Am. J. Clin. Nutr.* **33,** 2177 (1980).
624. L. Horlick, B. J. Kudchodkar, and H. S. Sodhi, *Circulation* **43,** 299 (1971).
625. W. D. Mitchell and L. E. Murchison, *Clin. Chim. Acta* **36,** 153 (1972).
626. H. S. Sodhi, L. Horlick, and B. J. Kudchodkar, *Clin. Res.* **17,** 663 (1969).
627. K. Einarsson, K. Hellström, and M. Kallner, *Eur. J. Clin. Invest.* **3,** 345 (1973).

628. L. A. Carlsson, A. G. Olsson, L. Oro, S. Rossner, and E. Walldius, *in* "Atherosclerosis III" (C. Schettler and A. Weizel, eds.), p. 768, Springer/Verlag, Berlin, Heidelberg, New York (1974).

629. H. T. I. Mok and S. M. Grundy, *Gastroenterology* **73**, 1235 (1977).

630. F. A. Abdurakhmaron, *Med. Zh. Uzb.* **12**, 32 (1965).

631. H. S. Sodhi, L. Horlick, and B. J. Kudchodkar, *Clin. Res.* **17**, 395 (1969).

632. I. Bekersky and E. H. Mosbach, *in* "The Bile Acids, Chemistry, Physiology and Metabolism" (P. P. Nair and D. Kritchevsky, eds.), Volume 2, p. 249, Plenum Press, New York (1973).

633. T. A. Miettinen, *J. Lab Clin. Med.* **71**, 537 (1968).

634. L. A. Simons and N. B. Myant, *Atherosclerosis* **19**, 103 (1974).

635. R. J. Winter, R. G. Thompson, and O. L. Green, *Metabolism* **28**, 1244 (1979).

636. P. R. Blackett, P. K. Weech, W. J. McConathy, and J. D. Fesmire, *Metabolism* **31**, 117 (1982).

637. J. R. Poley, J. D. Smith, J. B. Thompson, and J. F. Seely, *Pediatr. Res.* *12*, 1186 (1977).

638. J. E. Heubi, S. Burstein, M. A. Sperling, D. Gregg, M. T. Subbiah, and D. E. Matthews, *J. Clin. Endocrinol. Metab.* **57**, 885 (1983).

639. M. Friedman, S. O. Byers, R. H. Rosenman, C. H. Li, and R. Neuman, *Metabolism* **23**, 905 (1974).

640. T. A. Miettinen, *in* "Advances in Experimental Medicine and Biology" (G. W. Manning and M. D. Haust, eds.), Volume 82, p. 483, Plenum Press, New York (1977).

641. T. A. Miettinen, *Atherosclerosis* **15**, 163 (1972).

642. P. J. Nestel and T. Billington, *Atherosclerosis* **38**, 203 (1981).

643. Y. A. Kesaniemi and S. M. Grundy, *J. Lipid Res.* **25**, 780 (1984).

644. W. Bachrach and A. F. Hofmann, *Dig. Dis. sci.* **27**, 737 (1982).

645. W. Bachrach and A. F. Hofmann, *Dig. Dis. Sci.* **27**, 833 (1982).

646. A. F. Hofmann, *in* "Bile Acids and Cholesterol in Health and Disease" (G. Paumgartner, A. Stiehl, and W. Gerok, eds.), p. 301, MTP Press, Lancaster (1982).

647. G. P. Van Berge Henegouwen and A. F. Hofmann, *Gastroenterology* **73**, 300 (1977).

648. G. Paumgartner, A. Stiehl, and W. Gerok, "Bile Acids and Cholesterol in Health and Disease," MTP Press, Lancaster (1983).

649. E. D. Bell, B. Whitney, and R. H. Dowling, *Lancet* **2**, 1213 (1972).

650. R. G. Danzinger, A. F. Hofmann, L. J. Schoenfield, and J. L. Thistle, *N. Engl. J. Med.* **286**, 1 (1972).

651. Y. Nakano and K. Nakano, *J. Natl. Council Communal Hosp.* **70**, 25 (1973).

652. F. Sugata and M. Shimizu, *Jpn. J. Gastroenterol.* **71**, 75 (1974).

653. I. Makino, K. Shinozaki, K. Yoshimo, and S. Nakagawa, *Jpn. J. Gastroenterol.* **72**, 690 (1975).

654. I. Makino and S. Nakagawa, *Jpn. J. Gastroenterol.* **72**, 690 (1975).

655. T. Fedorowski, G. Salen, A. Colallilo, G. S. Tint, E. H. Mosbach, and J. C. Hall, *Gastroenterology* **73**, 1131 (1977).

656. J. Thistle, N. F. La Russo, A. F. Hofmann, J. Turcotte, G. L. Carlson, and B. J. Ott, *Dig. Dis. Sci.* **27**, 161 (1982).

657. T. Fedorowski, G. Salen, G. S. Tint, and E. H. Mosbach, *Gastroenterology* **77**, 1068 (1979).

658. H. Fromm, G. L. Carlson, A. F. Hofmann, S. Farivar, and P. Amin, *Am. J. Physiol.* **239**, G161 (1980).

659. G. Salen, D. Verga, A. K. Batta, G. S. Tint, and S. Shefer, *Gastroenterology* **82**, 341 (1982).

660. H. T. I. Mok, K. Von Bergmann and S. M. Grundy, *in* "Biological Effects of Bile Acids" (G. Paumgartner, G. Stiehl, and W. Gerok, eds.), p. 39, MTP Press, Lancaster (1978).

661. M. Tohma, R. Mahara, H. Takeshita, T. Kurosawa, S. Ikegawa, and H. Nittono, *Chem Pharm. Bull.* **33**, 3071 (1985).

662. J. M. Street, W. F. Balistreri, and K. D. R. Setchell, *Gastroenterology* **90**, 1773 (1986).

663. J. M. Street and K. D. R. Setchell, *Sem. Liver Dis.* **7**, 85 (1987).

APPENDIX

A.M. Lawson and K.D.R. Setchell

List of Retention Indices[a] for Methyl Ester-Trimethylsilyl (Me-TMS) Ether Derivatives of Bile Acids

Bile acid structure[b]	Iida et al.[c] 1987 t_r	MU	Iida et al.[d] 1983 t_r	Elliott[e] 1972 t_r	Almé et al.[f] 1977 t_r[h]	Setchell[g] and Lawson t_r	MU
None							
5α	0.53	2941				0.57	
5β	0.48	2906	0.42			0.52	
Monohydroxy							
3α-ol-5α	0.86	3141		0.72	0.82(0.76)		
3α-ol-5β	0.86	3141	0.85	0.85	0.80(0.74)	0.87	
3β-ol-5α	1.07	3232		0.82		1.06	
3β-ol-5β	0.85	3133	0.81	0.81		0.86	
3β-ol-Δ⁵					0.99(0.91)	0.99	
3α-ol-Δ⁶						0.81	
7α-ol-5α	0.60	2992					
7α-ol-5β	0.63	3012	0.59				
7β-ol-5α	0.82	3118					
7β-ol-5β	0.74	3079	0.70			0.75	
11β-ol-5β						0.78	
12α-ol-5α	0.64	3022					
12α-ol-5β	0.61	3000	0.56				
12β-ol-5α	0.66	3030					
12β-ol-5β	0.59	2981	0.53				
3β-ol-7α-OMe-Δ⁵							3251
3β-ol-7β-OMe-Δ⁵							3319
Dihydroxy							
3α,6α-ol-5α							
3α,6α-ol-5β	1.09	3238	1.14	1.12	1.11(1.02)	1.07	
3α,6β-ol-5β				1.10	1.05(0.97)	1.01	
3β,6β-ol-5α				1.33			
3α,7α-ol-5α	1.02	3210		0.84	1.00(0.92)		

(continued)

571

APPENDIX (*Continued*)

Bile acid structure[b]	Iida et al.[c] 1987 t_r	MU	Iida et al.[d] 1983 t_r	Elliott[e] 1972 t_r	Almé et al.[f] 1977 t_r^h	Setchell[g] and Lawson t_r	MU
3α,7α-ol-5β	1.05	3225	1.07	1.04	1.03(0.95)	1.04	3216
3α,7β-ol-5α	1.14	3259					
3α,7β-ol-5β	1.17	3268	1.23	1.21	1.15(1.06)		3252
3β,7α-ol-5α	1.04	3223		1.08			
3β,7α-ol-5β	0.96	3190	0.96	0.96	0.93(0.86)	0.96	
3β,7α-ol-Δ⁴							3200
3β,7α-ol-Δ⁵							3204
3β,7β-ol-5α	1.46	3360					
3β,7β-ol-5β	1.17	3268	1.23			1.14	
3α,12α-ol-5α	0.92	3173		1.18	0.91(0.84)	0.93	
3α,12α-ol-5β	1.00	3200	1.00	1.00	1.00(0.92)	1.00	
3α,12β-ol-5α	0.94	3181				0.95	
3α,12β-ol-5β	0.95	3186	0.96		0.92(0.85)	0.95	
3β,12α-ol-5α	1.16	3267		1.23	1.18(1.09)	1.19	
3β,12α-ol-5β	0.98	3198	0.99	1.00	1.05(0.88)	0.99	
3β,12β-ol-5α	1.27	3298				1.28	
3β,12β-ol-5β	1.00	3200	0.99			0.99	
3β,12α-ol-Δ⁵					1.14(1.05)		
7α,12α-ol-5α	0.65	3028					
7α,12α-ol-5β	0.90	3162	0.68			0.73	
7α,12β-ol-5α	0.67	3043					
7α,12β-ol-5β	0.71	3061	0.68				
7α,12α-ol-5α	0.89	3157					
7β,12α-ol-5β	0.86	3141	0.85				
7β,12β-ol-5α	0.97	3195					
7β,12β-ol-5β	0.88	3151	0.87				
Trihydroxy							
1,3,12-ol-5β					1.31(1.21)*		
3α,6α,7α-ol-5α							
3α,6α,7α-ol-5β	1.28	3302	1.39	1.15	1.38(1.27)		3320
3α,6β,7α-ol-5β				1.12			
3α,6β,7β-ol-5α				1.44			
3α,6β,7β-ol-5β				1.45			
3α,6α,7β-ol-5β				1.92			
3α,6β,12α-ol-5β					1.18(1.09)*		
3α,7α,12α-ol-5α	1.03	3211	1.09	1.05	1.03(0.95)		
3α,7α,12α-ol-5β	1.06	3228		1.09	1.09(1.00)	1.07	3227
3α,7α,12β-ol-5α	1.03	3211					
3α,7α,12β-ol-5β	1.02	3210	1.09			1.02	
3α,7β,12α-ol-5α	1.15	3262			1.18(1.09)		
3α,7β,12α-ol-5β	1.53	3375	1.31		1.25(1.15)		
3α,7β,12β-ol-5α	1.25	3289					

APPENDIX *(Continued)*

Bile acid structure[b]	Iida *et al.*[c] 1987 t_r	MU	Iida *et al.*[d] 1983 t_r	Elliott[e] 1972 t_r	Almé *et al.*[f] 1977 t_r^h	Setchell[g] and Lawson t_r	MU
3α,7β,12β-ol-5β	1.20	3274	1.34			1.17	
3β,7α,12α-ol-5α	1.07	3232		1.01			
3β,7α,12α-ol-5β	0.98	3198	1.00	0.96			
3β,7α,12α-ol-Δ⁴							3214
3β,7α,12β-ol-5α	1.18	3269					
3β,7α,12β-ol-5β	1.00	3200	1.06				
3β,7β,12α-ol-5α	1.50	3340			1.18(1.09)*		
3β,7β,12α-ol-5β	1.18	3269	1.31		1.22(1.13)		
3β,7β,12β-ol-5α	1.75	3429					
3β,7β,12β-ol-5β	1.29	3305	1.43				
3α,12α,16α-ol-5β						1.81	
3α,12α,23-ol-5β						1.33	
Tetrahydroxy							
1β,3α,7α,12α-ol-5β					1.36(1.25)		3325
1,3,7,12-ol-5χ					1.40(1.29)		
3,6,7,12-ol-5χ					1.42(1.31)		
3α,6α,7α,12α-ol-5β					1.36(1.25)		3313
3α,7α,12α,23-ol-5β					1.56(1.44)		
Oxo							
3-oxo-Δ⁴						1.27	
3-oxo-6α-ol-5β				1.32			
3-oxo-7α-ol-5α				1.08			
3-oxo-7α-ol-5β				1.17			
3-oxo-7α-ol-Δ⁴							3298
3-oxo-12α-ol-5α				1.17			
3-oxo-12α-ol-5β				1.09			
3-oxo-7α,12α-ol-5α				1.09			
3-oxo-7α,12α-ol-5β				1.35			
3-oxo-7α,12α-ol-Δ⁴							3360
6-oxo-3α-ol-5α							
7-oxo-3α-ol-5α				1.11			
7-oxo-3α-ol-5β				1.34		1.35	3315
7-oxo-3α,6α-diol-5β						1.42	
7-oxo-3α,12α-ol-5α				1.62			
7-oxo-3α,12α-ol-5β				1.59			
7-oxo-3β,12α-ol-5α				1.96			
12-oxo-3α-ol-5a						1.62	
12-oxo-3α-ol-5β				1.33	1.36(1.25)	1.65	
12-oxo-3β-ol-5β						1.31	
12-oxo-3α,7α-ol-5β				1.72			
12-oxo-3β-ol-5β-Δ⁹⁽¹¹⁾						1.32	

APPENDIX   (*Continued*)

Bile acid structure[b]	Iida et al.[c] 1987 t_r	Iida et al.[c] 1987 MU	Iida et al.[d] 1983 t_r	Elliott[e] 1972 t_r	Almé et al.[f] 1977 t_r^h	Setchell[g] and Lawson t_r	Setchell[g] and Lawson MU
Dioxo							
3,7-dioxo-12α-ol-5α				1.78			
3,7-dioxo-12α-ol-5β				1.56			
3,12-dioxo-5β						1.43	
7,12-dioxo-3α-ol-5β				1.88			
Nor-bile acids							
Nor-5β-cholan-23-oic						0.39	
Norlithocholic						0.66	
Nordeoxycholic						0.77	
Norchenodeoxycholic						0.79	
Norhyodeoxycholic						0.79	
Norcholic					0.76(0.70)	0.84	
Norursodeoxycholic						0.85	
C$_{27}$ bile acids							
3α,7α-ol-5β-C$_{27}$							3452
3α,7α,12α-ol-5β-C$_{27}$							3455
1β,3α,7α,12α-ol-5β-C$_{27}$							3550
3α,6α,7α12α-ol-5β							3540
3α,7α,12α,24-ol-5β[i]							3600
3α,7α,12α,24-ol-5β[i]							3612
3α,7α,12α,25-ol-5β							3628
3α,7α,12α,26-ol-5β							3715
C$_{29}$ bile acids							
3α,7α,12α-ol-5β-C$_{29}$-dioic							3930
C$_{20}$ bile acids							
3α-ol-5β						0.40	
3β-ol-5β						0.32	
3β-ol-Δ^5					0.38		

[a] Retention data are expressed for the Me-TMS ethers relative to the Me-TMS ether derivative of deoxycholic acid and/or as the methylene unit (MU) values relative to a homologous series of n-alkanes.

[b] Bile acids are listed according to the number of functional groups. For clarity unless indicated all bile acids have the cholanoic acid (C$_{24}$) nucleus and the presence of hydroxy groups is indicated by the term -ol (i.e., 3α-ol indicates 3α-hydroxy structure). The position of the hydrogen at C-5 is indicated as 5β or 5α and Δ signifies unsaturation at the position shown by the superscript. Nor-bile acids are indicated by their trivial names. C$_{27}$ indicates a cholestanoic acid nucleus. The C$_{20}$ bile acids have the androstane-17β-oic nucleus.

[c] Bile acid derivatives were chromatographed on an 18 meter OV-1 chemically bonded fused silica capillary column (HiCap-CBP-1; 0.2 mm internal diameter, 0.25 μm film thickness) at 270°C with helium as carrier gas (linear velocity 32 cm/sec)-[T. Iida *et al.*, *J. Chromatogr.* (1987)].

[d] Chromatography was performed on a 30 meter glass capillary column coated with OV-1 (Nihon Chromato Works Ltd, Tokyo, Japan) at 245°C using nitrogen (flow rate 1.7 ml/min) as carrier gas [Iida *et al.*, *J. Lipid Res.* **24**:211 (1983)].

[e] Chromatography was performed on conventional GLC column packed with 3% OV-1 liquid phase (see Ref. 56, Chapter 5).

[f] Chromatography was performed on 2.5 meter × 3.4 mm conventional GLC column packed with 1.5% SE-30 on Gas Chrom Q (80–100 ml) at 210–240°C (see ref. 13, Chapter 5).

[g] Chromatography was performed on a 30 meter DB-1 chemically bonded fused silica capillary column at 275°C using helium (flow rate 2ml/min) as carrier gas.

[h] Values were cited in the original article relative to the Me-TMS ether of cholic acid and these values are given in parentheses. Values have been recalculated and expressed relative to the Me-TMS ether of deoxycholic acid.

[i] These may be two C_{24} (*R* and *S*) or C_{25} isomers.

[*] This indicates a tentative identification.

INDEX